Electronics

Electronics originated as an applied field of physics. It is a subject of both science and technology that deals with electrical circuits with active devices, and their applications. In this course, students study various elements like amplification, rectification, communications, analog and digital system design, optoelectronics, etc.

This book is for physics honours/equivalent courses based on UGC CBCS curriculum and is also useful to AICTE recommended UG courses of engineering/technology. It aims to sensitise beginners to the subject, emphasizes on the theory and applications of the circuits and instruments and focusses on the fundamental concepts and the science involved in electronics.

The following are the specific features of the book.

- Each chapter contains solved examples of multiple-choice, reasoning and numerical types and past questions of national level competitive examinations to help students in self-learning and realizing their required level of preparation.
- The exercises are categorized as descriptive, conceptual and numerical questions.
- The book encourages raising questions. Distinct blocks entitled "Think a Bit" and "Note" are included along with the main text.
- End-chapter project work and computer simulations are given in most of the chapters with a view to inspire the independent thinking capability of the students and to buoy up the application-oriented knowledge.

Barun Raychaudhuri is a Professor at the Department of Physics, Presidency University, Kolkata. He is a senior member of IEEE. He has teaching experience of more than 25 years and his areas of research are atmospheric and hyperspectral remote sensing. Besides contributing to peer reviewed journals Raychaudhuri has written two books: *A Textbook of Digital Computer Electronics* (2005) and *A Textbook of Basic Electronics* (2012).

Electronics
Analog and Digital

Barun Raychaudhuri

CAMBRIDGE
UNIVERSITY PRESS

University Printing House, Cambridge CB2 8BS, United Kingdom

One Liberty Plaza, 20th Floor, New York, NY 10006, USA

477 Williamstown Road, Port Melbourne, VIC 3207, Australia

314–321, 3rd Floor, Plot 3, Splendor Forum, Jasola District Centre, New Delhi–110025, India

103 Penang Road, #05–06/07, Visioncrest Commercial, Singapore 238467

Cambridge University Press is part of the University of Cambridge.

It furthers the University's mission by disseminating knowledge in the pursuit of education, learning and research at the highest international levels of excellence.

www.cambridge.org
Information on this title: www.cambridge.org/9781009214230

© Barun Raychaudhuri 2022

This publication is in copyright. Subject to statutory exception and to the provisions of relevant collective licensing agreements, no reproduction of any part may take place without the written permission of Cambridge University Press.

First published 2022

Printed in India by Nutech Print Services, New Delhi 110020

A catalogue record for this publication is available from the British Library

Library of Congress Cataloging-in-Publication Data

Names: Raychaudhuri, Barun, author.
Title: Electronics : analog and digital / Barun Raychaudhuri.
Description: Cambridge, United Kingdom; New York, NY, USA: Cambridge University Press, 2022. | Includes bibliographical references and index.
Identifiers: LCCN 2022016938 (print) | LCCN 2022016939 (ebook) | ISBN 9781009214230 (paperback) | ISBN 9781009214247 (ebook)
Subjects: LCSH: Electronics. | Bisac: Technology & Engineering / Electronics / Circuits / General
Classification: LCC TK7816 .R375 2022 (print) | LCC TK7816 (ebook) | DDC 621.381--dc23/eng/20220601
LC record available at https://lccn.loc.gov/2022016938
LC ebook record available at https://lccn.loc.gov/2022016939

ISBN 978-1-009-21423-0 Paperback

Cambridge University Press has no responsibility for the persistence or accuracy of URLs for external or third-party internet websites referred to in this publication, and does not guarantee that any content on such websites is, or will remain, accurate or appropriate.

Contents

Preface   xvii

Chapter 1. Origin of Electronics   1
    1.1 What is Electronics   1
    1.2 Evolution of Electronics   2
       1.2.1 Revisiting the History   2
       1.2.2 Trends of Development   4
    1.3 Widespread Applications   5
    1.4 Electrons, Electricity and Electronics   6
       1.4.1 Electric Current   7
       1.4.2 Drift Velocity, Mobility and Conductivity   8
       1.4.3 Electron Emission from Metal   10
    1.5 Circuits and Sources   11
       1.5.1 Types of Circuits   11
       1.5.2 Voltage and Current Sources   12
    1.6 Active and Passive Device   14
    Multiple Choice-Type Questions and Answers   15
    Reasoning-Type Questions and Answers   16
    Solved Numerical Problems   16
    Exercise   18

Chapter 2. Semiconductor Fundamentals   20
    2.1 Crystalline Solids   20
       2.1.1 Lattice, Basis and Unit Cell   21
       2.1.2 Bravais Lattice and Miller Indices   22
    2.2. Energy Band   24
    2.3 Semiconductors   26
       2.3.1 Electron and Hole   27
       2.3.2 Intrinsic Semiconductor   29
       2.3.3 Doping and Extrinsic Semiconductor   30
       2.3.4 Doping in Compound Semiconductor   32
    2.4 Direct and Indirect Band Gap   32
    2.5 Effective Mass   33

2.6	Fermi Level, Energy Band and Semiconductors		35
	2.6.1 Energy Band of n-type Semiconductors		38
	2.6.2 Energy Band of p-type Semiconductors		39
2.7	Equilibrium Carrier Concentrations		40
2.8.	Drift and Diffusion of Carriers		43
	2.8.1 Drift Current Density		44
	2.8.2 Diffusion Current Density		45
	2.8.3 Semiconductor Current Density		45
	2.8.4 Einstein Relation		46
	2.8.5 Continuity Equation		47
2.9	Hall Effect		48
2.10	Resistivity and Four-Probe Technique		49
	Multiple Choice-Type Questions and Answers		50
	Reasoning-Type Questions and Answers		53
	Solved Numerical Problems		55
	Exercise		57
	Project Work on Chapter 2		60
Chapter 3.	**p–n Junction Diodes**		**61**
3.1	Fabrication of p–n Junction		61
3.2	Barrier Formation in p–n Junction		62
	3.2.1 Built-in Potential		64
	3.2.2 Fermi Level in p–n Junction		65
	3.2.3 Energy Band Diagram of p–n Junction		66
3.3	Forward and Reverse Bias		67
	3.3.1 Unbiased Diode		68
	3.3.2 Forward Biased Diode		69
	3.3.3 Reverse Biased Diode		69
3.4	Diode Current–voltage Characteristics		70
	3.4.1 Static and Dynamic Resistance		72
	3.4.2 Cut-in Voltage		74
3.5	Junction Capacitances		74
	3.5.1 Depletion Capacitance		74
	3.5.2 Diffusion Capacitance		75
3.6	Zener Diode		76
	3.6.1 Zener Breakdown		76
	3.6.2 Avalanche Breakdown		77
	3.6.3 Zener Diode Characteristics		79
	3.6.4 Zener Voltage Regulator		80
3.7	Light-Emitting Diode (LED)		82

	3.8	Photodiode and Solar Cell	84
	3.9	Metal–Semiconductor Contacts	86
	Multiple Choice-Type Questions and Answers		87
	Reasoning-Type Questions and Answers		91
	Solved Numerical Problems		92
	Exercise		95
	Project Work on Chapter 3		99
Chapter 4.	**Diode Applications**		**101**
	4.1	Piecewise Linear Model	101
	4.2	Load Line and Q-Point	102
	4.3	Rectifiers	104
		4.3.1 Half-Wave Rectifier	105
		4.3.2 Full-Wave Rectifier	111
		4.3.3 Bridge Rectifier	114
	4.4	Filters	117
		4.4.1 Capacitor Filter	117
		4.4.2 Inductor Filter	119
	4.5	Clippers	120
		4.5.1 Shunt Clipper	120
		4.5.2 Series Clipper	120
	4.6	Clamper	123
	4.7	Voltage Multiplier	125
	Multiple Choice-Type Questions and Answers		127
	Reasoning-Type Questions and Answers		131
	Solved Numerical Problems		132
	Exercise		134
	Project Work on Chapter 4		138
Chapter 5.	**Bipolar Junction Transistor (BJT)**		**141**
	5.1	Transistors: n–p–n and p–n–p	141
	5.2	Transistor Operating Principle	142
	5.3	Common-Emitter Configuration	143
		5.3.1 Current Amplification in Transistor	144
		5.3.2 Transistor Current Components	145
		5.3.3 Common-Emitter Output Characteristics	148
		5.3.4 Early Effect	150
		5.3.5 CE Input Characteristics	150
		5.3.6 CE Transfer Characteristics	150
	5.4	Common-Base Characteristics	151
	5.5	Common-Collector Configuration	153

		Multiple Choice-Type Questions and Answers	155
		Reasoning-Type Questions and Answers	157
		Solved Numerical Problems	159
		Exercise	160
		Project Work on Chapter 5	161
Chapter 6.	**Transistor Biasing and Amplification**		**163**
	6.1	Load Line and Q-Point	163
	6.2	Transistor Biasing and Stability	165
	6.3	Base Bias	167
	6.4	Emitter–Feedback Bias	169
	6.5	Collector–Feedback Bias	171
	6.6	Voltage-Divider Bias	172
	6.7	Load: DC and AC	175
	6.8	BJT Small Signal Voltage Amplifiers	177
		6.8.1 Common-Emitter (CE) Amplifier	178
		6.8.2 Common-Collector (CC) Amplifier	182
		6.8.3 Common-Base (CB) Amplifier	184
		Multiple Choice-type Questions and Answers	186
		Reasoning-Type Questions and Answers	188
		Solved Numerical Problems	189
		Exercise	194
		Project Work on Chapter 6	197
Chapter 7.	**Network Theorems and Transistor**		**200**
	7.1	Thevenin's Theorem	200
	7.2	Norton's Theorem	202
	7.3	Other Useful Theorems	203
		7.3.1 Superposition Theorem	203
		7.3.2 Maximum Power Transfer Theorem	204
	7.4	Two-Port Model and Hybrid Parameters	205
		7.4.1 Transistor as Two-port Network	207
		7.4.2 Significance of h Parameters	209
	7.5	Transistor Amplifier with h Parameters	211
	7.6	Simplified Hybrid Model	214
	7.7	r_e-Model and h-Model	215
	7.8	Transistor: Thevenin and Norton Equivalents	216
	7.9	Frequency Dependence of Gain	219
	7.10	Hybrid-Π Model	221
	7.11	Transistor Gain at High Frequency	224
	7.12	Gain and Decibel	227

	Multiple Choice-Type Questions and Answers	228
	Reasoning-Type Questions and Answers	229
	Solved Numerical Problems	229
	Exercise	235
	Project Work on Chapter 7	237
Chapter 8.	**Transistor Power and Multistage Amplifiers**	**239**
8.1	Need for Power Amplification	239
8.2	Conditions for Power Amplification	241
8.3	Distortions due to Nonlinearity	242
	8.3.1 Amplitude Distortion	243
	8.3.2 Harmonic Distortion	243
	8.3.3 Intermodulation Distortion	244
8.4	Amplifier Classes	245
8.5	Class A Amplifier	246
	8.5.1 Bias for Voltage Amplifier	246
	8.5.2 Resistive Load Power Amplifier	247
	8.5.3 Transformer Coupled Amplifier	249
8.6	Class B Amplifier	253
8.7	Push–Pull Amplifier	254
	8.7.1 Class A Push–Pull Amplifier	254
	8.7.2 Class B Push–Pull Amplifier	255
	8.7.3 Crossover Distortion	258
	8.7.4 Class AB Amplifier	259
	8.7.5 Complementary Symmetry Amplifier	260
8.8	Class C Amplifier	262
8.9	Multistage Amplifiers	264
	Multiple Choice-Type Questions and Answers	267
	Reasoning-Type Questions and Answers	269
	Solved Numerical Problems	270
	Exercise	274
Chapter 9.	**Field-Effect Transistor (FET)**	**276**
9.1	'Field-Effect' and 'Transistor'	276
9.2	Junction Field-Effect Transistor (JFET)	277
	9.2.1 JFET Current–Voltage Characteristics	278
	9.2.2 JFET Transfer Characteristics	282
9.3	FET Parameters	283
9.4	FET versus BJT	284
9.5	MOSFET	286
	9.5.1 n-Channel Depletion-Type MOSFET	286
	9.5.2 n-Channel Enhancement-Type MOSFET	289

9.6	FET Model		293
9.7	FET Biasing		294
	9.7.1	Self-Bias	294
	9.7.2	Drain-Feedback Bias	296
	9.7.3	Gate Bias	297
	9.7.4	Voltage-Divider Bias	298
9.8	FET Amplifiers		300
	9.8.1	Common-Source (CS) Amplifier	300
	9.8.2	Common-Drain (CD) Amplifier	301
	9.8.3	Common-Gate (CG) Amplifier	302

Multiple Choice-Type Questions and Answers — 304
Reasoning-Type Questions and Answers — 306
Solved Numerical Problems — 307
Exercise — 308
Project Work on Chapter 9 — 310

Chapter 10. Feedback Amplifiers and Oscillators — 311

10.1	Concept of Feedback		313
10.2	Types of Feedback		316
	10.2.1	Voltage–series Feedback	317
	10.2.2	Voltage–shunt Feedback	318
	10.2.3	Current–series Feedback	319
	10.2.4	Current–shunt Feedback	320
10.3	Advantages of Negative Feedback		321
	10.3.1	Stability Improvement	322
	10.3.2	Impedance Improvement	322
	10.3.3	Bandwidth Enhancement	324
	10.3.4	Noise Reduction	326
	10.3.5	Reduction of Nonlinear Distortion	326
10.4	Oscillators		327
	10.4.1	Positive Feedback and Oscillation	328
	10.4.2	Resonant Circuit Oscillators	329
	10.4.3	Colpitts Oscillator	331
	10.4.4	Hartley Oscillator	332
	10.4.5	Wien Bridge Oscillator	333
	10.4.6	Phase-Shift Oscillator	338
	10.4.7	Crystal Oscillator	341
10.5	Multivibrators		345
	10.5.1	Astable Multivibrator	345
	10.5.2	Monostable Multivibrator	348

	Multiple Choice-Type Questions and Answers		350
	Reasoning-Type Questions and Answers		352
	Solved Numerical Problems		353
	Exercise		357
	Project Work on Chapter 10		359

Chapter 11. Operational Amplifier — 362

- 11.1 A Review on Amplifiers — 362
- 11.2 Features of Op-Amp — 363
- 11.3 Differential Amplifier — 365
- 11.4 Common Mode Rejection Ratio — 368
- 11.5 Diff-Amp to Op-Amp — 369
- 11.6 Offset Parameters — 370
- 11.7 Slew Rate — 373
- 11.8 Linear Op-Amp Circuits — 374
  - 11.8.1 Inverting Amplifier — 375
  - 11.8.2 Noninverting Amplifier — 377
  - 11.8.3 Virtual Short and Virtual Ground — 382
  - 11.8.4 Voltage Follower — 383
  - 11.8.5 Op-Amp Adder (Inverting) — 384
  - 11.8.6 Op-Amp Adder (Noninverting) — 385
  - 11.8.7 Differential Amplifier — 386
  - 11.8.8 Instrumentation Amplifier — 388
  - 11.8.9 Passive and Active Filters — 389
  - 11.8.10 Active Low-Pass Filter — 392
  - 11.8.11 Active High-Pass Filter — 394
  - 11.8.12 Active Band-Pass and Band-Stop Filters — 395
- 11.9 Nonlinear Op-Amp Circuits — 396
  - 11.9.1 Integrator — 396
  - 11.9.2 Differentiator — 398
  - 11.9.3 Comparator — 400
  - 11.9.4 Schmitt Trigger — 401
  - 11.9.5 Logarithmic Amplifier — 402
  - 11.9.6 Solving Algebraic Equation — 403
  - 11.9.7 Solving Differential Equation — 404
  - 11.9.8 Precision Rectifier — 405
- 11.10 Op-Amp Waveform Generators — 406
  - 11.10.1 Square Wave Generator — 406
  - 11.10.2 Triangular Wave Generator — 407
  - 11.10.3 Sine Wave Generator — 409

		Multiple Choice-Type Questions and Answers	409
		Reasoning-Type Questions and Answers	414
		Solved Numerical Problems	415
		Exercise	422
		Projects on Chapter 11	428

Chapter 12. IC Technology and Instrumentation — 432

12.1	Integrated Circuit (IC)		432
12.2	IC Classification		433
12.3	IC Fabrication		436
12.4	IC Components: Active and Passive		437
12.5	Regulated Power Supply		439
12.6	Cathode Ray Oscilloscope (CRO)		442
	12.6.1	Construction of CRO	444
	12.6.2	Working Principle	445
	12.6.3	Electrostatic Focusing	446
	12.6.4	Electrostatic Deflection	447
	12.6.5	Waveform Display	449
	12.6.6	Applications of CRO	451
12.7	Digital Storage Oscilloscope		454
	Multiple Choice-Type Questions and Answers		456
	Reasoning-Type Questions and Answers		457
	Solved Numerical/Analytic Problems		458
	Exercise		459

Chapter 13. Digital Principles and Boolean Algebra — 460

13.1	The Digital System		460
	13.1.1	Analog and Digital	460
	13.1.2	Pros and Cons of Digital System	461
13.2	Number Systems and Conversions		462
	13.2.1	Binary Numbers	463
	13.2.2	Binary to Decimal Conversion	464
	13.2.3	Decimal to Binary Conversion	464
	13.2.4	Octal Numbers	467
	13.2.5	Hexadecimal Numbers	469
13.3	Digital Codes		470
13.4	Binary Arithmetic		472
	13.4.1	1's Complement and 2's Complement	473
	13.4.2	Radix Complements	477
	13.4.3	Signed Binary Numbers	478

13.5	Boolean Algebra		479
	13.5.1	OR Operation	482
	13.5.2	AND Operation	483
	13.5.3	NOT Operation	483
	13.5.4	De Morgan's Theorems	484
13.6	Boolean Simplification		485
13.7	Sum-of-Products and Product-of-Sums		488
	13.7.1	Sum-of-Products (SOP)	488
	13.7.2	Product-of-Sums (POS)	490
13.8	Karnaugh Map		493
	13.8.1	Don't Care Conditions	498
	13.8.2	Simplification is not Unique	499
	13.8.3	SOP and POS are Equivalent	499
Multiple Choice-Type Questions and Answers			500
Reasoning-Type Questions and Answers			503
Solved Numerical and Logical Problems			503
Exercise			507
Projects on Chapter 13			508
Chapter 14.	**Combinational Logic Circuits**		**510**
14.1	Boolean Algebra and Digital Electronics		510
	14.1.1	Combinational and Sequential Logic	511
	14.1.2	Positive and Negative Logic	512
14.2	Logic Gates		513
	14.2.1	OR Gate	513
	14.2.2	AND Gate	513
	14.2.3	NOT Gate	515
	14.2.4	NOR Gate (Universal Gate)	516
	14.2.5	NAND Gate (Universal Gate)	517
	14.2.6	Bubbled Gates	518
	14.2.7	Exclusive-OR (XOR) Gate	519
	14.2.8	Timing Diagram	522
14.3	Logic Families		522
	14.3.1	Transistor–Transistor Logic (TTL)	523
	14.3.2	MOS Logic	525
14.4	Arithmetic and Logic Circuits		527
	14.4.1	Half Adder	527
	14.4.2	Full Adder	528
	14.4.3	Half and Full Subtractors	530
	14.4.4	Adder–Subtractor	531
	14.4.5	Digital Comparators	532

14.5	Data Processing Circuits		533
	14.5.1	Multiplexer	533
	14.5.2	Demultiplexer	536
	14.5.3	Decoders	536
	14.5.4	Seven-Segment Display	537
	14.5.5	Encoders	540
	14.5.6	Parity Checker and Generator	541

Multiple Choice-Type Questions and Answers 542
Reasoning-Type Questions and Answers 546
Solved Logical Problems 547
Exercise 552
Project Work on Chapter 14 555

Chapter 15. Sequential Logic Circuits 557

- 15.1 Clock and Timer 557
  - 15.1.1 Clock Parameters 558
  - 15.1.2 Working Principles of IC 555 559
  - 15.1.3 Astable Multivibrator with IC 555 560
  - 15.1.4 Monostable Multivibrator with IC 555 562
- 15.2 Latch and Flip-Flop 563
  - 15.2.1 Bistable Multivibrator 564
  - 15.2.2 RS Flip-Flop with NOR Gates 565
  - 15.2.3 RS Flip-Flop with NAND Gates 567
  - 15.2.4 Clocked RS Flip-Flop 569
  - 15.2.5 D Flip-Flop 570
  - 15.2.6 JK Flip-Flop 571
  - 15.2.7 Racing and Propagation Delay 572
  - 15.2.8 Edge- and Pulse-Triggering 573
  - 15.2.9 JK Master–Slave Flip-Flop 575
  - 15.2.10 T Flip-Flop 577
- 15.3 Flip-Flop Characterization 578
  - 15.3.1 Characteristic Equation 579
  - 15.3.2 State Diagram 579
  - 15.3.3 Preset and Clear 580
- 15.4 Register 581
  - 15.4.1 Register with Series and Parallel Shifting 583
  - 15.4.2 Ring Counter 586
  - 15.4.3 Johnson Counter 587
  - 15.4.4 Register Applications 588
- 15.5 Counters 588
  - 15.5.1 Asynchronous Counter 589
  - 15.5.2 Synchronous Counter 591

15.6	Changing Counter Modulus		592
	15.6.1	Mod-3 Counter	593
	15.6.2	Mod-6 Counter	593
	15.6.3	Mod-5 Counter	594
	15.6.4	Decade (Mod-10) Counter	595
	15.6.5	Decade Counter using Preset-Clear	596
	15.6.6	Applications of the Counter	598

Multiple Choice-Type Questions and Answers — 598
Reasoning-Type Questions and Answers — 600
Solved Numerical Problems — 602
Exercise — 604

Chapter 16. Analog–Digital Conversion and Memory — 609

16.1	Why D/A and A/D Conversions		609
16.2	Binary Equivalent Weight		610
16.3	Digital-to-Analog (D/A) Conversion		611
	16.3.1	Weighted Resistor D/A Converter	611
	16.3.2	R-$2R$ ladder D/A Converter	614
	16.3.3	D/A Converter Performance	617
16.4	Analog-to-Digital (A/D) Conversion		618
	16.4.1	Flash A/D Converter	618
	16.4.2	Counter-Type A/D Converter	620
	16.4.3	Successive-Approximation A/D Converter	622
16.5	Memory		622
	16.5.1	Read-Only Memory (ROM)	625
	16.5.2	Random Access Memory (RAM)	627
	16.5.3	Memory Addressing	629
	16.5.4	Memory Read/Write	630

Multiple Choice-Type Questions and Answers — 631
Reasoning-Type Questions and Answers — 633
Solved Numerical Problems — 633
Exercise — 634
Project Work on Chapter 16 — 635

Chapter 17. Microcomputer and Microprocessor — 637

17.1	Evolution of Computer		637
	17.1.1	Historical Background	637
	17.1.2	Modern Computer	639
17.2	Computer, Microprocessor and Microcontroller		641
	17.2.1	Computer Organization	641
	17.2.2	Use of Microprocessor	643
	17.2.3	Use of Microcontroller	644

17.3	Hardware and Software		646
	17.3.1	Operating System	646
	17.3.2	Computer Languages	648
17.4	Microprocessor 8085		650
	17.4.1	Pin Configuration	652
	17.4.2	Operational Code (Opcode)	653
17.5	8085 Programming		654
17.6	Types of 8085 Instructions		668
17.7	Use of Subroutine		669
17.8	Arduino Programming		669
	17.8.1	Arduino Board	670
	17.8.2	Arduino IDE	670

Multiple Choice-Type Questions and Answers — 673
Reasoning-Type Questions and Answers — 674
Solved Examples — 675
Exercise — 677

Bibliography — 679

Index — 681

Preface

'Electronics' is quite a familiar word in our daily life. We see around us innumerable electronic gadgets for domestic use and many scientific, engineering, technical, and vocational teaching courses of different academic levels related to electronics. Actually electronics evolved as a part of modern physics almost a century ago and depicted significant contribution to modern science, technology, economy and society.

There is no dearth of excellent reference books on different aspects of electronics. Nevertheless, the University Grants Commission (UGC) guidelines of Choice-Based Credit System (CBCS) curriculum and Learning Outcomes based Curriculum Framework (LOCF), prescribing two full papers of electronics in the core course of undergraduate physics motivated this author to write something on electronics exclusively for young students who have just passed the high school and entered the higher studies. The outcome is this book; *Electronics: Analog and Digital*, prepared mainly for physics honours/equivalent courses.

The topics of electronic devices, circuits and systems are broadly classified into two categories: analog and digital, as is also specified in the CBCS curriculum. Maintaining the recommended topics, the subject matters are reorganized so as to facilitate the students.

The first introductory chapter outlines briefly the evolution, significance and widespread applications of electronics. Since semiconductors take major part in the fabrication of electronic devices, the subject learning starts with the basic properties and types of semiconductors, electrons and holes, concepts of energy band and effective mass and current transport phenomena (Chapter 2).

A fundamental structure in electronics is p–n junction. Chapter 3 explains its rectification property, forward and reverse biasing and the corresponding energy band diagrams. It also elucidates how the same p–n junction with constructional changes can give rise to different devices, such as rectifier diode, Zener diode and light-emitting diode. The important applications of diode as half-wave, full-wave and bridge rectifier, clipper and clamper are presented in Chapter 4.

The bipolar junction transistor (BJT) is a very remarkable device in electronics. It is the basic building block for many analog and digital circuits. So the book dedicates four chapters on different aspects of transistor. Chapter 5 contains the construction and working principle of n–p–n and p–n–p transistors, the amplifying action of transistor and detailed explanations of common-emitter, common-base and common-collector configurations.

Chapter 6 depicts different types of voltage biasing for transistor and the use of transistor as voltage amplifier. Chapter 7 includes the application of network theorems, such as Thevenin and Norton theorem and hybrid model for deriving the transistor equivalent circuits. The transistor of its own is a current amplifier and with proper biasing it can be made to act as a voltage amplifier. However, when it is required to amplify power, some specific construction and some special types of biasing are needed. These topics are clarified in Chapter 8. It also illustrates different classes of amplifier: A, B, AB and C and different types of coupling.

Chapter 9 consists of another type of transistor known as field-effect transistor (FET). The BJT discussed in Chapters 5 through 8 is basically a current-controlled device whereas the FET is a voltage-controlled device. Different types of FETs, such as JFET and MOSFET are introduced here.

Up to this, the characteristics of a single device with supporting circuits are discussed. The electronic devices, such as diode, BJT and FET are called *active device* and the other known electrical devices, such as resistor, capacitor and inductor are termed as *passive device*. Many useful electronic circuits can be constructed with suitable combination of active and passive elements. Chapter 10 deals with two important classes of electronic circuits, namely feedback amplifiers and oscillators. The concept of feedback, negative feedback and different oscillator circuits based on feedback, such as Hartley, Colpitts, Wien bridge and crystal oscillators and multivibrators are explained.

Another very important circuit known as operational amplifier aka op-amp is illustrated in Chapter 11. The fundamental properties and most popular applications of operational amplifier, such as inverting and noninverting amplifier, adder, integrator, differentiator, comparator and Schmitt trigger are narrated. These circuits comprise some or other type of application of feedback and can perform mathematical operations in terms of voltage.

Chapter 12 contains some discrete but important subjects of technology and instrumentation, namely integrated circuit (IC) and cathode ray oscilloscope (CRO). The devices and circuits in the previous chapters are analog in nature. The remaining chapters deal with digital systems.

Chapter 13 introduces digital electronics, different number systems and their interconversions, Boolean algebra and basic logic operations. Chapter 14 comprises the combinational logic circuits. It interprets the logic gates, such as AND, OR, NOT, universal gates, namely NAND and NOR and arithmetic and data processing circuits made of these, such as adder, subtractor, multiplexer and demultiplexer. Chapter 15 includes the sequential logic circuits, namely flip-flops, registers, counters and memories. Chapter 16 introduces the techniques of converting analog to digital and vice versa. Also different types of memory elements are outlined. Chapter 17 presents a brief sketch of the basic principles and architecture of a modern computer and explains in detail the programming technique for Microprocessor 8085. A brief idea is given on Arduino microcontroller system.

Since the book is for beginners, it illustrates the 'science' part of electronics and explains the fundamental concepts. With a view to help the students in self-learning, it has included lots of multiple-choice-type, short answer-type and numerical-type solved examples in every chapter. The separate blocks entitled "Think a Bit" and "Note" are introduced with the hope to encourage their independent thinking. The real world examples from past question papers of national level competitive examinations like National Eligibility Test (NET), Graduate Aptitude Test in Engineering (GATE), Joint Admission Test (JAM) and university examinations are expected to guide the students in realizing their required level of preparation. To enhance the application-oriented knowledge, relevant project work and computer simulations are given at the end of most of the chapters.

Apart from the above technical factors, this book is some sort of 'self-declaration'. It documents how I have been trying to learn and assimilate the subject and to present the same to the students for more than a quarter-century of my teaching profession. The nearly 450 diagrams of this book represent my own visualization of the phenomena.

I communicate heartfelt thanks to Cambridge University Press and all staff members for publishing the book and particularly I mention of Dr. Vaishali Thapliyal who have consistently encouraged me for writing. Thanks to Mr. Vikash Tiwari, Mr. Aniruddha De, and others for smooth publication of the book. I am also thankful to my student community. I have learnt a lot from them while interacting on the subject. The present efforts will be successful, if respected teachers and beloved student find this book up to their requirement. Any suggestion, criticism or other feedback is most welcome and the same may kindly be sent c/o the publisher's address.

Barun Raychaudhuri

1

Origin of Electronics

Electronics is a subject cultivated at different academic levels of undergraduate and postgraduate science and engineering curriculum. Beyond the classes, it has diversified applications in modern science, technology, economy, society, and daily life. The development of electronics throughout the last century may be treated as a distinct step along the progress of human civilization. This chapter sketches a brief outline of the background, evolution, and widespread applications of electronics. This also introduces the arrangement and relevance of topics in this book.

1.1 What Is Electronics

It is understood from our everyday experience that electronics is somehow related to the use of electricity. However, electricity is found in nature also, whereas electronic devices and the use of electricity in those devices are totally man-made. The techniques of electronic devices established several novel aspects in the use of electricity, which were never experienced earlier. Some salient features of electronics are mentioned in the following section.

- *Electrical Power Amplification*: An electronic device, such as a transistor can be made to amplify voltage and current simultaneously that cannot be achieved with other electrical gadgets, such as a transformer.
- *Nonlinear Current–Voltage Relationship*: According to Ohm's law, the steady current through a resistor, capacitor, or inductor varies linearly with voltage at a constant temperature. However, the current through electronic devices, such as a diode or a transistor undergoes nonlinear variation with voltage.
- *Impedance Transformation*: The same electronic device may exhibit different resistances across the input and the output terminals.

All the above characteristics were first realized with triode, a vacuum tube device invented by Lee de Forest in 1906. Therefore the invention of triode may be regarded as the foundation of electronics and we may feel that electronics has been a reliable companion of mankind for more than a century. The continuous research and development throughout this long period has enriched human civilization with innumerable equipment and gadgets like television, mobile phone, satellite communication, Internet, and a metamorphosis of computer from mechanical to electrically operated instrument. The researches on electronics and allied subjects have contributed to other branches of science and technology, and have given rise to new interdisciplinary fields. Throughout the 20th century, electronics has contributed to the civilization so much and influenced the society up to such a great extent that one may compare the advent of electronics with milestones like the inventions of wheel, printed letter, and steam engine.

1.2 Evolution of Electronics

Electronics originated as a part of modern physics after the discovery of electron by British physicist, Joseph John Thomson (1856–1940) in 1897. The studies on the dynamics of electrons brought about a turning point in the researches on current electricity that led to the development of electronics. The era of electronics, as is commonly reviewed, commenced with vacuum tube devices, such as triode and pentode. Later transistors and other semiconductor devices replaced those. However, it is less known that actually semiconductor devices were invented prior to vacuum tube devices. German engineer Karl Ferdinand Braun (1850–1918) noted in 1874 the dependence of the total resistance of a point metallic contact on metallic sulfide. Jagadish Chandra Bose of India studied systematically, during 1899–1904, the effect of oxidation of different metals on the radiation detecting property and patented (application in 1901, patented in 1904) lead sulfide point-contact rectifier, which was America's first publication in the area of solid state physics. It is also worth mentioning that he invented other instruments for wireless telecommunication, such as horn antenna, experimentally analyzed the properties of electromagnetic waves and was able to produce the first microwave radiation in laboratory.

Though point-contact rectifiers found applications in electromagnetic wave detectors, the primitive metal-semiconductor rectifying devices could not flourish only because there was no well-established theory to explain their properties. The working principle behind the rectifying action and the theory of current transport through the metal-semiconductor junction were not known in those days. Several decades later, British mathematician Alan Wilson (1906–1995) formulated the transport theory of semiconductor in 1931 that was applied to metal-semiconductor contacts. German physicist Walter Schottky (1886–1976) and British physicist Nevill Mott (1905–1996) independently suggested the development of potential barrier due to stable space charge in semiconductor. A brief idea of metal-semiconductor contacts is given in Chapter 3.

1.2.1 Revisiting the History

Electronics got maturity as a combination of two different trends of research: development of wireless communication and invention of vacuum tubes, and subsequent replacement

> **Note: J. C. Bose's Pioneering Contributions**
>
> Bose at Presidency College (now University) carried out ground breaking experiments on generating, transmitting, and receiving electromagnetic wave. He was the inventor of horn antenna and the first to generate microwave radiation in laboratory. As the detector, he devised a system of mercury contact with galena, the latter being identified today as Group II–VI semiconductor. Bose did his research almost a decade ago of the invention of triode when there was no trace of electronics and the concept of semiconductor was unknown. IEEE (Institute of Electrical and Electronics Engineers), the world's largest technical professional organization felicitated him posthumously with a plaque (preserved at the Physics Department, Presidency University) with the following inscription.
>
> IEEE MILESTONE IN ELECTRICAL ENGINEERING AND COMPUTING
>
> First Millimeter-wave Communication Experiments by J. C. Bose, 1894–1896
>
> Sir Jagadish Chandra Bose, in 1895, first demonstrated at Presidency College, Calcutta, India, transmission and reception of electromagnetic waves at 60 GHz, over a distance of 23 meters, through two intervening walls, by remotely ringing a bell and detonating gunpowder. For his communication system, Bose developed entire millimeter wave components such as a spark transmitter, coherer, dielectric lens, polarizer, horn antenna and cylindrical diffraction grating.
>
> September 2012   IEEE with logo

by semiconductor devices. Telecommunication was the first successful application of electronics where the new features, namely rectification and power amplification achieved with electronic devices were implemented.

The concept of wireless communication germinated when Scottish mathematician James Clerk Maxwell (1831–1879) put forward (1865) the theoretical explanations of the propagation of electromagnetic waves in space and German physicist Heinrich Hertz (1857–1894) could generate (1887) such waves in the laboratory. Jagadish Chandra Bose executed pioneering research during 1895–1904 on experimental verification of the nature of electromagnetic waves and invented equipment such as horn antenna for transmission of the wave. Italian inventor Guglielmo Marconi (1874–1937) succeeded in communicating electromagnetic wave over long distance across the Atlantic Ocean (1901) and could commercialize the wireless telecommunication system.

> **Note: A Missing Link in Wireless Telecommunication**
>
> The trail of development of telecommunication from the days of Maxwell to modern radio, television and mobile phone is not continuous. In 1912, a famous ship named Titanic sank in the Atlantic that Marconi's wireless signal could span in 1901. Titanic possessed on board wireless transmitter but that failed to contact the nearby ships. It was not a mere accident but this mishap revealed that the then existing knowledge on wireless transmission was insufficient to build up a universal 'grammar' of telecommunication that we have achieved today. Now one can easily access foreign radio broadcasting or can send email internationally. In those days, there was no device for continuous amplification of the weak electromagnetic wave propagating in space. The importance of triode (1906) was not realized at large. The wireless technology could have remained captivated in its primitive stage, if electronic devices were not there into the picture with the features of rectification, amplification, nonlinearity and oscillation.

Based on the electric bulb (1883) constructed by American inventor Thomas Edison (1847–1931), British engineer John Fleming (1849–1945) devised (1904) a vacuum tube equipment containing two electrodes and got a unidirectional current flow with it. The device was termed as 'diode' because of its two electrodes and was named 'valve' because its action was similar to the valve controlling liquid flow. The process of allowing current in a single direction through this device was called rectification. Very soon, American inventor Lee de Forest (1873–1961) incorporated (1906) a remarkable change in the vacuum tube equipment by introducing a third electrode. The tri-electrode vacuum tube device, popularly known as 'triode' could amplify electrical power. Thus a new dimension was added to the use of electricity.

1.2.2 Trends of Development

Jacob Millman (1911–1991), a renowned writer of reference books on electronics narrated (Jacob Millman and Arvin Grabel, *Microelectronics*, 2nd *Ed.*, Tata McGraw-Hill, India, 2009) the perpetuation of electronics in the group of four 'C's.

- C for Components
- C for Communications
- C for Computation, and
- C for Control

It is stated earlier that wireless telecommunication was the first significant application of electronics. Lee de Forest, the inventor of triode himself founded the first radio station (1916) in the USA and very rapidly radio broadcasting was commercialized.

After the invention of transistor (1948) by American physicists William Shockley (1910–1989), John Bardeen (1908–1991) and Walter Brattain (1902–1987), semiconductor devices replaced rapidly valve-made circuits. The nonlinear current–voltage relationship and other properties similar to those of vacuum devices were obtained with semiconductor junctions also (chapters 3, 5 and 9). Moreover, the conductivity of the material could be changed easily by adding other materials as impurity, the process being known as *doping* (Chapter 2). Thus semiconductor-made components were being developed for various purposes.

It is peculiar but a fact that the Second World War had a strong influence on the development of electronics. A lot of electronic components and controlling devices, such as radar, metal-oxide-semiconductor field-effect transistor (MOSFET) (Chapter 9), operational amplifier (Chapter 11), modern computer and other digital systems were developed during that period for the sake of war equipment. Efforts were being made all the way to fabricate miniaturized circuits suitable for missiles and artificial satellites. American engineer Jack Kilby (1923–2005) first achieved the successful fabrication of integrated circuit (IC) (Chapter 12) in 1958. The word 'microelectronics' became popular after the invention of IC. Today's unlimited series of control systems ranging from domestic remote controls to guided missiles cannot be imagined without ICs.

The concept of computation with instruments is centuries old. All the computing equipment, from the ancient *abacus* to the first computer named *difference engine* constructed by Charles Babbage were mechanical instruments. The perception of computer changed radically during the electronic era. Instead of mechanical structures, the computer appeared as an assemblage of electronic circuits, first of vacuum tube devices and then replaced with transistors and ICs. The development of electronics changed not only the computer but the whole concept of instrumentation, recording and measurement in scientific and technological applications and consumer products. The expansion of mercury in thermometer, photographic films with chemical emulsion, acoustic musical instruments and many other appliances have got an electronic alternative.

1.3 Widespread Applications

Electronics has developed as a combination of science and technology. The construction of electronic devices and circuits is based on the concept of electrical engineering while the background theories have originated from fundamental physical and chemical sciences, and mathematics. The theory and application of electronic devices have enriched other theoretical and experimental sciences and technologies and have given birth of many new ideas. Some are introduced here.

- *Negative Resistance*: The presence of resistance in a current carrying circuit causes some voltage drop across it. Therefore, one might expect that any 'negative resistance', if possible, would establish an amplification of voltage. Electronic devices have realized the concept of negative resistance. Some special electronic devices, such as Gunn diode, tunnel diode and IMPATT diode (all beyond the present scope of the book) can actually establish a negative resistance in a circuit. A few words on negative resistance will be stated in connection with the analysis of oscillators (Chapter 10).
- *Quantum Potential Well*: This is a theoretical model of quantum mechanics, which can be established experimentally with the technology of electronic devices.
- *Semiconductor Junctions*: In addition to the prevalent uses like rectification and voltage regulation and light emission (Chapter 3), the junction can be used as radiation detector in nuclear physics research. The presence of defects in the crystal structure can distort the simple energy band structure of a semiconductor (Chapter 2) and give rise to additional energy states, which can be investigated by studying the electrical properties of semiconductor and metal–semiconductor junctions.
- *Mathematical Modeling*: The studies on carrier transport in electronic devices have given rise to useful mathematical techniques, such as pseudopotential method, dangling bond treatment and filter transfer function, and many others, which are useful to other branches of physical science also.

Collaboration of electronics with other subjects have launched important interdisciplinary subjects. A few are elicited here.

- *Optoelectronics*: It is concerned with devices, such as light-emitting diode, laser, photodiode and solar cell that are meant for optical to electrical conversion and vice versa. Propagation of optical signal through glass or plastic fibers, known as fiber optical communication is a promising field of applications.
- *Information Technology*: It is a fusion of electronics, computer science, telecommunication, and statistics, and is concerned with data management, data mining, data-compression, cryptography, and many efficient transfer of information.
- *Medical Electronics*: Instruments, such as ECG, brain scanner, digital glucometer, digital sphygmomanometer, and laser eye surgery are electronic and optoelectronic instruments devoted to medical treatment and biodiagnostics.
- *Remote Sensing and Satellite Communication*: The techniques for observing the earth from space and spanning the continents through telephone and computer network acquired maturity with electronics.
- *Nanotechnology*: Collaborative researches on materials science and device fabrication technique have made it possible to construct electronic devices at molecular scale. Instead of 'microelectronics', the term 'nanoelectronics' is becoming popular. The nanoelectronic devices are produced with new fabrication techniques and these have size-dependent properties. The same material and its nanostructure counterpart, such as silicon and porous silicon, carbon and carbon nanotube are different. The nanostructure may exhibit novel electrical, optical or mechanical properties that are not available at macroscale.

As reviewed above, electronics has enriched science laboratories, technologies, and daily life. It has influenced human lifestyle and has become a significant component of today's economy, society, and human thought. It is not unreasonable to conclude as:

- if the invention of wheel and the discovery of fire have sublimed ape to man,
- if printed letter has given rise to renaissance,
- if steam engine has initiated industrial revolution,
- electronics has certainly created globalization of human civilization.

Perhaps one may add another 'C' to Millman's list to state 'C for Culture'.

1.4 Electrons, Electricity and Electronics

The history of science is often stranger than science fiction. Without knowing the fact that electric current through metallic wire is nothing but the flow of electrons, mankind had invented battery, dynamo and motor. Also the prime laws of current electricity including Ohm's law and Kirchhoff's laws were propounded prior to the discovery of electron. Nevertheless, the electronic artefacts were constructed consciously using the flow of electron, first with vacuum tube devices, such as triode, and then with semiconductor-made devices, such as the early transistors. The operation of electronic systems is somehow associated with electricity or movement of electrons, but it incorporates the flow of charge carriers through vacuum or semiconductor structures and emission of electrons from solids (Section 1.4.3), which are quite different phenomena in comparison with the simple flow of electrons in metals defined by Ohm's law.

1.4.1 Electric Current

Electric charge is the basic entity that assigns electrical property to an object. When two charged conductors are connected, charge flows from one to the other, and both objects acquire a common potential. This is an instantaneous and uncontrolled process. Rubbing a silk with a glass rod, discharging a capacitor, and thundering are such examples. When a constant potential difference is maintained between the conductors, a steady flow of charge termed as *electric current* takes place between them. The electric current is the rate of flow of charge through any area of a conducting material under the influence of a potential difference and it depends on

- the conducting property of the material medium quantified in terms of its *resistivity* or *conductivity*,
- the geometry of the object realized in terms of *resistance*,
- the motional property of the electric charges expressed in terms of *drift velocity* and *mobility*, and
- the potential difference and the surrounding parameters, such as temperature.

> **Think a Bit: Current, Current Density and Vector**
>
> Electric current has a direction but it is not vector because it does not obey the laws of vector addition. The sum or difference of 2 mA and 5 mA produces 7 mA or 3 mA only with the proper sign and nothing else. Current density is but a vector quantity because it describes how the charges flow at some point across a certain area. The vector direction informs the direction of charge flow at a point along a particular orientation of cross-sectional area. The current through a single conductor is the same at its all points but the current density increases and decreases according to the cross-sectional area of charge flow.

In the case of metallic conductors, the charge carriers are electrons only. Since the electrons are negatively charged, the direction of current is considered opposite to the direction of electron flow. In gases and electrolytes, both positive and negative ions contribute to the current but the net current occurs in a single direction; along the movement of positive charges. In vacuum, the electron flows in the form of a beam or stream that conducts current. Vacuum tube electronic devices utilize this process. Yet there is another material, named semiconductor, where the absence of electrons at a specific energy state is considered as positive charge and is termed as hole. Both electrons and holes contribute to the current in semiconductor devices, such as diodes and transistors. The salient features of electronics, namely electrical power amplification, nonlinear current–voltage relationship and impedance transformation within the device were first realized with vacuum tube devices and very soon the devices made of semiconductor devices replaced those made of vacuum tubes.

1.4.2 Drift Velocity, Mobility and Conductivity

In absence of any voltage and consequent electric field, the free charge carriers of a conductor undergo random motion in all directions due to thermal agitation, so that the resultant velocity is zero. When a voltage is established across the conductor, the electric field accelerates the charge carriers along it. Simultaneously the random motion and collision with atoms oppose the acceleration. As a combined effect, the carriers drift along the field with an average velocity called the *drift velocity*. In the case of metals, the charge carriers are electrons, which drift toward the higher potential, opposite to the direction of the electric field.

If an electron of mass m and charge e having initial thermal velocity $\vec{u}$ be accelerated by an electric field $\vec{E}$, its final velocity $\vec{v}$ just before a collision can be expressed as

$$\vec{v} = \vec{u} + \vec{a}\tau \tag{1.1}$$

where τ is the *relaxation time*, the short time for which a free electron accelerates before it undergoes a collision and $\vec{a} = e\vec{E}/m$ is the acceleration. Summing over all the electrons, the initial velocities are cancelled out and the drift velocity can be expressed as

$$v_d = \frac{eE}{m}\tau \tag{1.2}$$

Only the magnitudes are considered and the vector notations are avoided. Equation (1.2) indicates that the drift velocity is proportional to the electric field.

Let a conductor of length l and uniform cross-sectional area A contains n number of free electrons per unit volume. Then the total charge present in the conductor is $nelA$ and the time (t) taken in traversing the length of the conductor is l/v_d so that the current (charge/time) caused by these electrons is

$$I = neAv_d \tag{1.3}$$

Equation (1.3) expresses the current through a conductor in terms of electron drift velocity and the area of cross-section of the conductor.

The drift velocity acquired under unit electric field applied across the conductor is termed as *mobility* (μ) given by

$$\mu = \frac{v_d}{E} \tag{1.4}$$

where E is the magnitude of the electric field. The SI unit of mobility is m²V⁻¹s⁻¹. Using Equations (1.3) and (1.4),

$$I = ne\mu AE \qquad (1.5)$$

Equation (1.5) expresses the current in terms of carrier mobility. Using Equations (1.2), (1.4) and (1.5), we have

$$I = \frac{e^2 nAE\tau}{m} \qquad (1.6)$$

If V be the voltage applied across the conductor of length l, the electric field (E) is V/l. Then equating Ohm's law with Equation (1.6) the resistance (R) can be obtained as

$$R = \frac{V}{I} = \frac{m}{ne^2\tau} \frac{l}{A} \qquad (1.7)$$

From Equation (1.7), the resistivity (ρ) of the material (when $l = 1$, $A = 1$, $R = \rho$) can be identified as

$$\rho = \frac{m}{ne^2\tau} \qquad (1.8)$$

From Equations (1.6) and (1.8), the magnitude of the current density can be derived as

$$J = \frac{I}{A} = ne\mu E = \frac{E}{\rho} \qquad (1.9)$$

so that

$$\rho = \frac{1}{ne\mu} \qquad (1.10)$$

Also from Equations (1.4) and (1.9),

$$J = nev_d \qquad (1.11)$$

> **Think a Bit: Which Parameters Drift Velocity Depends On?**
>
> Equation (1.2) states that the drift velocity (v_d) of carriers in a conductor is proportional to the electric field (E) across it. However, using Equations (1.9)–(1.11) you will be able to derive the following expression for v_d.
>
> $$v_d = \frac{V}{ne\rho l}$$
>
> This implies that the drift velocity also depends on the geometry of the conductor. As the length (l) increases, the drift velocity decreases.

The SI unit of resistivity is Ωm. The inverse of resistivity is called conductivity (σ), which can be obtained from Equation (1.10) as

$$\sigma = ne\mu \tag{1.12}$$

The SI unit for conductivity is siemen (S).

Another useful relation can be derived from Equations (1.9) and (1.12) as

$$J = \sigma E \tag{1.13}$$

Equation (1.13) represents another form (vector form) of Ohm's law that has general validity. The familiar relationship $V = IR$ for bulk current is not valid for semiconductors. Equation (1.13) holds good at micro level. All the above concepts on drift velocity, mobility and conductivity are valid for semiconductors also. However, there exist the contributions of two types of carriers, electrons and holes, which have different values of mobility and other parameters. These topics are explained in detail in Chapter 2.

1.4.3 Electron Emission from Metal

Some electronic devices and related systems require the electrons to physically come out of the metal kept in vacuum. The outermost electrons in a metal are loosely associated with the ion core and can move freely within the metal boundary leading to its high conductivity. However, the electrons cannot escape the metal surface under ordinary conditions because of the net inward attraction of the positive ions. The electron must gain sufficient energy to overcome the attractive force. The highest energy level possessed by the electron at absolute zero is called the *Fermi energy*. The minimum amount of energy that must be given to the electron at the Fermi energy level so that it can escape the metal surface is called the *work function* of the metal. Such electron emission from metal surface can be caused by different energy sources, as stated below.

(i) *Thermionic Emission*: Thermal energy is supplied to the electrons by heating the metal in vacuum. The vacuum tube electronic devices like diode and triode utilize thermionic emission for generating electron flow in vacuum.

(ii) *Photoelectric Emission*: The metal is exposed to electromagnetic radiation of frequency (υ) greater than a certain critical value such that the photon energy (hυ) is greater than the work function. The emission is enhanced with the intensity of radiation. An optoelectronic device named photomultiplier tube emits electrons by this process.

(iii) *Field Emission*: This is also termed as cold cathode emission. At room temperature, electrostatic energy is supplied by applying a strong electric field caused by high positive voltage. Stronger electric field gives rise to greater emission. Electron microscope and carbon nanotube employ this technique.

(iv) *Secondary Emission*: Electrons or ions moving with high kinetic energy colliding with metal surface transfers the energy to the electrons inside the metal so that the electrons escape the metal surface. The scanning beam of electron microscope is an example of this process.

1.5 Circuits and Sources

Electrical circuits are indispensable part of electronics. The circuits with electronic components have some characteristics different from that with conventional electrical circuits. For instance,

- current is sometimes used as the independent variable in lieu of voltage (e.g. transistor operation in Chapter 5),
- different currents exist in different parts of the same closed loop (e.g. transistor biasing in Chapter 6),
- parameters of mixed dimensions are used in the same circuit (e.g. hybrid parameters in Chapter 7, feedback in Chapter 10) and
- one component functions as that of another; for instance, in the case of active pull-up (Chapter 14), a transistor might behave as a resistor.

Nevertheless, the basic concepts on circuits remain the same. This section introduces some circuit features relevant to electronic circuits used in this book.

1.5.1 Types of Circuits

An electric circuit is a closed conducting path through which electric current flows under the action of a source of electromotive force (emf). Any particular portion of the circuit may or may not contain voltage sources. An electronic circuit is also a category of electric circuits comprising components producing nonlinear current–voltage variation. Depending on the components and operations, circuits can be grouped into various categories. Few examples are cited here.

(i) *Resistive and Reactive Circuits*: Circuits containing resistors and power supply are called *resistive* or *dissipative* circuits. These dissipate energy across the resistors in the form of heat. The mathematical expressions describing the circuit performance are algebraic in nature. The transistor biasing circuits (Chapter 6) are of this category. A circuit containing capacitors and/or inductors is called *reactive* circuit. It can store energy and the mathematical description of the circuit operation involves integration and differentiation processes. The active filters in Chapter 11 are such examples.

(ii) *Linear and Nonlinear Circuits*: A linear circuit contains resistors, capacitors or inductors that have linear current–voltage relationship and obey the principle of superposition. In general, electronic circuits typically have nonlinear current–voltage variation. However, these can be made to act as linear under certain conditions. The diode rectifier for large ac input (Chapter 4) and the transistor amplifier for small ac input (Chapter 6) are two such examples.

(iii) *Active and Passive Circuits*: A circuit containing one or more sources of emf is called an *active* circuit because it can deliver energy. Diodes and transistors are called active components because these can be represented by equivalent voltage or current source. Section 1.6 discusses more on this matter. A circuit containing no emf source is called a *passive* circuit because it only consumes and dissipates energy. Resistors, capacitors and inductors are passive components.

(iv) *Time-Varying and Time-Invariant Circuits*: Time-varying circuits are those in which the resistance, capacitance, etc. get changed with time. For instance, the winding resistance of an electric motor changes with temperature. A time-invariant circuit does not show any change in voltage and current with time. The transistor with dc biasing (Chapter 6) is ideally a time-invariant circuit.

(v) *Bilateral and Unilateral Circuits*: A bilateral circuit works in the same manner in both directions. The input and output terminals can be interchanged. A Wheatstone bridge with four resistor is such an example. The unilateral circuit works in a single direction and its performance gets changed on changing the direction of supply voltage and input/output terminals. Electronic circuits containing diode, transistor, etc. are unilateral in nature.

(vi) *Lumped and Distributed Circuits*: The passive components, such as resistors and capacitors of common use are discrete elements and their properties are 'lumped' at a particular place of the circuit. These elements are joined with connecting wires and the lengths of the wires do not matter. All the circuits in this book are lumped circuits. However, in the case of long wires of transmission cable or high tension power line, the huge length does matter. In such a case, the resistance, capacitance or inductance is supposed to be distributed uniformly over the entire length of the wire. Here the concept of capacitance and conductance per unit length is adopted.

1.5.2 Voltage and Current Sources

A circuit requires an electrical power supply that delivers electrical energy to the circuit components for working. It may be a chemical battery or an electronic power supply. It is called an independent source because it does not depend on the circuit to which it supplies power. Each source has a certain upper limit of delivering power and it is of either of the following two types.

(i) *Voltage source*: It can maintain a constant voltage (DC or AC) across its terminals irrespective of the load or the current delivered. A good quality chemical cell having negligible internal resistance can be approximated as a voltage source. Also such sources are generated with electronic circuits (e.g. regulated power supply in Chapter 8).

(ii) *Current source*: It can supply a constant current (DC or AC) to any load irrespective of its value and the voltage across it. Such sources are realized with electronic circuits only.

In this book, all the power supplies are assumed to be voltage sources unless otherwise stated. The symbols of voltage and current sources are given in Figure 1.1 and a comparison of their characteristics are given in Table 1.1.

A practical voltage source, as indicated in Figure 1.2, possesses a small but finite internal resistance (R_s) appearing in series so that the output voltage (V_o) developed across the external load resistor (R_L) is

$$V_o = \frac{VR_L}{R_s + R_L} \quad (1.14)$$

Equation (1.14) implies that when $R_s \ll R_L$, $V_o \approx V$ and the voltage source approaches the ideal condition. Similarly, as shown in Figure 1.2, a practical current source possesses a very high but finite internal resistance (R_p) appearing in parallel. The output current (I_o) flowing through the external load resistor (R_L) is

$$I_o = \frac{IR_p}{R_p + R_L} \quad (1.15)$$

Equation (1.15) infers that when $R_p \gg R_L$, $I_o \approx I$ and the current source approximates the ideal state.

The voltage and current sources are mutually convertible. A constant voltage source can be converted to an equivalent constant current source and vice versa. When the voltage source (V) of Figure 1.2 is short circuited, it becomes a current source supplying a current (V/R_s) with parallel resistance R_s. Similarly, when the current source (I) of Figure 1.2 is open circuited, it becomes a voltage source supplying a voltage (IR_p) with series resistance R_p.

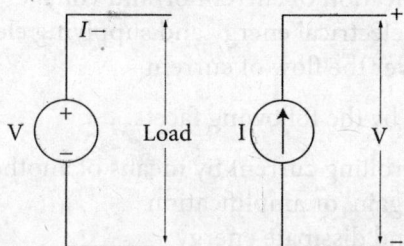

Figure 1.1 Symbolic representation of ideal voltage source (V) (left) and ideal current source (I) (right)

Table 1.1 Comparison of voltage source and current source

(i) An ideal voltage source has zero internal resistance.	(i) An ideal current source has infinite internal resistance.
(ii) It supplies a constant voltage independent of the current flowing out of it.	(ii) It supplies a constant current independent of the voltage across its terminals.
(iii) The current flowing out of the source is determined by the circuit connected to it.	(iii) The magnitude and polarity of the voltage across the current source is determined by the circuit connected to it.
(iv) Examples: a good quality chemical cell, a transistor used as emitter follower (Chapter 6).	(iv) Examples: a transistor in active region (Chapter 5), a junction field-effect transistor in saturation region (Chapter 9).

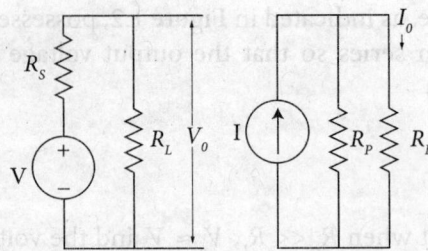

Figure 1.2 Practical voltage source (V) (left) with internal resistance (R_s) and external load resistance (R_L). Practical current source (I) (right) with internal resistance (R_p) and external load resistance (R_L).

1.6 Active and Passive Device

All the devices and components related to passage of electricity can be classified broadly into two categories: active and passive. This makes another demarcation between 'electric' and 'electronic'. An electronic circuit should contain at least one active component.

Active devices are those comprising one or more of the following features.

- Nonlinear current–voltage relationship under steady state
- Some sort of amplification of current or/and voltage
- Acting as source of electrical energy and supplying electrical power to a circuit
- Electrical control over the flow of current

Passive devices are identified by the following facets.

- Not capable of controlling current by means of another electrical signal
- Not providing any 'gain' or amplification
- Can receive, store and dissipate energy
- Exhibit linear current–voltage relationship when connected to steady DC/AC source.

Resistors, capacitors and inductors are passive components. These produce the same linear current–voltage variation on reversing the polarity of the voltage source. These cannot generate voltage and current and cause only a voltage drop proportional to the current. The resistor dissipates energy in the form of heat, the capacitor stores energy in the electric field and the inductor stores energy in the magnetic field.

Electronic components are different from all the above. Even the simplest diode, as will be seen in Chapter 3 is an active component because the current through it depends on the polarity of the power supply and the equivalent circuit of the diode consists of a voltage source. Not only the diode, most of the electronic devices and systems, namely bipolar transistors (Chapters 5–8), field-effect transistors (Chapter 9), operational amplifiers (Chapter 11) and logic gates in digital circuits (Chapters 14 and 15) comprising p-n junction as the basic building block are categorized as active devices.

> **Think a Bit: Is the Transformer a Passive Device?**
>
> Surprisingly yes! The transformer basically contains two coils (inductors). Though it raises the voltage or the current level, the power is kept constant and that too in the ideal case of no dissipation. Moreover the current is controlled magnetically. The primary and the secondary are electrically isolated.

Multiple Choice-Type Questions and Answers

1.1 The applied voltage across a variable resistor is constant. The graph between the current and the resistance is ....
   - (a) straight line with positive slope
   - (b) straight line with negative slope
   - (c) parabola
   - (d) rectangular hyperbola

1.2 The physical entity having the dimensions $[M^{-1}L^{-3}T^3A^2]$ is
   - (a) resistance
   - (b) resistivity
   - (c) conductivity
   - (d) mobility

1.3 The variation of the resistance (R) of a metallic wire with its diameter (D), keeping the length and temperature fixed is best represented by ....
   - (a) R proportional to $1/D$
   - (b) R proportional to $1/D^2$
   - (c) R proportional to $D/2$
   - (d) R proportional to D

1.4 A voltage source of 12 V and series resistance of 2 Ω can be converted to an equivalent current source of ....
   - (a) 6 A in parallel with a resistance of 2 Ω
   - (b) 6 A in series with a resistance of 2 Ω
   - (c) 6 A in parallel with a resistance of 24 Ω
   - (d) 6 A in parallel with a resistance of 24 Ω

1.5 A current source of 12 A and parallel resistance of 2 Ω can be converted to an equivalent voltage source of ....
   - (a) 24 V in series with a resistance of 6 Ω
   - (b) 24 V in parallel with a resistance of 6 Ω
   - (c) 24 V in series with a resistance of 2 Ω
   - (d) 24 V in parallel with a resistance of 2 Ω

1.6 The electrical resistance in a resistor originates from ....of electrons.
   - (a) thermal agitation
   - (b) random motion
   - (c) collisions
   - (d) all the above

1.7 The capacitance of a capacitor indicates its capacity of
   - (a) storing charge
   - (b) storing energy in electric field
   - (c) storing energy on the electrodes
   - (d) increasing the voltage

1.8 The inductance of an inductor is associated with ....
   - (a) inducing emf in a circuit due to change in current
   - (b) opposing the change in current
   - (c) storing energy in the magnetic field linked with the circuit
   - (d) all the above

1.9 In terms of fundamental units, 'ohm' can be expressed as ....
(a) $kgm^2s^{-3}A^{-2}$ (b) $kgm^{-2}s^3A^2$
(c) $kgm^2s^3A^{-2}$ (d) $kgm^{-2}s^{-2}A^{-2}$

1.10 A high tension supply (of the order of kV) should have .... internal resistance to restrict the current within a safe limit whereas a low voltage supply (say 5 V) should have .... internal resistance to increase the output current for a certain voltage.
(a) large, small (b) small, large
(c) moderate, large (d) small, moderate

Answers

1.1	(d)	1.2	(c)	1.3	(b)	1.4	(a)
1.5	(c)	1.6	(d)	1.7	(b)	1.8	(d)
1.9	(a)	1.10	(a)				

Reasoning-Type Questions and Answers

1.1 The drift velocity of electrons is very small. Indeed the electron moves less than a meter in an hour. Then how does an electric bulb at a long distance start glowing as soon as the switch is turned on?

Ans. The individual electrons move with the drift velocity but the electric field that drives the free electrons travel with a speed almost equal to that of light. Thus, on turning on the switch, all the electrons along the wire start moving at the same time and the current produced by the movement of the electrons nearest to the bulb turns it glowing.

1.2 How is a constant magnitude of current maintained through a wire of non-uniform cross-section?

Ans. The drift velocity and the current density adjust themselves so as to keep the current unchanged. If the cross-sectional area increases, the drift velocity decreases [Equation (1.3)] and when the drift velocity decreases, the current density decreases [Equation (1.11)].

Solved Numerical Problems

1.1 A 1.5 V rechargeable cell is labeled 1800 milliamp-hours. What is the maximum amount of electrical energy stored in it?

Soln. The charge stored in the cell is 1800×10^{-3} coulomb per second $\times 60 \times 60$ second
= 6480 coulomb

The total electrical energy is 6480 C × 1.5 V = 9720 J

[N.B. This is the maximum possible amount of electrical energy stored in the cell. Now it is up to the user whether the charge will be moved in the form of 1 mA steady current for 1800 hours or 1800 mA steady current for 1 hour or some intermediate values.]

1.2 A copper wire of diameter 1 mm carries a constant current of 5 A to a 100 W lamp. Calculate the magnitude of drift velocity of electrons. Copper has atomic mass of 63.5 unit and density of 9.0×10^3 kgm^{-3}.

Soln. The area of cross-section is $\pi \times (1 \times 10^{-3})^2/4 = 0.786 \times 10^{-6}$ m^2.

Assuming one free electron per atom of copper, the density of free electrons is

$$\frac{6.023 \times 10^{23} \times 9.0 \times 10^3 \times 10^3}{63.5} = 8.536 \times 10^{28} \text{ per cubic meter}$$

The current density is $\dfrac{5A}{0.786 \times 10^{-6} m^2} = 1.272 \times 10^6$ Am^{-2}

The drift velocity is [using Equation (1.11)] $\dfrac{1.272 \times 10^6}{8.536 \times 10^{28} \times 1.6 \times 10^{-19}} = 9.31 \times 10^{-5}$ ms^{-1}.

[Look how small it is!]

1.3 An electronic circuit with 10000 components performs its intended function successfully with a probability 0.99 if there are no faulty components in the circuit. The probability that there are faulty components is 0.05. If there are faulty components, the circuit performs successfully with a probability of 0.3. The probability that the circuit performs successfully is x/10000. What is x?

[JEST 2018]

Soln. Let Event A represents no faulty component, Event B represents that the circuit functions successfully and Event C represents that there are faulty components.

P(B|A) = 0.99
P(B|C) = 0.3
P(C) = 0.05
P(A) = 1 − P(C) = 0.95
P(B∩C) = P(C) P(B|C) = 0.05×0.3 = 0.015
P(B∩A) = P(A) P(B|A) = 0.95×0.99 = 0.9405
P(B) = 0.015 + 0.9405 = 0.9555 = x/10000
Therefore, x = 9555

1.4 In the circuit of Figure 1.3, each device D may be an insulator with probability p, or a conductor with probability (1 − p). The probability that a non-zero current flows through the circuit is

(a) $2 - p - p^3$
(b) $(1 - p)^4$
(c) $(1 - p)^2 p^2$
(d) $(1 - p)(1 - p^3)$

[NET June 2019]

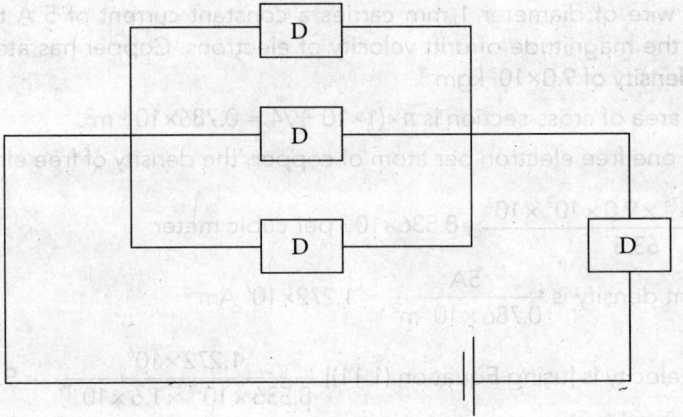

Figure 1.3

Soln. The current can flow, if the single right side device be a conductor with probability
$$p_1 = (1 - p)$$
and any one or more of the three parallel devices be conductor with probability
$$p_2 = (1 - p)(1 - p)(1 - p) + {}^3C_2 p(1 - p)(1 - p) + {}^3C_1 p^2(1 - p)$$
$$= (1 - p^3)$$
The total probability is
$$p_1 p_2 = (1 - p)(1 - p^3)$$

Exercise

Subjective Questions

1. Mention the significant new dimensions that electronics added to the regime of electricity.
2. Introduce how electronics has enriched the fields of science and technology during the last one century.
3. Show that the resistivity of a material decreases on increasing the free electron density in it.
4. Define the drift velocity of charge carriers. Derive a relation between the current flowing through a conductor and the drift velocity of free electrons in it.
5. What is the mobility of charge carrier? Find out a relationship of the mobility with the resistivity of the material.
6. Show that the units on the two sides of the equation $J = \sigma E$ are identical, symbols having their usual meaning.
7. Mention different techniques for causing electron emission from metal surfaces with the application in electronic devices/equipment.
8. Mention some notable features of the circuits with electronic components that make those distinct from the conventional electrical circuits.

Conceptual Test

1. Compare the flow of water from a tank through a tap with the flow of steady current.
2. Search two practical applications involving high voltage, low current and low voltage, high current.
3. Can electric current exist outside a material medium?
4. While cooking, metallic utensils are heated on oven. Does thermionic emission take place?

Numerical Problems

1. A steady current of 1 A is flowing through a conductor. How many electrons are passing through any cross-section in 16 seconds?
2. A rectangular block of length 5 cm, width 5 mm and thickness 0.5 mm carries a current of 2.5 mA. If the free electron density is 10^{15} cm^{-3}, find out the time taken by the electrons to traverse the full length of the sample.
3. Overhead power cable is required for a length of 10 km. If cables of the same length (l) and the same resistance (R) are manufactured with aluminum (Al) and copper (Cu), which one has less weight? [relative density of Al = 2.7 and that of Cu = 8.9, resistivity of Al = 2.63×10^{-8} Ωm and that of Cu = 1.72×10^{-8} Ωm]
4. For a given load resistance R_L = 4.7 ohm, the power transfer efficiencies ($\eta = P_{load}/P_{total}$) of a dc voltage source and a dc source of current with internal resistances R_1 and R_2, respectively, are equal. The product $R_1 R_2$ in units of ohm^2 (rounded off to one decimal place) is ....

[GATE 2019]

2
Semiconductor Fundamentals

This chapter introduces the types of semiconductors and their electrical properties. The origin of two types of charge carriers, namely electrons and holes in semiconductors are explained. The related theoretical concepts of energy band, Fermi energy and effective mass are introduced. The carrier transport in semiconductors causing drift and diffusion currents are discussed. Two important experimental techniques, namely Hall Effect and four-probe resistivity measurement are deliberated.

2.1 Crystalline Solids

A broad field of studies, known as *condensed matter physics* encompasses the macroscopic and microscopic physical properties of matter, mainly in the solid and liquid phases, being 'condensed' due to the electromagnetic forces between atoms. A part of this is *solid state physics* that studies the atomic level and bulk level properties of matter in its solid state. A solid is distinguished from liquid or gas by its definite shape and rigidity. This is because of the atoms or the molecules of a solid being closely packed and strongly bound. However, the atoms in a solid may or may not be in regular arrangement, which determines whether or not the solid is crystalline.

A *semiconductor* is generally a solid material having electrical conductivity level somewhere between that of conductors (e.g. metals) and insulators (e.g. ceramics). The conducting properties of semiconductors can be altered by application of external electrical fields, light and heat and by a process of incorporating impurities, known as *doping*. Devices made from semiconductors, such as diode, transistor and logic gates are the basic building blocks for modern electronic circuits. Before going through the properties of semiconductors, it is reasonable to get a brief idea on the fundamental properties of a crystalline solid.

A solid is called *crystalline*, if the atoms and the molecules in it are organized in symmetric arrays in three dimensions. A definite grouping of atoms is repeated periodically in three dimensions. If the crystal structure is perfect and the same regular arrangement is extended throughout the entire solid, it is called *single crystal* and if the periodicity is found to be intermittent and confined within a small region, it is called *polycrystalline* material. Solids not having such regular atomic arrangement are termed as *amorphous*. These have only a short range regularity among the nearest neighbor atoms. Metals, quartz, sodium chloride, calcium sulfate, etc. are some examples of crystalline solids, and calcium carbonate and glass are amorphous materials. Semiconductors are found in both single crystal and polycrystalline forms. Amorphous semiconductors also occur with limited use. A few technical terms related to crystals are introduced in the subsequent sections.

2.1.1 Lattice, Basis and Unit Cell

The study of crystal structures is known as crystallography. It has two major trends, direct observation through electron microscopy and indirect estimation with x-ray diffraction pattern. Such studies assume a model for the crystal structure consisting of the following terms.

Lattice: It is a schematic representation of the crystal when all the atoms are replaced by points so that a three-dimensional array of points is obtained.

Basis: It is the fundamental arrangement of atoms that is repeated in space along with the lattice. In fact adding atoms to the lattice is a basis that can be one atom or a group of atoms. In the lattice of an element, the basis is of single atom.

Unit Cell: This is the minimum geometric configuration of lattice points that is repeated in space so as to construct the whole lattice. The choice of unit cell is not unique. The cell containing just one lattice point is called 'primitive unit cell'.

Figure 2.1(a) explains the separate role of lattice and basis with the example of sodium chloride (NaCl) crystal. The atoms of both sodium (Na) and chlorine (Cl) are represented by points in the lattice at regular interval. At the same time, the combination of one Na atom and one Cl atom (more precisely two ions, sodium cation and chlorine anion) in the NaCl molecule is the basis, which is repeated throughout the crystal. The combination of the lattice and the basis constitute the actual crystal structure. Figure 2.1(b) interprets the significance of a unit cell. Two different repetitions of lattice points are indicated, one with the squares and the other with the parallelograms. Both are reasonable and each of these can represent the lattice independently.

In lieu of the pictorial representation, the lattice can be mathematically represented in the following way. Let vectors $\vec{a}$ and $\vec{b}$ join two lattice points to an arbitrary origin. Then any other lattice point can be defined as vector $\vec{c}$ with respect to the origin as

$$\vec{c} = n_1 \vec{a} + n_2 \vec{b} \tag{2.1}$$

where n_1 and n_2 are integers representing the extent of translations $\vec{a}$ and $\vec{b}$ directions, respectively. More realistic is the case of a three-dimensional lattice where one can represent an arbitrary lattice point $\vec{r}$ as

$$\vec{r} = n_1 \vec{a} + n_2 \vec{b} + n_3 \vec{c} \tag{2.2}$$

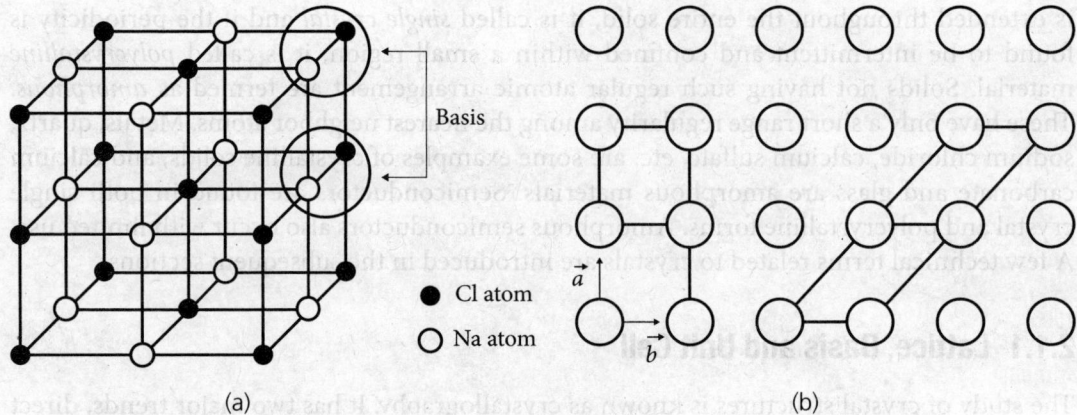

Figure 2.1 (a) Lattice representation of sodium chloride (NaCl) crystal with indication of the basis comprised of Na and Cl atoms; (b) Formation of two different unit cells in a two-dimensional lattice

in terms of integers, n_1, n_2 and n_3 representing lattice translation in three-dimension. The three vectors $\vec{a}$, $\vec{b}$ and $\vec{c}$ stand for the three sides of the unit cell and are termed as the *basis vectors* for the lattice.

It is understood from the above schematic representation that the crystal structure remains the same for translation through a vector like $\vec{r}$ that is the sum of integral multiples of the basis vectors. Therefore these are also called *crystal axes* and the angles between these axes are called *interfacial angles*. It is also obvious from the above discussion that there may be many choices for such sets of vectors and angles. Those are standardized as Bravais lattice introduced in the next section.

2.1.2 Bravais Lattice and Miller Indices

French physicist Auguste Bravais (1811–1863) introduced the concept of lattice and identified fourteen possible categories of lattice of seven different types based on the plane, axes and center of symmetry. These are summarized in Table 2.1. A cubic lattice is often a good model to understand the lattice arrangement. A *simple cubic* lattice structure has one atom at each of the eight corners of the cube. A *body-centered cubic* lattice has all these eight atoms plus one atom at the center of the cube. A *face-centered cubic* lattice has all these eight atoms plus one atom at the center of each of the six faces of the cube. Consequently the simple, body-centered and face-centered lattice has, respectively 1, 2 and 4 atoms per primitive cell. The NaCl lattice sketched in Figure 2.1(a) is an example of face-centered cubic lattice. Carefully observing the structure, it is understood that there exists an atom at each face and at each corner for either Na or Cl atoms.

Semiconductor Fundamentals

Table 2.1. Fourteen Bravais lattices for three-dimensional crystals

Lattice type	Specifications for crystal axes (a, b and c) and angles (α, β and γ)	Possible variations	Example
Cubic	$a = b = c$; $\alpha = \beta = \gamma = 90°$	3	NaCl
Trigonal or Rhombohedral	$a = b = c$; $\alpha = \beta = \gamma \neq 90°$	1	$CaSO_4$
Tetragonal	$a = b \neq c$; $\alpha = \beta = \gamma = 90°$	2	SnO_2
Hexagonal	$a = b \neq c$; $\alpha = \beta = 90°$, $\gamma = 120°$	1	ZnO
Orthorhombic	$a \neq b \neq c$; $\alpha = \beta = \gamma = 90°$	4	$BaSO_4$
Monoclinic	$a \neq b \neq c$; $\alpha = \beta = 90°$, $\neq \gamma$	2	$FeSO_4$
Triclinic	$a \neq b \neq c$; $\alpha \neq \beta \neq \gamma \neq 90°$	1	$K_2Cr_2O_7$

Similar to the crystal axes defined above, another important requisite in the study of crystals is defining a set of parallel crystal planes. British mineralogist W. H. Miller (1801–1880) put forward a notation system, now renowned as the *Miller indices*, to denote the orientation of crystal planes in terms of their intercepts with the three crystal axes. The method of assigning the Miller indices to a crystal is to take the reciprocals of the intercept units to the crystal axes and to convert those to the smallest whole numbers by using a common multiplier in the following steps.

1. Take any atom in the crystal as the origin.
2. Draw coordinate axes from this atom in the directions of the basis vectors.
3. Find the intercepts of the plane.
4. Take the reciprocals of these numbers and reduce to the smallest integers h, k, l, having the same ratio.

The orientation of crystal planes is specified by a set of the three whole numbers enclosed within brackets. Few examples of the Miller indices for simple cubic crystal of side length a:

(i) The crystal plane parallel to the yz plane having positive x-intercept at length a is denoted by (100). If the same plane intercepts at length a on the negative x-axis, it is ($\bar{1}$,0,0).

(ii) The crystal plane parallel to the xz plane and the xy plane having intercepts at length a on the y- and z-axis, respectively, (010) and (001).

(iii) Crystal planes having equal intercepts to the three perpendicular axes: (111)

Note: Symmetry Is Necessary in Lattice

It is implied from the Bravais lattices that there are many different ways of arranging atoms in three dimensions and the distances and orientation between the atoms can be of diverse forms. It is noteworthy that there is no pentagonal lattice whereas tetragonal and hexagonal lattices exist there. It is so because some symmetry is necessary in the translation and rotation of the lattice so that a basic structure can be repeated throughout the volume of the material.

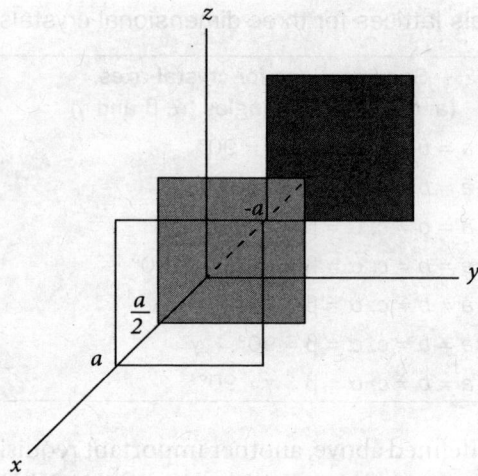

Figure 2.2 Example of determining Miller indices

> **Example 2.1**
>
> Figure 2.2 displays three planes, each parallel to the yz plane having intercepts of a (white), $a/2$ (light shaded) and $-a$ (deep shaded), respectively, on the x-axis. Determine the Miller indices for these three planes.
>
> **Solution**
>
> For x-intercept $= a$, the reciprocals are $1/a$, $1/\infty$, $1/\infty$ so that the Miller indices are $(1,0,0)$.
> For x-intercept $= a/2$, the reciprocals are $1/(a/2)$, $1/\infty$, $1/\infty$ and Miller indices $(2,0,0)$
> For x-intercept $= -a$, the reciprocals are $1/(-a)$, $1/\infty$, $1/\infty$, Miller indices $(\bar{1},0,0)$

2.2. Energy Band

According to Bohr's atomic model, the electrons in an isolated atom possess discrete energy levels separated by energy gaps. Such isolation of atoms may partly occur in gases but not in solids where the interatomic forces of ionic or covalent bonding hold the atoms closely spaced. The electrons in the outer orbit of an atom are influenced by the neighboring atoms. Subatomic particles like electrons have an associated wave aspect. So the electrons in solids get both their wave functions and energy states altered in presence of many interacting atoms. Following Pauli's exclusion principle, no two electrons in an interacting system of atoms may possess the same quantum states. Consequently the individual energy levels of the isolated atoms, instead of superimposing, co-exist very close to one another. Such a collection of closely spaced energy levels having very little difference ($\approx 10^{-19}$ eV) constitute a continuum of energy that is termed as *energy band*.

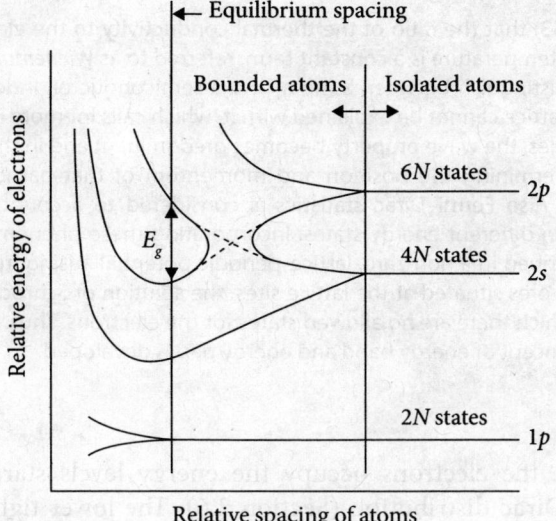

Figure 2.3 Formation of energy bands on bringing carbon atoms closer to one another

The process of energy band formation is outlined in Figure 2.3 with a simple model of carbon atoms. Let there be N number of isolated carbon atoms (atomic number = 6) in ground state. For each atom, the available energy levels are two $1s$ states, two $2s$ states and six $2p$ states so that there are total $2N$, $2N$ and $6N$ states, as sketched at the right side of Figure 2.3. Now if the interatomic separation is reduced, splitting of energy levels starts from the outer shell and the $2s$ and $2p$ bands get overlapped to provide with $2N + 6N = 8N$ available energy states. As the distance is further reduced and it reaches the equilibrium value of interatomic spacing for this particular crystal, the bands get separated into two bands, each having $4N$ states and separated by an energy gap. Among these three components,

- the higher energy band is called the *conduction band* (because the electrons must get accelerated and acquire higher energy for conduction),
- the lower energy band is called the *valence band* (because it represents the maximum obtainable energy by the valence electrons in absence of any external energy source) and
- the energy gap (E_g) is termed as the *band gap* or *forbidden gap* (because there is no available energy state for the electrons within this range for this particular crystal).

Note: Electrons and Energy Bands

Immediately after the discovery of electron (1897), German physicist Paul Drude propounded (1900) a model for electron transport in materials by assuming that the electrons in a solid behave like a classical ideal gas with no Coulomb interaction. The *Drude model* could explain a lot of properties of solid, such as conductivity, mobility, Ohm's law and Hall effect. Also it could partly address an important experimental finding established by German physicists Gustav Wiedemann and Rudolph Franz almost

> half a century ago (1853) that the ratio of the thermal conductivity to the electrical conductivity for most metals at a given temperature is a constant term, referred to as *Wiedemann–Franz law*. However, Drude model cannot distinguish between a metal and a semiconductor. Indeed the propagation of electrons through the lattice cannot be explained with it, which calls for more robust theories.
>
> For subatomic particles, the wave property becomes predominant and in that case, the mechanics, i.e. the method of determining the position and momentum of that particle requires the use of Schrodinger equation. Also Fermi–Dirac statistics is considered to account for the probability of residing the electrons in different energy states. Incorporating these phenomena, when the motion of the electrons is described in a non-zero, lattice-periodic potential arising from the periodic charge distribution of the ion cores situated at the lattice sites, the solution of Schrodinger equation yields a range of energies for which there are no allowed states for the electrons. Thus, instead of a continuum of energy states, the concept of energy band and energy gap is developed.

In the above system, the electrons occupy the energy levels starting from the lowest according to Fermi–Dirac distribution (Section 2.6). The lower tightly bound 1s band is filled with the $2N$ electrons originally residing in the 1s states of the isolated atoms. The $4N$ electrons residing originally in the 2s and 2p isolated states now occupy the states in either the valence band or in the conduction band depending on the amount of energy acquired. Two situations may occur.

(i) At absolute zero, all the $4N$ electrons occupy the $4N$ states of the valence band and the $4N$ available states in the conduction band remain empty.

(ii) At room temperature or at finite temperatures, some electrons achieve energy from the surroundings and get uplifted to the conduction band overcoming the band gap and leaving behind holes in the valence band. This explains why a semiconductor at normal temperatures contain both electrons and holes as charge carriers.

The amounts of all these energies are, of course, too small to joule so that these are measured in electron-volt (eV). The sign convention within the bands is that the electron energy upward from conduction band edge and the hole energy downward from valence band edge are defined to be positive. Each solid has its own characteristic energy band structure which determines the variation in the electrical properties. Metals, insulators and semiconductors can be distinguished on the basis of energy band theory. Use of energy band theory is continued with semiconductors in Section 2.6.

2.3 Semiconductors

The name itself implies that it is almost a conductor. There are certain elements like silicon (Si) and germanium (Ge) and compounds like gallium arsenide (GaAs) or indium phosphide (InP) that have electrical conductivities intermediate between that of metals and semiconductors. For instance, the resistivity of copper is of the order of 10^{-8} Ωm, while that of mica is around 10^{10} Ωm whereas the resistivity of silicon may vary over the range of 10^{-2} to 10^3 Ωm. Generally semiconductor atoms form covalent bonds. However, these are only qualitative definitions. The more important property associated with semiconductors, distinguishing those from metals and insulators is 'controlled variation of conductivity'. The conductivity of a semiconductor material can be changed over a wide range by the influence of temperature and optical excitation and by the process of doping. Thus the semiconductor has the following features.

(i) Generally the semiconductor is a solid having resistivity range of 10^{-2} to 10^3 Ωm.
(ii) It is manufactured in single crystal, polycrystalline and amorphous form.
(iii) Its electrical conductivity increases (resistivity decreases) with increase of temperature.
(iv) *Doping* is the process of adding small amount (e.g. 1 in 10^6) of other atoms as impurity, which can drastically change the conduction property of the semiconductor.

The above conductivity variations can be explained with band theory. Some examples of semiconductors are given in Table 2.2.

Table 2.2 Introduction to several elemental and compound semiconductor materials

Semiconductor	Type	Group(s)	Common applications
Silicon (Si)	element	IV	Majority of diode, transistor and IC
Germanium (Ge)	element	IV	Some transistors, nanodevices
Gallium arsenide (GaAs)	compound	III-V	Light-emitting diode (LED)
Gallium phosphide (GaP)	Compound	III-V	Light-emitting diode (LED)
Silicon carbide (SiC)	Compound	IV-IV	Light-emitting diode (LED)
Zinc sulfide (ZnS)	Compound	II-VI	Television screen, ultraviolet detector
AlGaAs	ternary compound	III-V	Heterojunction photodiode
GaInAsP	quaternary compound	III-V	Heterojunction photodiode

2.3.1 Electron and Hole

It may be recalled that the charge carriers in a metal are negatively charged electrons, whereas those in electrolytes are positive or negative ions, and insulators have no free charge carrier at all because the electrons are tightly bound to the ion core. Only a charge separation, known as *dielectric polarization* takes place under the action of an applied electric field. A semiconductor is quite different from all of these. It has two types of carriers, negatively charged electrons and positively charged holes, the latter being existent within the semiconductor only.

Since electrons are negatively charged, the removal of an electron from an isolated atom results in a positively charged ion. In a solid, atoms are closely spaced and the electrons of one atom is shared with others. The bound electrons in semiconductors can break the covalent bond by absorbing energy from the surroundings so that free electrons are generated. The removal of an electron from any atom acts as a local positive charge and is termed as *hole*. Another atom having deficiency of electron can accept the removed electron and lose its positive charge. This is represented as the movement of holes from one place to another within the semiconductor crystal.

The hole is not an actual particle like other subatomic particles—electron, proton and neutron. Moreover, it is not just the absence of electron. The energy state of a moving electron and that during its removal from a site are not the same. Consequently the hole exhibits some distinct features as follows.

- Energy band theory (Section 2.2) interprets the different energy states of electrons in a solid. An electron excited to the conduction band creates a hole in valence band, both having different energy states.
- The concept of effective mass (Section 2.5) allows to ignore the wave aspect of subatomic particles and treat them as classical particles. The effective mass of the hole is different from that of the electron.
- Mobility (Chapter 1) defines the ease of moving a charge carrier within a semiconductor under the action of electric field. The mobility of the hole is different from that of the electron.

Considering the above distinctive features, the hole is treated as a separate particle carrying positive charge in semiconductors.

Think a Bit: Can Holes Exist Outside the Semiconductor?

If a semiconductor bar is joined across a voltage source, electron–hole pairs are generated within the semiconductor due to cleavage of covalent bonds. The electrons and the holes get attracted toward the positive and the negative terminal, respectively, of the voltage supply. At the same time, electrons leave the voltage source through its negative terminal and neutralize the holes within the semiconductor. Thus the hole has no existence in the external circuit comprised of the metallic wire and the voltage source. The holes are generated, conducted, and recombined within the semiconductor material only. Looking from the external circuit, the current is due to electrons only and the semiconductor bar acts as a resistor.

Note: Pondering over Holes

The current density arising from a given band is $J = -n_o e \bar{v}$ where the equilibrium concentration of electrons is n_o and $\bar{v} = \frac{1}{n_o V} \sum_i v_i$ is the average velocity. The summation over all the velocities within volume V is considered. The current density is given by: $J = -\frac{e}{V} \sum_i v_i$. This current must be zero when summed over a full band because due to the symmetry of E–k curve about the k-axis [Equation (2.6)], for every state of positive velocity $\left(\frac{1}{\hbar}\frac{dE}{dk}\right)$, there exists a state of negative velocity of equal magnitude. If the band is nearly full and only a few empty states exist, then

$$J = -\frac{e}{V}\left(\sum_m v_m - \sum_n v_n\right) = +\frac{e}{V}\left(\sum_n v_n\right)$$

> The summation over m represents the sum over all the velocity states in the band and the summation over n is for all the unoccupied velocity states. The sum over m (nearly filled band) should vanish and the remaining sum over the unoccupied states, surprisingly, corresponds to a current that could be produced by positive charge carriers. Thus the current from an almost full band of electrons can be treated as a current from the motion of a small number of empty electronic states of opposite charges, which are designated as holes.

2.3.2 Intrinsic Semiconductor

A pure semiconductor material, either elemental or compound, contains only its own atoms and no other impurity atoms. This is called an *intrinsic semiconductor* because it possesses the inherent properties of that material only. At normal temperatures, it is very common phenomenon that the surrounding thermal, electrical, or optical energy breaks the covalent bonds of the intrinsic semiconductor freeing the electrons. These electrons can move within the crystal carrying negative charge and the corresponding positively charged holes can move in the opposite direction.

The above occurrences can be redefined in terms of energy bands as follows. The electrons get accelerated for movement, hence uplifted to a higher energy state of the conduction band leaving behind an empty, lower energy state in the valence band. The vacant state is treated as a positively charged hole. Thus there are equal number of electrons and holes in an intrinsic semiconductor. The number of electrons or holes per unit volume of the material is designated as the *intrinsic carrier concentration*.

At absolute zero, no electron can acquire energy to surmount the band gap and reach the conduction band so that the conduction band remains absolutely empty and the valence band remains totally filled. Consequently no charge carrier is available and the semiconductor behaves like an insulator at absolute zero. At finite temperatures, the electrical conduction in intrinsic semiconductors is determined by thermally generated electrons and holes.

The electrons in the conduction band get a large number of available energy states and can freely move around. The corresponding transport of holes in the valence band is not so smooth because most of the states are filled and only the few vacant states created by the transfer of electrons to the conduction band take part in carrier transport. Therefore the hole mobility is less than the electron mobility.

The *generation* of electron–hole pair takes place when an electron gains energy and gets transferred to the conduction band and a hole is created in the valence band. Complementary to this process, there occurs *recombination* of electron–hole pair when an electron in the conduction band releases energy and come down to the valence band filling up the empty state. An electron and a hole get annihilated together in this process. The rate of generation and recombination of carriers in semiconductors is temperature dependent and it is nil at absolute zero. At finite temperatures, the intrinsic semiconductor remains in equilibrium state under continuous generation and recombination processes.

> **Note: Usefulness of Silicon**
>
> Silicon has several beneficial features that have made it a very useful semiconductor material. It can be manufactured to a very high purity level, the impurity levels being reduced up to 1 part in 10^{10} atoms. This much purity is essential because doping is done of the order of 1 part in 10^6 atoms. Also as it will be discussed in Chapter 12 that silicon can be oxidized easily and the oxide can be removed precisely, which is very appropriate for the fabrication of integrated circuits.

2.3.3 Doping and Extrinsic Semiconductor

The intrinsic semiconductors have equal number of thermally generated charge carriers, which are generally not more than 10^{10} to 10^{13} cm^{-3} in number. So the conductivity of an intrinsic semiconductor is not high. It is possible to increase the number of electrons or holes up to 10^{15} to 10^{18} cm^{-3} or more and thereby to enhance the conductivity up to a great extent. This is achieved by introducing specific elements in controlled amount into the crystal as impurities. Such a process of varying the electrical conductivity of semiconductors by incorporating other atoms into it is known as *doping*. The dopant atoms can either replace the host crystal atoms or reside at vacancies or interstitials within the crystal. The following two types of doping are possible:

(i) n-type, where the number of electrons is increased (the letter 'n' stands for 'negative') and

(ii) p-type, where the number of holes is increased (the letter 'p' stands for 'positive').

Such a doped semiconductor, either n-type or p-type, where the concentration of charge carriers is much larger than that of pure semiconductor, is termed as *extrinsic semiconductor*. The dopant atoms are inserted into the intrinsic semiconductor crystal by any of the following two processes.

A. *Diffusion*: The semiconductor wafer is placed in a furnace at high temperature ($\approx 1000°C$) in a gaseous atmosphere of the dopant atoms, which by virtue of its thermal energy penetrates into the intrinsic semiconductor crystal. They sit into the lattice defects, vacancies and interstitials. The atoms of the host crystal also move out of their original lattice sites.

B. *Ion Implantation*: the ions of the dopant atom are imparted high kinetic energy ($\approx$ keV to MeV) applying accelerating potential and the ions penetrate into the intrinsic lattice. This method has more precise control on doping concentration and geometry.

The outcome of doping is sketched in Figure 2.4. Germanium and silicon, the two elemental semiconductors have four valence electrons. These tetravalent atoms can form covalent bonding with four adjoining atoms. Two different types of doping are explained using silicon (Si) lattice as model.

Semiconductor Fundamentals

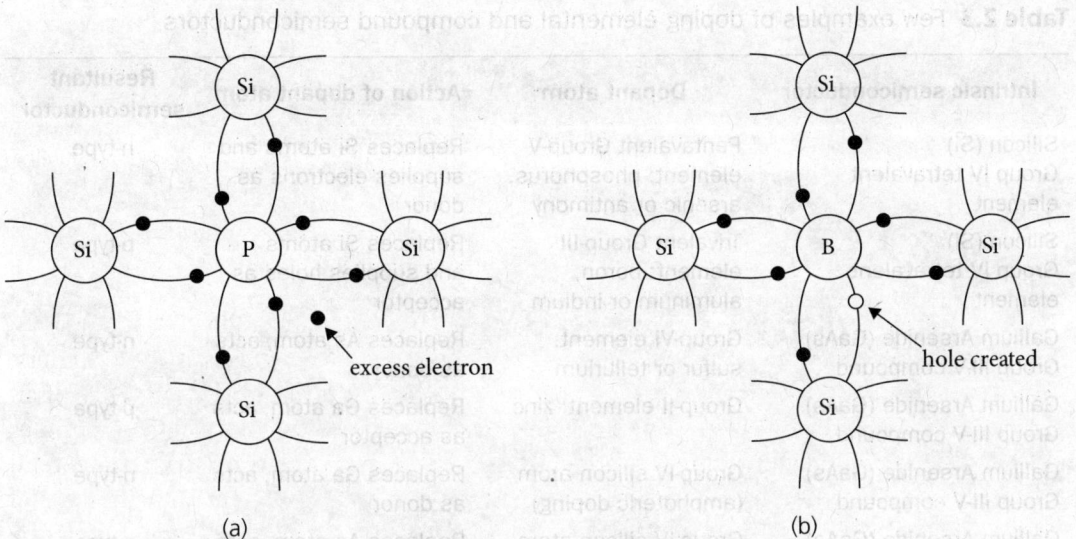

Figure 2.4 Tetravalent semiconductor (Si) doped with (a) pentavalent phosphorus (P) creating an excess electron and (b) trivalent boron (B) creating a hole

If the intrinsic, tetravalent silicon semiconductor is doped with pentavalent phosphorus (P) atoms, four out of its five valence electrons establish covalent bonds with four neighboring Si atoms [Figure 2.4(a)]. The fifth electron, being excess, remains loosely bound to the phosphorus atom and can easily free itself for contributing to current flow. Similar mechanism would take place, if the doping were done with any other pentavalent atom, such as antimony (Sb) or arsenic (As). The pentavalent atom is called the *donor atom* because it donates an electron. Such doping increases the electron content to a great extent and reduces the number of holes from that in intrinsic state because of recombination with electrons. Nonetheless, the product of the number of electrons and the number of holes remains fixed, which will be discussed further in section 2.6.3. Since the electrons outnumber the holes, the current flow in n-type semiconductor is mainly due to electrons. Therefore electrons are designated as the *majority carriers* in n-type semiconductor whereas the holes are termed as the *minority carriers*.

If the tetravalent silicon semiconductor is doped with trivalent boron (B) atoms [Figure 2.4(b)], the three valence electrons of boron form covalent bonds with the adjoining Si atoms but there remains deficiency of one electron in the bonding that creates a positively charged hole. Any moving electron can fill up the void, which is equivalent to a hole movement in the opposite direction. A similar situation would occur, if any other trivalent dopant atom, such as aluminum (Al) or gallium (Ga) were used. The trivalent dopant atom is called the *acceptor atom* because it accepts an electron for contributing to a hole in current flow. Doping with acceptor atoms increase the hole concentration to a large extent but the electron concentration is reduced from the intrinsic value because of recombination. The product of the electron and hole concentrations remains fixed. Since the number of holes dominates over that of electrons, the current flow in p-type semiconductor is mainly determined by the hole concentration. Now the holes are designated as the majority carriers and the electrons as the minority carriers.

Table 2.3 Few examples of doping elemental and compound semiconductors

Intrinsic semiconductor	Dopant atom	Action of dopant atom	Resultant semiconductor
Silicon (Si) Group IV tetravalent element	Pentavalent Group-V element: phosphorus, arsenic or antimony	Replaces Si atoms and supplies electrons as donor	n-type
Silicon (Si) Group IV tetravalent element	Trivalent Group-III element: boron, aluminum or indium	Replaces Si atoms and supplies holes as acceptor	p-type
Gallium Arsenide (GaAs) Group III-V compound	Group-VI element: sulfur or tellurium	Replaces As atom, acts as donor	n-type
Gallium Arsenide (GaAs) Group III-V compound	Group-II element: zinc	Replaces Ga atom, acts as acceptor	p-type
Gallium Arsenide (GaAs) Group III-V compound	Group-IV silicon atom (amphoteric doping)	Replaces Ga atom, acts as donor	n-type
Gallium Arsenide (GaAs) Group III-V compound	Group-IV silicon atom (amphoteric doping)	Replaces As atom, acts as acceptor	p-type

2.3.4 Doping in Compound Semiconductor

The sketches in Figure 2.4 with tetravalent semiconductors are just models for explaining the doping process. The process is more complicated in compound semiconductors like those mentioned in Table 2.2. There are a number of methods for preparing n- and p-type elemental and compound extrinsic semiconductors. The replacing of host atoms by the dopant atoms is the determining factor, particularly in the case of compound semiconductors. A few examples are given in Table 2.3.

In most cases, many types of impurity atoms may be present within a doped semiconductor. The predominance of donor or acceptor makes it n-type or p-type. However, the resultant concentration of electrons in n-type is the algebraic sum $(N_D - N_A)$ where N_D and N_A are the concentrations of the donor and acceptor atoms, respectively. Similarly the p-type semiconductor has hole concentration of $(N_A - N_D)$. This phenomenon of algebraic summation of charges in doped semiconductors is called *compensation*.

Another important parameter is the *charge neutrality*, which exists always at any space of the semiconductor taking into account the electrons, the holes and the donor and acceptor ions so as to make the material electrostatically neutral. The sum of the negative charges (electrons plus acceptor ions) must equal the sum of positive charges (holes plus donor ions).

2.4 Direct and Indirect Band Gap

Because of the periodic crystal structures, the charge carriers in a semiconductor move in a periodic potential. The calculation of the possible energy states of the charge carriers in such situation becomes a complicated quantum mechanical process. A simple model, known as *one-electron approximation* assumes the carriers to travel through a periodic potential, the wave function defining the motion being represented as

$$\psi(x) = u(\vec{k}, x)\exp(i\vec{k}x) \tag{2.3}$$

where the plane wave is modulated by the function $u(\vec{k}, x)$ depending on the periodicity of the lattice in x-direction. The vector $\vec{k}$ is called the *propagation constant* or *wave vector* and its magnitude is $k = 2\pi/\lambda$, expressed in terms of de Broglie wavelength (λ) of the charge carrier. The quantity $hk/2\pi = \hbar k$ indicates the momentum, h being the Plank constant and $\hbar = h/2\pi$.

> **Think a Bit: Is It Momentum?**
>
> The quantity $\hbar k$ is termed as the *crystal momentum*. It has the dimensions of that of momentum (Verify on your own!). Nevertheless, there are certain differences. The true momentum of an electron is not a constant of motion because of the periodic lattice potential whereas the crystal momentum is constant for a state of energy. The dynamical behavior of the electron in the lattice with respect to the crystal momentum is similar to that of a free electron with respect to real momentum.

Obviously the above mentioned periodicity depends on the direction of propagation. Even in a simple cubic lattice, the periodicity along the side is different from that along the face diagonal and that along the body diagonal. In an actual crystal, the variation is much more complicated and diversified. Consequently the variation of energy state (E) with $\vec{k}$ becomes different in different directions of the crystal. The actual variation in three dimensions is a complex phenomenon. Figure 2.5 gives an introductory idea about it with a simplified view of the energy band structure of silicon (Si) and gallium arsenide (GaAs) with reference to the magnitude of k considering different lattice directions. It indicates that in the case of Si, the energy variation in the conduction band has a minimum and that in the valence band has a maximum for different values of k. In contrast, the maximum and the minimum for GaAs occur at the same k value. Based on such k-alignment of bands, GaAs is called a *direct band gap semiconductor* and Si is called an *indirect band gap semiconductor*. In direct band gap materials where the band maxima and minima coincide in k-values, band-to-band transitions of electrons involve very small amount of energy changes whereas that in indirect band gap materials require a change in k-value also, hence a change in the momentum of carriers.

2.5 Effective Mass

Subatomic particles like electrons have a prominent associated wave characteristic, as propounded by French physicist Louis de Broglie (1892–1987). This wave aspect is considered in the wave mechanics of individual particles. However, if such concept is used to specify the motion of electrons and holes within a solid including a semiconductor crystal,

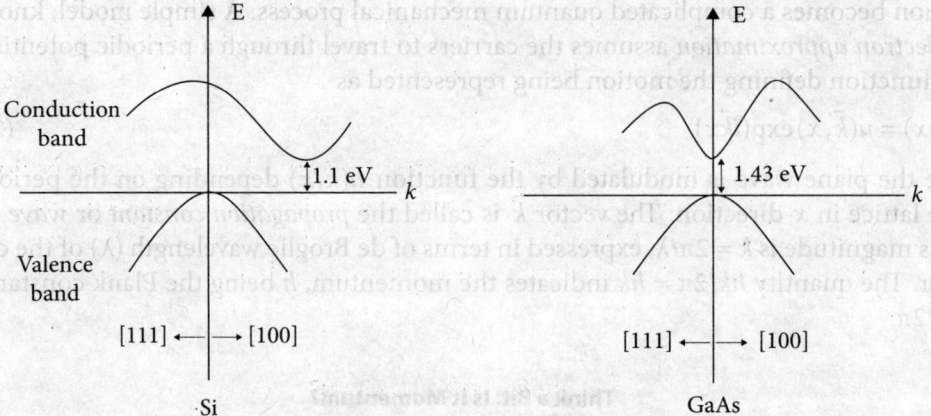

Figure 2.5 Simplified energy band diagrams for silicon (Si) and gallium arsenide (GaAs) with reference to [111] and [100] crystal orientations indicating the band maximum and minimum and the band gap (eV)

the influence of the periodic lattice, other electrons and ions cores are to be considered. That would make the mathematical analysis very complicated and unmanageable.

In order to simplify the theory, it is assumed that the rest mass of the particle has got changed under the combined influences of external forces in such a way that it can be treated as a free particle and the usual equations of electrodynamics can be applied to it. Such assumed altered mass is termed as the *effective mass* of the particle. This is expressed in terms of the rate of change of energy state with respect to momentum, as follows.

Let an electron of actual mass m moves within a crystal under the action of an external electric field ξ. The acceleration (f) gained by the electron will be such that

$$mf = -e\xi + x \tag{2.4}$$

where x is an unknown quantity representing the net force due to other electrons and ion cores. This unknown can be avoided by assuming that the effective mass of the electron under the present condition has changed to m^* so that

$$m^* f = -e\xi \tag{2.5}$$

Comparing Equations (2.4) and (2.5) it is understood that the concept of effective mass allows us to treat the charge carriers in solid as free particles and its energy (E) and momentum ($\hbar k$) can be related as

$$E = \frac{\hbar^2 k^2}{2m^*} \tag{2.6}$$

so that

$$\frac{dE}{dk} = \frac{\hbar^2 k}{m^*} \tag{2.7}$$

and

$$\frac{d^2E}{dk^2} = \frac{\hbar^2}{m^*} \qquad (2.8)$$

It can be stated from Equation (2.7) that the particles move with a group velocity of

$$v_g = \frac{1}{\hbar}\frac{dE}{dk} = \frac{\hbar k}{m^*} \qquad (2.9)$$

and the effective mass of the particle can be obtained from Equation (2.8) as

$$m^* = \frac{\hbar^2}{d^2E/dk^2} \qquad (2.10)$$

The effective mass of the electron and that of the hole are different even within the same semiconductor material because of difference in E–k relationship. Near the bottom of the conduction band (E_c) and the top of the valence band (E_v), the E–k relationship is approximated by Equation (2.6) where m^* is the effective mass of the charge carrier—electron or hole. For holes, the top of the valence band is denoted by either of two parabolic bands with different curvatures, termed as the *heavy hole band* with smaller d^2E/dk^2 and the *light hole band* with larger d^2E/dk^2. Electrons cannot be 'light' or 'heavy' but the holes can be so by virtue of its energy–momentum variation in the valence band.

> **Note: Experimental Determination of Effective Mass**
>
> The theoretical calculation of effective mass is a complicated process involving the energy states in three-dimensional space. Still, an average value of effective mass can be determined experimentally by allowing the electron to traverse a circular path of radius r with velocity v in a magnetic field of flux density B. Equating the magnetic flux qvB to the centripetal force, the angular frequency (ω) of rotation can be derived as $\omega = qB/m^*$. The electron has effective mass m^* and q magnitude of charge. An alternating electromagnetic field of variable frequency, usually of microwave range is applied to the sample. The maximum absorption of energy takes place at resonance when the frequency of the applied field matches the frequency of rotation. Noting the frequency at resonance and knowing q and B, m^* can be calculated. This technique is known as *cyclotron resonance*.

2.6 Fermi Level, Energy Band and Semiconductors

It is stated in Section 2.2 that the distribution of carriers over the available energy states in solids obey Fermi–Dirac statistics. According to the law of this statistics, the probability of an available energy state (E) being occupied by an electron in a solid at absolute temperature T is given by

$$f(E) = \frac{1}{1+\exp[(E-E_F)/kT]} \qquad (2.11)$$

The function $f(E)$ is termed as *Fermi–Dirac distribution function*, $k = 8.62 \times 10^{-5}$ eVK^{-1} denotes Boltzmann constant and E_F is a certain energy level, called *Fermi level*. All energy states are expressed in electron-volt (eV).

> **Think a Bit: The Term *kT/q* Is a Magic Figure!**
>
> The term kT, having the dimension of energy is sometimes written as kT/q to express in eV. In that case k would be 1.38×10^{-23} JK^{-1} and $q = 1.6 \times 10^{-19}$ C would represent the magnitude of charge of either an electron or a hole. At room temperature, kT/q comes out to be approximately 0.026 eV. It is termed as *volt-equivalent of temperature* and can be readily used in numerical calculations as a reasonable approximation.

The Fermi level is an important reference term in connection with the energy band structure of solids. The following conclusions about the distribution of electrons at different energy levels can be obtained from Equation (2.11).

(i) At absolute zero ($T = 0$ K),
  $f(E) = 0$ for $E > E_F$ and
  $f(E) = 1$ for $E < E_F$
(ii) For finite temperatures ($T > 0$ K),
  $f(E) = \frac{1}{2}$ for $E = E_F$
  $f(E) = 0$ for $E \gg E_F$ and
  $f(E) = 1$ for $E \ll E_F$

Thus at absolute zero, all the available energy states get filled up with electrons up to Fermi level and all the states above this level remain vacant. At any finite temperature above absolute zero, the probability of an energy state at Fermi level being filled up with electrons is ½. The graphical representation of the above statements is given in Figure 2.6(a). Solids including semiconductors undergo similar energy distribution of charge carriers.

The energy band diagram of an intrinsic semiconductor is schematically represented by the line diagram of Figure 2.6(b). It is not the actual band picture like that of Figure 2.5 but a simplified model for practical purpose. It is the convention to indicate the bottom of the conduction band (E_C) and the top of the valence band (E_V) by solid lines separated by band gap (E_g) and the intrinsic Fermi level (E_{Fi}) is indicated by dashed line. It implies that the Fermi level is not any actual energy level like the conduction or valence band edges but in represents an average level up to which the energy states are filled by charge carriers. The corresponding Fermi–Dirac energy distribution in the intrinsic semiconductor is also sketched in Figure 2.6(b) to indicate its relation with the intrinsic Fermi level. The band gap of a semiconductor is treated as a constant quantity. However, it varies slightly with temperature. For instance, changing from room temperature to absolute zero, the band gap of Si increases from 1.12 eV to 1.17 eV and that of GaAs increases from 1.42 eV to 1.52 eV.

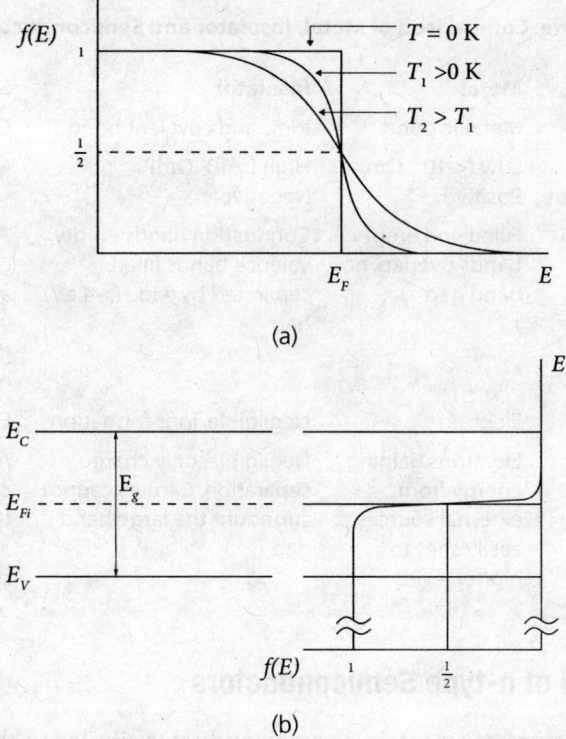

Figure 2.6 Graphical representation of (a) Fermi–Dirac energy distribution and (b) its correspondence with energy band diagram for intrinsic semiconductor

Example 2.2

In the energy band picture of a certain p-type semiconductor, the Fermi level is 0.36 eV above the valence band maximum. Calculate the probability of an energy state of (i) $(E_V - kT)$ and (ii) $(E_V + 16kT)$ being occupied at room temperature.

Solution

The room temperature is assumed to be 300 K so that $kT = 8.62 \times 10^{-5} \times 300 = 0.0259$ eV (approx.)

For case (i), $E = E_V - kT$ so that

$E - E_F = E_V - kT - E_V - 0.36$ (given) $= -0.3859$ eV

Putting this value of $(E - E_F)$ in Equation (2.11), $f(E) \approx 1$, which means the probability of occupying the state below E_V is 100%.

For case (ii), $E = E_V + 16kT$ so that

$E - E_F = E_V + 16 \times 0.0259 - E_V - 0.36 = 0.0544$ eV

Putting this value of $(E - E_F)$ in Equation (2.11), $f(E) = 0.109 = 11\%$ (approx.) It indicates that the probability of occupying the state above E_V decreases rapidly with increase in energy.

Note: Comparison of Metal, Insulator and Semiconductor

Parameter	Metal	Insulator	Semiconductor
Atomic bonding	Metallic bond	Ionic and covalent bond	Covalent bond
Resistivity and Temperature coefficient	Low (< 10^{-5} Ωm) Positive	High (> 10^5 Ωm) Negative	Moderate (>10^{-5} Ωm, < 10^5 Ωm) Negative
Energy bands and gap	Filled and empty bands overlap, no band gap	Conduction bands empty, valence bands filled, separated by wide (> 4 eV) gap	At room temperature, conduction bands almost empty, valence bands almost filled, separated by moderate (< 4 eV) gap
Free charge carriers	Electron	Negligible, ions formation	Electron and hole
Carrier transport	Electrons gaining energy from external sources easily shift to higher levels	Negligible, only charge separation. Carriers cannot surmount the large band gap	At room temperature, carriers surmount the band gap. At around absolute zero, behaves like insulator

2.6.1 Energy Band of n-type Semiconductors

The energy band structure of an intrinsic semiconductor similar to that of Figure 2.6(b) is constituted of the energy levels of atoms of that semiconductor only. When it is doped with other impurity atoms, the energy levels of those impurity atoms also take part in providing energy states to the carriers. Properly chosen dopant atoms can provide energy levels near the band edges. The levels of donor atoms take place near the conduction band edge (E_C) and that of acceptor atoms appear near the valence band edge (E_V).

Figure 2.7 displays the energy band diagram for an n-type semiconductor where tetravalent intrinsic semiconductor (say silicon) is doped with pentavalent donor atoms (say phosphorus).

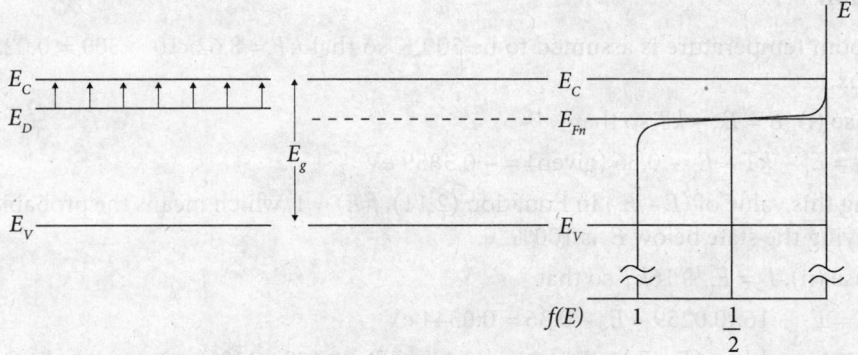

Figure 2.7 Energy band diagram of n-type semiconductor showing the positions of donor level (E_D) (left), donor Fermi level (E_{Fn}) (middle) and the corresponding Fermi–Dirac energy distribution (right)

The donor atoms have energy levels very close (0.01 to 0.04 eV) to the conduction band edge and it is termed as the *donor level* (E_D) because it 'donates' electrons to the conduction band. Such a level is sketched at the left side of Figure 2.7. The electrons in this level can acquire energy very easily to get uplifted to the conduction band (indicated by arrows) at room temperature thereby increasing the number of electrons there, hence increasing the electrical conductivity. Consequently the Fermi level for the n-type semiconductor (E_{Fn}), popularly known as the *donor Fermi level* rises upward, as shown in the middle of Figure 2.7. It indicates a higher energy level of electrons on the average. It may be noted that the donor level is an actual energy level and is indicated with solid line whereas the Fermi level, being a hypothetical level, is indicated with dashed line. The higher the donor doping concentration is, the closer the donor Fermi level shifts to the conduction band edge. The corresponding Fermi–Dirac distribution is shown at the right side of the figure. The nature of distribution is the same as that of the intrinsic semiconductor in Figure 2.6 but the average level has shifted up to the donor Fermi level (E_{Fn}). Conventionally the energy band diagram of n-type semiconductor is modeled with this Fermi level instead of showing the donor level.

2.6.2 Energy Band of p-type Semiconductors

Figure 2.8 presents a sketch for the energy bands in p-type semiconductors where a tetravalent intrinsic semiconductor (say, silicon) is doped with trivalent acceptor atoms (say, boron). The energy level of boron atoms, termed as the *acceptor level* (E_A) exists very close to the valence band edge of silicon, as indicated at the left side of the figure. It 'accepts' electrons from the valence band. Electrons can acquire energy at room temperature and shift easily from the valence band to this level (indicated by arrows). Consequently the number of holes in the valence band increases and the electrical conductivity is enhanced. The Fermi level for the p-type semiconductor (E_{Fp}), termed as the *acceptor Fermi level* shifts downward, as shown with dashed line in the middle of Figure 2.8. It denotes the average higher energy levels of holes. On increasing the acceptor doping concentration it comes closer to the valence band edge. The corresponding Fermi–Dirac energy distribution is shown at the right side where the average is centered about the acceptor Fermi level (E_{Fp}). Conventionally the energy band diagram of p-type semiconductor is sketched with this Fermi level instead of the actual acceptor level.

> **Think a Bit: Why 'Levels' for Donors and Acceptors**
>
> It has been elaborated that the closely spaced energy levels in solids including semiconductors constitute bands of energy. The same convention is followed for the conduction and valence bands of the semiconductor. Then why the donor and the acceptor atoms correspond to discrete energy levels? This is so because these energy levels are the characteristics of the dopants atoms and such atoms are actually discrete in the doped semiconductor. The amount of doping is very small, about 'one in a million' and each dopant atom find itself isolated within the crowd of the host semiconductor atoms.

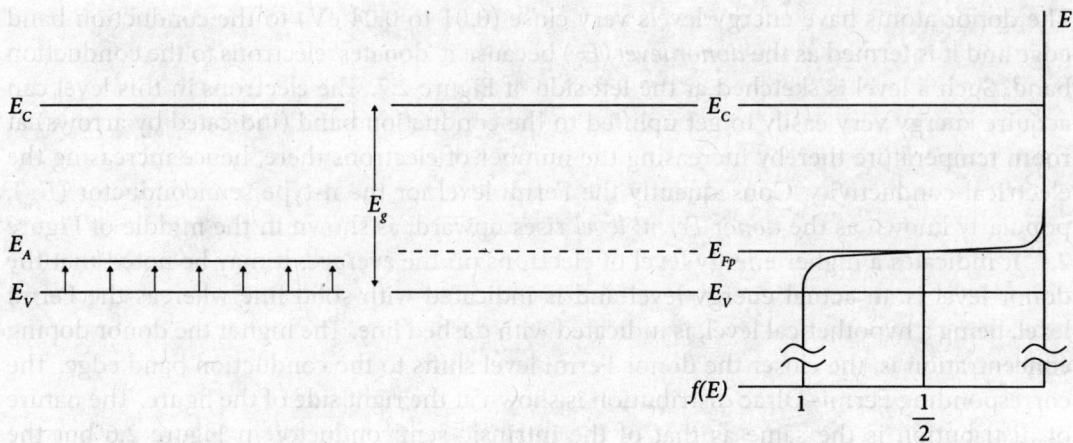

Figure 2.8 Energy band diagram of p-type semiconductor showing the positions of acceptor level (E_A) (left), acceptor Fermi level (E_{Fp}) (middle) and the corresponding Fermi–Dirac energy distribution (right)

2.7 Equilibrium Carrier Concentrations

The Fermi–Dirac distribution function expressed by Equation (2.11) when applied to conduction band edge (E_C), can be approximated as

$$f(E_C) = \exp[-(E_C - E_F)/kT] \tag{2.12}$$

because E_C and E_F have finite difference of energy (several kT/q) and the exponential term may be considered much greater than unity. Therefore the equilibrium concentration of electrons in the conduction band may be expressed as

$$n_o = N_C f(E_C) \tag{2.13}$$

where

$$N_C = 2\left(\frac{2\pi m_n^* kT}{h^2}\right)^{3/2} \tag{2.13a}$$

is the effective density of states at the conduction band edge, m_n^* is the effective mass of electrons there and the other symbols have the usual meaning. Thus the equilibrium concentration of electrons in conduction band is

$$n_o = N_C \exp[-(E_C - E_F)/kT] \tag{2.14}$$

Equation (2.14) is a general expression applicable to both the intrinsic and extrinsic semiconductors. As the probability of a state remaining occupied is denoted by $f(E)$, that of staying vacant may be articulated by $[1 - f(E)]$. Therefore the equilibrium concentration of holes in valence band edge (E_V) can be enunciated as

$$p_o = N_V[1 - f(E_V)] \tag{2.15}$$

where

$$N_V = 2\left(\frac{2\pi m_p^* kT}{h^2}\right)^{3/2} \tag{2.15a}$$

is the effective density of states at the valence band edge including m_p^* as the effective mass of holes there and other symbols with their standard meaning. Using Equation (2.11),

$$1 - f(E_V) = 1 - [1 + \exp(E_V - E_F)/kT]^{-1} \tag{2.16}$$

which can be approximated as

$$1 - f(E_V) = \exp[-(E_F - E_V)/kT] \tag{2.17}$$

Using Equations (2.15) and (2.17), the equilibrium concentration of holes in valence band is obtained as

$$p_o = N_V \exp[-(E_F - E_V)/kT] \tag{2.18}$$

Equation (2.18) is a general expression pertinent to both intrinsic and extrinsic semiconductors. The following useful relationships are derived using Equations (2.14) and (2.18).

A. Mass-Action Law

Equations (2.14) and (2.18) are multiplied and E_g is considered as $(E_C - E_V)$ to obtain

$$\begin{aligned} n_o p_o &= N_C N_V \exp[-(E_C - E_F)/kT]\exp[-(E_F - E_V)/kT] \\ &= N_C N_V \exp[-(E_g/kT)] \end{aligned} \tag{2.19}$$

Since the right side of Equation (2.19) is a constant quantity, it implies that the product of electron and hole concentrations in a semiconductor at equilibrium is constant irrespective of doping. This is a significant conclusion known as the *mass-action law*.

An intrinsic semiconductor has the same electron and hole concentration at equilibrium such that $n_o = p_o = n_i$, which is called the *intrinsic carrier concentration*. The product is thus

$$n_o p_o = n_i^2 \tag{2.20}$$

Equation (2.20) is the numerical form of the mass-action law.

B. Intrinsic Carrier Concentration

Using Equations (2.19) and (2.20), the intrinsic carrier concentration, in terms of semiconductor band gap and effective density of states at band edges, is obtained as

$$n_i = (N_C N_V)^{1/2} \exp[-\{E_g/(2kT)\}] \tag{2.21}$$

Equation (2.21) infers that the intrinsic Fermi level (E_{Fi}) lies in the middle of the band gap so that

$$E_C - E_{Fi} = E_{Fi} - E_V = \frac{E_g}{2} \tag{2.22}$$

Replacing the Fermi levels by the intrinsic Fermi level in Equations (2.14) and (2.18), the equilibrium concentration of intrinsic electrons (n_i) and intrinsic holes (p_i) are obtained as

$$n_i = N_C \exp[-(E_C - E_{Fi})/kT] \tag{2.23a}$$

$$p_i = N_V \exp[-(E_{Fi} - E_V)/kT] \tag{2.23b}$$

Actually $n_i = p_i$ because the carriers are created in pairs in intrinsic semiconductors.

C. Electron and Hole Concentrations

Equations (2.14) and (2.18), in combination with Equations (2.23a) and (2.23b) can lead to the following expression for the equilibrium electron and hole concentrations in terms of Fermi levels and intrinsic carrier concentration.

$$n_o = n_i \exp[(E_{Fn} - E_{Fi})/kT] \tag{2.24a}$$

$$p_o = n_i \exp[(E_{Fi} - E_{Fp})/kT] \tag{2.24b}$$

Equations (2.24a) and (2.24b) use $n_i = p_i$ and the Fermi levels E_{Fn} and E_{Fp} represent the Fermi level for n-type and p-type semiconductors, respectively. These also indicate that as the Fermi level moves away from the intrinsic level, either toward the conduction band or toward the valence band, the carrier concentration increases exponentially.

Example 2.3

A silicon sample is doped with 10^{16} phosphorus atoms cm^{-3}. Determine the equilibrium hole concentration at room temperature. Given: intrinsic carrier concentration = 1.5×10^{10} cm^{-3}.

Solution

It is assumed that each phosphorus atom contributes to one electron. Since the dopant atom is pentavalent, the resultant semiconductor becomes n-type. Considering $n_o = 10^{16}$ cm^{-3}, the hole concentration is calculated using Equation (2.20) as

$$p_o = (1.5 \times 10^{10})^2 / 10^{16} = 2.25 \times 10^4 \text{ cm}^{-3}$$

[It may be noted that the hole concentration in the n-type semiconductor becomes much lower than the intrinsic concentration.]

> **Example 2.4**
>
> Find out the Fermi level position with respect to the intrinsic Fermi level in the above case of Example 2.2.
>
> **Solution**
>
> Using Equation (2.24a), the energy difference is expressed as $E_{Fn} - E_{Fi} = kT \ln \frac{n_o}{n_i}$. Putting $k = 8.62 \times 10^{-5}$ eVK^{-1}, $T = 300$ K, $n_o = 10^{16}$ cm^{-3} and $n_i = 1.5 \times 10^{10}$ cm^{-3}, the Fermi level position (E_{Fn}) in the doped material comes out to be about 0.347 eV above the intrinsic Fermi level (E_{Fi}).

2.8. Drift and Diffusion of Carriers

The electric current in semiconductors is comprised of the movements of both electrons and holes that are not only of different charges, those have different mobility and effective mass as well. So the mechanisms of both of their transport are important. The following two processes of carrier transport in semiconductors give rise to two different current components.

(i) The influence of external electric field causes drift current.
(ii) The diffusion of carriers because of concentration gradient gives rise to diffusion current.

Chapter 1 has introduced the concepts of mobility, drift velocity and current density. The charge carriers in solids are always in the state of scattering with lattice, dopant atoms and other carriers so that there is no net motion along any specific direction and hence no net current flow happens. In presence of an external electric field, the carriers get accelerated to a particular direction and are retarded simultaneously because of the aforementioned collisions and scatterings. Ultimately the carriers under the acceleration and retardation attain a steady velocity, known as drift velocity (v_d), which is proportional to the electric field (E) and can be expressed [Equation (1.4)] as

$$v_d = \mu E \tag{2.25}$$

The mobility (μ) of a charge carrier is the drift velocity per unit electric field describing the easiness with which it is driven within the material under the action of an electric field. The drift current density arising from such carrier movement is given by [using Equation (1.9)]

$$J = nqv_d = nq\mu E = \sigma E \tag{2.26}$$

Here q denotes the charge magnitude, which may be of either the electron or the hole. The conductivity ($\sigma = nq\mu$) of the material depends on the carrier density and its mobility and the resistivity ($\rho = 1/\sigma$) is the inverse of the conductivity.

2.8.1 Drift Current Density

The above expressions for conductivity and current density are quite general. In the case of semiconductors, the holes drift toward the electric field and the electrons opposite to it with their individual mobility. The conductivity is the combined effect of these two carriers given by

$$\sigma = (nq\mu_n + pq\mu_p) \qquad (2.27)$$

where n and p represent the electron and hole concentration, respectively and μ_n and μ_p are their respective mobility. The *drift current density* in a semiconductor, from Equations (2.26) and (2.27) is

$$J = (nq\mu_n + pq\mu_p)E \qquad (2.28)$$

The individual contributions of electrons and holes to the total drift current are accounted in terms of their number density and mobility.

Example 2.5

An intrinsic silicon sample has carrier concentration of 1.5×10^{10} cm^{-3}, electron mobility of 1350 cm^{2}V^{-1}s^{-1} and hole mobility of 480 cm^{2}V^{-1}s^{-1}. Determine the resistivity of the sample.

Solution

The intrinsic semiconductor has the same electron and hole concentrations. Therefore, using Equation (2.27), the conductivity of the material is calculated as

$$\sigma = 1.6 \times 10^{-19} \times 1.5 \times 10^{10} \times (1350 + 480) = 4392 \times 10^{-9} \text{ S}.$$

The resistivity is $1/\sigma = 2.277 \times 10^5$ Ωcm

Example 2.6

How much drift current density would be obtained, if a voltage of 10 V were applied across 1 cm length of the sample in Example 2.5?

Solution

The electric field becomes 10 Vcm^{-1} so that using Equation (2.28) and the result of Example 2.5, the drift current density is calculated as: $4392 \times 10^{-9} \times 10 = 4.392 \times 10^{-5}$ Acm^{-2}.

2.8.2 Diffusion Current Density

The drift of charge carriers causing a current mentioned above may take place always whenever an external electric field is established across an intrinsic or extrinsic semiconductor sample. However, another important mechanism of current flow is initiated by diffusion of carriers when carriers in excess of thermal equilibrium quantities are imported by applying external voltage, optical excitation or other phenomena.

Non-uniform density of excess carriers in a semiconductor gives rise to spatial gradient of carrier concentration resulting in a net movement of carriers from the region of higher concentration to that of lower concentration through random thermal motion and scattering. This causes the diffusion current. The *diffusion current density* in a semiconductor are expressed separately for electrons and holes as

(i) For electrons: $J_n = qD_n \dfrac{dn}{dx}$ (2.29a)

(ii) For holes: $J_p = -qD_p \dfrac{dp}{dx}$ (2.29b)

The concentration gradients along x-direction for electrons and holes are denoted by dn/dx and dp/dx, respectively. The constants of proportionality D_n and D_p are called diffusion coefficients for electrons and holes, respectively, having the dimension of m²s⁻¹. The negative sign signifies that the net flow of carriers is in the direction of decreasing concentration. In the case of electrons, that is nullified by the negative charge.

2.8.3 Semiconductor Current Density

The most probable situation in a semiconductor is the existence of both the electric field due to potential gradient and the carrier diffusion due to concentration gradient. So, in general, the drift and diffusion components of current are added up and the total current density in the semiconductor are defined separately for electrons and holes as

(i) For electrons: $J_n = nq\mu_n E + qD_n \dfrac{dn}{dx}$ (2.30a)

(ii) For holes: $J_p = pq\mu_p E - qD_p \dfrac{dp}{dx}$ (2.30b)

All the quantities n, p, J_n and J_p are functions of distance (x) from a certain origin and there should not be any net current flow in a semiconductor at equilibrium. Therefore, any fluctuation in carrier concentration that initiates diffusion current should set up an electric field that redistributes the carriers by drift so as to maintain the condition of no net current flow. This implies that the two carrier transport processes, namely drift and diffusion are not independent. Indeed those are connected by the Einstein relation, as described in the following section.

2.8.4 Einstein Relation

Equations (2.29a) and (2.29b) indicate that the carrier diffusion coefficient is a constant term related to the carriers moving under unit concentration gradient. It should depend on both the scattering mechanism and the true energy distribution of the carriers. A relationship between the carrier density and the diffusion coefficient is established, as follows.

In a semiconductor, the potential energy (U) of electrons drifting under the action of an electric field (E) increases in the direction of the field because electrons drift is in the direction opposite to that of the field. The electrostatic potential (V) defined in terms of positive charge varies in the opposite direction. The convention is just opposite in the case of holes. Considering the above notion, one can write for electrons,

$$V = -\frac{U}{q} \tag{2.31a}$$

and

$$E = -\frac{dV}{dx} \tag{2.31b}$$

It is understood that both V and E in Equations (2.31a) and (2.31b) are functions of x. Using these two equations, the intrinsic Fermi level can be correlated with the applied electric field as

$$E = -\frac{d}{dx}\left(-\frac{E_{Fi}}{q}\right) = \frac{1}{q}\frac{dE_{Fi}}{dx} \tag{2.32}$$

Since there is no net current flow at equilibrium, putting $J_n = 0$ in Equation (2.30a) one obtains

$$E = -\frac{D_n}{n\mu_n}\frac{dn}{dx} \tag{2.33}$$

Here n represents the equilibrium electron concentration (function of distance) and can be represented by Equation (2.24a) so that

$$\frac{dn}{dx} = \frac{n}{kT}\left(\frac{dE_{Fn}}{dx} - \frac{dE_{Fi}}{dx}\right) \tag{2.34}$$

The term dE_{Fn}/dx may be assumed to be zero because the equilibrium Fermi level cannot vary with distance. Combining Equations (2.32), (2.33), and (2.34) we have

$$\frac{D_n}{\mu_n} = \frac{kT}{q} \tag{2.35a}$$

Similarly for holes one can have

$$\frac{D_p}{\mu_p} = \frac{kT}{q} \qquad (2.35b)$$

Equations (2.35a) and (2.35b) represent the same phenomenon, known as *Einstein relation* that states that there is a fixed ratio between the carrier diffusion coefficient and the carrier mobility at equilibrium.

2.8.5 Continuity Equation

This is a differential equation representing the change of electron and hole concentrations as function of time and distance within the semiconductor. To derive the equation, we consider an infinitesimal volume element (Adx) of length dx and cross-sectional area A within the semiconductor. The movement of holes is considered when the current flows in the same direction. Assuming average hole concentration p and one-dimensional flow of holes at time t in x-direction, let the hole current entering the volume at position x be I_p and that leaving the volume at position $(x+dx)$ be (I_p+dI_p). For a positive value of dI_p, the decrease in the number of holes per unit time within the volume is dI_p/q. Hence the decrease in the number of holes per unit volume per unit time can be defined in terms of the hole current density J_p $(= I_p/A)$ as

$$\frac{1}{Adx}\frac{dI_p}{q} = \frac{1}{q}\frac{dJ_p}{dx} \qquad (2.36)$$

The number of holes remains in equilibrium under the process of thermal generation and recombination with electrons. If p_o be the equilibrium value of hole concentration and τ_p be the mean lifetime for holes, an increase of p_o/τ_p number of holes per unit volume per unit time is estimated due to thermal generation. At the same time, there is a decrease of p/τ_p number of holes per unit volume per unit time due to recombination. Since the total charge must be conserved, the algebraic sum of the above mechanisms is given by

$$\frac{\partial p}{\partial t} = \frac{p_o - p}{\tau_p} - \frac{1}{q}\frac{\partial J_p}{\partial x} \qquad (2.37a)$$

Equation (2.37a) is called the continuity equation for holes in a semiconductor. Similarly the continuity equation for electrons can be derived in terms of electron current density (J_n), instantaneous concentration (n), equilibrium concentration (n_o) and lifetime (τ_n) as

$$\frac{\partial n}{\partial t} = \frac{n_o - n}{\tau_n} + \frac{1}{q}\frac{\partial J_n}{\partial x} \qquad (2.37b)$$

The partial derivative symbol indicates the dependence of p and n and J_p and J_n on both position (x) and time (t).

2.9 Hall Effect

American physicist Edwin Herbert Hall (1855–1938) discovered in 1879, even before the discovery of electron, that a voltage develops across a current carrying conductor placed in a magnetic field perpendicular to the direction of current flow. The voltage is perpendicular to both the current and the magnetic field. The above phenomenon is known as the famous *Hall effect*. It is a very useful experimental technique in

- determining the concentration of charge carrier,
- distinguishing positive and negative charges, and
- determining carrier mobility in a material knowing its resistivity.

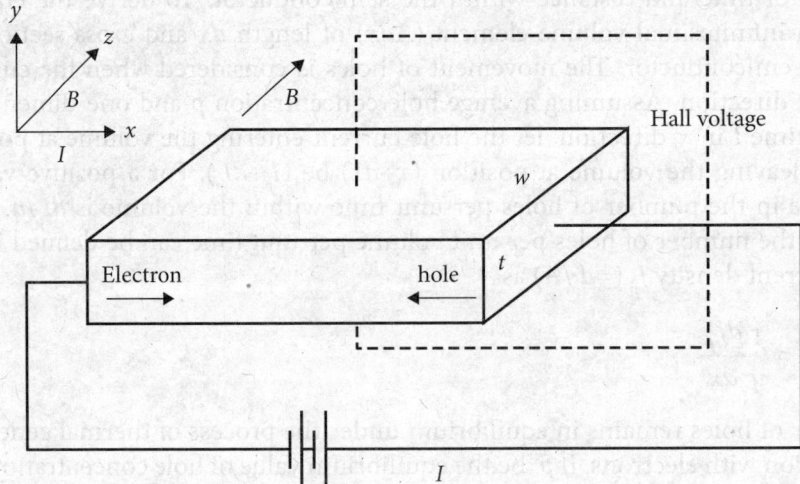

Figure 2.9 Hall effect set up

The Hall effect is indeed the real evidence that electric current in metals is caused by moving electrons and not protons. Figure 2.9 depicts a typical occurrence of Hall effect where a rectangular conductor (metal or semiconductor) slab of width w and thickness t carrying a steady current I along the x-axis is placed in a uniform magnetic field of magnitude B directed along the z-axis. The following three cases of the slab material may occur.

(i) In p-type semiconductor, the charge carriers are mainly holes with a small number of electrons. The directions of movement of both the carriers are indicated in the diagram. The Lorentz force deflects both the electrons and the holes along the negative y-axis. The holes, being the majority, form a positive charge at the bottom side of the slab and an electric field, referred to as the *Hall field* (E_H), is developed along the positive y-axis. The charge separation causes a potential difference, referred to as the Hall voltage (V_H) across the thickness of the slab. This voltage is measurable with external voltmeter.

(ii) In n-type semiconductor, the majority of charge carriers are electrons with few holes. This time also the Lorentz force pushes both the electrons and the holes toward the bottom side of the slab but the majority of electrons create a negative charge. Now E_H is directed toward the negative y-axis.

(iii) In metal, the charge carriers are electrons. The Lorentz force deflects the electrons toward the negative y-axis and the resultant E_H appears along the negative y-axis. However, in metals the Hall effect is negligible because of large carrier density and low drift velocity.

The deflection of charge carriers and the accumulation of charge take place at the same direction in all the above cases and it continues until the force on the carriers due to Hall field balances the magnetic force. At equilibrium,

$$qE_H = qv_d B \tag{2.38}$$

where v_d is the drift velocity of the carriers. Correlating it with the current density (J) using Equation (2.26), we obtain a constant known as the *Hall coefficient* (R_H) as

$$R_H = \frac{1}{qn} = \frac{E_H}{JB} \tag{2.39}$$

This is a characteristic of the material. In the case of p-type material, n is replaced by p. Using $E_H = V_H/t$ and $J = I/(wt)$, the Hall coefficient can be defined in terms of experimentally measurable quantities as

$$R_H = \frac{wV_H}{IB} \tag{2.40}$$

Observation and measurement of the Hall coefficient has the following useful applications.

(i) Since R_H is positive for p-type semiconductor (because q is +ve) and negative for n-type semiconductor (q is –ve), the type of semiconductor can be determined.

(ii) The carrier concentration (n or p) of the given sample can be obtained from Equation (2.39).

(iii) The Hall voltage is proportional to the magnetic field at a constant current [Equation (2.40)]. So it is a good probe for gauging an unknown magnetic field by measuring the current (I) and the Hall voltage (V_H) and knowing the sample dimension (w) and its Hall coefficient.

(iv) Knowing the conductivity or the resistivity of the material and measuring the carrier concentration, the carrier mobility can be determined.

2.10 Resistivity and Four-Probe Technique

The *resistivity* (SI units Ωm) of a material is the resistance offered by its sample of unit length and unit cross-sectional area. Knowing the resistivity of a material becomes essential for many experimental processes. If the sample be of long, wire-like geometrical shape, the resistivity [$\rho = (V/I)(A/l)$] can be determined by measuring the voltage drop (V) across its two ends for a known current (I) passing through it and knowing its length (l) and cross-sectional area (A). However, such simple *two-probe technique* is not suitable for samples of arbitrary shape, especially for semiconductors because of errors due to (i) contact resistance of measuring leads, (ii) hazard of material damage in soldering, and (iii) formation of rectifying Schottky contact (Chapter 3) due to metallic contact on semiconductor.

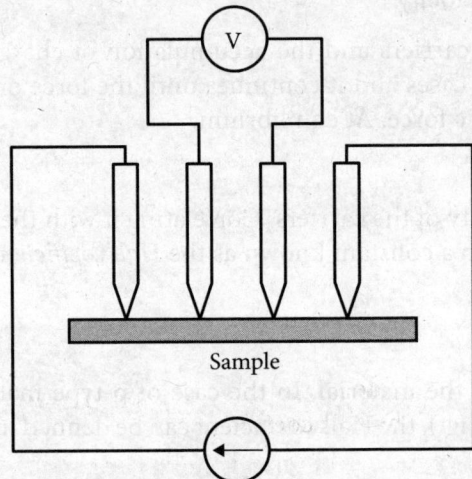

Figure 2.10 Four-probe technique

A *four-probe technique*, as sketched in Figure 2.10, is a generalized method for measuring resistivity. Four equally spaced ($\approx$ 1 mm) metal probes of pointed tips are connected with the sample by pressure contact. A high impedance current source supplies constant current (I) through the two outer probes and a high impedance voltmeter measures the voltage drop (V) across the two inner probes. The resistivity is obtained as $\rho = (V/I) \times$ constant, the constant term being dependent on probe spacing and sample thickness. Unlike the two-probe technique, where the voltage measuring contacts pass a current, the two inner probes of the four-probe technique draw negligible current (because of the high input impedance of the voltmeter) and thus avoids any unwanted voltage drop due to contact resistance.

Multiple Choice-Type Questions and Answers

2.1 Common semiconductor crystals have atoms bound mostly with
(a) ionic bond
(b) hydrogen bond
(c) covalent bond
(d) metallic bond

2.2 The band gap of semiconductors is of the order of
(a) $2-5 \times 10^{-19}$ J
(b) $20-50 \times 10^{-19}$ J
(c) $2-5 \times 10^{19}$ J
(d) $20-50 \times 10^{19}$ J

2.3 Semiconductors are stated to have .... temperature coefficient of resistance because their resistivity .... with increasing temperature.
(a) positive, decreases
(b) negative, decreases
(c) positive, increases
(d) positive, decreases

2.4 On increasing the temperature, the extrinsic semiconductor may act like
(a) n-type
(b) p-type
(c) amphoteric
(d) intrinsic

2.5 In an intrinsic semiconductor, the minority carrier concentration depends mainly on
(a) doping
(b) valency
(c) temperature
(d) conductivity

2.6 The density of the majority carriers in n-type semiconductor depends on the .... whereas that in p-type semiconductor depends on ....
(a) donor impurity ion concentration, acceptor impurity ion concentration
(b) donor impurity atom concentration, acceptor impurity atom concentration
(c) acceptor impurity atom concentration, donor impurity atom concentration
(d) acceptor impurity ion concentration, donor impurity ion concentration

2.7 On increasing the temperature, the Fermi level for n-type semiconductor moves .... the center of the band gap and the Fermi level for p-type semiconductor moves .... the center of the band gap.
(a) toward, away from
(b) away from, toward
(c) away from, away from
(d) toward, toward

2.8 The current flowing in an intrinsic semiconductor is mainly due to ....
(a) drift of conduction band electrons and valence band holes
(b) diffusion of conduction band electrons and valence band holes
(c) drift of conduction band electrons and diffusion of valence band holes
(d) diffusion of conduction band electrons and drift of valence band holes

2.9 The intrinsic carrier concentration of a semiconductor at 0 K is ....
(a) zero
(b) uncertain
(c) infinity
(d) material dependent

2.10 The carrier concentration in intrinsic semiconductor is function of ....
(a) pressure
(b) temperature
(c) both pressure and temperature
(d) neither pressure nor temperature

2.11 A semiconductor is said to be *fully compensated* when the ionized donor concentration is .... the ionized acceptor concentration and the semiconductor behaves as ....
(a) much greater than, extrinsic
(b) much less than, extrinsic
(c) equal to, intrinsic
(d) equal to, extrinsic

2.12 Which of the following parameters cannot be assessed directly from the measurement of the Hall voltage?
(a) Hall coefficient
(b) extrinsic type (n or p)
(c) magnetic field
(d) carrier mobility

2.13 When a potential difference is applied across a bar of intrinsic semiconductor, the total current through the bar is equal to the .... of electron and hole currents and the electron current is .... the hole current.
(a) difference, greater than
(b) sum, greater than
(c) sum, less than
(d) sum, equal to

2.14 The current through a p-type semiconductor flows due to ....
 (a) holes alone   (b) electrons alone
 (c) donor and acceptor ions   (d) both holes and electrons

 Hints. Though the majority of the carriers in the p-type semiconductor are holes, there is minor amount of electrons. Similarly there are holes as minority carriers in the n-type semiconductor.

2.15 The net charge of an n-type semiconductor is ....
 (a) positive   (b) zero
 (c) negative   (d) dependent on the dopant density
 [JEST 2012]

2.16 As shown in Figure 2.11, a conducting slab of copper is kept in the xy plane in a uniform magnetic field B along the x-axis. A steady current I flows through the cross-section of the slab along the y-axis.

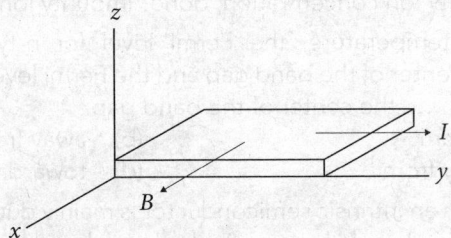

Figure 2.11

The direction of the electric field inside the slab, arising due to the applied magnetic field is along the:
 (a) negative y-direction   (b) positive y-direction
 (c) negative z-direction   (d) positive z-direction
 [JAM 2014]

 Hints. The electrons are pushed by the Lorentz force along the negative z-direction.

2.17 The experimentally measured transmission spectra of metal, insulator and semiconductor thin films are shown in Figure 2.12. It can be inferred that I, II and III correspond to

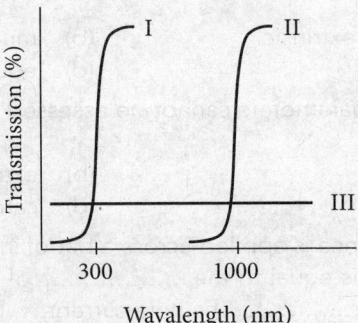

Figure 2.12

(a) insulator, semiconductor and metal (b) semiconductor, metal and insulator
(c) metal, semiconductor and insulator (d) insulator, metal and semiconductor

[NET June 2012]

Hints. Metals transmit very little. Both insulators and semiconductors absorb photon energy above a certain threshold. Insulators having higher band gap correspond to lower wavelength.

2.18 Consider a doped semiconductor having the electron and the hole mobilities μ_n and μ_p, respectively. Its intrinsic carrier density is n_i. The hole concentration p for which the conductivity is minimum at a given temperature is ....

(a) $n_i \sqrt{\dfrac{\mu_n}{\mu_p}}$ (b) $n_i \sqrt{\dfrac{\mu_p}{\mu_n}}$

(c) $n_i \dfrac{\mu_p}{\mu_n}$ (d) $n_i \dfrac{\mu_n}{\mu_p}$

[JAM 2008]

Hints. Put $n = \dfrac{n_i^2}{p}$ and equate $\dfrac{d\sigma}{dp}$ to zero.

Answers

2.1	(c)	2.2	(a)	2.3	(b)	2.4	(d)
2.5	(c)	2.6	(b)	2.7	(d)	2.8	(a)
2.9	(a)	2.10	(b)	2.11	(c)	2.12	(d)
2.13	(b)	2.14	(d)	2.15	(b)	2.16	(c)
2.17	(a)	2.18	(a)				

Reasoning-Type Questions and Answers

2.1 Are the energy and the momentum conserved in the transfer of carriers between the valence band and the conduction band?

Ans. When an electron in the valence band receives energy $h\nu$ in the form of photon (quantized electromagnetic radiation) greater than the band gap (E_g), the electron gets elevated to the conduction band consuming the energy equivalent to E_g and the remaining energy is dissipated as lattice vibration and heat. When an electron at the conduction band recombines with a hole in the valence band, the energy difference is released in the form of radiation, lattice vibration and heat depending on the type of semiconductor. For direct band gap semiconductors, the momentum (involving k) is conserved and the energy release is radiative. In the case of indirect band gap semiconductor, the momentum is changed (because of misalignment in k-space) and the energy released as lattice vibration and dissipation is non-radiative. Nevertheless the total energy and the total momentum in the interactions of carriers with photons and phonons (quantized lattice vibration) are always conserved.

2.2 How does the conductivity of (i) an intrinsic semiconductor and (ii) an extrinsic semiconductor change with temperature?

Ans. When the intrinsic semiconductor is raised to higher temperatures, more electrons get thermally excited to the conduction band so that the conductivity increases with temperature. In the case of extrinsic semiconductors, the picture is different. Let it be n-type semiconductor. The number of electrons contributed by the donor atoms remains unchanged with temperature whereas the concentration of the thermally generated electron–hole pairs increases with temperature. Ultimately the thermally generated electron–hole pairs may outnumber the free electron concentration due to donor atoms. This situation gives rise to almost equal concentrations of electrons and holes. Therefore, the n-type semiconductor behaves like an intrinsic semiconductor and its conductivity decreases with temperature.

2.3 Are the electrons and the holes 'particles'?

Ans. The electron is subatomic particle having an associated wave property. The hole is, after all a vacant space due to a missing electron. However, it is the difference in energy states that makes the distinction. The transport of electrons through the semiconductor is a complex process because of the interactions with ion cores and other electrons. The complexity is avoided by assuming an equivalent positively charged free particle of different effective mass (energy dependent). The same concept is applied to the electron also. Thus both the electron and the hole are 'quasi-particles'. Ignoring the wave aspect these are considered as Newtonian particles with different effective mass and different mobility.

2.4 What are 'light hole' and 'heavy hole'?

Ans. The valence band consists of mostly occupied energy states of the bonding electrons. The motion of holes in the almost filled valence band is associated with the rise of electronic energy. The hole movement occurs easily along the directions of bond between the neighboring atoms. So the holes along this direction have smaller effective mass and are referred to as *light holes*. In terms of wave vector, these have larger $\frac{\partial^2 E}{\partial k^2}$. Along any other direction, the movement is difficult and $\frac{\partial^2 E}{\partial k^2}$ is smaller, which corresponds to *heavy hole*.

2.5 What happens, if donor- and acceptor-type impurities are added in equal amount to an intrinsic semiconductor?

Ans. The resultant material continues to behave like intrinsic semiconductor. However, the conductivity gets reduced because of the growing lattice imperfections caused by the doping. A semiconductor of this category is known as *compensated semiconductor*.

2.6 What happens, if an n-type semiconductor is increasingly doped with acceptor-type impurities so that the concentration of the acceptor-type dopant atom becomes comparable with the donor concentration and ultimately exceeds that? What about the charge neutrality of the semiconductor during the above process?

Ans. The semiconductor tends to be intrinsic (actually compensated) and then acts as p-type semiconductor. This is indeed a common situation in IC technology. The semiconductor sample as a whole remains electrically neutral all the way. According to the charge neutrality condition, the total positive charge due to holes and donor ions must equal the total negative charge due to electrons and acceptor ions.

2.7 What is the physical significance of mass-action law?

Ans. It states that the product of the electron and hole concentrations in a semiconductor at a certain temperature is constant irrespective of the amount of donor or acceptor impurity atoms. It is so because the addition of donor impurities increases the number of free electrons but diminishes the number of holes by increased recombination with the excess electrons. Similar is the case of adding acceptor impurities and the charge neutrality condition is maintained always.

Solved Numerical Problems

2.1 Determine the Miller indices for the shaded crystal plane shown in Figure 2.13.

Soln. The intercepts are: twice the lattice distance a, thrice the lattice distance b and four times the lattice distance c.

The reciprocals are in proportion of 1/2, 1/3 and 1/4.

Multiplying each of the reciprocals by 12, the smallest possible triad of integers is 6, 4, 3.

Therefore, the Miller indices are (6,4,3)

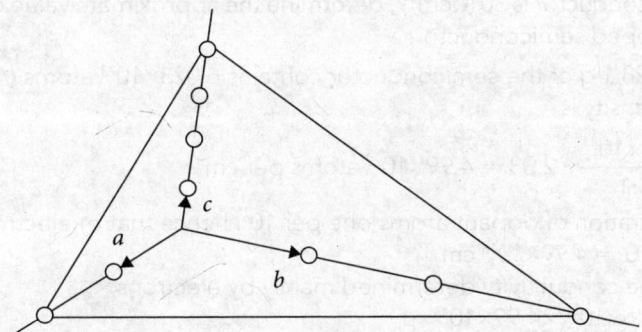

Figure 2.13

2.2 An intrinsic semiconductor sample has carrier concentration of 1.5×10^{10} cm^{-3} and electron and hole mobility of 0.135 m^{2}V^{-1}s^{-1} and 0.048 m^{2}V^{-1}s^{-1}, respectively. Calculate the conductivity of the semiconductor.

Soln. The intrinsic carrier concentration is the same for electrons and holes. Therefore, converting to the same system of units and using Equation (2.27),
the conductivity is calculated as: $1.6 \times 10^{-19} \times 1.5 \times 10^{-16} \times (0.135 + 0.048) = 0.439 \times 10^{-3}$ S

2.3 An experimentalist wishes to increase the conductivity of the above intrinsic semiconductor sample by 10^6 times through n-type doping. What would be the required concentration of donor atoms for this purpose?

Soln. The required conductivity is $0.439 \times 10^{-3} \times 10^6 = 439$ S.
Now the electrons are the dominant carriers and only the electron mobility (0.135 m^{2}V^{-1}s^{-1}) becomes significant. Therefore, the required doping concentration is
$439/(1.6 \times 10^{-19} \times 0.135) = 2.032 \times 10^{22}$ atoms per m^3.

2.4 An intrinsic semiconductor of atomic density 4.4×10^{28} m^{-3} is doped for n-type by adding donor atoms to the extent of one atom per 10^7 atoms of the host intrinsic material. Compare the conductivity of the resultant n-type material with that of the original sample. Given: electron mobility is 3800 cm^2 V^{-1}s^{-1}, hole mobility is 1800 cm^{2}V^{-1}s^{-1} and the intrinsic carrier concentration is 2.5×10^{13} cm^{-3}.

Soln. Converting to the same system of units, the atomic density is 4.4×10^{22} cm^{-3}.
Assuming each dopant atom per 10^7 semiconductor atoms contributes to one electron, the electron concentration in the n-type material is
$4.4 \times 10^{22}/10^7 = 4.4 \times 10^{15}$ cm^{-3}.
The hole concentration in the n-type sample, using mass-action law [Equation (2.20)] is
$(2.5 \times 10^{13})^2/4.4 \times 10^{15} = 1.42 \times 10^{11}$ cm^{-3}.
Using Equation (2.27), the conductivity of the resultant n-type sample is
$1.6 \times 10^{-19} \times (4.4 \times 10^{15} \times 3800 + 1.42 \times 10^{11} \times 1800)$ and
the conductivity of the original intrinsic sample is
$1.6 \times 10^{-19} \times (2.5 \times 10^{13} \times 3800 + 2.5 \times 10^{13} \times 1800)$
so that the ratio is about 119.43

2.5 A semiconductor element having atomic weight 28.1 g and density 2.33 g cm^{-3} is doped for n-type by adding impurity atoms one per million. If the conductivity of the resultant n-type semiconductor is 30 (Ωcm)$^{-1}$, determine the approximate value of electron mobility (μ_n) in the doped semiconductor.

Soln. Since 28.1 g of the semiconductor contains 6.023×10^{23} atoms (Avogadro number), its atomic density is
$$\frac{6.023 \times 10^{23}}{28.1} \times 2.33 = 4.99 \times 10^{22} \text{ atoms per cm}^3.$$
The concentration of dopant atoms one per 10^6, hence that of electrons is
$(4.99 \times 10^{22})/10^6 = 4.99 \times 10^{16}$ cm^{-3}
Assuming the conductivity determined mainly by electrons,
$30 = 1.6 \times 10^{-19} \times 4.99 \times 10^{16} \times \mu_n$
Hence, $\mu_n = 3758$ cm^{2}V^{-1}s^{-1} (approx.)

2.6 If a sample of silicon having intrinsic carrier concentration of 1.5×10^{16} m^{-3} is doped for n-type with donor atom concentration of 1×10^{23} m^{-3}, estimate the shift in Fermi level from its intrinsic position. What happens, if the doping concentration is increased up to 2.5×10^{25} m^{-3}?

Soln. The shift in Fermi level is estimated from Equation (2.24a) as
$E_{Fn} - E_{Fi} = 0.026 \times \ln(1 \times 10^{23}/1.5 \times 10^{16}) = 0.408$ eV (approx.)
Thus on doping for n-type, the Fermi level in the silicon sample shifts up toward the conduction band by 0.408 eV. [Here the approximation of $kT/q = 0.026$ eV has been used.]

In the second case, following the same process,
$E_{Fn} - E_{Fi} = 0.026 \times \ln(2.5 \times 10^{25}/1.5 \times 10^{16}) = 0.55$ eV (approx.)

[N.B. A shift of 0.55 eV from the mid-band position is equal to almost half of the silicon band gap implying that further doping might cause the Fermi level penetrate the conduction band. This is not unrealistic! Such a highly doped semiconductor is called *degenerate semiconductor* where the Fermi level pushes into the conduction and valence bands. Such materials have applications in fabricating novel semiconductor devices, such as tunnel diode and semiconductor laser.]

2.7 The energy required to break the covalent bonds in a semiconductor crystal and thus moving up an electron to the conduction band should be at least equal to the band gap energy. If the electrons of a semiconductor having band gap of 1.4 eV excited by a radiation are raised to the conduction band, what is the wavelength (λ) of the radiation?

Soln. In the limiting case, the energy per photon of the radiation equals the band gap energy so that
$$hc/\lambda = 1.4$$
where $h = 6.63 \times 10^{-34}$ Js $= 4.14 \times 10^{-15}$ eVs is Plank's constant and $c = 3 \times 10^8$ ms^{-1} is the speed of the radiation.

The wavelength is calculated as 887 nm. This is near-infrared radiation.

2.8 An intrinsic semiconductor sample is doped with donors from one side such that the doping concentration (N_D) decreases exponentially with distance (x) from the junction following the relation
$$N_D = N_{Do}\exp(-\alpha x)$$
where α is a constant and N_{Do} is the concentration at the initial position. Deduce an expression for the electric field $E(x)$ at equilibrium assuming N_D is much greater than the intrinsic concentration.

Soln. The electron current density is [Equation (2.30a)]
$$J_n = nq\mu_n E + qD_n \frac{dn}{dx}$$

At equilibrium, $J_n = 0$ so that electric field is
$$E(x) = -\frac{D_n}{n\mu_n}\frac{dn}{dx}$$

Here $n = N_D = N_{Do}\exp(-\alpha x)$ so that $\frac{dn}{dx} = -\alpha N_D$.

Using this expression for dn/dx, the above electric field can be formulated as
$$E(x) = \frac{D_n}{\mu_n}\alpha = \frac{kT}{q}\alpha \text{ [using Einstein relation Equation (2.35a)]}$$

Exercise

Subjective Questions

1. Explain *lattice* and *unit cell* in connection with crystal structure making necessary diagrams.
2. Discuss why a crystal does not possess a fivefold axis of rotational symmetry.
3. Sketch the orientations of (100) plane, (110) plane and (111) plane in a cubic unit cell.
4. Distinguish between electron and hole. Why the hole mobility is less than the electron mobility?
5. What is *effective mass* and what is its significance in semiconductor science and technology?
6. Show that $\hbar k$ has the dimension of momentum.

7. Show that N_C [Equation (2.13a)] and N_V [Equation (2.15a)] have the dimension of number per unit volume.
8. What is the Fermi level related to a solid? Explain the probability of energy distribution around the Fermi level (i) at 0 K and (ii) at 300 K.

Conceptual Test

1. Justify the statement, "The unit cell has all the structural properties of a given lattice".
2. Do you agree with the statement, "The lattice determines the mechanical properties of a crystal as well as its electrical properties"? Justify.
3. Assuming planes (hkl) and ($\overline{hkl}$) as identical, explain how many equivalent planes are there in the following cubic crystals.
   (a) (100),   (b) (110),   (c) and   (d) (123)
4. Can you make any reasoning for the naming of 'conduction band' and 'valence band'?
5. The energy (E) in a solid is a parabolic function of k given by $E = C(k-k_o)^2$, k_o and C are constants. Find the effective mass, hence justify that the dynamical behavior of the electron will be similar to that of a free particle with this effective mass.
6. Can a hole be compared with a positron?
7. The n-type semiconductor contains excess of electrons. Is it electrically neutral?
8. In multiple choice-type question No. 2.18, why does the method lead to the minimum hole concentration only whereas the general rule is that the derivative becomes zero for either minimum or maximum?
9. Can you determine the band gap (E_g) of an intrinsic semiconductor by measuring its conductivity (σ)?

   Hints. Using Equation (2.21), one can write $\sigma = \text{Const.} \times \exp[-E_g/(2kT)]$ where the constant term contains T but that has negligible contribution in comparison with the T within the exponential term for moderate temperatures (e.g. up to 100° C). Therefore, measuring the conductivity at different temperatures and plotting $\ln(\sigma)$ against $1/T$ one can estimate the band gap from the slope.

Numerical Problems

1. A unit cell has three orthogonal axes of lengths a = 1.21 Å, b = 1.84 Å and c = 1.97 Å, respectively. A plane of Miller indices ($23\overline{1}$) has an intercept of 1.21 Å along the x-axis. Find the lengths of intercept at y- and z-axes.
2. Sodium (Na) exhibits body-centered-cubic (BCC) crystal structure with atomic radius 0.186 nm. The lattice parameter of Na unit cell is .... nm. (Round off to 2 decimal places)

   [JAM 2019]
3. Determine the intrinsic carrier concentration of a semiconductor having resistivity of 0.47 Ωm. Given: electron mobility is 0.39 m²V⁻¹s⁻¹ and hole mobility is 0.19 m²V⁻¹s⁻¹.
4. A silicon sample having intrinsic carrier concentration of 1.5×10^{10} cm⁻³ is doped with 10^{21} phosphorus atoms per m³. What will be the hole concentration at equilibrium?
5. Determine the resistivity of the intrinsic semiconductor having μ_n = 1350 cm²V⁻¹s⁻¹, μ_p = 480 cm²V⁻¹s⁻¹ and $n_i = 1.5 \times 10^{10}$ cm⁻³, symbols having the usual meaning.

6. What should be the concentration of the donor atoms to be added to a sample of intrinsic silicon in order to produce n-type material of conductivity 0.432×10^3 $(\Omega m)^{-1}$? Given the electron mobility is 0.13 $m^2V^{-1}s^{-1}$.

7. A sample of direct band gap semiconductor has band gap of 1.99 eV. What will be the color of the radiation emitted by electron–hole recombination in the sample?

8. The band gap of an intrinsic semiconductor is 0.72 eV and the effective masses of hole and electron are related as $m_h^* = 6m_e^*$. At 300 K, the Fermi level with respect to the valence band edge is at .... Given: Boltzmann constant is 1.38×10^{-23} JK^{-1}.

[GATE 2015]

9. An intrinsic semiconductor of band gap 1.25 eV has electron concentration 10^{10} cm^{-3} at 300 K. Assume that the band gap is independent of temperature and the electron concentration depends only exponentially on temperature. If the electron concentration at 200 K is $Y \times 10^N$ cm^{-3} ($1 < Y < 10$, N = integer) then the value of N is ....

[JAM 2017]

10. A p-type semiconductor slab, as sketched in Figure 2.14, carries a current I = 100 mA in a magnetic field B = 0.2 T. If V_y = 0.25 mV and V_x = 2 mV, the mobility of holes in the semiconductor is .... $m^2V^{-1}s^{-1}$ up to two decimal places.

[GATE 2018]

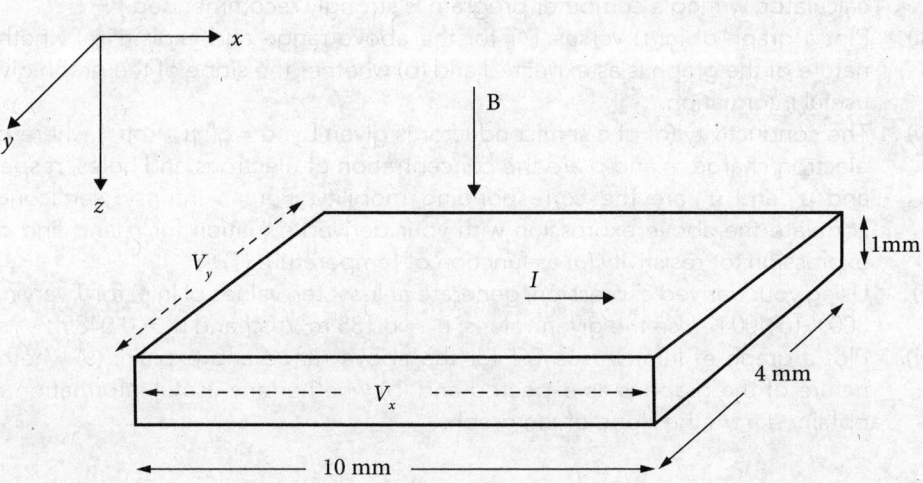

Figure 2.14

11. In an experiment, the resistance of a rectangular slab of a semiconductor is measured as a function of temperature. The semiconductor shows a resistance of 300 Ω at 200 K and 2 Ω at 250 K. Its energy band gap is .... [Given: ln(15) = 2.708, ln(10)= 2.303]
    (a) 0.138 eV (b) 0.431 eV (c) 0.690 eV (d) 0.862 eV

[JAM 2011]

Project Work on Chapter 2

At equilibrium state, the electron and hole concentration of an intrinsic semiconductor are given by
$$n_o = N_C \exp[-(E_C - E_F)/kT]$$
and
$$p_o = N_V \exp[-(E_F - E_V)/kT]$$
respectively, where

$$N_C = 2\left(\frac{2\pi m_n^* kT}{h^2}\right)^{3/2} \quad \text{and} \quad N_V = 2\left(\frac{2\pi m_p^* kT}{h^2}\right)^{3/2},$$

m_n^* and m_p^* being the effective mass electrons and holes, respectively. The other symbols have the usual meaning.

(i) Find out the above equations in this chapter and read the related discussions. Based on the above information, find out a suitable expression for intrinsic carrier concentration (n_i) of the semiconductor as function of temperature (T).

(ii) Using your derived expression, generate at least ten values of $\ln[n_i(T)]$ for T varying from 300 K to 400 K. Given: $m_o = 9.11 \times 10^{-31}$ kg, $m_n^* = 0.98 m_o$, $m_p^* = 0.16 m_o$, $E_g = 1.12$ eV, $k = 1.38 \times 10^{-23}$ JK^{-1} and $h = 6.62 \times 10^{-34}$ Js. [Though the work can be done with a calculator, writing a computer program is strongly recommended.]

(iii) Plot a graph of $\ln(n_i)$ versus $1/T$ for the above range and explain: (a) whether the nature of the graph is as expected and (b) whether the slope of the graph gives any useful information.

(iv) The conductivity (σ) of a semiconductor is given by: $\sigma = q(n\mu_n + p\mu_p)$ where q is the electron charge, n and p are the concentration of electrons and holes, respectively and μ_n and μ_p are the corresponding mobility. For an intrinsic semiconductor, correlate the above expression with your derived equation for n_i and find out the expression for resistivity (ρ) as function of temperature (T).

(v) Using your derived expression, generate at least ten values of $\ln(\rho)$ for T varying from 300 K to 400 K. Use the given values: $\mu_n = 0.135$ m^{2}V^{-1}s^{-1} and $\mu_p = 0.048$ m^{2}V^{-1}s^{-1}.

(vi) Plot a graph of $\ln(\rho)$ versus $1/T$ for the above range and explain: (a) whether the nature of the graph is as expected and (b) whether any useful information can be obtained from the slope of the graph.

3

p–n Junction Diodes

The semiconductor p–n junction is the most fundamental part of many electronic devices including diodes and transistors, and it is a basic element for understanding the working of other semiconductor devices. This chapter brings together the concepts on fabrication of a junction, barrier formation at the junction carrier and transport through it, forward and reverse biased conditions of a p–n junction and the corresponding energy band pictures and the wide variety of diodes that can be constructed using the p–n junction.

3.1 Fabrication of p–n Junction

Chapter 2 states that a piece of semiconductor, either n-type or p-type is a solid of variable electrical conductivity depending on the extent of doping. As soon as the combination of p- and n-type materials takes place, the electrical properties change radically. The junction acts a valve allowing current in a single direction. It is capable of converting an alternating input voltage or current to a pulsating direct voltage or current output, the phenomenon being referred to as the *rectification*.

Though it is called a 'junction', the p–n junction is not at all constructed by gluing or soldering two separate n-type and p-type pieces of semiconductors. In order to maintain the regularity of lattice structure, the same piece of intrinsic semiconductor crystal is doped for n- and p-type from two sides. Chapter 12 contains the detailed narration on the construction of p–n junction in connection with integrated circuit (IC) technology. The following are the common techniques of fabricating a p–n junction.

Grown Junction: This technique used to be adopted in the early days of semiconductor technology. In this process, the type of the dopant atom is abruptly changed during the crystal growth. For example, during crystal growth of silicon from melt, phosphorus is added for n-type doping and then boron atoms are added in larger concentration for p-type doping.

Alloyed Junction: This procedure is suitable for small scale production in laboratory. The doped semiconductor is alloyed with the material containing the opposite type of dopant.

For example, n-type germanium is heated with indium to form a molten alloy. On cooling, the germanium grows out of the alloy because of reduced solubility in solid state. At the interface of the alloy and the separated germanium, a region of germanium exists with high concentration of indium atoms and this germanium becomes p-type. Thus a p–n junction is formed.

Now-a-days, the planar technology is used for p–n junction and IC fabrication, which involves the growth of single-crystal semiconductor layers on another single-crystal semiconductor sheet, called the *substrate*. The processes is supported by periodic oxidation, selective doping by lithographic technique and formation of metallic connections. The following two techniques are popular for large scale production.

Diffused Junction: The semiconductor crystal, along with the dopant species is kept at high temperature ($\approx$ 1000°C) ambient when the dopant atoms acquire high kinetic energy and move into the semiconductor crystal.

Implantation Junction: A beam of impurity ions is accelerated to high energy ($\approx$ kV to MV) so that the ions get penetrated directly into the semiconductor crystal.

The latter two types of junction are formed with the widely used doping techniques of diffusion and ion implantation. These have the advantage of changing only the n- or p-type of doping that makes the junction. If the donor and the acceptor concentrations on the two sides of the junction are constant throughout, it is called an *abrupt junction* or a *step junction*. If the concentrations vary linearly with distance from the junction, it is called *linearly graded junction*.

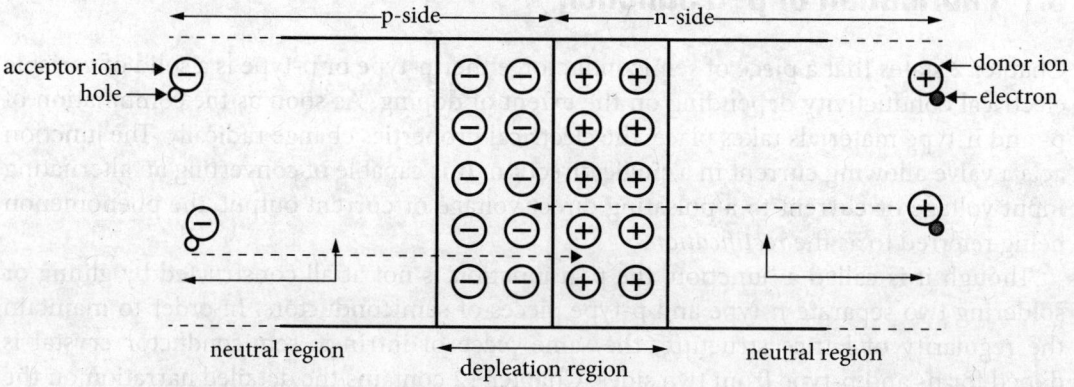

Figure 3.1 An ideal p–n junction and its components

3.2 Barrier Formation in p–n Junction

The physical properties and the corresponding energy band diagrams of n- and p-type semiconductors are illustrated separately in Chapter 2. Now the combined activities of these two types in a junction are elucidated. The mechanisms related to the formation of an ideal p–n junction with uniform doping are indicated in Figure 3.1.

The n-region contains plenty of electrons detached from the pentavalent dopant atoms, which become positive donor ions after donating the electrons. The p-region comprises holes in abundance because the trivalent dopant atoms have deficiency of electrons. These atoms convert to negative acceptor ions after accepting the electrons. The donor and the acceptor ions remain fixed in position by covalent bonding whereas the electrons and the holes are freely movable across the junction by the process of diffusion. Because of the concentration gradient, the electrons diffuse from the n-side, cross the junction and recombine with the holes at the p-side. Similarly the holes diffuse from the p-side, cross the junction and recombine with the electrons at the n-side. Thus a certain region on both sides of the junction contains neither electrons nor holes. This region has the following alternative nomenclatures:

- *Depletion region*, because this region is depleted of mobile charge carriers: electrons and holes,
- *Space-charge region*, because this space contains fixed immobile charges: positive donor ions at the n-side and negative acceptor ions at the p-side and
- *Transition region*, because it develops from the transition of charge carriers across this zone.

This book uses the name 'depletion region' throughout. The fixed ions in this space build altogether a net positive charge at the n-side and a net negative charge at the p-side and this charge separation on either side of the junction gives rise to the following two entities.

(i) An electric field is developed at the junction directed from the positive charge to the negative charge. It opposes the diffusion current of each type of carrier and creates a drift component of current from the n-side to the p-side, opposing the diffusion current. The current due to the drift of carriers cancels that due to diffusion and no net current flows across the junction at equilibrium. Any zone outside the depletion region is called the *neutral region* where the electric field is zero. Each of the neutral regions at the p-side and the n-side has a constant potential.

(ii) A potential difference, termed as *built-in potential* develops across the depletion region. This is also called *barrier potential* as it arrests further flow of electrons or holes through the junction. It is also known as *diffusion potential*, since the diffusing carriers must surmount this potential barrier in crossing the junction. The gradient of the potential is, as usual, in the direction opposite to the direction of the field. This book uses the name 'built-in potential' all over.

It is obvious that no further diffusion of carriers through the p–n junction can take place without applying any external source of energy, such as a voltage source. The process of applying such external voltage source to conduct current through a p–n junction is known as *biasing*, as discussed in the subsequent sections. The width of the depletion region decreases on increasing the doping concentration. For a p–n junction with a certain doping concentration, the depletion width can be changed by changing the biasing, as will be discussed later.

3.2.1 Built-in Potential

A suitable expression for the built-in potential can be derived in terms of doping concentrations as stated below. The built-in potential makes the net current through the p–n junction zero. So recalling Equation (2.30a) and equating the sum of drift and diffusion currents to zero,

$$n(x)q\mu_n E(x) = -qD_n \frac{dn(x)}{dx} \tag{3.1}$$

Both the electron concentration $n(x)$ and the electric field $E(x)$ are functions of distance (x) from a specific origin, the junction in the present case. Using Einstein relation [Equation (2.35a)] in Equation (3.1) and substituting for

$$E(x) = -\frac{dV(x)}{dx}, \tag{3.2}$$

Equation (3.1) can be converted as

$$\frac{dV(x)}{dx} = \frac{kT}{q} \frac{1}{n(x)} \frac{dn(x)}{dx} \tag{3.3}$$

Integrating over the depletion region,

$$\int_{V_p}^{V_n} dV(x) = \frac{kT}{q} \int_{n_p}^{n_n} \frac{dn(x)}{n(x)} \tag{3.4}$$

where V_p and V_n are the potential at the end of the depletion region at the p-side and the n-side, respectively. The equilibrium concentration of electrons outside the depletion region at the p-side and the n-side are denoted by n_p and n_n, respectively. Integrating Equation (3.4) with the given limits,

$$V_n - V_p = \frac{kT}{q} \ln \frac{n_n}{n_p} \tag{3.5}$$

The difference $(V_n - V_p)$ is the built-in potential (V_o). Approximating n_n as the donor atom concentration (N_D) and approximating

$$n_p \approx \frac{n_i^2}{N_A}, \tag{3.6}$$

N_A being the acceptor atom concentration and n_i the intrinsic carrier concentration, Equation (3.5) becomes

$$V_o = \frac{kT}{q} \ln \frac{N_A N_D}{n_i^2} \tag{3.7}$$

Equation (3.7) expresses the built-in potential in terms of doping concentrations. The same expression can be obtained from the current transport equation [Equation (2.30b)] for holes.

p–n Junction Diodes

The built-in potential (V_o) actually appears as a barricade of qV_o (expressed in electron-volt). The electrons at the n-side must acquire this much energy to surmount the barrier. The hurdle for the holes is also of the same amount but the direction is opposite.

> **Note: Temperature Dependence of Built-in Potential**
>
> The potential barrier at the junction depends on doping density, temperature and intrinsic carrier concentration. For a certain junction, increase of temperature causes more drift of minority carriers across the junction so that the barrier of potential is reduced. Example: For silicon, the reduction of barrier potential is about 2 mV per °C rise in temperature. If the built-in potential of a silicon p–n junction be 0.7 eV at 25°C, at 100°C it would be 0.7 – 0.002(100 – 25) = 0.55 eV.

Example 3.1

A p–n junction diode has $N_D = 10^{15}$ cm^{-3}, $N_A = 10^{17}$ cm^{-3} and $n_i = 10^{10}$ cm^{-3}, symbols having the usual meaning. Determine the value of the built-in potential at the junction.

Solution

Using Equation (3.7), the built-in potential is

$$0.026 \ln \frac{10^{15} \times 10^{17}}{\left(10^{2}\right)^{2}} = 0.718 \text{ eV (approx.)}$$

3.2.2 Fermi Level in p–n Junction

It may be recollected from Chapter 2 that the energy bands of a solid actually represent the energy–momentum (E–k) relationship for its charge carriers and it depends on the crystal periodicity, hence on the orientation of crystal planes in three dimensions. The energy bands and the interim gaps are different in [100] and [111] directions and the maxima and minima of the bands may or may not occur at the same value of k depending on the material. For simple interpretations, it is customary to indicate the energy band diagram of a semiconductor just by two horizontal lines denoting the conduction and valence band edges and the Fermi level is indicated by dotted line within the gap at the middle, above or below depending on intrinsic, n-type or p-type material.

Extending the same simplification, the energy band diagram for a p–n junction is constructed by joining the individual band pictures for n- and p-type, as indicated in Figure 3.2. Here the vertical direction, of course represents energy but the horizontal direction corresponds to distance. The foremost condition for drawing an energy band diagram for a p–n junction is a Fermi level (E_F) aligned throughout the n- and p-type regions. If it were not so, one side would indicate higher average energy than the other without applying any external voltage source, which is impossible. The job is still incomplete and it carries forward with a question of how to join the conduction and valence band edges of these two sides: straight line or any curve?

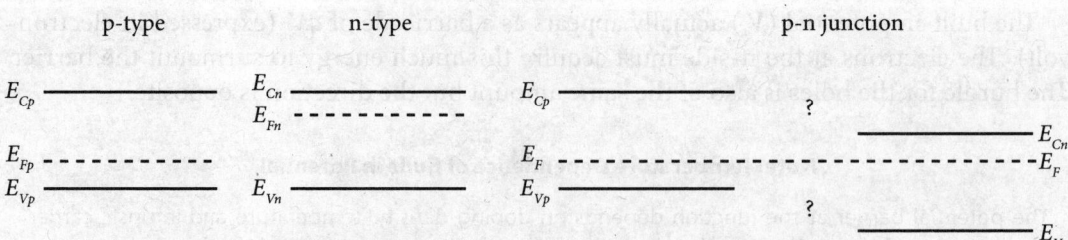

Figure 3.2 Alignment of Fermi level while constructing the band diagram for p–n junction

3.2.3 Energy Band Diagram of p–n Junction

To complete the picture of Figure 3.2, we need explicit information on the change of potential (V) with distance (x) that can be correlated with the charge density (ρ) distribution using Poisson's equation as

$$\frac{d^2V}{dx^2} = -\frac{\rho}{\varepsilon} \qquad (3.8)$$

The permittivity of the semiconductor is denoted by ε. The electric field obtained by integrating Equation (3.8) is given by

$$E = -\frac{dV}{dx} = -\int \frac{\rho}{\varepsilon} dx \qquad (3.9)$$

A p–n junction with depletion region extending from x_p to x_n is shown in Figure 3.3(a). The variation of electric field with distance, as apparent from Equation (3.9) is linear for constant charge density. The field is zero at x_p and at x_n, the two ends of the depletion region and is maximum (E_o) at the junction, as indicated in Figure 3.3(b). The potential is obtained by integrating the field expressed by Equation (3.9) as

$$V = -\int E dx = -\int \frac{\rho}{\varepsilon} x dx \qquad (3.10)$$

Equation (3.10) implies that the potential varies in proportion of x^2 and this nonlinear variation is sketched in Figure 3.3(c). The potential energy is the product of potential and charge, hence the potential energy variation, as denoted in Figure 3.3(d), is just opposite to that of Figure 3.3(c). Following these concepts, the energy band diagram for the whole p–n junction is drawn as that in Figure 3.3(e). The conduction band edges and the valence band edges of the two sides are now joined by nonlinear curves composed of two second-order polynomials, each one starting from the junction.

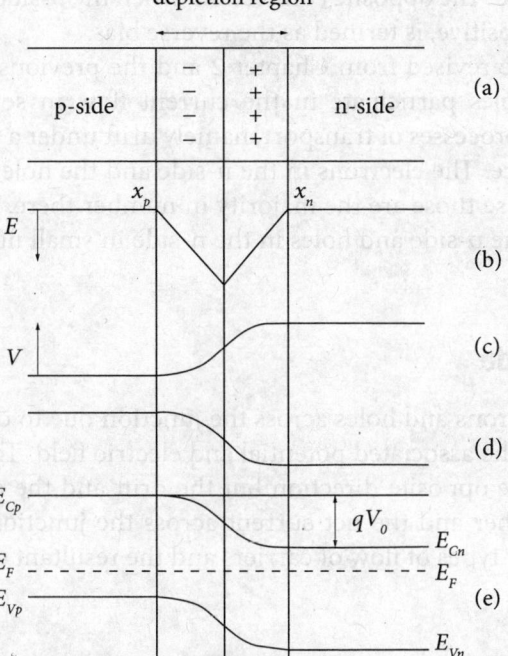

Figure 3.3 Properties of p–n junction: (a) depletion region, (b) variation of electric field (E) over it, (c) corresponding electrostatic potential, and (d) potential energy for electrons; (e) Energy band diagram for p–n junction indicating the band edges at the p-side (E_{Cp} and E_{Vp}) and the n-side (E_{Cn} and E_{Vn}) and the aligned Fermi level (E_F)

Note: Importance of Band Gap

In Chapter 2, we have seen that the magnitude of the energy gap between the conduction and the valence band distinguishes among metals, insulators and semiconductors. The conductivity of the intrinsic semiconductor depends on it. In this chapter we will see that many other device properties, such as diode forward cut-in voltage and fabrication of light-emitting diodes of different wavelengths, photodiodes and solar cells take into consideration the semiconductor band gap. Also junctions of dissimilar band gap materials, such as GaAs and AlAs can create *heterojunction* and *quantum well* structures.

3.3 Forward and Reverse Bias

A semiconductor p–n junction with metallic connections taken out from the n- and p-sides, is available commercially as p–n junction diode. It has the property of *rectification* or allowing current in a single direction. 'Biasing a diode' means connecting an external voltage source across its two metallic leads. When the p-side is joined to the positive terminal of the source and the n-side with the negative terminal, it is called the forward bias applied to the diode. The opposite connection, when the p-side is joined to the negative and the n-side to the positive, is termed as the reverse bias.

A few words may be revised from Chapter 2 and the previous portion of this chapter. Both electrons and holes participate in the current flow in semiconductors and both undergo two separate processes of transport, namely drift under a field and diffusion out of concentration difference. The electrons in the n-side and the holes in the p-side are called majority carriers because those are the majority in number there. Thermal excitations also generate electrons in the p-side and holes in the n-side in small numbers, which are called minority carriers.

3.3.1 Unbiased Diode

The initial flow of electrons and holes across the junction due to diffusion gives rise to the depletion region with the associated potential and electric field. The field tends to drift the minority carriers in the opposite direction but the drift and the diffusion components of current cancel each other and the net current across the junction of the unbiased diode becomes zero. The four types of flow of carriers and the resultant currents are summarized in Table 3.1.

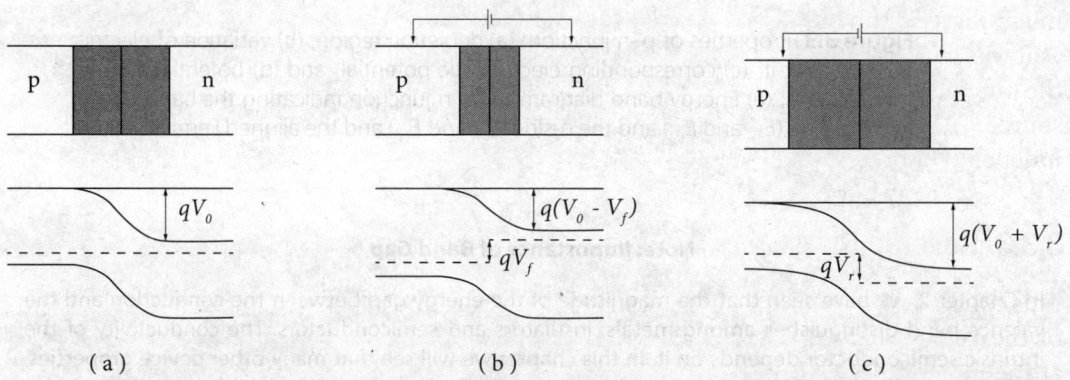

Figure 3.4 Depletion region and energy band diagram for (a) unbiased diode, (b) forward biased diode, and (c) reverse biased diode

Table 3.1 Summary of the carrier flow and the resultant current directions across a p–n junction

Carrier flow	causing current	Net current direction
Hole (majority) diffusion from p to n	from p to n	Total diffusion current from p to n
Electron (majority) diffusion from n to p	from p to n	
Hole (minority) drift from n to p	from n to p	Total drift current from n to p
Electron (minority) drift from p to n	from n to p	

The depletion region of the unbiased diode attains a steady width, as sketched in Figure 3.4(a). This also interprets the corresponding energy band characteristics. The built-in potential (qV_o) poses a barricade to the carriers and the Fermi level gets aligned for both the sides.

3.3.2 Forward Biased Diode

The working principle of the diode under forward bias is explained with Figure 3.4(b), both in terms of the carrier flow and in connection with the energy band picture. As indicated, the positive terminal of the external voltage source is joined to the p-side. It pushes the positive holes of the p-side toward the junction. Similarly the negative terminal of the source repels the negative electrons of the n-side toward the junction. Thus the carriers continue to gather in the direction of the junction from both sides and the depletion width decreases with forward bias. If these electrons and holes can cross the junction, the negative and the positive terminal of the voltage source attract the holes and the electrons, respectively and a net current flows in the external circuit directed from the negative to the positive terminal. This current increases rapidly with the forward bias voltage.

In terms of the energy band diagram, as indicated in Figure 3.4(b), it is inferred that the applied forward bias (V_f) lowers the potential barrier to $q(V_o - V_f)$ that facilitates the majority carriers to cross the junction. So the diffusion of both electrons and holes increase much thereby enhancing the diffusion current much more while the drift current as small as it was in the unbiased condition. The Fermi level remains no longer lined up throughout the junction but gets separated by an energy difference of qV_f caused by the external voltage source. The minority carriers participating in the drift component of current is almost independent of the applied bias and potential barrier.

3.3.3 Reverse Biased Diode

This condition is defined with Figure 3.4(c). This time the positive and the negative terminal of the voltage source is joined to the n-side and the p-side, respectively and because of attraction, both the electrons and the holes get away from the junction. So the width of the depletion region increases with reverse bias. Negligible diffusion of carriers comes about across the junction from either side leading to virtually no diffusion current. Just the drift of the minority carriers causes a very feeble current through the junction under reverse bias.

The potential barrier for an applied reverse bias V_r, as depicted in Figure 3.4(c), increases up to $q(V_o + V_r)$, which erects a larger potential barricade to the carriers for diffusing across the junction. The Fermi level is misaligned in the opposite direction by an amount of qV_r.

3.4 Diode Current–voltage Characteristics

The above demonstrations make it obvious that the p–n junction diode allows the current through it in a single direction. It conducts under forward bias and does not conduct under reverse bias. Thus it acts analogous to a valve, which governs the flow of fluids in a single direction. The property of causing unidirectional current flow, known as rectification was first realized with vacuum tube diode, which earned it the nickname 'valve diode'. Though the semiconductor diode has similar characteristic, it is not called 'valve'. The mechanisms of current conduction are quite different in the two devices. The present section deals with the current–voltage characteristics of p–n junction diode only.

Figure 3.5(a) gives the circuit symbol for p–n junction diode. The p- and n-sides are indicated for convenience. These are not written in the circuit diagram. Figure 3.5(b) and Figure 3.5(c) represent the forward and reverse biased conditions, respectively. Figure 3.5(d) suggests that the diode can be represented by an equivalent circuit containing a voltage (V_D) due to the built-in potential at the junction and a series resistance (R_s) owing to the bulk semiconductor. It is reasonable, since the current through the diode depends on the polarity of the power supply. It does not amplify but controls the direction of current. In this sense, the diode is quite different from passive elements, such as resistors, capacitors and inductors. Therefore, the diode is treated as an *active device* though it does not deliver energy but dissipates energy. There was a short discussion on active and passive devices in Chapter 1. The current (I) through a diode under forward bias voltage (V) can be expressed as

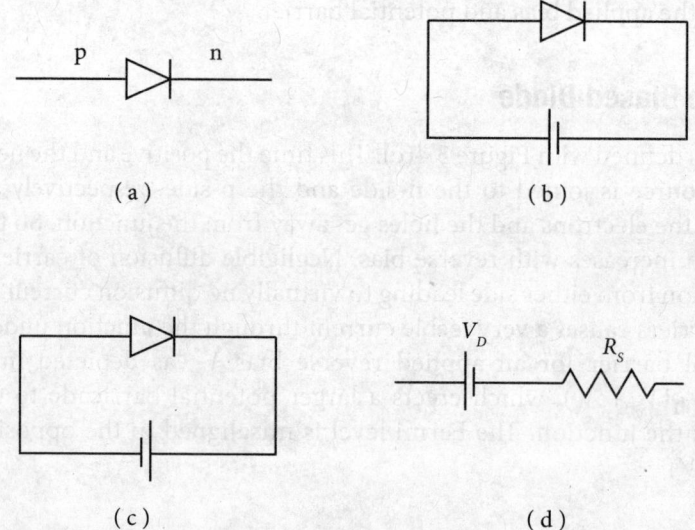

Figure 3.5 The p–n junction diode: (a) symbol, (b) under forward bias, (c) under reverse bias, and (d) its equivalent circuit showing the voltage (V_D) across the diode and resistance (R_s) in series

$$I = I_o[\exp(qV/kT) - 1] \tag{3.11}$$

The mathematical derivation of Equation (3.11) is a lengthy process and we adopt this widely used equation in this readymade form. To be more specific, the exponential term in Equation (3.11) is actually $[\exp(\text{constant} \times qV/kT)]$ where the constant term can be considered as unity for higher currents, relevant to all practical purpose. The quantity I_o is called the *reverse saturation current*, which arises from the drift of thermally generated carriers and is almost independent of the bias voltage. This is also termed as the *generation current* because the participant carriers are thermally generated.

The current through the diode under reverse bias is almost I_o, a negligibly small quantity independent of bias. Typically it is of the order of microamperes (μA) for Ge diodes and nanoamperes (nA) for Si diodes. It increases with temperature because the thermal generation of carriers increases with temperature. More carriers can be generated in presence of external sources, such as optical excitation, as will be elaborated in Section 3.7. A significant feature of a reverse biased diode is sudden, rapid increase of current beyond a certain large bias voltage, which is known as *breakdown*. The breakdown phenomena and related devices will be discoursed on later, in Section 3.6.

Figure 3.6 displays the current–voltage (*I–V*) characteristics of a typical p–n junction diode under both forward and reverse bias. If the theoretical values were generated using Equation (3.11), the forward characteristic would look like that of the first quadrant in this figure. The reverse current is so small that it becomes almost zero in comparison with the forward current. Let us summarize what 'characteristics' of the diode are revealed from Figure 3.6.

- First of all, the *I–V* relationship is nonlinear, a salient feature of electronic devices.
- The rectifying property of the diode is prominent from the graph, as the forward current is large and the reverse current is much smaller.
- The reverse current undergoes abrupt increase at a large reverse voltage, the happening being termed as the *breakdown* of the diode. This is discussed further in connection with Zener diode.
- The forward current appears to be zero (actually negligible) up to a certain voltage and then increase rapidly. This threshold voltage is referred to as the *cut-in voltage*, indicated by V_c in Figure 3.6.
- Since the *I–V* plot is not linear, the resistance of the device is not a constant quantity. The rate of change of forward current with forward voltage, hence their ratio, varies continuously. The voltage-to-current ratio is defined as the *dynamic resistance*.

The above mentioned terms defined in connection with the diode *I–V* characteristics are quite useful parameters and are illustrated here.

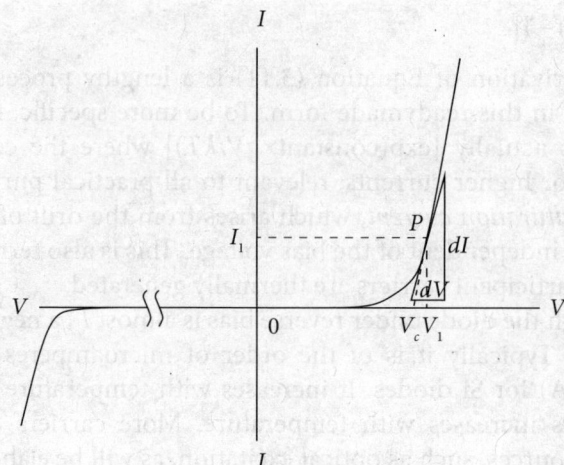

Figure 3.6 Forward and reverse current–voltage (I–V) characteristics of p–n junction diode indicating the reverse breakdown and the determination of forward static and dynamic resistance at point P

3.4.1 Static and Dynamic Resistance

Figure 3.6 indicates that one cannot estimate a fixed value of resistance from the slope (dV/dI) of the forward I–V curve because the quantity changes continuously from point to point. So the resistance of a forward biased diode is defined in the following two ways.

Static Resistance: This is the ratio of any spot value of voltage to the corresponding current. In Figure 3.6, the forward static resistance at point P is

$$r_s = \frac{V_1}{I_1} \tag{3.12}$$

Dynamic Resistance: This is ratio of the small change of forward voltage (dV) to the corresponding change of forward current (dI), as indicated in Figure 3.6. For larger values of V in Equation (3.11), the '1' in the expression for current can be neglected in comparison with the exponential term and the dynamic resistance can be simply expressed as

$$r_d = \frac{dV}{dI} = \frac{kT}{qI} \tag{3.13}$$

Using the standard approximation of 0.026 for kT/q, as explained in Chapter 2,

$$r_d = \frac{26}{I(\mathrm{mA})} \tag{3.14}$$

Equation (3.14) is quite useful formula for forward biased diodes. Under reverse bias, except for the breakdown condition, the diode resistance is very large because the current is very small.

Example 3.2

A p–n junction diode is forward biased with 0.26 volt. If the reverse saturation current at room temperature be 2 μA, find the values of (i) the forward current, (ii) the static resistance, and (iii) the dynamic resistance of the diode.

Solution

(i) Using Equation (3.11), the forward current at the given voltage is

$$2 \times 10^{-6} \left[\exp\left(\frac{0.26}{0.026}\right) - 1 \right] = 44 \text{ mA}$$

(ii) Using Equation (3.12), the static resistance is

$$\frac{0.26}{44 \times 10^{-3}} = 5.9 \text{ }\Omega$$

(iii) Using Equation (3.14), the dynamic resistance is

$$\frac{26}{44} = 0.59 \text{ }\Omega$$

Think a Bit: Digging into the Origin of Diode Forward I–V Characteristic

Equation (3.11) implies that for higher values of forward bias voltage, when the exponential term becomes much larger than unity, the diode forward I–V relationship assumes the shape of x versus e^x plot (Curve 1). The question is how the I–V curve would look like, if the lower current region of Figure 3.6, say that below I_1 were expanded?

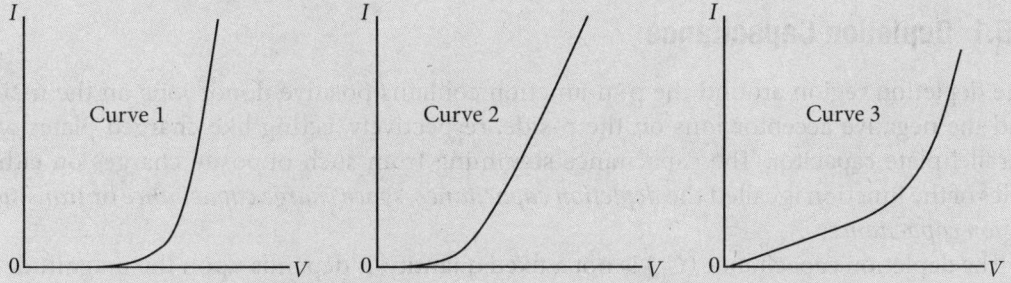

The answer is also fascinating: the curve would look almost the same (Curve 2) except for being more slanted (thereby increasing the dynamic resistance). This is the basic nature of exponential variation that e^x increases more rapidly as x increases. So the 'zero current' in Figure 3.6 for forward bias voltages lower than V_c is not actually zero but much lower than that at higher voltages. Now think in the reverse way. If we go on decreasing the forward bias voltage, the '1' in Equation (3.11) no longer remains negligible. For a very small voltage, the exponential term can be approximated by $(1 + qV/kT)$ and the I–V plot becomes almost liner (Curve 3). Ultimately at zero voltage, the exponential term becomes unity making the current zero. Now it is your job to verify the above conditions putting suitable numeric values.

3.4.2 Cut-in Voltage

The current–voltage relationship of a forward biased diode represented by Equation (3.11) implies that some current flows always through the diode even for very small, 'zero-plus' values of forward bias voltage. However, too small a current value is not of practical use. An appreciable bias voltage, when the exponential term becomes much greater than unity is of practical significance. Such a voltage, indicated by V_c in Figure 3.6 is referred to as the *cut-in voltage*. This voltage corresponds to the current that:

- reaches the order of milliampere,
- becomes several percent of the maximum rating, and
- increases rapidly with bias voltage.

The cut-in voltage is considered as a threshold. Above this voltage, both the static and the dynamic resistance yield vary small values, less than ohms and the diode acts as a closed switch. Below this voltage, the current is negligibly small and the diode performs like an open switch.

3.5 Junction Capacitances

The p–n junction involves the presence of immobile opposite charges on either side of the junction, and the flow of mobile charge carriers of opposite polarity across the junction. As a result it gives rise to two types of capacitances, namely depletion capacitance and diffusion capacitance, predominant at reverse and forward bias, respectively. These capacitances are quite important for applications with time-varying signals and the studies on the structure of the junction.

3.5.1 Depletion Capacitance

The depletion region around the p–n junction contains positive donor ions on the n-side and the negative acceptor ions on the p-side, respectively, acting like charged plates of a parallel plate capacitor. The capacitance stemming from such opposite charges on either sides of the junction is called the *depletion capacitance, space charge capacitance* or *transition region capacitance*.

The depletion capacitance (C_D) is not a fixed quantity, it depends upon the magnitude of the bias voltage (V) and is defined as

$$C_D = \left| \frac{dQ}{dV} \right| \tag{3.15}$$

Indicating the ratio of the change in immobile space charge (Q) to the corresponding change in bias voltage. The depletion capacitance per unit area is quantified as

$$C_D = \frac{\varepsilon A}{W} \qquad (3.16)$$

The expression in Equation (3.16) is similar to that of a parallel plate capacitor with ε being the permittivity of the semiconductor, A the cross-sectional area of the junction, and W the width of the depletion region. The depletion capacitance originates from the immobile dipoles across the depletion region and it is the dominating junction capacitance at reverse biased condition of the diode when mobile carriers are negligible. The capacitance decreases with the increase in reverse bias as the depletion width decreases.

3.5.2 Diffusion Capacitance

This type of junction capacitance dominates at the forward biased condition of the diode when the mobile charge carriers, i.e., the electrons and the holes diffuse through the junction and both become minority carriers after crossing the junction. The density of these excess minority carriers decreases rapidly with distance from the junction due to recombination. However, momentarily a large amount of minority carriers gets stored near the junction and gives rise to a capacitance termed as the *diffusion capacitance*. This is also named as the *storage capacitance* because this capacitance originates from the 'charge storage effects' caused by the charges lagging behind the voltage as the current changes.

The charge carriers appear at the junction at the same rate as that at which these disappear due to recombination. It is a quasi-steady process resulting in a constant forward current (I) through the junction for a certain forward bias (V) given by Equation (3.11). Let the minority carrier storage develops a charge Q and τ be the mean lifetime of the carrier. The diffusion capacitance can be articulated as

$$C_d = \frac{dQ}{dV} = \tau \frac{dI}{dV} \qquad (3.17)$$

It is interesting to note that Equation (3.17) dimensionally represents a time interval. Using Equations (3.11) and (3.17), the diffusion capacitance can be approximated as

$$C_d = \tau \frac{I}{kT/q} \qquad (3.18)$$

Equation (3.18) implies that the diffusion capacitance is proportional to the forward current. Consequently it increases rapidly with forward bias and masks the effect of depletion capacitance. In reverse biased condition, dI/dV is very small and the diffusion capacitance becomes negligible compared to the depletion capacitance.

The diffusion capacitance has an opposing effect to the rapid switching of a p–n junction from forward to reverse biased condition. If the large forward current through a forward biased junction is stopped suddenly by reverse biasing the junction, the minority charge carriers left in the depletion region take some time to get flushed out of the junction. This charge storage effect is equivalent to a capacitance that prevents the junction from quick change under high frequency ac signals.

3.6 Zener Diode

It is stated earlier that the reverse saturation current due to thermally generated minority carriers is almost voltage independent and it increases slightly with temperature. However, on increasing the reverse voltage beyond a critical value, the reverse current increases abruptly, the occurrence being known as 'breakdown'. There is a special category of p–n junction diode, known as Zener diode that is operated in this breakdown mode only under reverse bias. Moreover, the Zener diode is used not as a rectifier but as a reference voltage source. This device is named after American physicist Clarence Melvin Zener (1905–1993) who first interpreted (1934) the breakdown of electrical insulator. First let us get acquainted with the following two types of breakdown mechanisms because the operating principle of a Zener diode is based on either of these two.

3.6.1 Zener Breakdown

The width of the depletion region of a p–n junction decreases with increasing doping concentration. When the doping concentration on both the n- and p-sides become very large, of the order of 10^{18} cm^{-3} or more, the depletion region becomes very thin even under reverse bias. Because of the narrowing of the depletion region width, a large electric field appears across the junction at moderate reverse bias. The high field exerts a large force on the electrons and tears those out of the covalent bonds. Thus electron–hole pairs are produced in profusion and the reverse current increases rapidly without further increase in the reverse voltage. This phenomenon of producing large number of electron–hole pairs because of the rupture of covalent bonds under the action of high electric field across the reverse biased p–n junction is referred to as the *field ionization* or *Zener breakdown* after its discoverer. The following conditions are necessary for the occurrence of Zener breakdown of a reverse biased p–n junction.

(i) Sharp, abrupt junction
(ii) High ($\geq 10^{18}$ cm^{-3}) doping on either side
(iii) Thin depletion region (< 100 nm)
(iv) High electric field ($\geq 10^6$ Vcm^{-1})
(v) Moderate reverse bias, 5–6 V or less

The Zener breakdown phenomenon can be described in terms of the energy band picture also, as interpreted with Figure 3.7. Following Equations (2.24a) and (2.24b), it is inferred that because of the high doping, the aligned Fermi level stays very close to the conduction and valence band edges [Figure 3.7(a)]. Applying a small reverse bias brings the conduction band at the n-side to the same level of the valence band at the p-side [Figure 3.7(b)]. As a result, the large number of empty, available energy states in the n-side conduction band gets aligned with the large number of filled energy states (indicated with shade) in the p-side valence band. At the same time, the potential barrier separating these two is very narrow because the depletion region is very thin. Consequently many electrons from the p-side valence band can tunnel through the potential barrier and reach the n-side conduction band causing a large reverse current from the n- to the p-side.

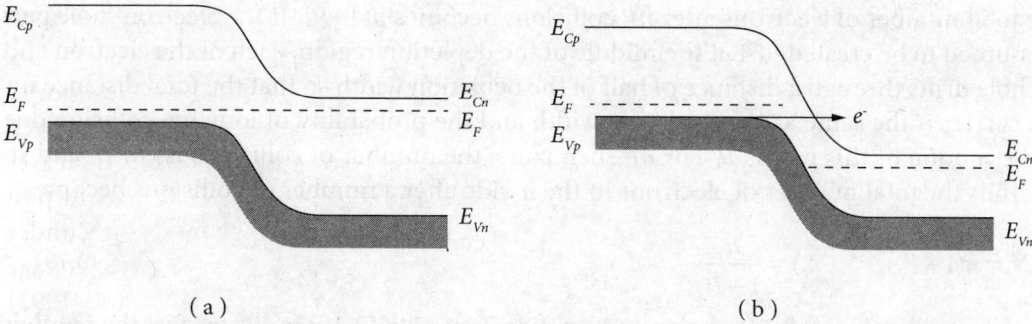

Figure 3.7 Heavily doped p–n junction (a) at equilibrium showing Fermi level closer to the band edges and (b) at reverse bias indicating the electron (e⁻) tunneling from the p-side to the n-side

3.6.2 Avalanche Breakdown

If the doping concentration is not so high, the depletion region at reverse bias does not narrow down as in the previous case and the electric field developed across it does not become so high. Nevertheless, if the reverse bias is increased further, the minority carriers can acquire high kinetic energy and collide with the lattice to knock out the electrons from the covalent bonds. Such a phenomenon of creating electron–hole pairs by the process of impact of carriers having high kinetic energy at moderately doped p–n junction under large reverse bias is termed as *impact ionization*.

The electrons and the holes generated by impact ionization get swept to the n-side and the p-side, respectively, by the electric field. These carriers also possess high kinetic energy to cause further ionizing collisions with the lattice. As a result, more and more electron–hole pairs are generated like a chain reaction. Such collision-induced carrier multiplication is popularly named as the *avalanche breakdown*, as a metaphor of avalanche, a natural calamity when large mass of ice and rocks fall down the mountains. The followings are the essential conditions for avalanche breakdown in a p–n junction.

(i) Moderate doping ($\approx 10^{16}$ cm^{-3} or less)
(ii) High reverse bias, more than 6 V and may be up to 100 V or more
(iii) The critical value of reverse bias voltage at which the avalanche breakdown initiates increases with semiconductor band gap and decreases with increase in doping concentration.

The large number of carriers generated by avalanche breakdown causes a large reverse current from the n- to the p-side without further increase of reverse bias.

An elementary estimate of the number of carriers generated in avalanche breakdown can be obtained in the following way. Let P be the probability of an electron or a hole making an ionizing collision with the lattice while being accelerated through the depletion region. For n number of carriers, either electrons from the p-side or holes from the n-side, there will be nP ionizing collisions, each causing an electron–hole pair. Let us consider electrons only.

The total number of electrons after nP collisions becomes $n(1+P)$. If the electron–hole pair is assumed to be created just at the middle of the depletion region, each of the electron and the hole drifts through a distance of half of the depletion width so that the total distance for any carrier is the same as the depletion width and the probability of ionizing collision due to the motion of this pair is P. For nP such pairs, the number of collisions is $(nP)P$ and so on. Thus the total number of electrons in the n-side after a number of collisions becomes

$$N = n(1 + P + P^2 + \ldots) = \frac{n}{1-P} \tag{3.19}$$

A similar expression holds for holes in the p-side. Equation (3.19) indicates that the number of carriers becomes infinitely large when the ionizing probability approaches unity.

Table 3.2 presents a comparative view of avalanche and Zener breakdown mechanisms in p–n junctions under reverse bias. The similarity is that both of the processes give rise to a huge number of carriers, hence a large current after a certain threshold voltage and no further increase of voltage is required to maintain this current. So the reverse voltage across the p–n junction remains constant that can serve the purpose of voltage regulation.

Table 3.2 Comparison of the Zener and avalanche breakdown processes

Zener Breakdown	Avalanche Breakdown
It occurs with thin depletion region at small reverse voltage.	It occurs with wide depletion region at large reverse voltage.
It is caused by high electric field across the depletion region that tears out the carriers of the bands. The process is called the *field ionization*.	It is caused by highly energetic charge carriers accelerating through the depletion region and knocking out the carriers of the bonds. The operation is called the *impact ionization*.
It does not involve collisions of carriers with the crystal ions. The initially available carriers do not possess sufficient energy to break the covalent bond. It is the strong electric field that exerts the force on the bound electrons and breaks the bond.	The initially available carriers as well as the newly generated carriers pick up sufficient energy from the applied reverse bias to collide with crystal ions and create more electron–hole pairs. Each new carrier, in turn, produces more carriers through collision and disruption of bonds. Thus the process is 'cumulative'.
The temperature coefficient is negative.	The temperature coefficient is positive.
The consequence is large reverse current at small (5–6 V) reverse bias.	The consequence is large reverse current (not as steep as that of Zener breakdown) at large (more than 6 V) reverse bias.

Another significant entity is the temperature of the device. Semiconductor devices have a general property of temperature sensitivity. In the case of p–n junctions, it is quantified in terms of the *temperature coefficient*, which is the percentage change in the reference voltage across the junction for unit change in temperature of the device. In the case of narrow depletion region, when the Zener effect is likely to occur, an increase in temperature increases the kinetic energy of the valence electrons so that these can escape at less applied voltage from the bonds. As a result, the Zener breakdown voltage decreases on increasing the temperature and it is said to have the *negative temperature coefficient*. In the case of wide depletion region, when avalanche multiplication is likely to take place, an increase in temperature enhances the vibrational displacements of atoms in the crystal. It increases the probability of collisions of the intrinsic electrons and holes while crossing the depletion width. Thus they lose energy and require higher voltage to gain sufficient energy to take part in the impact ionization process. So the avalanche breakdown has *positive temperature coefficient*. Generally a temperature coefficient of ± 0.1% per °C is noted.

3.6.3 Zener Diode Characteristics

The Zener diode is basically a p–n junction diode and its forward current–voltage characteristic is similar to other p–n junction diodes. Even so unlike common diodes, it is used in the breakdown condition under reverse bias. The mechanism of breakdown may be either Zener or avalanche but the device is called the 'Zener diode' because both the processes cause similar outcomes of current and voltage. On increasing the reverse bias, the device starts conducting heavily at a certain breakdown voltage and the current increases rapidly without further increase of voltage. Thus after the breakdown, the Zener diode maintains a constant voltage across itself, referred to as the *Zener voltage* and acts as a *voltage regulator*. The value of the reverse breakdown voltage can be adjusted

- by changing the material so that the band gap changes, and
- by changing the doping concentration.

A wide range of Zener diodes is commercially available having regulating voltage from less than a volt to hundreds of volt. Generally the Zener diodes of breakdown voltage less than 6 V get operated by Zener breakdown and the Zener diodes of higher breakdown voltages get acted by avalanche breakdown.

> **Think a Bit: Breakdown Is Not So Destructive!**
>
> The term 'breakdown' is a bit misleading. It is true that a high current takes place all on a sudden at the breakdown condition of reverse bias. It is also grim fact that a high current may damage the diode because of high power dissipation across it. But such a situation may happen during the forward bias too. There is no hazard on either side unless the maximum power rating of the device is exceeded. That is why a current limiting resistance is always present in the circuit. This chapter explains the use of resistors in reverse biased Zener diodes. The rectifier circuits of Chapter 4 consist of external resistor in the cases of forward biased p–n junction diodes. Zener diodes are specially fabricated by tailoring the construction and doping for operating exclusively in breakdown mode. Ordinary p–n junction diodes are meant for rectification and the breakdown voltage occurs at too high voltage (say several hundreds of volt) when the current would become unmanageable.

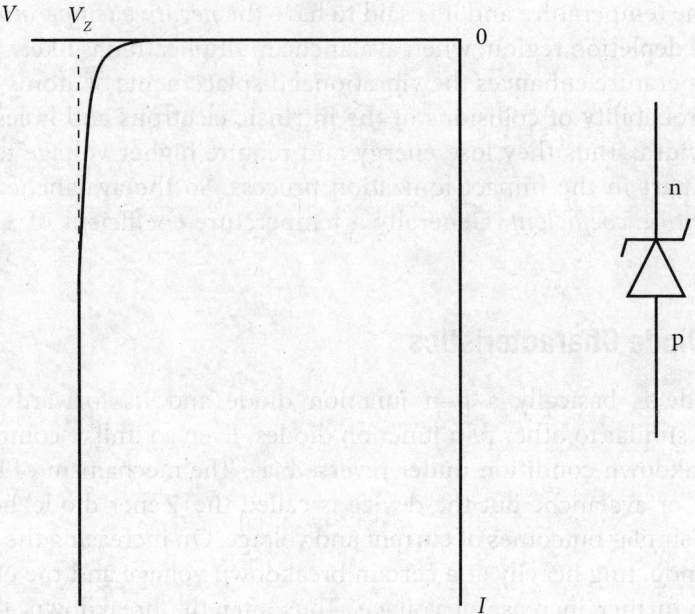

Figure 3.8 Reverse current–voltage characteristic of the Zener diode (left) and its symbol (right)

The forward current–voltage characteristic curve of a Zener diode is similar to that of a typical p–n junction diode (Figure 3.6). The reverse current–voltage characteristic curve is somewhat different, as sketched in Figure 3.8. This also introduces the circuit symbol of Zener diode given at the right side. It is slightly different from that of p–n junction diode. Figure 3.8 reveals the following characteristics of the Zener diode.

- A constant voltage (V_Z) is maintained across the Zener diode after breakdown that enables it to act as voltage regulator.
- A minimum current is also to be maintained through the Zener diode during the breakdown.
- The current after breakdown increases very rapidly with slight change of voltage implying very small dynamic resistance of the Zener diode after breakdown.

3.6.4 Zener Voltage Regulator

The practical use of Zener diode as voltage regulator is demonstrated with the circuit diagram given in Figure 3.9. The supply voltage is V and the constant Zener voltage across the Zener diode after breakdown is V_Z. This voltage appears across the load resistor (R_L) joined in parallel with the Zener diode. The total current (I) divides into two components: I_Z through the Zener diode and I_L through the load resistance. The resistance R is referred to as current limiting resistor that controls the current through the Zener diode so as to keep the power dissipation well below the upper limit. The following two conditions may arise.

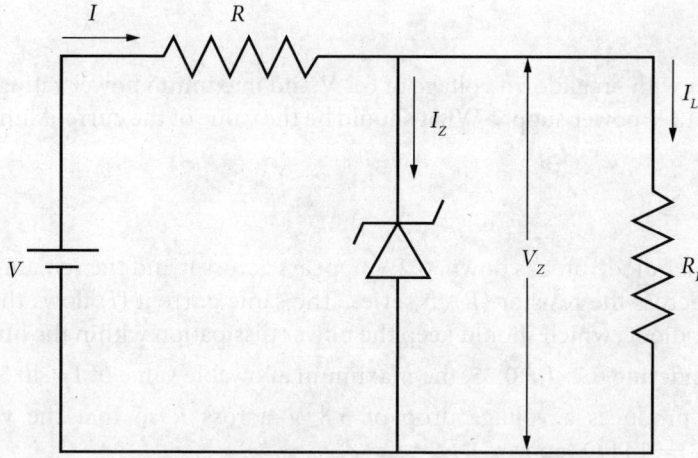

Figure 3.9 Zener diode as voltage regulator

Condition-I: If the supply voltage (V) remains constant but the resistance (R_L) of the load resistor varies, the regulation property of the Zener diode is called the *load regulation*. The voltage across R_L is given by

$$V_Z = \frac{VR_L}{R+R_L} \tag{3.20}$$

Equation (3.20) does not present the full picture. If the load resistance becomes too small, almost all the current flows through that and the current through the Zener diode becomes so low that it does not reach the breakdown region and the voltage regulation property gets lost. The minimum allowable value of load resistance is

$$R_{L\min} = \frac{RV_Z}{V-V_Z} \tag{3.21}$$

Combining Equations (3.20) and (3.21) it is inferred that the Zener diode can act as voltage regulator provided the load resistance is high (light load) or it is at open circuited (no load) condition. If the load resistance becomes very small (heavy load), the load regulation fails.

Condition-II: If the load resistance (R_L) remains constant but the supply voltage (V) fluctuates, the regulation property of the Zener diode is termed as *line regulation*. From Equation (3.20), the minimum allowable supply voltage is

$$V_{\min} = \frac{(R+R_L)V_Z}{R_L} \tag{3.22}$$

Equation (3.22) denotes that the supply voltage must be well above the breakdown voltage of the Zener diode in order to hold its line regulation.

> **Example 3.3**
>
> A Zener diode with breakdown voltage of 6.2 V and maximum power rating of 0.25 W is joined across 12 V power supply. What should be the value of the current limiting resistor in series?
>
> **Solution**
>
> When the Zener diode breaks down, 6.2 V appears across it and the remaining 5.8 V (12 − 6.2) drops across the resistor (R) in series. The same current (I) flows through this R and the Zener diode, which should keep the power dissipation within the limit of 0.25 W.
>
> Considering $6.2 \times I = 0.25$, the maximum allowable value of $I = 40.3$ mA.
>
> This current produces a voltage drop of 5.8 V across R so that the value of R is $5.8/(40.3 \times 10^{-3}) = 144\ \Omega$ (approx.)

3.7 Light-Emitting Diode (LED)

This device has turned out to be one of the most popular light sources in domestic, scientific, and commercial appliances, and has replaced the tungsten bulb in most cases by virtue of its much lower power consumption, lower cost, and varieties of color of light. It should be clarified at the very beginning that the color of the emitted light does not depend on the colored encapsulation but originates from the device properties.

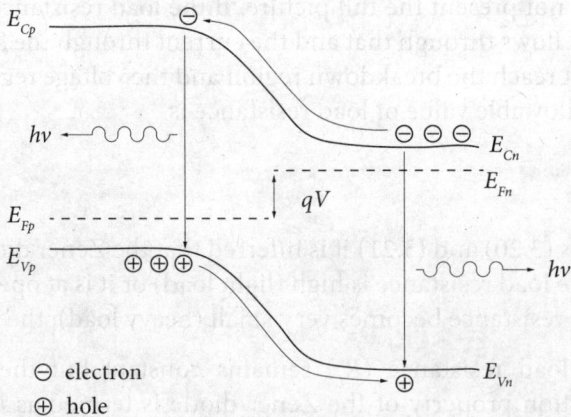

Figure 3.10 Energy band diagram of light-emitting diode under forward bias (V) showing the recombination of electrons and holes and the emission of the consequent radiation ($h\nu$)

The light-emitting diode (LED) is basically a forward biased p–n junction where the electrons and the holes are injected across the junction, as usual, by the external bias voltage. However, the carriers do not traverse long distance but recombine within and around the depletion region and the energy is released in the form of visible or infrared radiation of different wavelengths depending on the band gap and other properties of the material forming the junction. The carrier injection, possible recombination, and the release of radiation ($h\nu$) are outlined in Figure 3.10. The forward bias voltage (V) has split the Fermi level at the n- and p-sides by the amount of qV.

The LED is different from a conventional forward biased p–n junction diode because the semiconductor materials are different. Let us recapitulate direct and indirect band gap semiconductors (Section 2.4). A direct band gap material, such as GaAs or GaAsP is suitable for LED fabrication since the recombination is associated with radiative transition yielding electromagnetic radiation in the form of visible light or infrared or ultraviolet radiation. In the case of indirect band gap semiconductor, such as Si or Ge, the transition is non-radiative because some energy is lost in the momentum change involved and the energy released in the recombination is released as heat to the lattice.

> **Note: LED with Indirect Band Gap Semiconductor**
>
> It is true that direct band gap semiconductors are quite suitable for LED because the momentum is conserved in recombination and there is high probability of inter-band transitions resulting in the emission of radiation. The recombination in indirect band gap materials involves momentum change and phonons (quantized lattice vibration) and other scattering agents participate in the process of conserving the momentum. The probability for inter-band transition is very small.
>
> However, the available direct band gap semiconductor materials can emit radiation of only few selective wavelengths corresponding to their band gaps. In order to get diversities in emission wavelength and varieties of color of the emitted light, wider choice of material is essential. The research on novel LED materials finds importance in this regard. There are many avenues of achieving suitable materials for emitting radiation. For instance, the ternary compound semiconductor AlGaAs exhibits regular variation of band gap, hence of the emission wavelength with the relative proportion of As and Ga. Indirect band gap materials are also involved by using a special doping technique. Some impurities, such as nitrogen atoms in GaAsP compound semiconductor act as highly localized recombination centers. Consequently the electrons in such centers remain tightly bound in physical space (x) and their momentum (p) can spread out in momentum space following the fundamental rule $\Delta x \Delta p$ = constant. Thus the problem of momentum change in indirect band gap semiconductor is overcome.

Table 3.3 Some semiconductor materials for fabricating light-emitting diode (LED)

Material	Band gap	Emission in LED
$In_xGa_{1-x}N$	2.7 to 3.6 eV	Blue and violet
GaAs	1.4 eV	Infrared
$GaAs_{1-x}P_x$	1.424 to 1.977 eV direct band gap for $0 < x < 0.45$ and indirect band gap for $x > 0.45$	Red for $x = 0.4$ Orange for $x = 0.65$ Yellow for $x = 0.85$ Green for $x = 1.0$

The forward and reverse current–voltage characteristics of an LED are similar to those of a conventional p–n junction diode. The LED starts glowing when the current enters the cut-in voltage region. A resistance is joined in series to keep the current within safe limit. Few examples of novel LED materials are given in Table 3.3. The demand for wide variation of emission wavelength over the ultraviolet, visible and infrared range has enhanced the research on this subject.

Table 3.4 Comparison of photodiode and solar cell

Parameter	Photodiode	Solar Cell
Working principle	It is an optoelectronic device where the conduction of current occurs due to generation of electrons and holes under illumination.	It is an optoelectronic device based on the same principle where the voltage generated across the junction is more important. So it is also called *photovoltaic cell*.
Construction	It may be wavelength selective. It may have an optional reverse bias.	It is designed to uniform response over large wavelength range in order to utilize the maximum of solar radiation. It has no external bias. The device itself delivers power to the external circuit.
Application	It is used for the purpose of detecting optical illumination level. It converts the optical signal to proportional electric signal. An important application is in optical communication.	It is employed to convert solar radiation to useful electrical power. The most important application is as an alternative source of energy. A collection of a number of individual solar cells, known as solar panel is used to generate a large amount of photovoltaic power.

3.8 Photodiode and Solar Cell

These devices act just the opposite of an LED. Both the photodiode and the solar cell consist of a p–n junction that is exposed to optical radiation. The absorption of energy gives rise to electron–hole pairs that, in turn, generate current or voltage. These two devices have differences in the construction, mode of operation and application, as summarized in Table 3.4 but the basic operation of a p–n junction under incident radiation is a common phenomenon and the same is explained here.

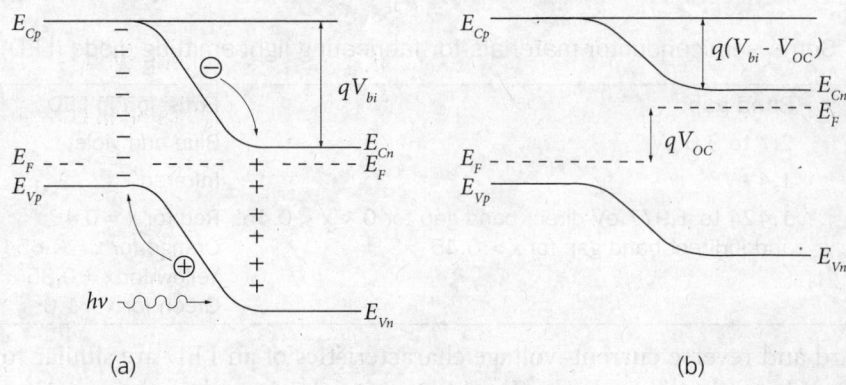

Figure 3.11 Illuminated p–n junction: (a) Generation and drift of electrons and holes at the depletion region under optical excitation, (b) band bending and Fermi energy positions of the illuminated junction

In the case of an ordinary p–n junction under reverse bias, the minority carriers are generated thermally in the vicinity of the junction and are swept away by the external field to cause a weak current. This is referred to as the *dark current* since it can take place without any external illumination. If the junction is illuminated, the generation of carriers enhances much more and occurs even without any bias. This process is indicated in Figure 3.11(a). It denotes the positive immobile ions at the n-side, the negative acceptor ions at the p-side and the aligned Fermi level throughout the p–n junction.

If the junction is exposed to optical excitation (denoted by $h\nu$), electrons and holes are generated at around the junction. These electrons and holes are drifted away immediately to the n- and the p-side, respectively, because of the immobile ions, as indicated in the figure. These electrons and holes decrease the positive charge on the n-side and the negative charge on the p-side so that an open circuit voltage (V_{oc}) appears across the junction having the polarity opposite to that of the built-in potential (V_{bi}), as sketched in Figure 3.11(b). The situation is similar to that of a forward biased diode but the carrier flow is in the opposite direction.

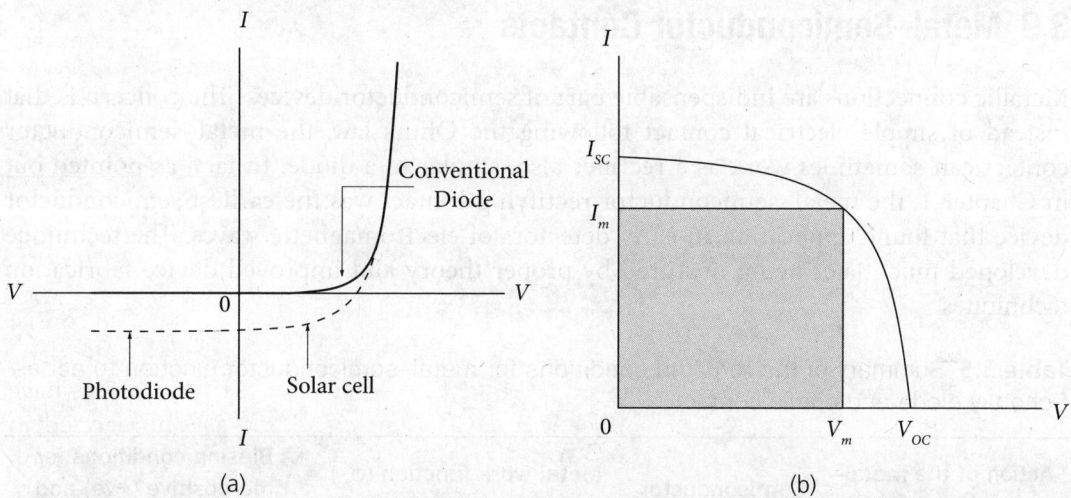

Figure 3.12 Current–voltage characteristics of an illuminated p–n junction: (a) full view and (b) magnified and inverted view of the fourth quadrant

Think a Bit: When the Illuminated Junction Acts as a Battery?

Let us recollect that in a conventional diode under forward bias, power is delivered to the device from an external circuit, both the current (*I*) and the voltage (*V*) across the junction are positive and the *I–V* plot is drawn in the first quadrant of graph. Under reverse bias, both the current and the voltage are negative, the *I–V* plot is in the third quadrant and the current is negligible except for breakdown.

When this reverse biased diode is a photodiode having its junction illuminated under optical source, the optical excitation enhances the current but the *I–V* plot is still in the third quadrant and power is delivered to the device from external source.

In the case of a solar cell, the scenario changes abruptly. The current flows within the device from the negative to the positive side of the voltage, as in a battery and the *I–V* plot is drawn in the fourth quadrant when the junction voltage is positive and the current is negative. Now power is delivered from the device to the external circuit, similar to a battery.

Since the current due to optically generated carriers is directed from the n- to the p-side, opposite to the conventional current, the reverse current–voltage characteristic of the illuminated junction gets changed, as depicted in Figure 3.12(a). The fourth quadrant is magnified and inverted in Figure 3.12(b), which is useful to estimate the performance of a solar cell. The curve at Figure 3.12(b) has two extremes,

- Open circuit voltage (V_{oc}) when the current becomes zero, and
- Short circuit current (I_{sc}) when the voltage is zero.

None of these two conditions is suitable for delivering the maximum power to the load because the voltage-current product becomes zero in either case. Somewhere in the middle, as shown by the shaded rectangle in Figure 3.12(b), the solar cell delivers the maximum power ($V_m I_m$). The ratio ($V_m I_m$)/($V_{oc} I_{sc}$) is called the *fill factor*, a figure of merit of the solar cell.

3.9 Metal–Semiconductor Contacts

Metallic connections are indispensable part of semiconductor devices. The concern is that instead of simple electrical contact following the Ohm's law, the metal–semiconductor contact can sometimes work as a rectifier also, similar to a diode. In fact, as pointed out in Chapter 1, the metal–semiconductor rectifying contact was the earliest semiconductor device that found application in early detectors of electromagnetic waves. The technique developed much later being nurtured by proper theory and improved device fabrication techniques.

Table 3.5 Summary of the required conditions for metal–semiconductor junction to act as Schottky diode and ohmic contact

Action of the metal–semiconductor junction	Semiconductor involved	Metal work function (φ_m) and semiconductor work function (φ_s) relationship	Biasing conditions for the positive (+ve) and negative (–ve) terminals of the power supply
Schottky diode	n-type	$\varphi_m > \varphi_s$	Forward bias: metal +ve, semiconductor –ve. Reverse bias: metal –ve, semiconductor +ve.
	p-type	$\varphi_m < \varphi_s$	Forward bias: metal –ve, semiconductor +ve. Reverse bias: metal +ve, semiconductor –ve.
Ohmic contact	n-type	$\varphi_m < \varphi_s$	Linear current–voltage characteristic in both biasing conditions.
	p-type	$\varphi_m > \varphi_s$	

The metal–semiconductor junction acting as rectifier is called Schottky diode. It has important applications in microwave devices, photodiodes, solar cells, and field-effect transistors. As ohmic contact, the metallic contact over the semiconductor provides with an electrical connection. All semiconductor devices need external metallic connections, which are provided by the ohmic contact. The distinction of acting as either a rectifier or an electrical connection originates from the choice of metal–semiconductor pair, their work functions and the semiconductor doping type. When the junction acts as a rectifier, the difference of the metal work function and the semiconductor electron affinity gives rise to the potential barrier at the junction. A brief account of the required combinations is present in Table 3.5. For practical purpose, it is sometimes difficult to provide with the above combinations of metal and semiconductor work functions for ohmic contacts. A practical ohmic contact is fabricated by doping the semiconductor heavily in the contact region. That makes the depletion width very thin so as to allow the carriers to tunnel through the barrier at the junction.

Note: The Diode 'Zoo'

The p–n junction has been one of the most explored and the best exploited devices in electronics. Just the changing of the doping level can convert a p–n junction from rectifier to voltage regulator (Zener diode). Some other novel uses of the p–n junction, such as light source (light-emitting diode), light detector (photodiode) and power source (solar cell) are illustrated in this chapter. Such radical changes in the device operation are achieved with
- proper choice of semiconductor material,
- constructional specifications and
- change of doping concentration.

Even a metal–semiconductor junction, instead of semiconductor p–n junction can act as a rectifier (Schottky diode). There are many novel devices acting as diode with the p–n junction as ingredient. For instance:

Tunnel diode: It is made of very highly doped ($\approx 10^{20}$ cm^{-3}) p–n junction and exhibits negative resistance. This device has application in the amplification of microwave signals.

Varactor diode: It is a reverse biased p–n junction diode that acts as a voltage-variable capacitor. The junction capacitance decreases with increasing depletion width, hence with increasing reverse bias. The varactor diode, with the use of proper fabrication technology, exhibits this phenomenon more prominently and is used in telecommunication circuits.

Multiple Choice-Type Questions and Answers

3.1 The n-side and the p-side of the depletion region of a semiconductor p–n junction contain .... and ...., respectively.
 (a) positive immobile donor ions, positive mobile acceptor ions
 (b) negative immobile donor ions, negative immobile acceptor ions
 (c) positive immobile donor ions, negative immobile acceptor ions
 (d) negative immobile donor ions, negative mobile acceptor ions

3.2 The electric field across an unbiased p–n junction is ....
   (a) directed from p- to n-side
   (b) directed from n- to p-side
   (c) bidirectional
   (d) directed along the direction of current flow

3.3 In an unbiased diode, electrons .... and ....
   (a) diffuse from n- to p-side, drift from p- to n-side
   (b) diffuse from p- to n-side, drift from n- to p-side
   (c) diffuse from n- to p-side, do not drift at all
   (d) do not diffuse, drift from p- to n-side

3.4 The current (I) values through a forward biased diode are measured for bias voltage (V) values above the cut-in. The $\log_e I$ vs V plot is expected to be ....
   (a) straight line with positive slope
   (b) straight line with negative slope
   (c) straight line with positive slope and passing through the origin
   (d) logarithmic curve

3.5 A p–n junction has n-side at .... potential than the p-side and there is an electric field at the junction directed from .... side.
   (a) lower, p- to n-   (b) lower, n- to p-
   (c) higher, p- to n-   (d) higher, n- to p-

3.6 The electrostatic potential at the depletion region of a p–n junction ....
   (a) increases monotonically toward p-region
   (b) increases monotonically toward n-region
   (c) is uniform throughout the junction
   (d) has fixed negative value at p-side and fixed positive value at n-side

3.7 *Statement-I*: In p–n junction, there is an electric field across the junction directed from the n-side toward the p-side.
   Statement-II: The n-side has positively charged immobile ions and the p-side has negatively charged immobile ions.
   (a) *Statement-I* is correct but *Statement-II* is incorrect.
   (b) *Statement-II* is correct but *Statement-I* is incorrect.
   (c) Both statements are correct and *Statement-II* is the reason for *Statement-I*.
   (d) Both statements *I* and *II* are correct but these two have no relation.

3.8 A p–n junction was formed with a heavily doped (10^{18} cm^{-3}) p-region and lightly doped 10^{14} cm^{-3}) n-region. Which of the following statement(s) is (are) correct?
   (a) The width of the depletion layer will be more in the n-side of the junction.
   (b) The width of the depletion layer will be more in the p-side of the junction.
   (c) The width of the depletion layer will be the same on both sides.
   (d) If the p–n junction is reverse biased, then the width of the depletion region increases.
   [JAM 2016]

3.9 In reverse biased condition, the junction capacitance of a step graded p–n junction diode varies proportionally with bias voltage (V) as ....
   (a) $V^{-1/4}$ (b) $V^{-1/3}$ (c) $V^{-1/2}$ (d) V^{-1}

3.10 The temperature coefficient of Zener breakdown voltage is ....
 (a) zero
 (b) negative
 (c) positive
 (d) positive at room temperature

3.11 The high doping of Zener diode .... the depletion width thereby .... the electric field across it and providing with large number of carriers .... the breakdown.
 (a) decreases, increasing, after
 (b) increases, decreasing, before
 (c) increases, increasing, after
 (d) decreases, decreasing, after

3.12 If the load resistance decreases in a Zener regulator, the current through its series resistance ....
 (a) increases
 (b) decreases
 (c) emains the same
 (d) becomes independent of the breakdown voltage

3.13 The Zener diode shown in Figure 3.13 has an input voltage in the range 15 V – 20 V and a load current in the range of 5 mA – 20 mA. If the Zener voltage is 6.8 V, the value of the series resistance should be
 (a) 390 Ω
 (b) 420 Ω
 (c) 440 Ω
 (d) 460 Ω

[JAM 2015]

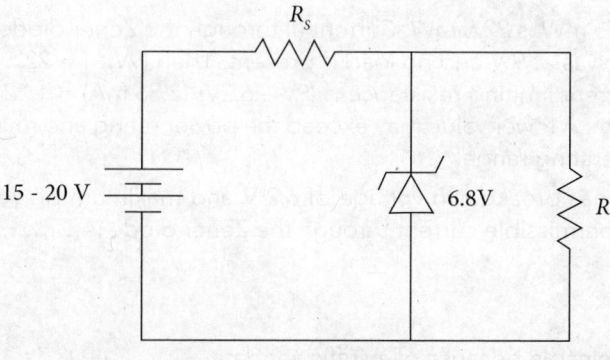

Figure 3.13

Hints. The worst condition is that when the Zener diode has to maintain 6.8 V across the load resistor with input voltage 15 V (minimum) and load current 20 mA (maximum). In that case,

$R_s = (15 - 6.8)/20$ mA $= 410$ Ω

Actually R_s should be less than 410 Ω.

3.14 In Figure 3.14, the knee current of the ideal Zener diode is 10 mA. To maintain 5 V across R_L, the minimum value of R_L in Ω and the minimum power rating of the Zener diode in mW are ....
 (a) 125 and 125
 (b) 125 and 250
 (c) 250 and 125
 (d) 250 and 250

[GATE 2013]

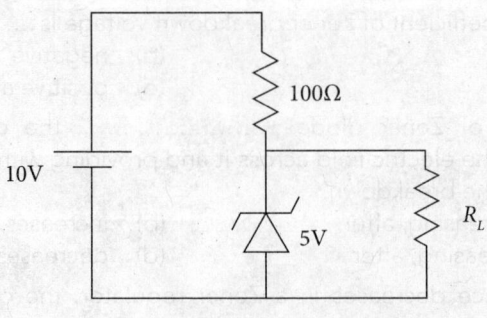

Figure 3.14

Hints. The knee means breakdown. The current through 100 Ω is $(10 - 5)/100 = 50$ mA. The current through R_L is $50 - 10 = 40$ mA. Therefore, $R_L = (5\text{ V})/(40\text{ mA}) = 125$ Ω.

At open circuit condition ($R_L = 0$), the whole of 50 mA passes through the Zener diode and it must withstand that. So the minimum power rating should be $(5\text{ V}) \times (50\text{ mA}) = 250$ mW.

3.15 A Zener diode of 25 mW has breakdown voltage of 6.20 V. It is connected with a power supply of 0–9 V range. A maximum of 90 % power consumption is allowed in the Zener diode. The minimum value of the current limiting resistor ....
(a) 1 kΩ (b) 500 Ω (c) 2.12 kΩ (d) 1.5 kΩ

Hints. 90 % of 25 mW is 22.5 mW. Current (I) through the Zener diode is maximum when the power supply is at 9 V and no load is present. Then $(9\text{V}) \times I = 22.5$ mW so that $I = 2.5$ mA and the current limiting resistance is $(9\text{V} - 6.2\text{V})/(2.36\text{ mA}) = 1.12$ kΩ. Slightly higher value is optimum. A lower value may exceed the power rating and much higher value will shorten the operating range.

3.16 A Zener diode has breakdown voltage of 6.2 V and maximum power rating of 0.25 W. The maximum permissible current through the Zener diode is ....
(a) 0.04 mA
(b) 40 mA
(c) 1.55 mA
(d) not determinable because of insufficient data

Answers

3.1	(c)	3.2	(b)	3.3	(a)	3.4	(a)
3.5	(d)	3.6	(b)	3.7	(c)	3.8	(a) and (d)
3.9	(c)	3.10	(b)	3.11	(a)	3.12	(c)
3.13	(a)	3.14	(b)	3.15	(d)	3.16	(b)

Reasoning-Type Questions and Answers

3.1 Justify the statement: "The potential barrier across a p–n junction prevents the flow of minority carrier".

Ans. The electrons in the n-side and the holes in the p-side of the p–n junction are majority carriers and the potential barrier does not oppose their motions. It prevents the electrons from moving into the p-side and the holes from moving into the n-side where those are treated as minority carriers.

3.2 Justify the statement: "The forward bias on p–n junction diode causes *minority carrier injection*".

Ans. The forward bias causes steady injection of holes into the n-region and electrons into the p-region where both become minority carriers.

3.3 A normal diode has a typical doping concentration of 10^{15} to 10^{16} cm^{-3}. A Zener diode has doping concentration of around 10^{18} cm^{-3} and that of a tunnel diode is about 10^{20} cm^{-3}. What may happen, if the doping concentration is increased further?

Ans. Obviously the upper limit is Avogadro number 6.023×10^{23} cm^{-3}, beyond which atoms per unit volume cannot be accommodated any more. However, much before this condition, the material gets converted into an alloy.

The objective of doping is to keep the number of dopant atoms per unit volume much less than that of the surrounding host material atoms so that discrete energy levels can be assumed for the carriers contributed by the dopant atoms. Such condition would be violated on increasing the doping concentration abnormally and the carriers from the dopant atoms would also enter the bands for transport similar to the host atoms. The semiconductor in such a condition is referred to as *degenerate*.

3.4 Is the development of built-in potential in unbiased diode creation of energy?

Ans. No, it is just the maintenance of equilibrium at the junction. The neutral region at the n-side, far from the junction, has a constant potential. Similarly, the neutral p-region also has a constant potential. The difference of these two appears as an electrostatic potential at the junction, higher at the n-side and lower at the p-side, the gradient being in the direction opposite to the electric field. The electrons or the holes must overcome the potential barrier, which is possible only by applying an external energy source in the form of biasing.

3.5 Can one measure the built-in potential across a p–n junction using a voltmeter? Explain.

Ans. No, it cannot be done. First let us understand the actual meaning of 'measuring a voltage with voltmeter'. A classic analog voltmeter is based on the principle of charge flow through a coil in magnetic field and the resultant galvanic deflection. A digital voltmeter is calibrated in terms of an amplified voltage difference between the inputs of a differential amplifier (Chapter 11). Whatever may be the process, the fundamental requirement is some source of electromotive force causing the movement of charge. In the present case, the barrier potential arising from the immovable ions arrests the flow of movable charge carriers and only that much potential is developed across the junction that can nullify the diffusion of carriers and maintain the equilibrium. So there is no agent for doing the work for further charge flow. If one still attempted to place an ideal voltmeter across the device, new contact potentials would be formed at each terminal of the voltmeter thereby cancelling the built-in potential.

If some external potential were imposed by supplying energy from outside, then the resultant voltage could be measured joyfully. For instance, the open-circuit voltage of a solar cell can be measured with a voltmeter.

3.6 The process of light emission in LED is called the *spontaneous emission*. Can you state why so?

Ans. The forward bias injects minority carriers, electrons in the p-side and holes in the n-side. The energy of the carriers is raised by the external bias. The excited electrons, after staying a certain time in the higher energy state give up the energy naturally as photon ($h\nu$) of the order of the semiconductor band gap (E_g) and recombine with the holes in the valence band. Since the excitation and the recombination are random, spontaneous processes, the light emission is called spontaneous emission.

[N.B. There is another process of light emission known as the *stimulated emission* where the electron at the higher energy state is further excited by a photon. The radiation in this case is much more coherent. This principle is used in laser.]

3.7 Blue LED was invented in 1990s and Isamu Akasaki, Hiroshi Amano and Shuji Nakamura of Japan were awarded Nobel Prize in 2014 for the achievement. Has this invention got any special significance?

Ans. White light is composed by proper mixing of red, green and blue lights. The red and the green LEDs had been there in existence for a long time (1960s) and used to be disbursed as decorative lighting. It was the blue LED that remained a challenge over a long time and its invention paved the way for white LED light for domestic and commercial lighting, smartphone and computer screen. It largely replaced the conventional sources, such as tungsten bulb and fluorescent tube.

3.8 What are *excess carriers* in semiconductor devices?

Ans. Generally the operation of a semiconductor device is determined by the charge carriers produced in excess of thermal equilibrium by some external agent. For example, a p–n junction diode at equilibrium has no net current flow because the drift of carriers under the electric field at the junction just cancels the diffusion of carriers across the junction. An external forward bias injects more carriers to both sides crossing the junction causing the current. In a light-emitting diode, similar external forward bias injects excess minority carriers into the regions where they recombine with the majority carriers giving rise to both the current and the radiation emission. In an illuminated junction, such as in photodiode or solar cell, the optical excitation from outside generates electrons and holes that cause the current or the voltage. All these are examples of excess carrier conduction.

Solved Numerical Problems

3.1 The reverse saturation current of a p–n junction diode is 1 pA at room temperature. Estimate at what forward bias the diode current would reach 10 mA.

Soln. Let the required bias voltage be V. Then using Equation (3.11),

$$10 \times 10^{-3} = 1 \times 10^{-12} \times \left[\exp\left(\frac{V}{0.026}\right) - 1 \right]$$

Hence, $V = 0.598$ V.

[N.B. In practice, a larger voltage is required across the diode because of the voltage drop at the bulk semiconductor resistance and the contact resistances.]

3.2 Calculate the range of forward bias voltage after cut-in, over which the dynamic resistance of the diode changes by a factor of 100.

Soln. Let the dynamic resistances corresponding to voltage V_1 and V_2 be denoted by r_{d1} and r_{d2}, respectively. Following Equation (3.13),

$$\frac{r_{d1}}{r_{d2}} = 100 = \exp[(V_1 - V_2)/(kT/q)]$$

Hence, $V_1 - V_2 = (kT/q)\ln(10^2) = 0.12$ V approx.

[N.B. The result implies that after cut-in, the dynamic resistance of the diode decreases very rapidly with voltage so that the forward current increases quickly over a short range of bias voltage.]

3.3 In Figure 3.9, the Zener diode has breakdown voltage of 12 V, the load resistor is of 1 kΩ and the current limiting resistor is of 150 Ω. If the maximum allowable current through the Zener diode be 20 mA, then assuming the minimum Zener current to be zero, estimate the required range of the input voltage (V).

Soln. When zero current passes through the Zener diode, the total current through the 150 Ω resistor is

$$0 + (12 \text{ V})/(1 \text{ k}\Omega) = 12 \text{ mA}$$

In this case, the input voltage will be

$$12 + 12 \times 10^{-3} \times 150 = 13.8 \text{ V}$$

When 20 mA passes through the Zener diode, the total current through the 150 Ω resistor is

$$20 + (12 \text{ V})/(1 \text{ k}\Omega) = 32 \text{ mA}$$

This time the input voltage will be

$$12 + 32 \times 10^{-3} \times 150 = 16.8 \text{ V}$$

Thus the operating range of the input voltage should be 13.8 V to 16.8 V

3.4 A diode at room temperature ($kT = 0.025$ eV) with a current of 1 μA has a forward bias voltage $V_F = 0.4$ V. For $V_F = 0.5$ V, find out the diode current in μA.

[JAM 2015]

Soln. Let the required diode current be I. Using Equation (3.11) and considering the given value kT as kT/q,

$$I = I_o \exp(0.5/0.025) \quad (1)$$

and

$$1 \times 10^{-6} = I_o \exp(0.4/0.025) \quad (2)$$

Solving for I_o from (2) and putting the same in (1),

$$I = 54.598 \times 10^{-6} = 54.6 \text{ μA (approx.)}$$

3.5 A variable power supply (5 V – 20 V) is connected to a Zener diode (Figure 3.15) specified by a breakdown voltage of 10 V. Determine the ratio of the maximum power to the minimum power dissipated across the load resistance.

[JAM 2013]

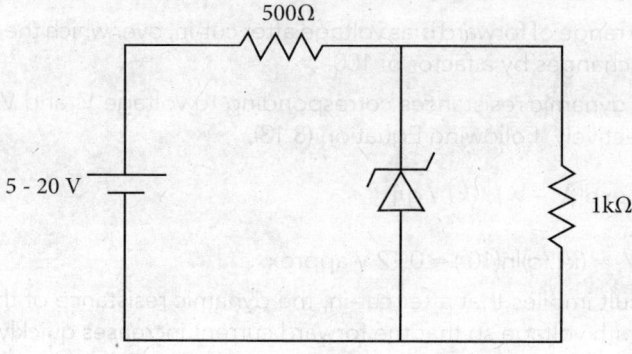

Figure 3.15

Soln. When the supply voltage becomes 5 V, the minimum value, the Zener diode does not break down and acts as open circuit. The total current through the load resistor is

$$\frac{5V}{(500+1000)\Omega} = \frac{10}{3} mA$$

Therefore, the (minimum) power dissipated across the load is

$$\left(\frac{10}{3}\right)^2 \times 10^{-6} \times 1 \times 10^3 = \frac{100}{9} mW$$

When the supply voltage becomes 10 V or more, the Zener diode breaks down and holds a fixed voltage of 10 V across the load resistor and the (maximum) power dissipated across the load is

(10×10)/1000 = 100 mW

Thus the ratio of the maximum to the minimum power dissipated is 9:1.

3.6 A 0.25 W Zener diode has breakdown voltage of 6 V and a dynamic resistance of 5 Ω. If the current limiting resistance in series be 500 Ω, show that the Zener diode acts as voltage regulator satisfactorily for the range of variation of 12 to 15 V of the supply voltage.

Soln. When the supply voltage is 12 V, the Zener current is

$$\frac{12-6}{500+5} = 11.88 mA$$

The output voltage is the Zener breakdown voltage plus the voltage drop across the dynamic resistance and is given by

$$6 + 5 \times 11.88 \times 10^{-3} = 6.059 V$$

Similarly, when the supply voltage is 15 V, the Zener current is

$$\frac{15-6}{500+5} = 17.82 \, \text{mA}$$

The output voltage is

$$6 + 5 \times 17.82 \times 10^{-3} = 6.089 \, \text{V}$$

Thus in both cases, the output voltage remains almost constant.

The maximum power dissipated across the Zener diode is

$$17.82 \times 10^{-3} \times 6 + (17.82)^2 \times 10^{-6} \times 5 = 0.108 \, \text{W},$$

which is well below the given maximum power rating of 0.25 W.

From the above results it can be concluded that the Zener diode acts as voltage regulator satisfactorily for the given range of variation of the supply voltage.

Exercise

Subjective Questions

1. Explain the formation of built-in potential at the p–n junction. What are the minority carriers in a p–n junction diode and what are their effects?
2. Derive an expression for the built-in potential in terms of doping concentrations. Mention the factors on which the width of the depletion region depends.
3. Explain with diagram, how the space charge, the electric field and the electrostatic potential vary with distance across an unbiased p–n junction.
4. Explain the drift and diffusion current across the p–n junction mentioning the carriers involved and their direction of movement. How do the drift and the diffusion current change with bias?
5. Draw a graph of the forward and reverse current–voltage characteristics of a p–n junction diode and discuss what properties of the diode are reflected from the graph. What is the significance of the cut-in voltage of a diode?
6. What is Zener diode? Compare the Zener breakdown and the avalanche mechanisms of a reverse biased p–n junction.
7. Make a comparison between the features of a Zener diode and that of a conventional rectifier diode.
8. What is the origin of the junction capacitance of a p–n junction diode? What types of capacitances are formed at the junction? What happens in the case of a Zener diode?
9. Explain the basic principles of light-emitting diode (LED). Is silicon a suitable semiconductor for fabricating LED?
10. Why any arbitrary semiconductor material may not be suitable for the fabrication of light-emitting diode (LED)? What are the main uses of LED?

Conceptual Test

1. An intrinsic semiconductor, when doped for n-type, gets much increased number of electrons. Does the material remain electrically neutral?
2. A circuit consists of a series combination of a voltage source (V), a forward biased p–n junction diode and a resistor R. Experimentally one can measure the current (I) flowing in the circuit and the voltage drop across the diode. Can these parameters be estimated theoretically?

   Hints. The forward current becomes $I = I_o \left[\exp\left(\dfrac{q(V-IR)}{kT} \right) - 1 \right]$, which is different from Equation (3.11). Both sides of the present equation contain I and it cannot be determined analytically. Such equations are called *transcendental equations* that are solved by *iterative* method. Trial values of I are put on both sides and the differences are verified. The value of I corresponding to the minimum difference is considered.
3. A toy car has solar panel on its roof. Absorbing energy from the sun, the car starts moving without using any other fuel. Are the first and the second law of thermodynamics obeyed?
4. Can the illuminated junction of a solar cell and the light-emitting junction of an LED be treated as reversible devices, such as heat engine and refrigerator? Explain.
5. Compare the action of photoelectric effect with that of a photodiode. (**Hints.** The former one takes place in metal or semimetal and the electron, acquiring energy, comes out of the material surface. The latter one occurs in semiconductor and the electrons transfer from one energy level to another within the material and do not leave the material.)

Numerical Problems

1. The reverse saturation current of a silicon or germanium diode gets doubled for every 10°C rise in temperature. Base on this information, determine the increase in temperature for 100 times increase in the current.
2. Assuming the diodes in the circuit of Figure 3.16 to be ideal find the voltage V_o.

   [GATE 2010]

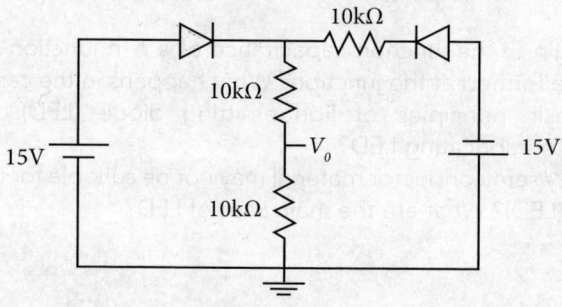

Figure 3.16

3. In Figure 3.9, the supply voltage is 20 V, the Zener breakdown voltage is 10 V and the device requires minimum 25 mA of current for maintaining a constant breakdown voltage across the load resistor of 100 Ω. Determine the resistance value of R for proper voltage regulation.

4. Someone has made the following changes in the Zener diode circuit of Figure 3.15: Supply voltage 10 V, Zener breakdown voltage 6 V, load resistance 10 kΩ and the current limiting resistance 1 kΩ. Under such a condition, calculate (i) the load current, (ii) the Zener current and (iii) the load current, if the load resistance is reduced to 600 Ω.

5. Each of the diodes D_1 and D_2 in Figure 3.17 have forward cut-in voltage of 0.7 V. The Zener diode has reverse breakdown voltage of 5.6 V. The input voltage is $v_i = 10\sin 100\pi t$ volt. What are the maximum and minimum values of the output voltage (v_o)?

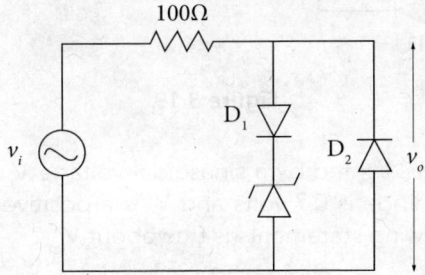

Figure 3.17

6. An LED operates at 1.5 V and 5 mA in forward bias. Assuming 80% external efficiency of the LED, how many photons are emitted per second?
   (a) 5×10^{16} (b) 1.5×10^{16} (c) 0.8×10^{16} (d) 2.5×10^{16}
   [NET Jun. 2012]

7. The forward diode current is given by $I = \kappa T^\alpha e^{-E_g/k_B T}(\exp(eV/k_B T) - 1)$, where E_g is the band gap of the semiconductor, V is the voltage drop across the diode, T is the temperature of the diode operating near room temperature and α and κ are constants. A diode is used as a thermal sensor in the circuit shown in Figure 3.18. If V is measured using an ideal voltmeter to estimate T, the variation of the voltage V as a function of T is best approximated by (in the following a and b are constants)
   (a) $aT^2 + b$ (b) $aT + b$ (c) $aT^3 + b$ (d) $aT + bT^2$
   [NET Jun. 2019]

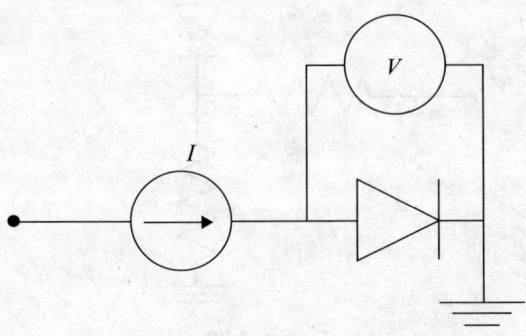

Figure 3.18

8. Figure 3.19 shows a circuit containing two diodes D_1 and D_2 with threshold voltages V_{TH} of 0.7 V and 0.3 V, respectively. Considering the simplified diode model, which assumes diode I–V characteristic as shown in the plot on the right, the current through the resistor R is .... μA.

[JAM 2020]

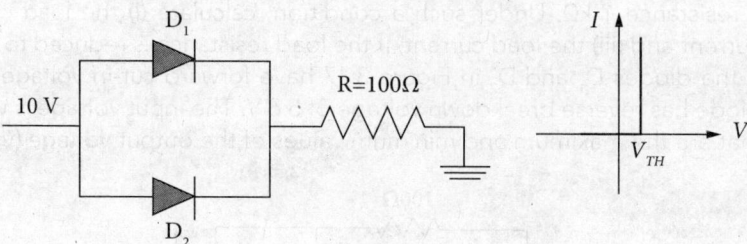

Figure 3.19

9. The circuit of Figure 3.20 is fed by a sinusoidal voltage $V_{in} = V_o \sin\omega t$. Assume that the cut-in voltage of the diode is 0.7 volts and V_1 is a positive dc voltage smaller than V_o. Which one of the following statements is true about V_{out}?

(a) Positive part of V_{out} is restricted to a maximum voltage of $0.7 + \dfrac{R_2}{R_1 + R_2} V_1$

(b) Negative part of V_{out} is restricted to a maximum voltage of $0.7 + \dfrac{R_2}{R_1 + R_2} V_1$

(c) Positive part of V_{out} is restricted to a maximum voltage of $0.7 + \dfrac{R_1}{R_1 + R_2} V_1$

(d) Negative part of V_{out} is restricted to a maximum voltage of $0.7 + \dfrac{R_1}{R_1 + R_2} V_1$

[JEST 2019]

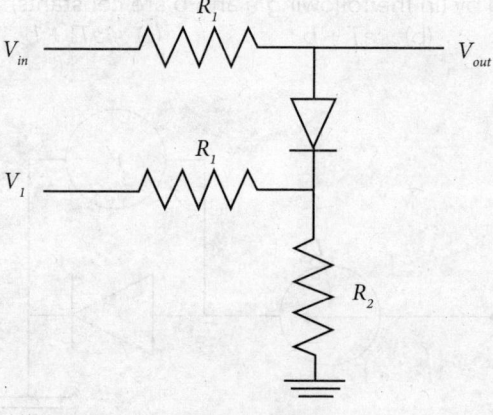

Figure 3.20

Project Work on Chapter 3

Study the forward and reverse current–voltage characteristics of a p–n junction diode and determine the cut-in voltage, static resistance and dynamic resistance of the diode.

Theory:
The p–n junction diode is *forward biased* when the positive terminal of the external voltage source is joined to the p-side and the negative terminal of the voltage source to the n-side. This pushes the holes and the electrons toward the junction and causes a high current to flow in the external circuit. The current (I) through a diode under forward bias voltage (V) can be expressed as [Equation (3.11)]

$$I = I_o[\exp(qV/kT) - 1]$$

I_o is the small reverse saturation current, almost independent of the bias voltage. The forward current increases rapidly and reaches the order of mA at the *cut-in voltage*.

The forward I–V plot is not linear and the resistance calculated from the slope (dV/dI) varies continuously, which is defined in the following ways.

Static Resistance is the ratio of any spot value of voltage to the corresponding current.

Dynamic Resistance is the ratio of the small change of forward voltage (dV) to the corresponding change of forward current (dI). For higher values of forward current, the dynamic resistance (r_d) can be approximated as [Equation (3.14)]

$$r_d = \frac{26}{I(mA)}$$

The diode is *reverse biased* when the positive and the negative terminal of the voltage source is joined to the n-side and the p-side, respectively. This time a negligible current, of the order of I_o flows through the diode up to the breakdown region of reverse bias voltage.

Computer analysis: The experiment is simulated with LTspice open software. The assembled circuit for the forward biased diode is shown in Figure P3.1. The following are the specifications for simulation.

The diode MURS320 was selected from the given choices. From the 'Simulate' menu, the simulation command 'DC sweep' was selected. The single voltage source, treated as the 1st source was named V. Linear sweep was implemented with start value of 0, stop value of 2 (volt) and increment of 0.01 (volt). The forward output characteristic curve for the diode was obtained on the screen by Clicking on the 'Run' command followed by right clicking on the viewer and selecting 'Add Traces' option. The actual set of simulated data was obtained by right clicking on the viewer and selecting 'File' and 'Export data as text'.

The reverse biased condition of the diode was simulated by reversing the voltage source, as shown in Figure P3.2. The above simulation process was repeated for the reverse biased condition and the current was found to be much less (of the order of 10^{-9} A) than that under forward bias and was almost independent of the bias voltage up to 5 V.

Results and Discussion: Using the simulated data, the forward current–voltage characteristic curve of the diode is plotted with Origin graph-plotting software, as shown in Figure P3.3. The forward cut-in voltage of the diode, as indicated by dashed line in the figure is about 0.55 V. The reverse current values plotted in the same scale were essentially zero and are not shown here.

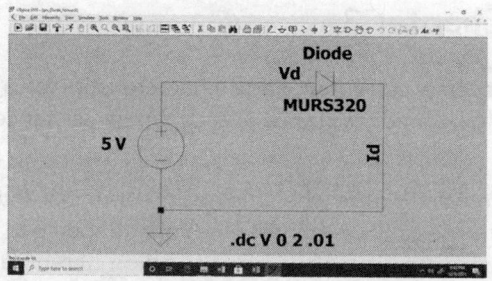

Figure P3.1 LTspice simulated diode under forward bias

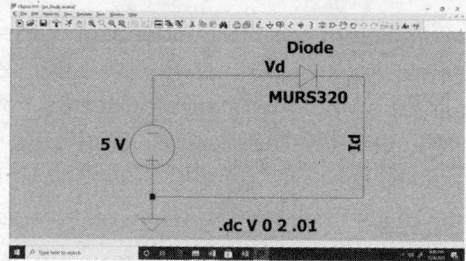

Figure P3.2 LTspice simulated diode under reverse bias

For the forward voltage variation from 0.4 V to 0.6 V, the calculated static resistance reduced from 4608 Ω to 24.39 Ω and the calculated dynamic resistance reduced from 4032 Ω to 1.43 Ω.

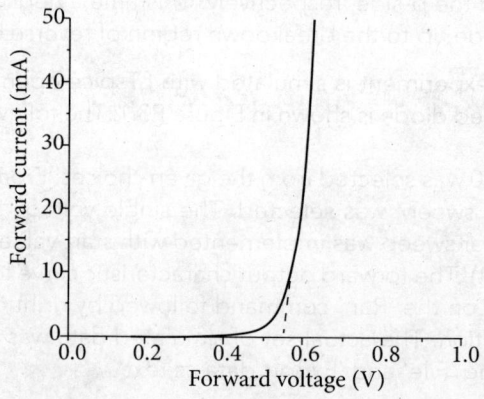

Figure P3.3 LTspice simulated forward current–voltage characteristic curve for the diode; the cut-in voltage is indicated by dashed line

Discuss the following points.

- Rectifying property of the diode revealed from the experiment
- Nonlinear current–voltage variation and its consequences on the device resistance
- The order of forward current around the cut-in voltage
- The trend of variation of static and dynamic resistances for different current–voltage values around the cut-in voltage

4

Diode Applications

The most significant property of the p–n junction diode, as introduced in Chapter 3, is *rectification* or converting alternating current (ac) to direct current (dc) because the diode conducts under forward bias only. This property is utilized for fabricating rectifier circuits. The three major categories of rectifier namely half-wave, full-wave, and bridge are illustrated in this chapter. The rectifying property of a diode is also implemented to the transmission of a portion of an alternating voltage waveform. Such circuits, known as the *clippers* are elaborated. Two other important circuits, namely the *clamper* and the *voltage multiplier* that use diodes with capacitors are brought together.

4.1 Piecewise Linear Model

The diode is, of course, a nonlinear device because the current through it undergoes nonlinear change with voltage. Yet the diode can be approximated as a linear element part-by-part under certain conditions. The concept of such 'linearizing the diode', as explained with Figures 4.1(a) and 4.1(b), is quite useful to the analysis of the circuits containing diodes. Both the diagrams symbolize the forward current–voltage characteristic curve of a typical diode but at different scales. Figure 4.1(a) represents the enlarged view for the condition just after cut in. The current is now determined by the junction property and the nonlinearity of the current–voltage curve is quite prominent. The same diode at a forward voltage much higher than the cut-in voltage behaves like that of Figure 4.1(b) where the current is dominated by the bulk resistance of the semiconductor, the nonlinearity below 0.7 V gets squeezed into a small region of the characteristic curve and the major portion of the curve becomes linear.

The above demonstration implies that when the bias voltage becomes much larger than the forward cut-in voltage of the diode, the equivalent circuit can be represented by the combination of the following three pieces of linear circuit elements in series.

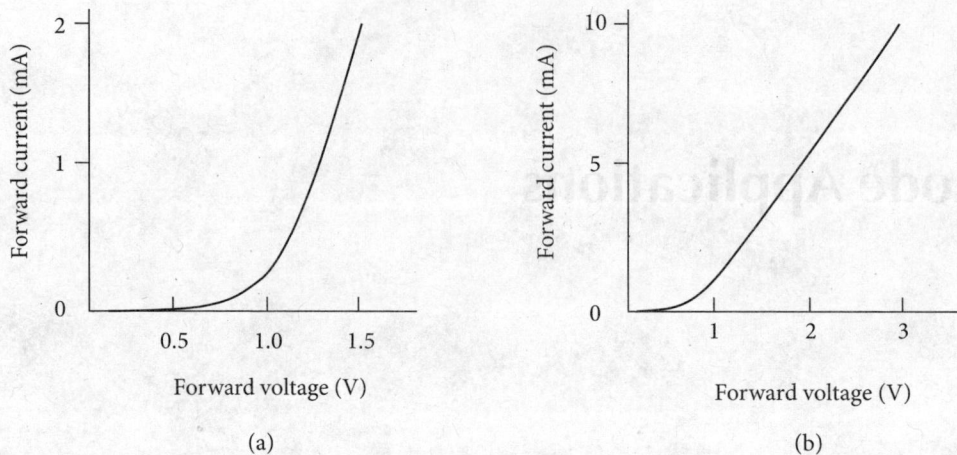

Figure 4.1 Schematic representation of the forward current–voltage characteristic of a diode, (a) immediately after cut-in and (b) at a high forward voltage

- An on/off switch representing the forward/reverse biased condition of the diode.
- A voltage source in lieu of the voltage drop across the diode.
- A resistor denoting the semiconductor bulk resistance.

Therefore, such an equivalent circuit of the diode is termed as the piecewise linear model of the diode. Figure 4.2 illustrates the action of this model. The switch (S) is the proxy of the forward and reverse-biased conditions of the diode. The voltage source (V_D) denotes the voltage drop across the diode when it is turned on. The resistor (R) stands for the resistance of the bulk semiconductor in the diode. Figure 4.2(a) indicates that when the diode is forward-biased, the switch (S) is assumed to be closed. The diode turns on when the forward voltage across it reaches V_D. The forward current–voltage curve after this voltage becomes straight line with the slope determined by resistance R. This linearity turns out to be more predominant, if an external resistor is joined in series. If the external resistance becomes much larger than the bulk semiconductor resistance of the diode, the latter one can be neglected and the equivalent circuit can be further simplified with omission of R [Figure 4.2(b)]. If the applied voltage is much larger than V_D, this voltage drop is also omitted from the equivalent circuit [Figure 4.2(c)] and the diode is approximated by simply an on/off switch under forward and reverse bias, respectively.

4.2 Load Line and Q-Point

If an external resistor is joined in series with the diode, the situation becomes like that of Figure 4.3(a). The supply voltage (V_S) forward biases the diode and the same forward current (I_D) flows through both the diode and the resistor (R_L). This is called *load resistor* because the supply has to deliver current through it and it appears as a 'load' or 'burden' to

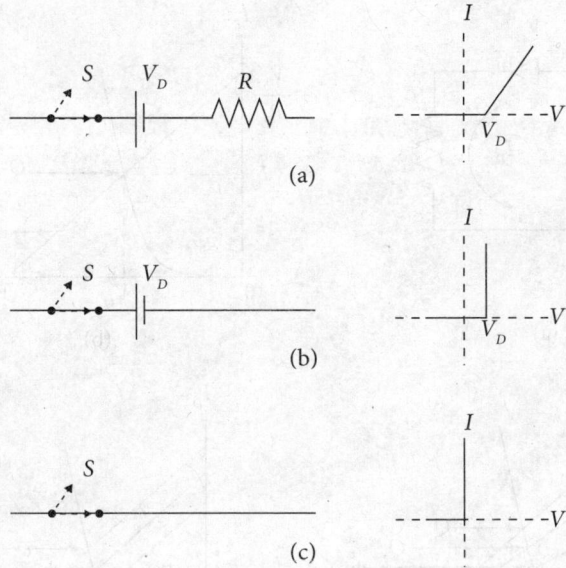

Figure 4.2 Piecewise linear model of the diode (left) and the corresponding current–voltage variation (right): (a) series combination of switch (S), voltage source (V_D) and resistor (R) in general, (b) approximated as series combination of S and V_D when the externally joined resistance is much larger than R, and (c) approximated as a switch (S) only when the applied voltage is much larger than V_D.

the voltage source. The supply voltage drops partly across the diode as V_D and partly across the load resistor as $I_D R_L$ so that the above parameters can be related as

$$V_S = V_D + I_D R_L \tag{4.1}$$

The expression for the current can be derived from Equation (4.1) as

$$I_D = -\frac{V_D}{R_L} + \frac{V_S}{R_L} \tag{4.2}$$

Equation (4.2) signifies that the forward current (I_D) through the diode–resistor series combination and the voltage drop (V_D) across the diode are related by a straight line whose slope depends on the resistance value of the load (R_L). Therefore, Equation (4.2) is designated as the *load line* of the diode. Figure 4.3(b) shows that the load line intercepts the voltage-axis and the current-axis at V_S and (V_S/R_L), respectively. When the forward current–voltage (I–V) characteristic is plotted on the same graph, as shown in Figure 4.3(b), the point of its intersection with the load line is termed as *operating point* or *quiescent point* or, more popularly, Q-point of the diode. The voltage (V_Q) and the current (I_Q) corresponding to the Q-point represent the voltage and current values at which the diode of Figure 4.3(a) is operating with a load resistance of R_L. The load line, hence the Q-point gets changed with the variations of the load resistance and the supply voltage in two different ways, as sketched in figures 4.3(c) and 4.3(d), respectively.

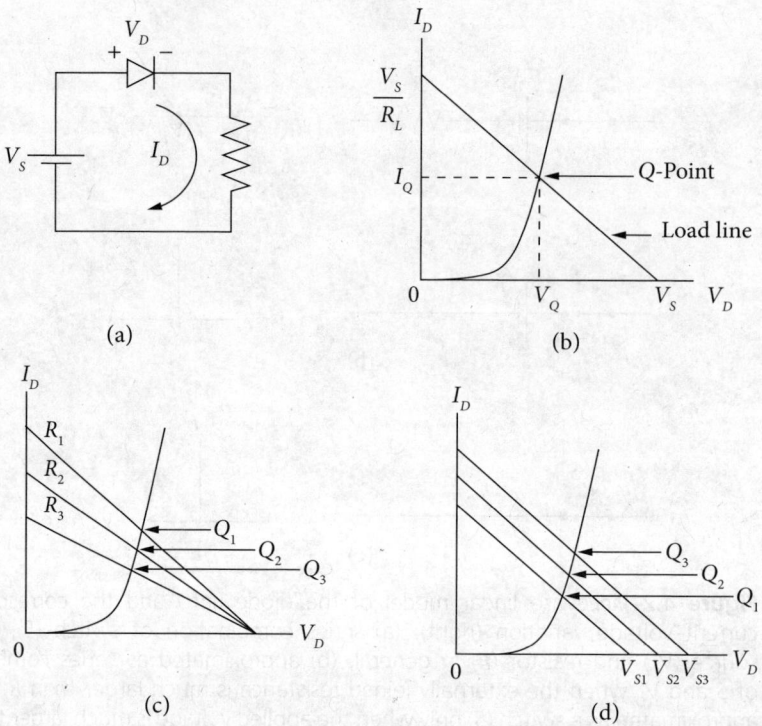

Figure 4.3 Diode with load resistance in series, (a) circuit diagram, (b) graphical representation of load line and Q-point, (c) variation of load line and Q-point with load resistance ($R_1 < R_2 < R_3$), and (d) variation of load line and Q-point with supply voltage ($V_{S1} < V_{S2} < V_{S3}$).

> **Note: Significance of Load Line and Q-Point**
>
> The load line and the Q-point are two interrelated parameters of an electronic device, which determine the operating values of the current and the voltage for a specific value of load resistance and supply voltage. This determination is executed by intersecting the load line with the output current–voltage characteristic curve. The present discussion is on the Q-point of a diode. Similar concepts are applied to other electronic devices, such as the bipolar junction transistor (Chapter 6).

4.3 Rectifiers

Household gadgets, such as light, fan, and refrigerator etc., run on 220 V, 50 Hz AC main supply. The same 220 V AC is used for charging the batteries of electronic gadgets, such as mobile phone and laptop computer. However, the electronic circuits within these equipment do not get in direct connection with the main ac supply. The high value of alternating voltage is stepped down to a suitable lower value with a transformer and is then passed through an electronic circuit made of diodes, known as the *rectifier*, which converts the alternating current (ac) to direct current (dc). This book adopts the symbols 'ac' and 'dc' when used as adjective denoting the property of a source. When used as noun or at the beginning of a sentence, the symbols 'AC' and 'DC' are used. Though the diode performs the main job of rectification, the rectifier refers to the whole circuit comprised of the diode, the step-down transformer and the possible load resistance. It is called a 'rectifier' because it straightens the current in a single direction from a state of periodically reversing direction. Three major categories of rectifier circuits are referred to as half-wave, full-wave and bridge rectifiers, depending on their mode of operation. This section presents an overview of all these rectifiers. It is always assumed that the input voltage amplitude is much larger than the forward cut-in voltages of the diodes involved in the rectifier.

4.3.1 Half-Wave Rectifier

It consists of only one diode, as shown in the circuit diagram of Figure 4.4(a) and it can utilize only one-half of the input ac cycle. The step-down reduces the amplitude of the main supply of alternating voltage but the voltage waveform and its frequency remain unchanged. Let this stepped down alternating voltage be sinusoidal given by

$$v = V_m \sin \omega t \tag{4.3}$$

The amplitude (V_m) is much larger than the forward cut-in voltage of the diode. The changes of the phase (ωt) and time period (T) are indicated in the sketch of Figure 4.4(b). During the positive half-cycle of input, when $0 \leq \omega t \leq \pi$ or $0 \leq t \leq T/2$, the diode is forward-biased and the input alternating current flows through the diode and the load resistor (R_L). The voltage variation across R_L is, therefore, similar to the input waveform. Through the negative half-cycle of the input, when $\pi \leq \omega t \leq 2\pi$ or $T/2 \leq t \leq T$, the diode gets reverse-biased so that the input alternating current cannot flow through the load resistor. Thus only the positive half of the input ac waveform is obtained across the load resistor, as drawn in Figure 4.4(b). The variation of current and voltage across the load is sinusoidal but in only one direction, when the diode is forward-biased. The following parameters are defined for the purpose of assessing the rectifier performance.

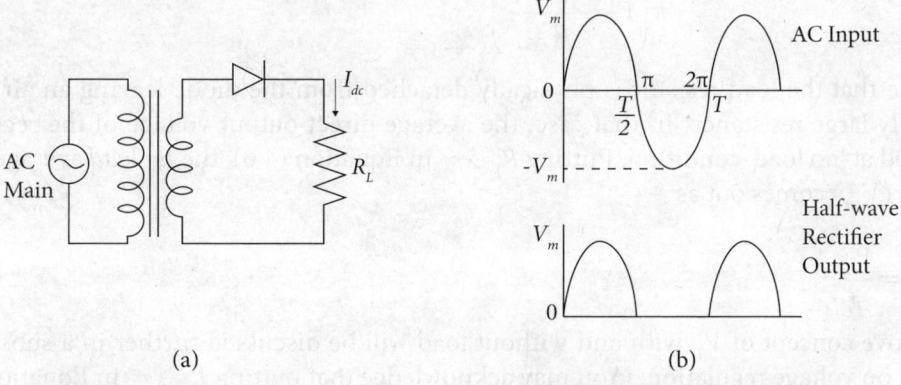

Figure 4.4 Half-wave rectifier (a) circuit diagram and (b) input and output voltage waveforms

Average Direct Current (I_{dc}): The output current of the half-wave rectifier flowing through the load resistor varies as sinusoidal in one direction for one half of the input and remains zero for the other half. Consequently, it has some finite dc value on the average, which can be estimated in the following way.

$$I_{dc} = \frac{1}{2\pi}\left[\int_0^{\pi} I_m \sin \omega t \, d(\omega t) + \int_{\pi}^{2\pi} 0.d(\omega t)\right] = \frac{I_m}{\pi} \qquad (4.4)$$

In Equation (4.4), the maximum value of the input alternating current is

$$I_m = \frac{V_m}{R_L} \qquad (4.4a)$$

Strictly speaking, the forward resistance (R_f) of the diode and the Ohmic resistance (R_s) of the transformer secondary coil are also added and the accurate expression for I_m should be

$$I_m = \frac{V_m}{R_L + R_f + R_s} \qquad (4.4b)$$

However, R_f and R_s are negligibly small in practice and Equation (4.4a) is a reasonable approximation for I_m.

Average Direct Voltage (V_{dc}): The average direct voltage obtained across the load resistor (R_L) during the full ac cycle is given by

$$V_{dc} = I_{dc} R_L = \frac{I_m R_L}{\pi} \qquad (4.5)$$

Shortly before this, we have discarded R_f and R_s. Would you mind, if those were reconsidered? Indeed that would provide with a very useful definition of load condition. Repeating Equation (4.5) with the inclusion of R_f and R_s we have

$$V_{dc} = \frac{V_m R_L}{\pi(R_L + R_f + R_s)} = \frac{V_m}{\pi\left(1 + \dfrac{R_f + R_s}{R_L}\right)} \qquad (4.6)$$

Imagine that the load resistor is physically detached from the diode leaving an air gap of infinitely large resistance. In that case, the average direct output voltage of the rectifier is obtained at 'no load' condition. Putting $R_L \to \infty$ in Equation (4.6), the *no load* average direct voltage (V_{NL}) comes out as

$$V_{NL} = \frac{V_m}{\pi} \qquad (4.7)$$

The above concept of V_{dc} with and without load will be discussed further in a subsequent section on voltage regulation. (You may acknowledge that putting $R_L \to \infty$ in Equation (4.5) would not allow us for this mathematical manipulation.)

Voltage Regulation: The rectifier cannot produce a constant direct voltage across the load resistor irrespective of its resistance value because the current through the load goes on changing with the resistance. The question is up to what extent the rectifier can approach the ideal situation of constant load voltage independent of the load current. This quality of a rectifier is defined in terms of *voltage regulation*, a figure of merit indicating up to what percentage the ideal condition is deviated from.

A formal expression for the voltage regulation of a rectifier is derived as follows. This is derived with the parameters of half-wave rectifier but is valid for the other rectifiers as well. From Equations (4.4) and (4.4b), the average direct current through the load is expressed as

$$I_{dc} = \frac{V_m}{\pi(R_L + R_f + R_s)} \tag{4.8}$$

Rearranging Equation (4.8) and considering $V_{dc} = I_{dc}R_L$ from Equation (4.5),

$$V_{dc} = \frac{V_m}{\pi} - I_{dc}(R_f + R_s) \tag{4.9}$$

Using Equation (4.7) and designating the V_{dc} value for an arbitrary load resistance as V_L, Equation (4.9) can be rewritten as

$$V_L = V_{NL} - I_{dc}(R_f + R_s) \tag{4.10}$$

Equation (4.10) denotes that for fixed values of R_f and R_s, the voltage (V_L) across the load resistor decreases linearly with increasing load current. The occurrence is outlined graphically in Figure 4.5. The voltage regulation is expressed in percentage as

$$\text{Regulation}(\%) = \frac{V_{NL} - V_L}{V_L} \times 100 \tag{4.11}$$

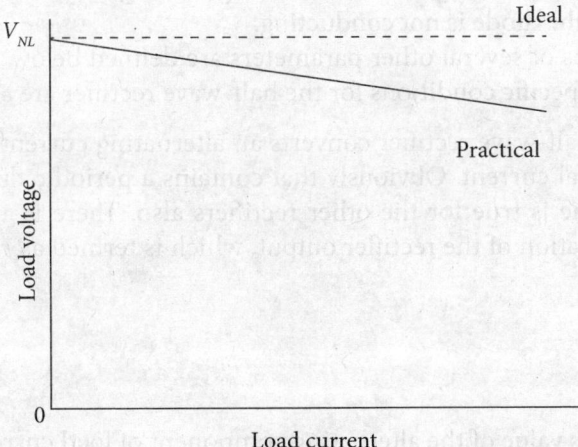

Figure 4.5. Variation of load voltage with load current for a practical rectifier compared with the ideal condition.

At no load condition, the load current is zero and $V_L \approx V_{NL}$ so that the regulation is 0%. An ideal regulation remains zero percent for all values of load current, as indicated by the dotted line in Figure 4.5. Actually the load voltage decreases steadily with increasing load current and the regulation percentage goes on increasing. The larger is the regular percentage, the poorer is the regulation.

RMS Current: An ac ammeter, if joined in series with the load resistance would read the *root-mean-square* (*rms*) value of the pulsating current. This is estimated as

$$I_{rms} = \left[\frac{1}{2\pi}\int_0^\pi I_m^2 \sin^2 \omega t\, d(\omega t)\right]^{1/2} = \frac{I_m}{2} \tag{4.12}$$

It is remarkable to note that the rms current defined by Equation (4.12) is quite different from the conventional value of $I_m/\sqrt{2}$ in the case of continuous alternating current.

Peak Inverse Voltage: All the above mentioned parameters of the rectifier are somehow associated with the conducting state of the diode at forward-biased condition during one half of the input voltage. The non-conducting state of the diode at the other half should also be considered. The maximum of the input alternating voltage appears across the diode under reverse bias and it must be well below the reverse breakdown voltage of the diode. The maximum allowable limit of reverse voltage that the rectifier circuit can withstand without reaching the breakdown region is defined in terms of a *peak inverse voltage*. This quantity is determined by

- the material and construction of the diode itself,
- the number of diodes involved in the rectification process and
- the arrangement of the diode(s) in the rectifier circuit.

A half-wave rectifier has a peak inverse voltage of V_m, the amplitude of the input alternating voltage. It is to be noted that the peak inverse voltage has significance during the part of the input ac cycle when the diode is not conducting.

General definitions of several other parameters are defined below, which are applicable to all rectifiers. The specific conditions for the half-wave rectifier are also mentioned.

Ripple Factor: The half-wave rectifier converts an alternating current into a pulsating and varying unidirectional current. Obviously that contains a periodic fluctuation, something like ripples. The same is true for the other rectifiers also. There is a general formula for estimating the fluctuation of the rectifier output, which is termed as *ripple factor* (*r*) and is defined as

$$r = \frac{I_{r(rms)}}{I_{dc}} \tag{4.13}$$

where $I_{r(rms)}$ is the rms value of the alternating component of load current expressed as

$$I_{rms}^2 = I_{dc}^2 + I_{r(rms)}^2 \tag{4.14}$$

Using Equations (4.13) and (4.14) and substituting for I_{dc} and I_{rms} from Equations (4.8) and (4.12), respectively, the ripple factor is obtained as

$$r = \left[\left(\frac{I_{rms}}{I_{dc}}\right)^2 - 1\right]^{1/2} \tag{4.15}$$

Proceeding in the same way as above, the ripple factor can be expressed in terms of rms and dc values of voltage also, as given by Equation (4.15a).

$$r = \left[\left(\frac{V_{rms}}{V_{dc}}\right)^2 - 1\right]^{1/2} \tag{4.15a}$$

For the half-wave rectifier, $I_{rms} = I_m/2$ [Equation (4.12)] and $I_{dc} = I_m/\pi$ [Equation (4.4)]. Putting these values in Equation (4.15), the numerical value of r comes out as 1.21 or 121%, which indicates that the half-wave rectifier does not perform efficient rectification. It is further clarified in the next paragraph.

> **Note: Significance of Ripple Factor**
>
> The effectiveness of the rectification process is judged by the ripple factor. The rectifier output should ideally give dc only but in practice it contains dc as well as ac component. The ripple factor defines the ratio of the rms value of the ac component to the rms value of the dc component within the rectified output.

Efficiency: The efficiency (η) of a rectifier, also called the *ratio of rectification* is defined as the ratio of the dc output power (P_{dc}) across the load resistor to the ac input power (P_{ac}) supplied to the rectifier from the transformer secondary terminals.

$$\eta = \frac{P_{dc}}{P_{ac}} \times 100\% \tag{4.16}$$

These two power parameters are expressed in terms of the corresponding currents as

$$P_{dc} = I_{dc}^2 R_L \tag{4.16a}$$

$$P_{ac} = I_{rms}^2 R_L \tag{4.16b}$$

For practical purpose, the forward resistance of the diode (R_f) and the ohmic resistance of the transformer secondary coil (R_s) are assumed to be negligibly small. Putting the expressions for $I_{dc} = I_m/\pi$ and $I_{rms} = I_m/2$ in Equation (4.16), the efficiency of the half-wave rectifier is found to be 40.6%.

Transformer Utilization Factor (TUF): Most of the diode rectifier circuits involve a transformer for stepping down the ac main supply. Therefore the transformer rating is also included in the assessing parameters of a rectifier in terms of a transformer utilization factor (*TUF*) given by

$$TUF = \frac{\text{DC power output}}{\text{effective volt-ampere rating of the transformer}} \quad (4.17)$$

The denominator of Equation (4.17) can be expressed as the product $V_{rms}I_{rms}$ for the secondary coil. For the half-wave rectifier, the average dc power is, of course, $I_{dc}^2 R_L = (I_m/\pi)^2/R_L$. The rated alternating voltage at the transformer secondary is $(V_m/\sqrt{2})$, where V_m can be approximated as $I_m R_L$. The current is $I_m/2$ so that the ac power (voltage × current) is $I_m^2 R_L/(2\sqrt{2})$. Putting these values in Equation (4.17), TUF = 0.287 for the half-wave rectifier.

> **Note: Rectifier Efficiency versus *TUF***
>
> The transformer utilization factor (*TUF*) is a performance evaluation parameter of the rectifier. It is different from the efficiency of rectification. The efficiency denotes how competent is the rectification process in converting the ac input power to the dc output power whereas the *TUF* signifies how much the transformer is utilized in the rectification process. The higher is the efficiency, the more is the conversion of ac into dc and the higher is the *TUF*, the lesser is the volt–ampere rating (hence the bulk size) of the transformer.

Example 4.1

A silicon diode with forward resistance of 20 Ω is employed in half-wave rectifier. If the input voltage is 20sin314*t* and the load resistor is of 1 kΩ, determine the average direct output current. Neglect the resistance of the transformer secondary coil.

Solution

From the given expression, the maximum input voltage is 20 V. Using Equation (4.4b), the maximum current is

$$20/(1000 + 20) = 19.608 \text{ mA}$$

Using Equation (4.4), the average direct output current is 6.239 mA.

Example 4.2

Determine (i) the average direct voltage across the load, (ii) the ac input power and (iii) the power efficiency of the rectifier in Example 4.1 above.

Solution

(i) The average direct voltage across the load is

$(6.239 \text{ mA}) \times (1 \text{ k}\Omega) = 6.239 \text{ V}$

[Note that this could be determined from Equation (4.6) assuming $R_s = 0$]

(ii) The rms current is [Equation (4.12)] $19.608/2 = 9.804$ mA so that the ac input power is

$(9.804 \times 10^{-3})^2 \times (1000 + 20) = 98.04 \text{ mW}$

(iii) The dc output power across the load resistance is

$(6.239 \times 10^{-3})^2 \times 1000 = 38.93 \text{ mW}$

Therefore, the efficiency is $(38.93/98.04) \times 100 = 39.71$ %

[The deviation from the ideally expected value of 40.6 is due to the existence of diode resistance.]

4.3.2 Full-Wave Rectifier

This rectifier utilizes the full of the input alternating voltage waveform to convert in a unidirectional output. The circuit diagram is given in Figure 4.6(a) and the input–output voltage waveforms are illustrated in Figure 4.6(b). It comprises two diodes D_1 and D_2 connected opposite to each other in series with the transformer secondary coil. As shown in the figure, the transformer is center-tapped; a connection is taken out from just the middle of the secondary coil. As a result, the secondary voltage is divided into two equal parts v_1 and v_2, each having the peak value of V_m and each being 180° out of phase with respect to each other. When v_1 forward biases D_1, v_2 reverse biases D_2 and vice versa. Tracing the current path on the circuit, it is understood that in both cases, the current (I_{dc}) through the load resistor flows in the same direction, as indicated by solid and dashed arrows. Thus both the halves of the input alternating voltage and current get rectified. Similar to the half-wave rectifier, the following parameters are defined for the full-wave rectifier assuming a sine wave voltage as the input.

Average Direct Current (I_{dc}): The input is sinusoidal for both the halves but opposite in direction so that the average value of the unidirectional variable current across the load is estimated as

$$I_{dc} = \frac{1}{2\pi} \left[\int_0^{\pi} I_m \sin \omega t \, d(\omega t) + \int_{\pi}^{2\pi} (-I_m \sin \omega t) \, d(\omega t) \right] = \frac{2I_m}{\pi} \qquad (4.18)$$

The peak value of input current (I_m) in each half can be approximated by V_m/R_L.

Average Direct Voltage (V_{dc}): Using Equation (4.18), the average direct voltage obtained across the load resistor during the full-cycle is calculated, for practical purpose, as

$$V_{dc} = I_{dc} R_L = \frac{2 I_m R_L}{\pi} \tag{4.19}$$

If we go for rigorous accuracy, it would be

$$V_{dc} = \frac{2 V_m}{\pi \left(1 + \dfrac{R_f + R_s}{R_L} \right)} \tag{4.19a}$$

The parameters are already defined earlier, in connection with Equation (4.4b).

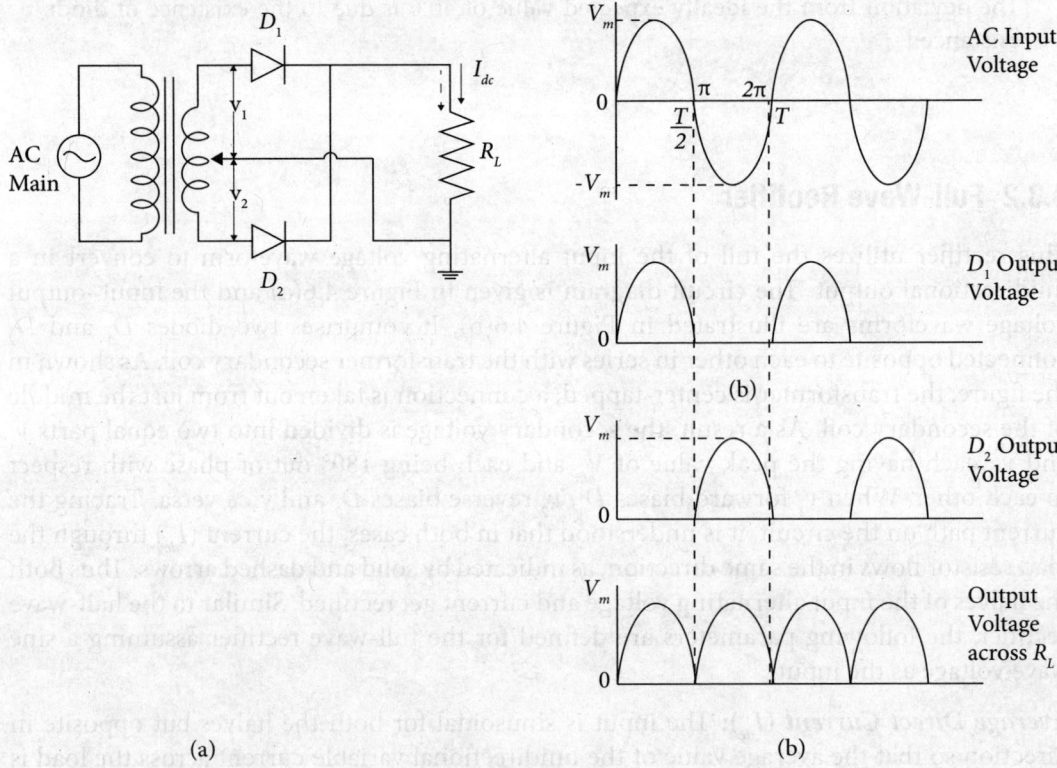

Figure 4.6 Full-wave rectifier (a) circuit diagram and (b) input and output voltage waveforms

Voltage Regulation: The concept of regulation is similar to that of the half-wave rectifier but the numerical value is different. Proceeding in the same way as that of the half-wave rectifier, the load voltage of the full-wave rectifier can be defined as

$$V_L = V_{NL} - I_{dc}(R_f + R_s) \tag{4.20}$$

Equation (4.20) appears to be the same as Equation (4.10) but the expressions for the no-load voltage (V_{NL}) and the average direct current (I_{dc}) are different in Equation (4.20) are different, as follows.

$$V_{NL} = \frac{2V_m}{\pi} \tag{4.20a}$$

$$I_{dc} = \frac{2V_m}{\pi(R_L + R_f + R_s)} \tag{4.20b}$$

Equation (4.11) defining the regulation percentage is applicable to this full-wave rectifier also.

RMS Current: Integrating over both the halves of the input alternating current and averaging over the full-cycle, the rms current is obtained as

$$I_{rms} = \left[\frac{1}{2\pi} \left(\int_0^\pi I_m^2 \sin^2 \omega t \, d(\omega t) + \int_\pi^{2\pi} I_m^2 \sin^2 \omega t \, d(\omega t) \right) \right]^{1/2} = \frac{I_m}{\sqrt{2}} \tag{4.21}$$

It is noted that the rms value of the full-wave rectified current is the same as that of the conventional alternating current.

Peak Inverse Voltage: The two halves, v_1 and v_2, of the secondary alternating voltage of the full-wave rectifier are indicated in Figure 4.6(a). When one half of the secondary alternating voltage forward biases a diode, the other half reverse biases the other diode and this sequence is continued. Figure 4.6(b) sketches the corresponding output waveforms where each half has a peak value of V_m. Since the forward-biased diode has very small resistance, almost $V_m + V_m = 2V_m$ appears across the reverse-biased diode and the diode must not break down under this voltage. Thus each of the diodes in a full-wave rectifier should have reverse breakdown voltage greater than $2V_m$. In other words, the peak inverse voltage of the full-wave rectifier is $2V_m$, which is the maximum amplitude of the ac input that the rectifier can withstand.

Ripple Factor: The concept is the same as that of the half-wave rectifier and Equation (4.15) is applicable to define the ripple factor of full-wave rectifier also. However, the expressions for I_{dc} and I_{rms} are different. Substituting for those values from (4.18) and (4.21), respectively, the value of ripple factor for full-wave rectifier is 0.48 approximately. Note that this is much smaller than that of the half-wave rectifier implying that the drawback of ac fluctuation is much improved in the present case.

Efficiency: Equation (4.16) can be used to define the efficiency of full-wave rectifier also. Obviously, the numerical values of P_{dc} and P_{ac} come out to be different from those of the half-wave rectifier because the I_{dc} and I_{rms} are different. Putting the appropriate values, the efficiency (η) is found to be 81.2%, much higher than that of the half-wave rectifier.

Transformer Utilization Factor (TUF): If one observes from the primary coil side, the current is found to flow in any one half of the secondary coil for both the positive and the negative half-cycle of the input and the *TUF* for the primary can be calculated by using Equation (4.17) as

$$TUF_p = \frac{\left(\frac{2I_m}{\pi}\right)^2 R_L}{\left(\frac{V_m}{\sqrt{2}}\right)\left(\frac{I_m}{\sqrt{2}}\right)} = 0.81 \tag{4.22a}$$

However, the picture becomes different on looking from the secondary coil side. Each half of the secondary coil draws power from the main supply in one half of the ac cycle. Therefore, the effective TUF for the secondary is

$$TUF_s = 0.287 + 0.287 = 0.574 \tag{4.22b}$$

Considering both the primary and the secondary of the transformer, the average TUF for the whole full-wave rectifier is

$$TUF = (TUF_p + TUF_s)/2 = 0.693 \tag{4.22c}$$

4.3.3 Bridge Rectifier

This is also a kind of full-wave rectifier but different from the conventional full-wave rectifier discussed above. The bridge rectifier has the following distinguished features.

- The transformer for stepping down the input is not center-tapped.
- Two diodes are employed in series for each half of the ac input.
- It has a bridge-like construction, the ac input and the load being connected to diagonally opposite terminals.

Figure 4.7(a) displays the circuit diagram for a bridge rectifier. Figure 4.7(b) demonstrates that during one half of the secondary alternating voltage, diodes D_1 and D_2 get forward-biased and the other two, being reverse-biased, act as open circuits. Those two diodes, namely D_3 and D_4, get forward-biased during the other half of the input, D_1 and D_2 become reverse-biased and act as open circuits [Figure 4.7(c)]. Tracing the current path, it is found that in both the above cases the current flows through the load resistor (R_L) in the same direction. Figure 4.7(d) compares the output voltage waveform obtained across R_L with the input alternating voltage waveform.

Though the designation of 'full-wave' is not assigned, the bridge rectifier utilizes the full of the alternating input voltage in producing the rectified output voltage across the load resistor. It has the following similarities with the conventional full-wave rectifier.

- The average direct current (I_{dc}) and the rms current (I_{rms}) through the load and the average direct voltage (V_{dc}) across it have the same values and can be expressed by the same equations.
- The ripple factor and the efficiency are the same and can be expressed by the same equations.
- The voltage regulation is also the same as that of the conventional full-wave rectifier.

There are several dissimilarities too, as mentioned below that have given the bridge rectifier a distinct identity.

- The peak inverse voltage is V_m (and not $2V_m$), appearing across each reverse-biased diode.
- The TUF need not be averaged. It is calculated simply with Equation (4.22a) and is found to be 0.81, a much larger value than that obtained with a full-wave rectifier. Indeed this is the main advantage of bridge rectifier that makes it suitable for deriving large dc power.

The three types of rectifiers discussed above have their own merits and limitations, as compared in Table 4.1 and Table 4.2.

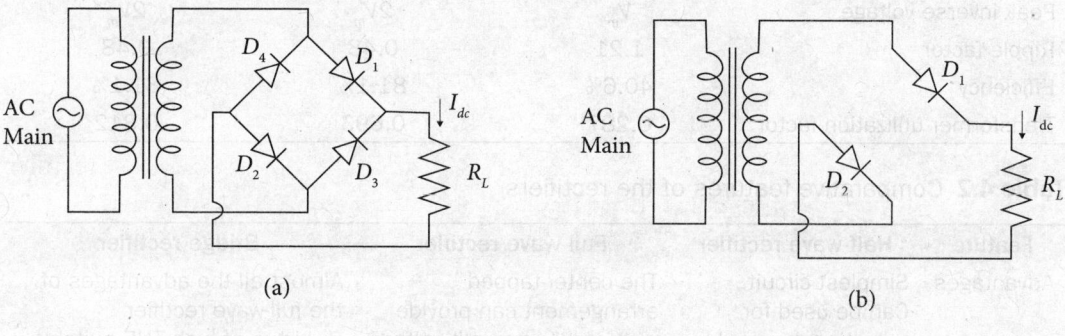

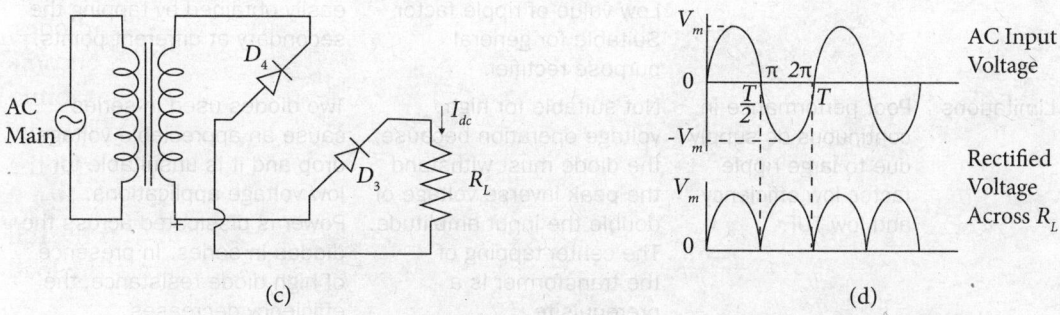

Figure 4.7 Bridge rectifier, (a) circuit diagram, (b) diodes D_1 and D_2 get forward-biased during one half of the input, (c) diodes D_3 and D_4 are forward-biased during the other half, (d) the output voltage waveform obtained across the load resistor (R_L) is compared with the input alternating voltage waveform

Table 4.1 Comparison of rectifier circuit parameters

Parameter	Half-wave rectifier	Full-wave rectifier	Bridge rectifier
Average direct current (I_{dc})	$\dfrac{I_m}{\pi}$	$\dfrac{2I_m}{\pi}$	$\dfrac{2I_m}{\pi}$
Average direct voltage (V_{dc})	$\dfrac{I_m R_L}{\pi}$	$\dfrac{2I_m R_L}{\pi}$	$\dfrac{2I_m R_L}{\pi}$
No load voltage (V_{NL})	$\dfrac{V_m}{\pi}$	$\dfrac{2V_m}{\pi}$	$\dfrac{2V_m}{\pi}$
RMS current (I_{rms})	$\dfrac{I_m}{2}$	$\dfrac{I_m}{\sqrt{2}}$	$\dfrac{I_m}{\sqrt{2}}$
Peak inverse voltage	V_m	$2V_m$	$2V_m$
Ripple factor	1.21	0.48	0.48
Efficiency	40.6%	81.1%	81.1%
Transformer utilization factor	0.287	0.693	0.812

Table 4.2 Comparative features of the rectifiers

Feature	Half-wave rectifier	Full-wave rectifier	Bridge rectifier
Advantages	Simplest circuit. Can be used for pulsating dc supply.	The center-tapped arrangement can provide with +ve & –ve split voltage with respect to the ground. High values of efficiency and TUF. Low value of ripple factor. Suitable for general purpose rectifier.	Almost all the advantages of the full-wave rectifier. By virtue of high TUF, suitable for high dc power operations. Center-tapping is not needed. Variable voltage output is easily obtained by tapping the secondary at different points.
Limitations	Poor performance in continuous dc supply due to large ripple factor, low efficiency and low TUF.	Not suitable for high voltage operation because the diode must withstand the peak inverse voltage of double the input amplitude. The center-tapping of the transformer is a prerequisite.	Two diodes used in series cause an appreciable voltage drop and it is unsuitable for low voltage applications. Power is dissipated across the diodes in series. In presence of high diode resistance, the efficiency decreases.

Think a Bit: Are the Efficiencies Universal?

It is mentioned in Table 4.1 that the efficiency of a half-wave rectifier is 40.6% and that of a full-wave rectifier is 81.1%. Are these values always true? These are actually the ideal values. In the case of a practical diode, if the forward resistance (R_f) becomes significant, the efficiency fluctuates. For instance, if R_f is considered in the derivation of efficiency (η) for half-wave rectifier using (4.16), we have $\eta = 40.6/(1 + R_f/R_L)$, which is dependent on R_f. Similar situation occurs for the full-wave rectifier.

4.4 Filters

It is now quite apparent that none of the rectifiers talked about so far—half-wave, full-wave and bridge—resembles a good dc source because all of those produce undulating current or voltage that are unidirectional but fluctuate periodically between zero and a certain maximum. An ideal dc source should produce a constant direct voltage across the load all the time. A good quality chemical cell fulfils such requirement satisfactorily. If the extent of variation of the rectifier output could be reduced, that would resemble the supply of a constant direct voltage similar to a chemical battery. The filter is such a circuit aid to the rectifier for this purpose. Two different types of filter are introduced here.

> **Note: A Filter Is a Frequency-Selective Circuit**
>
> The word 'filter' in connection with electronic circuits has a broad range of meaning. A filter is a circuit including some reactive elements, such as capacitor, inductor or both. It may contain passive as well as active devices. Depending on the change of reactance with frequency, the filter can pass or block electrical signals of a certain frequency range. In this sense, the present filters are 'low-pass' because they allow the dc component (low or zero frequency) and block the ac components of the rectified output. However, the present objective of employing filters to the rectifiers is to achieve a steady, non-fluctuating direct voltage. More deliberation on filters will be continued in Chapter 11.

4.4.1 Capacitor Filter

The simplest and the most widely used filter with rectifiers is simply a high value capacitor (C), as depicted in Figure 4.8. The capacitor filter is placed in parallel with the load resistor (R_L) and the capacitance (C) is so chosen that its reactance ($1/\omega C$) becomes much smaller than the load resistance at the frequency (ω) of the input voltage.

The graphs of Figure 4.8 are showing the voltage variation across the load resistor. Both for the half-wave and the full-wave rectifier, the filter action takes place in the same way. The capacitor charges up to V_m, the peak value of the rectified output during its increasing phase. Then the rectifier output voltage starts decreasing (dashed line) and the capacitor starts discharging through R_L with the time constant (CR_L) so that the voltage across it also decreases (solid line). The value of C is selected in such a way that for a range of R_L, the product CR_L remains much larger than the time period of the input voltage. As a result, the rate of discharge becomes lower than the rate of decrease of the rectified voltage and before the capacitor gets fully discharged, the next growing phase of charges it again up to V_m. Such a sequence is continued for either type of rectifier, as outlined in Figure 4.8. Comparing the solid and dashed lines, it is obvious that the capacitor filter smoothens the fluctuation of the rectifier output up to a great extent and the result is even better with the full-wave rectifier.

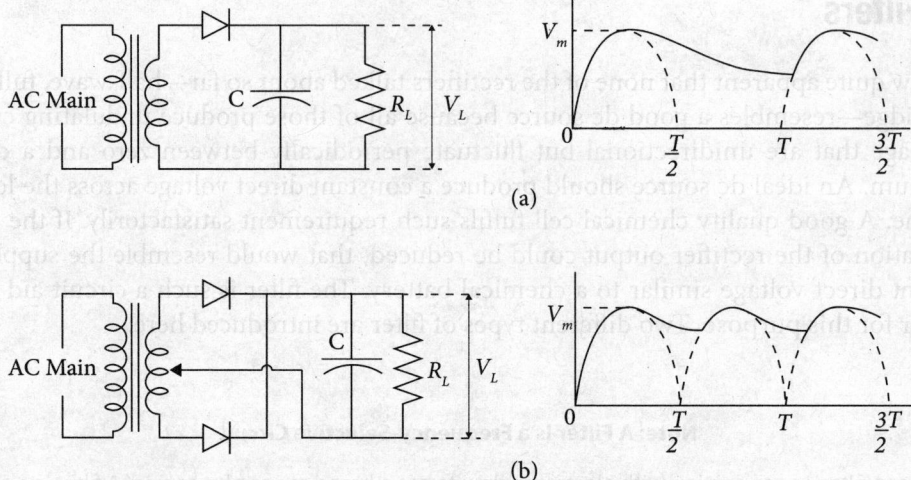

Figure 4.8 Capacitor filter with (a) half-wave and (b) full wave rectifiers. The circuit implementations of the capacitor (C) in the rectifiers are presented (left) and the corresponding time variations of the output voltage (V_L) across the load resistor (R_L) (solid lines) and the original rectified waveforms (dashed lines) are indicated (right)

Think a Bit: What Is the Peak Inverse Voltage with Filter?

An interesting point is that the full-wave rectifier with filter has peak inverse voltage of $2V_m$, same as that without filter whereas the half-wave rectifier with filter also has the same peak inverse voltage of $2V_m$, instead of the usual value of V_m without filter. This is so because both the peak value of the alternating input voltage and the peak value of the charged capacitor add together across the diode.

Example 4.3

The main ac supply of 220 V, 50 Hz is stepped down with center-tapped transformer and rectified with full-wave rectifier. Now the circuit designer intends to filter the rectified output voltage with a capacitor of 1000 µF. The required range of variation of the load resistance is from 1 kΩ to 100 Ω. Is the chosen value of capacitance appropriate for this filter?

Solution

The time period of the input alternating voltage is 1/50 = 0.02 s

When the load resistance becomes 1 kΩ, the time constant for the capacitor filter is

$10^3 \times 10^{-6} \times 10^3 = 1$ s

When the load resistance reached 100 Ω, the time constant is

$10^3 \times 10^{-6} \times 10^2 = 0.1$ s

For both the extremes of the load resistance, the time constant remains much larger than the time period of the input voltage. So the capacitor choice is appropriate.

4.4.2 Inductor Filter

An inductor opposes any change of the rectified current through the load and can be used as a filter. The inductor filter is joined in series with the load resistor. Figure 4.9 outlines the use of inductor filter and the resultant output variations in half-wave and full-wave rectifiers. In the half-wave rectifier, the output current persists beyond the phase angle π and reduces to zero at a later phase angle so that the average current fluctuates less than that without the filter. In the case of full-wave rectifier, the output current does not reduce to zero and assumes a wavy nature resulting in lesser fluctuation of the average current in comparison with that of half-wave rectifier.

Table 4.3 presents a comparison between capacitive and inductive filters. Combinations of capacitor (C) and inductor (L) are also used as filters. The series combination of L with R_L and C in parallel is called T-filter. The ripple can be made almost independent of R_L for some critical value of L. The combination of two capacitors in parallel on either side of an inductor is termed as Π-filter. It reduces the higher harmonics very sharply. Thus each filter has its own merits, limitations and fields of applications.

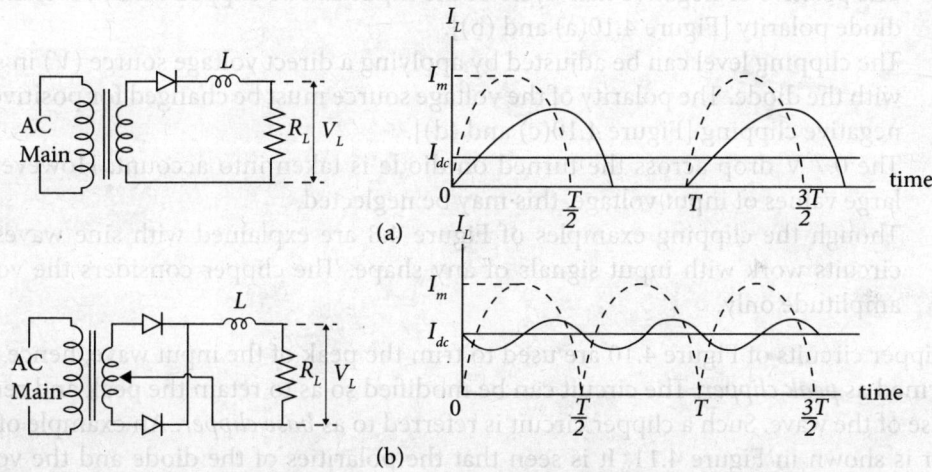

Figure 4.9 Inductor filter with (a) half-wave and (b) full wave rectifiers. The circuit implementations of the inductor (L) in the rectifiers are presented (left) and the corresponding time variations of the output current (I_L) across the load resistor (R_L) (solid lines) and the original rectified waveforms (dashed lines) are indicated (right)

Table 4.3 Comparison of capacitor (C) and inductor (L) filters

Feature	C-filter	L-filter
Mode of use	In parallel with load resistor (R_L).	In series with load resistor (R_L).
Action	The capacitor stores electrical energy during the increase of the rectified voltage and compensates by discharging the same during the decrease of the rectified voltage.	The inductor stores magnetic energy when the rectified current is high and releases the same to compensate for the decrease of the rectified current.
Application	Suitable for high R_L, hence small load current.	Suitable for small R_L, hence high load current.

4.5 Clippers

A *clipper*, also called *limiter*, is an electronic circuit using diode that can trim or cut off a portion of an alternating voltage waveform above or below a reference level, and transmit the remaining portion to the output. In this sense, a half-wave rectifier also could be designated as a 'clipper', since it removes one half of the input voltage. Howbeit, the actual clipper is more precise and flexible in tailoring the wave. The purpose of clipper is to confine an arbitrary waveform within some specified limit. Two largely distinguished categories of clippers are series and shunt, depending on the position of the diode with reference to the load.

4.5.1 Shunt Clipper

Several examples of shunt clipper are shown in Figure 4.10. The common feature is that the diode is always joined in parallel with the load resistor. Some other characteristics of the circuits are also noted.

- The positive or negative half-cycle of the input can be clipped off by reversing the diode polarity [Figure 4.10(a) and (b)].
- The clipping level can be adjusted by applying a direct voltage source (V) in series with the diode. The polarity of the voltage source must be changed for positive and negative clipping [Figure 4.10(c) and (d)].
- The 0.7 V drop across the turned on diode is taken into account. However, for large values of input voltage, this may be neglected.
- Though the clipping examples of Figure 4.8 are explained with sine waves, the circuits work with input signals of any shape. The clipper considers the voltage amplitude only.

The clipper circuits of Figure 4.10 are used to trim the peak of the input wave; hence these are termed as *peak clipper*. The circuit can be modified so as to retain the peak and remove the base of the wave. Such a clipper circuit is referred to as *base clipper*. An example of base clipper is shown in Figure 4.11. It is seen that the polarities of the diode and the voltage source are inverted. So long as the input voltage (v_i) is less than the source voltage (V), the output voltage remains constant at V because the diode is forward-biased at this voltage. When v_i exceeds V, the time variation is noted at the output. If V is large, the addition of 0.7 V can be neglected, as is done in the graph of Figure 4.11.

4.5.2 Series Clipper

In this category, the diode appears as a series element with respect to the load. Some examples of series clipper are given in Figure 4.12. The circuit of Figure 4.12(a) acts similar to a half-wave rectifier. In presence of the voltage source (V), the diode turns on when the input exceeds ($V + 0.7$) volt. Then the input variation is replicated at the output so that only the peak portion of the positive half-cycle is obtained. During the whole negative cycle of the input, the diode remains off and no output is obtained. The circuit of Figure 4.12(b) works similar to that of Figure 4.10(c). The circuit of Figure 4.12(c) acts like that of Figure 4.11. In both cases, the diode and the resistor are interchanged the position and the diode polarity is also changed.

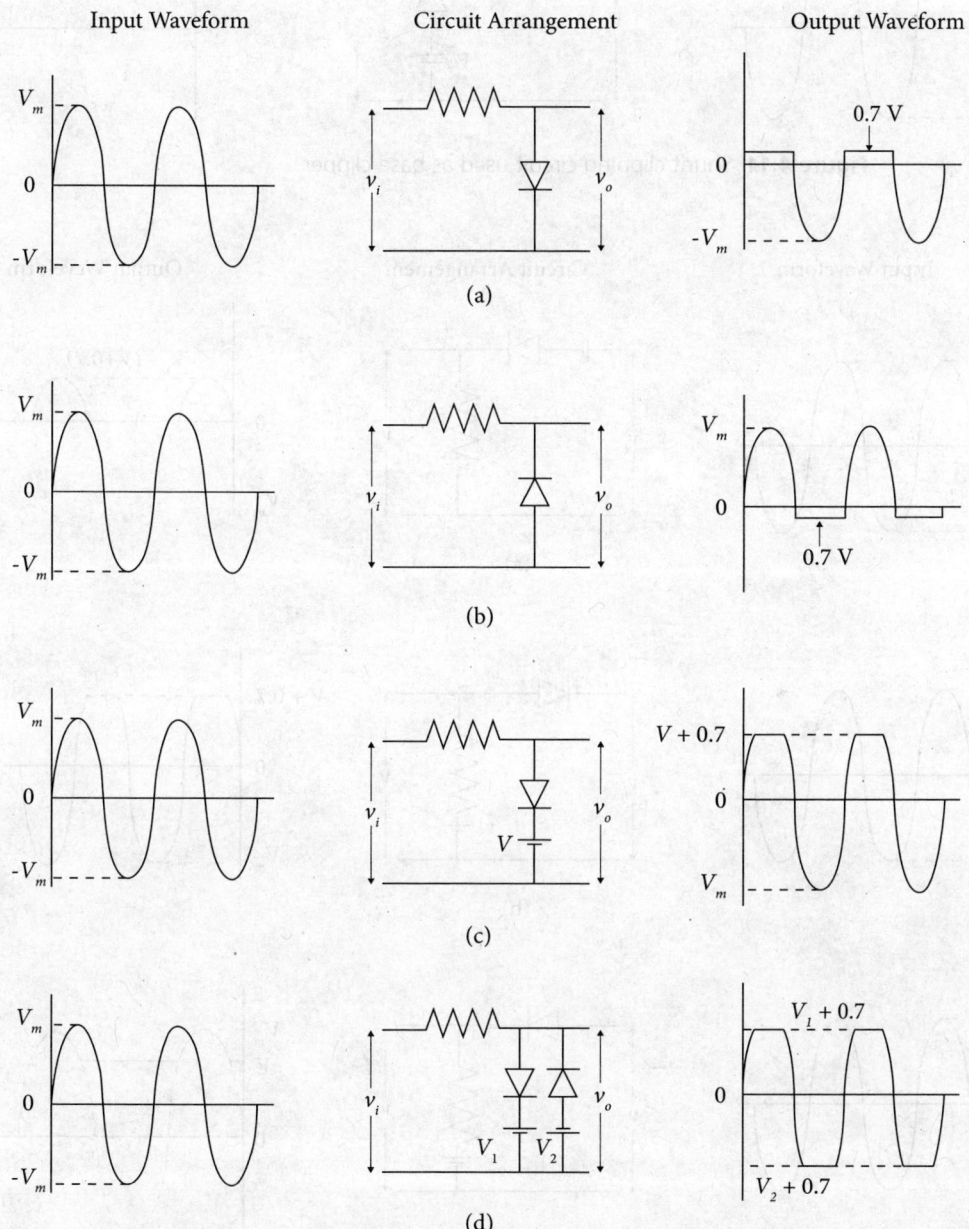

Figure 4.10 Shunt clipper circuits: (a) positive clipping, (b) negative clipping, (c) adjustable positive clipping, and (d) positive and negative clipping with adjustable levels

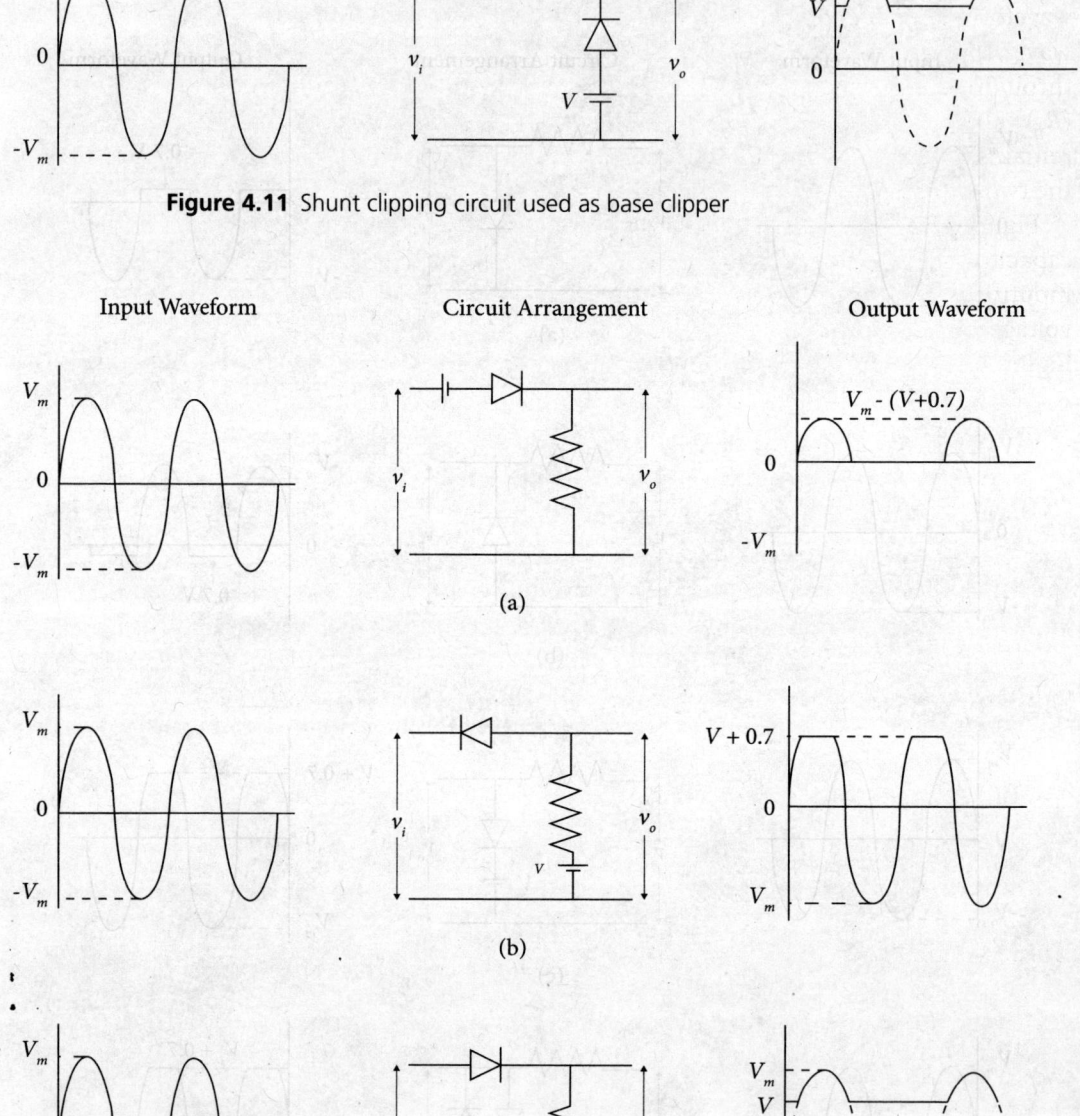

Figure 4.11 Shunt clipping circuit used as base clipper

Figure 4.12 Series clipper circuit used as (a) negative clipper, (b) positive adjustable clipper and (c) positive base clipper

4.6 Clamper

It is a useful combination of diode and capacitor that cause the peak value of the voltage waveform to 'clamp' or get fixed at a specified value without changing the shape of the wave. The working principle is to add some positive or negative dc value to the waveform through the charging and discharging of a capacitor. For a certain range of load resistance (R_L) values, the capacitance (C) value is selected in such a way that the time constant ($R_L C$) remains much larger than the time period of the voltage waveform under consideration for the two extremes of the load resistance variation.

Figure 4.13 illustrates the working principle of a clamper comprised of a diode (D), a capacitor (C) and a load resistor (R_L). A symmetric square wave voltage is applied as the input. It is assumed that the input amplitude (V) is much larger than the forward cut-in voltage of the diode. Both the circuit diagram and the input waveform are sketched in Figure 4.13(a). The following operations are executed in sequence.

(i) During the positive half-cycle of the input voltage, the diode gets forward-biased and acts as a short circuit [Figure 4.13(b)]. It makes the capacitor to charge promptly through it up to $+V$, the positive peak value of the input voltage. The voltage polarities of the figure. The voltage $+V$ across the charged capacitor opposes the input so that the output becomes zero.

(ii) During the negative half-cycle of the input voltage, the diode becomes reverse-biased and acts like an open circuit [Figure 4.13(c)]. Now the capacitor starts discharging through R_L. Since the time constant ($R_L C$) is much larger than the input time period, the capacitor retains almost the whole of voltage V during the negative half-cycle. The input voltage, the polarities being indicated, gets added to the capacitor voltage thereby making the output voltage $-2V$.

(iii) The resultant output voltage waveform is shown in Figure 4.13(d). Comparing with the input waveform, it is found that the wave shape is unchanged but the whole waveform is shifted in the voltage axis through a value of $-V$. It appears that someone has clamped the waveform to a shifted voltage value keeping the square shape intact.

(iv) If the diode polarity were reversed in the circuit of Figure 4.13(a), the output waveform would be clamped at another value, as shown in Figure 4.13(e).

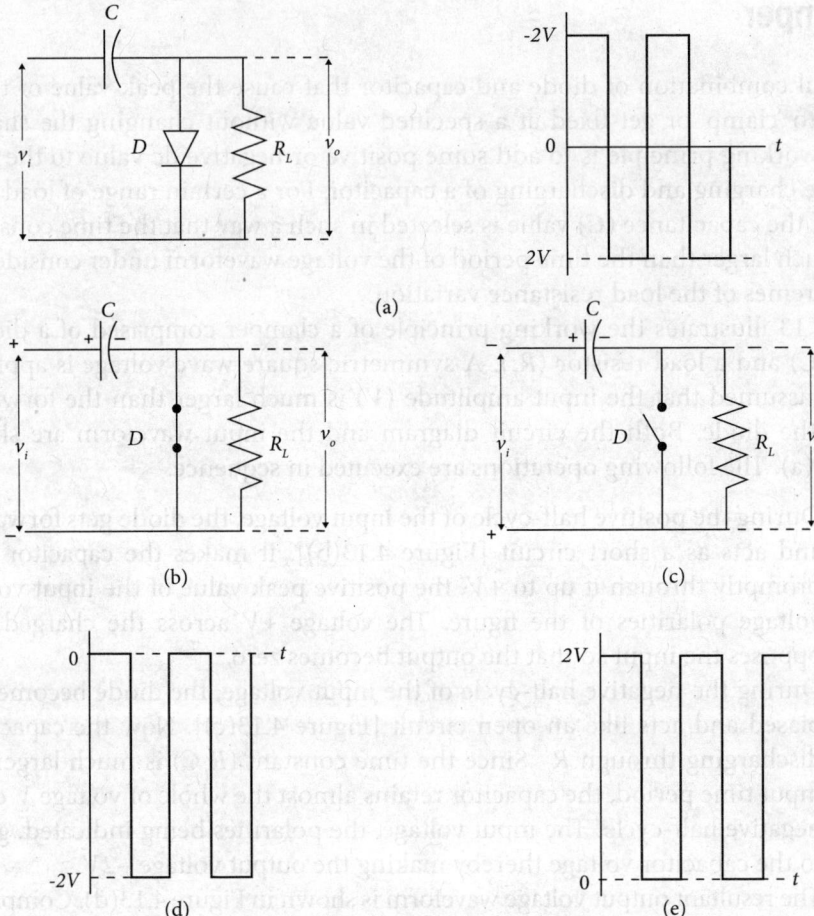

Figure 4.13 Clamper circuit constructed with diode (D), capacitor (C) and resistor (R_L). (a) Circuit diagram (left) and input waveform (right), circuit conditions during, (b) the positive, (c) the negative half-cycle of the input and the output voltage waveforms, (d) with the present circuit condition, and (e) with the diode polarity reversed

Example 4.4

A sine wave, as sketched in Figure 4.14, is applied to the input of the clamper circuit of Figure 4.13(a). Sketch the output waveform (i) at the present condition and (ii) at the condition when the diode polarity is changed.

Solution

The output waveforms are also sketched in Figure 4.14. The explanation is the same as that given above. In all the cases, the forward resistance and the cut-in voltage of the diode are ignored.

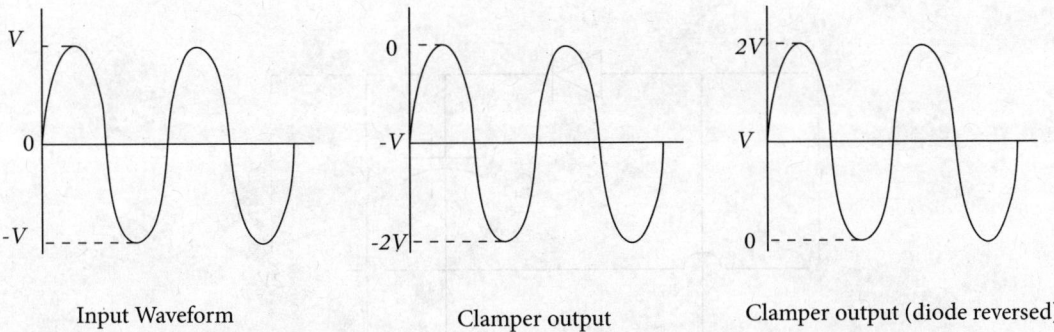

Input Waveform   Clamper output   Clamper output (diode reversed)

Figure 4.14 Clamper circuit operation with sinusoidal input voltage

4.7 Voltage Multiplier

This is another useful combination of diodes and capacitors. The voltage multiplier circuit can produce a direct voltage equal to a multiple of the peak value of the alternating input voltage. Figure 4.15(a) represents such a basic multiplier circuit, which can double the input voltage and is named the *voltage doubler*. It consists of two diodes, D_1 and D_2, and two capacitors C_1 and C_2. Let the input be a sinusoidal voltage with an amplitude of V. For the sake of convenience, first we consider the negative half of the input cycle. At the peak of the negative half, D_1 is forward-biased and D_2 is reverse-biased [Figure 4.15(b)]. Therefore, C_1 charges up to V with polarity indicated in the figure. At the peak of the positive half-cycle, D_2 is forward-biased [Figure 4.15(c)]. Consequently the capacitor C_2 finds two voltage sources in series: C_1 charged up to V and the input voltage reaching up to the peak value of V. Thus C_2 gets charged up to $2V$ and the voltage appearing across C_2 becomes a unidirectional fixed value of twice the input peak value. Since the capacitors are charged during only one half of the input cycle, the circuit is termed as the *half-wave voltage doubler*.

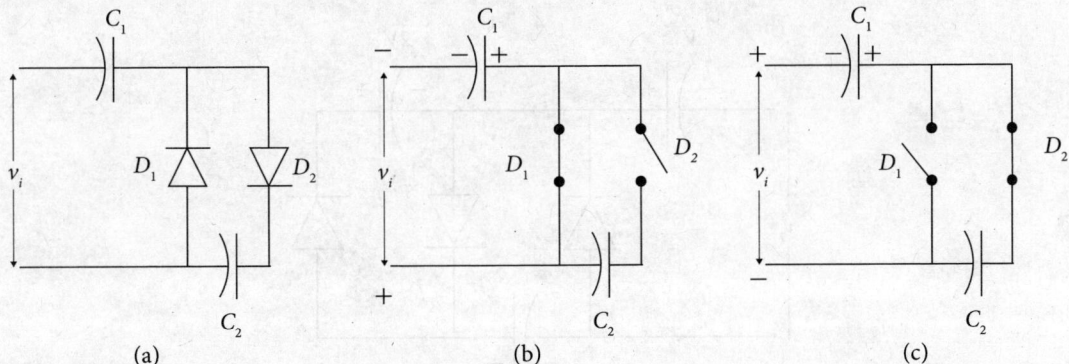

Figure 4.15 Half-wave voltage doubler constructed with diodes and capacitors, (a) circuit diagram and circuit conditions (b) at the negative half-cycle, and (c) at the positive half-cycle of the input voltage

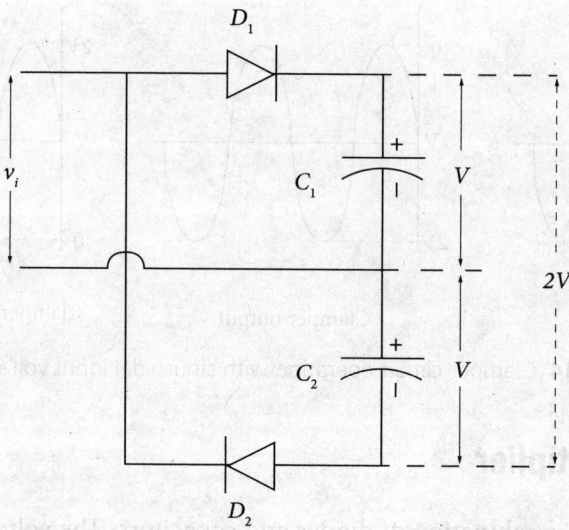

Figure 4.16 Full-wave voltage doubler

There also exists a *full-wave voltage doubler*, as introduced in Figure 4.16. During the positive half-cycle, C_1 charges through D_1 up to voltage V and during the negative half-cycle, C_2 charges through D_2 up to voltage V and a resultant voltage of $2V$ is obtained at the output. The number of diode–capacitor combination can be increased to get the output voltage multiplied further. For example, a *voltage tripler* is given in Figure 4.17. Adding more such combinations, it is possible to increase the voltage level up to a high value, of the order of kilovolts. Such multiplier circuits are used in television, cathode ray oscilloscope, and other instruments.

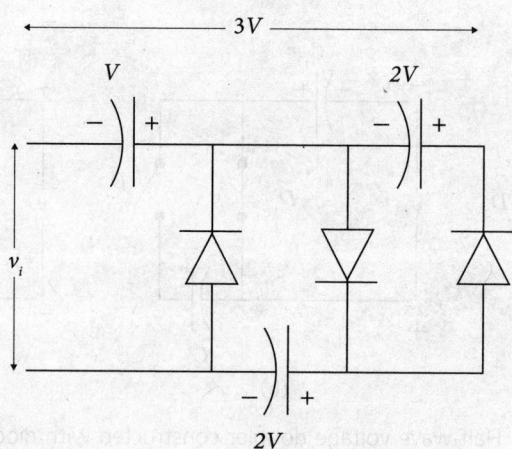

Figure 4.17 Voltage tripler

Multiple Choice-Type Questions and Answers

4.1 If the RMS value of an ac input voltage is 10 V, the peak inverse voltage of a full-wave rectifier is ....
(a) 7.07 V (b) 14.14 V (c) 28.28 V (d) 10 V

4.2 In a center-tapped full-wave rectifier, the transformer secondary peak voltage at each half-cycle is V_m. The maximum voltage across each diode is
(a) $2V_m$ (b) V_m (c) $V_m/2$ (d) $V_m\sqrt{2}$

4.3 A full-wave rectifier: "(i) uses two diodes in opposite connection, (ii) utilizes both the halves of input cycle and (iii) produces higher dc value of current than that of a half-wave rectifier for a given input."

Based on the above statements, the full-wave rectifier is considered more efficient than the half-wave rectifier because ....
(a) statements (i) and (ii) are true
(b) statements (i), (ii) and (iii) are true
(c) statements (ii) and (iii) are the reasons
(d) statement (ii) is the reason

4.4 The ac main supply voltage is not constant; the rms value fluctuates between 200 V and 220 V. This main supply voltage is stepped down by a transformer of 10:1 primary-to-secondary ratio and is rectified by a half-wave rectifier. The peak output voltage of the rectifier will ....
(a) not change due to this fluctuation
(b) fluctuate between 20 V and 22 V approximately
(c) fluctuate between 28.28 V and 31.11 V approximately
(d) fluctuate between 14.14 V and 15.56 V approximately

4.5 A 220 V, 50 Hz AC main supply is stepped down, half-wave rectified and applied to a tungsten bulb. The bulb would appear to ....
(a) glow with the same brightness as that with the original ac supply
(b) glow with less brightness than that with the original ac supply
(c) glow with more brightness than that with the original ac supply
(d) blink rapidly

Hints. Because of intermittent dc, the bulb would actually blink but human eye would not be able to detect that. The average brightness would get reduced due to the less output dc value than the input rms value.

4.6 Reversing the polarity of the diode in a half-wave rectifier reverses the direction of ....
(a) load current
(b) load voltage
(c) the current through the transformer secondary
(d) all the above

4.7 The ac main supply voltage has a frequency of 50 Hz. This is stepped down and rectified: first with a half-wave rectifier and then with a bridge rectifier. The output frequency of the half-wave rectifier is .... and that of the bridge rectifier .... is
(a) 50 Hz, 100 Hz
(b) 25 Hz, 100 Hz
(c) 50 Hz, 50 Hz
(d) $50\sqrt{2}$ Hz, $100\sqrt{2}$

4.8 If the transformer secondary voltage for a half-wave rectifier is 12 V RMS, the voltage across the load resistor of 1 kΩ is .... approximately.
(a) 3.82 V DC (b) 5.4 V DC (c) 16.97 V DC (d) 8.49 V DC

4.9 In a full-wave rectifier, accidentally one of the diode has got open circuit due to loose connection. Which of the following statements seems to be most logical?
(a) The circuit can be used as a half-wave rectifier with the same peak value of load voltage as the original value.
(b) The circuit can be used as a half-wave rectifier but with a peak value of load voltage half of the original value.
(c) The circuit can be used as a half-wave rectifier but with a peak value of load voltage 0.7 times the original value.
(d) The circuit is spoiled and is of no further use.

4.10 A designer wishes to build-up a rectifier that can be operated directly from 220 V AC main supply without using any transformer. Which of the following statements is most reasonable?
(a) The designer is wrong; any rectifier must involve a transformer.
(b) A full-wave rectifier can be constructed without a transformer.
(c) A capacitor can be used in lieu of a transformer.
(d) A half-wave rectifier and a bridge-rectifier can work without any transformer.

4.11 The peak inverse voltage of a half-wave rectifier with capacitor filter and input voltage $V\sin\omega t$ is ....
(a) V (b) $V\sqrt{2}$ (c) $V/\sqrt{2}$ (d) 2V

4.12 The peak inverse voltage of a full-wave rectifier with capacitor filter and input voltage $10\sin 350t$ is ....
(a) 10 (b) 20 (c) $10\sqrt{2}$ (d) $20\sqrt{2}$

input voltage waveform 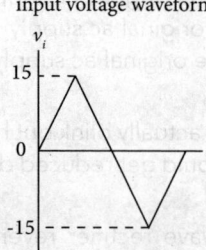 diode circuit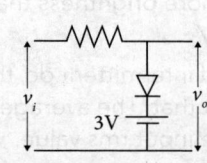

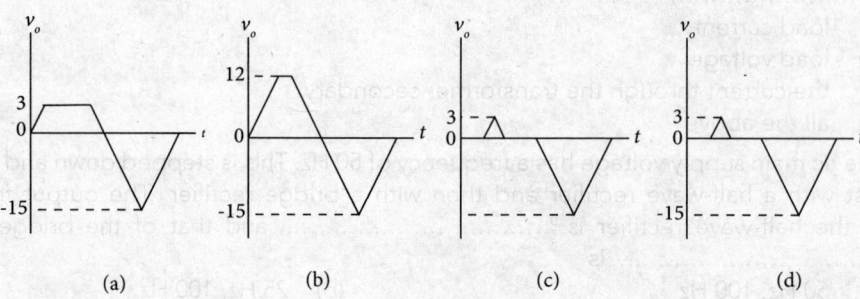

(a) (b) (c) (d)

Figure 4.18

4.13 The average direct current through each diode of a center-tapped full-wave rectifier is .... the average direct current through the load.
 (a) equal to (b) greater than
 (c) $1/\sqrt{2}$ times (d) half of

4.14 Figure 4.18 shows an input voltage (v_i) waveform applied to a diode circuit and all the circuit components are supposed to be ideal. Which of the following sketches represents the output voltage (v_o) waveform?

4.15 In the above question, which of the sketches given in Figure 4.19 represents the current (i_o) through the diode?

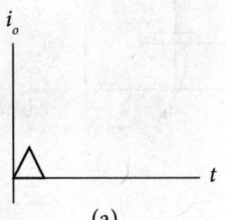
(a)

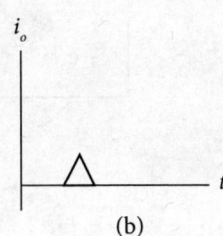
(b)

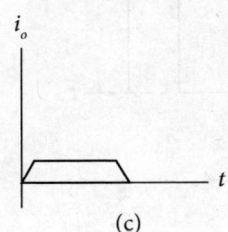
(c)

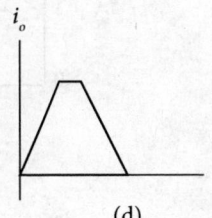
(d)

Figure 4.19

4.16 Figure 4.20 depicts an input voltage (v_i) waveform applied to a circuit comprised of diode and capacitor. Assuming all the circuit components to be ideal, state which of the following sketches represents the output waveform (v_o).

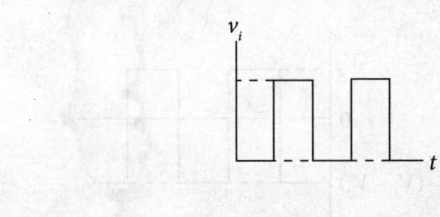

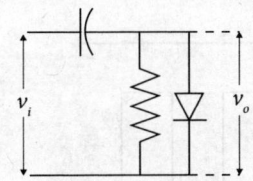

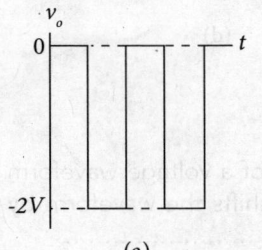
(a)

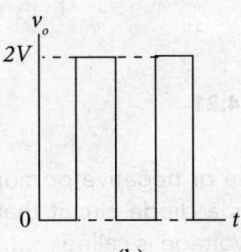
(b)

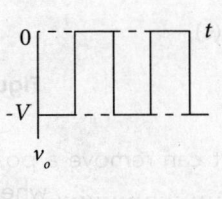
(c)

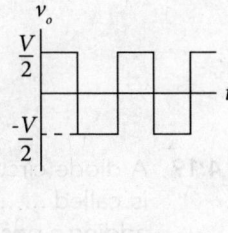
(d)

Figure 4.20

4.17 Figure 4.21 consists of a circuit and an input voltage (v_i) waveform. Which of the following sketches correctly represents the output voltage (v_o)?

4.18 In a full-wave rectifier with capacitor filter, the load voltage .... on increasing the secondary voltage of the step-down transformer and the ripple voltage .... on increasing the filter capacitance.
(a) increases, decreases
(b) remains constant, increases
(c) increases, increases
(d) increases, remains constant

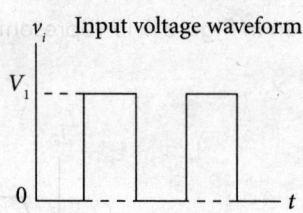
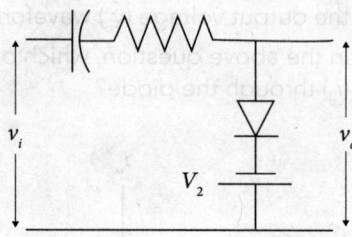
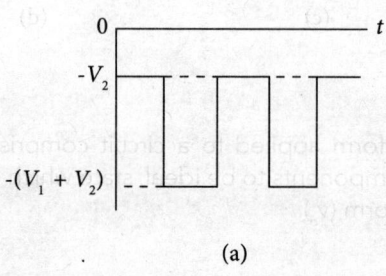
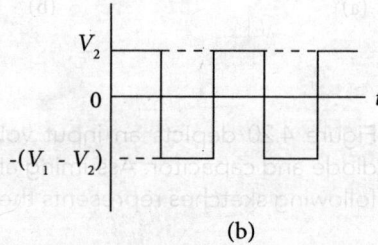
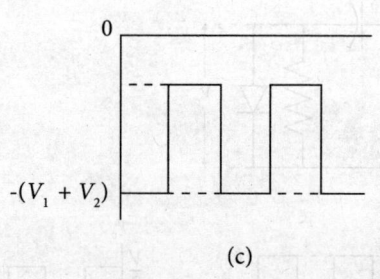
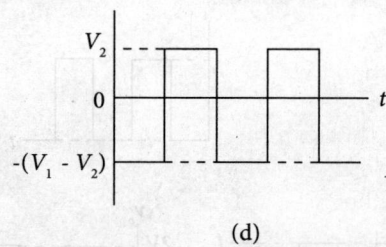

Figure 4.21

4.19 A diode circuit that can remove a positive or negative portion of a voltage waveform is called .... whereas a diode circuit that shifts the waveform by adding a positive or negative dc value of voltage is called ....
(a) clamper, clipper
(b) clipper, clamper
(c) limiter, shifter
(d) shaper, limiter

4.20 The voltage multiplier circuit is suitable for producing a voltage .... the input voltage and a current .... the input current.
(a) higher than, equal to
(b) higher than, less than
(c) higher than, greater than
(d) higher than, slightly greater than

Answers

4.1	(c)	4.2	(a)	4.3	(c)	4.4	(c)
4.5	(b)	4.6	(d)	4.7	(a)	4.8	(b)
4.9	(a)	4.10	(d)	4.11	(d)	4.12	(b)
4.13	(d)	4.14	(a)	4.15	(b)	4.16	(c)
4.17	(a)	4.18	(a)	4.19	(b)	4.20	(b)

Reasoning-Type Questions and Answers

4.1 It appears from Table 4.1 that the half-wave rectifier has poor regulation characteristics. Does it have any practical application?

Ans. In spite of the low dc values, poor efficiency and high ripple factor, the half-wave rectifier is the cheapest and the simplest rectifier circuit. If a pulsating DC serves the purpose, e.g. in the case of a calling bell, the half-wave rectifier fits well. It has two very important applications: as detector in AM radio and in power diode applications, such as battery charger.

4.2 Is the transformer an essential component of the rectifier?

Ans. The transformer accounts for attenuating the ac main input voltage amplitude up to a required level. In most cases, that is essential. However, if the original input is permitted without any reduction of amplitude, the half-wave and bridge rectifiers can be used without any transformer. In the case of a full-wave rectifier, the transformer is compulsory to divide the input voltage into two equal parts.

4.3 Both the full-wave rectifier and the bridge rectifier produce the same output voltage waveform. Is there any difference in their characteristics?

Ans. A subtle feature makes the difference. In a full-wave rectifier, each diode conducts during the half-cycle and the combination is obtained at the output. In the case of a bridge rectifier, the whole of the secondary voltage is used as the input to the rectifier. Therefore, using a transformer of fewer (ideally half) turns in the secondary coil (hence less weight, size and cost), the bridge rectifier can produce the same output voltage as the conventional full-wave rectifier.

4.4 Let the diodes and the transformer in a rectifier be ideal and the provision for center-tapping of the transformer be also available. Are the full-wave and bridge rectifiers interchangeable in such a case?

Ans. No, the exact field of application should be considered. If the requirement is high dc power, the bridge rectifier is the better choice because of (i) higher utilization of transformer and (ii) the peak inverse voltage across each diode being one-half of that for a full-wave rectifier. In contrast, if the secondary voltage is small and the power dissipation across the diodes is vital, a full-wave rectifier works better than the bridge rectifier.

4.5 The diode is a device of nonlinear current–voltage relationship. Does this property distort the output waveform of a rectifier?

Ans. No, the rectifier circuit generally acts with an ac input voltage (say 6 V or 12 V) much larger than the diode cut-in voltage (≈ 0.7 V) and the diode current is dominated by its bulk resistance with the load resistance in series. The piecewise linear model of the diode is quite linear in such a case. If the input voltage were too small, comparable to the diode cut-in voltage, the nonlinearity would be a real problem.

[N.B. For a very small input voltage, the process of rectification adopts a different technique. An operational amplifier is used as precision rectifier that will be introduced in Chapter 11.]

Solved Numerical Problems

4.1 A center-tapped step-down transformer having primary-to-secondary turns ratio 10:2 is used for constructing full-wave rectifier with two diodes, each having forward resistance of 30 Ω. The ac main supply is of 220 V rms, 50 Hz and the load resistance applied to the rectifier is of 1 kΩ. Determine (i) the ripple factor and (ii) the efficiency of the full-wave rectifier. Neglect the resistance of the transformer secondary coil.

Soln. The secondary voltage is

$$2 \times 220/10 = 44 \text{ V rms}$$

Each half of the center-tapped secondary produces peak voltage of $22\sqrt{2}$ volt and the peak current is

$$\frac{22\sqrt{2}}{(1000+30)} = 30.2 \text{ mA}$$

The rms current is [Equation (4.21)]

$$\frac{30.2}{\sqrt{2}} = 21.35 \text{ mA}$$

The direct current through the load is [Equation (4.20b)]

$$\frac{44\sqrt{2}}{\pi(1000+30)} = 19.23 \text{ mA}$$

(i) The ripple factor is [Equation (4.15)]

$$\left[\left(\frac{21.35}{19.23}\right)^2 - 1\right]^{1/2} = 0.482$$

(ii) The efficiency is [Equation (4.16)]

$$\eta = \frac{(19.23)^2 \times 1000}{(21.35)^2 \times (1000+30)} \times 100 = 78.76\%$$

[N.B. The deviations from the ideal values of Table 4.1 are due to the existence of the diode resistance.]

4.2 Determine the percentage regulation for (i) full-wave rectifier and (ii) bridge rectifier fabricated with diodes, each having forward resistance of 50 Ω. The load resistance is of 1 kΩ and neglect the resistance of the transformer secondary coil.

Soln.

(i) In the case of full-wave rectifier, $V_L = V_{NL} - I_{dc}R_f$ [Equation (4.20)] so that

$$\frac{V_{NL} - V_L}{V_L} = \frac{I_{dc}R_f}{I_{dc}R_L} \times 100$$

Thus the regulation is

$$\frac{50}{1000} \times 100 = 5\%$$

(ii) The bridge rectifier contains two diodes in series so that $V_L = V_{NL} - 2I_{dc}R_f$ and the regulation is

$$\frac{2 \times 50}{1000} \times 100 = 10\%$$

[N.B. This solution presents a useful expression for percentage regulation in terms of resistances only. When $R_f \to 0$, the regulation approached the ideal condition of 0%. It also indicates that the regulation property of the bridge rectifier degrades when the diode resistance becomes significant.]

4.3 Calculate the efficiency (η) of the two rectifiers mentioned in Q. 4.2 above.

Soln. Using Equation (4.16) and considering the existence of the diode forward resistance (R_f), the efficiency of full-wave rectifier can be derived as

$$\eta = \left(\frac{I_{dc}}{I_{rms}}\right)^2 \frac{1}{1 + R_f/R_L} \times 100$$

Putting the values of I_{dc} and I_{rms} for full-wave rectifier,

$$\eta = \frac{81.1}{1 + R_f/R_L} \times 100 = 77.24\%$$

Following the same logic as above, the expression for the efficiency of the bridge rectifier can be obtained as

$$\eta = \left(\frac{I_{dc}}{I_{rms}}\right)^2 \frac{1}{1 + 2R_f/R_L} \times 100 = 77.24\%$$

[N.B. The above solution provides with a useful expression for rectifier efficiency in presence of diode forward resistance. It also demonstrates that despite the same value of I_{dc} and I_{rms}, the bridge rectifier retards slightly in efficiency because of the forward resistances of two diodes being added. When that resistance approaches zero, both the full-wave and the bridge rectifier exhibit the same efficiency.]

4.4 The main ac supply of 220 V, 50 Hz is stepped down and rectified with a full-wave rectifier including a capacitor filter of 1000 µF. Let the current remains constant at a value of 6 mA during the discharge of the capacitor. Determine the peak-to-peak ripple voltage of the rectifier with filter.

Soln. The peak-to-peak ripple voltage can be defined as the change of voltage (dV) during the time (dt) of one cycle of the full-wave rectified output. The current (I) can be related to the capacitance (C) and the rate of change of voltage as

$$I = C\frac{dV}{dt}$$

In the present case, $C = 1000\ \mu F$, $I = 6\ mA$ and $dt = 1/(100\ Hz) = 10\ ms$. Therefore,

$dV = (6\times10^{-3}\times10\times10^{-3})/(1000\times10^{-6}) = 0.06\ V$

4.5 Can you derive an expression for the ripple factor for the full-wave rectifier with capacitor filter given in Figure 4.8?

Soln. For large values of CR_L, the exponential discharge of C through R_L is approximated by a linear decrease and the mean output voltage can be expressed as

$$V_{dc} = V_m - \frac{\Delta V}{2} \tag{i}$$

Where ΔV is the total reduction of capacitor voltage from its peak value (V_m) during discharge. Considering voltage = (current × time) ÷ (capacitance),

$$\Delta V = \frac{V_{dc}}{R_L}\frac{T}{2}\frac{1}{C} \tag{ii}$$

T is the time period of the input voltage and $f = 1/T$ is the frequency of the input voltage. From (i) and (ii),

$$V_{dc} = V_m - \frac{V_{dc}}{4fCR_L} \tag{iii}$$

Assuming the ripple varying from $\Delta V/2$ to $-\Delta V/2$ over the time span of 0 to $T/2$, the RMS value of the ripple can be expressed as

$$V_{rms} = \left[\frac{2}{T}\int_0^{T/2}\left(\frac{\Delta V}{2} - \frac{2\Delta Vt}{T}\right)^2 dt\right]^{1/2} = \frac{\Delta V}{2\sqrt{3}} \tag{iv}$$

Using (ii) and (iv), the ripple factor is determined as

$$r = \frac{1}{4\sqrt{3}fCR_L} \tag{v}$$

[N.B. From (v), it is apparent that higher value of C reduces the ripple factor. Practically electrolytic capacitors of the order of 100 to 1000 μF are used as filter. Also it is implied that larger values of the load resistance reduce the ripple factor.]

Exercise

Subjective Questions

1. Explain the load line and Q-point related to the performance of a diode.
2. Interpret with diagrams, how the load line of a diode varies with (i) the load resistance value and (ii) the supply voltage value.

3. Draw the circuit diagram of a full-wave rectifier, briefly introduce its working principle and trace the current path with red and blue ink, respectively, for the two halves of the input.
4. Repeat the above jobs with a bridge rectifier.
5. Derive expressions for (i) the efficiency and (ii) the ripple factor of a rectifier. Are the same expressions applicable to half-wave, full wave and bridge rectifiers? If so, wherein lie the differences?
6. Derive the efficiency of a full-wave rectifier when the forward resistance of the diode and the transformer secondary coil resistance are no longer negligible.
7. What is load regulation of a rectifier? Find out some expression to quantify the extent of regulation and state how it is different for the three rectifiers.
8. Compare the merits and limitations of a full-wave rectifier and a bridge rectifier.
9. Explain, with necessary diagrams, how the capacitor filter improves the performance of a rectifier.
10. Draw the circuit diagram of a shunt clipper and outline its working principle. How the series clipper is different from it?
11. Draw a clamper circuit diagram and draw the output waveforms for (i) positive and (ii) negative clamping with a triangular wave as input.
12. Briefly introduce the principle of voltage doubling with diode and capacitor using a circuit diagram.

Conceptual Test

1. Does the load line of a diode vary with the type of the diode? Depict the diagram, if necessary.
2. What happens, if the diode polarity is reversed in a half-wave rectifier? Draw the output voltage variation across (i) the load resistor and (ii) the diode.
3. What happens, if in a full-wave rectifier, (i) the two diodes are not identical, (ii) the secondary is not exactly center-tapped?
4. Compare the voltage regulation of the rectifier with the voltage regulation of a Zener diode.
5. Compare a voltage multiplier with a step-up transformer.

Numerical Problems

1. A half-wave rectifier has load resistance of 1 kΩ, diode forward resistance of 100 Ω and transformer secondary voltage of 12 V (rms). Calculate (i) the average direct current through the load, (ii) the AC component of the load current and (iii) the efficiency of the rectifier.
2. The voltage $1000\sin\omega t$ is applied across YZ in Figure 4.22. Assuming ideal diode, the voltage measured across WX is
   (a) $\sin\omega t$
   (b) $(\sin\omega t + |\sin\omega t|)/2$
   (c) $(\sin\omega t - |\sin\omega t|)/2$
   (d) 0 for all t

   [GATE 2013]

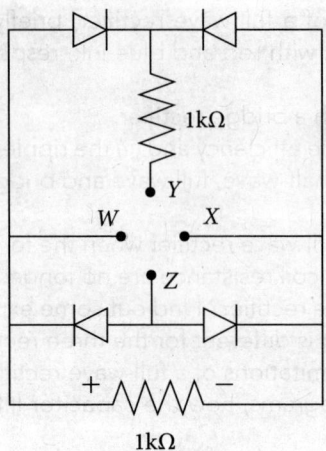

Figure 4.22

3. A sine wave of 5V amplitude is applied at the input of the circuit shown in Figure 4.23. Which of the following waveforms represents the output most closely?

[JAM 2014]

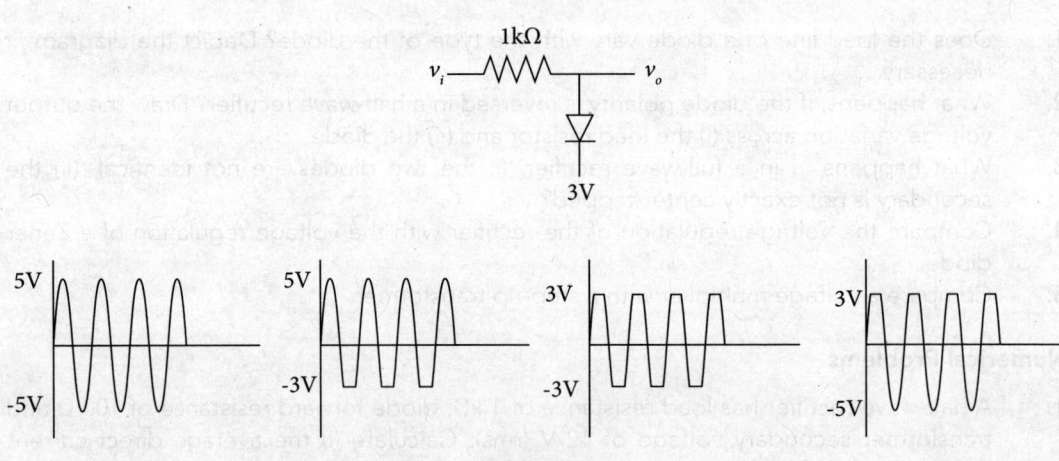

Figure 4.23

4. Which of the following correctly describes the output waveform of the circuit of Figure 4.24?

[JAM 2005]

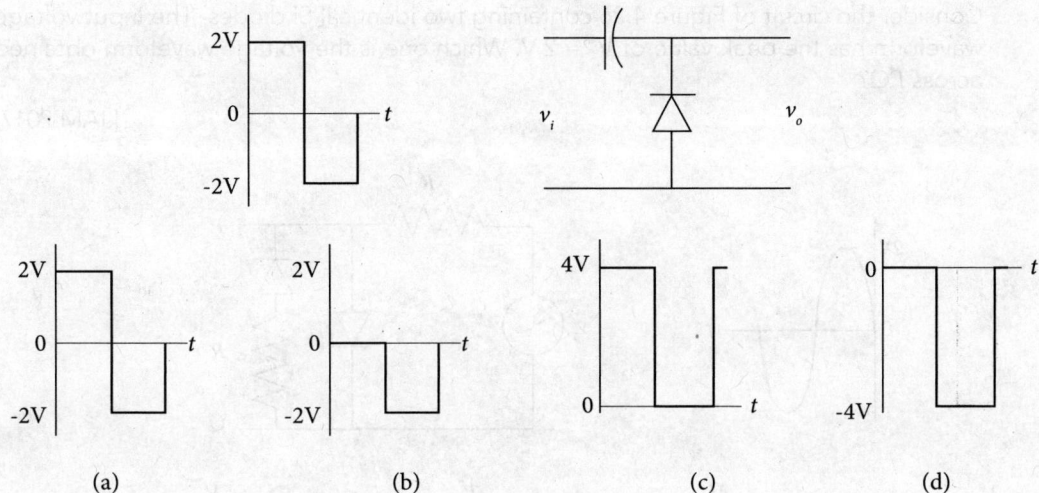

Figure 4.24

5. A sinusoidal voltage having a peak value of V_p is input to the circuit of Figure 4.25 where the dc voltage is V_b. Assuming ideal diode, which one of the following graphs best describes the output waveform?

[NET Dec. 2018]

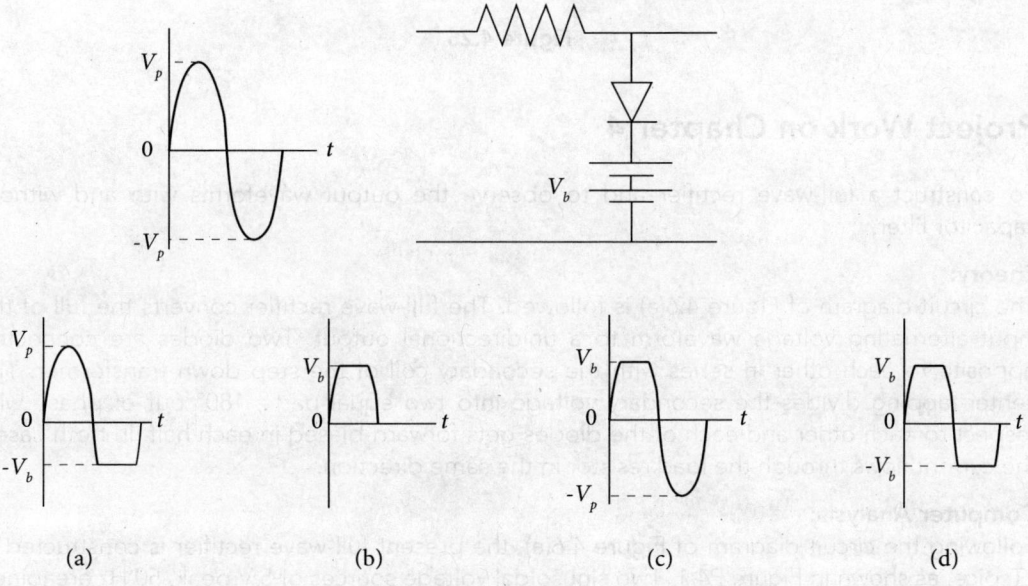

Figure 4.25

6. Consider the circuit of Figure 4.26 containing two identical Si diodes. The input voltage waveform has the peak value of V_p = 2 V. Which one is the voltage waveform obtained across PQ?

[JAM 2017]

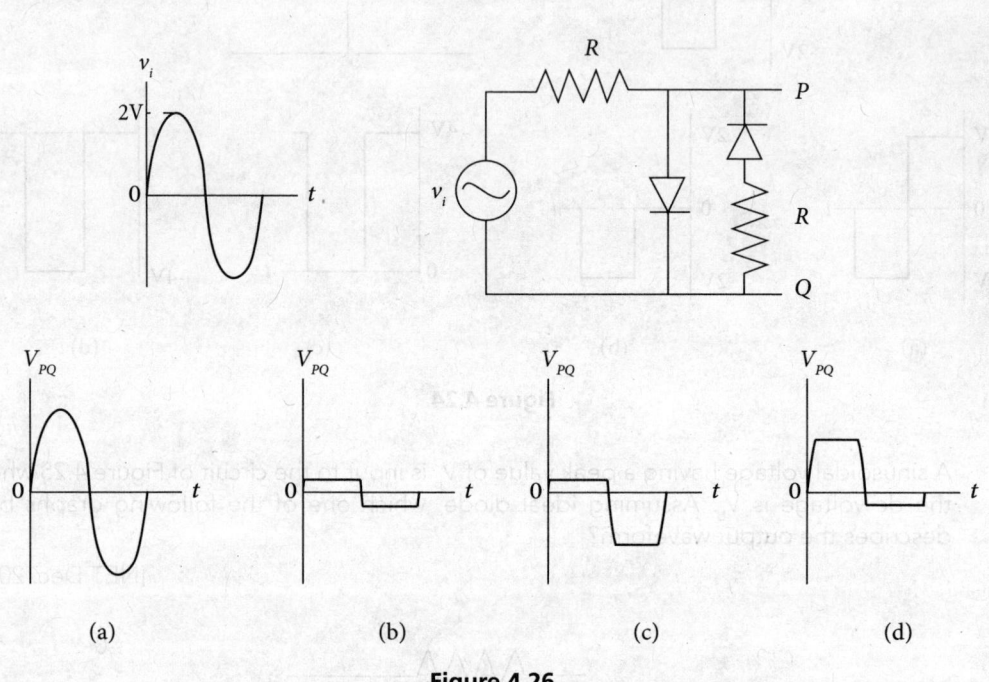

Figure 4.26

Project Work on Chapter 4

To construct a full-wave rectifier and to observe the output waveforms with and without capacitor filter.

Theory:
The circuit diagram of Figure 4.6(a) is followed. The full-wave rectifier converts the full of the input alternating voltage waveform to a unidirectional output. Two diodes are connected opposite to each other in series with the secondary coil of the step down transformer. The center-tapping divides the secondary voltage into two equal parts, 180° out of phase with respect to each other and each of the diodes gets forward-biased in each half. In both cases, the current flows through the load resistor in the same direction.

Computer Analysis:
Following the circuit diagram of Figure 4.6(a), the present full-wave rectifier is constructed in LTspice, as shown in Figure P4.1. Two sinusoidal voltage sources of 5 V peak, 50 Hz are joined together to simulate center-tapped step down condition of the ac input and the two diodes D_1 and D_2 are connected to this composite ac source, as shown in the figure. The load resistor (R_L) is of 1 kΩ. The voltage developed across it is denoted by V_{out}.

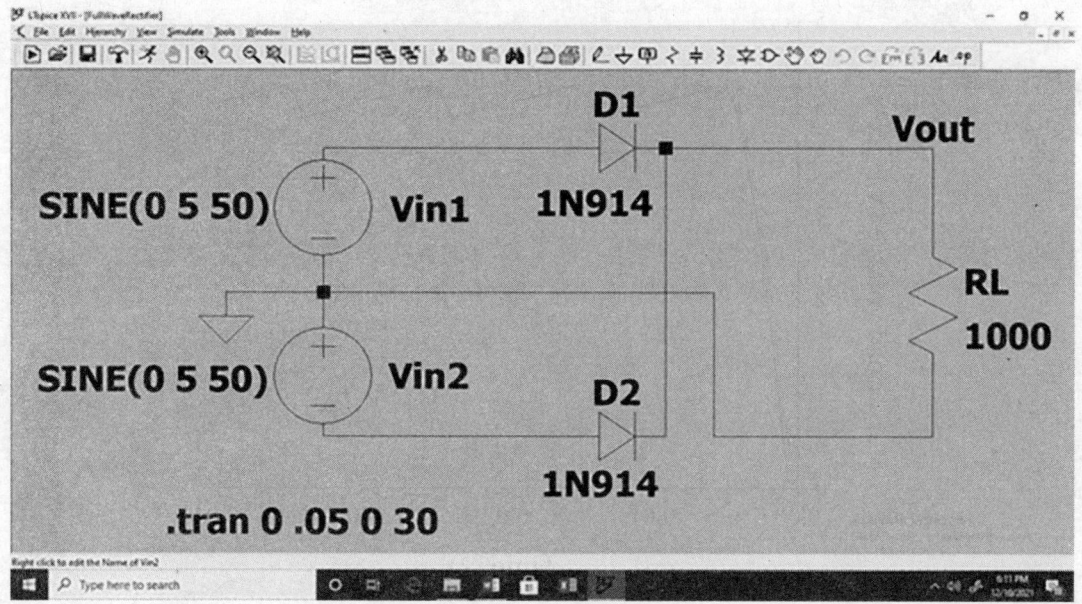

Figure P4.1 LTspice simulated full-wave rectifier

In Figure P4.2, a capacitor of 100 µF is joined in parallel to the load resistor (R_L) and this capacitor acts as filter. The output voltage (V_{out}) data generated for both with and without filter are compared in Figure P4.3.

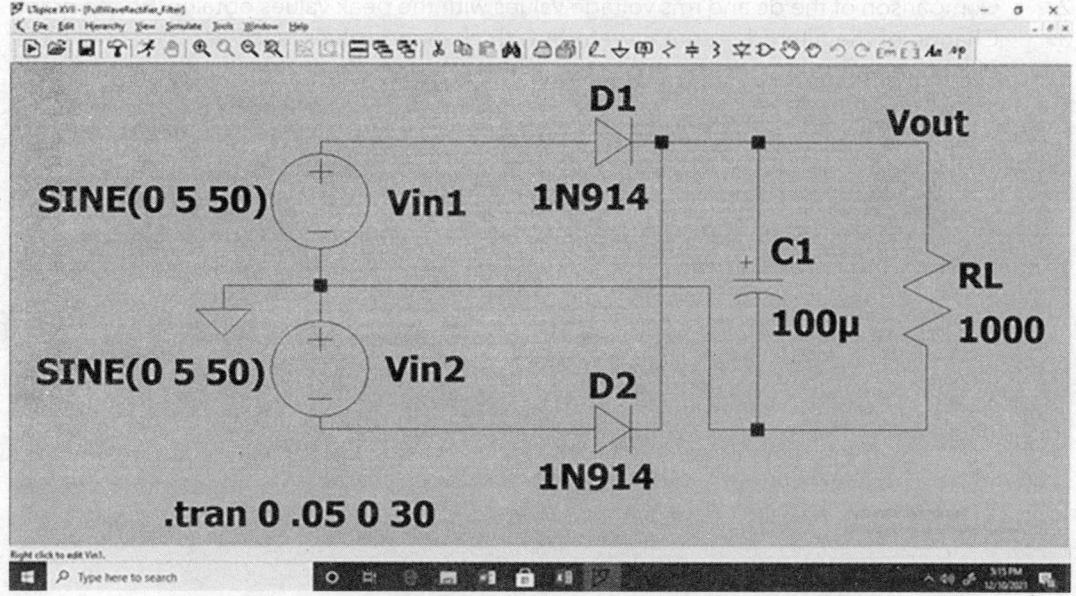

Figure P4.2 LTspice simulated full-wave rectifier with capacitor filter

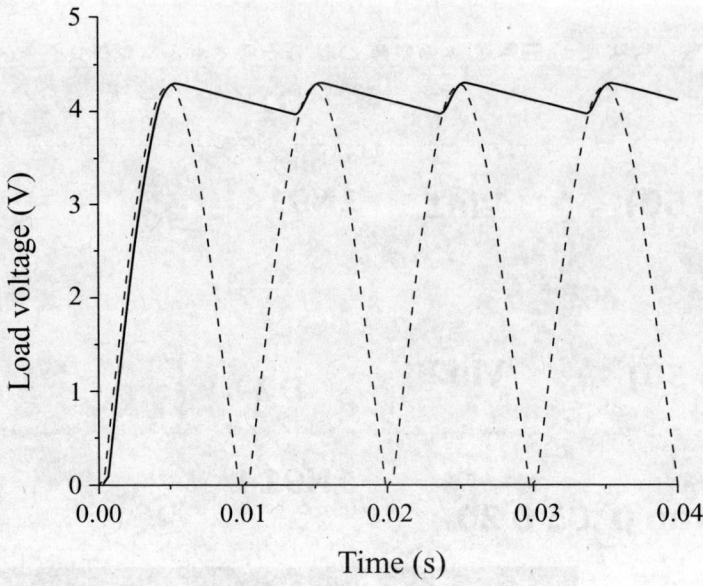

Figure P4.3 Output voltage waveform across the load with capacitor filter (solid line) and without filter (dotted line)

The following studies should be carried out with the circuits.

1. Change of filtered output waveform with changed values, such as 1000 μF and 2000 μF of the capacitor
2. Comparison of the dc and rms voltage values with the peak values obtained in the graph
3. Change of the filtered output waveform with changed values, such as 500 Ω and 100 Ω of load resistance.

5

Bipolar Junction Transistor (BJT)

This chapter introduces the construction and working principle of n–p–n and p–n–p transistors with proper discussion on common-emitter, common-base, and common-collector configurations. The amplifying action of the device is explained. The device is formally termed as *bipolar junction transistor* (BJT) because both electrons and holes, the charges of two opposite polarities take part in current conduction. There is yet another type of transistor, known as *field-effect transistor* (FET) that is *unipolar*; consisting of charge carriers of a single polarity, electrons or holes. Such devices will be illustrated in Chapter 9. Unless otherwise stated, the word 'transistor' refers to the BJT throughout this book.

5.1 Transistors: n–p–n and p–n–p

A transistor is made of semiconductor, mostly of silicon in present days. Two types of constructions are possible, either a p-region sandwiched between two n-regions or vice versa, as sketched in Figures 5.1 (a) and (b). These devices are termed as n–p–n transistor and p–n–p transistor, respectively. The corresponding circuit symbols are also given in Figure 5.1. The construction of both types of transistors includes the following stipulations.

(i) The transistor has two p–n junctions and three distinct regions, namely emitter, base, and collector. These regions have different sizes and doping concentrations.
(ii) The emitter is heavily doped. It injects carriers into the base: electrons for n–p–n transistors and holes for p–n–p transistors.
(iii) The base is very thin (a few percent of the total device width) and lightly doped. It allows the recombination of a small fraction of the carriers injected from the emitter and the passage of most of the carriers through it toward the collector.
(iv) The collector is the largest in size and of moderate doping; often the lightest among the three regions. The carriers crossing the base are collected in this region. The collector–base junction area is made to be maximized.
(v) A metallic ohmic contact is made with each region for applying external bias voltages.

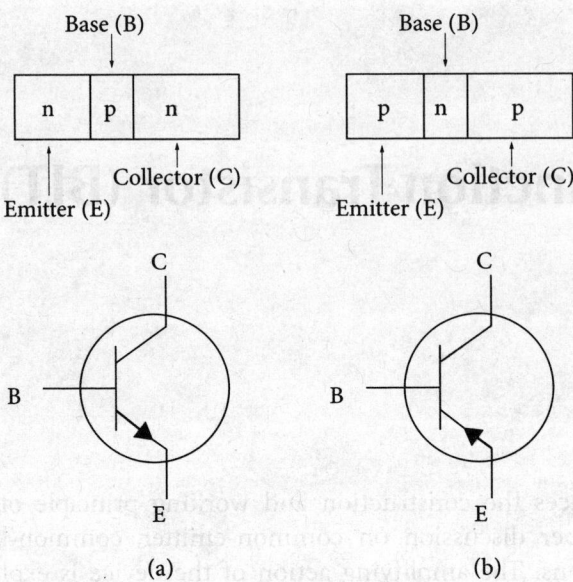

Figure 5.1 Transistor construction and circuit symbol (a) n–p–n and (b) p–n–p

5.2 Transistor Operating Principle

The transistor came forth as the first semiconductor device that could reproduce the major characteristics of electronic devices obtained earlier with vacuum tube devices (such as triode and pentode), namely

- electrical power amplification,
- nonlinear current–voltage relationship, and
- different input and output impedances.

The real advantages of semiconductor devices in comparison with vacuum tube devices, such as lower operating voltage, higher efficiency and smaller size were achieved with the transistor. The BJT is a *current amplifier* on its own. It can be made to amplify voltage (Chapters 6 and 7) and power (Chapter 8) also with the use of other circuit components. The transistor can also be made to act as on/off switch. Such operations are generally useful to digital circuits (Chapter 14).

An unbiased transistor does not allow any current to flow through it. The emitter–base and the collector–base junctions are to be biased with external voltage sources. The voltages at different terminals of a transistor are indicated by the single- and double-subscripted notations given in Table 5.1. In course of biasing the two junctions, any one of the emitter—base or collector—becomes common to both the biasing voltage sources. Thus three configurations of transistor operation are possible: common-emitter (CE), common-base (CB) and common-collector (CC). For the CE and CB biasing, there exist the following three modes of current flow through the transistor.

A. *Saturation Mode*: If both the emitter–base junction and the collector–base junction are forward-biased, large emitter current and large collector current flow through the transistor and the device acts as a closed switch. Such condition of transistor operation is called the *saturation mode*.
B. *Cutoff Mode*: If both the emitter–base and collector–base junctions are reverse-biased, negligible current flows through the emitter and the collector terminals and the transistor acts as an open switch. Such condition of transistor operation is termed as the *cutoff mode*.
C. *Active Mode*: If the emitter–base junction is forward-biased and the collector–base junction is reverse-biased, a unique feature is obtained. The current gets either amplified or attenuated. Generally the amplification of current is preferred. When the transistor acts like this, it is said to be in the *active mode*.

The above conditions of transistor operation are explained in the following sections in connection with common-emitter, common-base, and common-collector connections of transistor.

Table 5.1 Conventional symbols for the voltages at different terminals of a transistor

V_E, V_B and V_C	Emitter, base and collector voltage, respectively, with reference to ground
$V_{CE} = V_C - V_E$	Collector-to-emitter voltage difference (collector voltage larger in magnitude)
$V_{BE} = V_B - V_E$	Base-to-emitter voltage difference (base voltage larger in magnitude)
$V_{EB} = V_E - V_B$	Emitter-to-base voltage difference (emitter voltage larger in magnitude)
$V_{CB} = V_C - V_B$	Collector-to-base voltage difference
V_{CC}	Supply voltage at the collector
V_{BB}	Supply voltage at the base
V_{EE}	Supply voltage at the emitter

5.3 Common-Emitter Configuration

The common-emitter (CE) configuration is the most widely used mode of operation of a transistor because it has an inherent property of current amplification and by applying suitable techniques (Chapters 6–8), it can be made to amplify voltage and power also. The biasing of n–p–n transistor in CE mode is demonstrated in Figure 5.2. The voltage source V_{BB} forward-biases the base–emitter junction. The voltage source V_{CC} is greater than V_{BB} so that the collector–base junction gets reverse-biased. The grounded emitter terminal is common to both the input voltage (V_{BE}) and the output voltage (V_{CE}). A similar biasing process is applicable to p–n–p transistors with the polarities of the supply voltages reversed. This is the simplest bias configuration where V_{BB} and V_{BE} or V_{CC} and V_{CE} become numerically equal. However, in a practical circuit, as will be discussed in the next chapter, these become different due to the presence of resistors and other passive elements in the circuit. The current components through the transistor and the amplification process are now explained.

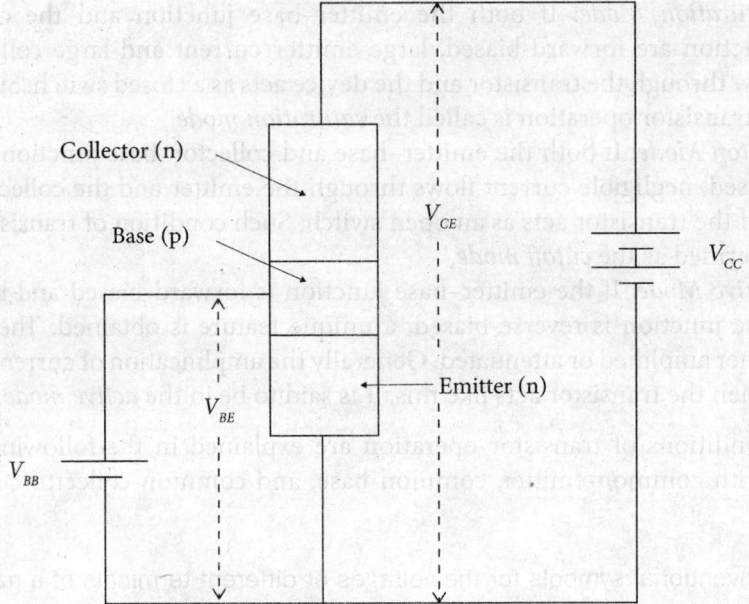

Figure 5.2 Transistor biasing in common-emitter mode

5.3.1 Current Amplification in Transistor

We are considering the n–p–n transistor. If the forward bias at the base–emitter junction becomes greater than its built-in potential, a large number of electrons are injected to the base from the emitter. The base region being very thin and lightly doped, the following two phenomena take place.

(i) Only a small fraction of the incoming electrons recombines with the holes in the base. To compensate for this charge loss, more holes enter the base from the external biasing source so that a small base current, opposite in direction to the emitter current is established.

> **Think a Bit: Do the Holes Exist outside the Semiconductors?**
>
> The above narration of recombination can be viewed from a different aspect. A new electron produced by breaking of covalent bond within the base enters the positive terminal of the base voltage source and at the same moment, an electron leaving the negative terminal of the base voltage source enters the emitter. In the case of p–n–p transistor, electrons enter the base from the base voltage source to compensate for the electrons that have filled the covalent bond within the base. Thus the holes exist within the semiconductor only and there is no hole from the point of view of the external circuit.

(ii) The majority of the electrons passes through the base without being recombined and diffuses into the collector–base depletion region. The electric field across this region drifts the electrons into the collector region where the positive V_{CB} attracts the electrons to cause a large collector current, almost equal to the emitter current.

Similar explanations as above hold in the case of p–n–p transistors with the reversal of the carrier polarities and the biasing terminals.

The base current is much smaller than the collector current. Therefore, a small change in the base current due to change in base–emitter forward bias causes a large change in the collector current. This is the origin of the current amplification mechanism executed by a transistor operating in common-emitter mode. We will see later that similar current amplification occurs in common-collector but not in common-base mode.

> **Note: What Does 'Transistor' Mean**
>
> The word 'transistor' is a combination of 'transfer' and 'resistor'. The output junction of a transistor is reverse-biased offering high resistance but the input current is transferred toward the output. Thus the transistor provides with the transfer of current from low resistance to high resistance. Effectively it offers such a resistance that can amplify an electrical signal transmitted through it.
>
> John R. Pierce, scientist and writer coined the name 'transistor' to account for its transresistance similar to the transconductance of vacuum tubes. The name fitted well with the names of other devices involving current transport, such as varistor and thermistor.
>
> In earlier days, the transistor was also named 'semiconductor triode' because it reproduced similar features of nonlinear current–voltage relationship and electrical power amplification obtained with the triode for the first time. The operating principles are, however, quite different. The BJT is a current-controlled device whereas the triode is a voltage-controlled device. In this regard, the FET of Chapter 9 can be treated as the actual semiconductor equivalent of the triode.

5.3.2 Transistor Current Components

Referring to Figure 5.2, the following current components can be obtained through the emitter, the base, and the collector terminal with the participation of electrons and holes.

(i) *Emitter Current* (I_E): It consists of two components: that due to many electrons diffusing from the emitter to the base and that due to few holes diffusing from the base to the emitter. Both these current components have the same direction. However, the latter one is negligibly small and for practical purpose, the entire emitter current is assumed to be that due to the flow of electrons from the emitter to the base.

(ii) *Base Current* (I_B): As explained above, the small base current stems from the recombination of few electrons and holes within the base region. This current is opposite to I_E in direction.

(iii) *Collector Current* (I_C): This current is caused by the electrons reaching the collector after diffusing through the base region. Since the flow of charge carriers from the emitter is the origin of both the base and emitter currents, the emitter current is the sum of the base and collector currents irrespective of n–p–n or p–n–p transistors. Therefore,

$$I_E = I_C + I_B \tag{5.1}$$

(iv) *Reverse Saturation Current* (I_{CO}): The drift of minority carriers across the reverse-biased collector–base junction gives rise to a feeble current in the direction of I_C. It has two components: the current due to the electrons moving from the p-type base to the n-type collector and that due to the movement of holes from the n-type collector to the p-type base. The reverse saturation current remains present always, even if the input forward bias is removed by opening the emitter terminal and it flows in the direction of I_C. That is why it is indicated by the symbol I_{CO} where the *o*-subscript stands for 'open'.

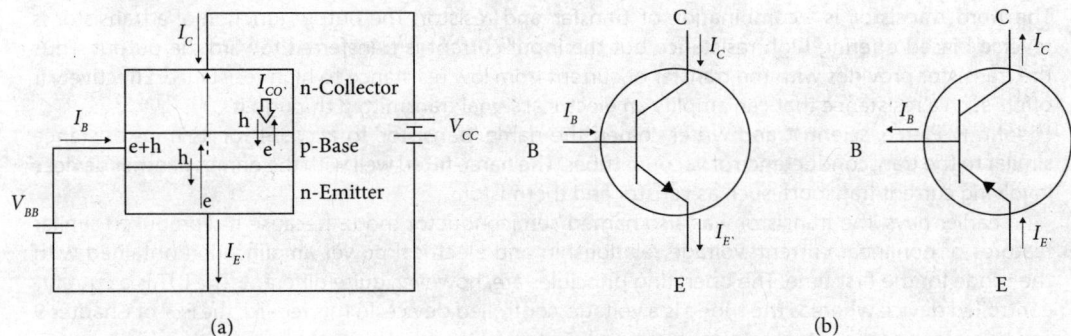

Figure 5.3 (a) components of emitter current (I_E), base current (I_B) and collector current (I_C) for n–p–n transistor in CE configuration, the amount and direction of flow for electrons (e) and holes (h) being indicated by long and small arrows and (b) circuit symbols with current directions for (b) n–p–n and (c) p–n–p transistor

Table 5.2 Current direction and voltage polarity for different terminals of a transistor

Transistor type	I_E	I_B	I_C	V_{BE}	V_{CB}	V_{CE}
n–p–n	negative	positive	positive	positive	positive	positive
p–n–p	positive	negative	negative	negative	negative	negative

Equation (5.1) has considered only the levels or magnitudes of the currents. If the directions of flow of electrons and holes are taken into account, it is found that the collector and the base currents have the same polarity, which is opposite to that of the emitter current. This feature is obtained with both n–p–n and p–n–p transistors. Table 5.2 summarizes the signs for the currents and the voltages through different terminals of a transistor and Figure 5.3 illustrates the directions of these currents and their components. Also the following conventions of current and voltage are maintained.

- *Current Convention*: The current flowing into the transistor is assumed to be positive and that flowing out of the device is considered negative.
- *Voltage Convention*: The voltage polarity for transistor is considered positive when the terminal marked by the first subscript is positive with respect to the second one.

At the cutoff condition of the input forward bias, $I_E = 0$ but $I_C = I_{CO}$. From this state, the ratio of the collector current increment up to an arbitrary value (I_C) to the corresponding increment of emitter current (I_E) is denoted by a symbol α (alpha) such that

$$\alpha = \frac{I_C - I_{CO}}{I_E - 0} \tag{5.2}$$

If I_{CO} is neglected, Equation (5.2) reduces to

$$\alpha = \frac{I_C}{I_E} \tag{5.3}$$

Strictly speaking, α is not a constant; it varies with emitter current, collector voltage and temperature. However, for the practical purpose, α is assumed to be a constant term close to unity representing the ratio of the collector current to the emitter current. If I_{CO} is taken into consideration, combining Equations (5.1) and (5.2) and rearranging one obtains

$$I_C = \frac{\alpha I_B}{1-\alpha} + \frac{I_{CO}}{1-\alpha} \tag{5.4}$$

Introducing another constant (beta)

$$\beta = \frac{\alpha}{1-\alpha} \tag{5.5}$$

and consequently

$$\alpha = \frac{\beta}{\beta+1}, \tag{5.6}$$

Equation (5.4) changes as

$$I_C = \beta I_B + (\beta+1)I_{CO} \tag{5.7}$$

In most practical cases of transistors operating in active region, $I_{CO} \ll I_B$ so that Equation (5.7) yields

$$\beta = \frac{I_C}{I_B} \tag{5.8}$$

In Equation (5.8), β represents the ratio of the collector current to the base current and indicates the extent of current amplification in a transistor. It is obvious from Equation (5.5) that a small change in α gives rise to a large change in β. So keeping the amplification constant is a matter of considerable importance. Here lies the importance of transistor biasing that will be discussed in Chapter 6.

5.3.3 Common-Emitter Output Characteristics

The transistor in CE mode is drawn with circuit symbol in Figure 5.4(a). This is actually a practical form of Figure 5.2. In practice, some circuit conventions are followed that we will discuss in Chapter 6. The base current (I_B) is the input current and the collector current (I_C) is the output current. The collector-to-emitter voltage (V_{CE}) is the output voltage. The variation of I_C against V_{CE} with I_B as a parameter is graphically represented by a set of curves called the *static output characteristics* of the transistor in CE mode, an example being given in Figure 5.4(b). The graph area has three distinct regions, namely the saturation region, the active region, and the cutoff region, indicating the activity of the transistor at three different domains of current–voltage variation.

A. *Saturation Region*: When V_{CE} ($= V_{BE} - V_{BC}$) attains a small positive value (*e. g.* few tenths of a volt), both the base–emitter and the collector–base junctions become forward-biased because both the junctions have cut-in voltage less than a volt. Consequently I_C increases very rapidly with voltage (exponentially, according to the diode relationship) and attains a saturation value for a fixed I_B. The saturation occurs within a very small voltage (V_{CEsat}) less than diode cut-in voltage and the saturation resistance given by the ratio V_{CEsat}/I_C becomes very small, less than a few ohms. Because of these two properties, the transistor acts like a closed switch in the saturation region.

B. *Active Region*: When V_{CE} is further increased keeping I_B fixed, the collector–base junction gets reverse-biased. Then I_C remains almost constant irrespective of the change in V_{CE}. On increasing I_B, the saturation of I_C takes place at higher values and then again remains almost independent of V_{CE}. Thus the whole area to the right of $V_{CE} \approx$ cut-in voltage and above $I_B \approx 0$ in Figure 5.4(b) is the active region of the transistor. Here the base–emitter junction is forward-biased and the collector–base junction is reverse-biased. Two important phenomena are noted in the active region.

(i) It is linear region because the change in input current (I_B) produces a proportional change in the output current (I_C).
(ii) The transistor amplifies current in this region, the current gain being defined by β [Equation (5.8)]. Also, to be discussed later, the ratio of the corresponding changes of the collector and base currents ($\Delta I_C/\Delta I_B$) expresses current amplification of the transistor for ac operations.

The current gain of a transistor depends on (i) the transistor itself, (ii) collector current, and (iii) temperature. The current gain has no effect on the base current. For a certain situation, the base current is fixed. Depending on the current gain, the collector current varies. When the current gain increases, the collector current increases.

C. *Cutoff Region*: This is the region close to the V_{CE} axis when $I_B \approx 0$. In this case, both the base–emitter and the collector–base junctions are reverse-biased. It appears from Figure 5.4(b) that just making $I_B = 0$ one may attain the situation of cutoff. However, there lies a subtle paradox.

Let the base–emitter junction be open-circuited thereby making $I_B = 0$. Now putting $I_B = 0$ in Equation (5.7), I_C comes out to be

$$I_{CEO} = (\beta + 1)I_{CO} = I_{CO}/(1-\alpha) \tag{5.9}$$

The symbol I_{CEO} in Equation (5.9) denotes the reverse current through the collector–base junction in CE mode when the base is open circuited. Depending on α (which, in turn depends on the semiconductor material), this current may attain an appreciable value even at zero base current. Thus simply making $I_B = 0$ is not the sufficient condition for cutoff. One has to slightly reverse bias the base–emitter junction. Any way with silicon transistors, α is much smaller than unity for small values of collector current and cutoff can be achieved at $V_{BE} \approx 0$, which is equivalent to shorting the base and the emitter. The collector–base junction gets already reverse-biased, as we increase V_{CE}. The collector current at cutoff is not zero but of the order of I_{CO}.

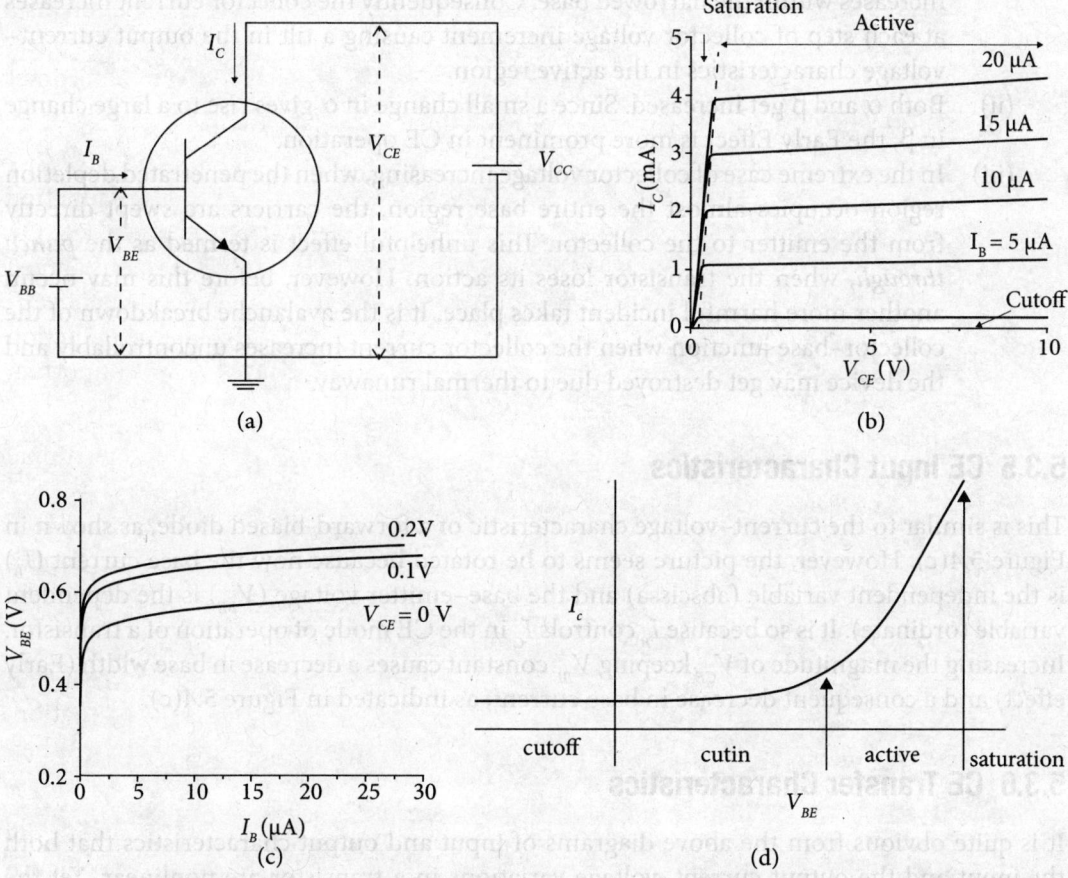

Figure 5.4 n–p–n transistor in common-emitter configuration: (a) circuit diagram, (b) sample output characteristics for different base currents, (c) sample input characteristics, and (d) schematic transfer characteristics

5.3.4 Early Effect

As apparent from Figure 5.4(b), the variation of I_C with V_{CE} in the active region is not horizontal but a bit tilted, which is more prominent at higher values of I_B. Such increase of I_C with V_{CE} in the active region originates from *Early Effect* propounded by American electrical engineer J. M. Early (1922–2004). The reason is the reverse bias applied across the collector–base junction of the transistor in active mode that influences the base width. Since the base is thin and lightly doped, the depletion region penetrates significantly into the base thereby reducing the effective base width. The depletion width goes on increasing with collector bias thereby making the base narrower. This phenomenon is termed as the *Early Effect* or *base-width modulation*. It has the following effects on the transistor output characteristics.

(i) Because of the effective reduction of base width, the recombination rate within the base decreases and the concentration gradient of the minority carriers increases within the narrowed base. Consequently the collector current increases at each step of collector voltage increment causing a tilt in the output current–voltage characteristics in the active region.

(ii) Both α and β get increased. Since a small change in α gives rise to a large change in β, the Early Effect is more prominent in CE operation.

(iii) In the extreme case of collector voltage increasing, when the penetrated depletion region occupies almost the entire base region, the carriers are swept directly from the emitter to the collector. This unhelpful effect is termed as the *punch through*, when the transistor loses its action. However, before this may occur, another more harmful incident takes place. It is the avalanche breakdown of the collector–base junction when the collector current increases uncontrollably and the device may get destroyed due to thermal runaway.

5.3.5 CE Input Characteristics

This is similar to the current–voltage characteristic of a forward-biased diode, as shown in Figure 5.4(c). However, the picture seems to be rotated because now the base current (I_B) is the independent variable (abscissa) and the base–emitter voltage (V_{BE}) is the dependent variable (ordinate). It is so because I_B controls I_C in the CE mode of operation of a transistor. Increasing the magnitude of V_{CE} keeping V_{BE} constant causes a decrease in base width (Early effect) and a consequent decrease in base current, as indicated in Figure 5.4(c).

5.3.6 CE Transfer Characteristics

It is quite obvious from the above diagrams of input and output characteristics that both the input and the output current–voltage variations in a transistor are nonlinear. Yet the true nonlinearity of a transistor is realized with its *transfer characteristic*; the variation of its output current with respect to the input voltage. The transfer characteristic of a transistor represents the mutual relationship between the input and the output that an input signal is likely to undergo while getting transferred through the device. The nature of the transfer characteristic of a silicon transistor in CE mode is sketched in Figure 5.4(d). The overall variation of the output current (I_C) with the input voltage (V_{BE}) is nonlinear. It is noteworthy that a small segment of the curve may be approximated as a piece of straight line. This approximation is the key factor to small signal linear amplification of a transistor, as will be discussed in Chapters 6 and 7.

> **Example 5.1**
>
> A transistor operating in CE mode has current gain 120. Determine the collector current, if the emitter current is 5 mA.
>
> **Solution**
>
> Using Equation (5.6), the transistor has α = 120/(120 + 1) = 0.9917
>
> Using Equation (5.3), the collector current (magnitude) is calculated as
>
> I_C = 0.9917×5 = 4.9585 mA

5.4 Common-Base Characteristics

The transistor circuit diagram in common-base (CB) mode is shown in Figure 5.5(a). The explanations for electron flow from the emitter, recombination within the base and collection at the collector in this mode are similar to that of CE configuration explained in section 5.3.1. However, there are the following marked differences.

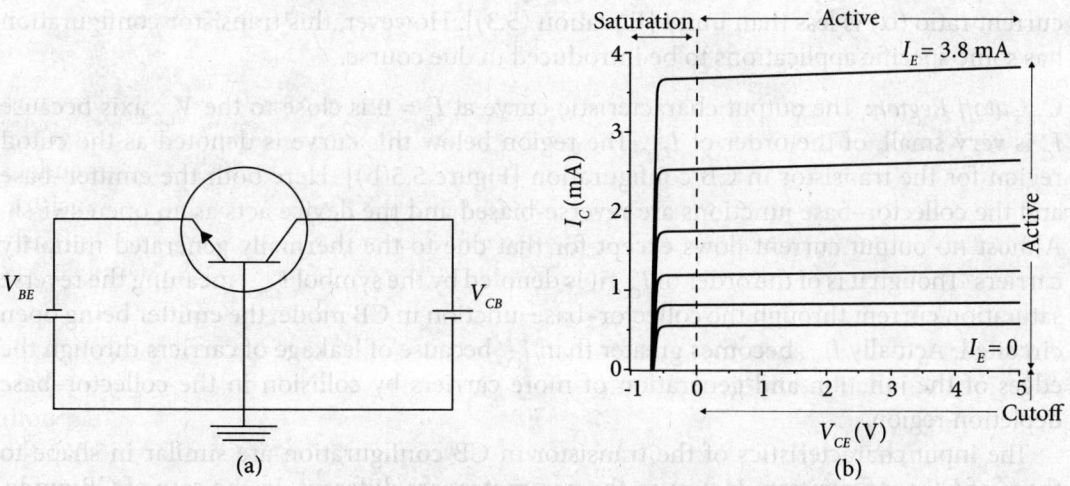

Figure 5.5 n–p–n transistor in common-base configuration: (a) circuit diagram and (b) sample output characteristics for different emitter currents

(i) Here the input current is the emitter current (I_E) and the output current is the collector current (I_C). The output voltage is the collector–base voltage (V_{CB}).

(ii) Figure 5.5(b) shows an example of the *static output characteristics* for a transistor in CB configuration where the output current (I_C) is determined by collector–base voltage (V_{CB}) with the emitter current (I_E) as a parameter.

(iii) The locations and the properties of the saturation, active and cutoff regions are different from those of CE configuration, as mentioned below.

A. *Saturation Region*: The condition is the same; both the junctions are to be forward-biased. However, that occurs at the region slightly left of $V_{CB} = 0$ and above $I_E = 0$, as indicated in Figure 5.5(b). It is so because the voltage source V_{BE} forward biases the emitter–base junction and the electrons flow in abundance into the base. The positive terminal of V_{BE} attracts the electrons crossing the base and reaching the collector. Thus the collector current can reach the saturation value even when V_{CB} is made zero by short-circuiting the collector and the base. In order to resist the electron flow to the external circuit, a small negative voltage at the collector is needed. That is essentially forward-biasing the collector–base junction. It is indicated in Figure 5.5(b) that the collector current reduces to zero not at zero V_{CB} but at its negative value.

B. *Active Region*: Now the collector–base junction is to be reverse-biased and all $V_{CB} > 0$ satisfy the condition. The emitter–base junction remains forward-biased. So the whole area of $V_{CB} > 0$ and $I_E > 0$ in Figure 5.5(b) represents the active region for the transistor operating in CB mode. The output current (I_C) is almost independent of the output voltage (V_{CB}) and depends only on the input current (I_E). Here the Early effect is not so predominant like that in CE mode because the input (emitter) and the output (collector) currents are of the same order. Therefore, as apparent from the figure, the variation of I_C with V_{CB} is almost horizontal, particularly for lower values of I_E.

The CB mode of operation does not amplify current because the output (I_C) to input (I_E) current ratio (α) is less than unity [Equation (5.3)]. However, this transistor configuration has some specific applications to be introduced in due course.

C. *Cutoff Region*: The output characteristic curve at $I_E = 0$ is close to the V_{CE} axis because I_C is very small, of the order of I_{CO}. The region below this curve is denoted as the cutoff region for the transistor in CB configuration [Figure 5.5(b)]. Here both the emitter–base and the collector–base junctions are reverse-biased and the device acts as an open switch. Almost no output current flows except for that due to the thermally generated minority carriers. Though it is of the order of I_{CO}, it is denoted by the symbol I_{CBO}, meaning the reverse saturation current through the collector–base junction in CB mode, the emitter being open circuited. Actually I_{CBO} becomes greater than I_{CO} because of leakage of carriers through the edges of the junction and generation of more carriers by collision in the collector–base depletion region.

The input characteristics of the transistor in CB configuration are similar in shape to those of CE configuration. However, the parameters are different. In the case of CB mode, V_{BE} is plotted against I_E with V_{CB} as parameter.

> **Example 5.2**
>
> An n–p–n transistor operating in CB mode has $\alpha = 0.99$. If the emitter current is 10 mA and the reverse saturation current is 2 µA, calculate the collector and base currents.
>
> **Solution**
>
> Using Equation (5.2), the collector current is: $0.99 \times 10 \times 10^{-3} + 2 \times 10^{-6} = 9.902$ mA
>
> Using Equation (5.1), the base current is: $10 - 9.902 = 0.098$ mA or 98 µA

Note: Voltage and Current Variables in Transistor Characteristics

In all the three transistor configurations, the input current and the output voltage are considered as the independent variables whereas the output current and the input voltage are treated as the dependent variables. However, the specific parameters are different, as summarized below.

Configuration	input current	output voltage	output current	input voltage
common-emitter	base current (I_B)	collector-emitter voltage (V_{CE})	collector current (I_C)	base–emitter voltage (V_{BE})
common-base	emitter current (I_E)	collector–base voltage (V_{CB})	collector current (I_C)	emitter–base voltage (V_{EB})
common-collector	base current (I_B)	emitter-collector voltage (V_{EC})	emitter current (I_E)	base-collector voltage (V_{BC})

5.5 Common-Collector Configuration

The base and the emitter terminals of a transistor can serve as input terminal and also as ground terminal. The collector, unlike these two, cannot be used as the input of a transistor. However, with suitable circuit configuration, it can act as a common point for both the input and the output. That is the common-collector (CC) mode of transistor operation shown in Figure 5.6(a). It has the following features.

(i) The collector terminal is directly connected to the supply voltage V_{CC} having magnitude greater than the base supply (V_{BB}) with the polarities indicated.
(ii) The CC circuit contains a load resistor R between the emitter and the ground.
(iii) The use of such a resistor in addition to the voltage supply is essential in CC mode because the output voltage (V_o) is obtained across this resistor and is given by

$$V_o = V_{BB} - V_{BE} \tag{5.10}$$

For a constant V_{BE} (≈ 0.7 V for silicon transistors), the change in V_o is in phase with the change in V_{BB}. It appears from Equation (5.10) that the output voltage across the emitter 'follows' the input voltage at the base. Therefore, the common-collector configuration of transistor is also termed as the *emitter follower*.

Though the 'common' collector is not so noticeable in Figure 5.6(a), yet we can imagine the collector-to-emitter (V_{CE}) and the collector-to-base (V_{CB}) voltage differences as two biasing sources and draw the transistor upside down so that it looks like that of Figure 5.6(b). Now it is clear that the base is the input, the emitter is the output and the collector is common to both base and emitter circuits. The load resistor (R) receives currents from both the base and the collector circuits.

Figures 5.6(c) represents the output characteristics of the transistor in CC operation. It shows the variation of emitter (output) current (I_E) with output voltage (V_{CE}) for fixed base (input) currents (I_B). The output current becomes zero when I_B equals to I_{CO}. On increasing I_B, the transistor passes through active and saturation regions. The characteristic curves indicate that the CC configuration of transistor can cause larger change in emitter current with small change in base current, hence current amplification takes place similar to the CE configuration.

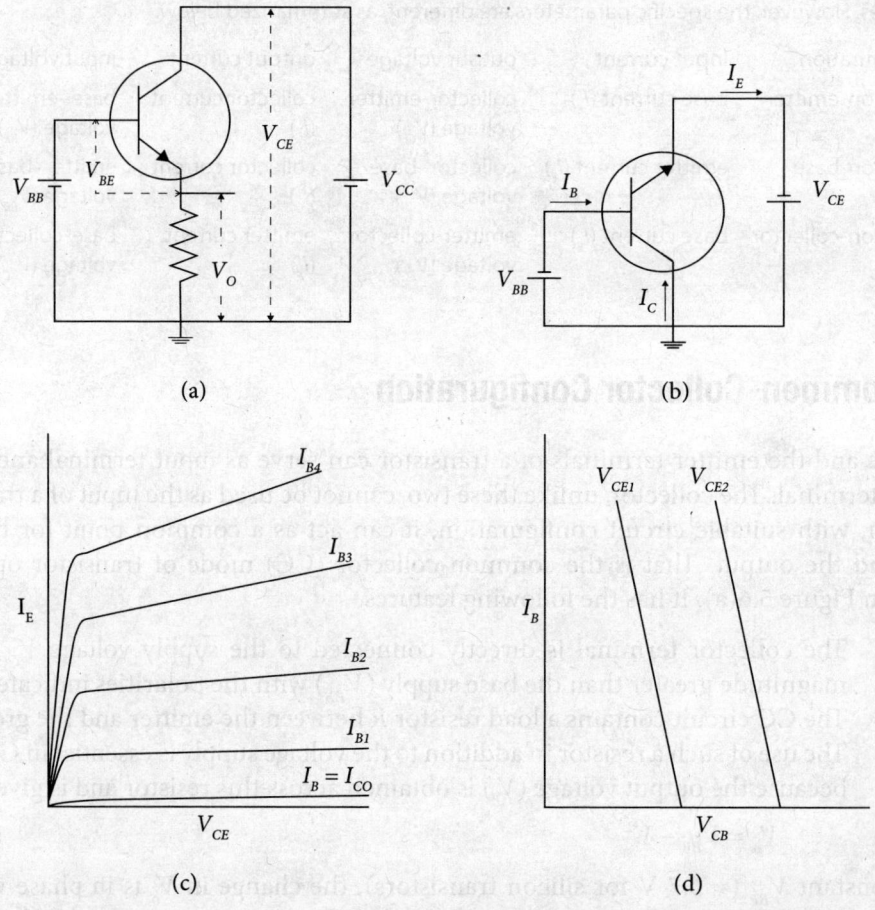

Figure 5.6 n–p–n transistor in common-collector configuration: (a) circuit diagram, (b) equivalent circuit showing the common connection of the collector, (c) output characteristics, and (d) input characteristics

Figure 5.6(d) sketches the input characteristics of the transistor in CC mode showing the relationship between the base (input) current (I_B) and the collector–base (input) voltage (V_{CB}) for fixed output voltage (V_{CE}). The input voltage is mainly determined by the level of the output voltage so that

$$V_{CB} = V_{CE} - V_{BE} \qquad (5.11)$$

Multiple Choice-Type Questions and Answers

5.1 In the active region of common-emitter configuration, the collector terminal is ………………… with respect to the base in the case of n–p–n transistor and that is ………………… in p–n–p transistors.
    (a) positive, negative   (b) negative, positive
    (c) positive, positive   (d) negative, negative

5.2 The following(s) can be an output terminal of a transistor.
    (a) emitter   (b) base   (c) collector   (d) all the three

5.3 The following terminal(s) ………………… cannot be an input terminal of a transistor.
    (a) emitter   (b) base   (c) collector   (d) ground

5.4 The independent variables related to the transistor characteristics are
    (a) input voltage and input current   (b) input current and output voltage
    (c) output voltage and output current   (d) output current and input voltage

5.5 The dependent variables related to the transistor characteristics are
    (a) input voltage and input current   (b) input current and output voltage
    (c) output voltage and output current   (d) output current and input voltage

5.6 A small change in α produces a ………………… change in β.
    (a) smaller   (b) large   (c) small   (d) negligible

5.7 If the α of a transistor is 0.99, then its β is …………………
    (a) 0.01   (b) 1.01   (c) 99   (d) 1.0

5.8 Which of the following is incorrect?
    (a) $\alpha = \beta/(1 + \beta)$ (b) $\beta = \alpha/(1 - \beta)$
    (c) $1 - \alpha = 1/(1 + \beta)$ (d) $\alpha = \beta/(1 - \beta)$

5.9 The ratio $I_E : I_B : I_C$ in a transistor is approximately …………………
    (a) $1 : (1 - \alpha) : \alpha$ (b) $\alpha : (1 - \alpha) : 1$
    (c) $1 : (1 + \beta) : \beta$ (d) $\beta : (1 + \beta) : 1$

5.10 The following two statements are given in connection with the Early effect.
Statement-I: It reduces the base width.
Statement-II: The base is lightly doped.
The following conclusion can be made based on the above.
    (a) Both Statement-I and Statement-II are true but Statement-II has nothing to do with Statement-I.
    (b) Statement-I is true, Statement-II is also true and it is the reason for Statement-I.
    (c) Both Statement-I and Statement-II are true.
    (d) Statement-I is false but Statement-II is true.

5.11 In a common-collector amplifier, the .... is at ac ground and the input voltage is coupled to the ....
   (a) collector, base
   (b) base, emitter
   (c) emitter, collector
   (d) base, collector

5.12 A small ac input signal is applied to a common-collector amplifier. The collector voltage is equal to ....
   (a) the dc biasing voltage of the base plus the input voltage waveform.
   (b) the dc biasing voltage of the base plus the input voltage multiplied by β.
   (c) a constant voltage of zero volt.
   (d) a constant voltage of V_{CC}.
   Hints. The collector is at ac ground, hence no ac signal is there.

5.13 The α of a transistor is defined for ....
   (a) the minority carriers only
   (b) the majority carriers only
   (c) the majority carriers for n–p–n transistor only
   (d) the minority carriers for n–p–n transistor only

5.14 The minority carrier component of the collector current can flow when the emitter current is ....
   (a) zero
   (b) greater than the collector current
   (c) either of (a) and (b)
   (d) none of (a) and (b)

5.15 For an unbiased Silicon n–p–n transistor in thermal equilibrium, which one of the following electronic energy band diagrams is correct? (E_C = conduction band minimum, E_V = valence band maximum, E_F = Fermi level).

[JAM 2020]

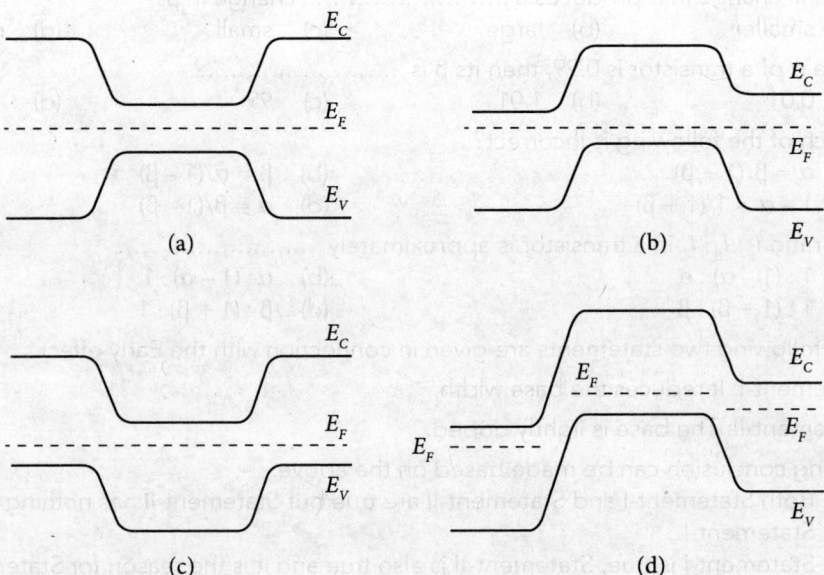

Figure 5.7

Answers

5.1	(a)	5.2	(d)	5.3	(c)	5.4	(b)
5.5	(d)	5.6	(b)	5.7	(c)	5.8	(d)
5.9	(a)	5.10	(c)	5.11	(a)	5.12	(d)
5.13	(b)	5.14	(c)	5.15	(b)		

Reasoning-Type Questions and Answers

5.1 The p–n–p or the n–p–n structure of a transistor appears to be symmetric with respect to the base. So can the emitter and the collector be interchanged?

Ans. No, the emitter and the collector regions have different amounts of doping and surface area.

5.2 There are two depletion regions in a transistor. Are those similar?

Ans. No, the depletion region at the emitter–base junction is thinner than that of the collector–base junction because the emitter is the most heavily doped region. The depletion region penetrates the minimum within the emitter and the maximum within the base region because the base is lightly doped.

5.3 The transistor resembles two p-n junction diodes back-to-back, comprising the emitter–base and the collector–base junction, respectively. Can a transistor be actually fabricated this way by joining two diodes?

Ans. No, the necessary conditions for transistor action are thin base width and less doping concentration in comparison with that of the emitter. The collector width is much more than that of the base. None of the above conditions is satisfied with two oppositely connected diodes.

5.4 When is the transistor operated as amplifier and when does it act as switch?

Ans. The transistor in active mode acts as an amplifier capable of amplifying current. In saturation mode and in cutoff mode, it works as closed and open switch, respectively.

5.5 Does the transistor in active region act as a current source?

Ans. Yes, the output current (I_C) in active region is almost independent of the output voltage (V_{CE} in CE mode and V_{CB} in CB mode). The input current (I_B in CE mode and I_E in CB mode) controls this fixed value of the output current. So the transistor in active region acts as a current-controlled current source.

5.6 How will the static output characteristic curves change, if a transistor is replaced by another of the same category?

Ans. The nature of the characteristic curves will remain the same showing the saturation, active and cutoff regions. However, the value of β changes randomly from device to device so that the output voltage value for the transition from saturation to active mode will be different and the output current value for the same input current will be different.

[N.B. Actually the effect of such variations is overcome by biasing with passive components (Chapter 6).]

5.7 What happens, if the output voltage of the transistor is increased indefinitely?

Ans. After reaching the active region, the collector current remains almost fixed independent of the output voltage (V_{CE} for CE mode and V_{CB} for CB mode). However, it remains so provided the reverse bias across the collector–base junction stays well below the breakdown voltage. On increasing the output voltage indefinitely, the collector–base junction breaks down and the collector current increases uncontrollably. This may cause self-heating of the device due to the power dissipation at the junction. The increase of temperature enhances the reverse saturation current through the junction that, in turn, further increases the collector current. This cumulative process goes on and ultimately causes permanent thermal damage to the transistor. Such destruction of the device is termed as the *thermal runaway*.

5.8 Draw the energy band diagram corresponding to the emitter, base and collector regions of a transistor operating in active mode.

Ans. Figure 5.8 indicates the energy band picture of a transistor operating in active region. Forward-biasing the emitter–base junction raises the Fermi level of the emitter region and causes the electrons from emitter conduction band to enter the base conduction band. The electrons become minority carrier in the p-type base. A few recombine with holes and the major portion fall down to the collector conduction band. The energy released is dissipated in the form of heat and lattice vibration.

[N.B. The power rating of a BJT is determined by the collector–base junction area.]

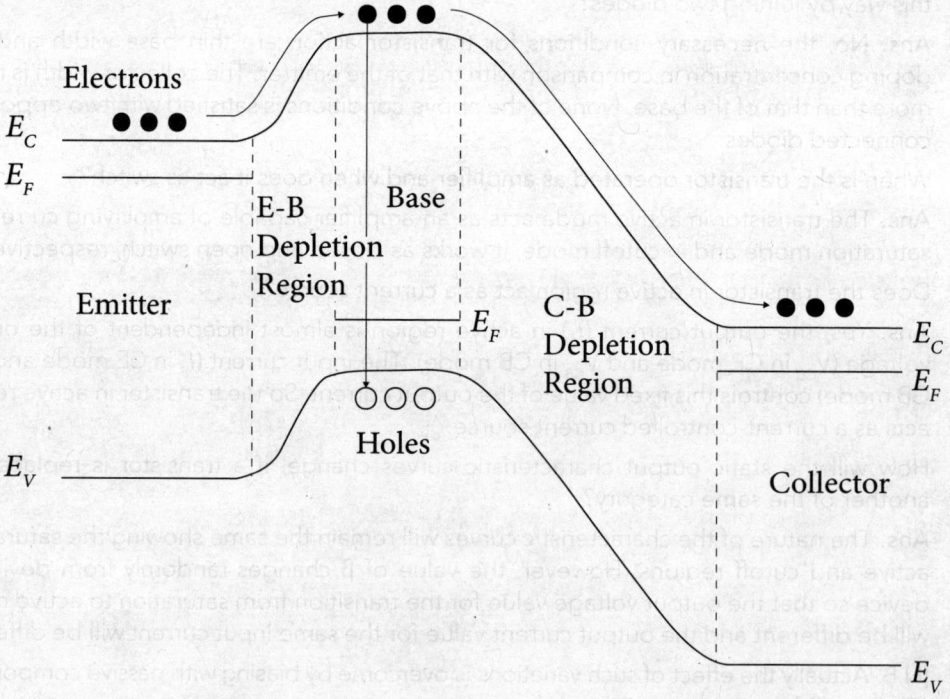

Figure 5.8

Solved Numerical Problems

5.1 A BJT operating in CE mode has $\beta = 100$ and its collector–base leakage current is 10 nA. What will be the collector current when (i) the base terminal is kept open and (ii) there is a base current of 10 µA.

Soln. In Case (i), following Equation (5.9), the collector current is

$(100+1) \times 10 \times 10^{-9} = 1.01$ µA.

In Case (ii), following Equation (5.7), the collector current is

$100 \times 10 \times 10^{-6} + (100+1) \times 10 \times 10^{-9} = 1.001$ mA

5.2 A transistor operating in CE mode has base current of 20 µA, reverse saturation current of 0.5 µA and $\alpha = 0.99$. Calculate the collector and emitter currents.

Soln. Using Equation (5.5), the β of the transistor is $0.99/(1 - 0.99) = 99$

Then using Equation (1), the collector current is

$99 \times 20 \times 10^{-6} + (99+1) \times 0.5 \times 10^{-6} = 2.03$ mA

The emitter current is

2.03 mA $+ 20$ µA $= 2.05$ mA

5.3 A BJT operating in CB mode has collector current of 5 mA, base current of 15 µA and reverse saturation current of 1 µA. What should be the value of the approximate value of the base current so as to make the collector current 25 mA, assuming other parameters remain unchanged?

Soln. Assuming $\beta + 1 \approx \beta$ in Equation (5.7),

$5 \times 10^{-3} = \beta \times (15 \times 10^{-6} + 1 \times 10^{-6})$, hence $\beta = 312.5$

Using this value of β, the base current (I_B) required to make the collector current of 25 mA can be approximated as

$25 \times 10^{-3} = 312.5 \times I_B + 313.5 \times 1 \times 10^{-6}$

Therefore, the required base current is 79 µA (approx.)

5.4 The base current and the emitter current of a transistor are 20 µA and 5 mA, respectively. Calculate the approximate values of α, β and I_C, symbols having their usual meaning.

Soln. Neglecting the effect of I_{CO} in Equation (5.7), we have $I_C = \beta I_B$ and putting the same in Equation (5.1),

$I_E = I_B + \beta I_B$ (Only the magnitude is considered.)

$\Rightarrow \beta + 1 = I_E/I_B = 250$

Hence, $\beta = 249$

Using Equation (5.6), $\alpha = 249/(249+1) = 0.996$

Using Equation (5.3), $I_C = 0.996 \times 5 \times 10^{-3} = 4.98$ mA (Magnitude only)

5.5 For a transistor in CE configuration, the reverse saturation current is 35 µA. When the same transistor is connected in CB configuration, the current reduces to 0.25 µA. Determine the approximate values of α and β for the transistor.

Soln. Given: $I_{CEO} = 35$ µA and $I_{CBO} = 0.25$ µA is assumed to be equal to I_{CO}. Then using Equation (5.9),

$\beta + 1 = 140$ so that $\beta = 139$

Using Equation (5.6), $\alpha = 139/(139+1) = 0.993$ (approx.)

5.6 In a certain transistor, α is 0.98 and it increases by 1% due to Early effect, as V_{CE} changes from 1 V to 10 V. Calculate the percentage change in β.

Soln. Originally β was (Equation (5.)] 0.98/(1 − 0.98) = 49

Now α has become 0.98 + 0.0098 = 0.9898 so that

β becomes 0.9898/(1 − 0.9898) = 97

Thus 1% change in α causes almost 98% change in β.

[N.B. Such large effect on β due to small change in α may cause wide variation among the output characteristics of transistors of the same type. In order to minimize the effect of variation of β and to maintain stability in operation, different biasing techniques are used. These will be discussed in the subsequent chapters.]

Exercise

Subjective Questions

1. Draw the circuit diagram of a p–n–p transistor in CB configuration and draw the output characteristics for different emitter currents properly indicating the active, saturation and cutoff regions.
2. Draw the circuit diagram of a p–n–p transistor in CE configuration and sketch the output characteristics for different base currents. Indicate the active, saturation and cutoff regions on the diagram. Can you state that the transistor acts as a current source in active region?
3. Drawing a block diagram for p–n–p transistor, indicate the different components of emitter current, base current and collector current for the transistor being in active region. Indicate the proper directions of the currents.
4. Introduce the α and β of a transistor and establish a relation between these two parameters. Explain the significance of the reverse saturation current and derive a relation between the collector current and the base current including the effect of the reverse saturation current.
5. Considering the forward-biased emitter–base junction and the reverse-biased collector–base junction, explain the output characteristics of a p–n–p transistor. Define β from this characteristics.
6. What is the Early effect? What are its effects on the transistor output characteristics?
7. Compare the output characteristics of a transistor in CE and CB modes.

Conceptual Test

1. Discuss whether the transistor acts as active element in any of saturation, active and cutoff mode.
2. What would happen, if the emitter, base and collector regions of a bipolar junction transistor had the same doping concentration?
3. What is the significance of the term 'bipolar' in connection with transistors?
4. The bipolar junction transistor has contributions of both electrons and holes in the current flow. So has the p-n junction diode but it is not named 'bipolar diode'. Do you find any justification for this?

5. It is fact that the intention of transistor construction is to somehow minimize α and maximize β. Is the thin and lightly doped base in conformity with that?
6. Who does saturate in the saturation region and how?
7. For a certain doping concentration, the saturation current is proportional to the junction area, hence the device area. Therefore it is also called scale current. For two transistors having different junction areas, a larger collector will be obtained with the larger area device for the same base–emitter voltage. Now, based on the above fact, what would happen to α and β, if someone attempted to use a transistor interchanging the emitter and the collector?

Numerical Problems

1. A transistor has $\beta = 100$, $I_{CEO} = 1$ μA and $I_C = 1$ mA. Determine I_B and I_E. The symbols have the usual meaning.
2. In a transistor, 99.5% of the carriers injected from the emitter into the base cross the collector–base junction. If the collector current is 10 mA and the collector–base leakage current is 5 μA, determine the emitter current.
3. In an n–p–n transistor, 10^{11} electrons flow from the emitter to the base during each microsecond and 10^9 holes move from the base to the emitter during each microsecond. If the base current be 20 μA, determine the collector current. Assume that each electron and hole possesses the charge of 1.6×10^{-19} C in magnitude. The number of electrons and the number of holes flowing per unit time are not the same; is it justified?
4. A transistor having $\alpha = 0.98$ operates in CB mode. If the emitter current is 5 mA and the reverse saturation current is 10 μA, calculate the collector current and the base current.
5. The CE current gain of a transistor is 100. If the emitter current be 5 mA, calculate the collector current.
6. A transistor is operating in CE mode. A 560 Ω resistor is joined between the collector and the power supply and a voltage drop of 0.6 V appears across it. If $\alpha = 0.97$, determine the base current.

Project Work on Chapter 5

Draw the output characteristics of a bipolar junction transistor in common-emitter configuration and determine the current gain.

Theory:
The transistor is made to act in active region by making the emitter–base junction forward-biased and the collector–base junction reverse-biased. The emitter terminal is common to both the input and the output circuits. For a fixed base current (I_B), the collector current (I_C) is plotted as function of the collector-to-emitter voltage (V_{CE}). The same is repeated for several I_B values. The current gain is given by

$$\beta = \frac{\Delta I_C}{\Delta I_B}$$

I_C and I_B are measured in mA and μA, respectively.

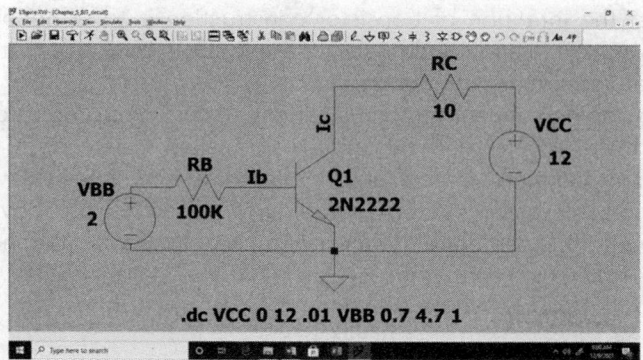

Figure P5.1 LTspice circuit diagram for generating transistor output characteristics

Computer analysis:

The experiment was simulated with LTspice open software. The assembled circuit is shown in Figure P5.1 and the specifications for simulation are given below.

The high base resistance (100 kΩ) keeps the base current at microampere level. A small resistance of 10 Ω is inserted to the collector circuit.

From the given choices of transistors, 2N2222 was selected as an example. From the 'Simulate' menu, the simulation command 'DC sweep' was selected. The 1st source was named V_{CC}. Linear sweep was implemented with start value of 0, stop value of 12 (volt) and increment of 0.01 (volt). The 2nd source was named V_{BB}. Linear sweep was implemented with start value of 0.7, stop value of 4.7 (volt) and increment of 1 (volt). Such uneven values were chosen to make the I_B values at round figures.

The output characteristic curves were made visible on the screen by Clicking on the 'Run' command, right clicking on the viewer and selecting 'Add Traces' option. Right clicking, selecting 'File' and 'Export data as text', the actual dataset was accessed and saved in a separate folder. The use of some efficient graph-plotting software, such as Origin can produce a better output. A set of output characteristics prepared with Origin is given in Figure P5.2.

Results and Discussion:

Some sample data on the variation of I_C with V_{CE} are given in Table P5.1.

Table P5.1 I_C versus V_{CE} data with I_B as a parameter

I_B = ... µA		I_B = ... µA		I_B = ... µA		I_B = ... µA		I_B = ... µA	
V_{CE} (V)	I_C (mA)	V_{CE} (V)	I_C (mA)	V_{CE} (V)	I_C (mA)	V_{CE} (V)	I_C (mA)	V_{CE} (V)	I_C (mA)

Several curves are generated and β values are calculated from adjacent curves, as summarized in Table P5.2.

Table P5.2 Calculation of β at different operating points

V_{CE}	I_{C2}	I_{C1}	I_{B2}	I_{B1}	$\beta = \dfrac{I_{C2} - I_{C1}}{I_{B2} - I_{B1}}$

6

Transistor Biasing and Amplification

As understood from Chapter 5, the bipolar junction transistor is a current amplifier of its own. However, it can be made to amplify voltage also, as will be discussed in this chapter. The key factor for voltage amplification with transistor is the use of external passive elements such as resistors and capacitors in the transistor circuit driven by a single voltage supply. The process is termed as *biasing*. This chapter illustrates several popular biasing techniques for the common-emitter (CE), common-base (CB) and common-collector (CC) configurations of the transistor.

6.1 Load Line and *Q*-Point

The concept of operating point and load line has been introduced in Chapter 4 in connection with p–n junction diodes. Similar requirements of tracing load line and establishing operating point are realized with the transistor also. The convention remains the same, but the parameters get changed in the case of a transistor.

The operating point for a transistor is commonly named as the Q-point or the *quiescent point* because it specifies the output voltage and the output current at which the transistor operates steadily. The Q-point of a particular transistor circuit is determined by establishing proper voltage levels at the three terminals using the voltage drops across external resistors, capacitors or inductors and a single electrical power supply. This process is known as biasing, to be discussed in the next section. The fixation of Q-point depends on several factors, such as

- the dc and dc loads of the amplifier,
- the range of the power supply,
- the maximum power rating of the given transistor and
- the allowable distortion in amplification.

Based on the above, there are several different types of classifications of amplifiers, namely class A, class B, class AB, and class C, to be discussed in Chapter 8. In practice, the Q-point of a transistor is determined with reference to output characteristics in the following way.

Figure 6.1(a) shows an n–p–n transistor in CE configuration with an external resistor (R_C) in series with the collector terminal. The output characteristics for different base current are sketched in Figure 6.1(b). These are similar to Figures 5.4 (a) and (b), respectively, of Chapter 5 but there is an important difference. The transistor output characteristics of Chapter 5 are *static characteristics* because those are the properties of the transistor itself and under the same condition of supply voltage and ambient temperature, the same characteristic curves should be obtained. In the present case, the collector current (I_C) is given by

$$I_C = \frac{V_{CC} - V_{CE}}{R_C} \tag{6.1}$$

so that it gets changed on changing R_C thereby changing the output characteristics. So in presence of external resistor in collector circuit, as in Figure 6.2, the output characteristics are termed as *dynamic output characteristics* of the transistor. Equation (6.1) can be modified as

$$I_C = -\frac{1}{R_C}V_{CE} + \frac{V_{CC}}{R_C} \tag{6.2}$$

Equation (6.2) is termed as *load line* because it represents the equation of a straight line having the slope (negative) dependent on the load resistance (R_C) at the collector circuit. It has voltage-axis intercept of V_{CC} and current-axis intercept of V_{CC}/I_C, as denoted in Figure 6.1(b).

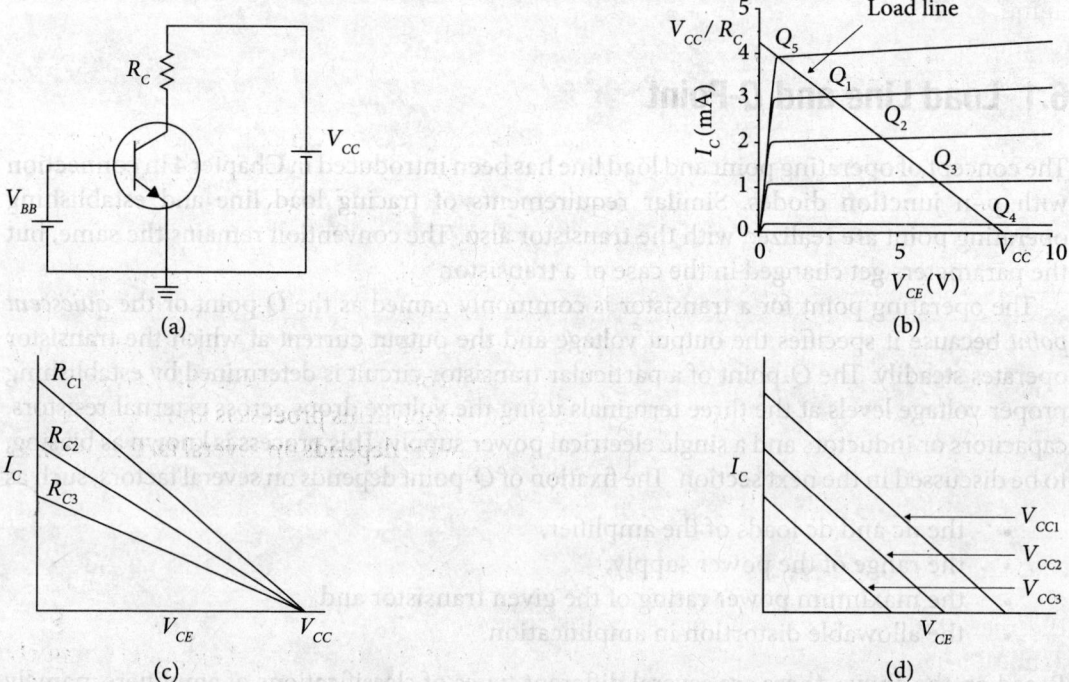

Figure 6.1 The transistor in CE configuration with resistance R_C in series with the collector: (a) circuit diagram, (b) output characteristics with load line and Q-points, (c) variation of load line with resistance ($R_{C1} < R_{C2} < R_{C3}$) in collector circuit, and (d) variation of load line with supply voltage ($V_{CC1} > V_{CC2} > V_{CC3}$)

The intersection of the load line with any output characteristic curve is designated as the Q-point of the transistor. As the input (base) current is changed, the output (collector) current and the output voltage (V_{CE}) also change and the point of intersection with the load line gets shifted. Thus the Q-point of the transistor moves on the load line depending on the transistor operation. In Figure 6.1(b), the points Q_1, Q_2 and Q_3 indicate the amount of current flowing through R_C and the amount of voltage drop across the collector and the emitter terminals while the transistor operates in different states of the active region. The points Q_4 and Q_5 drive the transistor to cutoff and saturation, respectively. The load line changes its position and orientation with both the value of R_C and the supply voltage (V_{CC}), as sketched in Figures 6.1(c) and (d), respectively. On increasing the resistance value, the slope of the line decreases. The slope remains unchanged with the change of supply voltage but the intercepts get changed with reduction of intercepts with smaller voltage.

6.2 Transistor Biasing and Stability

The transistor circuits in Chapter 5 include two power supplies, each one for the input and the output part. However, those are only for the sake of easy interpretation. Any practical system is expected to be energized by a single power supply. A practical transistor circuit is also driven by a single supply voltage and that is achieved by biasing. What is more significant, it stabilizes the transistor operation. Thus the biasing of a transistor is important for the following reasons.

- It provides with proper voltage levels at the three terminals of the transistor from a single supply using passive components.
- Proper biasing allows the transistor to amplify voltage and power.
- The Q-point can be fixed at a certain position on the load line with proper biasing.
- It provides with stability in operation of the transistor.

The current and voltage values at the operating point should be stable during the circuit operation whereas there are several agents, mainly the followings, to fluctuate it.

- Replacement of a transistor by another of the same type may cause variability in electrical parameters.
- Even with the same transistor, the reverse saturation current changes significantly with temperature (almost doubles for every 10°C rise) resulting in change in the collector current.
- The temperature variation also causes significant change in the base–emitter voltage, hence in the stability.

The main objective of biasing is to minimize the above fluctuations thereby imparting stability in the circuit operation. The process is termed as *bias stability*, which keeps the collector current relatively constant with variations in the reverse saturation current and the current gain (β). The stabilization process actually utilizes a technique, known as *negative feedback*, to be explained in Chapter 10.

The bias stability is quantified in terms of the *stability factor* (S); the rate of change of collector current (I_C) with respect to the reverse saturation current (I_{CO}) given by

$$S = \frac{\partial I_C}{\partial I_{CO}} \qquad (6.3)$$

The relationship between I_C and I_{CO} is established with Equation (5.7), which is repeated here as Equation (6.4).

$$I_C = \beta I_B + (\beta+1)I_{CO} \qquad (6.4)$$

It indicates that a small change in I_{CO} produces a large, $(1+\beta)$ times change in I_C. So the effect of temperature is significant and that is quantified in terms of the stability factor (S). Using Equations (6.3) and (6.4), one can obtain

$$\beta \frac{dI_B}{dI_C} = 1 - (1+\beta)\frac{1}{S}$$

so that

$$S = \frac{1+\beta}{1-\beta\dfrac{dI_B}{dI_C}} \qquad (6.5)$$

assuming β to be constant. The larger is the value of S, the greater is the thermal stability.

Note: Definitions of Stability Factor

Since the fluctuation in collector current may be caused due to variation in current gain (β) and base-to-emitter voltage (V_{BE}) with change in temperature, the stability factor may also be defined as

$$S_1 = \frac{\partial I_C}{\partial \beta} \text{ or } S_2 = \frac{\partial I_C}{\partial V_{BE}}$$

However, Equation (6.5) is the most popular form of expressing the stability factor.

The inclusion of a resistor in the collector circuit [Figure 6.1(a)] may be considered as the first step toward biasing. Many biasing techniques, especially in CE mode, retain this process. The following sections present several standard methods of transistor biasing. Except for the base bias, which is suitable for switching operations, all the other biasing methods are suitable for linear amplification by setting up the Q-point around the middle of the load line. These are based on the stabilization technique of making the circuit insensitive to the fluctuation of β. The general expression of Equation (6.5) can be applied to estimate the stability for all these circuits.

6.3 Base Bias

This is also termed as *fixed bias*. The circuit for base biasing is shown in Figure 6.2. The supply voltage V_{CC} biases both the collector and the base terminals with the help of two resistors R_C and R_B, respectively. The emitter terminal is grounded. The base current can be expressed as

$$I_B = \frac{V_{CC} - V_{BE}}{R_B}, \tag{6.6}$$

which is independent of I_C. Therefore, from Equation (6.5), the stability factor is obtained as

$$S = 1 + \beta \tag{6.7}$$

Thus I_C increases $(1+\beta)$ times as fast as I_{CO} and there is no provision for checking the change in β. So the Q-point is not stable in base bias. This biasing technique is simple and is suitable for digital circuits where the transistor switches between saturation and cutoff conditions. When the transistor is intended to be used as an amplifier, other biasing techniques are more appropriate, as illustrated in Figure below.

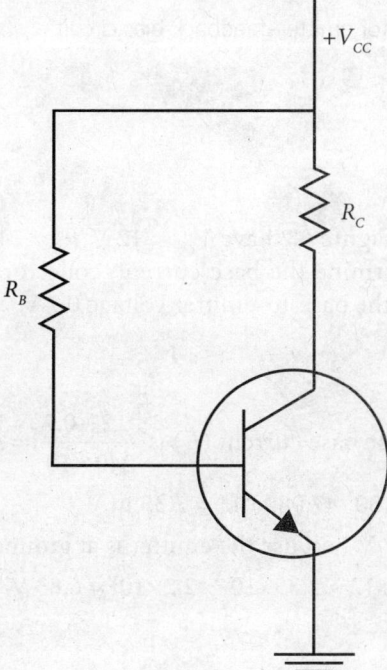

Figure 6.2 Transistor base bias circuit

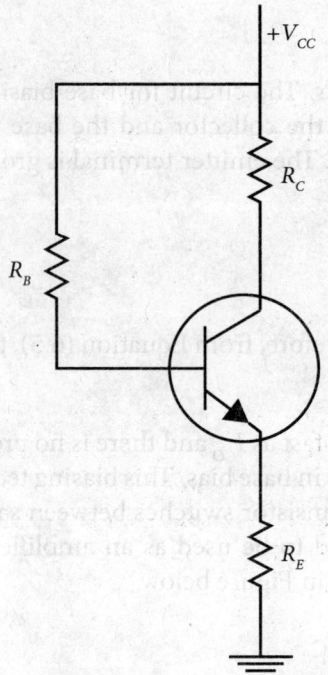

Figure 6.3 Transistor emitter–feedback bias circuit

Example 6.1

Let the base bias circuit of Figure 6.2 have $V_{CC} = 12$ V, $R_B = 240$ kΩ, $R_C = 2.2$ kΩ and the transistor has β = 50. Determine the base current, collector current, base voltage and collector voltage assuming the base-to-emitter voltage 0.7 V.

Solution

Following Equation (6.6), the base current (I_B) is $\dfrac{12 - 0.7}{240 \times 10^3} = 47.08$ μA

The collector current (I_C) is $50 \times 47.08 \times 10^{-6} = 2.35$ mA

The base voltage (V_{BE}) is 0.7 V because the emitter is at ground potential.

The collector voltage (V_C) is $12 - 2.35 \times 10^{-3} \times 2.2 \times 10^3 = 6.83$ V

6.4 Emitter–Feedback Bias

It is, as presented in Figure 6.3, an improved version of base bias with the inclusion of a resistor R_E in series with the emitter. The resistor R_E is called *feedback resistor* because it is common to both the input and the output and a portion of the output is returned to the input through it. Please note two important features in regard of Figures 6.2 and 6.3.

(i) The way of drawing the supply voltage and the ground in these circuit diagrams is different from that of Chapter 5. The present one is more professional method of indicating the power supply in circuit diagrams and we will be continuing this mode of drawing.

(ii) Both the circuits, and also the circuits in the subsequent figures contain a single power supply. Try to maintain this feature in transistor biasing.

In Figure 6.3, the emitter is no longer at ground potential as in base bias. Instead, the emitter voltage varies in proportion with the base voltage because the base–emitter junction has a fixed voltage drop (around 0.7 V for silicon transistors) across it. Such a proportional variation of the emitter voltage with respect to the base voltage is termed as the bootstrapping.

The presence of the feedback resistor (R_E) at the emitter controls any fluctuation in the collector current (I_C) in the following way. If I_C increases for some reason, the emitter voltage ($I_E R_E$) also increases. Consequently the voltage across the base resistor ($I_B R_B$) decreases because the supply voltage (V_{CC}) is constant. The decreased base voltage reduces the base current (I_B), which, in turn, reduces the original increase in I_C. Following the same reason for possible decrease in I_C, it can be justified that the voltage drop across R_E fed back to the base can modify the base current in a direction opposite to the change in the collector current.

The expression for the stability factor with this emitter–feedback biasing circuit is derived as follows. Applying the Kirchhoff voltage law around the collector circuit, we have

$$V_{CC} = I_E R_E + R_C I_C + V_{CE} \tag{6.8}$$

where V_{CE} is the transistor collector-to-emitter voltage drop. Assuming $I_E = I_C + I_B$ (magnitude only) and $(1+\beta) \approx \beta$, the collector current can be expressed as

$$I_C = \frac{V_{CC} - V_{CE}}{R_C + R_E} + \frac{R_E}{R_C + R_E} I_{CO} \tag{6.9}$$

where Equation (6.4) is used for eliminating I_B. Equation (6.9) represents the dc load line in the case of emitter–feedback bias circuit, which is dependent on I_{CO}.

Now applying the Kirchhoff voltage law around the base circuit and taking into account the base-to-emitter voltage drop (V_{BE}),

$$V_{CC} = I_E R_E + V_{BE} + I_B R_B \tag{6.10}$$

Assuming $I_E = I_C + I_B$ (magnitude only), the base current can be expressed as

$$I_B = \frac{V_{CC} - V_{BE}}{R_B + R_E} - \frac{R_E}{R_B + R_E} I_C \tag{6.11}$$

Since V_{BE} and V_{CC} are independent of I_C, one can obtain from Equation (6.11)

$$\frac{dI_B}{dI_C} = -\frac{R_E}{R_B + R_E} \tag{6.12}$$

Substituting for dI_B/dI_C from Equation (6.12) in Equation (6.5), the stability factor for emitter–feedback bias is obtained as

$$S = \frac{1 + \beta}{1 + \dfrac{\beta R_E}{R_B + R_E}} \tag{6.13}$$

Using Equations (6.4) and (6.11) and assuming $(1+\beta) \approx \beta$, the collector current can be approximated as

$$I_C \approx \frac{V_{CC} - V_{BE} + (R_E + R_B) I_{CO}}{R_E + R_B/\beta} \tag{6.14}$$

The following points may be noted concerning the above stability.

(i) The denominator in Equation (6.13) is greater than unity so that $S < (1+\beta)$, a condition better than that of the base bias circuit.

(ii) If $R_E \gg R_B$, $S \approx 1$, the apparently best condition. However, in practice, R_E cannot be so large because that would lead the transistor to saturation.

(iii) Equation (6.14) indicates that $R_E \gg R_B/\beta$ can make the collector current insensitive to β. However, this is also unrealistic.

(iv) If $R_E = R_B/\beta$, $S \approx \beta/2$, which is the suitable practical condition for circuit designing.

Example 6.2

Let the emitter–feedback bias circuit of Figure 6.3 have V_{CC} = 20 V, R_B = 430 kΩ, R_C = 2 kΩ, R_E = 1 kΩ and the transistor has β = 50. Determine the base current, collector current, emitter voltage, base voltage and collector voltage assuming the base-to-emitter voltage 0.7 V.

Solution

From the base circuit of Figure 6.3, $V_{CC} = I_B R_B + V_{BE} + I_E R_E$. Assuming $I_C = \beta I_B$ and $I_E = (\beta+1) I_B$ and putting the given numerical values, the base current is calculated as

$$I_B = \frac{20 - 0.7}{430 \times 10^3 + (50+1) \times 1 \times 10^3} = 40.12 \ \mu A$$

> The collector current (I_C) is 50×40.12×10⁻⁶ = 2.006 mA
>
> The emitter voltage (V_E) is 51×40.12×10⁻⁶×1×10³ = 2.046 V
>
> The base voltage (V_B) is 2.046 + 0.7 = 2.746 V
>
> The collector voltage (V_C) is 20 − 2.006×10⁻³×2×10³ = 15.988 V
>
> [Note that the base voltage is greater than the emitter voltage but less than the collector voltage so that the collector–base junction is reverse-biased and the transistor acts in active region.]

6.5 Collector–Feedback Bias

The circuit diagram is given in Figure 6.4. The base resistor (R_B) is joined to the collector instead of the supply voltage. If somehow I_C increases, the collector-to-emitter voltage (V_{CE}) decreases because of larger voltage drop across R_C. Consequently I_B decreases and I_C tends to reduce to its original value. Similar stability is expected for sudden decrease in I_C.

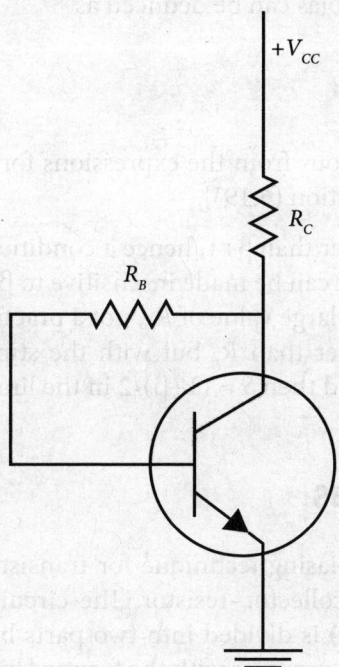

Figure 6.4 Transistor collector–feedback bias circuit

Applying Kirchhoff voltage law around the collector circuit of Figure 6.4,

$$V_{CC} = (I_B + I_C)R_C + I_B R_B + V_{BE} \tag{6.15}$$

so that the base current is

$$I_B = \frac{V_{CC} - I_C R_C - V_{BE}}{R_C + R_B} \tag{6.16}$$

and

$$\frac{dI_B}{dI_C} = -\frac{R_C}{R_C + R_B} \tag{6.17}$$

Because V_{BE} is almost independent of collector current. Putting the value of dI_B/dI_C from Equation (6.17) in Equation (6.5), the stability factor for collector–feedback bias is obtained as

$$S = \frac{1+\beta}{1+\dfrac{\beta R_C}{R_C + R_B}} \tag{6.18}$$

Using Equations (6.4) and (6.15) and assuming $\beta \gg 1$, hence $(\beta + 1) \approx \beta$, the collector current for collector–feedback bias can be deduced as

$$I_C = \frac{V_{CC} - V_{BE} + (R_C + R_B)I_{CO}}{R_C + R_B/\beta} \tag{6.19}$$

The following features are obvious from the expressions for the stability [Equation (6.18)] and the collector current [Equation (6.19)].

(i) The stability is smaller than $\beta+1$, hence a condition better than that of base bias.
(ii) The collector current can be made insensitive to β, if $\beta R_C \gg R_B$. However, such a prerequisite calls for large value of R_C, not a practicable condition.
(iii) If R_C becomes smaller than R_B but with the stipulation of $\beta R_C \geq R_B$, that is a feasible condition and then $S \approx (1+\beta)/2$ in the limiting case.

6.6 Voltage-Divider Bias

This is the most widely used biasing technique for transistor linear amplifier and can be used even in absence of any collector–resistor. The circuit diagram is shown in Figure 6.5(a). The supply voltage (V_{CC}) is divided into two parts by the series combination of R_1 and R_2 and the voltage across R_2 provides with the forward bias to the base–emitter junction of the transistor. Using the Thevenin's theorem (Chapter 7), the circuit can be simplified to the equivalent circuit of Figure 6.5(b) where

Transistor Biasing and Amplification

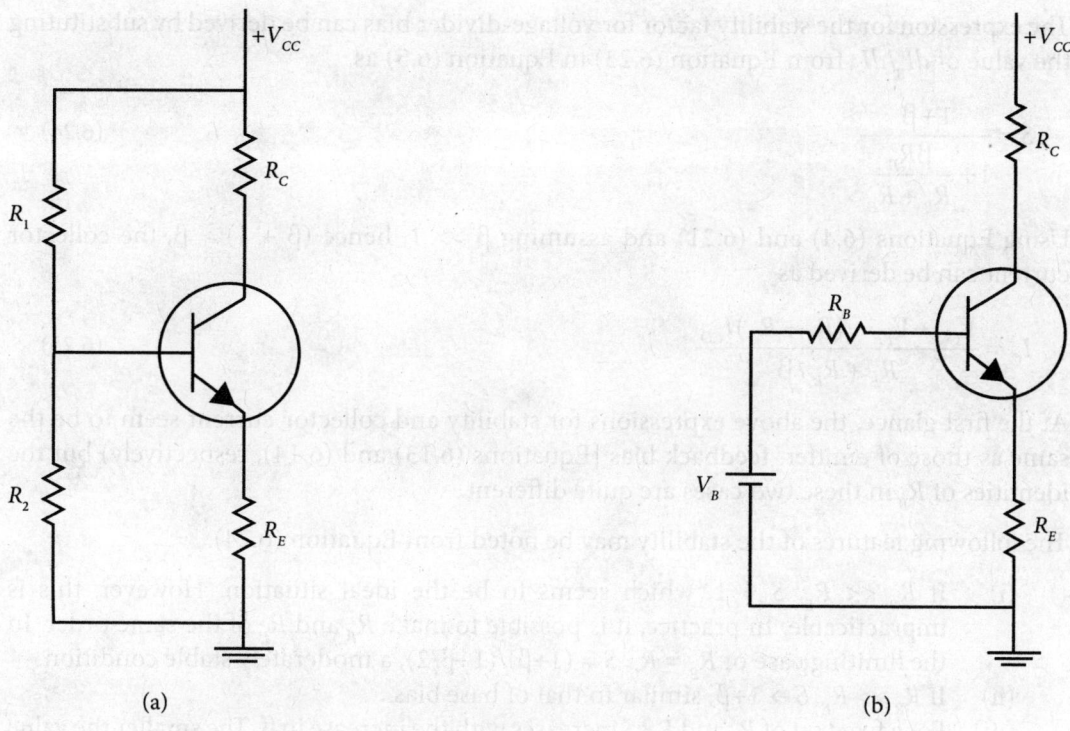

Figure 6.5 Transistor voltage-divider bias (a) actual circuit and (b) equivalent circuit

$$V_B = \frac{R_2 V_{CC}}{R_1 + R_2} \tag{6.20a}$$

and

$$R_B = \frac{R_1 R_2}{R_1 + R_2} \tag{6.20b}$$

From the circuit of Figure 6.5(b),

$$V_B = I_B R_B + V_{BE} + (I_B + I_C) R_E \tag{6.21}$$

so that

$$I_B = \frac{V_B - V_{BE} - I_C R_E}{R_E + R_B} \tag{6.22}$$

and

$$\frac{dI_B}{dI_C} = -\frac{R_E}{R_E + R_B} \tag{6.23}$$

The expression for the stability factor for voltage-divider bias can be derived by substituting the value of dI_B/dI_C from Equation (6.23) in Equation (6.5) as

$$S = \frac{1+\beta}{1 + \dfrac{\beta R_E}{R_B + R_E}} \qquad (6.24)$$

Using Equations (6.4) and (6.21) and assuming $\beta \gg 1$, hence $(\beta + 1) \approx \beta$, the collector current can be derived as

$$I_C = \frac{V_{CC} - V_{BE} + (R_E + R_B)I_{CO}}{R_E + R_B/\beta} \qquad (6.25)$$

At the first glance, the above expressions for stability and collector current seem to be the same as those of emitter–feedback bias [Equations (6.13) and (6.14), respectively] but the identities of R_B in these two cases are quite different.

The following features of the stability may be noted from Equation (6.24).

(i) If $R_B \ll R_E$, $S \approx 1$, which seems to be the ideal situation. However, this is impracticable. In practice, it is possible to make R_B and R_E of the same order. In the limiting case of $R_B = R_E$, $S = (1+\beta)/(1+\beta/2)$, a moderately stable condition.
(ii) If $R_B \gg R_E$, $S \to 1+\beta$, similar to that of base bias.
(iii) For a fixed set of R_B and R_E, S increases with the increase in β. The smaller the value of R_B, the better is the stabilization. However, that cannot be made indefinitely because that would drive the transistor into saturation.
(iv) The condition of R_B and R_E of the same order and $R_B > R_E$ is the best compromise.

Note: Self Bias

The voltage-divider bias is also known as *self bias* or *emitter bias* because the variation of the input signal itself, applied between the base and the ground, causes a proportional change in the emitter bias voltage. The combination of the emitter biasing resistor and the voltage-divider resistors build-up a convenient voltage at the base and at the emitter, which work even in absence of any dc resistance in series with the collector. The transformer-coupled amplifier of Chapter 8 is one such example. The base current and the stability factor can be specified independently.

Example 6.3

Let the voltage-divider bias circuit of Figure 6.5(a) have the following parameters: V_{CC} = 12 V, R_C = 2.2 kΩ, R_E = 1 kΩ, R_1 = 82 kΩ, R_2 = 22 kΩ and the β of the transistor is 100. Determine the Q-point of the circuit.

Solution

The Q-point is the combination of I_C and V_{CE}. Using Equation (6.20a), the base voltage is calculated as

$$V_B = \frac{22 \times 12 \times 10^3}{(82+22) \times 10^3} = 2.538 \text{ V, approximated as } 2.54 \text{ V}$$

The resistance appearing at the base is, using Equation (6.20b), $R_B = (82 \times 22)/(82+22) = 17.35$ kΩ

Using Equation (6.21), the base current (I_B) is calculated as $\dfrac{2.54 - 0.7}{(17.35 + 101 \times 1) \times 10^3} = 15.55$ μA (approx.)

The collector current (I_C) is $100 \times 15.55 \times 10^{-6} = 1.55$ mA

The collector-to-emitter voltage (V_{CE}) is $12 - (1.55 \times 2.2 + 101 \times 15.55 \times 10^{-6} \times 1 \times 10^3) = 7.02$ V

Thus the Q-point is denoted by $I_C = 1.55$ mA and $V_{CE} = 7.02$ V

[In most cases, a good biasing requires the Q-point to be located at the middle point of the load line so as to make $V_{CE} \approx V_{CC}/2$.]

6.7 Load: DC and AC

So far the properties of the bipolar transistor have been cultivated under dc conditions only. The transistor with the above mentioned different types of biasing can act as amplifier for an alternating voltage signal also. The application of the ac input signal in addition to the dc bias causes the Q-point to fluctuate continuously around its fixed position on the load line consistent with the amplitude and frequency of the ac input and a magnified version of this variation is produced at the output. Not only that, the load line itself gets changed depending on whether the input is dc or ac.

The above statement is explained with Figure 6.6. This is actually the base bias circuit of Figure 6.2 with two capacitors joined at the input and the output. A separate load resistance of R_L also has been joined. The circuit acts in different ways with dc and ac, as explained below.

For dc operations, the capacitors act as open circuits, R_L gets detached and the load line is similar to that of Figure 6.1(b).

For ac input, the capacitors act as short circuits and the effect load is the parallel combination of R_C and R_L. These two come in parallel because for ac, both the $+V_{CC}$ and the ground terminals are same. This is a very useful concept and we will use it in many occasions. Since the load resistance changes with ac, the load line also changes for the ac operation. This is exemplified with the following numerical exercise.

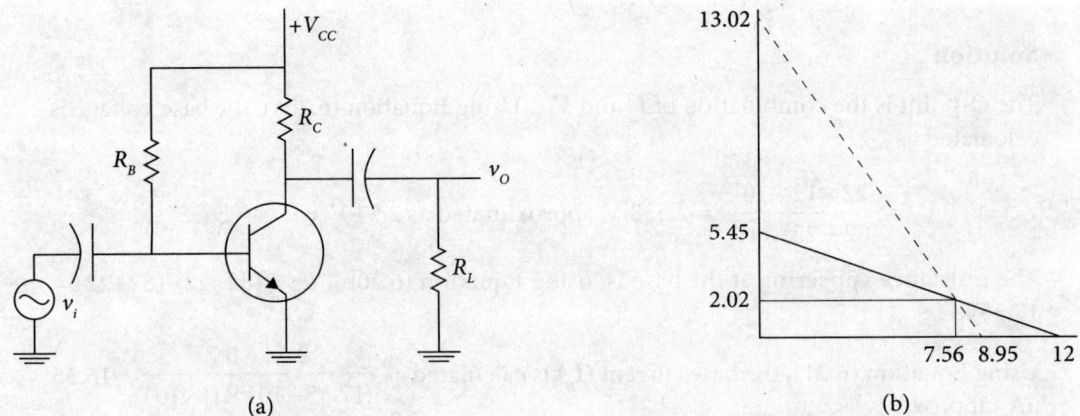

Figure 6.6 (a) Transistor amplifier with base bias for accepting ac input and (b) its dc load line (solid line), ac load line (dashed line), and operating point (thin lines) with numerical values

---

Note: A Brief Recapitulation on Transistor Operation

The bipolar junction transistor itself can act as a current amplifier. In common-emitter connection, a small change in the base (input) current gives rise to a large change in the collector (output) current and the amplification is measured in terms of the current gain (β) [Equation (5.8)]. In the case of common-base connection, the change in input (emitter) current causes a smaller change in output (collector) current quantified in terms of α [Equation (5.4)] and the amplification can be stated to be is less than unity.

The transistor operation is made stable by suitable biasing with passive elements. It fixes the operating point of output current and voltage defined in terms of load line and Q-point.

In presence of a small ac signal in addition to the dc bias, the effective value of the load resistance may change causing change in the load line and the Q-point.

There are broadly two types of transistor circuits: amplifying and switching. For amplifying circuits, the Q-point must remain within the active region under all operating conditions. Otherwise the output signal gets distorted whenever saturation or cutoff occurs. For switching circuits (mainly used in digital applications), the Q-point switches between saturation and cutoff.

It will be discussed in the subsequent section that the amplifiers dealing with ac input signals have both dc and ac equivalent circuits, hence dc and ac load lines. The dc load line contains the Q-point for dc biasing and the ac load line allows the Q-point to swing over with the ac input. For small signals, the difference is not so critical but in the case of large signals, the location of the Q-point location on the ac load line should be considered carefully for allowing maximum swing. More deliberations on large signal operations are presented in Chapter 8.

Example 6.4

Let the transistor circuit of Figure 6.6(a) has the parameters $\beta = 100$, $V_{CC} = 12$ V, $R_C = 2.2$ kΩ, $R_B = 560$ kΩ and $R_L = 1$ kΩ, symbols having the usual meaning. Determine the dc and ac load lines.

Solution

The base current is $I_B = \dfrac{V_{CC} - V_{BE}}{R_B} = \dfrac{12 - 0.7}{560 \times 10^2} = 20.2$ µA

The collector current is $I_C = \beta I_B = 100 \times 20.2 \times 10^{-6} = 2.02$ mA

The collector-to-emitter voltage is $V_{CE} = V_{CC} - I_C R_C = 7.56$ V

Thus the Q-point with the load of 2.2 kΩ is the intersection of $I_C = 2.02$ mA and $V_{CE} = 7.56$ V on the load line, which cuts the voltage-axis at $V_{CC} = 12$ V and the current-axis at $V_{CC}/R_C = 5.45$ mA. All these are indicated (not drawn up to scale) in Figure 6.6(b).

Under the ac condition, the effective load becomes $R_C // R_L = 0.687$ kΩ. Any way the load line must pass through the Q-point calculated above. So the equation for the ac load line is

$$I_C - 2.02 \times 10^{-3} = -\dfrac{V_{CE} - 7.56}{0.687 \times 10^3}$$

Putting $I_C = 0$, $V_{CE} = 8.95$ V and putting $V_{CE} = 0$, $I_C = 13.02$ mA. The ac load line with these intercepts is sketched (not up to scale) in Figure 6.6(b).

6.8 BJT Small Signal Voltage Amplifiers

The previous section has introduced how a properly biased transistor circuit accepts the input of alternating voltage for amplification and changes the load line accordingly keeping the dc position of the Q-point unchanged. The Q-point swings on the ac load line in accordance with the amplitude and frequency of the input signal. This fluctuation should not reach either the saturation end or the cutoff for the purpose of linear amplification. Several more features are to be considered, as follows.

- The amplitude of the ac input voltage must be very small, well below V_{BE} (0.7 V for silicon transistors) of the transistor.
- The frequency of the input should not be too high ($\approx$ MHz or more) lest the effects of the junction capacitances become significant.
- Though the transistor has nonlinear current–voltage relationship, linearity may be assumed in the transfer characteristic [e.g. Figure 5.4(d)] at the active region for small input signals.
- For small signals, the dc and the ac analysis of the transistor circuit can be treated as linear combinations and linear theories, such as the superposition theorem, Thevenin theorem and Norton theorem can be utilized (Chapter 7).

One of the important features of an electronic device, as introduced in Chapter 1, is the capability of amplifying both voltage and current simultaneously. These amplifying properties are defined in terms of voltage gain and current gain mentioned below. Another salient characteristic of an electronic device is the ability of transforming impedance within the device itself. So the input and the output impedances are defined separately. Thus a transistor amplifier has the following four characteristic parameters for handling voltage and current.

- A. *Current Gain*: This is the most fundamental quantity of a transistor defined in terms of the ratio of the output current to the input current. The dc values of β and α are parameters of this category but a more generalized form of current gain is the ratio of the rate of change in output current to the corresponding rate of change in the input current, as derived in the following sections.
- B. *Voltage Gain*: This is defined as the ratio of the ac output voltage to the ac input voltage and is obtained in terms of the voltage drop across the output resistance and its change.
- C. *Input Impedance*: This is the impedance obtained on looking into the input terminals of the transistor and is defined by the ratio of the input voltage to the input current. In the case of low frequency amplifiers, as in the present cases, the impedance is approximated by resistance.
- D. *Output Impedance*: This is the impedance appearing between the output terminals and can be approximated by resistance in the case of low frequency amplifiers.

6.8.1 Common-Emitter (CE) Amplifier

The input signal causes fluctuation of the dc Q-point, hence variability in the collector current. This is much larger than the input variation of the base current. Therefore, the ac gain is high, which leads to current amplification. This amplified current increases the voltage drop across the collector–resistor, hence a voltage amplification is achieved. For the positive half of the input cycle, the base current increases causing an increase of the collector current. Consequently, the voltage drop across the collector–resistor increases causing a decrease in the collector voltage. For the negative half of the input cycle, the opposite sequence takes place and the collector voltage increases. Thus the CE amplifier output at the collector terminal produces a magnified replica of the input signal but with inverted phase.

Figure 6.7(a) represents the circuit diagram of a small signal, moderate frequency CE amplifier with voltage-divider biasing formed with resistors R_1 and R_2. A source of alternating input voltage (v_i) signal is connected between the base and the ground through a capacitor (C_B), termed as *blocking capacitor*, which blocks the dc bias voltage from reaching the ac signal source and at the same time acts as short circuit to the ac input signal. The dc equivalent circuit of this amplifier is similar to the voltage-divider bias circuit of Figure 6.5(a) where all capacitors act as open circuits.

The amplified output voltage (v_o) appears across the collector and the ground and the collector–resistor (R_C) acts as the load. It is so because both the positive V_{CC} and the ground terminal look the same to the ac input as it changes polarity continuously. Therefore, R_1 and R_2 appear in parallel and R_C appears between the collector and the ground in the ac equivalent circuit of the amplifier, as sketched in Figure 6.7(b). For the same reason, if any additional load resistance (R_L) is joined as shown by dotted line, it would come in parallel with R_C [Figure 6.7(b)].

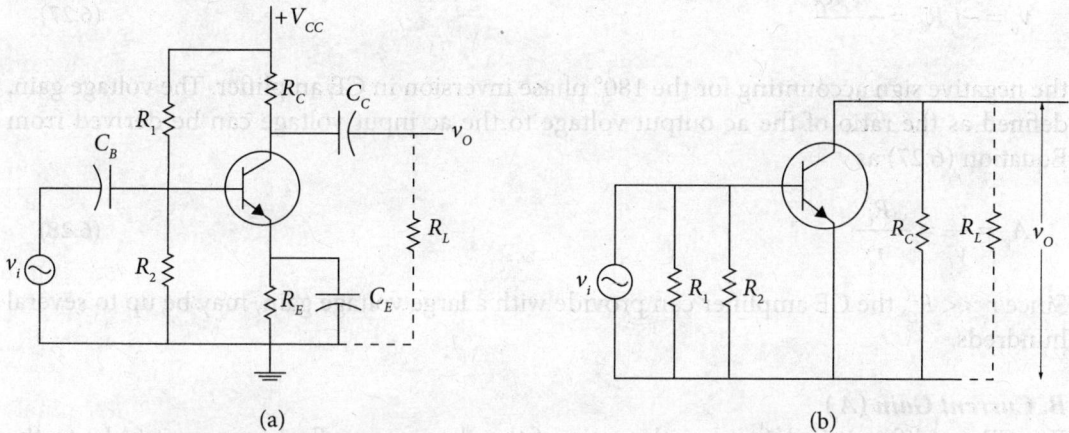

Figure 6.7 BJT small signal common-emitter (CE) amplifier: (a) circuit diagram and (b) AC equivalent circuit

There are two more capacitors in the circuit. The capacitor (C_C) in the output circuit, called the *coupling capacitor*, couples only the amplified ac part of the signal to the output terminals and blocks the dc part. The capacitor (C_E) in parallel with the emitter resistor is named the *emitter bypass capacitor*. It acts as an ac-short for the emitter terminal to the ground and at the same time preserves the dc bias across R_E. This way it saves for the loss due to the signal voltage drop across R_E by providing with a 'bypass path' for the ac only.

Thus the amplifier has a dc equivalent circuit taking care of the biasing voltages and an ac equivalent circuit looking after the amplifying properties. The gain and impedance parameters for the CE amplifier are derived as follows.

> **Think a Bit: What about the Input and the Output Phase in CE Amplifier?**
>
> The assumption of small input ac signal allows superposition, hence linear combination of dc and ac. The total current and voltage in any branch of the circuit is the algebraic sum of the direct current and voltage and the alternating current and voltage through that branch. As the input ac signal voltage increases during the positive half cycle, the collector current also increases so that the voltage drop across the collector–resistor (R_C) increases. Consequently the collector voltage decreases. The opposite phenomenon takes place when the input current decreases. The same is repeated for the negative half cycle with the opposite polarity. As a whole, a phase inversion of 180° takes place between the ac input and the amplified ac output in CE amplifier.

A. Voltage Gain (A_v)
The emitter current at ac condition is given by

$$i_e = \frac{v_i}{r_e} \qquad (6.26)$$

where $r_e = (26 \text{ mV})/(I_C \text{ in mA})$ is the ac emitter resistance offered by the bypassed emitter. The output voltage is

$$v_o = -i_e R_C = -\frac{v_i R_C}{r_e} \qquad (6.27)$$

the negative sign accounting for the 180° phase inversion in CE amplifier. The voltage gain, defined as the ratio of the ac output voltage to the ac input voltage can be derived from Equation (6.27) as

$$A_v = \frac{v_o}{v_i} = -\frac{R_C}{r_e} \qquad (6.28)$$

Since $r_e \ll R_C$, the CE amplifier can provide with a large voltage gain, may be up to several hundreds.

B. Current Gain (A_i)
For CE amplifier, it is defined as the ratio of the change in collector current (ΔI_C) to the corresponding change in base current (ΔI_B) and is expressed as

$$A_i = \frac{\Delta I_C}{\Delta I_B} \qquad (6.29)$$

The current gain of the transistor given by Equation (6.29) may be compared with its dc ratio (β) given by Equation (5.8). The present one is sometimes called 'ac beta' and that of Equation (5.8) is 'dc beta'. In practice both are of the same order of numerical values.

It is important to note that the current gain of a transistor is the upper limit for the voltage amplification with it. Equations (6.28) and (6.29) together indicate that the CE amplifier can produce both high voltage gain and high current gain, hence high power gain. More discussion on power amplification is presented in Chapter 8.

C. Input Impedance (z_i)
The impedance appearing between the input terminals of the amplifier is defined as

$$z_i = \frac{v_i}{i_b} = \frac{i_e r_e}{i_b} = \beta r_e \text{ (assuming } i_e \approx i_c\text{)} \qquad (6.30)$$

In our present context of small signal, moderate frequency amplification, the impedance can be stated simply as *input resistance*. On looking into the base–emitter terminals, the external biasing resistors R_1 and R_2 come in parallel with this input resistance for ac inputs. However, these resistances are made much larger than βr_e so that these are ignored in parallel connection.

> **Note: Expression for AC Emitter Resistance**
>
> The current through a forward biased p–n junction is (Chapter 3) expressed as $I = I_o \left[\exp\left(\frac{qV}{kT}\right) - 1 \right]$ so that the ac resistance of the junction is obtained as $r = \frac{dV}{dI} = \frac{kT}{q(I+I_o)} \approx \frac{kT}{qI}$, since $I \gg I_o$.
>
> For the forward biased emitter–base junction of the transistor, $r = r_e$ and $I = I_E \approx I_C$ so that $r_e = \frac{kT}{qI_C}$.
>
> Putting the numerical values of Boltzmann constant (k), electron charge (q) and assuming room temperature (T), the expression comes out to be approximately $r_e = \frac{26 \text{mV}}{I_C}$, where I_C is in mA.

D. Output Impedance (z_o)

The output impedance of the transistor between the collector and the ac ground is very high because the collector–base junction is reverse-biased. Therefore the resistance of the collector-resistor (R_C) parallel to the output terminals acts as the effective output impedance. If any additional load resistor (R_L) is joined between the collector and the ground, the effective output impedance becomes

$$z_o = R_C // R_L \tag{6.31}$$

The output impedance may also be expressed as $\Delta V_{CE}/\Delta I_C$, if the results are available.

> **Note: Concept of Power Gain**
>
> The power gain of a transistor, in general, is the ratio of the output power to the input power. Since the emitter is part of the input in a CE amplifier, the power gain can be expressed as $\frac{(\Delta I_C)^2 R_L}{(\Delta I_E)^2 r_e} = \frac{\Delta I_C}{\Delta I_E} \frac{\Delta I_C R_L}{\Delta I_E r_e}$
>
> where ΔI_E represents small change in input (emitter) current and ΔI_C the corresponding change in the output (collector) current. The load resistance between the collector and the ac ground is R_L and the dynamic resistance of the base–emitter junction is r_e.
>
> The current gain is $A_i = \frac{\Delta I_C}{\Delta I_E}$ and the voltage gain is $A_v = \frac{\Delta I_C R_L}{\Delta I_E r_e} = \frac{\Delta v_o}{\Delta v_i}$. Thus the power gain is the product ($A_v A_i$) of the voltage gain and the current gain. Another feature is revealed here. The advantage of large I_C/I_B ratio is not received directly. Also, if we assume $\Delta I_C \approx \Delta I_E$, power gain $\approx A_v$, which does not exceed β. Thus the power gain of a CE amplifier does not seem to surpass the numerical limit prescribed by β of the transistor. In an actual power amplifier (Chapter 8) the output power is increased by other techniques, such as changing the effective value of impedance and using more than one amplifier stage.

6.8.2 Common-Collector (CC) Amplifier

Figure 6.8 represents the circuit diagram of a transistor CC amplifier. This is similar to the CC configuration of Figure 5.6(a) with the addition of the ac input voltage (v_i). This, after amplification through the amplifier, appears as output voltage (v_o) across the un-bypassed emitter resistor (R_E). As explained earlier, the $+V_{CC}$ terminal of the power supply acts as the ground for the ac input and bring the collector terminal in common with respect to the input and the output circuits.

The circuit has some similarity with the voltage-divider biasing of CE amplifier [Figure 6.7(a)] but there are the following important differences.

- There is no resistor in series with the collector,
- the collector is directly connected with the positive terminal of the power supply,
- the emitter resistor is not capacitor bypassed, and
- instead of the collector, the output is taken at the un-bypassed emitter.

The common-collector amplifier is also known as the *emitter follower* because the ac output voltage across the emitter resistor is in phase with the input voltage and is almost equal to it. As if the emitter voltage follows the input voltage. The followings are its gain and impedance properties.

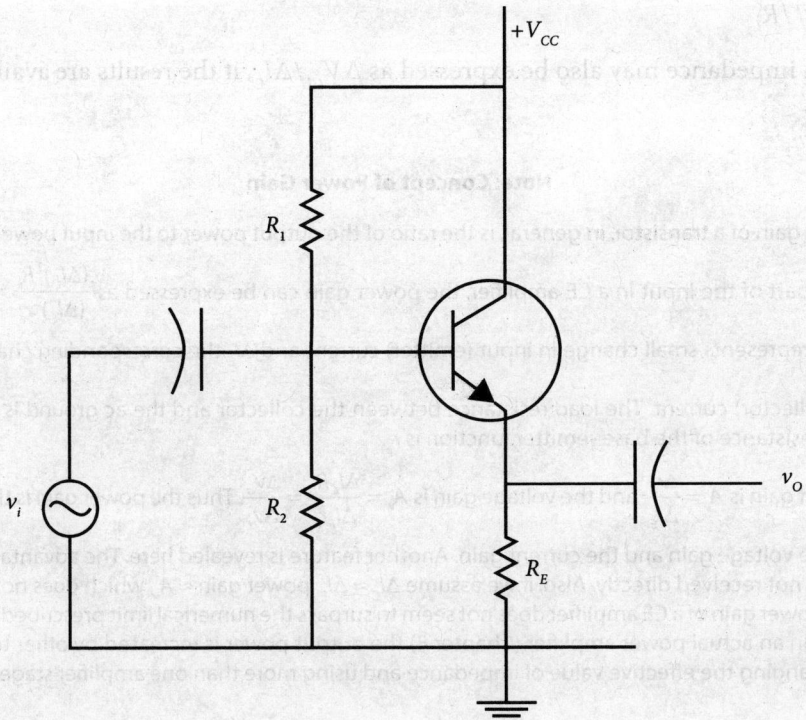

Figure 6.8 BJT common-collector (CC) amplifier

A. Voltage Gain (A_v)

The ac output voltage of the CC amplifier in Figure 6.8 is

$$v_o = i_e R_E \tag{6.32}$$

and the ac input voltage is

$$v_i = i_e R_E + \beta r_e i_b = i_e (R_E + r_e) \tag{6.33}$$

so that the voltage gain is

$$A_v = \frac{v_o}{v_i} = \frac{R_E}{R_E + r_e} \tag{6.34}$$

Since $R_E \gg r_e$, Equation (6.34) indicates that the voltage gain of CC amplifier is almost unity, unlike that of the CE amplifier.

B. Current Gain (A_i)

The emitter current (i_e) is the ac output current and the base current (i_b) is the ac input current so that the current gain is

$$A_i = \frac{i_e}{i_b} \approx \beta \tag{6.35}$$

Thus the CC amplifier has the same high current gain as that of CE amplifier.

C. Input Impedance (z_i)

It is defined as the ratio of the ac input voltage to the ac input current so that

$$z_i = \frac{v_i}{i_b} = \frac{i_e(R_E + r_e)}{i_b} \approx \beta R_E \tag{6.36}$$

because $i_e \approx i_c = \beta i_b$ and $R_E \gg r_e$. The quantity βR_E is numerically large so that we conclude that the input resistance of CC amplifier is high.

D. Output Impedance (z_o)

Looking into the output port of the CC amplifier between the emitter and the ground, the output impedance is obtained as

$$z_o = r_e + \frac{R_1 // R_2}{\beta} \tag{6.37}$$

where r_e is the ac resistance due to the emitter–base junction and the parallel combination ($R_1//R_2$) divided by the current gain (β) accounts for the fact that the input (base) current is almost $1/\beta$ times the output (emitter) current. This second term in Equation (6.37) is practically negligible and the effective output resistance is just r_e. Thus we conclude that the output impedance of CC amplifier is very small.

> **Think a Bit: CC Amplifier Can Be Used as Buffer**
>
> If an amplifier has low input impedance (e.g. transistor CB amplifier) and a signal source is connected to its input like the diagram given below, the amplifier itself acts as short circuit and the output, instead of being amplified, gets attenuated. This undesired effect is called *loading*.
>
>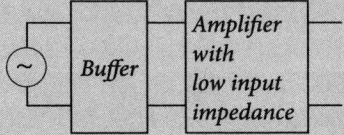
>
> To avoid this, a CC amplifier can be joined as a *buffer* or shield between the amplifier and the source. By virtue of high input impedance, the buffer brings the source voltage to the amplifier input without attenuation and at the same time the low output impedance of the buffer matches the low input impedance of the amplifier.

6.8.3 Common-Base (CB) Amplifier

The circuit diagram is given in Figure 6.9. It uses a negative supply voltage $(-V_{EE})$ at the emitter terminal through the emitter resistor (R_E). The base is grounded and the ac input signal is applied between the emitter and the ground. The output is obtained across the collector-resistor (R_C). It has the following gain and impedance parameters.

A. Voltage Gain (A_v)
The voltage gain of CB amplifier can be expressed as

$$A_v = \frac{v_o}{v_i} = \frac{i_c R_C}{i_e r_e} \tag{6.38}$$

Assuming $i_e \approx i_c$ in Equation (6.38), A_v can also be simplified as R_C/r_e. Any way it is clear that the CB amplifier has high voltage gain similar to that of CE amplifier. However, unlike the CE amplifier, it has output in the same phase with that of the input.

B. Current Gain (A_i)
The collector current (i_c) is the ac output current and the emitter current (i_e) is the ac input current. Thus the current gain can be expressed as

$$A_i = \frac{i_c}{i_e} \tag{6.39}$$

which is almost equal to unity.

C. Input Impedance (z_i)
It is obvious from the circuit of Figure 6.8 that the input impedance of CB amplifier is

$$z_i \approx r_e \tag{6.40}$$

because the base is grounded and the input voltage is applied to the emitter. Thus the CB amplifier has very low input impedance, which is indeed a weak point of this circuit.

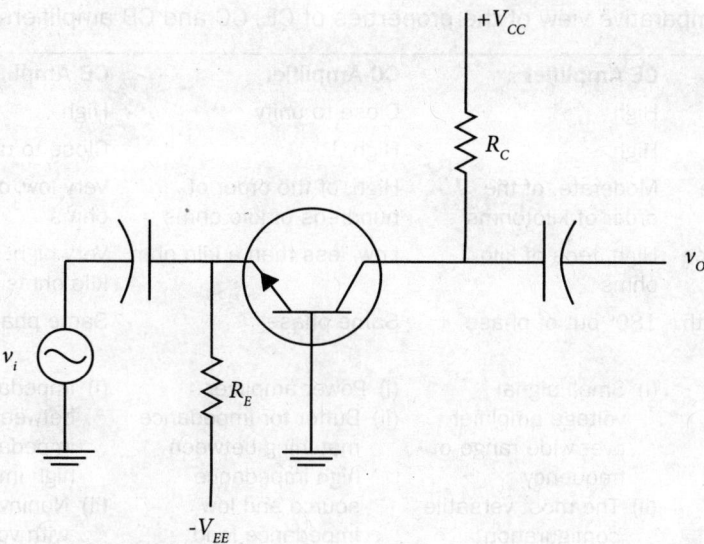

Figure 6.9 BJT common-base (CB) amplifier

D. Output Impedance (z_o)

The output impedance of the transistor is very high because the collector–base junction is at reverse bias. The resistance of the collector-resistor (R_C) joined in parallel acts as the effective output resistance so that

$$z_o \approx R_C \tag{6.41}$$

Example 6.5

A CB amplifier with n–p–n transistor has load resistor of 1 kΩ. If the current gain be 0.97 and the input resistance be 194 Ω, calculate the voltage gain and the power gain of the amplifier.

Solution

Using Equations (6.38), (6.39) and (6.40), the voltage gain is $A_v = 0.97 \times 1 \text{ k}\Omega / 194 \text{ }\Omega = 5$

The power gain ($A_v A_i$) is $5 \times 0.97 = 4.85$

Section 6.8 above present in detail the individual properties of transistor voltage amplifiers fabricated in common-emitter (CE), common-collector (CC) and common-base (CB) mode. It has been illustrated how the gain and impedance features change abruptly with the change in biasing configuration of the same transistor. The main characteristics of these three categories are summarized in Table 6.1 below.

Table 6.1 A comparative view of the properties of CE, CC and CB amplifiers

Parameter	CE Amplifier	CC Amplifier	CB Amplifier
Voltage gain	High	Close to unity	High
Current gain	High	High	Close to unity
Input impedance	Moderate, of the order of kilo ohms	High, of the order of hundreds of kilo ohms	Very low, of the order of ohms
Output impedance	High, tens of kilo ohms	Low, less than a kilo ohm	Very high, hundreds of kilo ohms
Output phase with respect to input	180° out of phase	Same phase	Same phase
Main use	(i) Small signal voltage amplifier over wide range of frequency (ii) The most versatile configuration among the three	(i) Power amplifier (ii) Buffer for impedance matching between high impedance source and low impedance load (iii) Enhancing the input impedance of circuits like operational amplifier	(i) Impedance matching between low impedance source and high impedance load (ii) Noninverting amplifier with voltage gain greater than unity (iii) Constant current source

Multiple Choice-type Questions and Answers

6.1 The operating point is to be established within the .... region when the transistor functions most ....
 (a) active, linearly (b) active, nonlinearly
 (c) cutoff, linearly (d) saturation, linearly

6.2 The load line shifts parallel to itself when the .... is changed.
 (a) emitter resistance (b) collector resistance
 (c) supply voltage (d) load resistance

6.3 The voltage drop across the emitter resistor .... biases the base–emitter junction in .... mode.
 (a) forward, active (b) reverse, active
 (c) forward, saturation (d) forward, cutoff

6.4 The choice of Q-point depends on – Statement-I: The supply voltage, Statement-II: The load resistance, Statement-III: The input signal amplitude. The correct statements are:
 (a) I and II (b) I and III
 (c) II and III (d) all the above

6.5 The stability of the collector current may be disturbed due to – Statement-I: Transistor replacement, Statement-II: change of reverse saturation current with temperature, Statement-III: Statement-III: change of base–emitter voltage due to temperature. The correct statement(s) is(are):
 (a) only II (b) I and II
 (c) all the above (d) II and III

6.6 The .... is the stability factor, the .... is the stability of transistor operation within certain limits.
   (a) smaller, worse
   (b) smaller, better
   (c) larger, better
   (d) larger, worse

6.7 The load line of a transistor in CE configuration has .... slope and it relates graphically ...., symbols having usual meaning.
   (a) positive, I_C and V_{BE}
   (b) negative, I_C and V_{BE}
   (c) negative, I_C and V_{CE}
   (d) negative, I_C and V_{CC}

6.8 The input resistance is maximum in .... amplifier whereas the output resistance is maximum in .... amplifier.
   (a) CC, CB
   (b) CE, CB
   (c) CE, CC
   (d) CB, CE

6.9 The output resistance is minimum in .... amplifier whereas the input resistance is minimum in .... amplifier.
   (a) CB, CE
   (b) CC, CB
   (c) CC, CE
   (d) CE, CC

6.10 A CE amplifier operates at I_C = 1.3 mA and the collector–resistor is 1 kΩ. The ac emitter resistance is ....
   (a) 10 kΩ
   (b) 20 kΩ
   (c) 20 Ω
   (d) cannot be determined because data are insufficient

6.11 An emitter follower can be used as
   (a) voltage amplifier
   (b) current amplifier
   (c) power amplifier
   (d) impedance matcher

6.12 The largest power gain is obtainable with .... amplifier.
   (a) CB
   (b) CC
   (c) CE
   (d) voltage

6.13 If a CE amplifier has β = 100 and I_C = 2.6 mA, the input resistance of the amplifier is about ....
   (a) 10 kΩ
   (b) 1 kΩ
   (c) 2.6 kΩ
   (d) 260 Ω

6.14 The load line is almost horizontal when load resistance is .... and is almost vertical when load resistance is ....
   (a) of the order of kΩ, of the order of kΩ
   (b) of the order of Ω, of the order of kΩ
   (c) very large, very small
   (d) very small, very large

6.15 The voltage-divider bias gives better stability for .... values of emitter resistance and .... values of current gain.

6.16 The dc and ac load lines .... intersect at the Q-point in absence of ac input.
   (a) sometimes
   (b) always
   (c) never
   (d) occasionally

6.17 The current axis intercept and the voltage axis intercept of a load line corresponds to .... and ...., respectively.
   (a) cutoff, saturation
   (b) saturation, active
   (c) saturation, cutoff
   (d) active, cutoff

6.18 The following two statements are under consideration.

Statement-I: The ac load line is steeper than the dc load line.
Statement-II: The ac collector and emitter resistances are lower than the corresponding dc resistances.
(a) Both statements are correct and *Statement-II* is the proper reason for *Statement-I*.
(b) Both statements are correct but *Statement-II* is not the reason for *Statement-I*.
(c) *Statement-I* is not correct but *Statement-II* is correct.
(d) *Statement-I* is correct but *Statement-II* is incorrect.

6.19 Consider the following two statements.

Statement-I: The CE amplifier is not suitable for small load resistances.
Statement-II: It is difficult to obtain a small output resistance with the CE amplifier.
(a) Both statements are correct and *Statement-II* is a valid reason for *Statement-I*.
(b) Both statements are correct but not related any way.
(c) *Statement-I* is correct but *Statement-II* is incorrect.
(d) *Statement-II* is correct but *Statement-I* is incorrect.

6.20 An emitter follower circuit has output voltage taken across the emitter resistor of 1 kΩ and the ground. If the emitter current is 1.03 mA, the total ac emitter resistance of the circuit is .... approximately.
(a) 25 Ω (b) 1 kΩ (c) 1025 kΩ (d) 1025 Ω

Hints. Assuming $I_C \approx I_E$, the total ac emitter resistance including the unbypassed emitter resistor is 26/1.03 + 1000 = 1025 Ω approximately.

Answers

6.1	(a)	6.2	(c)	6.3	(b)	6.4	(d)
6.5	(c)	6.6	(b) & (d)	6.7	(c)	6.8	(a)
6.9	(b)	6.10	(c)	6.11	(b) & (d)	6.12	(c)
6.13	(b)	6.14	(c)	6.15	(d)	6.16	(b)
6.17	(c)	6.18	(a)	6.19	(a)	6.20	(d)

Reasoning-Type Questions and Answers

6.1 We know that transistor biasing is the process of maintaining proper voltage levels at the three terminals of a transistor but that could be achieved with two good voltage sources or two sets of battery. Why are the external resistors, capacitors, etc. essential?

Ans. First of all, a single voltage source can serve the purpose with the use of external passive components. The main objective of biasing is to achieve stable operation. The biasing network can provide with suitable conditions for less fluctuation of the output current with temperature and other parameters. Moreover, the biasing can establish a suitable operating point and allows the transistor to act as voltage amplifier, current amplifier, power amplifier and on/off switch.

6.2 What are the criteria of positioning the Q-point on the load line?

Ans. There are several factors in different contexts that are to be taken into consideration while assigning the operating point of a transistor. For instance, voltage amplification over a wide range of frequency needs the Q-point to be fixed at the middle of the load line. The switching operation fixes the Q-point at the upper (saturation) or lower (cutoff) end of the load line. The ratings of the transistor, such as the maximum allowable power dissipation must not be exceeded. The dc and ac loads, as applicable, should be considered in fixing the operating point.

6.3 It is noted that the collector current has increased whereas the biasing components and the ambient temperature are fixed. What might be the reasons?

Ans. The device temperature might have increased during prolonged operation thereby increasing the reverse current. The transistor replaced by a similar one with larger β may cause the same result.

6.4 Is collector–feedback bias suitable for transistor circuits with small values of load resistance?

Ans. No, in that case the stability factor becomes almost (1+β) [The parallel combination of load and R_C tends to zero in Equation (6.18)], similar to that of base bias.

6.5 This chapter has illustrated the stability in different biasing modes but all in CE configuration. What about CB configuration?

Ans. The requirement of bias stability is most prominent in CE configuration because of the fluctuation of β due to various reasons. In the case of CB configuration, the output (collector) current is $I_C = \alpha I_E + I_{CO}$ so that $\partial I_C / \partial I_{CO}$ is almost unity. Thus the CB configuration does not face the problem of bias stabilization.

Solved Numerical Problems

6.1 The transistor in Figure 6.10 has β = 100. Assuming V_{BE} = 0.7 V, V_{CEsat} = 0.2 V, determine whether the transistor is working in cutoff, saturation or active mode. Find the value of the output voltage (V_o).

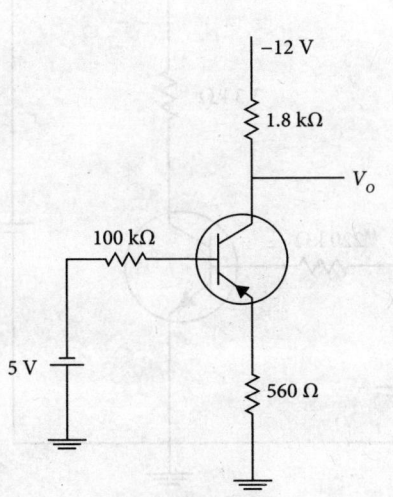

Figure 6.10

Soln. [The transistor is p–n–p, hence the polarities of the supply voltages are opposite to that of n–p–n transistor. However, for being in active region, the magnitude of the collector voltage should be greater than the magnitude of the base voltage in this case also.]

Assuming the transistor in active region and considering only the magnitude of the voltage, the base current is calculated as

$$I_B = \frac{5 - 0.7}{100 \times 10^3 + 101 \times 560} = 27.46 \text{ μA}$$

The collector current (I_C) is $100 \times 27.46 \times 10^{-6} = 2.75$ mA (approx.)

The base voltage (V_B) is $101 \times 27.46 \times 10^{-6} \times 560 + 0.7 = 2.25$ V

The collector voltage (V_C) is $12 - 1.8 \times 10^3 \times 2.75 \times 10^{-3} = 7.05$ V

Since $|V_C| > |V_B|$, the transistor is working in active region and the output voltage (V_o) is −7.05 V (the collector voltage itself with proper sign).

6.2 Calculate the values of I_B, I_C and I_E in the circuit of Figure 6.11 assuming $\beta = 100$, $I_{CO} = 20$ nA and $V_{BE} = 0.7$ V, symbols having the usual meaning. Repeat the calculation, if a resistor of 2.2 kΩ is added in between the emitter and the ground. What are your comments about this change?

Soln. Applying the Kirchhoff's voltage law to the base circuit, the base current is calculated as

$$I_B = \frac{5 - 0.7}{220 \times 10^3} = 19.54 \text{ μA}$$

The collector current is calculated [using Equation (6.4)] as

$$I_C = 100 \times 19.54 \times 10^{-6} + 101 \times 20 \times 10^{-9} = 1.96 \text{ mA (approx.)}$$

The emitter current is (in magnitude) $I_E = I_C + I_B = 1.98$ (approx.)

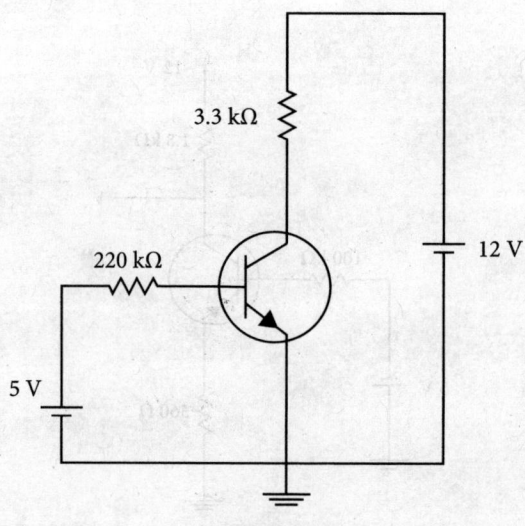

Figure 6.11

In presence of the resistor of 2.2 kΩ in the emitter circuit, the base voltage can be expressed as

$$V_B = I_B R_B + V_{BE} + (\beta+1)I_B R_E + (\beta+1)R_E I_{CO}$$

so that putting the given numerical values, the base current comes out as 9.71 µA and consequently the collector current is calculated as $100 \times 9.71 \times 10^{-6} + 101 \times 20 \times 10^{-9} = 0.973$ mA. Therefore, the emitter current becomes

0.973 mA + 9.71 µA = 0.98 mA

The currents, particularly the output currents have reduced in presence of the emitter resistor. However, this imparts more gain stability to the circuit (emitter–feedback bias, section 6.4).

6.3 Determine whether the transistor in the circuit of Figure 6.12 is working in cutoff, saturation or active mode. If not active, then estimate the minimum value of R_E for which the transistor remains in active region. Given $\beta = 100$, $V_{BEsat} = 0.8$ V and $V_{CEsat} = 0.2$ V, symbols having the usual meaning.

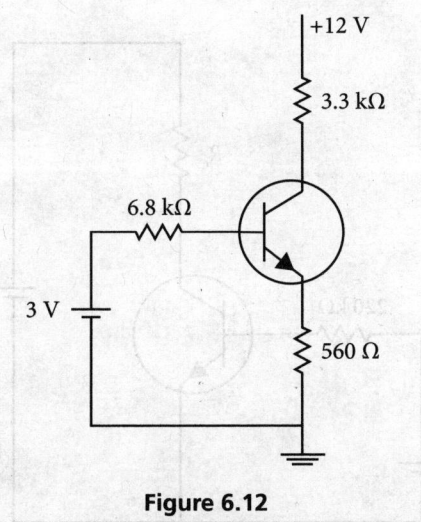

Figure 6.12

Soln. Let us assume that the transistor operates in active region, $I_E \approx I_C$ and $I_C \approx \beta I_B$, symbols having the usual meaning. Though the given value of base-to-emitter voltage is specified for saturation, it is assumed to be applicable to the limiting case of active region. In that case, applying the Kirchhoff's voltage law to the base–emitter circuit,

$$6.8 \times 10^3 \times I_B + 0.8 + 560 \times 100 \times I_B = 3$$

so that the base current $(I_B) = 35.0$ µA

and the collector current (I_C) is 100×35.03 µA = 3.50 mA.

In the collector–emitter circuit,

$$(3.3 \times 10^3 + 560) \times 3.5 \times 10^{-3} + V_{CE} = 12$$

so that $V_{CE} = -1.51$ V

Thus, for the given values, the collector-to-emitter voltage comes out to be negative, which is not possible in active region. So the first conclusion is that the transistor is not in active region.

The base voltage is

$$V_B = 3 - 35.03 \times 10^{-6} \times 6.8 \times 10^3 = 2.76 \text{ V}$$

The collector voltage is

$$V_C = 10 - 3.50 \times 10^{-3} \times 3.3 \times 10^3 = -1.55 \text{ V}$$

Since the base voltage is greater than the collector voltage, the transistor is in saturation.

The limiting condition for remaining in active region is that the base voltage should be equal to the collector voltage for a given minimum value of resistance of the emitter resistor (R_E) when

$$V_{BB} - R_B I_B = V_{CE} + \beta I_B R_E$$

Using the given value of 0.2 V for the collector-to-emitter voltage at saturation and putting the other numerical values, R_E results in 731 Ω, which is the minimum resistance to be maintained in the emitter circuit to keep the transistor in active mode.

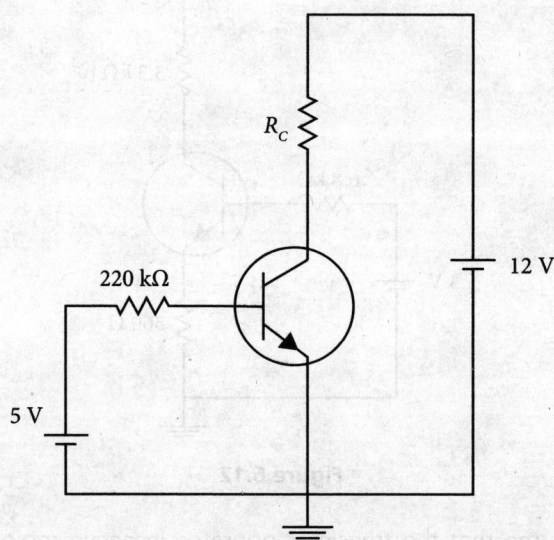

Figure 6.13

6.4 The transistor in Figure 6.13 has $\beta = 100$, $V_{BEsat} = 0.8$ V and $V_{CEsat} = 0.2$ V. Find the minimum value of R_C for which the transistor remains in saturation.

Soln. In the base–emitter circuit,

$$220 \times 10^3 \times I_B + 0.8 = 5$$

so that the base current (I_B) is 19.09 µA and
the collector current (I_C) is $100 \times 19.09 \times 10^{-6} = 1.91$ mA (approx.)

In the collector–emitter circuit,
$$1.91 \times 10^{-3} \times R_C = 12 - 0.2$$
so that $R_C = 6.18$ kΩ (approx.)

6.5 Determine the voltage gain for the CE amplifier of Figure 6.7(a), if $R_1 = 10$ kΩ, $R_2 = 2.2$ kΩ, $R_C = 4.7$ kΩ, $R_E = 1$ kΩ and the capacitor values are chosen in such a way that they serve their purpose well.

Soln. The base voltage [using Equation (6.20a)] is $(10 \times 2.2)/(10 + 2.2) = 1.8$ V.

Therefore, the emitter voltage is $1.8 - 0.7 = 1.1$ V.

The emitter current is $(1.1\ \text{V})/(1\ \text{k}\Omega) = 1.1$ mA.

Let the collector current also be 1.1 mA.

The ac emitter resistance [defined with Equation (6.26)] is $26/1.1 = 23.04\ \Omega$.

The voltage gain [using Equation (6.28)] is $4700/23.04 = 199$ (approx.)

6.6 In the collector–feedback circuit shown in Figure 6.14, the base–emitter voltage $V_{BE} = 0.7$ V and current gain $\beta = I_C/I_B = 100$ for the transistor. The value of the base current I_B is
(a) 20 µA (b) 40 µA (c) 10 µA (d) 100 µA

[NET Dec. 2019]

Soln. In the given circuit,
$$V_{CC} = I_C R_C + I_B R_B + V_{BE}$$
The symbols have the usual meaning. Therefore, putting the numerical values,
$$I_B = (20.7 - 0.7)/[(100 \times 5 + 500) \times 10^3]$$
$$= 20\ \mu A$$

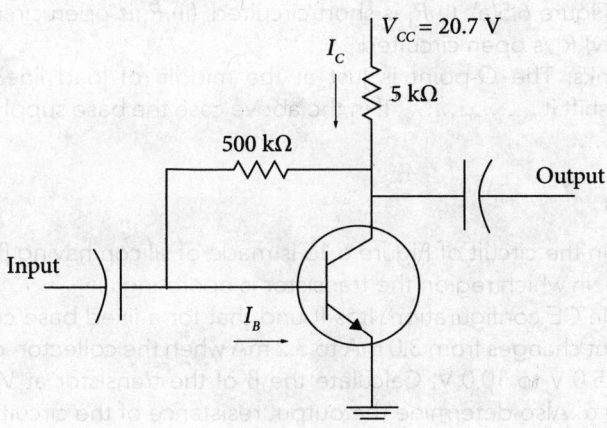

Figure 6.14

Exercise

Subjective Questions

1. What is transistor biasing and why is it needed?
2. Explain the *load line* of a transistor and mention its significance.
3. Elucidate, with diagrams, how the load line of a transistor in CE configuration changes with (i) the load resistance and (ii) the supply voltage.
4. What are the main causes of disturbing the bias stability of a transistor? Introduce the *stability factor* and obtain a suitable expression for it.
5. Draw the circuit diagram for a fixed bias and discuss its use and limitations. Explain the use of a resistor in between the emitter and the ground for stabilizing transistor operation.
6. Mention the utilities of (i) the coupling capacitors and (ii) the bypass capacitor in a CE amplifier.
7. Clarify the significance of the name 'emitter follower'. Derive the expressions for the voltage gain, current gain, input impedance and output impedance of this circuit and mention some useful applications of the circuit.

Conceptual Test

1. In the solved numerical problem 6.5, the current gain β was never required. Is the voltage gain independent of current gain? If not, put up your logic.
2. A common-collector amplifier has voltage gain of unity so that the output voltage would be a mere replica of the input. Then how is this circuit proved to be useful?
3. Collector–feedback bias is not suitable for small values of load resistance. Has this biasing any merit too?
4. Interpret the consequences, if the following changes are made in the voltage-divider bias circuit of Figure 6.5(a): (i) R_1 is short circuited, (ii) R_1 is open circuited, (iii) R_2 is short circuited and (iv) R_2 is open circuited.
5. Fill in the blanks: The Q-point is just at the middle of load line. Decrease of base resistance will shift it .... If in the above case the base supply increases, Q-point shifts ....

Numerical Problems

1. The transistor in the circuit of Figure 6.15 is made of silicon having β = 100 and I_{CO} = 20 nA. Determine, in which region the transistor is operating.
2. A transistor is in CE configuration. It is found that for a fixed base current of 25 μA, the collector current changes from 3.0 mA to 3.2 mA when the collector–emitter voltage (V_{CE}) changes from 5.0 V to 10.0 V. Calculate the β of the transistor at V_{CE} = 10.0 V and the corresponding α. Also determine the output resistance of the circuit.
3. Determine the Q-point of the voltage-divider bias circuit of Figure 6.5(a), if the supply voltage be of 20 V, R_1 = 10 kΩ, R_2 = 2.2 kΩ, R_C = 3.3 kΩ, R_E = 1 kΩ and the transistor has β = 150.

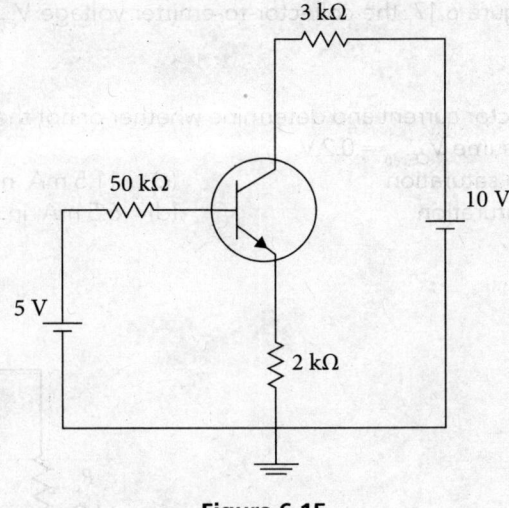

Figure 6.15

4. A transistor operating in CB mode has $\alpha = 0.99$ and dynamic resistance of the base–emitter junction is 25 Ω. If a load resistance of 5 kΩ be applied between the collector and the base, calculate the voltage gain and the power gain.
5. A transistor connected in CE configuration has supply voltage of 10 V and the drop across a resistance of 820 Ω connected in the collector circuit is 2 V. If $\alpha = 0.99$, determine (i) the collector voltage and (ii) the base current.
6. The transistor in Figure 6.16 has $\alpha = 0.98$ and $V_{BE} = 0.7$ V. Determine the value of R_1 for an emitter current of 2 mA.

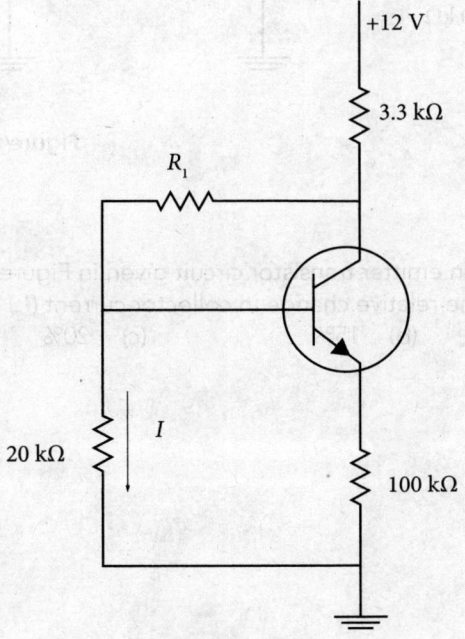

Figure 6.16

7. In the circuit of Figure 6.17, the collector-to-emitter voltage V_{CE} is .... Neglect V_{BE}, take $\beta = 100$.

[JAM 2016]

8. Calculate the collector current and determine whether or not the transistor in Figure 6.18 is in saturation. Assume $V_{CE(sat)} = 0.2$ V.
   (a) 6.5 mA, not in saturation
   (b) 11.5 mA, in saturation
   (c) 11.5 mA, not in saturation
   (d) 6.5 mA, in saturation

[JEST 2020]

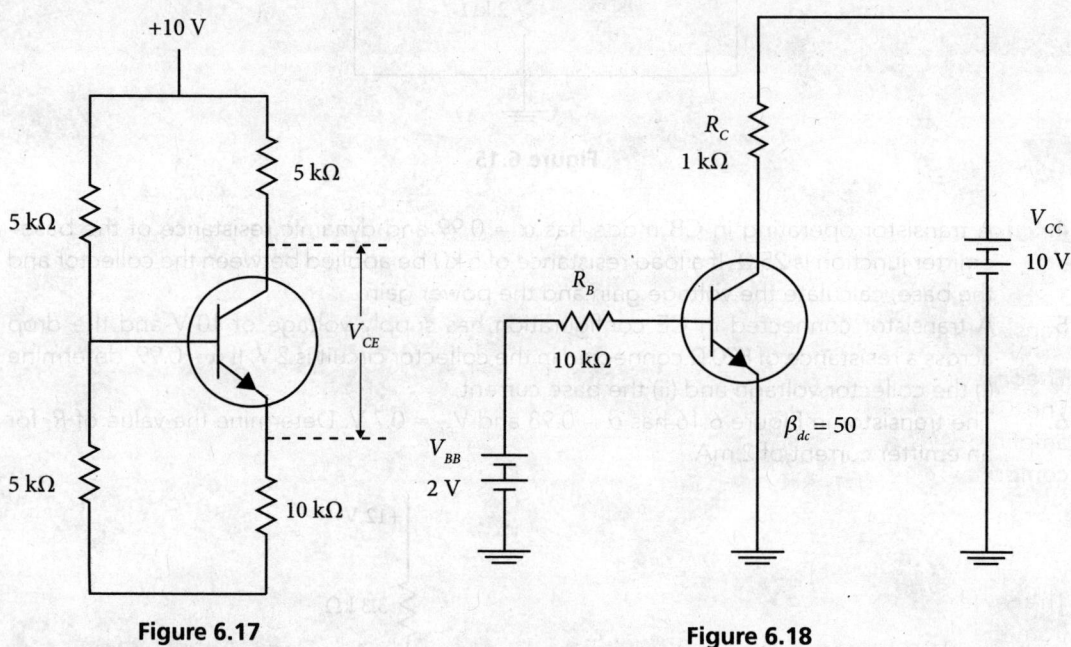

Figure 6.17   Figure 6.18

9. Analyze the common emitter transistor circuit given in Figure 6.19. If the current gain (β) increases by 50%, the relative change in collector current (I_C) is approximately
   (a) 5%   (b) 15%   (c) 20%   (d) 25%

[JEST 2020]

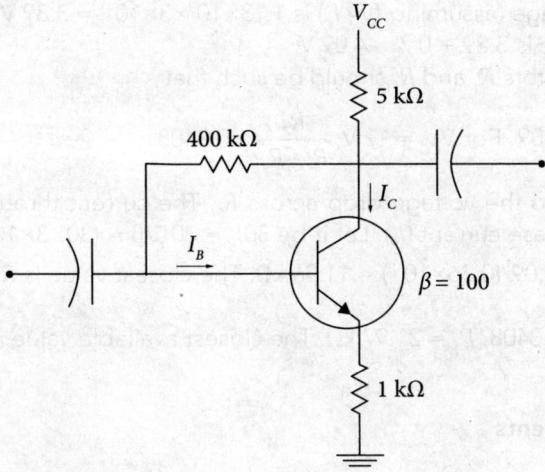

Figure 6.19

Project Work on Chapter 6

To construct a common-emitter (CE) voltage amplifier of given gain with suitable biasing of the transistor.

Theory:
The CE amplifier of Figure 6.7(a) with voltage-divider biasing is chosen for designing the amplifier of fixed voltage gain with realistic values of components. The amplifier design has two components:

(i) DC bias for fixing the Q-point and
(ii) AC bias for passing the input signal properly.

The key factor is the h_{fe} or β of the transistor and it must be known beforehand. It can be estimated from the dc output characteristics ($\Delta I_C / \Delta I_B$) or can be measured with digital multimeter. A large h_{fe} transistor is useful to voltage amplifier.

Biasing for dc components

1. It is assumed that the h_{fe} of the transistor is 250. The supply voltage is assumed to be 12 V. The designing starts with fixing the collector current (I_C) at a moderate value. Let I_C = 3 mA.
2. The ac emitter resistance is $r_e = 26/I_C$ (mA) = 26/3 = 8.67 Ω
3. The input resistance is $h_{ie} = \beta r_e = 250 \times 8.67 = 2.168$ kΩ
4. Let the targeted voltage gain (A_v) be 100. The collector resistance (R_C) acts as load. Using the approximate formula $A_v = R_C / r_e$, R_C = 867 Ω. Choosing the nearest available value, R_C = 820 Ω.
5. To fix the Q-point near the middle of the load line, $V_{CE} = V_{CC}/2$. In the collector circuit, $V_{CC} = I_E R_E + V_{CE} + I_C R_C$. Assuming $I_E \approx I_C$, considering V_{CC} = 12 V and putting $V_{CE} = V_{CC}/2$, R_E = 6000/3 − 867 = 1.13 kΩ. The nearest available value is R_E = 1 kΩ.

6. The emitter voltage (assuming $I_E \approx I_C$) is $1.13 \times 10^3 \times 3 \times 10^{-3}$ = 3.39 V.
7. The base voltage is 3.39 + 0.7 = 4.09 V.
8. The biasing resistors R_1 and R_2 should be such that

$$\frac{R_2 V_{CC}}{R_1 + R_2} = 4.09. \text{ For } V_{CC} = 12 \text{ V}, \frac{R_2}{R_1 + R_2} = 0.3408$$

9. This 4.09 V is also the voltage drop across R_2. The current through R_2 should be much larger than the base current (I_B). Let it be $30 I_B = 30 I_C/\beta = (30 \times 3 \times 10^{-3})/250 = 0.36$ mA
Therefore, $R_2 = 4.09/(0.36 \times 10^{-3}) = 11.36$ kΩ. The closest value is $R_2 = 10$ kΩ.

10. Using $\frac{R_2}{R_1 + R_2} = 0.3408$, $R_1 = 21.97$ kΩ. The closest available value is $R_1 = 22$ kΩ.

Biasing for ac components

1. The lower and the upper cutoff frequencies are to be fixed. Let the lower cutoff frequency = 500 Hz. The emitter bypass capacitor (C_E) should be such that $1/\omega C_E \ll R_E$ at 500 Hz. Considering $1/\omega C_E = 0.1 R_E$,

$$\frac{1}{2\pi \times 500 \times C_E} = 0.1 \times 1.13 \times 10^3 \Rightarrow C_E = 2.8 \mu F.$$

The next higher available value is $C_E = 10 \mu F$ (electrolytic)

2. In the AC equivalent circuit, h_{ie}, R_1 and R_2 appear in parallel to produce an equivalent resistance $R_{eq} = h_{ie} // R_1 // R_2$. Considering $h_{ie} = 2.168$ kΩ, $R_1 = 21.97$ kΩ and $R_2 = 11.36$ kΩ, $R_{eq} = 1.68$ kΩ.

The input blocking capacitor (C_B) is such that in the limiting case at $f = 500$ Hz,

$$C_B = 1/(2\pi f R_{eq}) = 0.189 \mu F.$$

A larger value of 0.47 μF or 1 μF is chosen for C_B (non-electrolytic).

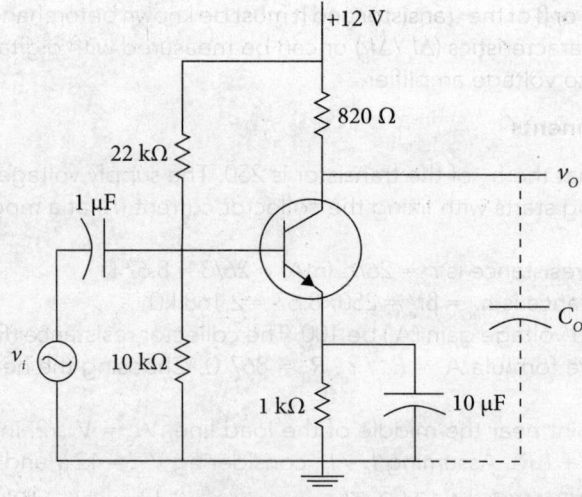

Figure P6.1 Transistor CE amplifier designed for a specific gain of 100

After joining the components according to the above designing, the CE amplifier looks like that of Figure P6.1. The input voltage should be very small, such as less than 40 mV. The amplitude remains constant over a large range of frequency above the lower cutoff frequency of the input signal. At around 1 MHz, the output amplitude starts decreasing when the junction capacitances become significant. In an actual experiment, the coupling capacitor can be eliminated and the amplified output voltage can be observed directly on cathode ray oscilloscope, introduced in Chapter 12.

The output may be made to decrease before the above, at a desired cutoff frequency (f_c) by joining a capacitor (C_o) across the output, as indicated in Figure P6.1. This provides a method for adjusting the bandwidth of the amplifier. For instance, if the desired cutoff frequency (f_c) be 20 kHz, C_o can be calculated as

$$C_o = 1/(2\pi f_c R_C) = \text{around 10 nF}.$$

7

Network Theorems and Transistor

The bipolar transistor is, no doubt, a nonlinear device. However, under certain conditions, it may be made to act as a linear device, as mentioned in Chapter 6. The present chapter explains how the well-known circuit theorems, such as Thevenin's theorem and Norton's theorem, meant for linear circuits can be applied to transistor operations with the assumption of a two-port device model. A suitable set of circuit parameters, known as hybrid parameters, are introduced in order to derive more accurate expressions for the voltage and current gains of a transistor and its other parameters. Also the frequency dependence of the transistor gain is explained in terms of the hybrid parameters.

7.1 Thevenin's Theorem

It is a very useful theorem for simplifying electrical circuits. It states that any two terminals of a linear network can be substituted by a single voltage source equal to the open circuit voltage between those two terminals in series with a single resistance equal to the equivalent resistance appearing between the two open terminals when all voltage and current sources related to the circuit are replaced by their internal resistances.

The above mentioned open circuit voltage is termed as *Thevenin voltage* and the equivalent resistance is called *Thevenin resistance*. An ideal voltage source is replaced by a short circuit and an ideal current source is replaced by an open circuit. The properties of voltage and current sources may be recalled from Chapter 1. Now let us have some practical idea about this theorem using the circuit of Figure 7.1(a). Suppose the current through the load resistor (R_L) is to be determined. It is not too difficult to do that with the present circuit. However, one can simplify it further through the following steps using Thevenin's theorem.

Step-1: Terminals P and Q are made open to detach R_L and the circuit looks like that of Figure 7.1(b). We proceed with this circuit.

Step-2: The current through R_2 is $V/(R_1 + R_2)$ so that the Thevenin voltage across R_2 is

$$V_T = \frac{VR_2}{R_1 + R_2} \tag{7.1}$$

Step-3: The voltage source (V), assumed to be ideal is replaced by short circuit so that the circuit gets modified to that of Figure 7.1(c). The Thevenin resistance between terminals P and Q is

$$R_T = \frac{R_1 R_2}{R_1 + R_2} \tag{7.2}$$

Step-4: The simplified circuit between P and Q is similar to that of Figure 7.1(d) having the Thevenin equivalent voltage source in series with the Thevenin resistance. Now R_L is joined again and the current through it is calculated easily as $V_T/(R_T + R_L)$.

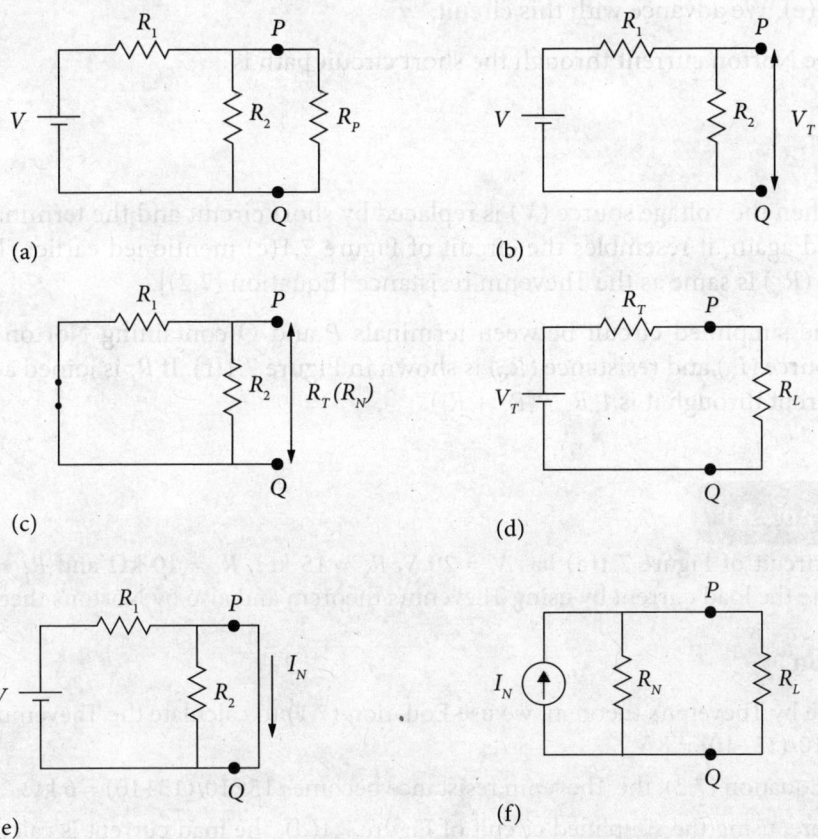

Figure 7.1 Circuit simplification with Thevenin's and Norton's theorems: (a) the original circuit, (b) terminals P and Q are open circuited, (c) the voltage source is replaced by short circuit and the Thevenin or Norton equivalent resistance (R_T or R_N) are indicated, (d) Thevenin equivalent circuit across R_L, (e) terminals P and Q are open circuited, and (f) Norton equivalent circuit across R_L

7.2 Norton's Theorem

This is another useful theorem for circuit simplification. According to it, any two terminals of a linear network can be replaced by a single current source equal to the short circuit current between those two terminals in parallel with a single resistance equal to the equivalent resistance acting between the two terminals when all voltage and current sources related to the circuit are exchanged by their internal resistances. Ideally the voltage source and the current source are replaced by short circuit and open circuit, respectively.

The aforementioned short circuit current supplied by the current source is called *Norton current* and the equivalent resistance in parallel is referred to as *Norton resistance*. The use of Norton's theorem is complementary to that of Thevenin's theorem and vice versa. The same circuit of Figure 7.1(a) is now being simplified using Norton's theorem.

Step-1: Terminals P and Q are short circuited and the modified circuit looks like that of Figure 7.1(e). We advance with this circuit.

Step-2: The Norton current through the short circuit path is

$$I_N = \frac{V}{R_1} \tag{7.3}$$

Step-3: When the voltage source (V) is replaced by short circuit and the terminals P and Q are opened again, it resembles the circuit of Figure 7.1(c) mentioned earlier. The Norton resistance (R_N) is same as the Thevenin resistance [Equation (7.2)].

Step-4: The simplified circuit between terminals P and Q containing Norton equivalent current source (I_N) and resistance (R_N) is shown in Figure 7.1(f). If R_L is joined across P and Q, the current through it is $I_N R_N / (R_N + R_L)$.

Example 7.1

If the circuit of Figure 7.1(a) has $V = 20$ V, $R_1 = 15$ kΩ, $R_2 = 10$ kΩ and $R_L = 10$ kΩ, calculate the load current by using Thevenin's theorem and also by Norton's theorem.

Solution

To solve by Thevenin's theorem, we use Equation (7.1) to calculate the Thevenin voltage as $20 \times 10/(15+10) = 8$ V.

Using Equation (7.2), the Thevenin resistance becomes $15 \times 10/(15+10) = 6$ kΩ.

Therefore, using the simplified circuit of Figure 7.1(d), the load current is calculated as $8/(6+10) = 0.5$ mA.

To solve the same problem by Norton's theorem, Equation (10.3) is used to determine the Norton current of $(20 \text{ V})/(15 \text{ k}\Omega) = 4/3$ mA.

The Norton resistance is the same as the Thevenin resistance (6 kΩ).

Therefore, using the simplified circuit of Figure 7.1(f), the load current comes out to be 6×(4/3)/(6+10) = 0.5 mA

[This example indicates that the simplification of a circuit with either Thevenin's theorem or Norton's theorem leads to the same result.]

7.3 Other Useful Theorems

There are many important circuit theorems that are beyond the present scope. Just a few are introduced here very briefly.

7.3.1 Superposition Theorem

When a linear circuit contains more than one source of electromotive force (emf), the current in any branch of the circuit is the algebraic sum of the individual currents produced by each source replacing the other sources by their respective internal resistances. The following example explains the use of this theorem.

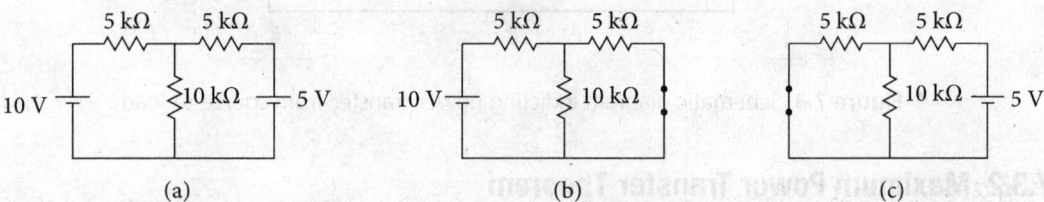

Figure 7.2 Explanation of superposition theorem: (a) the original circuit and short circuit replacing the voltage source of (b) 5 V and (c) 10 V, respectively.

Example 7.2

Using the superposition theorem, determine the current flowing through the 10 kΩ resistor in Figure 7.2(a).

Solution

Figure 7.2(b) shows the condition when the 5 V source is replaced by its internal resistance (ideally short circuited). In this case, the total current supplied by the 10 V source is

$$10/(5 + 50/15) = 6/5 \text{ mA}$$

Consequently, the current through the 10 kΩ resistor is

$$(6/5) \times 5/(10 + 5) = 2/5 \text{ mA} = I_1 \text{ (say)}$$

Similarly, when the 10 V source is shorted [Figure 7.2(c)], the total current due to the 5 V source is

$$5/(5 + 50/15) = 3/5 \text{ mA}$$

The current through the 10 kΩ resistor is

$$(3/5) \times 5/(10 + 5) = 1/5 \text{ mA} = I_2 \text{ (say)}$$

Therefore, the total current through the 10 kΩ resistor due to the two voltage sources is

$$I_1 + I_2 = 3/5 \text{ mA}$$

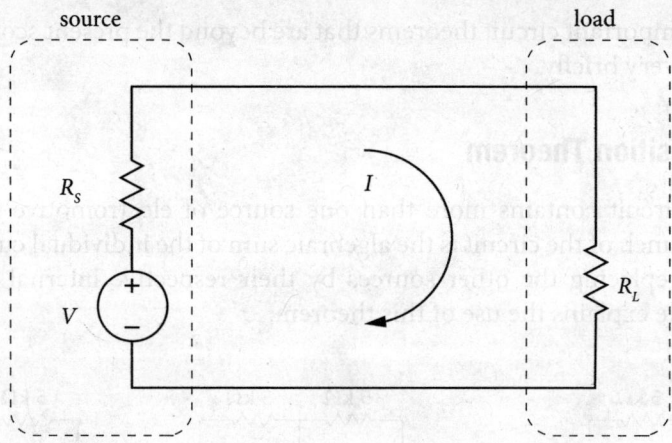

Figure 7.3 Schematic diagram indicting power transfer from source to load

7.3.2 Maximum Power Transfer Theorem

It ascertains that when the load resistance becomes equal to the source resistance, the amount of power extracted by the load from the source becomes maximum. This is true to both dc and ac sources and to circuits containing reactive elements like inductors and capacitors. In that case the word 'impedance' is more appropriate. Impedance matching is necessary between the source and the load for maximum power transfer.

Figure 7.3 outlines the power transfer from a source to a load. The current flowing in the circuit is

$$I = \frac{V}{R_s + R_L} \tag{7.4}$$

Using this expression for the current, the power delivered to the load can be determined as

$$P_L = I^2 R_L = \frac{V^2 R_L}{R_s^2 + R_L^2 + 2R_s R_L} \tag{7.5}$$

The power delivered to the load would be maximum when $dP_L/dR_L = 0$ leading to $R_s = R_L$. Putting this condition in Equation (7.5), the maximum power delivered to the load is given by

$$P_{L\max} = \frac{V^2}{4R_L} \tag{7.6}$$

The input power at this condition can be calculated from Equation (7.3), by putting $R_s = R_L$, as

$$P_i = \frac{V^2}{2R_L} \tag{7.6a}$$

> **Think a Bit: Where Does the Other Half Go?**
>
> Comparing the conditions of Equations (7.6) and (7.6a) it is revealed that
> (i) the maximum power is transferred to the load when the source resistance equals the load resistance and
> (ii) the maximum power transferred to the load can be only half of the input power.
>
> Then what about the remaining half? It is lost in the components in the form of heat dissipation.

7.4 Two-Port Model and Hybrid Parameters

The transistor is a three-terminal nonlinear device. In contrast, the circuit theorems discussed so far are applicable to two terminals of a linear circuit. Then how are those relevant to the transistor? It is so because we have read in Chapters 5 and 6 that any one of the emitter, base, or collector is used in common with both the input and the output terminals. If the input and output terminals are considered separately, these theorems are found to be useful for simplifying the circuit. Indeed this is the significance of two-port network and modeling the transistor like that.

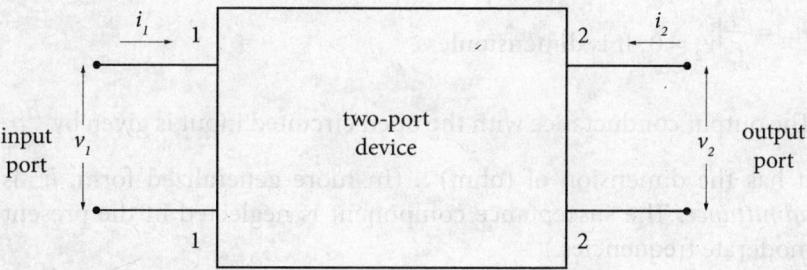

Figure 7.4 Two-port network

Figure 7.4 introduces the concept of two-port network that considers the input and the output separately. The exclusivity is that the analysis can be made in terms of several electrical parameters without knowing the actual interior view of the circuit construction. It is sufficient to assume that within the 'black box', there are:

(i) linear circuit elements and
(ii) no reactive element.

Whatever may be within the actual circuit, it possesses two input terminals (1, 1), two output terminals (2, 2) and four electrical parameters, namely the input voltage (v_1), the input current (i_1), the output voltage (v_2), and the output current (i_2). All these parameters are denoted by lowercase letters to imply both direct current (dc) and alternating current (ac) operations. Out of the four parameters, two are selected as independent variables and the other two are defined in terms of these independent variables.

If the input current (i_1) and the output voltage (v_2) are considered as the independent variables, the four parameters can be related by the following equations.

$$v_1 = h_{11} i_1 + h_{12} v_2 \qquad (7.7a)$$

$$i_2 = h_{21} i_1 + h_{22} v_2 \qquad (7.7b)$$

The quantities h_{11}, h_{12}, h_{21} and h_{22} are the constants of proportionality, known as the *hybrid parameters* or more popularly as the *h parameters*. The parameters are called 'hybrid' because these are originated from the mixing of voltage and current in each equation and are not all alike dimensionally, as explained below.

(i) The input resistance with the output short circuited is given by $h_{11} = \frac{v_1}{i_1}|_{v_2 = 0}$.

It has the dimension of ohms. (In general, h_{11} should be treated as *impedance*. However, the present book considers only low and medium frequencies of the input signal when the reactive part can be neglected.)

(ii) The fraction of the output voltage at the open circuited input is given by $h_{12} = \frac{v_1}{v_2}|_{i_1 = 0}$. It is dimensionless.

(iii) The forward current transfer ratio with the output short circuited is given by $h_{21} = \frac{i_2}{i_1}|_{v_2 = 0}$. It is dimensionless.

(iv) The output conductance with the open circuited input is given by $h_{22} = \frac{i_2}{v_2}|_{i_1 = 0}$.

It has the dimension of (ohm)$^{-1}$. (In more generalized form, h_{22} is treated as *admittance*. The susceptance component is neglected in the present context of moderate frequencies.)

Based on the above inferences, the hybrid circuit model for the two-port network of Figure 7.4 can be constructed, as given in Figure 7.5. It contains a dependent voltage generator $h_{12} v_2$ and a dependent current generator $h_{21} i_1$. These voltage and current sources are designated as 'dependent' because unlike those of external sources, these are created within the circuit and depend (linearly) on the voltage and current at some other point within the same circuit.

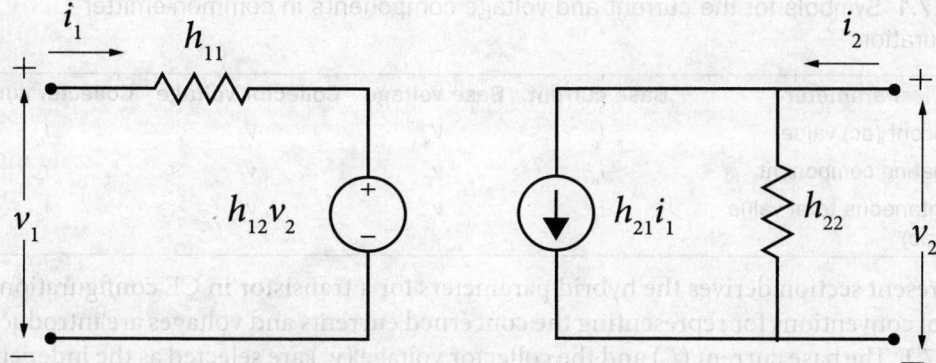

Figure 7.5 Hybrid circuit model for the two-port network of Figure 7.4. The input and output ports are treated separately.

7.4.1 Transistor as Two-port Network

The two-port network model can be useful to explain the operation of various devices, such as transistors, transformers, filters, and many others. Various electrical quantities, such as impedance, admittance, resistance, conductance, or a mix of these can be taken into consideration. There is also wide diversity in the choice of dependent and independent variables and the methodology of analysis. In the present context, we will consider the two-port model for the transistor where the hybrid parameters are quite useful. Generally the following stipulations are maintained.

(i) The electrical signal should be small (much smaller than the V_{BE} of the transistor) to ensure linear operation.
(ii) The frequency of the signal should be low (around audio range) to avoid the effect of reactance.
(iii) There are both dc bias to maintain the quiescent condition and small ac input superimposed on it ensuring linearity of operation.

The following symbols and the notations are used for the circuit parameters. In lieu of the 11, 12, 21 and 22 subscripts mentioned above, the following set of subscripts is used for the hybrid parameters.

11 → i (input)
12 → r (reverse transfer)
21 → f (forward transfer)
22 → o (output)

A second subscript is used to indicate the type of configuration: common-emitter (CE), common-base (CB) or common-collector (CC) by specifying the common terminal: emitter (e), base (b) or collector (c). For instance, h_{ie} denotes the input resistance in common-emitter configuration following the above notations.

Table 7.1 Symbols for the current and voltage components in common-emitter configuration

Parameter	Base current	Base voltage	Collector voltage	Collector current
Quiescent (dc) value	I_B	V_B	V_C	I_C
Alternating component	i_b	v_b	v_c	i_c
Instantaneous total value (dc + ac)	i_B	v_B	v_C	i_C

The present section derives the hybrid parameters for a transistor in CE configuration. The symbol conventions for representing the concerned currents and voltages are introduced in Table 7.1. The base current (i_B) and the collector voltage (v_C) are selected as the independent variables. The other variables, namely the base voltage (v_B) and the collector current (i_C) are functions of the above two given by

$$v_B = f_1(i_B, v_C) \tag{7.8a}$$

$$i_C = f_2(i_B, v_C) \tag{7.8b}$$

Making Taylor series expansion around the quiescent point (I_B, V_C) and neglecting higher order terms (to maintain linearity),

$$\Delta v_B = \frac{\partial f_1}{\partial i_B}\bigg|_{V_C} \Delta i_B + \frac{\partial f_1}{\partial v_C}\bigg|_{I_B} \Delta v_C \tag{7.9a}$$

$$\Delta i_C = \frac{\partial f_2}{\partial i_B}\bigg|_{V_C} \Delta i_B + \frac{\partial f_2}{\partial v_C}\bigg|_{I_B} \Delta v_C \tag{7.9b}$$

Substituting the small incremental quantities Δv_B by v_b, Δi_B by i_b, Δv_C by v_c, and Δi_C by i_c from Table 7.1, Equations (7.9a) and (7.9b) can be written as

$$v_b = h_{ie} i_b + h_{re} v_c \tag{7.10a}$$

$$i_c = h_{fe} i_b + h_{oe} v_c \tag{7.10b}$$

Equations (7.10a) and (7.10b) include the following hybrid parameters for the transistor in CE configuration.

$h_{ie} = \frac{\partial f_1}{\partial i_B}\bigg|_{V_C} = \frac{\partial v_B}{\partial i_B}\bigg|_{V_C}$ is the input resistance,

$h_{re} = \frac{\partial f_1}{\partial v_C}\bigg|_{I_B} = \frac{\partial v_B}{\partial v_C}\bigg|_{I_B}$ is the reverse open-circuit voltage amplification,

$h_{fe} = \frac{\partial f_2}{\partial i_B}\bigg|_{V_C} = \frac{\partial i_C}{\partial i_B}\bigg|_{V_C}$ is the forward short-circuit current gain, and

$h_{oe} = \frac{\partial f_2}{\partial v_C}\bigg|_{I_B} = \frac{\partial i_C}{\partial v_C}\bigg|_{I_B}$ is the output conductance.

Similar equations can be obtained for CB and CC configurations. A summary is given in Figure 7.6. Since the ac signal continuously switches between positive and negative values, an instantaneous situation is shown in the figures. Comparing with the general two-port structure of Figure 7.5, it is noted that the input and output ports are not isolated but connected through the common (ground) terminal in the case of transistors. The three terminals for each case are indicated in Figure 7.6.

The output conductance (h_{oe}) is sometimes represented in terms of a resistance ($1/h_{oe}$), as mentioned in the subsequent section. There are specific conversion formulae for h parameters in different configurations that are summarized in Table 7.2. It is apparent from Table 7.2 that the h parameters for CE configuration can be obtained in terms of those of CB configuration and vice versa just by interchanging the 'b' and 'e' subscripts.

Table 7.2 Conversion of hybrid parameters for common-emitter (CE), common-collector (CC), and common-base (CB) configurations

Parameter	CE	CC	CB
Input impedance with output shorted	$h_{ie} = \dfrac{h_{ib}}{1+h_{fb}}$	$h_{ic} = h_{ie}$	$h_{ib} = \dfrac{h_{ie}}{1+h_{fe}}$
Reverse voltage gain with input opened	$h_{re} = \dfrac{h_{ib}h_{ob}}{1+h_{fb}} - h_{rb}$	$h_{rc} \approx 1$	$h_{rb} = \dfrac{h_{ie}h_{oe}}{1+h_{fe}} - h_{re}$
Forward current gain with output shorted	$h_{fe} = \dfrac{-h_{fb}}{1+h_{fb}}$	$h_{fc} = -(1+h_{fe})$	$h_{fb} = \dfrac{-h_{fe}}{1+h_{fe}}$
Output admittance with input opened	$h_{oe} = \dfrac{h_{ob}}{1+h_{fb}}$	$h_{oc} = h_{oe}$	$h_{ob} = \dfrac{h_{oe}}{1+h_{fe}}$

7.4.2 Significance of *h* Parameters

At audio frequencies, the *h* parameters are real numbers. These can be measured easily even from the static characteristic curves. The hybrid parameters have the following useful features in connection with transistor circuit analyses.

- *Transistor Performance*: The *h* parameters provide with suitable expressions for transistor performance quantities, such as voltage and current gains and input and output resistances.
- *Circuit Design*: These parameters offer an easy method for circuit analysis and design. Manufacturers generally specify a set of hybrid parameters for a specific transistor.

- *Construction Independence*: These parameters work with the terminal (input/output) characteristics rather than the internal construction and individual components, namely the emitter, base, collector and the junctions.
- *Range of Operation*: The hybrid parameters do not specify any actual operating condition like the Q-point. Instead, a range of each parameter is provided.

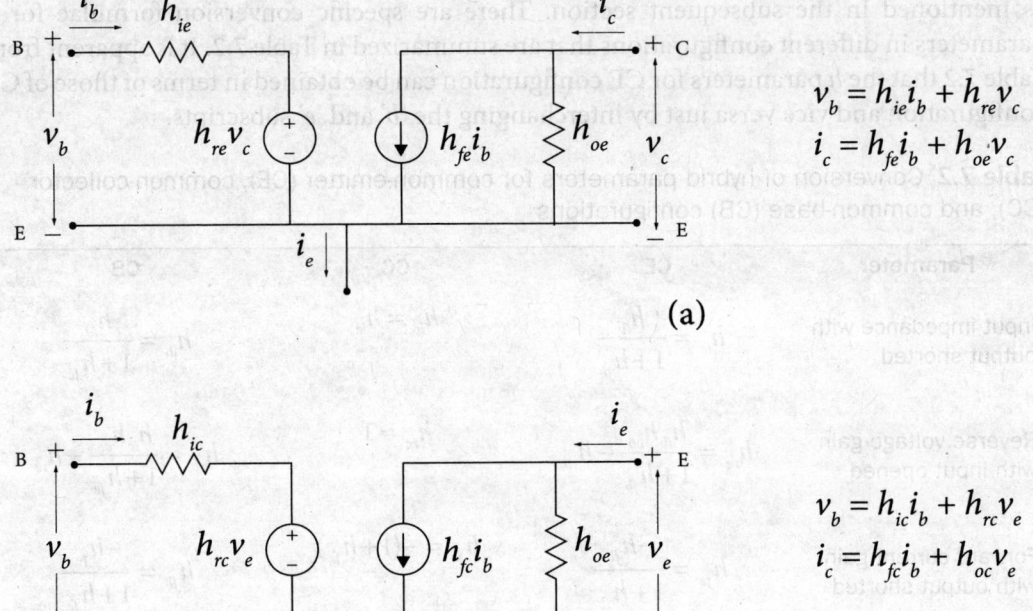

$$v_b = h_{ie}i_b + h_{re}v_c$$
$$i_c = h_{fe}i_b + h_{oe}v_c$$

$$v_b = h_{ic}i_b + h_{rc}v_e$$
$$i_c = h_{fc}i_b + h_{oc}v_e$$

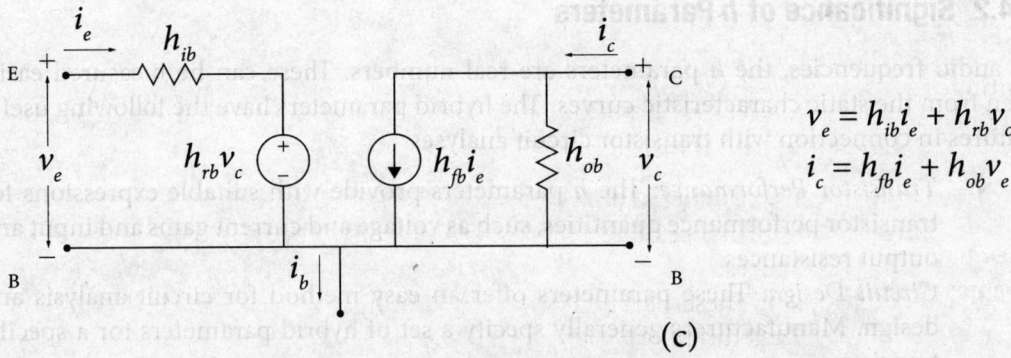

$$v_e = h_{ib}i_e + h_{rb}v_c$$
$$i_c = h_{fb}i_e + h_{ob}v_c$$

Figure 7.6 Hybrid circuit model and current–voltage equations for (a) common-emitter, (b) common-collector, and (c) common-base configurations

7.5 Transistor Amplifier with *h* Parameters

The previous section illustrated that the hybrid equivalent model holds good for all the three configurations (CE, CC and CB) of a transistor. We choose one of those, namely the CE configuration given in Figure 7.6(a). The transistor amplifier can be realized by joining the following three parameters to that hybrid model.

(i) Proper dc biasing with passive components
(ii) An external input ac source (v_s)
(iii) An external load (R_L)

Figure 7.7 represents such two-port hybrid model for the transistor CE amplifier. Though the circuit is valid for both pure resistance and impedance, it is assumed that no reactance is involved. The reader is encouraged to revisit the simplified explanation for CE amplifier given in section 6.8.1. It was explained there that the effect of dc biasing can be ignored under certain conditions. The present interpretation is meant for the same amplifier but the gain and impedance parameters are derived from a different point of view.

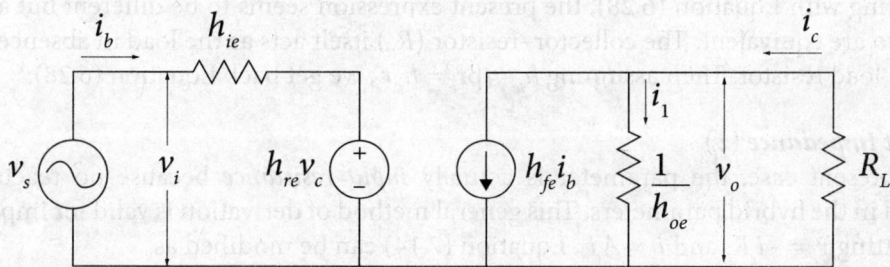

Figure 7.7 Two-port hybrid model for transistor amplifier in CE configuration

A. Current Gain (A_i)

The output voltage (v_o) across R_L is essentially v_c with the opposite sign. Let the current through the output resistance ($1/h_{oe}$) be $i_1 = v_c h_{oe}$. The output current (i_c) can be written as

$$i_c = i_1 + h_{fe} i_b \tag{7.11}$$

Putting the value of i_1 and using $v_c = -i_c R_L$ in Equation (7.11), the current gain can be obtained as

$$A_i = \frac{i_c}{i_b} = \frac{h_{fe}}{1 + h_{oe} R_L} \tag{7.12}$$

The quantity h_{oe} is very small so that $h_{oe} R_L \ll 1$ in Equation (7.12) and the current gain, can be approximated for practical purposes by

$$A_i = h_{fe} \tag{7.13}$$

Comparing with Equation (6.29), one can conclude that h_{fe} and the transistor current gain (β) are equivalent.

B. Voltage Gain (A_v)

The input voltage (v_i) in the circuit of Figure 7.7 is essentially v_b, and is given by

$$v_b = i_b h_{ie} + h_{re} v_c \tag{7.14}$$

Substituting for i_b from Equation (7.12), putting $i_c = -v_c/R_L$ and simplifying, the voltage gain is derived as

$$A_v = \frac{v_c}{v_b} = -\frac{h_{fe} R_L}{h_{ie} + R_L(h_{ie} h_{oe} - h_{re} h_{fe})} \tag{7.15}$$

Since h_{oe} and h_{re} are very small quantities, $(h_{ie} h_{oe} - h_{fe} h_{re}) \ll h_{ie}$ and can be ignored in Equation (7.15) thereby approximating the voltage gain by

$$A_v = -\frac{h_{fe} R_L}{h_{ie}} \tag{7.16}$$

Equation (7.16) is useful to represent the voltage gain of a CE amplifier for practical purpose. The negative sign indicates a phase inversion of 180° between the input and the output. Comparing with Equation (6.28), the present expression seems to be different but actually these two are equivalent. The collector–resistor (R_C) itself acts as the load in absence of any external load resistor. Then assuming $h_{ie} = \beta r_e = h_{fe} r_e$, we get back Equation (6.28).

C. Input Impedance (z_i)

In the present case, the parameter is actually *input resistance* because no reactance is involved in the hybrid parameters. This general method of derivation is valid for impedance also. Putting $v_c = -i_c R_L$ and $i_c = A_i i_b$, Equation (7.14) can be modified as

$$v_b = h_{ie} i_b - h_{re} A_i i_b R_L \tag{7.17}$$

Therefore, the input impedance is

$$z_i = \frac{v_b}{i_b} = h_{ie} - h_{re} A_i R_L \tag{7.18}$$

The quantity h_{re} being too small, the input impedance can be approximated, for practical purposes by

$$z_i = h_{ie} \tag{7.19}$$

D. Output Impedance (z_o)

This also can be treated as *output resistance* in the present case but the generalized term 'impedance' is retained. The output impedance of an amplifier is defined as the ratio of the output voltage to the output current when the input voltage source is short circuited. Applying this condition to v_s in Figure 7.7, the input current becomes

$$i_b = -\frac{h_{re} v_c}{h_{ie}} \tag{7.20}$$

This current flows in a direction opposite to that shown in the figure, under the action of the reverse voltage source $h_{re}v_c$. From Equations (7.11) and (7.20), the output impedance can be derived as

$$z_o = \frac{v_c}{i_c} = \frac{1}{h_{oe} - h_{fe}h_{re}/h_{ie}} \qquad (7.21)$$

Now h_{re} is very small and that divided by h_{ie} is even much smaller a quantity that can be neglected in Equation (7.21). For practical purpose, the output impedance is given by

$$z_o = \frac{1}{h_{oe}} \qquad (7.22)$$

The above analysis assumed the voltage source (v_s) to be ideal with zero internal resistance. If the voltage source had finite internal resistance (R_s), still all the above interpretations would be valid just h_{ie} being replaced by ($h_{ie} + R_s$) in Equations (7.15) and (7.21).

Example 7.3

A transistor CE amplifier has $h_{fe} = 100$ and $h_{oe} = 30 \times 10^{-6}$ S. If the load resistor be of 3 kΩ, determine the current gain.

Solution

Using Equation (7.12), the current gain is calculated as
$100/(1 + 30 \times 10^{-6} \times 3 \times 10^3) = 91.74$

Example 7.4

A CE amplifier has $h_{ie} = 2$ kΩ, $h_{re} = 2\times10^{-4}$, $h_{fe} = 50$ and $h_{oe} = 20\times10^{-6}$ A/V. If the load resistance is 4 kΩ and the source resistance is 200 Ω, determine the input resistance, the output resistance, the voltage gain and the power gain of the amplifier.

Solution

The current gain will be needed. Using Equation (7.12), it is calculated as
$50/(1 + 20\times10^{-6}\times4\times10^3) = 46.3$

Using Equation (7.18) and the above value of current gain, the input resistance comes out to be
$2000 - 2\times10^{-4}\times46.3\times4\times10^3 = 1.963$ kΩ.

To calculate the output resistance, Equation (7.21) is used. Since the source resistance comes in series with the input resistance, h_{ie} in Equation (7.21) is replaced by ($h_{ie} + 200$). Then the output resistance value becomes 64.7 kΩ.

The voltage gain is, using Equation (7.15),

$$\frac{-50 \times 4 \times 10^3}{2 \times 10^3 + 4 \times 10^3 \times (2 \times 10^3 \times 20 \times 10^{-6} - 50 \times 2 \times 10^{-4})} = -94.34$$

The power gain (in magnitude) is the product of the current gain and the voltage gain given by

$$46.3 \times 94.34 = 4.368 \times 10^3$$

N. B. The following two points are noteworthy in connection with the above calculation of voltage gain.

If one calculates the approximate voltage gain using Equation (7.16), it becomes −100, which is 6% greater than the accurate value.

If the source resistance of 200 Ω is included in the amplifier, h_{ie} in Equation (7.15) is replaced by (h_{ie} + 200) and the voltage gain becomes −85.61, which is 9.25% less than the accurate calculation.

Both of the deviations are of the order of experimental errors.

7.6 Simplified Hybrid Model

Two of the hybrid parameters, namely the reverse voltage amplification (h_{re} for CE mode) and the output conductance (h_{oe} for CE mode) are negligibly small. Therefore, assuming $h_{re} v_c \approx 0$ and $1/h_{oe} \to \infty$ in the hybrid equivalent circuit, one can replace these two components by short circuit and open circuit, respectively. That makes the transistor hybrid model further simplified, similar to that shown in Figure 7.8. The input voltage source (v_s) and the load resistor (R_L) are joined as usual. Referring to Table 7.2, similar simplifications can be made with a transistor in CB mode. This approximation is reasonable and convenient for all practical transistor circuits with moderate frequencies and small amplitudes of the input signal.

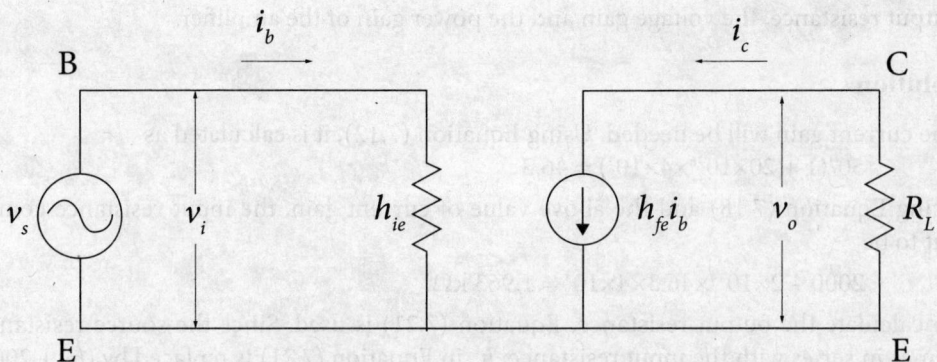

Figure 7.8 Simplified hybrid model for transistor in CE configuration comprising the voltage source (v_s) and the load resistor (R_L).

7.7 r_e-Model and h-Model

Let us first recapitulate our earlier concept of ac emitter resistance (r_e) discussed in Chapter 6. The base–emitter junction of an n–p–n transistor in CE configuration acts as a forward biased diode with dynamic resistance r_e and the output terminal is approximated as a current source of β times the input current (i_b). So the equivalent circuit for the transistor in CE mode can be drawn like that of Figure 7.9(a). Following that equivalent circuit, the input impedance is given by

$$z_i = \frac{v_b}{i_b} = \frac{i_e r_e}{i_b} \qquad (7.23)$$

Putting $i_e = i_c + i_b$, $i_c = \beta i_b$ and $\beta + 1 \approx \beta$ in Equation (7.23), symbols having their usual meaning,

$$z_i = \beta r_e \qquad (7.24)$$

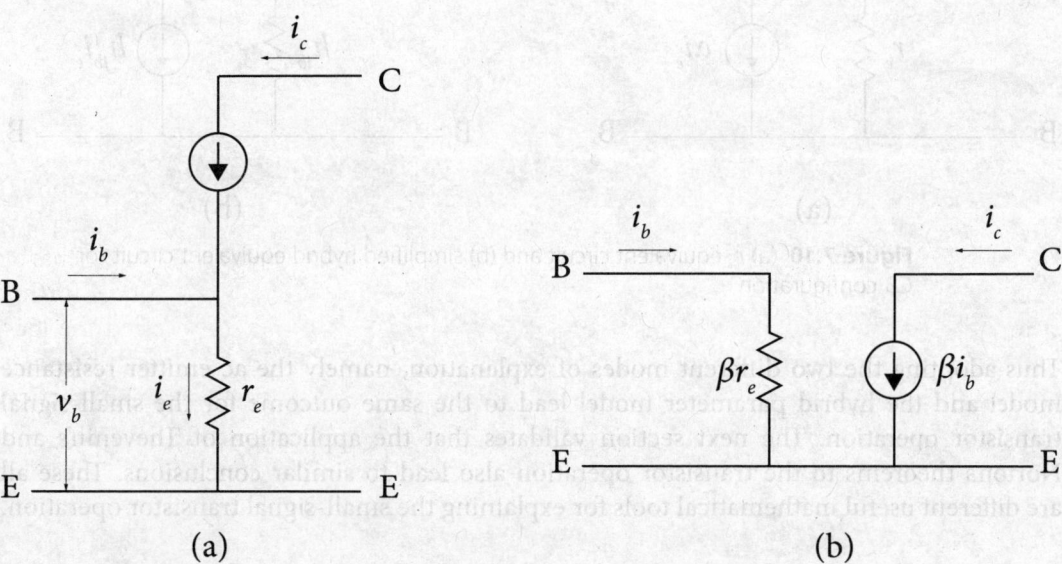

Figure 7.9 r_e-Model for CE configuration, (a) T-model, (b) Π-model

Thus the impedance of the transistor looking into the base is βr_e so that the equivalent circuit can be represented as that of Figure 7.9(b). It is actually an improved version of the circuit of Figure 7.9(a) because both the input and the output terminals are considered separately. The equivalent circuits of Figure 7.9 (a) and (b) are known as the r_e-model for CE configuration. Sometimes these two equivalent circuits are called the *T-model* and Π-model, respectively. The T-model is actually one of the earliest transistor ac models, known as *Ebers–Moll* model.

Now comparing the above model with the simplified hybrid model of Figure 7.8, the following similarities can be noted.

$$h_{ie} = \beta r_e \qquad (7.25a)$$

$$h_{fe} = \beta \qquad (7.25b)$$

Similarly, in the case of CB configuration, one can come up with the equivalent circuits like those of Figure 7.10(a) and (b) and obtain the following two equivalences.

$$h_{ib} = r_e \qquad (7.26a)$$

$$h_{fb} = \alpha \qquad (7.26b)$$

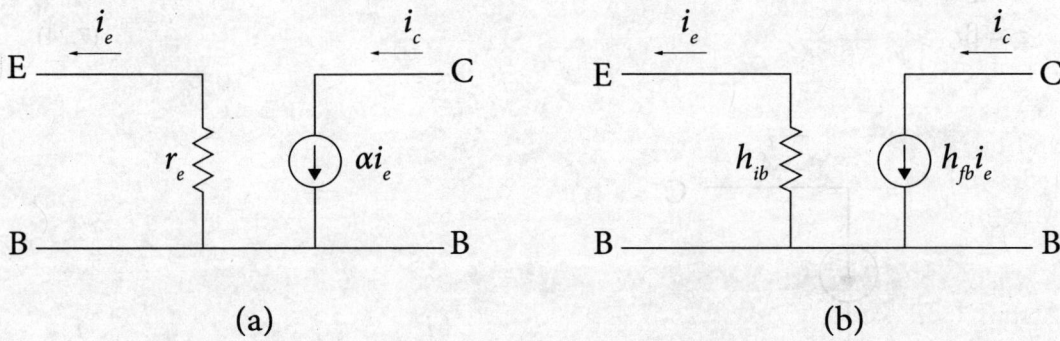

Figure 7.10 (a) r_e-equivalent circuit and (b) simplified hybrid equivalent circuit for CB configuration

Thus adopting the two different modes of explanation, namely the ac emitter resistance model and the hybrid parameter model lead to the same outcome for the small-signal transistor operation. The next section validates that the application of Thevenin's and Norton's theorems to the transistor operation also lead to similar conclusions. These all are different useful mathematical tools for explaining the small-signal transistor operation.

7.8 Transistor: Thevenin and Norton Equivalents

The previous sections have interpreted the gain and impedance properties of the transistor with the use of hybrid parameters. The possible simplifications are pointed out and the results are compared with those obtained from other models, such as ac emitter resistance and collector current source assumptions. Now the effect of the biasing components on the hybrid model is being explored. This time the applicability of Thevenin's and Norton's circuit theorems to the transistor are also validated.

Both Thevenin's and Norton's theorems (sections 7.1 and 7.2) are applicable to two-terminal linear networks. The transistor is neither 'two-terminal' nor 'linear'. Nevertheless the discussions on the hybrid model (sections 7.4 and 7.5) establish that the two input terminals and the two output terminals of the transistor can be treated separately as linear devices with the stipulation of small amplitude and low frequency of the input signal. The two-port hybrid model for transistor amplifier in CE configuration (Figure 7.7) has the following properties.

(i) The input is represented as a voltage source $h_{re}v_c$ in series with a single resistance h_{ie}. So it is the *Thevenin equivalent* of the input terminal. Similar conclusions can be made using the hybrid models for the other configurations given in Figure 7.6.

(ii) The output is represented as a current source $h_{fe}i_b$ in parallel with a single resistance $1/h_{oe}$. Thus it is the *Norton equivalent* of the output terminal. Similar logic holds to the other configurations of Figure 7.6.

(iii) The application of both Thevenin's theorem and Norton's theorem to the same circuit is another reason for calling it a 'hybrid'.

The transistor CE amplifier with voltage-divider bias for dc and with the blocking, coupling and bypass capacitor for ac was introduced in Chapter 6 [Figure 6.7(a)]. The circuit is redrawn here as Figure 7.11(a) and the dc and the ac components are considered separately with the following steps.

A. *Thevenin's theorem is applied to the two input terminals, indicate by 'in'.*
Assuming the circuit to be disconnected at these points, the open circuit voltage between these two terminals can be derived with simple circuit concept as

$$V_T = \frac{R_2 V_{CC}}{R_1 + R_2} \qquad (7.27a)$$

This is the *Thevenin voltage* between the two input terminals that act as the base supply of the transistor. This concept was already introduced with Equation 6.20(a). The corresponding *Thevenin resistance*, determined by replacing V_{CC} by short circuit is

$$R_T = \frac{R_1 R_2}{R_1 + R_2} \qquad (7.27b)$$

This was introduced in Chapter 6 [Equation 6.20(b)] also. The dc equivalent circuit looks like that of Figure 7.11(b).

B. *Norton's theorem is applied to the two output terminals, indicate by 'out'.*
Assuming the circuit to be short circuited between these two points, the Norton current becomes $h_{fe}i_b$. Assuming V_{CC} to be short circuited, the Norton resistance becomes the parallel combination of the transistor output resistance ($1/h_{oe}$) and the collector biasing resistance (R_C). Since $1/h_{oe} \gg R_C$, the effective output resistance is R_C alone.

C. The ac part has the following components.

The blocking capacitor (C_B), the coupling capacitor (C_C) and the emitter bypass capacitor (C_E) are supposed to act as open circuit for the dc bias and as short circuit for the ac signal of a certain given frequency range. The ac input voltage (v_i) joined across the input terminals ('in') through C_B appears across the Thevenin voltage (V_T) mentioned above. The external load resistor (R_L) is joined through C_C comes in parallel with R_C to form the effective load.

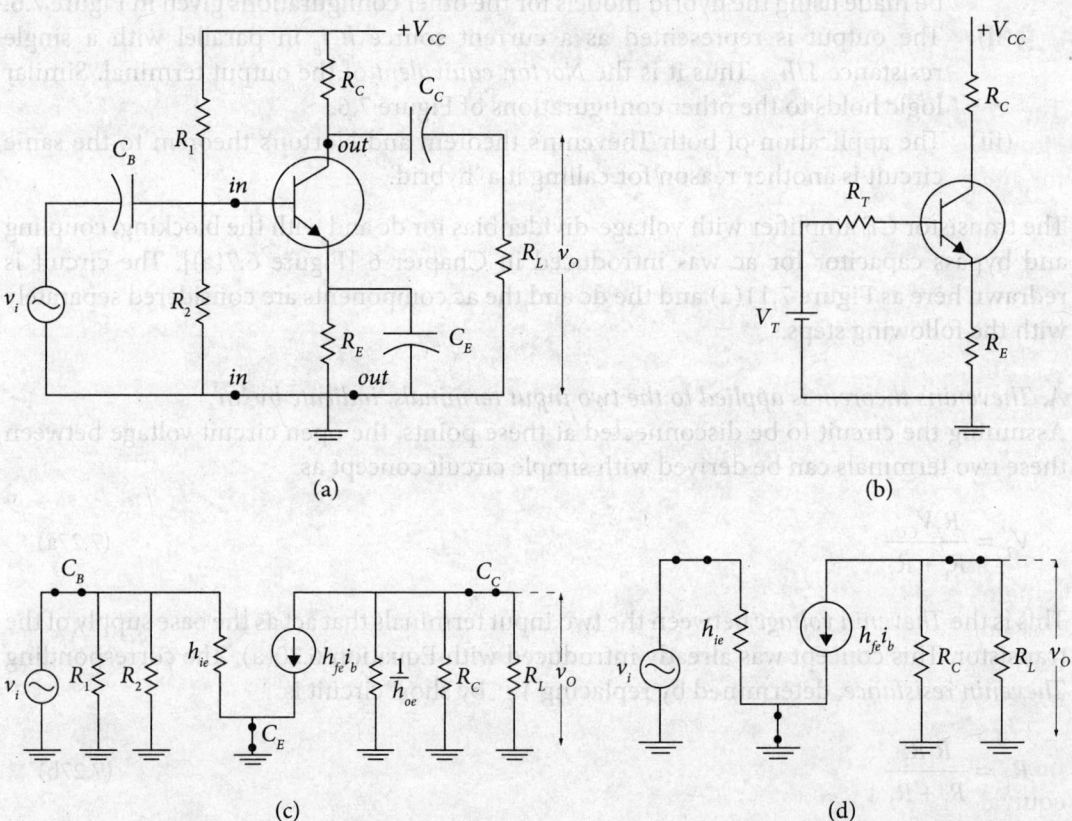

Figure 7.11 Transistor CE amplifier with voltage divider bias: (a) circuit diagram, (b) dc equivalent circuit, (c) ac equivalent circuit, and (d) simplified ac equivalent circuit

Figure 7.11(c) depicts the ac equivalent circuit of the transistor amplifier. The hybrid parameters are the same as those of the simplified hybrid model of Figure 7.8. The external biasing resistors R_1, R_2 and R_C are added in parallel keeping the hybrid model intact. The biasing capacitors C_B, C_C and C_E act as short circuits, as indicated in the figure. One can make the following simplifications.

(i) The resistance $1/h_{oe}$, being much larger than R_C and R_L can be ignored in the parallel combination.
(ii) Similarly, the biasing resistors R_1 and R_2 can be neglected in the parallel combination, if the resistance of their parallel combination is much greater than h_{ie}.

The ac equivalent made easier with the above considerations regains the simplified hybrid equivalent circuit of Figure 7.11(d). Comparing with Figure 7.8 it is understood that the original hybrid model meant for the transistor alone can be kept almost undisturbed in the case of the transistor amplifier with proper choice of the biasing components.

7.9 Frequency Dependence of Gain

The voltage gain of a transistor discussed so far is supposed to be a constant quantity. However, in reality, it remains constant over a wide range of input signal frequency but not for any frequency. It is so because of the following reasons.

(i) At lower frequencies (e.g. below several hundreds of hertz), the coupling capacitors may not act as perfect short circuit.
(ii) At higher frequencies (e.g. above 1 MHz), the junction capacitances of the transistor, though very small ($\approx$ picofarads) may have remarkable effect.
(iii) Particularly for the CE amplifier, which has the output amplified as well as inverted, the effect of the collector junction capacitance at high frequency has a magnified effect at the input. This is known as *Miller effect*.
(iv) Also the stray capacitances due to wiring and other metallic structures creep into the operating of the transistor at high frequencies.

Consequently the voltage gain of a transistor becomes a function of signal frequency. It remains constant over a range, called the *mid-frequency* range. Depending on the biasing, this range may vary from about 1 kHz to several hundreds of kHz. Below this range, the gain decreases with decreasing frequency and above this range, the gain decreases with increasing frequency.

For detailed illustration of the above phenomena, we start with the simplified hybrid model of Figure 7.11(d). For further convenience, the current source is converted to the equivalent voltage source. The resultant circuit is shown in Figure 7.12(a). The circuit performance at the following three ranges of signal frequency are considered.

A. Gain at mid-frequency

In the circuit of Figure 7.12(a), the coupling capacitor is expected to act as perfect short circuit and the output current is given by

$$i_o = -\frac{h_{fe} i_b R_C}{R_C + R_L} \tag{7.28}$$

The output voltage is, therefore,

$$v_o = -\frac{h_{fe} i_b R_C R_L}{R_C + R_L} \quad (7.29)$$

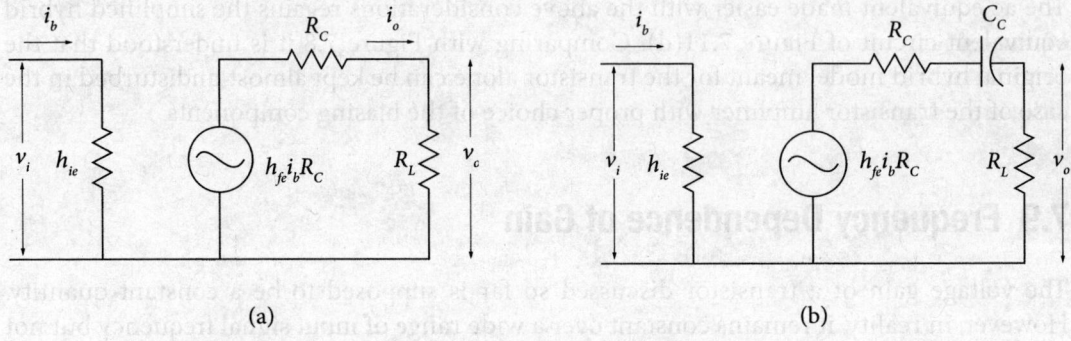

(a) (b)

Figure 7.12 Simplified hybrid model for transistor CE amplifier modified for voltage source (a) at mid-frequency, and (b) at low frequency

The input voltage is, obviously,

$$v_i = h_{ie} i_b \quad (7.30)$$

Using Equations (7.29) and (7.30), the voltage gain of the transistor CE amplifier at mid-frequency is given by

$$A_v = \frac{v_o}{v_i} = -\frac{h_{fe} R_C R_L}{h_{ie}(R_C + R_L)} \quad (7.31)$$

If R_C is made comparable to h_{ie}, the expression for the voltage gain becomes

$$A_v = -\frac{h_{fe} R_L}{(R_C + R_L)} \quad (7.32)$$

If, in addition, R_L becomes much larger than R_C, Equation (7.32) yields $A_v \approx -h_{fe}$, the magnitude of which is the maximum obtainable gain from a transistor.

B. Gain at low frequency
This time the coupling capacitor (C_C) cannot be treated as short circuit. Considering its presence in the circuit, as shown in Figure 7.12(b), the output current is given by

$$i_o = -\frac{h_{fe} i_b R_C}{R_C + R_L - j/\omega C_C} \quad (7.33)$$

Hence the output voltage is

$$v_o = -\frac{h_{fe} i_b R_C R_L}{R_C + R_L - j/\omega C_C} \quad (7.34)$$

The input voltage is already given by Equation (7.30). Therefore, the voltage gain of the transistor CE amplifier at low frequency is given by

$$A_{vl} = \frac{v_o}{v_i} = -\frac{h_{fe} R_C R_L}{h_{ie}(R_C + R_L - j/\omega C_C)} \qquad (7.35)$$

Making R_C comparable to h_{ie} in Equation (7.35), dividing both the numerator and the denominator by $(R_C + R_L)$ and substituting for A_v from Equation (7.32) one can derive the following expression for the voltage gain at low frequency (A_{vl}) in terms of that at mid-frequency (A_v).

$$A_{vl} = \frac{A_v}{1 - j\dfrac{f_{cl}}{f}} \qquad (7.36)$$

The frequency

$$f_{cl} = \frac{1}{2\pi C_C (R_C + R_L)} \qquad (7.36a)$$

in Equation (7.36) is defined as the *lower cutoff frequency* of the transistor CE amplifier biased with the collector series resistor R_C and joined to the external load resistor R_L. If the signal frequency (f) becomes just equal to this f_{cl}, the gain reduces to $1/\sqrt{2}$ times the constant gain at mid-frequency. Therefore, in terms of decibel (to be discussed later), the lower cutoff frequency is also termed as *lower 3dB frequency*.

> **Think a Bit: Reactance Responsible for Diminishing the Low-frequency Gain**
>
> The derivation of Equation (7.36) has considered the reactance effect of the coupling capacitor alone on the reduction of the amplifier gain at low frequencies. The blocking and bypass capacitors are assumed to be acting as ideal short circuits at the frequency under consideration. In practice, these capacitances also contribute significantly to the reactance and thereby reduction of the gain at low frequencies. A numerical problem in this chapter has addressed this phenomenon that particularly the emitter bypass capacitor has remarkable effect at a low frequency.

7.10 Hybrid-Π Model

The transistor gain and its variation at middle and low ranges of the signal frequency are discussed in the previous section. The next inevitable question is what about the high frequency. Indeed the whole thing is to be recast in the high frequency region because the conventional hybrid model itself does not hold in this case. The transistor analysis at high frequencies adopts a different circuit archetype, known as *hybrid-Π model* ($\Pi \equiv pi$). That model for an n–p–n transistor in CE configuration is sketched in Figure 7.13. It has the following parameters.

A. Transconductance

When the time taken by the diffusion of charge carriers from one region of the transistor to another becomes comparable to the switching time of the polarity of the high frequency signal, the immediate responding of the transistor to the instantaneous changes of input voltage or current is no longer valid. Consequently, if the CE configuration is considered, the assumption of the transistor output current source $h_{fe}i_b$ for input current i_b is not reasonable. However, the concentration of the minority carriers injected into the base is still proportional to the input voltage v_{be} across the base–emitter junction so that the output current source can be represented by $g_m v_{be}$ where g_m, the *transconductance* represents the mutual relationship between the input and the output ports of the device. At room temperature, this transconductance can be approximated by

$$g_m = \frac{I_C(\text{mA})}{26} \qquad (7.37)$$

This may be treated as the inverse of the dynamic resistance of a p–n junction [Equation (3.14)]. The number 26 in Equation (7.37) comes from the standard value of kT/q at room temperature and it is obvious that the transconductance is inversely proportional to temperature.

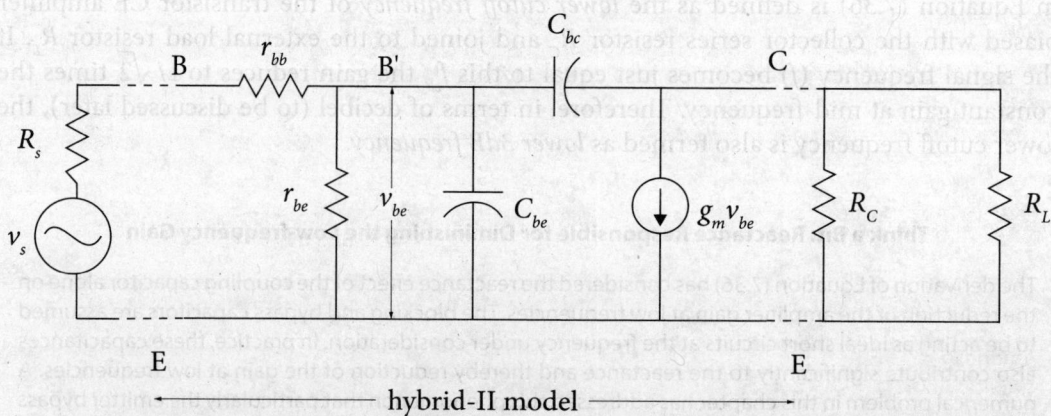

Figure 7.13 Hybrid-Π model for a transistor in CE configuration showing the resistive (r_{bb} and r_{be}) and capacitive (C_{be} and C_{bc}) parameters of the device, the transistor output current source ($g_m v_{be}$) and the positions for the external signal source (v_s) with internal resistance (R_s) and the parallel combination of collector series resistor (R_C) and external load resistor (R_L).

B. Other Conductance

The minority carriers injected into the base cause an enhanced recombination current, which is addressed by a conductance g_{be} between the base and the emitter. The inverse of this conductance is treated as a resistance r_{be} ($= 1/g_{be}$). Another resistance, known as the *base-spreading resistance* (r_{bb}) arises at the input due to the bulk resistance of the base material. In practice, $r_{bb} \ll r_{be}$.

The collector voltage causes base-width modulation due to the Early effect (Chapter 5) that makes change in the emitter current, hence in the collector current. This feedback effect is accounted for by a conductance g_{ce} between the collector and the emitter. The reciprocal of this conductance is considered as a resistance r_{ce} (= $1/g_{ce}$). A conductance is created between the base and the collector also but that is negligibly small.

C. Junction Capacitances

The capacitances of the two junctions become significant at high frequencies. The capacitance of the base–emitter junction and that of the collector–base junction are indicated by C_{be} and C_{bc}, respectively. The lowercase subscripts signify that these take place for high frequency ac signals only and have no permanent existence in the circuit.

Due to *Miller effect*, the base-to-collector capacitance (C_{bc}) is divided into an input and an output part, the input component (C_i) being much more prominent than the output component and appearing in parallel with C_{be}.

D. Comparison with Low-Frequency Hybrid Model

The input, transfer and output parameters of the hybrid-Π model have the following relationships with the corresponding low-frequency hybrid parameters.

$$r_{bb} = h_{ie} - \frac{1}{g_{be}} = h_{ie} - r_{be} \qquad (7.38a)$$

$$g_m = h_{fe} g_{be} = \frac{h_{fe}}{r_{be}} \qquad (7.38b)$$

$$g_{ce} \approx h_{oe} \qquad (7.38c)$$

Note: Miller Effect

American engineer John Miller (1882–1962) propounded some theorem for creating equivalent circuits. An inference of Miller's theorem is the Miller effect, which states that the equivalent input capacitance of an inverting voltage amplifier can get increased due to the voltage gain. The enhanced input capacitance, termed as Miller capacitance (C_M) is given by $C_M = C_f(1 + |A_v|)$ where $-A_v$ is the voltage gain of the inverting amplifier and C_f is the feedback capacitance. Not only capacitance, any impedance connected between the input and a gain-producing terminal can modify the amplifier input impedance through this effect. In the case of BJT, the Miller effect can drastically increase the junction capacitance between the base and the collector. Here the base-to-collector capacitance (C_{bc}) stands for C_f and appears in enhanced form between the input and the ground. The CE amplifier has the most prominent Miller effect. The CC and CB amplifiers have already one side directly grounded and the Miller effect becomes negligible. For CC mode, V_{cc} is ac grounded so that the input capacitance equals C_{bc}. For CB mode, C_{bc} is grounded at the base end and does not affect the input capacitance (C_{be}).

> **Example 7.5**
>
> A transistor used in CE circuit has $h_{fe} = 100$, $h_{ie} = 2$ kΩ, $C_{bc} = 4$ pF and $C_{be} = 10$ pF. If the collector resistance is 2 kΩ, calculate the effective input capacitance including the Miller effect.
>
> [C.U. 2009]
>
> **Solution**
>
> The voltage gain is $-(100 \times 2k\Omega)/2k\Omega = -100$
>
> The effective input capacitance is $10 \text{ pF} + (1 + 100) \times 4 \text{ pF} = 414 \text{ pF}$

7.11 Transistor Gain at High Frequency

Considering the above features of hybrid-Π model, the circuit of Figure 7.13 is simplified further as shown in Figure 7.14. Let the voltage source (v_s) have a small internal resistance R_s. Applying Thevenin's theorem between B' and E terminals, the Thevenin resistance at the input becomes

$$R_{Ti} = r_{be} // (r_{bb} + R_s) \tag{7.39}$$

The Thevenin voltage at the input becomes

$$v_{Ti} = \frac{v_s r_{be}}{r_{be} + r_{bb} + R_s} = kv_s \tag{7.40}$$

where

$$k = \frac{r_{be}}{r_{be} + r_{bb} + R_s} \tag{7.40a}$$

is a dimensionless constant fraction. For an ideal voltage source, $R_s \approx 0$, hence $k \approx 1$.

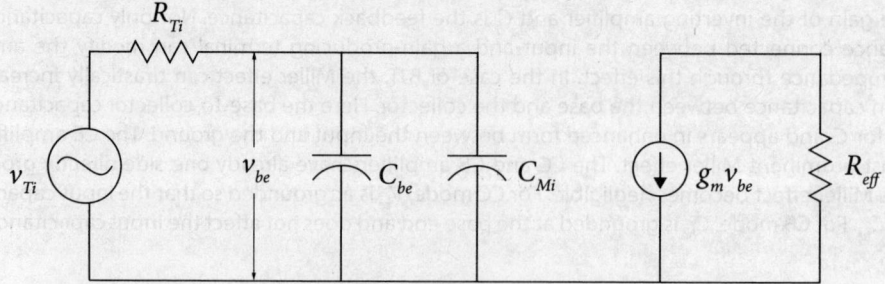

Figure 7.14 Simplified hybrid-Π model for the transistor in CE configuration showing the Thevenin input voltage (v_{Ti}), the Thevenin input resistance (R_{Ti}), the input Miller capacitance (C_{Mi}), and the effective load resistance (R_{eff})

Following the Miller effect, the input component of C_{bc} appearing in parallel with C_{be} is represented by C_{Mi} so that the equivalent capacitance at the input is

$$C_{eq} = C_{be} + C_{Mi} \qquad (7.41)$$

The effective load (R_{eff}) at the output is the parallel combination of r_{ce}, R_C and R_L. The conductance g_{ce} is very small so that r_{ce} can be treated as open circuit and the effective load resistance can be approximated as

$$R_{eff} = \frac{R_C R_L}{R_C + R_L} \qquad (7.42)$$

The input voltage is given by

$$v_{be} = \frac{1/j\omega C_{eq}}{R_{Ti} + 1/j\omega C_{eq}} k v_s = \frac{k v_s}{1 + j\omega C_{eq} R_{Ti}} \qquad (7.43)$$

Therefore,

$$v_s = \frac{v_{be}(1 + j\omega C_{eq} R_{Ti})}{k} \qquad (7.44)$$

The output voltage is, obviously,

$$v_o = -g_m v_{be} R_{eff} \qquad (7.45)$$

From Equations (7.44) and (7.45), the voltage gain of the CE amplifier at high frequency can be defined as

$$A_{vh} = \frac{v_o}{v_s} = \frac{-k g_m R_{eff}}{1 + j\omega C_{eq} R_{Ti}} \qquad (7.46)$$

The numerator ($-k g_m R_{eff}$) of Equation (7.46) can be considered almost equivalent to the mid-frequency gain (A_v) stated by Equation (7.32) with the following reasonable approximations. For an ideal voltage source, $R_s \to 0$ and in practice, $r_{bb} \ll r_{be}$ so that $k \to 1$ in Equation (7.40a). Using Equations (7.38a) and (7.38b), if $r_{be} \approx h_{ie}$, then $g_m \approx h_{fe}/h_{ie}$. The effective load R_{eff} is defined by Equation (7.42) where R_C can be made of the order of h_{ie}. Considering all the above, Equation (7.46) can be modified in the following way to represent the frequency dependent voltage gain of the CE amplifier at high frequency.

$$A_{vh} = \frac{A_v}{1 + j\dfrac{f}{f_{cu}}} \qquad (7.47)$$

where the frequency

$$f_{cu} = \frac{1}{2\pi C_{eq} R_{Ti}} \qquad (7.47a)$$

is defined as the *upper cutoff frequency* of the transistor CE amplifier. In terms of decibel, this is called the *upper 3dB frequency* because when the signal frequency (f) becomes equal to this f_{cu}, the voltage gain reduces to $1/\sqrt{2}$ times the constant gain at mid-frequency.

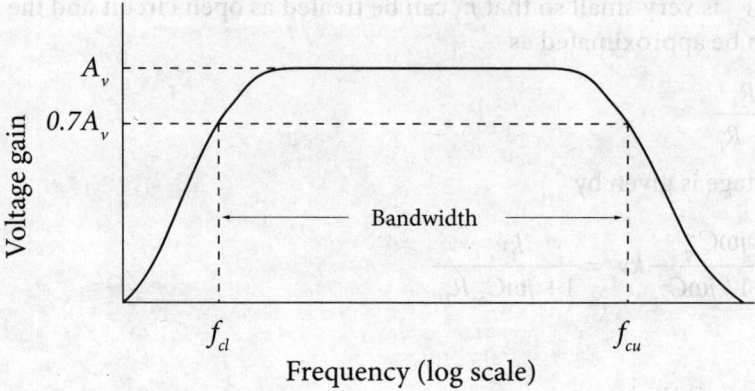

Figure 7.15 Gain-frequency plot for the transistor CE amplifier showing the constant voltage gain (A_v) over a range of frequency, its reduction at lower cutoff (f_{cl}) and upper cutoff (f_{cu}) frequencies and the bandwidth

> **Note: Gain–Bandwidth Product**
>
> The physical interpretation of the gain–bandwidth product can be obtained from the hybrid-Π model parameters. For an ideal signal source, the source resistance (R_s) is zero and the Thevenin resistance in Equation (7.39) becomes almost $r_{bb'}$, which is a small fraction ($k \approx 0.1$) of r_{be}. Assuming $r_{bb} = kr_{be}$ and using $r_{be} = h_{fe}/g_m$ from Equation (7.38b), the upper cutoff frequency in Equation (7.47a) becomes
>
> $$f_{cu} = \frac{g_m}{2\pi k C_{eq} h_{fe}} \text{ that can be rearranged as } f_{cu} \times h_{fe} = \frac{g_m}{2\pi k C_{eq}}$$
>
> The bandwidth is ($f_{cu} - f_{cl}$) but $f_{cu} \gg f_{cl}$, hence bandwidth $\approx f_{cu}$. Also the maximum possible voltage gain is h_{fe}. Thus the above expression indicates that the product of the voltage gain and the bandwidth of the amplifier is a constant quantity because the right side terms are constants, particularly for a fixed dc operating point when the dc value of the collector current is unchanged. A few words more can be concluded. The right side has the dimension of frequency. So, it is a measure of the high frequency when the gain becomes unity.

Assembling the conditions of Equations (7.32), (7.36) and (7.47), the frequency variation of the voltage gain of the CE amplifier can be obtained. A curve similar to that of Figure 7.15 is acquired by continuously varying the frequency of the input signal and plotting the corresponding voltage gain. The frequency is plotted in logarithmic scale because the gain remains constant (A_v) over a large range of frequency. Such a plot is termed as 'semilog plot'.

The lower and the upper cutoff frequencies (f_{cl} and f_{cu}, respectively) are indicated in the figure when the gain reduces to $1/\sqrt{2} (\approx 0.7)$ times the maximum gain. The frequency span between these two limits is known as the *bandwidth* of the amplifier. It quantifies the range of input signal frequency over which the gain of the amplifier remains almost constant. The product of gain and bandwidth of an amplifier, known as *gain–bandwidth product* is a constant quantity.

7.12 Gain and Decibel

The voltage gain or the current gain of an amplifier is a ratio of similar quantities and these should not have any unit or dimension. However, sometimes two levels of voltage, current or power are compared in terms of their logarithmic ratio, termed as *decibel*. Strictly speaking, it is not any 'unit' of any standard system but it serves the purpose of quantifying a physical parameter.

Let there be two power levels P_1 and P_2. The P_2 level with respect to P_1 level is expressed in decibel (dB) by log of ratio as $10\log_{10}(P_2/P_1)$. For example, if the output power of an amplifier be 100 times the input power, the power gain in decibel is written as $10\log_{10}(100/1) = 20$ dB. It could also be expressed as $\log_{10}(100/1) = 20$ bel but the 'bel' unit is too large a quantity for practical purpose and the decibel (= bel/10) unit is the most optimum scale. Since power is proportional to the square of voltage or current, it is customary to express the decibel voltage and current gains in terms of the input and output quantities in the following way.

- Voltage gain in decibel is $20\log_{10} \dfrac{v_{out}}{v_{in}}$

- Current gain in decibel is $20\log_{10} \dfrac{i_{out}}{i_{in}}$

The decibel system can be used for any form of energy, such as sound level or light intensity level. The decibel system is actually based on the fact that the human response to seeing or listening is nonlinear and can be approximated by logarithmic variation.

Following the above concept of decibel, the voltage gain at the lower cutoff frequency [Equation (7.36)] or at the upper cutoff frequency [Equation (7.47)] can be expressed with respect to the mid-frequency constant voltage gain [Equation (7.32)] as

$$20\log_{10}\left(\frac{A_v}{\sqrt{2}A_v}\right) \approx -3 \text{ dB} \tag{7.48}$$

Therefore, f_{cl} in Equation (7.36a) is designated as *lower 3 dB frequency* and f_{cu} in Equation (7.47a) is labeled as *upper 3 dB frequency*. Since power is square of voltage, the power at these two frequencies are half of that at mid-frequency. So these two frequencies are also called *half power frequency*.

Multiple Choice-Type Questions and Answers

7.1 Thevenin's theorem and Norton's theorem are applicable to circuits comprising
  (i) linear elements   (ii) only ac voltage sources
  (iii) only dc voltage sources (iv) any of dc or ac voltage sources
  (a) only (iii) is true   (b) only (i) is true
  (c) only (iv) is true   (d) both (i) and (iv) are true

7.2 Thevenin's theorem cannot be applied to networks containing..........................elements.
  (a) linear   (b) nonlinear   (c) active   (d) passive

7.3 The hybrid model is applicable to .... transistors.
  (a) only n–p–n   (b) only power
  (c) both n–p–n and –p–n–p (d) only switching

7.4 The h parameters obtainable from the CE static output characteristics are ....
  (a) h_{fe} and h_{re} (b) h_{re} and h_{oe}
  (c) h_{fe} and h_{oe} (d) h_{ie} and $h_{fe10.5}$

7.5 The h parameters obtainable from the CE static input characteristics are ....
  (a) h_{fe} and h_{ie} (b) h_{ie} and h_{re} (c) h_{re} and h_{oe} (d) h_{fe} and h_{re}

7.6 A CE amplifier has $h_{fe} = 100$ and $I_C = 5.2$ mA. The h_{ie} is approximately ....
  (a) 0.5 kΩ   (b) 1.0 kΩ   (c) 2.6 kΩ   (d) 5.2 kΩ
  (Hints. $h_{ie} = \beta r_e$, $r_e = 26/5.2$)

7.7 A CE amplifier has collector current of 1.3 mA and the resistance joined in series with the collector is 1 kΩ. If the transistor has $h_{fe} = 100$, the voltage gain is approximately ....
  (a) 100   (b) 50   (c) 2.6   (d) 26

7.8 The hybrid parameters of a transistor are
  (i) fixed quantities
  (ii) dependent on temperature
  (iii) dependent on the position of the quiescent point on the output characteristic curve
  (iv) dependent on the position of the quiescent point on the input characteristic curve
  (a) statements (ii) and (iii) are true   (b) statements (ii) and (iv) are true
  (c) statements (i) and (ii) are true   (d) statements (ii), (iii) and (iv) are true

7.9 The parameter h_{fe} depends on I_C and V_{CE} in the following way.
  (a) h_{fe} depends on both I_C and V_{CE} and is more sensitive to I_C than to V_{CE}.
  (b) h_{fe} depends on both and more sensitive to V_{CE} than to I_C.
  (c) h_{fe} depends on both and is equally sensitive to both I_C and V_{CE}.
  (d) h_{fe} depends on I_C only and not on V_{CE}.

7.10 Only two of the four hybrid parameters, namely .... and ....
  are sufficient to the approximate analysis of transistor operations at ....
  provided the load resistance is .... the $1/h_{oe}$ of the transistor.
  (a) h_{fe}, h_{oe}, low, much less than   (b) h_{ie}, h_{fe}, low, much less than
  (c) h_{fe}, h_{ie}, low, comparable to   (d) h_{ie}, h_{oe}, low, greater than

7.11 The output voltage across the load resistor of a capacitor coupled CE amplifier is
....

(a) dc only
(b) ac only
(c) a mixture of dc and ac
(d) dc and ac half-half

Answers

7.1	(d)	7.2	(b)	7.3	(c)	7.4	(c)
7.5	(b)	7.6	(a)	7.7	(b)	7.8	(d)
7.9	(a)	7.10	(b)	7.11	(b)		

Reasoning-Type Questions and Answers

7.1 Can the transformer be represented by two-port network? Can one assign the primary and secondary voltages as the independent variables?

Ans. Yes, the working of a transformer can be modeled with two-port network; one port for the primary and the other port for the secondary of the transformer. However, the primary and the secondary voltages are related by a constant (turns ratio). So one cannot consider these two as the independent variables.

7.2 Can it be stated that the frequency dependence of the gain of a BJT amplifier is a capacitive effect?

Ans. It is not too unreasonable to say so. The different external and internal capacitances of the device play important role in influencing the gain at different frequencies of the input signal.

At low frequencies, the external biasing capacitors, namely the blocking, bypass and coupling capacitors offer significant reactance that reduce the gain. The internal capacitances of the two junctions act as open circuits.

At medium range of frequencies, all the external capacitors act as short circuit and the internal (junction) capacitances act as open circuit.

At high frequencies, all the external capacitors act as short circuit and the junction capacitances offer notable reactance that diminish the gain.

Solved Numerical Problems

7.1 If a transistor with current gain 100 has emitter current of 2 mA and output resistance of 50 kΩ, draw the approximate hybrid equivalent circuit.

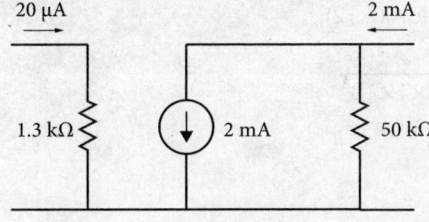

Figure 7.16

Soln. The ac emitter resistance (r_e) is

(26 mV)/(2 mA) = 13 Ω

It is given that $\beta = h_{fe} = 100$

Therefore, the input resistance (h_{ie}) is $\beta r_e = 1.3$ kΩ

Assuming the emitter current almost equal to the collector current, the base current is

(2 mA)/100 = 20 μA

The output resistance ($1/h_{oe}$) is 50 kΩ (given).

Using all the above values, the approximate hybrid equivalent circuit can be sketched as that given in Figure 7.16.

7.2 Draw the approximate hybrid equivalent circuit and calculate the numerical values of the parameters for the fixed bias transistor circuit of Figure 7.17(a). Given: base current = 40 μA, current gain = 50, R_B = 470 kΩ, R_C = 2.2 kΩ and the transistor output resistance = 50 kΩ.

Soln. The required hybrid equivalent circuit is presented in Figure 7.17(b). The collector current ($h_{fe}i_b$) is

50×40 μA = 2 mA

The ac emitter resistance (r_e) is

(26 mV)/(2 mA) = 13 Ω

The input resistance (h_{ie}) is

50×13 = 650 Ω

The total input resistance of the hybrid equivalent circuit is

$R_B // h_{ie} \approx h_{ie} = 650$ Ω because $R_B \gg h_{ie}$

Similarly, the output resistance is

$(1/h_{oe}) // R_C \approx R_C = 2.2$ kΩ because $1/h_{oe} \gg R_C$

7.3 Consider a CE amplifier that has a load resistance of 1 kΩ. The h parameters are h_{ie} = 1.2 kΩ, $h_{re} = 7\times10^{-4}$, h_{fe} = 135 and h_{oe} = 50 μS. Determine the voltage gain and the current gain of the amplifier.

Soln. Using Equation (7.15), the generalized expression for the voltage gain (A_v) in CE mode,

$$A_v = \frac{-135 \times 1 \times 10^3}{1.2 \times 10^3 + (1.2 \times 10^3 \times 50 \times 10^{-6} - 135 \times 7 \times 10^{-4}) \times 1 \times 10^3}$$
$$= -109.36$$

Let us consider the magnitude only. Similarly, using the expression for the current gain (A_i) in CE mode [Equation (7.12)],

$$A_i = \frac{135}{1 + 50 \times 10^{-6} \times 1 \times 10^3}$$
$$= 128.57$$

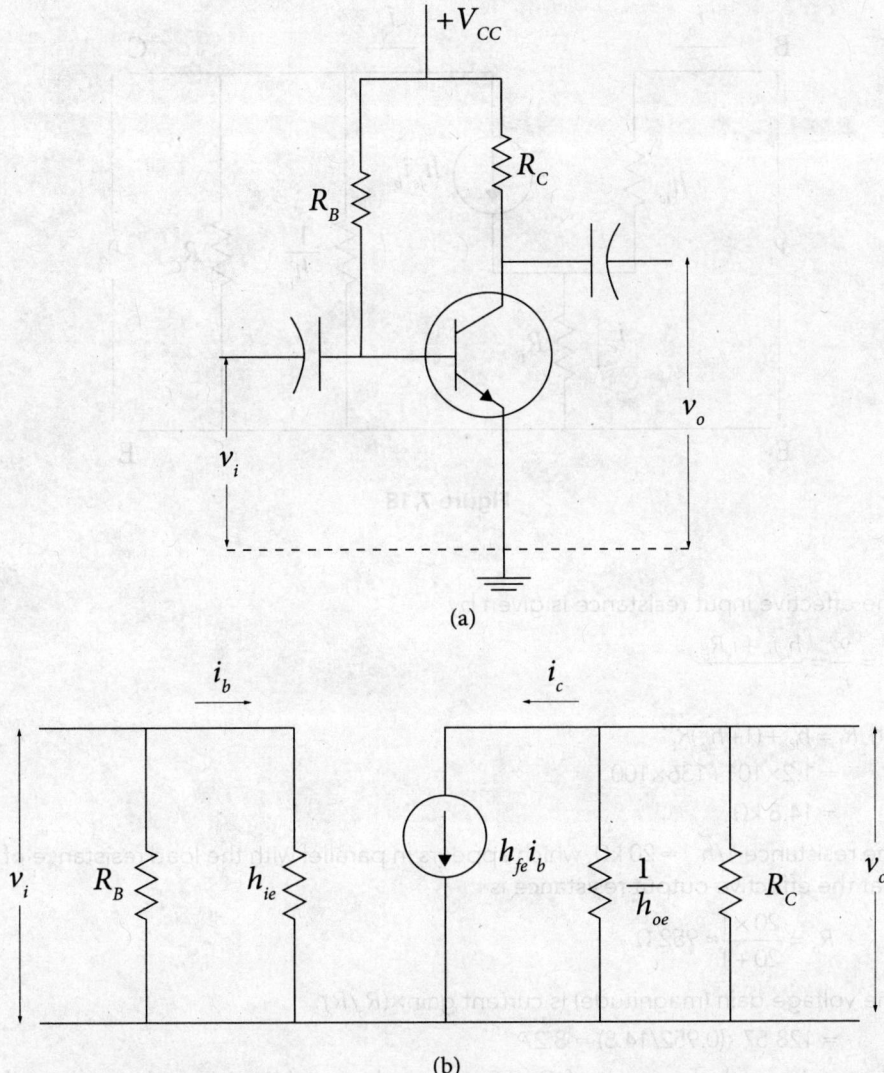

Figure 7.17

7.4 In the above example (7.3), the emitter resistor was supposed to be perfectly bypassed. Now suppose that the emitter resistance is of 100 Ω and that is not bypassed. In such a condition, draw the hybrid equivalent circuit and determine the voltage and current gains.

Soln. The approximate hybrid model for the CE amplifier with unbypassed emitter is given in Figure 7.18. The current gain remains the same as that calculated in the previous example (Q. 7.3), i.e. 128.57. The voltage gain, however, differs because the input resistance gets changed.

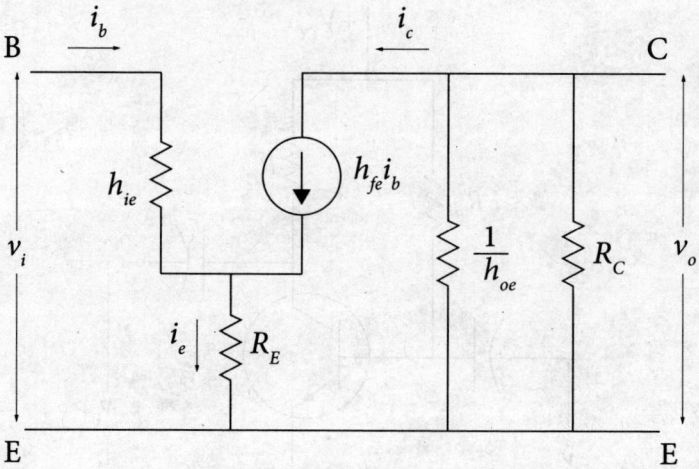

Figure 7.18

The effective input resistance is given by

$$R_i = \frac{v_i}{i_b} = \frac{h_{ie}i_b + i_e R_E}{i_b}$$

Or, $R_i = h_{ie} + (1+h_{fe})R_E$
$= 1.2 \times 10^3 + 136 \times 100$
$= 14.8 \text{ k}\Omega$

The resistance $1/h_{oe} = 20$ kΩ, which appears in parallel with the load resistance of 1 kΩ so that the effective output resistance is

$$R_o = \frac{20 \times 1}{20 + 1} \approx 952 \text{ }\Omega$$

The voltage gain (magnitude) is current gain×(R_o/R_i)
$= 128.57 \times (0.952/14.8) = 8.27$

[Comparing with the result of Q. 7.3, it is understood that the voltage gain of the CE amplifier diminishes severely in absence of bypassing the emitter resistor.]

7.5 The transistor amplifier circuit of Figure 7.19(a) corresponds to the following parameters: $h_{fe} = 100$, $R_C = 2$ kΩ, $h_{ie} = 560$ Ω, $R_E = 1$ kΩ and $R_B = 600$ kΩ, symbols having their usual meaning. Determine the simplified hybrid model for this amplifier and calculate (i) the input resistance, (ii) the output resistance and (iii) the voltage gain of the amplifier.

Soln. The required hybrid model is given in Figure 7.19(b) where the parameters h_{re} and h_{oe} are neglected. For the ac signal, the base circuit resistor (R_B) appears to take place between the base and the ground and the collector circuit resistor (R_C) acts as the load.

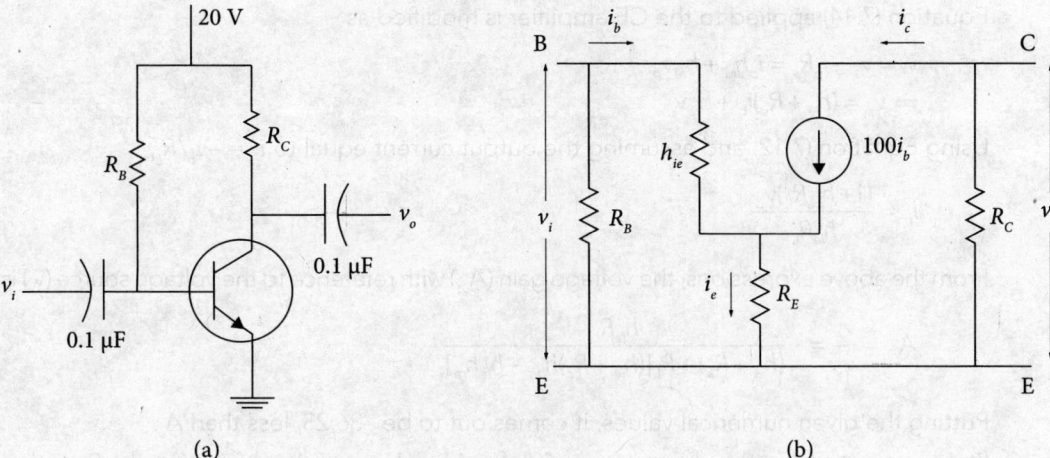

Figure 7.19

(i) The input resistance between the input terminals is (as derived in Q. 7.4)

$R_i = h_{ie} + (1 + h_{fe})R_E$

$= 560 + 101 \times 1 \times 10^3$

$= 101.56 \text{ k}\Omega$

This occurs in parallel with R_B (= 600 kΩ) so that the ultimate input resistance becomes

(600×101.56)/(600 + 101.56) = 86.86 kΩ

(ii) The output resistance is effectively R_C (= 2 kΩ)

(iii) The voltage gain is [modified form of Equation (7.16)]

$$-\frac{h_{fe}R_C}{R_i} = -\frac{100 \times 2}{86.86} = -2.30$$

[Again it is noted that a CE amplifier with unbypassed emitter results in a very small voltage gain.]

7.6 Compute the voltage gain, the input impedance and the output impedance of a transistor amplifier connected in CE configuration given that h_{ie} = 2 kΩ, h_{re} = 10^{-3}, h_{fe} = 100, h_{oe} = 10^{-4} mho, R_L = 1 kΩ and R_s (output impedance of the source) = 600 Ω.

[C.U. 2000]

Soln. The voltage gain (magnitude) of the amplifier is [using Equation (7.15)]

$$A_v = -\frac{100 \times 1000}{2000 + 1000 \times (2000 \times 10^{-4} - 100 \times 10^{-3})} = -47.62$$

When the source impedance (R_s) is included in the amplifier, the voltage gain (A_{v1}), as seen by the voltage source (v_s) is derived in the following way.

Equation (7.14) applied to the CE amplifier is modified as

$$v_i = v_s - i_b R_s = i_b h_{ie} + h_{re} v_o$$
$$\Rightarrow v_s = (h_{ie} + R_s)i_b + h_{re} v_o$$

Using Equation (7.12) and assuming the output current equal to $i_o = -v_o/R_L$,

$$i_b = \frac{(1+h_{oe}R_L)v_o}{h_{fe}R_L}$$

From the above expressions, the voltage gain (A_{v1}) with reference to the voltage source (v_s) is

$$A_{v1} = \frac{v_o}{v_s} = -\frac{h_{fe}R_L}{(h_{ie}+R_s)+R_L[(h_{ie}+R_s)h_{oe}-h_{fe}h_{re}]}$$

Putting the given numerical values, it comes out to be −36.23, less than A_v.

[It may be noted that the above expression for A_{v1} reduces to that of A_v given by Equation (7.15) when $R_s \to 0$ and $v_s \to v_o$.]

The current gain is given by [Equation (7.12)]

$$\frac{100}{1+10^{-4} \times 1 \times 10^3} = 90.9$$

The input impedance z_i (actually resistance) of the amplifier in presence of source impedance (R_s) can be derived in the following way.

$$v_i = v_s - R_s i_b = (h_{ie} - h_{re}A_i R_L)i_b$$
$$\Rightarrow z_i = \frac{v_s}{i_b} = h_{ie} + R_s - h_{re}A_i R_L$$

Putting the given numerical values, z_i = 2.51 kΩ

The expression for the output impedance (z_o) given by Equation (7.21) can be modified in presence of source impedance (R_s) as follows.

$$z_o = \frac{1}{h_{oe} - \dfrac{h_{fe}h_{re}}{h_{ie}+R_s}}$$

Putting the given numerical values, $z_o = 1.62 \times 10^{-4}$ Ω

7.7 In the case of a CE amplifier with unbypassed emitter, derive an expression for the input resistance using the hybrid model.

Soln. [The required expression has already been derived in connection with solved numerical problem 7.4. Another form of that expression is derived here.]

For any input signal (v_i) entering the transistor,

$$v_i = \beta r_e i_b + i_e R_E$$

where $h_{ie} = \beta r_e$ and $i_e = (\beta+1)i_b$.

Considering $\beta+1 \approx \beta$,

$$v_i = \beta r_e i_b + \beta i_b R_E$$

Therefore, the input resistance is

$$\frac{v_i}{i_b} = \beta(r_e + R_E)$$

It may have two extreme cases.

(i) When R_E is bypassed (as in the CE amplifier), $R_E \to 0$ and the input resistance is βr_e.
(ii) When R_E is not bypassed (as in CC amplifier and $R_E \gg r_e$), the input resistance is βR_E.

7.8 The mid-band gain of an RC coupled amplifier is 120. At frequencies of 100 Hz and 100 kHz, the gain falls to 60. Determine the bandwidth.

[C.U. 2013]

Soln. Following Equation (7.36), the magnitude of the gain at low frequency (= 100 Hz) can be expressed in terms of the mid-band gain (= 100) and the lower cutoff frequency (f_{cl}) as

$$60 = \frac{120}{\sqrt{1+\left(\frac{f_{cl}}{100}\right)^2}} \text{ so that } f_{cl} = 173.2 \text{ Hz}$$

Similarly, following Equation (7.47), the magnitude of the gain at high frequency (= 100 kHz) is expressed in terms of the mid-band gain (= 100) and the upper cutoff frequency (f_{cu}) as

$$60 = \frac{120}{\sqrt{1+\left(\frac{100 \times 10^3}{f_{cu}}\right)^2}}, \text{ hence } f_{cu} = 57.73 \text{ kHz}$$

Therefore, the bandwidth is $f_{cu} - f_{cl} = 57.56$ kHz

Exercise

Subjective Questions

1. What do you mean by two-port model for a device/circuit? What are hybrid parameters in this connection?
2. Explain under what conditions the two-port model and the hybrid parameters can be applied to the operation of a transistor. How can the transistor hybrid model be further simplified?
3. What are the useful features of h parameters in relation to the transistor operation?
4. Draw the hybrid models for n–p–n and p–n–p transistors and point out the differences.
5. Draw the hybrid model for a CE amplifier and derive the expressions for its current gain, voltage gain, input impedance and output impedance.
6. Why does the voltage gain of a transistor become frequency dependent? Derive expressions for the voltage gain at medium and low frequencies of the input voltage.
7. Draw the hybrid-Π model for a transistor operating at high frequencies and explain the parameters associated with it.
8. Define the lower and upper cutoff frequencies and the bandwidth for a transistor.

Conceptual Test

1. Express the h parameters of two-port network as a square matrix of order 2×2 relating the voltage and current of the input and output ports.
2. The method of calculation of the h parameters is also 'hybrid'; sometimes open circuited and sometimes short circuited. Clarify.
3. Why is small amplitude of the input signal essential in hybrid models?
4. The small-signal ac response of a transistor can be described by both hybrid model and r_e model. However, the transistor datasheet often mentions the h parameters. Justify.
5. Justify that $h_{fe}i_b$ has not only the dimension of current, it is actually a current source.
6. The ac component $i_e r_e$ is equal to the ac base voltage. Justify the statement.

Numerical Problems

1. The current read by the ammeter (A) in the circuit of Figure 7.20 is
    (a) 27.3 mA   (b) 100.0 mA   (c) 54.5 mA   (d) 50.0 mA

    [TIFR 2011]

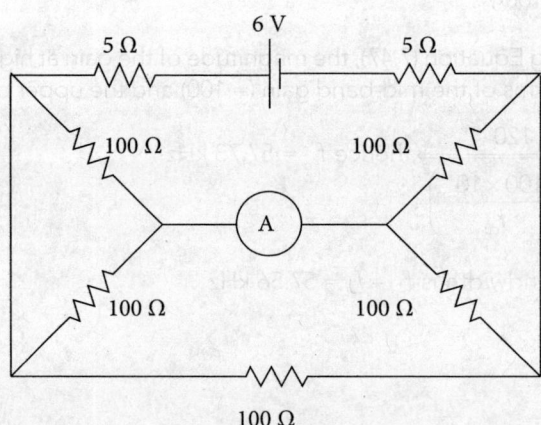

Figure 7.20

2. A small signal transistor amplifier with base bias has R_B = 470 kΩ, R_C = 1 kΩ, I_B = 40 μA and β = 40, symbols having their usual meaning. Draw the approximate hybrid equivalent circuit for the amplifier.
3. Calculate the input impedance, output impedance and voltage gain of the transistor amplifier given in Figure 7.21. Given: β = 200, V_{BE} = 0.7 V and r_e = (26 mV)/I_E.

    [JAM 2013]
4. A CE amplifier is designed with an n–p–n transistor having β = 100. The supply voltage (V_{CC}) is 20 V and the biasing resistors are R_1 = 40 kΩ, R_2 = 10 kΩ, R_C = 4 kΩ and R_E = 2 kΩ, notations having the usual meaning. The input blocking capacitor (C_B) is of 10 μF, the output coupling capacitor (C_C) is of 1 μF and the emitter bypass capacitor (C_E) is of 20 μF. Draw the approximate hybrid equivalent circuit for this amplifier and determine its lower cutoff frequency.

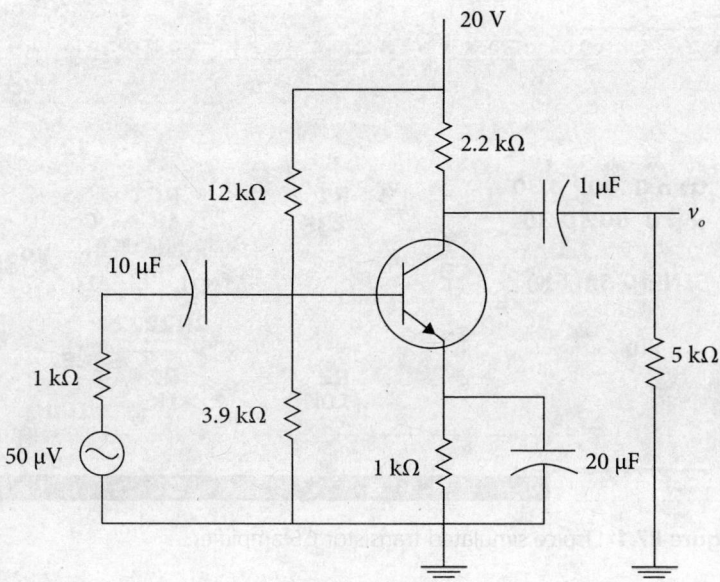

Figure 7.21

5. An amplifier having output resistance of R_o can amplify an input by 5 times. If an external load resistance of 10 kΩ is joined, the amplification becomes 2 times the input. From these data, determine the value of R_o.

Project Work on Chapter 7

To study the linearity and frequency response of a transistor CE amplifier of a given voltage gain.

Theory:
The voltage gain of the transistor CE amplifier remains constant over a wide range of input signal frequency and reduces at very low and very high frequencies characterized by lower cutoff and upper cutoff frequencies. The linearity of the amplifier is maintained over the frequency range of constant gain where the output voltage is proportional to the input voltage.

Computer Analysis:
A circuit similar to that fabricated in the project of Chapter 6 (Figure P6.1) was simulated with LTspice, as presented in Figure P7.1. The component values are slightly changed.

Results and Discussion:
The simulated data for the input and output waveforms are plotted in Figure P7.2. The following features are noted.

- The output voltage amplitude is amplified by almost 100 times. (The exact increase can be obtained from the dataset).
- The output waveform is 180° out of phase with respect to the input waveform.
- The sinusoidal waveform is retained and is not distorted at the output.

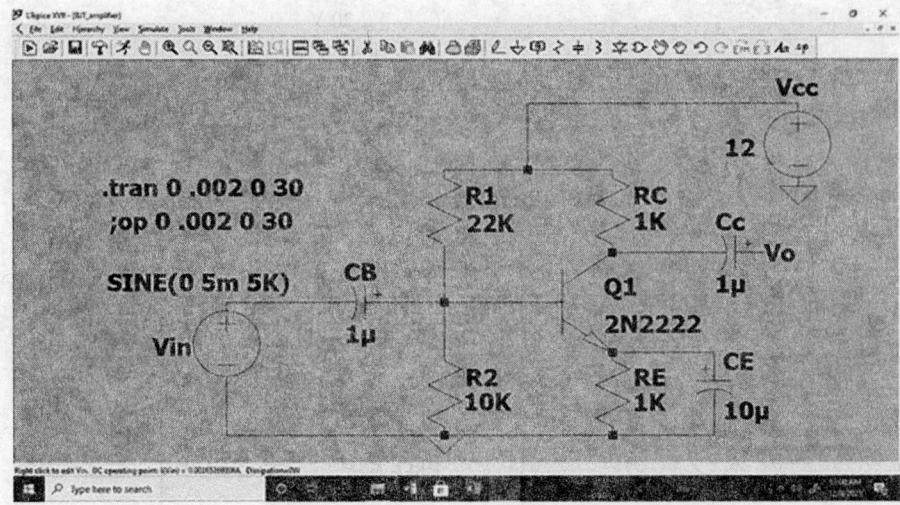

Figure P7.1 LTspice simulated transistor CE amplifier

The following extensions of the work should be made.

1. For a moderate frequency, such as 5 kHz of the input signal, increase the input amplitude and note the corresponding output amplitude. The graph of input versus output amplitude graph would be straight line up to a certain extent. This determines the range of linearity of the amplifier and the input amplitude should be well within this range.
2. For a reasonable, constant amplitude of the input signal, change its frequency and note the output amplitude. The amplitude remains constant over a wide range of frequency and decreases at very low and very high values.
3. Repeat the above frequency variation by joining different values of C_o shown in Figure P6.1 and note the change in bandwidth.

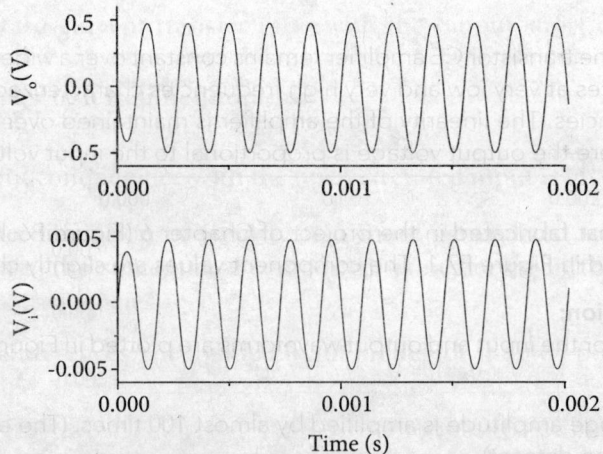

Figure P7.2 LTspice simulated input and output voltage waveforms for the CE amplifier

# 8

Transistor Power and Multistage Amplifiers

The essential conditions for a bipolar junction transistor to work as power amplifier are illustrated in this chapter. Sometimes the output of an amplifier is used as the input of another amplifier in order to produce larger amplification. Such cascading of two or more amplifier stages is called coupling, which may be done with resistor, capacitor transformer, or with direct connection. Different types of coupling and the resultant multistage amplifiers are discussed. Several classes of amplifier operation, such as A, B, AB and C are introduced. Some specific types of biasing techniques suitable for transistor power amplification, such as push–pull and tuned amplifier are explained.

8.1 Need for Power Amplification

In Chapters 5 through 7, we have come across different types of transistor configurations and biasing circuits acting as voltage or current amplifiers. The amplifier converts a portion of the electrical energy obtained from the dc power supply into the energy obtained at the output in proportion to the input. The input signal just controls the mode of conversion. There are wide varieties of amplifier depending on the requirement of

- ac or dc amplification,
- voltage, current, or power amplification,
- amplification over a wide range of frequency of the input signal (wideband amplifier), and
- amplification around only a fixed frequency of the input signal (narrowband or tuned amplifier) and many others.

The loudspeaker in a public address system is a very common example where high level amplification is required for a weak electrical signal. Servomechanisms, such as the movement of the motor in a printer connected to a computer and signal transmissions in radio/television broadcasting are other popular examples where high level amplifications are compulsory. Such large extent of amplification cannot be achieved with the BJT amplifiers discussed in Chapters 6 and 7 because of the following constraints.

(i) Achieving voltage gain is not possible for large signals because the positive and negative swings of a large (≥ 0.7 V) ac input would drive the Q-point to saturation and cutoff, respectively.

(ii) The current gain can still be achieved because it is the fundamental property of the transistor.

(iii) The nonlinearity in the transistor transfer characteristic becomes predominant for large input signals.

Considering the above features simultaneously, one can speculate that amplification of large input signals can be achieved with neither voltage, nor current but with the product of the voltage and the amplified current, i.e. the electrical power, provided that the effects of nonlinearity can be minimized. Table 8.1 presents a comparative view of the amplification conditions relevant to small and large input signals. It is apparent that the requirements for power amplification are somewhat different from those for voltage and current amplification. That calls for special types of transistors with special biasing techniques, as outlined in the next section.

Table 8.1 Transistor amplifiers for small and large input signals

BJT Small Signal Amplifier	BJT Large Signal Amplifier
It aims at gain and linearity in amplification. The objective is to increase the voltage/current level of the input signal.	It aims at high power efficiency. Nonlinearity is unavoidable but can be minimized with suitable biasing circuits.
It amplifies voltage or current and the power efficiency is not so important here.	It provides sufficient power to the output load. Therefore, not only the power efficiency, the power handling capability of the circuit and impedance matching to the output load are also important.
The specified current gain (β) of the transistor for voltage amplification is large (> 100). So the amplifier provides with high voltage gain.	The β of the transistor meant for power amplification is relatively small (< 50). So this amplifier can provide high power at the output but not much voltage gain.

Table 8.2 Comparison of the requirements for voltage and power amplifiers

Voltage Amplifier	Power Amplifier
It is designed for achieving the maximum voltage amplification.	It is planned for attaining the maximum output power.
Both bipolar junction transistors and field-effect transistors are used for constructing voltage amplifiers.	Bipolar junction transistors are more suitable for power amplifiers.
The resistance in the collector circuit is high, of the order of kilo ohms.	The collector circuit resistance is low, of the order of ohms.
Generally RC coupling and direct coupling are used in joining the load or cascading the amplifier stages.	Generally transformer coupling is used.
Class A operation is quite appropriate.	Class B and Class C operations are more suitable.

8.2 Conditions for Power Amplification

The concept of electrical power amplification was first realized with the triode, the first *electronic* device in true sense. Later the bipolar junction transistor was found to exhibit similar properties of amplification. As recalled from the previous chapters, the transistor exhibits the following two characteristics.

(i) Fundamentally it is a dc device and its dc level of operation depends on its operating point (Q-point) that may oscillate with the ac signal at the input.

(ii) The dc and the ac components of the input follow the superposition principle for small values of input signals. Though the transistor is a nonlinear device, small fluctuations in its transfer characteristic act as linear change.

None of these features can be preserved in the case of large input signals when the nonlinearity of the transfer characteristic becomes predominant. A comparison of the requirements of a voltage amplifier and a power amplifier is made in Table 8.2. Considering all these factors, the amplification of large input signals with a transistor, popularly known as *power amplification* should fulfil the following conditions.

Maximizing the Power: Instead of the voltage or the current alone, the product of these two is targeted to be maximized.

Novel Biasing Techniques: The biasing techniques suitable for voltage amplification, as discussed in Chapter 6 are not appropriate to power amplification. Special biasing techniques, such as class B or class AB operation and push–pull configuration, are implemented to minimize the distortion due to nonlinearity.

Transistor Construction: All transistors are not suitable for power amplification. Manufacturers construct different versions of bipolar junction transistor meant for voltage, current or power amplification processes. The transistors fit for handling high power have the following constructional features.

- Both the emitter and the base regions have higher doping concentrations than those of conventional transistors.
- The emitter–base junction area is made large in order to minimize the junction resistance, hence to reduce the I^2R power loss.
- The collector area is made large and it is connected to a metallic heat sink to facilitate heat dissipation.

Power Rating: The transistor output characteristics graph introduced in Chapter 5 is repeated in Figure 8.1 with a certain portion of the active area shaded where the product of the collector current (I_C) and the collector-to-emitter voltage (V_{CE}) exceeds the maximum power rating for the transistor prescribed by the manufacturer. This shaded region must be avoided and the transistor biasing should be such that the power ($I_C V_{CE}$) dissipating at the collector at any moment remains below the maximum specified limit of power rating. In other words, the Q-point must not enter the shaded region.

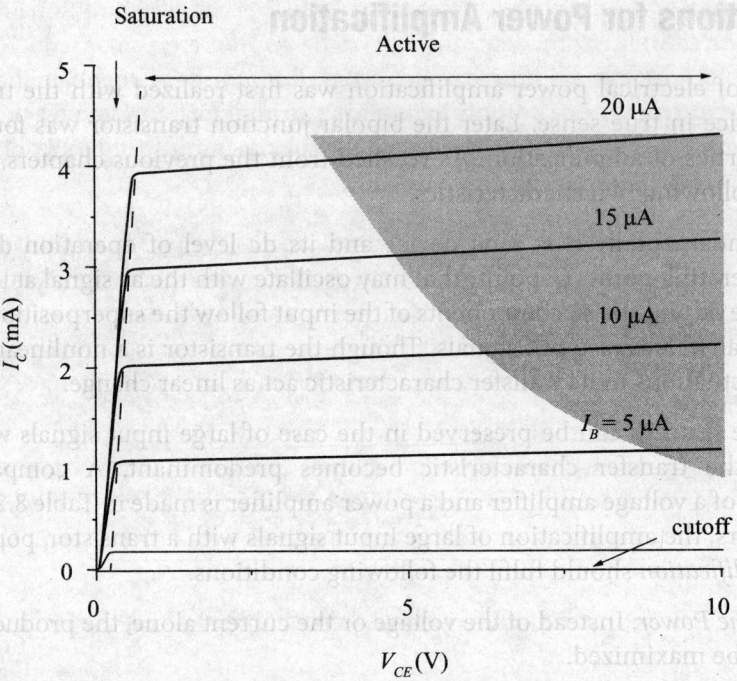

Figure 8.1 Transistor output characteristics showing the maximum allowable power dissipation for different base currents. Any value of $I_C V_{CE}$ should not reach the shaded area, which indicates the maximum specified limit of power rating and beyond

> **Note: Dissipation Hyperbola**
>
> The dc power drawn from the supply is partly utilized as ac power across the load and the remaining part is lost in the circuit. Dissipation across the resistors and the transistor itself cause the power loss. When the load draws the minimum power, the dissipation at the transistor is maximum. The maximum possible power dissipation ($I_C V_{CE}$) without damaging the transistor should always be less than the maximum limit specified by the manufacturer. The locus of the product $I_C V_{CE}$ for different base current is almost hyperbola, known as dissipation hyperbola, as shown by the edge of the shaded area in Figure 8.1.

8.3 Distortions due to Nonlinearity

It is discussed in Chapter 5 that the transfer-characteristic curve of a transistor accounting for the variation of the collector current (I_C) with the base-to-emitter voltage (V_{BE}) is nonlinear as a whole. Several types of distortion of the input signal may stem from this nonlinearity, particularly for large amplitudes of the signal and the transistor power amplifier must take into account such undesired effects. The followings are the major distortions arising from the nonlinearity of the device.

8.3.1 Amplitude Distortion

The left side plot of Figure 8.2 indicates that for small input ac signals, the variation through the transfer curve may be approximated as linear. In the case of larger amplitude of the input, the nonlinearity of the transfer curve becomes predominant, as sketched in the right side plot of Figure 8.2. The disproportionate variation across the nonlinear curve causes distortion in the output waveform.

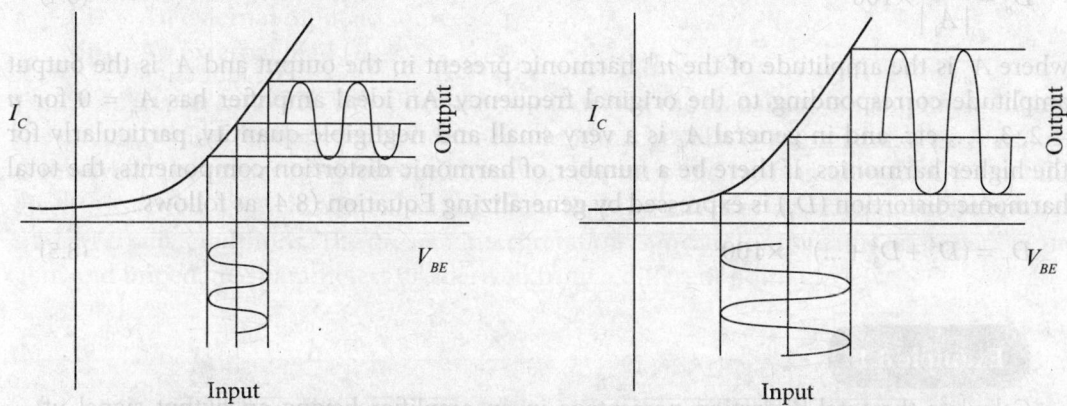

Figure 8.2 Amplitude distortion of the input signal due to the nonlinear transfer curve of the transistor: a small input signal causes negligible distortion (left) and a large input signal causes significant distortion (right) in the output waveform

8.3.2 Harmonic Distortion

Let the nonlinear collector current (i_c) as function of the base current (i_b) be expressed by the following second order polynomial.

$$i_c = k_1 i_b + k_2 i_b^2 \tag{8.1}$$

The higher terms are neglected. The constants k_1 and k_2 depend on the device. If the variation of the base current be sinusoidal of the form

$$i_b = I_b \cos \omega t \tag{8.2}$$

I_b being the maximum value, the expression for i_b can be substituted in Equation (8.1) to obtain

$$i_c = k_3 + k_3 \cos 2\omega t + k_4 \cos \omega t \tag{8.3}$$

where $k_3 = \dfrac{k_2 i_b^2}{2}$ and $k_4 = k_1 I_b$ are constants. Equation (8.3) indicates that a new frequency 2ω that was not present originally at the input has been produced at the output. This is a harmonic of the input frequency. If the nonlinearity of Equation (8.1) were represented by higher order polynomials, more harmonics could be obtained. Such generation of harmonic frequencies due to the nonlinearity in device transfer characteristic is referred to as *harmonic distortion*.

The harmonic distortion of an amplifier output for the nth harmonic is expressed in percentage by

$$D_n = \frac{|A_n|}{|A_1|} \times 100 \qquad (8.4)$$

where A_n is the amplitude of the nth harmonic present in the output and A_1 is the output amplitude corresponding to the original frequency. An ideal amplifier has $A_n = 0$ for $n = 2, 3, \ldots$, etc. and in general A_n is a very small and negligible quantity, particularly for the higher harmonics. If there be a number of harmonic distortion components, the total harmonic distortion (D_T) is expressed by generalizing Equation (8.4) as follows.

$$D_T = (D_2^2 + D_3^2 + \ldots)^{1/2} \times 100 \qquad (8.5)$$

Example 8.1

Calculate the total distortion percentage in an amplifier having an output signal of fundamental amplitude of 2.5 V, second harmonic amplitude of 0.25 V and third harmonic amplitude of 0.1 V.

Solution

Following Equation (8.5), the total distortion (D_T) is

$$\left[\left(\frac{0.25}{2.5}\right)^2 + \left(\frac{0.1}{2.5}\right)^2\right]^{1/2} \times 100 = 10.77\%$$

8.3.3 Intermodulation Distortion

If two or more signals of different frequencies are applied simultaneously to the input, the output contains sum and difference frequencies of those two. The following example will clarify the phenomenon. Let two sinusoidal input signals of angular frequency ω_1 and ω_2 be given by

$$v_1 = V_1 \sin \omega_1 t \qquad (8.6a)$$

and

$$v_2 = V_2 \sin \omega_2 t \qquad (8.6b)$$

The symbols have their usual meaning. If these two signals are applied at a time to a transistor amplifier having nonlinear transfer characteristic, the instantaneous collector current as function of input voltage would be

$$i_C = I_C + k_1(V_1 \sin\omega_1 t + V_2 \sin\omega_2 t) + k_2(V_1 \sin\omega_1 t + V_2 \sin\omega_2 t)^2 \tag{8.7}$$

neglecting the higher terms. Here I_C is the dc value of the collector current and k_1 and k_2 are constants depending on the device. It is obvious that the squared term $(V_1 \sin\omega_1 t + V_2 \sin\omega_2 t)^2$ in Equation (8.7) yields terms like $\cos(\omega_1+\omega_2)t$ and $\cos(\omega_1-\omega_2)t$. Such sum/difference frequencies $(\omega_1 \pm \omega_2)$ were not originally present at the input. These are generated because of the nonlinearity of the device characteristic. Such appearance of sum and difference frequencies of the original input signal frequencies at the output due to the nonlinearity is termed as the *intermodulation distortion*.

> **Note: Amplitude Modulation**
>
> 'Modulation' is the fundamental phenomenon involved in any sort of wireless transmission, such as radio, television, mobile phone, or satellite communication, and the nonlinearity of the electronic device is somehow involved there. The basic of modulation is to modify the amplitude, frequency or phase of a high frequency signal, known as *carrier* by a low frequency *message* signal. In the case of amplitude modulation, which is widely used in radio broadcasting, the amplitude of the carrier signal is modulated by the message signal so that sum and difference frequencies are produced. The mechanism is somewhat similar to that of the aforementioned intermodulation distortion.

8.4 Amplifier Classes

Different types of classification of amplifiers, such as dc and ac, low frequency and high frequency, wideband and narrowband, etc. are possible. However, on the basis of the mode of operation and the location of the Q-point, the following classes of amplifiers are available.

Class A: The output current flows throughout the 360° phase change of the ac input cycle. The Q-point is assigned near the middle of the load line. Generally this class of amplifier is suitable for small signal, high gain voltage amplification over a wide range of frequency. As power amplifier, the class A amplifier has low efficiency.

Class B: The output current flows through only 180° phase change that is half of the input cycle. The Q-point is located near the cutoff end of the load line. The class B amplifier is suitable to large signal amplification or power amplification. The efficiency is higher than that of class A.

Class C: The output current flows through less than 180° phase change of the input cycle so that a continuous input signal results in output pulses of the same frequency. The Q-point is fixed below the cutoff and a small forward bias of the transistor base–emitter junction is required to pull up the Q-point to the active region. The class of amplifier is appropriate to tuned amplifier capable of amplifying around a single frequency. Telecommunication circuits require this type of amplifier. It is of high efficiency.

In addition, there are some other classes. For instance, class AB has some properties intermediate between those of the class A and the class B amplifier. The different classes are discussed in details in the subsequent sections.

8.5 Class A Amplifier

As stated above, the class A amplifier circuit is biased in such a way that the dc operating point remains at the middle of the load line, halfway between the voltage values of zero and V_{CC}. The ac input causes the Q-point swing up and down along the load line but it never reaches the saturation or the cutoff region. Consequently the output current flows during the entire cycle of the input signal. In fact the CE amplifier discussed in Chapters 6 and 7 is a class A amplifier. The usefulness of such an amplifier as power amplifier is now the topic of interest.

8.5.1 Bias for Voltage Amplifier

This is to explore whether the CE voltage amplifier circuit with voltage divider bias (e.g. that of Figure 6.7a or Figure 7.11a) is useful to power amplification process. The dc power consumed from the supply is $V_{CC}I_C$ and the power consumed by the transistor is $V_{CE}I_C$, symbols having the usual meaning. The following conditions are considered.

(i) For dc bias only, the algebraic sum of the bias voltage is

$$V_{CC} = V_{CE} + I_C(R_C + R_E) \tag{8.8}$$

Multiplying both sides of Equation (8.8) by I_C and rearranging,

$$V_{CC}I_C - V_{CE}I_C = I_C^2(R_C + R_E) \tag{8.9}$$

The right side of Equation (8.9) represents the amount of power that is of no use and is unnecessarily dissipated across the resistors.

(ii) In presence of ac signal, R_E is bypassed and the effective load resistance becomes $R_{L1} = R_C // R_L$ so that

$$V_{CC}I_C = I_C^2 R_{L1} + V_{rms}I_{rms} + P_D \tag{8.10}$$

In Equation (8.10), V_{rms} and I_{rms} represent the root mean square values of output voltage and current, respectively and P_D is the average power dissipated across the transistor. Putting $V_{CC} = V_C + I_C R_{L1}$ in Equation (8.10) and simplifying, we obtain

$$P_D = V_C I_C - V_{rms} I_{rms} \tag{8.11}$$

Equation (8.11) implies that the power dissipation is maximum when no ac signal is applied and the ac power ($V_{rms}I_{rms}$) is zero. It concludes that

- the biasing designed for voltage amplification is not suitable for power amplification, and
- it results in wastage of power in the form of dissipation.

(iii) Can the power dissipation be nullified by making the resistors zero in Equation (8.9)? R_E cannot be made zero because it is a part of stabilizing the operating point. However, R_C can be replaced by an inductor having negligible dc resistance. This is the principle of transformer coupling discussed later. Another way out is to use base bias, as mentioned in the next section.

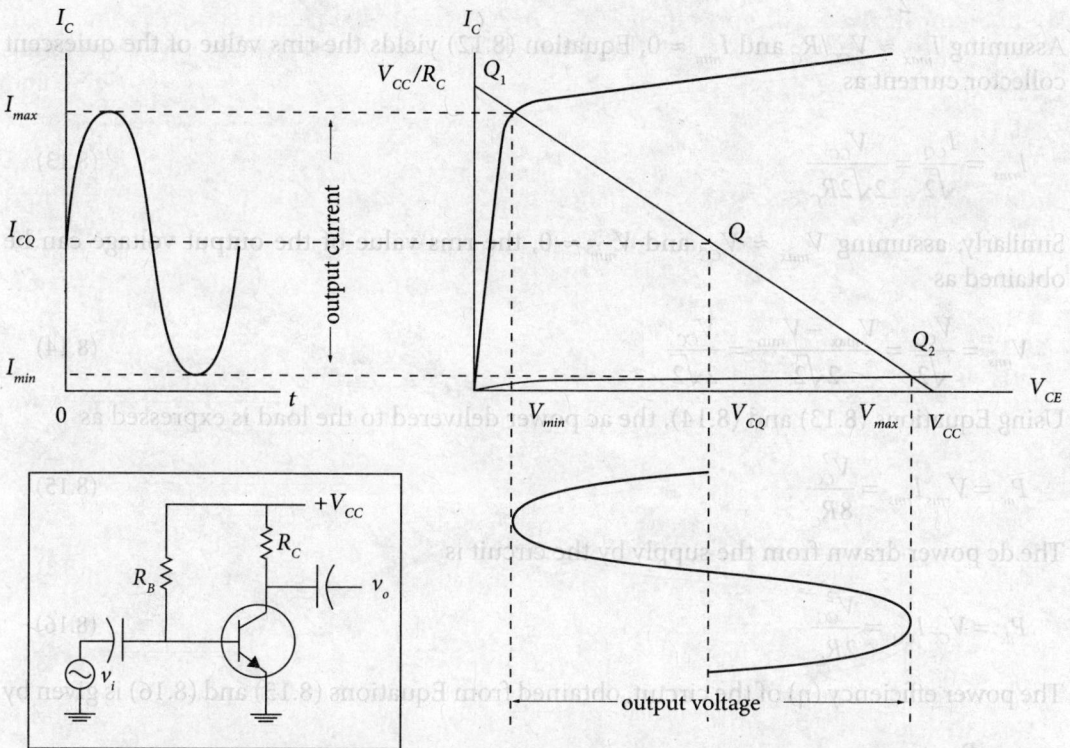

Figure 8.3 Class A resistive load power amplifier (inset) showing the variation of the quiescent collector current (I_{CQ}) and the corresponding Q-point and output voltage due to sinusoidal input signal

8.5.2 Resistive Load Power Amplifier

The CE amplifier with base bias in connection with voltage amplification (e.g. Figure 6.6a) may be recalled. A similar circuit is displayed at the inset of Figure 8.3. This is also class A amplifier. Now it is acting as power amplifier and it has the following distinguished features compared to the base bias for voltage amplification.

- This transistor has low value of β (< 50).
- The resistance R_C is small, of the order of ohms.
- The base and collector currents are much larger than those of voltage amplifier.
- The capacitors usually act as short circuits at the input signal frequency.

The variation of the collector current (I_C) and the corresponding swing of the Q-point from Q_1 to Q_2 for a sinusoidal input are sketch in Figure 8.3. The maximum possible variation of the collector current without driving the transistor into saturation or cutoff corresponds to the quiescent value of collector current

$$I_{CQ} = \frac{I_{max} - I_{min}}{2} \qquad (8.12)$$

Assuming $I_{max} \approx V_{CC}/R_C$ and $I_{min} \approx 0$, Equation (8.12) yields the rms value of the quiescent collector current as

$$I_{rms} = \frac{I_{CQ}}{\sqrt{2}} = \frac{V_{CC}}{2\sqrt{2}R_C} \qquad (8.13)$$

Similarly, assuming $V_{max} \approx V_{CC}$ and $V_{min} \approx 0$, the rms value of the output voltage can be obtained as

$$V_{rms} = \frac{V_{CQ}}{\sqrt{2}} = \frac{V_{max} - V_{min}}{2\sqrt{2}} = \frac{V_{CC}}{2\sqrt{2}} \qquad (8.14)$$

Using Equations (8.13) and (8.14), the ac power delivered to the load is expressed as

$$P_{ac} = V_{rms} I_{rms} = \frac{V_{CC}^2}{8R_C} \qquad (8.15)$$

The dc power drawn from the supply by the circuit is

$$P_{dc} = V_{CC} I_{CQ} = \frac{V_{CC}^2}{2R_C} \qquad (8.16)$$

The power efficiency (η) of the circuit, obtained from Equations (8.15) and (8.16) is given by

$$\eta = \frac{P_{ac}}{P_{dc}} \times 100 = 25\% \qquad (8.17)$$

Equation (8.17) states that the efficiency of power amplifier with resistive load is not so good. Indeed the 25% is the ideal value. A practical circuit yields much less efficiency, as exemplified below.

Example 8.2

The power amplifier of Figure 8.3 has $V_{CC} = 24$ V, $R_C = 24$ Ω and quiescent collector current (I_{CQ}) of 500 mA. When the input signal is applied at the base, the collector current varies by ± 200 mA. Determine the overall power efficiency of the amplifier. Is the amplifier biased for class A?

Solution

The dc power drawn by the circuit from the power supply is

$$24 \times 500 \times 10^{-3} = 12 \text{ W}$$

The collector current increases and decreases by ± 0.2 A with respect to the quiescent value of 0.5 A. Therefore, the ac power delivered at the load is

$$(0.2/\sqrt{2})^2 \times 24 = 0.48 \text{ W}$$

The power efficiency is

$$(0.48/12) \times 100 = 4\%$$

The collector-to-emitter voltage is

$$V_{CE} = V_{CC} - I_C R_C = 24 - 0.5 \times 24 = 12 \text{ V}$$

Since V_{CE} is found to be $V_{CC}/2$, the Q-point is at the middle of the load line. Therefore the amplifier is biased for class A and will continue to act in that way until the large swing due to the input signal drives it into saturation or cutoff.

Example 8.3

What will be the efficiency in the above example, if we consider the transistor alone?

Solution

The dc power dissipated at the collector load is

$$P_{diss} = (0.5)2 \times 24 = 6 \text{ W}$$

The power delivered to the transistor is

$$P_T = P_{dc} - P_{diss} = 12 - 6 = 6 \text{ W}$$

The efficiency of the transistor alone, known as *collector efficiency* is

$$(0.48/6) \times 100 = 8\%$$

The worsening of efficiency of the circuit as a whole is due to the dissipation across the load.

8.5.3 Transformer Coupled Amplifier

Figure 8.6(a) depicts the circuit diagram of transformer coupled class A power amplifier. The resistor in the collector circuit is replaced by the primary coil of a transformer. The load resistor (R_L) is very small, of the order of few ohms and is connected to the secondary coil of the transformer. The transformer coupling serves the following purposes.

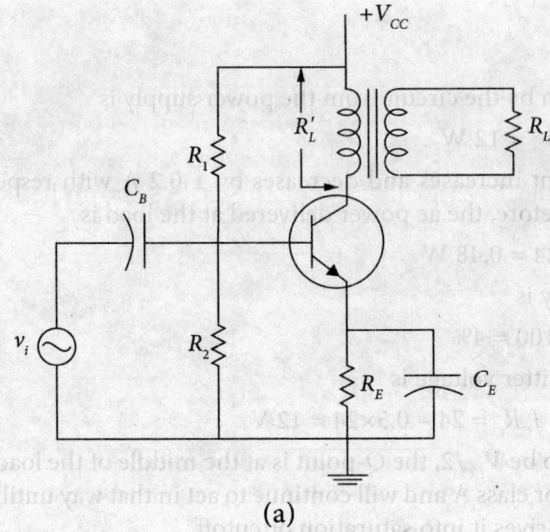

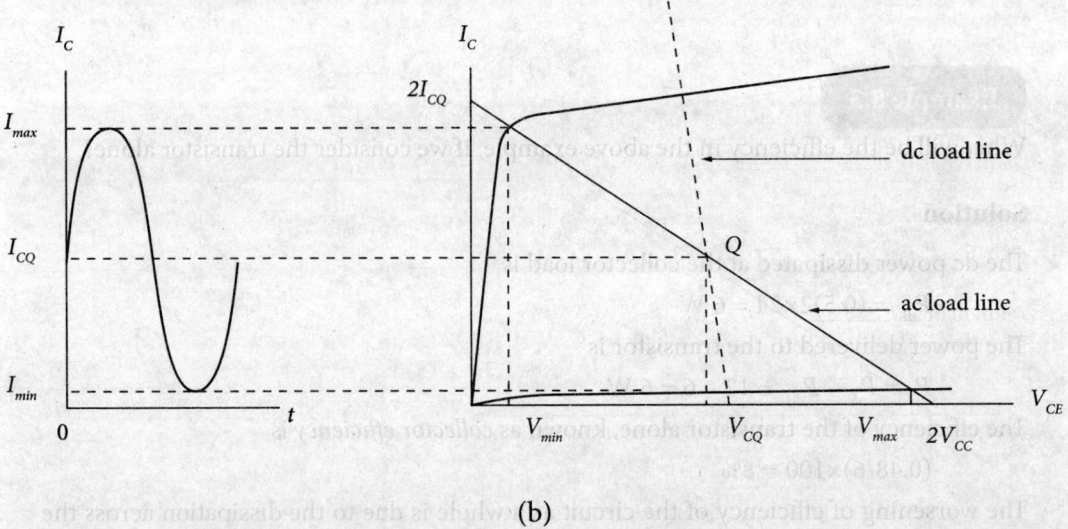

Figure 8.4 Class A transformer coupled amplifier (a) circuit diagram and (b) variation of quiescent collector current (I_{CQ}) and Q-point due to change in sinusoidal input signal

(i) It minimizes the power dissipation in the collector circuit because the ohmic resistance of the primary coil is negligible.

(ii) It provides with impedance matching between the transistor output and the load resistance. The reflected load (R'_L) into the collector circuit is $n^2 R_L$ where $n:1$ is the primary–secondary turn ratio. Thus the transformer provides with a high impedance primary to the high impedance collector and a low impedance secondary to the low resistance load.

The dc load line of the transformer coupled amplifier is ideally vertical because of zero resistance at the collector. Practically it becomes slightly tilted, as indicated in Figure 8.4(a), due to R_E and finite ohmic resistance of the coil. The ac load line has the slope of $-1/R'_L$ and the Q-point swings along this line on varying the sinusoidal input signal. Consequently the collector current also undergoes sinusoidal variation. This is another class A amplifier.

The dc power drawn from the supply can be expressed as

$$P_{dc} = V_{CC} I_{CQ} = \left(\frac{V_{max} + V_{min}}{2}\right)\left(\frac{I_{max} + I_{min}}{2}\right) \tag{8.18}$$

The ac power delivered across the transformer primary is given by

$$P_{ac} = V_{rms} I_{rms} = \left(\frac{V_{max} - V_{min}}{2\sqrt{2}}\right)\left(\frac{I_{max} - I_{min}}{2\sqrt{2}}\right) \tag{8.19}$$

Using Equations (8.18) and (8.19), the efficiency can be estimated as

$$\eta = \frac{P_{ac}}{P_{dc}} = \frac{1}{2}\frac{(V_{max} - V_{min})(I_{max} - I_{min})}{(V_{max} + V_{min})(I_{max} + I_{min})} \times 100\% \tag{8.20}$$

The following salient features are noted for the transformer coupled power amplifier.

Improved Power Efficiency: If $V_{min} \approx 0$ and $I_{min} \approx 0$, Equation (8.20) yields $\eta = 50\%$, which is double the efficiency of the resistive load power amplifier. Though it is the ideal case, the power efficiency undoubtedly increases to a great extent on coupling the load resistor to the amplifier by means of a transformer because the power dissipation across the collector circuit can be minimized.

Impedance Matching: It is known that the maximum power transfer occurs from the source to the load when both are of comparable resistance. In the case of a load of few ohms, such as speaker or motor, the transformer coupling can match the load to the amplifier output resistance of more than kiloohms.

Suitable for Low Resistance Loads: A load of low resistance drawing a large amount of current is appropriate for transformer coupling. It prevents the large direct current from flowing directly in the load and thus avoiding magnetic saturation.

Example 8.4

The transformer coupled amplifier of Figure 8.4(a) has supply voltage of 24 V, primary-to-secondary turns ratio of 10 and the load connected to the secondary is 5 Ω. What will be the maximum ac power output, if the quiescent collector current with no signal is 50 mA? Is the amplifier matched properly for maximum power transfer to the load?

Solution

Since the turn ratio is 10, the reflected resistance at the transformer primary due to the load of 5 Ω at the secondary is

$(10)^2 \times 5 = 500$ Ω

The rms value of the given collector current of 50 mA is

$50 \times 10^{-3}/(\sqrt{2}) = 0.05/\sqrt{2}$ A

The maximum ac power output is

$(0.05/\sqrt{2})^2 \times 500 = 625$ mW

The resistance (V_{CC}/I_{CQ}) across the primary due to biasing is

$24/(50 \times 10^{-3}) = 480$ Ω

Since the resistance across the primary due to biasing (480 Ω) is comparable to the transformer reflected resistance at the primary (500 Ω), the amplifier is reasonably matched for power transfer.

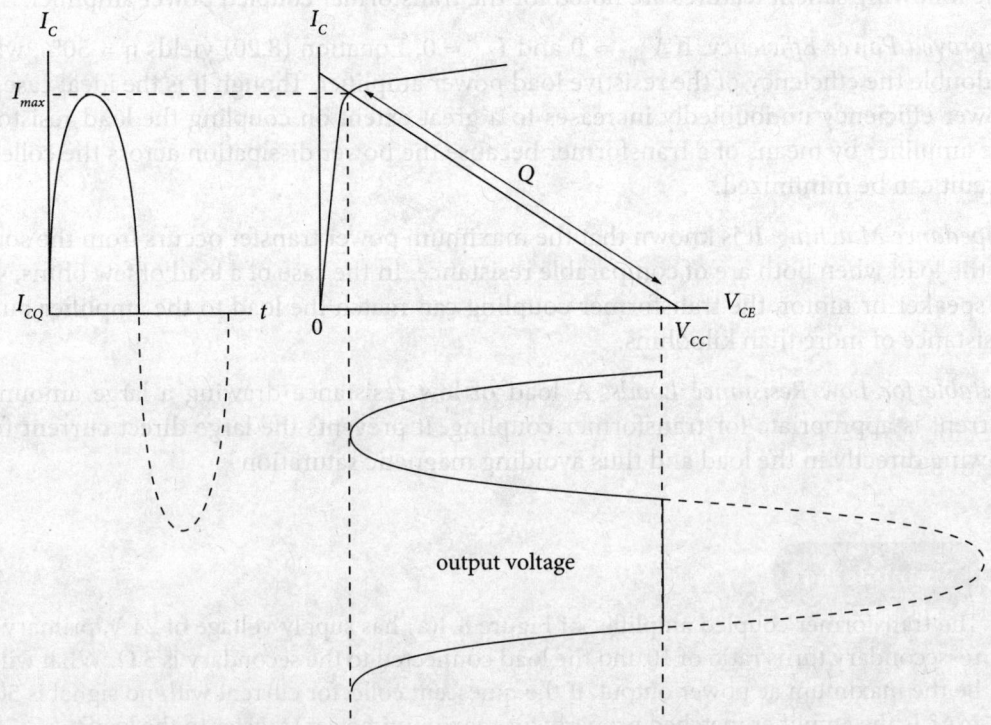

Figure 8.5 Shifting of the Q-point and the corresponding variations of the collector current and the output voltage in class B amplifier

8.6 Class B Amplifier

The biasing of this class of amplifier is such that the Q-point of the transistor is fixed at cutoff point, the lower extreme end of the load line. For the positive half cycle of the input, the Q-point shifts up along the load line and can move up to the saturation point, as the transistor conducts with higher collector current. Throughout the negative half cycle of the input, the current at the Q-point remains zero because the transistor does not conduct at all. Thus the output current of the transistor amplifier flows for only one half of the input cycle. Figure 8.5 explains the nature of shifting of the Q-point and the corresponding variations of the collector current and the output voltage. The power efficiency of the class B amplifier can be estimated in the following way.

The dc quiescent value of the collector current is estimated as

$$I_{CQ} = \frac{1}{2\pi}\int_0^\pi I_{max} \sin\theta d\theta = \frac{I_{max}}{\pi} \qquad (8.21)$$

Using the expression for I_{CQ} from Equation (8.21), the dc input power derived from the supply comes out to be

$$P_{dc} = V_{CC}I_{CQ} = \frac{V_{CC}I_{max}}{\pi} \qquad (8.22)$$

The output power obtained during one half of the input cycle is

$$P_{ac} = \frac{V_{CC}V_{min}}{\sqrt{2}}\frac{I_{max}}{\sqrt{2}} \qquad (8.23)$$

No ac output power is obtained during the other half of the input cycle. Therefore, the mean ac power (P_{av}) attained at the output is $(1/2)P_{ac}$ where P_{ac} is given by Equation (8.23). Neglecting V_{min} in Equation (8.23) (because $V_{min} \ll V_{CC}$), the expression for the mean ac power can be simplified as

$$P_{av} = \frac{V_{CC}I_{max}}{4} \qquad (8.24)$$

Using Equations (8.22) and (8.24), the following power efficiency of the class B amplifier can be obtained.

$$\eta = \frac{P_{av}}{P_{dc}} \times 100 = \frac{\pi}{4} \times 100 = 78.5\% \qquad (8.25)$$

The class B amplifier has the following advantageous features.

(i) Comparing the value of power efficiency obtained from Equation (8.25) with those obtained from Equation (8.17) and Equation (8.20), it is quite apparent that the efficiency of class B amplifier is much higher than that of class A amplifiers.

(ii) The span of swing of the Q-point in a class B amplifier is the whole length of the load line, about double the span for class A where the swing is possible through half of the length (middle to one end) only. Therefore, the class B amplifier can accommodate for larger amplitude signals and is more suitable for power amplification.

(iii) When there is no signal, the transistor also ceases to conduct. So the condition of power dissipation during no signal is eliminated.

The class B amplifier has the following limitations too.

(i) Since the transistor conducts for only one half of the input cycle, it cannot complete the full signal amplification. It requires another matching transistor for the remaining half.
(ii) The class B biasing is not suitable to all active device configurations. For example, self-bias cannot be implemented to field-effect transistor circuit for class B.
(iii) The supply voltage must be well regulated because the average direct current increases with the signal.
(iv) There exists a small range of input voltage around zero where the transistor remains cutoff and no output is produced. This results in crossover distortion of the output.

8.7 Push–Pull Amplifier

Class A power amplifiers with either resistive load or transformer coupled load can be constructed with a single transistor. The class B operation with a single transistor cannot produce output power for the whole input cycle but it can amplify signals of larger amplitudes. Both the advantages of these two classes, namely amplifying larger amplitude signals and extracting power at the load throughout the whole input cycle can be achieved with a special combination of two identical transistors known as *push–pull* connection.

The push–pull amplifier is a type of power amplifier, generally transformer coupled where two identical transistors are involved in the amplification process, each being activated to one half of the input cycle. Both the transistors are of the same type, either n-p-n or p-n-p and are closely matched. The ac input signal is divided into two halves of 180° phase difference with a transformer. Each part is applied to a transistor and each transistor performs amplification during one half of the ac input. An output transformer combines the collector currents of the two transistors and produces an output signal of the same waveform as that of the original input. The push–pull combination can be made for class A, class B and other classes of amplifier. However, class B push–pull is the most appropriate amplifier for power amplification over a range of frequencies.

8.7.1 Class A Push–Pull Amplifier

The circuit diagram for class A power amplifier with push–pull connection is given in Figure 8.6(a). The resistor combination R_1-R_2 provides with voltage divider bias for each transistor, as interpreted in Figure 8.6(b). The coils of the transformers have negligible ohmic resistance. Each transistor acts as a transformer coupled power amplifier for each half of the input cycle, the Q-point being at the middle of the ac load line.

During one half of the input voltage (v_i), the upper end of the secondary of the input transformer (T_1) attains positive polarity with respect to the center and the upper transistor (Q_1) conducts in this half. The bottom transistor (Q_2) remains cutoff because the lower end of the transformer secondary attains negative polarity. During the next half of the input signal, the polarities of T_1 get reversed and Q_2 now starts conducting, Q_1 remaining cutoff. The output current (i_1 and i_2) of the two transistors flowing through the primary of the output transformer (T_2) are 180° out of phase with respect to each other. Considering the presence of harmonic distortion, the two currents can be expressed as

$$i_1 = I_C + k_1 \cos \omega t + k_2 \cos 2\omega t + k_3 \cos 3\omega t + \ldots \tag{8.26a}$$

and

$$i_2 = I_C + k_1 \cos(\omega t + \pi) + k_2 \cos 2(\omega t + \pi) + k_3 \cos 3(\omega t + \pi) + \ldots \tag{8.26b}$$

I_C being the dc value of the collector current, ω the original frequency of the input voltage and k_1, k_2, etc. constants. The net output current becomes

$$i = i_1 - i_2 = 2(k_1 \cos \omega t + k_3 \cos 3\omega t + \ldots) \tag{8.27}$$

Equation (8.27) reveals that the even harmonics are balanced out in the push–pull amplifier. It is definitely an advantage. Only the third harmonic remains as the source of harmonic distortion. The fifth, seventh or higher terms are of quite negligible amplitudes. Thus the push–pull amplifier can minimize the harmonic distortion.

Another advantage of push–pull connection, as apparent from Equation (8.27) is that the dc components of the collector current oppose the magnetic effect of each other and thus the tendency of magnetic saturation of the transformer core can be avoided.

8.7.2 Class B Push–Pull Amplifier

This is more efficient form of power amplifier with push–pull configuration. The beneficial feature of reducing harmonic distortion, as mentioned above in connection with class A operation is available with this case also. Moreover, the higher power efficiency of class B operation [Equation (8.25)] is availed of. Also the signals of higher amplitude can be amplified. As a whole, the class B transformer coupled power amplifier is a very suitable circuit for audio frequency signal amplification.

Figure 8.7(a) displays the circuit for class B push–pull amplifier with transformer coupled load where R_1 is opened and R_2 is almost zero. Comparing it with the class A push–pull amplifier of Figure 8.6(a), it is understood that the mechanism of circuit action is similar in the two cases with the difference in the transistor biasing. The two resistors R_1 and R_2 are absent here. Therefore, both the transistors Q_1 and Q_2 remain cutoff in absence of signal voltage. During each half cycle of the input, Q_1 and Q_2 conduct alternately showing the property of Figure 8.5. The collector current of each transistor is half sinusoidal so that the total load current becomes a full sine wave.

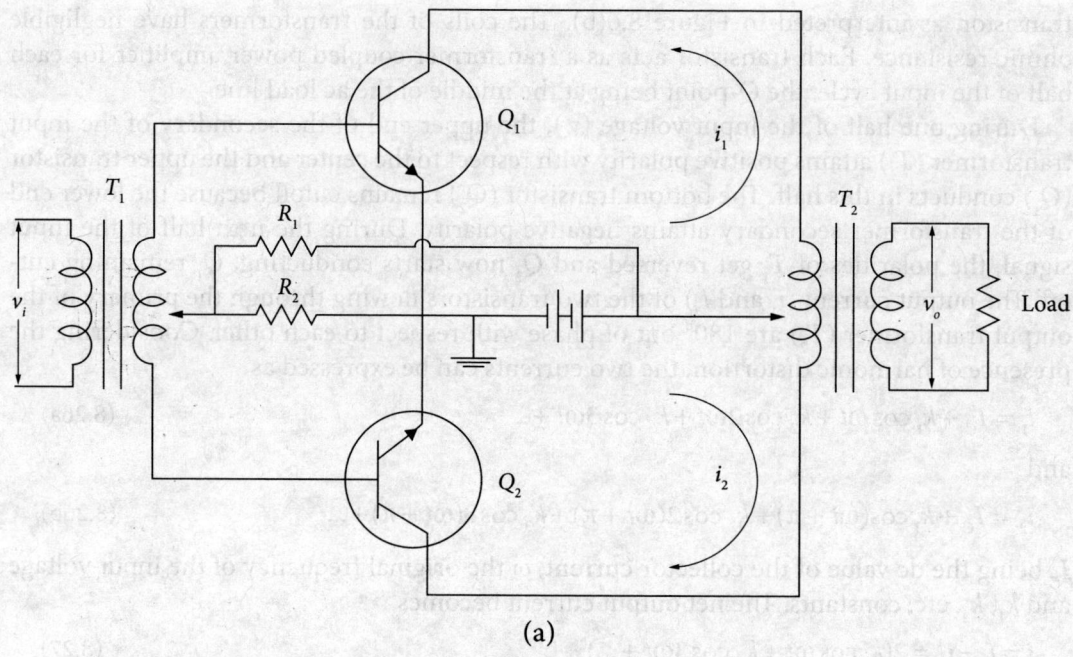

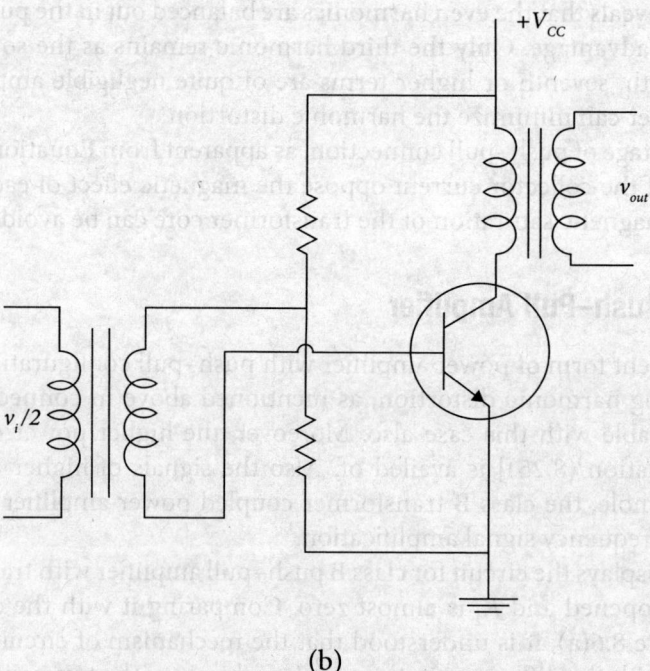

Figure 8.6 Class A push–pull amplifier: (a) actual circuit diagram and (b) equivalent circuit for each transistor

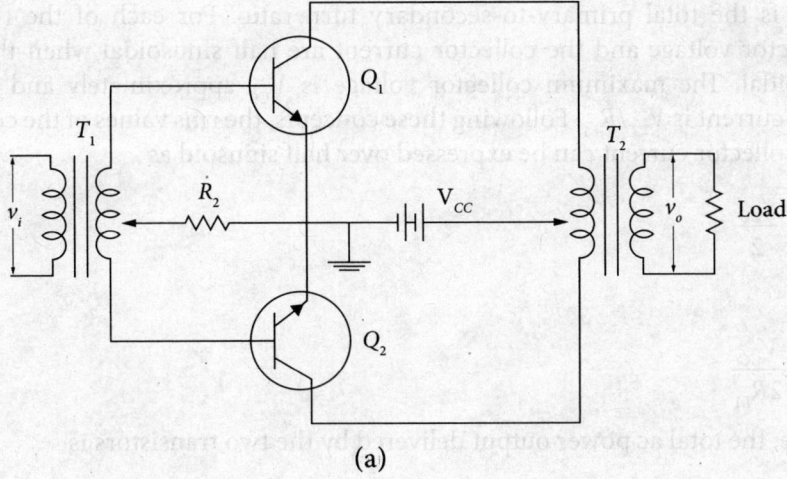

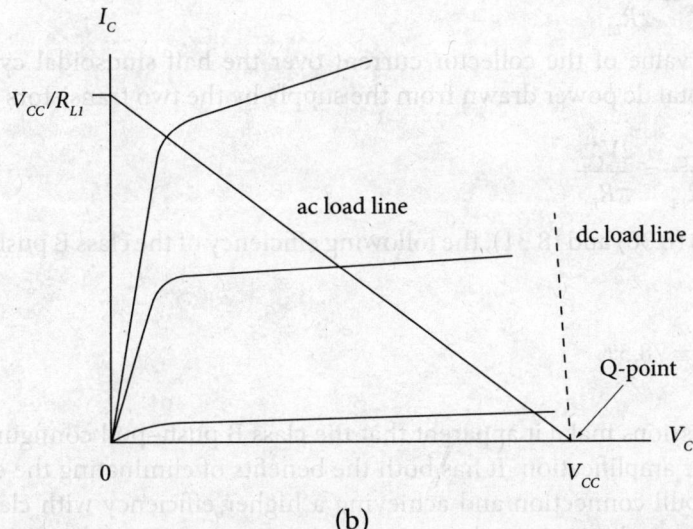

Figure 8.7 Class B push–pull amplifier: (a) circuit diagram, (b) dc and ac load lines

Figure 8.7(b) shows the positions of the Q-point and the dc and ac load lines of the class B push–pull amplifier with respect to the output characteristics. The ohmic resistance of the transformer primary coil is very small so that the dc load line is almost vertical at the cutoff end, as shown in the figure. The Q-point fixed here swings along the ac load line throughout the positive half of the input cycle.

If the actual load at the transformer secondary be R_L, the reflected load resistance at the transformer primary between the central tap and one end would be

$$R_{L1} = \left(\frac{n}{2}\right)^2 R_L \tag{8.28}$$

where n is the total primary-to-secondary turn ratio. For each of the two transistors, the collector voltage and the collector current are half sinusoidal when the input signal is sinusoidal. The maximum collector voltage is V_{CC} approximately and the maximum collector current is V_{CC}/R_{L1}. Following these concepts, the rms values of the collector voltage and the collector current can be expressed over half sinusoid as

$$V_{rms} = \frac{V_{CC}}{2} \tag{8.29a}$$

and

$$I_{rms} = \frac{V_{CC}}{2R_{L1}} \tag{8.29b}$$

Therefore, the total ac power output delivered by the two transistors is

$$P_{ac} = 2V_{rms}I_{rms} = \frac{V_{CC}^2}{2R_{L1}} \tag{8.30}$$

The average dc value of the collector current over the half sinusoidal cycle is $V_{CC}/\pi R_{L1}$. Therefore, the total dc power drawn from the supply by the two transistors is

$$P_{dc} = 2V_{CC}\frac{V_{CC}}{\pi R_{L1}} = \frac{2V_{CC}^2}{\pi R_{L1}} \tag{8.31}$$

Using Equations (8.30) and (8.31), the following efficiency of the class B push–pull amplifier is estimated.

$$\eta = \frac{P_{ac}}{P_{dc}} \times 100 = 78.5\% \tag{8.32}$$

The above discussions make it apparent that the class B push–pull configuration is a good design for power amplification. It has both the benefits of eliminating the even harmonics with the push–pull connection and achieving a higher efficiency with class B operation. However, it has some limitations also.

The use of push–pull mode can discard the even harmonics in the output of both class A and class B amplifiers only if the two transistors are properly matched. If not, as in a practical case, some harmonics may remain within the output signal. Another limitation of class B push–pull operation is distortion at small input voltages, as mentioned in the next section. This problem can be overcome by the use of a hybrid of class A and class B, known as class AB of amplifier configuration.

8.7.3 Crossover Distortion

The input characteristic of a transistor is nonlinear, similar to that of a p–n junction diode. No appreciable base current flows until the base–emitter junction gets forward biased by a cut-in voltage of about 0.7 V. Now it is clear from Figure 8.7(a) that the transistors in a class B push–pull amplifier are switched on alternately by the input alternating voltage itself; one transistor in each half cycle. If the input voltage becomes small and remains below the threshold of 0.7 V, no output signal is obtained. In other words, there exists a small range of input voltage centered around zero when both the transistors remain cutoff. The corresponding range of the output voltage remains absent thereby causing a distortion in the output waveform. This distortion is referred to as the *crossover distortion* because this originates from the crossover of the nonlinear regions of the two transistors.

Figure 8.8 explains pictorially the formation of crossover distortion in the output waveform due to the nonlinearity in both of the transistor input characteristics (i_{B1} vs v_{B1} and i_{B2} vs v_{B2}) indicated by the shaded region. This type of distortion clips off some portion of the output signal between the half cycles. It can be eliminated by the use of class AB amplifier discussed in the next section.

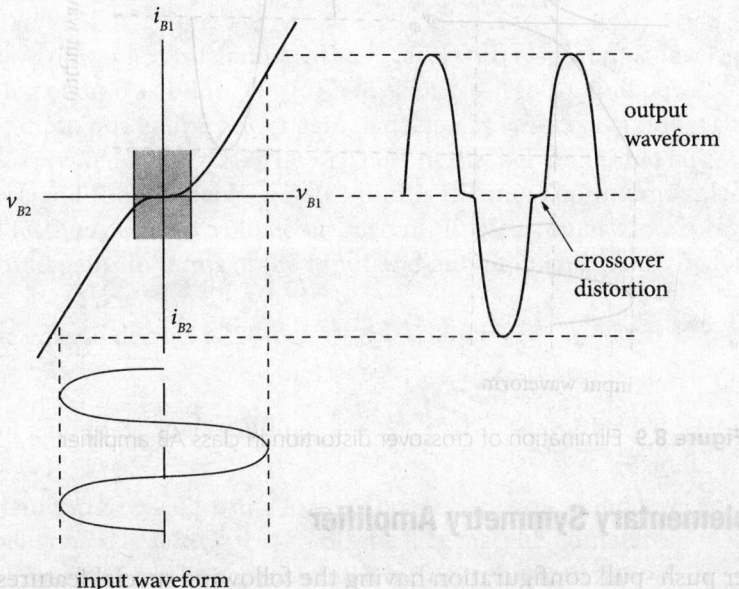

Figure 8.8 Crossover distortion due to nonlinear input characteristics of transistors (shaded region) and the resultant distorted output waveform

8.7.4 Class AB Amplifier

The crossover distortion would not occur, if the transistors were of linear characteristics all the way. The same is virtually accomplished with class AB amplifier by making the two transistors slightly forward biased so that a small current can flow at zero signal.

Figure 8.6(a) may be revisited. That circuit was made for class A operation. In the case of class B operation, both the resistors R_1 and R_2 were removed [Figure 8.7(a)]. Now, in class AB operation, R_2 is made not zero but of a small value such that the biasing voltage drop across R_2 just meets the limiting condition of forward biasing a transistor. This adjustment of the resistor R_2 is the key factor of class AB amplifier and Figure 8.9 explains the working principle of eliminating the crossover distortion with class AB operation. The small forward bias across R_2 shifts the zero base current region slightly and the overall transfer characteristic becomes linear. A finite base current flows on just crossing the zero condition of the input signal and the full, undistorted waveform is obtained as the amplified output. The output current in each transistor flows for slightly more than 180° of the input cycle. Since the circuit operation is a hybrid of class A and class B, it is termed as class AB.

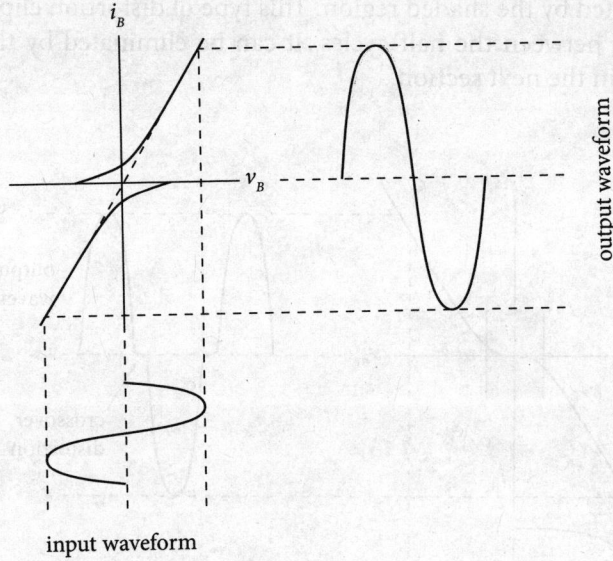

Figure 8.9 Elimination of crossover distortion in class AB amplifier

8.7.5 Complementary Symmetry Amplifier

This is another push–pull configuration having the following special features.

(i) It includes two closely matched but opposite types of transistors; one n–p–n and the other p–n–p.
(ii) It does not require any transformer.

The circuit diagram of a complementary symmetry power amplifier is shown in Figure 8.10. The input signal is coupled to the transistor through a CR network. The voltage appears across the resistor (R) and the capacitor (C) allows the ac signal only. The two transistors Q_1 and Q_2 turn on during alternate half cycles of the input in the following sequence. When no input signal is present, both the transistors remain off. During the positive half cycle of the input, transistor Q_1 (n–p–n) conducts because its base–emitter junction gets forward biased. Transistor Q_2 (p–n–p) remains cutoff and acts as open circuit. During the negative half cycle of the input, Q_1 remains cutoff and Q_2 conducts.

It is apparent from the circuit that the conducting transistor (either Q_1 or Q_2 in the respective half cycle) acts as an emitter follower thereby providing unity voltage gain, high current gain and high input impedance. The collector currents of the two transistors in the two half cycles flow in opposite directions through the load resistor (R_L) and the total current variation through R_L produces the output voltage waveform, which is an amplified version of the input signal in the same phase. Thus the complementary symmetry push–pull amplifier has the following advantageous features.

(i) It has high input impedance, a favorable quality for voltage amplification.
(ii) The elimination of transformer makes the circuit simple, cost effective and suitable for both high and low frequencies.
(iii) There is no phase inversion at the output.
(iv) It allows direct coupling of load that provides the suitability for both dc and ac amplification.

Despite the above benefits, the complementary symmetry amplifier has some major limitations also.

(i) The power supplies are required with respect to the ground, as shown in Figure 8.10, facing inconvenience in matching other circuits.
(ii) It is very difficult to obtain a perfectly matched pair of transistors of two different types. If not, the harmonic and crossover distortions cannot be avoided.

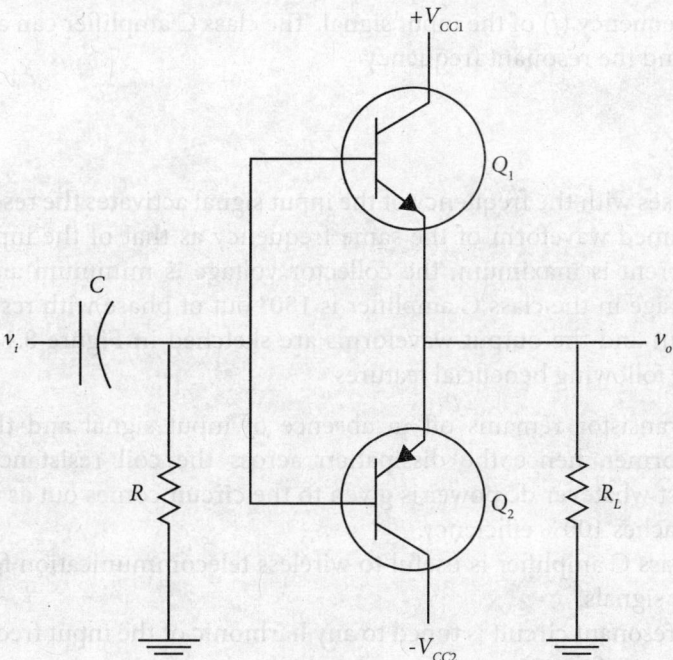

Figure 8.10 Complementary symmetry push–pull power amplifier

8.8 Class C Amplifier

It is mentioned at the beginning of this chapter that a tuned or narrowband amplifier is that category of amplifier which, instead of a wide range of frequency, amplifies selective input signals having frequency around a fixed, predetermined value only. That frequency is determined by a resonant circuit associated with the load. Such tuned amplifiers are used in radio frequency range of signals, mainly in telecommunication related circuits. A tuned amplifier can be constructed with class A or class B operations but another category, known as class C is more appropriate for this purpose.

The circuit diagram of a class C tuned power amplifier is presented with Figure 8.11(a). There is no dc bias to the input port from the supply voltage ($+V_{CC}$). Moreover, a small negative voltage V_{BB} reverse biases the base–emitter junction so that the transistor is biased with the Q-point positioned below cutoff, as indicated in Figure 8.11(b). When the input signal (v_i) grows up in the positive half cycle and exceeds V_{BB} + 0.7 V (for silicon transistors), the transistor gets forward biased and produces a proportionally variable collector current. Throughout the negative half of the input cycle, the transistor remains cutoff and no collector current is produced. Thus the collector current is produced in the form of a series of pulses, each pulse corresponding to one input cycle. Since the duration of each pulse is less than half input cycle, the output current at the collector flows for less than 180° of the input cycle.

The load resistor (R_L) in the class C amplifier is connected to the secondary of the output transformer (T). The primary of the transformer acts as an inductor of inductance L and a capacitor of capacitance C is joined in parallel so that an LC resonant circuit is formed that is tuned to the frequency (f) of the input signal. The class C amplifier can amplify signals of frequencies around the resonant frequency

$$f = \frac{1}{2\pi\sqrt{LC}} \tag{8.33}$$

The collector pulses with the frequency of the input signal activates the resonant circuit and produces a sustained waveform of the same frequency as that of the input signal. When the collector current is maximum, the collector voltage is minimum and vice versa. So the collector voltage in the class C amplifier is 180° out of phase with respect to the input voltage. The input and the output waveforms are sketched in Figure 8.11(c). The class C amplifier has the following beneficial features.

(i) The transistor remains off in absence of input signal and there is no input transformer, hence the dissipation across the coil resistance is minimized. Almost whatever dc power is given to the circuit comes out as load power. So it approaches 100% efficiency.

(ii) The class C amplifier is useful to wireless telecommunication for handling high power signals.

(iii) If the resonant circuit is tuned to any harmonic of the input frequency, amplified waveform of that frequency can be obtained.

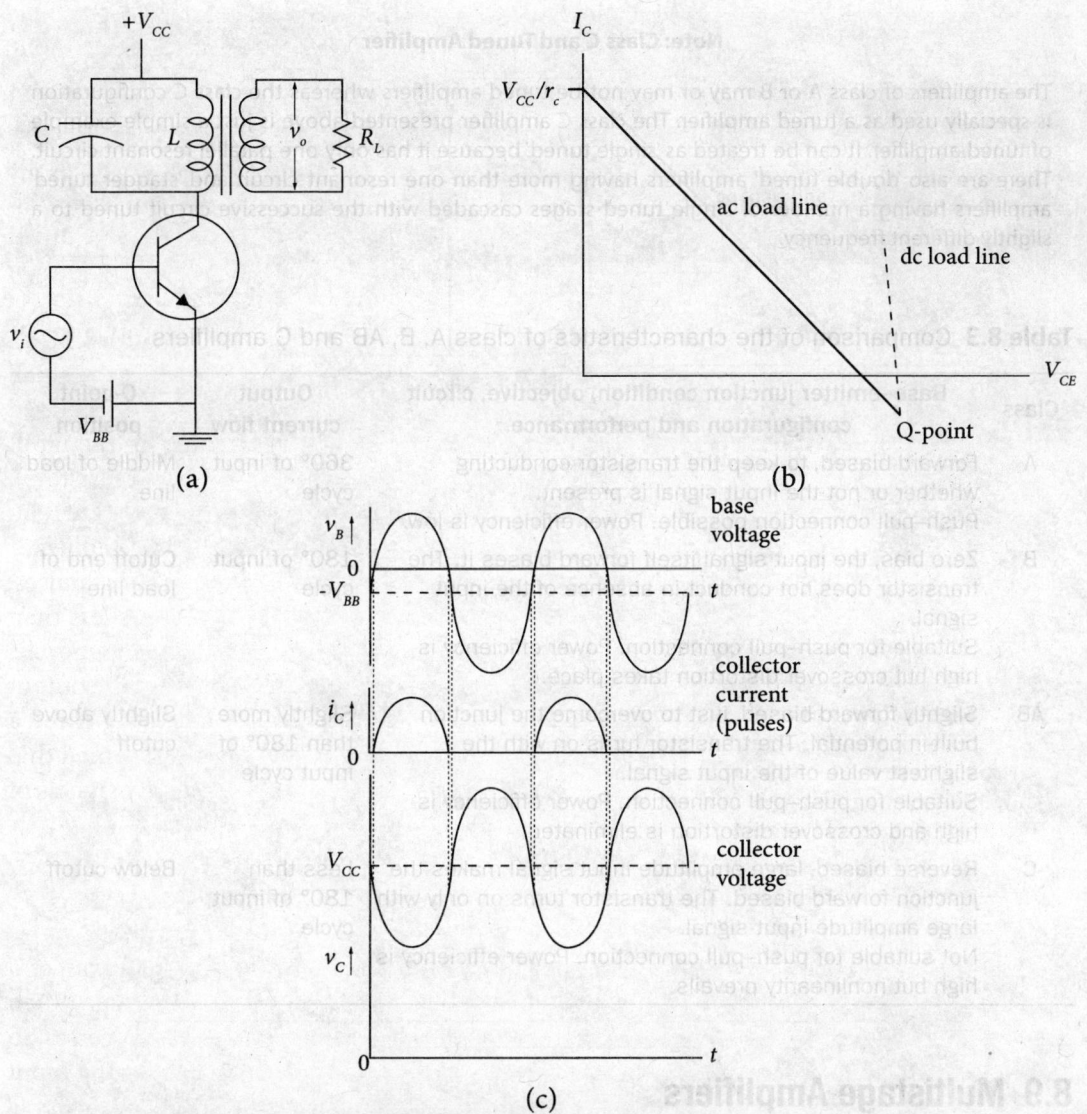

Figure 8.11 Class C tuned power amplifier: (a) circuit diagram, (b) dc and ac load lines, and (c) input and output waveforms (not drawn up to scale)

The class C amplifier has the following limitations.

(i) The amplifier is highly nonlinear and is not suitable for small signal amplification.
(ii) The harmonics at the output are eliminated by the use of resonant circuit tuned to the fundamental frequency of the input signal but the existence of harmonics is the unavoidable component of the amplifier.

Now having several classes of amplifier operations illustrated in details, Table 8.3 presents a brief comparative view of the main characteristics of the amplifiers of class A, B, AB and C.

> **Note: Class C and Tuned Amplifier**
>
> The amplifiers of class A or B may or may not be tuned amplifiers whereas the class C configuration is specially used as a tuned amplifier. The class C amplifier presented above is just a simple example of tuned amplifier. It can be treated as 'single tuned' because it has only one parallel resonant circuit. There are also 'double tuned' amplifiers having more than one resonant circuit and 'stagger tuned' amplifiers having a number of single tuned stages cascaded with the successive circuit tuned to a slightly different frequency.

Table 8.3 Comparison of the characteristics of class A, B, AB and C amplifiers

Class	Base–emitter junction condition, objective, circuit configuration and performance.	Output current flow	Q-point position
A	Forward biased, to keep the transistor conducting whether or not the input signal is present. Push–pull connection possible. Power efficiency is low.	360° of input cycle	Middle of load line
B	Zero bias, the input signal itself forward biases it. The transistor does not conduct in absence of the input signal. Suitable for push–pull connection. Power efficiency is high but crossover distortion takes place.	180° of input cycle	Cutoff end of load line
AB	Slightly forward biased, just to overcome the junction built-in potential. The transistor turns on with the slightest value of the input signal. Suitable for push–pull connection. Power efficiency is high and crossover distortion is eliminated.	Slightly more than 180° of input cycle	Slightly above cutoff
C	Reverse biased, large amplitude input signal makes the junction forward biased. The transistor turns on only with large amplitude input signal. Not suitable for push–pull connection. Power efficiency is high but nonlinearity prevails.	Less than 180° of input cycle	Below cutoff

8.9 Multistage Amplifiers

All the bipolar junction transistor (BJT) amplifiers discussed so far in this chapter and in Chapters 5 through 7, have only one amplifying unit. Even in the case of push–pull amplifiers, each transistor amplifier works individually in a separate half cycle. The maximum amplification obtainable with a single transistor amplifier, as learnt earlier, is limited up to the current gain (β) that is not more than few hundreds. However, we will find in Chapter 11 that an amplifier named operational amplifier can amplify the input voltage by 10^6 times or more. An input of microvolt may result in several volts at the output. Such enormous amplification is achieved by not a single amplifier but a number of amplifiers at a time. The output of one amplifier is connected to the input of another amplifier and this way a small input voltage is increased in steps.

The process of joining one amplifier at the output of another is called cascading. The principle of amplifier cascading is outlined in Figure 8.12. The two amplifiers have voltage gain A_1 and A_2, respectively. The input voltage (v_i) is applied to the first amplifier and its output $(v_1 = A_1 v_i)$ serves as the input to the second amplifier. The final output voltage (v_o) becomes $A_2 A_1 v_i$, which is amplified by the product $(A_1 A_2)$ of the individual gains. Extending this concept to a number of amplifying stages, the overall gain of an n-stage amplifier is given by the product

$$A = A_1 A_2 \ldots A_n \tag{8.34}$$

where $A_1, A_2, \ldots A_n$ are the gains of the individual amplifying stages.

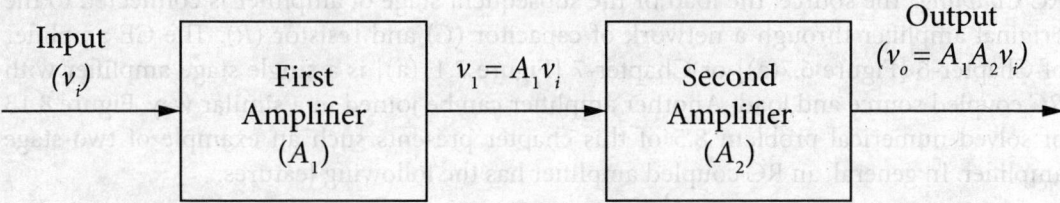

Figure 8.12 Block diagram for amplifier cascading

Note: Darlington Pair

Sometimes CC amplifier cascading is also done by using direct connections of BJT. The combination of two BJTs, named Darlington pair after the inventor is one such example. The circuit is shown in the adjacent figure. Two identical transistors are directly coupled in such a way that the emitter of the first BJT (Q_1) is the connected to the base of the second BJT (Q_2) and the collectors of these two transistors are connected together. The three-terminal device combination, indicated by dotted boundary can be used as a single device. It acts like an emitter follower, as shown in the figure. If each transistor has current gain of β, the overall current gain of the Darlington pair is β^2, a large quantity. The input resistance also increases β times but the overall voltage gain remains unity.

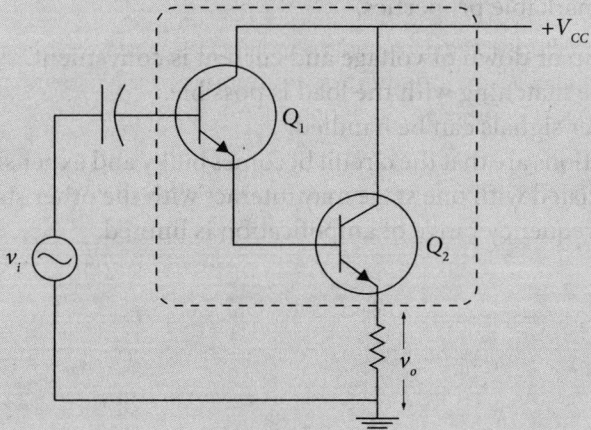

The significant points of consideration are: which amplifiers are suitable for cascading and how to cascade. For instance, a combination of common-collector (CC) amplifiers is not suitable for obtaining large voltage gain because an individual CC amplifier has voltage gain almost unity. However, it can be cascaded for high current gain. A common-base (CB) amplifier is not appropriate for high current gain because the current gain is almost unity. The common-emitter (CE) amplifier having high current gain is quite suitable for cascading. In a multistage amplifier designed for high voltage gain, the intermediate stages should be CE amplifiers. The input and output stages depend on the type of the source and load. An amplifier stage is made connected to another amplifier or a source or a load through passive elements and this process is known as coupling. The following types of coupling are generally used.

RC Coupling: The source, the load or the subsequent stage of amplifier is connected to the original amplifier through a network of capacitor (C) and resistor (R). The CE amplifier of Chapter-6 [Figure 6.7(a)] or Chapter-7 [Figure 7.11(a)] is a single stage amplifier with *RC* coupled source and load. Another amplifier can be joined in a similar way. Figure 8.13 of solved numerical problem 8.5 of this chapter presents such an example of two-stage amplifier. In general, an RC coupled amplifier has the following features.

- (i) It is generally used for small signal voltage amplification.
- (ii) High voltage gain can be achieved.
- (iii) A constant voltage gain over a wide range of frequency of input signal can be obtained.
- (iv) The circuit is compact in size, cost effective and low power consuming.
- (v) The limitation of RC coupling is that the amplification falls at low frequency and is quite incompatible to dc amplification.

Transformer Coupling: The push–pull amplifiers of this chapter are examples where the load is coupled to the amplifier through transformer. Two or more such amplifiers also can be coupled through transformers. Generally transformer coupling is used for power amplification and in the construction of tuned amplifiers. The use of transformer coupling has the following remarkable properties.

- (i) Stepping up or down of voltage and current is convenient.
- (ii) Impedance matching with the load is possible.
- (iii) High power signals can be handled.
- (iv) The limitations are that the circuit becomes bulky and expensive and the magnetic field associated with one stage may interact with the other stages of the amplifier. Also the frequency range of amplification is limited.

Direct Coupling: The load or another amplifier stage is connected directly with the amplifier using connecting wires without employing any passive component. The complementary symmetry amplifier (Figure 8.10) is one such example. The internal circuit of an operational amplifier (Chapter 11) also involves direct coupling. The followings are the salient features of this type of coupling.

(i) The gain remains constant down to dc (zero frequency) and up to a high range of frequency of the input signal.
(ii) The circuit is simple and less expensive.
(iii) However, direct coupling can be applied to limited cases where the dc component of the output does not disturb the normal operation of the load device.

The purpose of coupling is not just electrical connection; it also isolates the dc biasing voltage of one stage from the input of the next stage. None of the coupling techniques is absolutely good or bad. Each type of coupling has its own merit and drawback and specific field of application. The amplifiers may be of different purposes with different modes of operations. The load may be a simple resistor, speaker, motor, relay or any optoelectronic device. The choice of coupling depends on all these factors.

Think a Bit: Decibel Gain for Cascaded Amplifiers?

If the individual voltage gains of the amplifiers are $A_1, A_2, ..., A_n$ and the individual power gains are $P_1, P_2, ..., P_n$ then the overall voltage gain in decibel system is

$20\log_{10}(A_1 A_2 ... A_n) = 20\log_{10} A_1 + 20\log_{10} A_2 + ... + 20\log_{10} A_n$

and the overall power gain in decibel system is

$10\log_{10}(P_1 P_2 ... P_n) = 10\log_{10} P_1 + 10\log_{10} P_2 + ... + 10\log_{10} P_n$

Thus the total decibel gain of the cascaded amplifier is the sum of the decibel gains of the individual amplifiers.

Multiple Choice-Type Questions and Answers

8.1 The voltage amplifier biasing is not suitable for power amplification because of ...., hence ....
 (a) low power efficiency, large power dissipation
 (b) large power dissipation, low power efficiency
 (c) low voltage gain, high power efficiency
 (d) high voltage gain, high power dissipation

8.2 The output signal is opposite in phase with respect to the input signal in ....
 (a) CB and CE amplifiers   (b) CE amplifiers only
 (c) CB and CC amplifier   (d) CE and CC amplifiers

8.3 The highest current gain is obtained with .... amplifiers.
 (a) CB   (b) CE   (c) CC   (d) CB and CE

8.4 The highest input impedance is obtained with .... amplifiers whereas the output impedance is achieved with .... amplifiers.
(a) CB, CC (b) CE, CC (c) CC, CE (d) CC, CB

8.5 The class A amplifier with transformer coupling can achieve the maximum overall efficiency of .... percent.
(a) 25 (b) 50 (c) 78.5 (d) 100

8.6 Zero input signal represents the ------------ condition in class A amplifier and the .... condition in class B amplifier.
(a) worst, best (b) best, worst
(c) uncertain, worst (d) best, uncertain

8.7 The average current in class B operation over the whole input cycle is .... that in class A operation.
(a) half of (b) greater than
(c) equal to (d) less than

8.8 The following statements are given about the occurrence of crossover distortion.
(i) It takes place when the current conduction is crossed over from one transistor to the other in the push–pull amplifier.
(ii) It occurs with both class A and class B push–pull amplifier.
(iii) It is observed most predominantly in the case of low amplitude input signals.
(a) only (i) is true (b) only (ii) is true
(c) both (i) and (ii) are true (d) all (i), (ii) and (iii) are true

8.9 The following reasons are given for transformer coupling having the highest efficiency.
(i) It provides with negligible dc resistance.
(ii) It does not need any coupling capacitor.
(iii) It provides with impedance matching for maximum power transfer.
(a) (i) and (iii) are correct (b) only (i) is correct
(c) only (iii) is correct (d) (i), (ii) and (iii) are correct

8.10 The output of a class B amplifier ....
(a) consists of positive half cycles only with respect to the ground.
(b) acts like a full-wave rectifier.
(c) appears for less than 180° of input cycle.
(d) is free from even order components of harmonic distortion.

8.11 The overall decibel gain of a cascaded amplifier is equal to the .... of the gains of the individual amplifiers.
(a) product (b) ratio (c) sum (d) difference

8.12 Direct coupled amplifiers are suitable for ....
(a) DC signals only (b) low frequency signals only
(c) DC and low frequency signals only (d) small amplitude signals only

8.13 Doubling the power level of an amplifier is expressed in decibel scale as ....
(a) 2 dB (b) –3 dB (c) 20 dB (d) 3 dB

8.14 A power amplifier having efficiency of 75% is most likely to be ....
(a) transformer coupled amplifier (b) class AB push–pull amplifier
(c) class C amplifier (d) RC coupled amplifier

8.15 In a tuned amplifier, the load impedance is supplied by
(a) capacitor
(b) inductor
(c) LC parallel resonant circuit
(d) LC series resonant circuit

8.16 What should be the primary-to-secondary turns ratio for a transformer coupling a load of 8 Ω so that it appears as a 5 kΩ effective load at the primary?
(a) 5:1
(b) 1:5
(c) 25:1
(d) 1:25

8.17 A gain of 60 dB means a power ratio of .... and a voltage ratio of ....
(a) $10^6, 10^3$
(b) $10^3, 10^6$
(c) $10^3, 10^3$
(d) $10^6, 10^6$

8.18 The voltage gain of an amplifier is 10^5. It corresponds to .... dB.
(a) 5
(b) 10
(c) 20
(d) 100

8.19 Maximum power transfer occurs in an amplifier when the input and output resistances become equal. In that case, the power gain in dB is .... the current gain in dB and the voltage gain in dB.
(a) less than
(b) equal to
(c) greater than
(d) double of

8.20 A high gain voltage amplifier, known as operational amplifier has voltage gain ideally infinity. In a practical class, the ratio of such gain to that of a typical CE amplifier is best represented as
(a) 10^{100}
(b) 10^{10}
(c) 100 dB
(d) 10^5 dB

Answers

5.1	(b)	5.2	(b)	5.3	(c)	5.4	(d)
5.5	(b)	5.6	(a)	5.7	(d)	5.8	(d)
5.9	(b)	5.10	(a)	5.11	(c)	5.12	(c)
5.13	(d)	5.14	(b)	5.15	(c)	5.16	(c)
5.17	(a)	5.18	(d)	5.19	(b)	5.20	(c)

Reasoning-Type Questions and Answers

8.1 What is the *frequency response* of an amplifier?

Ans. The variation of the voltage gain of an amplifier with the input signal frequency is termed as the frequency response of that amplifier. It quantifies the activity of an amplifier in response to the input signal of a specific frequency and this property depends on the construction of the amplifier. For example, an RC coupled class A amplifier can amplify signals over a wide range of frequency whereas a transformer coupled class C amplifier can amplify the signals only around a fixed frequency.

8.2 What are the factors that determine the frequency response of an amplifier?

Ans. Mainly the following three entities are involved in deciding the frequency response of an amplifier.
- The coupling elements, such as capacitor or transformer coil
- The resonant circuit, if any
- The property of the device itself, such as the junction capacitance of a transistor.

8.3 How is a wideband amplifier characterized?

Ans. The amplifier having a stable gain over a wide range of frequency of input electrical signal is called *wideband amplifier* or *video amplifier*. The RC coupled amplifier is such an example. The lower and upper cutoff frequencies, where the voltage gain reduces to $1/\sqrt{2}$ times the maximum and the power gain reduces to half of the maximum are separated by a wide range (e.g. tens of kHz or more) of frequency. In other words, the bandwidth of the wideband amplifier is large.

8.4 How is a narrowband amplifier characterized?

Ans. The narrowband amplifier amplifies electrical signals around a given frequency only and attenuates the others. A resonant circuit at the output tuned to a certain frequency is an essential component of this type of amplifier. So these are also called *tuned amplifier*. The bandwidth is too narrow and is sometimes defined in terms of full width at half maximum (FWHM). It is the small frequency span, at the end of which the gain reduces to half of the maximum.

8.5 The *gain-bandwidth product* of an amplifier was discussed in Chapter 7 with the conclusion that it is a constant quantity. Can you provide any alternative interpretation?

Ans. The mid-frequency gain (A_m) of an amplifier is a constant quantity for a particular load and the bandwidth is the difference of the upper cutoff frequency (f_{cu}) and the lower cutoff frequency (f_{cl}). Since $f_{cu} \gg f_{cl}$, the bandwidth can be approximated by f_{cu} and the gain-bandwidth product can be expressed as $A_m f_{cu}$, a product of two constant terms. This matter can also be treated from a practical point of view. As the frequency of the input signal goes higher and higher, the transistors have to shift the Q-point around the mean position at a faster rate. If, at the same time, the same large extent of fluctuation had to be maintained, that would need more and more energy and violate the basic laws of physics. So the extent of fluctuation (hence the gain) gets reduced with the faster rate of change of Q-point.

[The gain-bandwidth product is a fundamental property of an amplifier and will be discussed further in Chapter 10 in connection with feedback.]

Solved Numerical Problems

8.1 Calculate the power efficiency of a class A power amplifier with base bias and resistive load, if V_{CC} = 12 V, R_C = 10 Ω, R_B = 1 kΩ and β = 50, symbols having the usual meaning. The input voltage causes maximum ±10 mA change in base current.

Soln. The base current is calculated as

$$I_B = (V_{CC} - 0.7)/R_B = 11.3 \text{ mA}$$

At the Q-point, the collector current is

$$I_C = \beta I_B = 50 \times 11.3 \times 10^{-3} = 0.565 \text{ A}$$

The dc power consumed from the power supply is

$$P_{dc} = V_{CC} I_C = 12 \times 0.565 = 6.78 \text{ W}$$

Since the input voltage causes a change of 10 mA in base current, the maximum change appearing at collector current is

$50 \times 10 \times 10^{-3} = 0.5$ A

The ac power delivered at the load (R_c) is

$$P_{ac} = \left(\frac{0.5}{\sqrt{2}}\right)^2 \times 10 = 1.25 \text{ W}$$

The power efficiency is

$$\frac{P_{ac}}{P_{dc}} \times 100 = \frac{1.25}{6.78} \times 100$$
$$= 18.44\%$$

8.2 A class A power amplifier has supply voltage of 12.5 V and the maximum collector current change is 100 mA. A loudspeaker of 5 Ω resistance is used as load. Calculate the power transferred to the speaker when it is (i) directly connected to the collector circuit and (ii) coupled through transformer for maximum power transfer.

Soln.

(i) The maximum ac voltage developed across the 5-Ω loudspeaker is

$100 \times 10^{-3} \times 5 = 0.5$ V

Therefore, the AC power delivered across it is

$0.5 \times 100 \times 10^{-3} = 50$ mW

(ii) With transformer coupling, the output impedance looking into the collector circuit is

$12.5/(100 \times 10^{-3}) = 125$ Ω

Considering the ideal case of maximum power transfer, the reflected primary resistance should be 125 Ω when the actual resistance across the secondary is 5 Ω. If the primary-to-secondary turn ratio be n,

$125 = n^2 \times 5$, hence $n = 5$

The secondary voltage (primary voltage divided by turn ratio) is

$12.5/5 = 2.5$ V

The load current is

$2.5/5 = 0.5$ A

Therefore, the power delivered across the load is

$(0.5)^2 \times 5 = 1.25$ W

8.3 A class B push–pull amplifier has supply voltage of 12 V, the output transformer has primary-to-secondary turn ratio of 5:1 and the load resistance joined across the transformer secondary is of 4 Ω. Calculate the dc power drawn from the supply, the ac power delivered to the load and the efficiency of the amplifier.

Soln. The reflected load resistance at the primary of the transformer for each transistor is [using Equation (8.28)]

$(6/2)^2 \times 4 = 36$ Ω

The total ac power drawn from the supply is [using Equation (8.31)]

$$\frac{2\times(12)^2}{\pi\times 36} = 2.546 \text{ W}$$

The ac output power is [using Equation (8.30)]

$$\frac{(12)^2}{2\times 36} = 2 \text{ W}$$

The efficiency is

$$(2/2.546)\times 100 = 78.55 \%$$

8.4 A multistage amplifier employs five stages, each of which has a power gain of 30. What is the gain of the amplifier in dB?

[C.U. 2013]

Soln. The overall power gain in decibel is

$10\log_{10}(30/1) + 10\log_{10}(30/1) + 10\log_{10}(30/1) + 10\log_{10}(30/1) + 10\log_{10}(30/1)$

$= 5\times 14.77$

$= 73.86$ dB

8.5 Figure 8.13 shows the circuit diagram of a two-stage RC coupled amplifier. Let the two transistors be identical. The input voltage (v_i) is applied to the first stage and its output is applied to the second stage through a coupling capacitor (C). The other biasing components are understood as usual. Derive an expression for the voltage gain of this cascaded amplifier at the optimum range of medium frequencies when the coupling capacitor acts as short circuit.

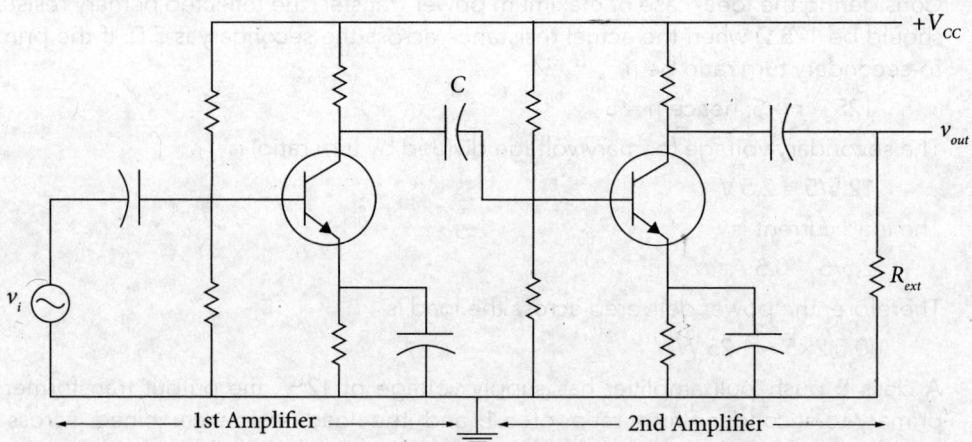

Figure 8.13 Two-stage RC coupled transistor CE amplifier. The biasing resistors and capacitors are usual, for example, similar to those of Figure 6.7(a) or Figure 7.11(a)

Soln. The voltage gain of the first amplifier is determined as follows. The effective resistance for the RC coupling is the input resistance (h_{ie}) of the second amplifier. The biasing resistors are supposed to be neglected in the parallel combination. The capacitor (C) acts as short circuit at the optimum frequency range. Therefore, the hybrid model for the first amplifier can be drawn similar to that of Figure 8.14(a). For convenience, it is modified for voltage source in Figure 8.14(b). (The logics are similar to those in Figure 7.12 except for the capacitor acting as short circuit.)

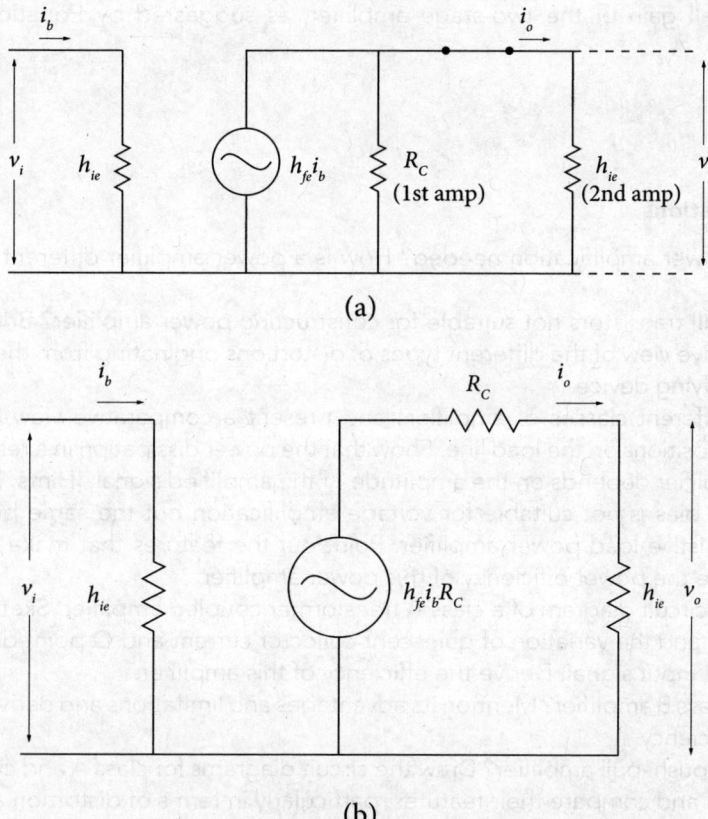

Figure 8.14

The output current is given by

$$i_o = -\frac{h_{fe} i_b R_C}{R_C + h_{ie}} \qquad (1)$$

Therefore, the current gain is given by

$$A_i = \frac{i_o}{i_b} = -\frac{h_{fe} R_C}{R_C + h_{ie}} \qquad (2)$$

Considering the output voltage (v_o) as $i_o h_{ie}$ and the input voltage (v_i) as $h_{ie} i_b$, the voltage gain of the first amplifier at mid-frequency is given by

$$A_{v1} = \frac{v_o}{v_i} = -\frac{h_{fe} R_C}{R_C + h_{ie}} \qquad (3)$$

For the second amplifier, the effective load resistance (R_L) is the parallel combination of the external load resistor (R_{ext}) and the resistor in its collector circuit (R_C). The voltage gain (A_{v2}) of the second amplifier is similar to that derived in the previous chapter for the circuit of Figure 7.12 and can be expressed by Equation (7.31).

The overall gain of the two-stage amplifier, as suggested by Equation (8.34), is the product $A_{v1} A_{v2}$.

Exercise

Subjective Questions

1. Why is power amplification needed? How is a power amplifier different from a voltage amplifier?
2. Why are all transistors not suitable for constructing power amplifier? Briefly introduce a comparative view of the different types of distortions originating from the nonlinearity of the amplifying device.
3. Define different classes of amplifiers and present a comparative view by drawing the Q-point positions on the load line. Show that the power dissipation in a resistive load class A CE amplifier depends on the amplitude of the amplified signal. [**Hints.** Equation (8.11)]
4. The base bias is not suitable for voltage amplification but the same bias can provide with a resistive load power amplifier. Point out the features that make this distinction. Determine the power efficiency of this power amplifier.
5. Draw the circuit diagram of a class A transformer coupled amplifier. Sketch its dc and ac load lines and the variation of quiescent collector current and Q-point due to change in sinusoidal input signal. Derive the efficiency of this amplifier.
6. What is class B amplifier? Mention its advantages and limitations and derive an expression for its efficiency.
7. What is a push–pull amplifier? Draw the circuit diagrams for class A and class B push–pull amplifiers and compare their features, particularly in terms of distortion and efficiency.
8. State about the origin of crossover distortion in an amplifier, its effect on the output and a suitable method of eliminating it.
9. Draw the circuit diagram of a complementary symmetry amplifier and mention its benefits and limitations.
10. Briefly explain the working principle of a class C amplifier with suitable circuit diagram and state about its specific uses.
11. Present a brief comparison among transformer coupling, capacitive coupling and direct coupling in multistage amplifiers. Write down the expression for the overall power gain in decibel system for a multistage amplifier.
12. Derive an expression for the voltage gain of a two-stage RC coupled amplifier cascaded amplifier at the optimum range of medium frequencies. (**Hints.** See solved numerical problem 8.5)

Conceptual Test

1. Can the hydraulic press be treated as a mechanical amplifier?
2. A car engine produces a magnified version of the torque exerted to the stirring wheel. Is the engine a mechanical amplifier?
3. The step-up transformer can increase the level of either voltage or current. Is it an amplifier?
4. An amplifier produces a larger voltage at the output with a smaller voltage at the input. Does it violate the laws of thermodynamics?
5. It is a fact that the class C amplifier is ideally 100% efficient whereas a Carnot engine is far below 100% efficiency. Wherein lies the paradox?
6. What type of amplifier is suitable for tuning to a particular radio station or specific television channel?
7. Do you think CE amplifier are more suitable for cascading than CC and CB amplifiers? If so, why?

Numerical Problems

1. The output voltage of an amplifier is 10 V at 5 kHz and 7.07 V at 25 kHz. What is the decibel change in output power level?
2. The mid-band voltage gain of an RC coupled amplifier is 150. It is found that the values of the gain are 50 at the frequency of 40 Hz and 60 at the frequency of 500 kHz. Find the bandwidth of the amplifier.

   [C.U. 2008]
3. A resistive load class A power amplifier has V_{CC} = 24 V, I_{CQ} = 500 mA and R_L = 20 Ω, the symbols having the usual meaning. If the maximum output current swing is 250 mA, calculate the efficiency of the amplifier.
4. Determine the magnitude of voltage gain of an RC coupled amplifier in mid-frequency range, if it has current gain of 150, collector current of 4 mA and load resistance of 5 kΩ.
5. A transformer coupled class A power amplifier has load resistance of 10 Ω. If the primary-to-secondary turn ratio be 6:1 and the output current has peak-to-peak swing of 250 mA, calculate the output power. Neglect the resistance of transformer windings.
6. The rms output voltage in the mid-band region of an amplifier is 2 V and the power gain is 42 dB. The power output at the lower cutoff frequency (100 Hz) is 0.4 W. Find (i) the rms output voltage at 100 kHz, (ii) the output power in the mid-band region and (iii) the rms input voltage, if the input resistance be 1 kΩ.

   [C.U. 2005]
7. A class B amplifier has supply voltage of 20 V and it provides a peak voltage of 12 V to a load of 10 Ω. Determine the amplifier efficiency.
8. The two-stage RC coupled CE amplifier of Figure 8.13 has the following parameters. Each of the two amplifiers has R_1 = 30 kΩ, R_2 = 10 kΩ, R_C = 2 kΩ and R_E = 1 kΩ. The usual positions of these resistors in the circuit are understood. The external load resistor (R_{ext}) is of 5 kΩ. Each transistor has β = 100 and V_{BE} = 0.7 V, symbols having the usual meaning. All the capacitors are serving the purpose in ideal manner. Determine the overall voltage gain in decibel.

9

Field-Effect Transistor (FET)

This is a separate category of transistor where the output current is controlled by the input voltage. There are two broad categories of such transistors, namely junction field-effect transistor (JFET) and metal–semiconductor field-effect transistor (MOSFET). The device construction, electrical characteristics and the related parameters of JFET and MOSFET are illustrated in this chapter. Different types of amplifier configurations realized with these devices are introduced.

9.1 'Field-Effect' and 'Transistor'

Chapters 5 to 8 have gone through various properties of the three-terminal bipolar junction transistor (BJT) where the output current is controlled by another current at the input. The present electronic device is also three-terminal and it can amplify voltage and switch electrical signals. So this is also a transistor. However, unlike the bipolar transistor, the current through the output terminals is controlled by an electric field resulting from the voltage at the third terminal. Indeed the electric field may control the current conducting path of the output circuit without any direct electrical contact between the causing and the resultant parameters. That is why it is named *field-effect transistor*, abbreviated as FET.

The FET is contemplated in different forms. There is a category of junction FET (JFET) where the input voltage varies the depletion width of a reverse-biased p–n junction. Another group of FET having a metal electrode separated from the semiconductor by an insulator is known as *metal-oxide-semiconductor* FET (MOSFET). A remarkable difference from BJT is that unlike the participation of both electrons and holes, the output current in a FET is composed of either electrons or holes and accordingly the FET is referred to as *unipolar device* and *majority carrier device*. Both the JFET and the MOSFET have the following three terminals.

Source: It is the point through which the majority carriers enter the current carrying semiconductor bar, called *channel*. The source terminal of the FET may be compared to the emitter of the BJT.

Drain: Through this the terminal, the majority carriers flow out of the channel. The drain is analogous to the collector of a BJT.

Gate: It is the terminal with either a metal (for MOSFET) or a heavily doped region of opposite type semiconductor (p^+ for n-channel and vice versa) (for JFET) meant for controlling the drain current. The gate is comparable to the base of the BJT.

In addition to the three metallic leads corresponding to the above three terminals, a practical FET sometimes contains a fourth metallic lead connected to the semiconductor substrate over which the whole device is fabricated. The FET has the following advantageous features.

(i) It is less noisy by virtue of being unipolar.
(ii) It has high switching speed because of being free from minority carriers.
(iii) The input resistance is very high due to reverse bias in JFET and insulating layer in MOSFET. This is very suitable for property for voltage amplification.
(iv) The FET has good thermal stability. JFETs are generally more temperature stable than BJTs.
(v) The fabrication technique is simple and FETs are much smaller in size than BJTs, which is a suitable property for IC fabrication.

Nevertheless there are some limitations of FETs.

(i) Small range of input signal frequency with stable amplification
(ii) Less voltage gain than BJT
(iii) Large resistance at on-state.

9.2 Junction Field-Effect Transistor (JFET)

The construction of this device is outlined in Figure 9.1(a). It consists of a uniformly doped semiconductor (e.g. Si) bar, known as *channel* with metallic connections at the two ends. The two ends are called the *source* and *drain*, respectively, as indicated in the figure. The present interpretation considers an n-type channel. Two heavily doped p-regions, denoted by p^+ are fabricated on both sides of the n-channel. A similar explanation holds for p-channel device also where two n^+ regions are grown on both sides of a p-type semiconductor bar. A third metallic contact, namely gate is made with both the p^+ regions in common. The present structure of JFET is actually a model adopted for easy explanation of the working principle. An actual JFET contains a single p^+–n junction on one side of the channel and the whole structure is fabricated on a semiconductor plate, known as the *substrate*. The symbols for n-channel and p-channel JFET are given in Figure 9.1(b) and 9.1(c), respectively. The arrow in the gate terminal of the symbol indicates the direction of current flow, if the gate-channel junction were forward-biased.

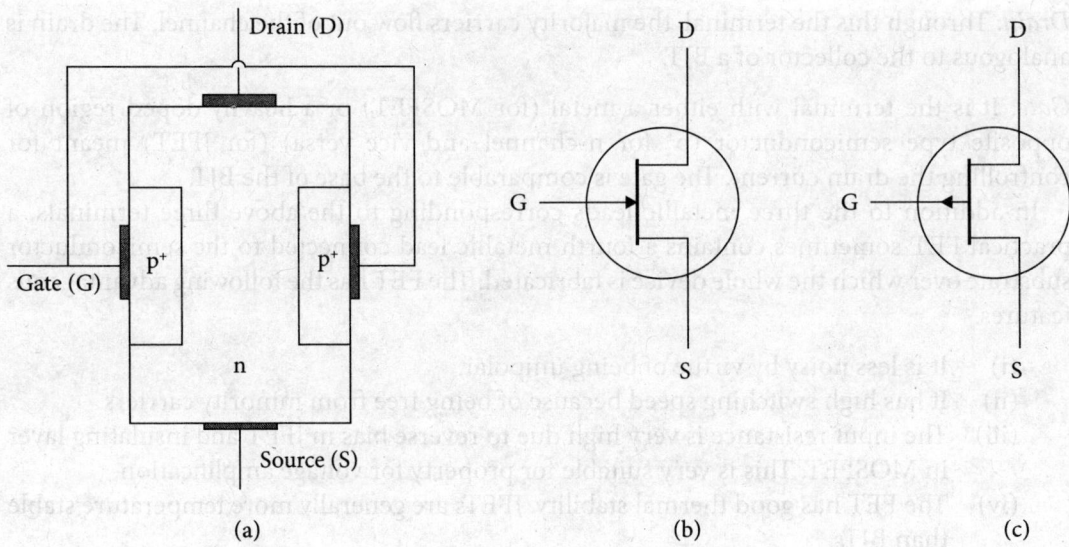

Figure 9.1 Junction field-effect transistor (JFET): (a) construction of n-channel and symbol of (b) n-channel JFET, and (c) p-channel JFET

9.2.1 JFET Current–Voltage Characteristics

The output current–voltage characteristics of JFET are interpreted for an n-channel device drawn in Figure 9.2(a). The mechanisms are the same in the case of p-channel device, but the voltage polarities are opposite to the present one. The following device features are notable in connection with the current–voltage characteristics [Figure 9.2(b)].

Drain and Gate Biasing: Figure 9.2(a) shows that a positive bias voltage $+V_{DD}$ is applied to the drain (D) terminal of an n-channel JFET. A negative bias voltage $-V_{GG}$ is optionally applied to the gate terminal. The voltage and current notations related to a JFET (and MOSFET also) are summarized in Table 9.1. The biasing conditions are presented in Table 9.2.

Table 9.1 Voltage and current notations for JFET and MOSFET

Symbol	Meaning
V_{DD}	Drain-to-source supply voltage, +ve for n-channel and –ve for p-channel device
V_{DS}	Direct voltage, as measured between the drain and the source. It is different from V_{DD} in presence of any external biasing resistor.
V_{GG}	Gate-to-source supply voltage (reverse bias)
V_{GS}	Direct voltage, as measured between the gate and the source
I_D	General representation for drain current

Field-Effect Transistor (FET)

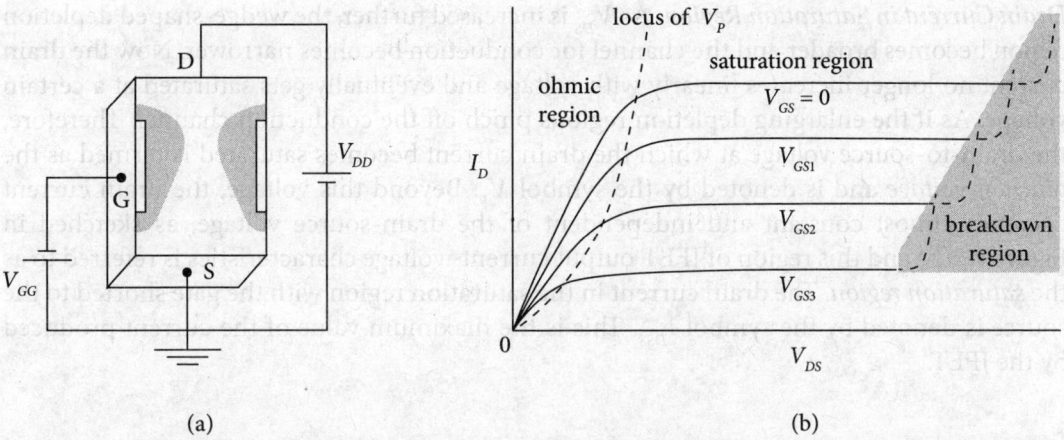

Figure 9.2 n-channel JFET: (a) biasing and formation of wedge-shaped depletion region (shaded) and (b) variation of drain current (I_D) with drain-to-source voltage (V_{DS}) for different gate voltages ($V_{GS1} < V_{GS2} < V_{GS3}$). Generally V_{DS} is in volts and I_D is in mA

Table 9.2 Biasing conditions for JFET

JFET condition	I_D	V_{DS}	V_{GS}
n-channel	positive	positive	negative
p-channel	negative	negative	positive

Drain Current in Ohmic Region: Let us consider the simplest case when no gate bias is applied and consequently $V_{GS} = 0$. The electrons move from source to drain under the influence of the positive drain voltage (V_{DS}) causing a drain current (I_D). Initially, for small values of V_{GS}, the channel of n-type semiconductor works like a simple resistor and I_D increases with V_{DS}. This region of JFET output current–voltage characteristics is referred to as *ohmic region* or *linear region*, as denoted in Figure 9.2(b).

Formation of Depletion Region: As the drain bias is increased, an increasing voltage drop is established throughout the channel between the source and the drain with the voltage increasing in magnitude toward the drain. This voltage drop reverse biases the gate p⁺–n junctions due to zero voltage at p-side and positive voltage at n-side. Consequently a depletion region is created within the channel having the following two features.

(i) Since the n-type channel is lesser doped than the p⁺ region, the depletion region spreads more into the channel.

(ii) Due to the voltage gradient from the source to the drain, the depletion becomes wedge-shaped with the wider side toward the drain, as shown by the shaded region in Figure 9.2(a).

Drain Current in Saturation Region: As V_{DS} is increased further, the wedge-shaped depletion region becomes broader and the channel for conduction becomes narrower. Now the drain current no longer increases linearly with voltage and eventually gets saturated at a certain voltage. As if the enlarging depletion regions pinch off the conduction channel. Therefore, the drain-to-source voltage at which the drain current becomes saturated is termed as the *pinchoff voltage* and is denoted by the symbol V_p. Beyond this voltage, the drain current becomes almost constant and independent of the drain-source voltage, as sketched in Figure 9.2(b) and this region of JFET output current–voltage characteristics is referred to as the *saturation region*. The drain current in the saturation region with the gate shorted to the source is denoted by the symbol I_{DSS}. This is the maximum value of the current produced by the JFET.

Think a Bit: Why Is I_D Constant at Saturation?

It is a fact that the pinchoff condition narrows down the channel so that the drain current gets saturated. But why does the channel not get blocked completely to make the drain current zero? One may think that if the channel were choked completely making the current zero, the consequences would be physically contradictory. The voltage drop along the channel length causing the reverse bias itself would come down to zero and there would be none to produce the pinchoff voltage. The current density (current divided by zero area) would rise up to infinity. Anyway no one has tutored the drain current about these predicted consequences. Then how does it adjust so nicely? Actually we have to go back to the phrase 'field effect'. The increase in V_{DS} causes higher electric field when the carrier mobility decreases and the drift velocity saturates so as to retard the drain current. There is also a repulsion among the immobile ions of the depletion regions, which encounter closely with each other and on increasing the drain voltage, the region of close proximity spreads along the channel. Ultimately a balanced condition is reached when a small but nonzero channel width is maintained. Also the value of V_{DS} cannot be increased indefinitely because that would result in breakdown of the gate junction.

Thus the variation of I_D with V_{DS} has two distinct regions: ohmic region where I_D varies linearly with small values of V_{DS} and saturation region where I_D is independent of V_{DS} up to a large range. Though the two regions are not so sharply demarcated, the pinchoff voltage occurs somewhere between these two.

Note: Saturation Region or Active Region

The saturation region of the JFET output current–voltage characteristics is also sometimes mentioned as 'active region' in literature, similar to that of BJT. There may be some logic in support because the action of JFET in the saturation region is like a current source similar to the BJT. There is, obviously a marked difference: the JFET is a voltage-controlled current source whereas the BJT is a current-controlled current source. This book, however, adheres to the term 'saturation region' for the horizontal region of drain current.

Voltage Amplification: In the above discussion, there was no gate voltage. If now a reverse gate-source voltage (V_{GS}) is applied, that produces an additional reverse bias to the gate p$^+$–n junction and all the above mentioned features of depletion region formation, pinchoff mechanism and current flows in ohmic and saturation regions are exhibited. As V_{GS} is increased, saturation occurs at lower values of I_D and the pinchoff condition takes place at lower values of V_{DS}, as outlined in Figure 9.2(b).

The above phenomenon of adding a reverse gate bias gives rise to a voltage amplification. A small change in V_{GS} produces the same change in I_D as that obtained with a larger change in V_{DS}. Thus the JFET and also the MOSFET introduced later performs as a voltage amplifier. This property is quantified in terms of an amplification factor, introduced later. Indeed the FET is a solid-state equivalent of triode, the vacuum tube device that put forward the first concept of voltage amplification.

The pinchoff voltage (V_P) of a FET is defined for zero gate voltage. So the drain voltages causing saturation in drain current in presence of gate voltages are, of course, not 'truly' pinchoff voltages but act in the same manner. These voltages continue to drop in almost parabolic path with increasing gate voltages, as indicated in Figure 9.2(b). Eventually, when $V_{GS} = -V_P$, the saturation of drain current takes place at almost zero value and the device becomes turned off.

> **Note: Cutoff and Pinchoff**
>
> The gate-source cutoff voltage, denoted by the symbol $V_{GS(off)}$ is that value of gate-source voltage, which makes the drain current (I_D) cutoff to almost zero. This voltage is related to the gate bias whereas the pinchoff voltage is related to the drain bias. The pinchoff voltage (V_P) is that value of drain-source voltage (V_{DS}), which levels off I_D at a fixed value. Above this voltage, the drain current is almost constant and independent of the drain voltage unless the latter reaches the breakdown limit. $V_{GS(off)}$ is numerically equal to V_P and opposite in sign. These two voltages lead the device to the same situation where the depletion regions of gate junctions almost touch each other.

Breakdown Region: In addition to the ohmic and saturation regions, a third zone, namely breakdown region may take place on the JFET drain characteristic curves, as indicated in Figure 9.2(b). If V_{DS} is increased indefinitely, avalanche breakdown happens across the gate p+–n junction and the drain current suddenly increases in an uncontrollable manner. This condition should always be avoided in JFET operation.

JFET Ohmic Region: As mentioned earlier, the region left to the pinchoff in Figure 9.2(b) is called ohmic region because the drain current (I_D) varies almost linearly with the drain voltage (V_{DS}) over this region and the slope of the linear variation changes with the gate voltage (V_{GS}). Thus the JFET performs as a *voltage-controlled resistor*. The I_D versus V_{DS} curve is almost straight line and V_{GS} is the controlling agent for the slope of the curve, hence the resistance. This is a unique feature of JFET having no BJT counterpart. Though the variable resistance property is realized with the output characteristic curves, the input (gate) is an indispensable component because the device resistance varies with the input voltage. This resistance in the ohmic region can be defined approximately by

$$r_{(ohm)} = \frac{r_{o(ohm)}}{\left(1 - V_{GS}/V_P\right)^2} \qquad (9.1)$$

This is different from the drain resistance obtained in the saturation region. In Equation (9.1), when V_{GS} is zero, the resistance reached its minimum value of $r_{o(ohm)}$, which is equal to V_P/I_{DSS}. As the magnitude of the gate voltage is increased, the resistance also increases.

9.2.2 JFET Transfer Characteristics

The JFET as amplifier is generally used in the saturation region of drain current. The transfer characteristic of the JFET is the relationship between the output current (I_D) and the input voltage (V_{GS}) given by

$$I_D = I_{DSS}\left(1 - \frac{V_{GS}}{V_P}\right)^2 \qquad (9.2)$$

As defined earlier, I_{DSS} represents the saturation drain current at zero V_{GS} and V_P is the pinchoff voltage. Equation (9.2) implies that when the gate-source voltage becomes equal to the pinchoff voltage, the drain current becomes zero. That value of gate-source voltage is denoted by $V_{GS(off)}$. The derivation of Equation (9.2), also known as Shockley equation after the founder is beyond the present scope and we will be using it in this ready form. It has the following properties.

(i) Equation (9.2) is valid for the transfer characteristic of the MOSFET also.
(ii) The equation indicates that the drain current varies in proportion with the square of the gate voltage, hence the FET is also termed as square-law device.
(iii) When the gate voltage (V_{GS}) becomes equal to V_P, the drain current becomes zero.
(iv) Figure 9.3 pictorially represents the typical shape of transfer characteristic curve and its association with different states of drain characteristics.

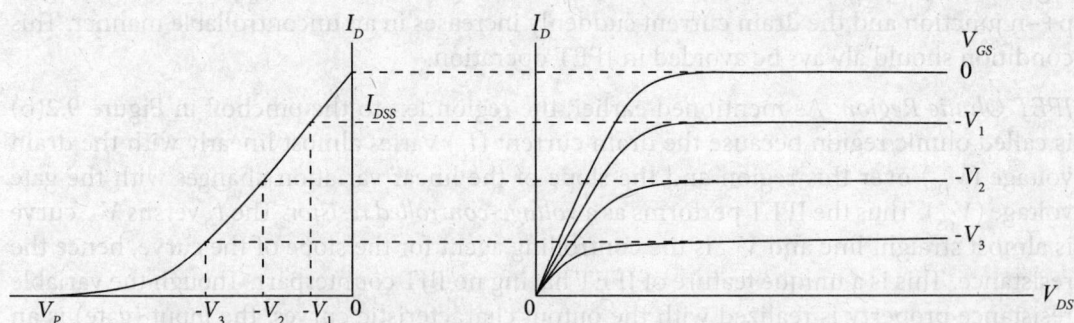

Figure 9.3 A typical transfer characteristic of n-channel JFET (left) and its comparison with different drain characteristics (right)

9.3 FET Parameters

The linear small signal equivalent of FET can be obtained by expressing the drain current as function of both the drain and the gate voltage given by

$$i_D = f(v_{DS}, v_{GS}) \tag{9.3}$$

Equation (9.3) represents the generalized varying quantities of drain current (i_D), drain voltage (v_{DS}) and gate voltage (v_{GS}) in presence of both dc bias and ac signal. If both v_{GS} and v_{DS} are varied, the change in drain current is

$$\Delta i_D = \frac{\partial i_D}{\partial v_{GS}}\Big|_{V_{DS}} \Delta v_{GS} + \frac{\partial i_D}{\partial v_{DS}}\Big|_{V_{GS}} \Delta v_{DS} \tag{9.4}$$

Because of small signals, the higher terms in the Taylor series expansion are ignored. For only the ac components, Equation (9.3) can be written as

$$i_d = g_m v_{gs} + \frac{1}{r_d} v_{ds} \tag{9.5}$$

It defines the following three important device parameters applicable to both JFET and MOSFET.

- Mutual Conductance or Transconductance: $g_m = \dfrac{\partial i_D}{\partial v_{GS}}\Big|_{V_{DS}}$ (9.6a)

- Drain Resistance: $r_d = \dfrac{\partial v_{DS}}{\partial i_D}\Big|_{V_{GS}}$ (9.6b)

- Amplification Factor: $\mu = g_m r_d = \dfrac{\partial v_{DS}}{\partial v_{GS}}\Big|_{I_D}$ (9.6c)

The *mutual conductance* is the ratio of the change in drain current to the corresponding change in gate voltage for a fixed drain voltage. It has the following properties.

(i) It represents the mutual relationship between the input (gate) voltage and the output (drain) current.
(ii) It is also named *transconductance* because it represents the transfer property of the device. It quantifies how efficiently the input voltage effect is transferred to the output current.
(iii) The dimension of g_m is that of conductance (siemens).

The *drain resistance* is the ratio of the change in drain voltage to the corresponding change in drain current for a fixed gate voltage.

The *amplification factor* represents the ratio of the change in drain voltage to the corresponding change in gate voltage maintaining a constant drain current. The μ of a FET quantifies its voltage amplification property and may be compared to the β of a BJT.

9.4 FET versus BJT

Table 9.3 presents a brief comparison of the overall properties and applications of FETs and BJTs. The general term 'FET' is used because the comparison holds for both the JFET discussed so far and the MOSFET introduced in the next section.

Table 9.3 Comparison of bipolar junction transistor (BJT) and field-effect transistor (FET)

BJT	FET
It is bipolar device. Both electrons and holes equally take part in current conduction.	It is unipolar device. Either electrons (in n-channel) or holes (in p-channel) make the current conduction.
Minority carriers are involved in the current flow.	It is majority carrier device.
It has three terminals: emitter, base and collector. One of these is connected in common to both the input and the output port.	It has three terminals: source, gate and drain, one of these being common to both the input and the output port.
It is current-controlled device. The base current in CE mode or the emitter current in CB mode controls the collector current.	It is voltage-controlled device. The gate voltage controls the drain current.
The carriers move through the base by the process of diffusion.	The carriers move through the channel by the process of drift.
The electric field across the junction is along the direction of carrier flow and the field indirectly influences the collector current through Early effect.	The electric field across the junction is transverse to the direction of carrier flow and the electric field along the channel is aligned to the direction of carrier movement. The electric fields have direct effect on the drain current causing its saturation.
It has low input resistance, of the order of kilo ohm or less.	It has very high input resistance, of the order of $10^8\ \Omega$ or more due to the reverse gate bias in JFET and the presence of insulating layer in MOSFET.
The minority carriers make BJT a noisy device.	The FET is less noisy than BJT because only one type of charge carriers is involved.
If the small reverse saturation current is ignored, a linear relationship exists between I_C and I_B ($I_C = \beta I_B$). So it is suitable for linear amplification, particularly power amplification.	For FETs, the input-output relationship (between I_D and V_{GS}) is nonlinear [Equation (9.1)]. FET is preferred to BJT for high speed switching applications because the former has no stored charge due to minority carriers, hence no time is wasted in removing the charge from the junction area. Also a JFET can be used as a voltage-variable resistor.

Example 9.1

In a JFET, I_{DSS} is 10 mA and V_p is 5 V, symbols having the usual meaning. Calculate the drain current in saturation region for gate voltage of –1 V.

Solution

Using Equation (9.2), the required drain current is

$$10 \times 10^{-3} \times (1 - 1/5)^2 = 6.4 \text{ mA}$$

[Only the magnitude of the gate voltage is to be considered.]

Example 9.2

In Example 9.1, determine the device resistance in the ohmic region.

Solution

The minimum value of the device resistance in the ohmic region is

$$(5 \text{ V})/(10 \text{ mA}) = 500 \text{ }\Omega$$

Following Equation (9.1), the resistance in the present case is

$$500/(1 - 1/5)^2 = 781.25 \text{ }\Omega$$

Example 9.3

If, in Example 9.1, the drain voltage is changed to 4.5 V keeping the gate voltage unchanged, determine the drain resistance.

Solution

Using Equation (9.2), the drain current with $V_{DS} = 4.5$ V is

$$10 \times 10^{-3} \times (1 - 1/4.5)^2 = 6.05 \text{ mA}$$

The drain resistance, using Equation (9.6b) is calculated as

$$(5.0 - 4.5)/[(6.4 - 6.05) \times 10^{-3}] = 1.43 \text{ k}\Omega \text{ (approx.)}$$

[It may be noted that the usual drain resistance, defined in the saturation region is much larger than the voltage-variable resistance in the ohmic region for comparable values of other parameters.]

9.5 MOSFET

The metal-oxide-semiconductor field-effect transistor (MOSFET) is another category of FET that has gate, source, and drain terminals similar to a JFET and the output and transfer characteristics have similarities with those of the JFET. However, the working principle of MOSFET is different from that of JFET. The gate terminal in a MOSFET remains insulated from the channel by a silicon dioxide layer. Because of this insulating layer, the input resistance of a MOSFET is very high ($> 10^{10}\ \Omega$), even higher than that of the JFET ($10^8\ \Omega$). Since the gate in the MOSFET remains insulated from the channel, the device is also termed as *insulated gate field-effect transistor* (IGFET). There are two distinct categories of the MOSFET, namely depletion-type and enhancement-type. Both may be of n-channel and p-channel type. Here the n-channel version of these two devices are interpreted. The p-channel versions are similar in operation with the bias and the charge polarities reversed.

9.5.1 n-Channel Depletion-Type MOSFET

The construction of the device is illustrated in Figure 9.4(a). It consists of two heavily doped n⁺ regions, namely source and drain within a p-type substrate. A moderately doped n-channel is fabricated between the source and the drain. The gate terminal is connected to the metal layer over the oxide, which isolates the n-channel from the gate. The substrate is either internally connected to the source terminal or an external substrate terminal is provided, which may be grounded separately. The drain voltage is made positive with respect to the source and the gate voltage can be made zero, positive or negative optionally. Based on the gate bias, the depletion-type MOSFET can execute two different modes of working: *depletion mode* and *enhancement mode*, respectively. Figure 9.4(b) is a pictorial representation of the drain characteristics in these two modes.

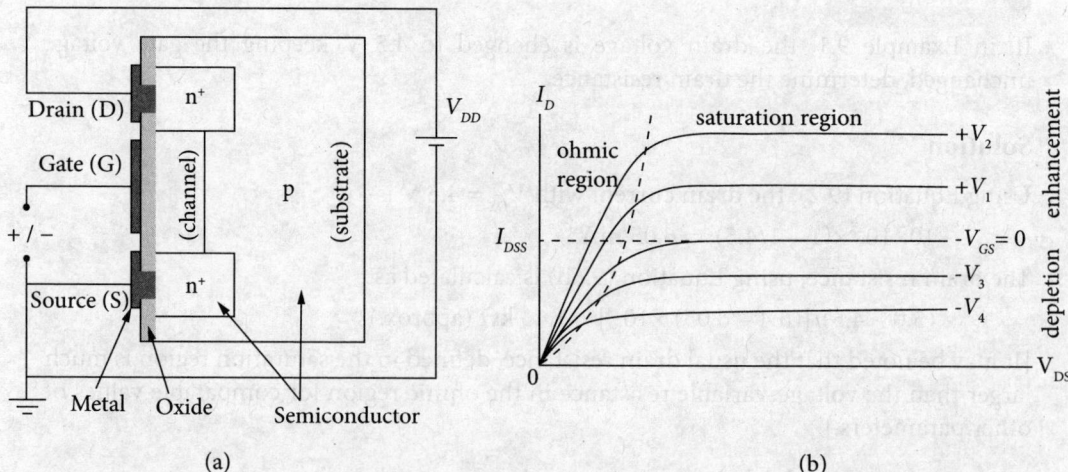

Figure 9.4 n-channel depletion-type MOSFET (a) construction and (b) drain characteristics in depletion and enhancement modes

A. Depletion-Mode Performance

When $V_{GS} = 0$, an appreciable number of electrons flow from the source to the drain through the channel under the influence of positive drain supply voltage ($+V_{DD}$) so that the drain current (I_D) increases linearly. However, the increasing drain-source voltage (V_{DS}) causes a depletion region across the reverse-biased channel–substrate junction so that the channel width gets reduced. Eventually, at higher values of drain voltage, pinchoff condition occurs and I_D becomes constant and independent of V_{DS}. Similar to the JFET, the saturation level of the drain current for $V_{GS} = 0$ is denoted by I_{DSS}, as given in Figure 9.4(b).

When V_{DS} is negative, the depletion of carriers is more pronounced. The negative gate voltage repels the electrons of the n-channel toward the p-substrate and attracts the holes of the p-substrate to the n-channel. Recombination of electrons and holes take place, the channel gets depleted thereby diminishing the drain current. The saturation of I_D occurs at lower values for larger magnitude of negative V_{GS}, as indicated in Figure 9.4(b). Ultimately, the drain current reduces to zero for $V_{GS} = V_P$ and Equation (9.2) is valid for this MOSFET also.

> **Note: 'Normally On' Device**
>
> Both the depletion-mode MOSFET and the JFET are considered *normally on* device because both devices can yield a large amount of drain (output) current when the gate (input) voltage is zero. However, there is a basic difference. For the JFET, the maximum possible drain current is I_{DSS} whereas in the depletion-mode MOSFET, the drain current can exceed I_{DSS} on applying the gate voltage of the appropriate polarity. For n-channel device, the drain current can exceed I_{DSS} with positive V_{GS}.

B. Enhancement-Mode Performance

This is a new feature of the MOSFET not obtainable with the JFET. In this case, the gate is made positive or of the same polarity as that of the drain. The positive gate voltage enhances the number of carriers by the following two ways.

(i) It attracts additional electrons from the p-substrate.
(ii) Collision of accelerated particles break the bonds and cause new carriers.

Consequently the drain current increases more rapidly and the saturation level occurs at a value higher than I_{DSS}, as sketched in Figure 9.14(b).

Thus the depletion-type MOSFET can operate in both depletion and enhancement modes depending on the gate bias polarity. For depletion mode, the gate bias is of opposite polarity and for enhancement mode, it is of the same polarity as compared with the drain bias. This convention is maintained for both n-channel and p-channel devices. The p-channel depletion-type MOSFET acts in a similar manner as that explained above In that case, the gate-source voltage is either zero or positive for depletion mode and negative for enhancement mode, the drain voltage being negative all the way.

The symbols of n-channel and p-channel depletion-type MOSFETs are introduced in Figure 9.5. The symbols are self-explanatory. The gap between the gate and the other two terminals imply that due to the presence of the oxide layer, there is no direct electrical connection between the gate and the channel. The arrow direction, specifying the n- or p-channel category indicates the direction of current (opposite to that of electron flow) from source to drain through the channel.

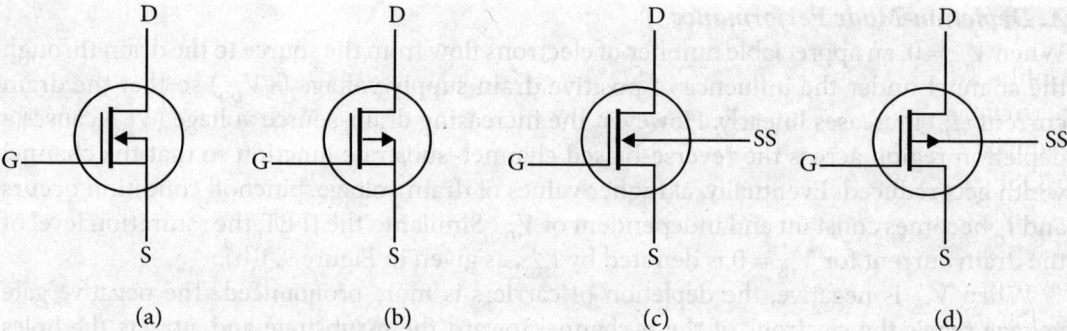

Figure 9.5 Symbols for depletion-type MOSFET with internally connected source (S) and substrate (SS) for (a) n-channel and (b) p-channel devices and with separate external substrate (SS) connection for (c) n-channel, and (d) p-channel devices

C. Transfer Characteristic

Figure 9.6 (left side) displays the transfer characteristic curve of an n-channel MOSFET showing the typical nature of variation of I_D with V_{GS}. It is of the similar parabolic nature as that of a JFET. However, comparing with the corresponding drain characteristics for different values of V_{GS}, as shown in Figure 9.6 (right side), it is found that the V_{GS} values can be both negative and positive and the I_D values can extend up to the positive side of V_{GS} axis. For p-channel depletion-type MOSFET, the transfer characteristic curve would resemble the mirror image of that of the n-channel device. Equation (9.2) is valid for both the depletion and enhancement regions of both n- and p-channel devices just with the proper polarity of V_{GS} incorporated.

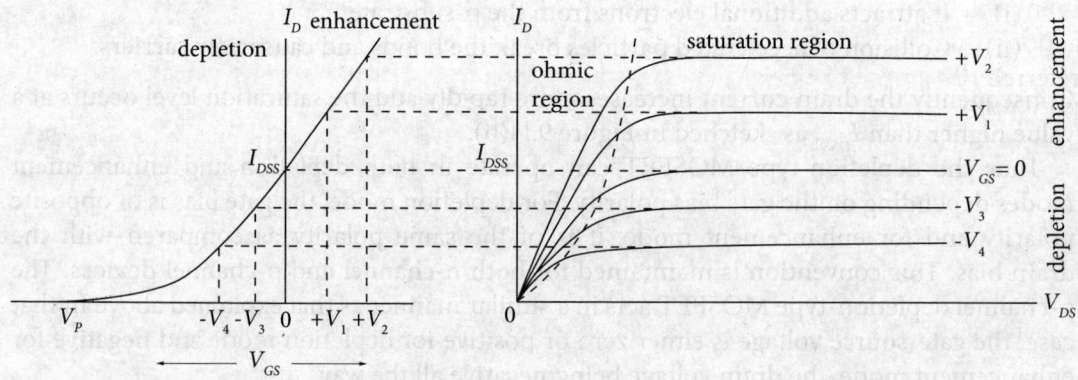

Figure 9.6 Transfer characteristic of n-channel depletion-type MOSFET compared with its drain characteristics

9.5.2 n-Channel Enhancement-Type MOSFET

Unlike the previously mentioned depletion-type device, this category of MOSFET has no permanent channel between the source and the drain. The channel is created by enhancing induced charges with gate bias of the same polarity as that of the drain bias. This type of MOSFET was invented later than the depletion-type MOSFET.

Figure 9.7(a) outlines the architecture of a typical n-channel enhancement-type MOSFET. It has several similarities with the construction of depletion-type MOSFET. The thin oxide layer grown over the semiconductor surface is overlaid by the usual metallic connection of the gate. A lightly doped p-type substrate contains highly doped n^+ regions for source and drain with metallic connection. Similar to the depletion-type MOSFET, the substrate is sometimes internally connected to the source and also sometimes a separate external connection is made available for the substrate. However, as a remarkable difference, no channel exists between the source and the drain as a constructed component. Also the drain voltage and the gate voltage are biased with the same polarity. For the n-channel MOSFET similar to that of Figure 9.7(a), both the gate-source voltage (V_{GS}) and the drain-source voltage (V_{DS}) increase in the positive direction and for a p-channel device, both V_{DS} and V_{GS} are negative. The following biasing conditions are considered in connection with the n-channel device of Figure 9.7(a).

When $V_{GS} = 0$, no drain current flows in spite of positive V_{DS} being applied. This situation is analogous to two reverse-biased p–n junctions, drain-substrate and substrate-source. When V_{GS} is made slightly positive, the V_{DS} remaining unchanged, the positive gate potential repels the holes of the substrate adjacent to the oxide layer. At the same time, the electrons (minority carriers) of the substrate are attracted toward the gate and are accumulated in the vicinity of the substrate edge. Because of the oxide layer, these electrons cannot drift into the metallic gate and build up an *inversion layer* of charge carriers of opposite polarity within the p-substrate. On further enhancing the positive V_{GS}, the electron concentration is enhanced and a current flow is commenced between the source and the drain under the action of the positive V_{DS}. The limiting value of V_{GS} that initiates this drain current (I_D) is referred to as *threshold voltage* (V_T). Below V_T, the drain current remains zero. This type of MOSFET functions in enhancement mode only and there is no scope for depletion mode.

Note: 'Normally Off' Device

When the gate voltage of the enhancement-type MOSFET is made zero, no current flows between the source and the drain. When the gate voltage becomes larger than the threshold voltage, the device switches from cutoff to saturation. Since the device yields no output (drain) current when the input (gate) voltage is zero, this MOSFET is called *normally off* device.

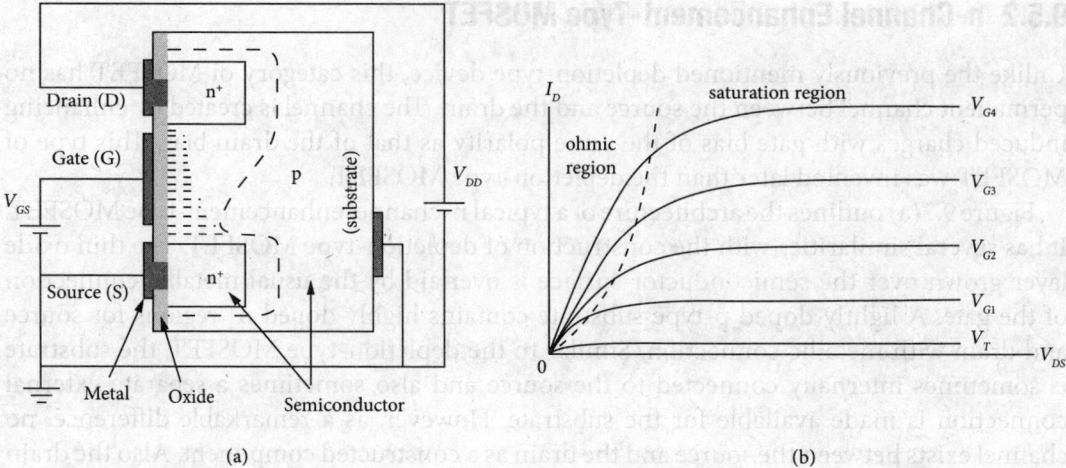

Figure 9.7 n-channel enhancement-type MOSFET, (a) construction and (b) drain characteristics for positive gate voltages ($V_{G4} > V_{G3} > V_{G2} > V_{G1} > V_T$), V_T being the threshold voltage

Figure 9.7(b) represents the drain current–voltage characteristics of the n-channel enhancement-type MOSFET. The following three distinct regions of the characteristic curves may be noted.

Cutoff Region: When $V_{GS} < V_T$, there is no drain current.

Ohmic or Linear Region: For lower values of V_{DS}, I_D increases linearly and the channel acts like a simple resistor. The stipulations for the device operating in ohmic region are: $V_{GS} > V_T$ but $V_{DS} \leq (V_{GS} - V_T)$. In this region, the drain current is expressed by (derivation is beyond the present scope)

$$I_D = k(V_{GS} - V_T - V_{DS}/2)V_{DS} \tag{9.7}$$

The constant k, depending on the device architecture is given by

$$k = \frac{\mu C w}{L} \tag{9.7a}$$

The involved parameters are: μ the electron mobility, C the capacitance per unit area of the gate-oxide structure, w the channel width and L the channel length. It is interesting to note that the current depends not only on the device structure but also on the type of carriers because electrons and holes have different values of mobility for a specific semiconductor.

Saturation Region: For a certain V_{GS}, if V_{DS} is increased, the induced channel becomes narrower, especially at the drain-end because of the depletion region occurring between the reverse-biased, drain-substrate region indicated by dashed line in Figure 9.7(a). Thus a pinchoff process takes place and I_D reaches a saturation level [Figure 9.7(b)] similar to that of JFET and depletion-type MOSFET. The saturation occurs for $V_{DS} \geq (V_{GS} - V_T)$. Putting the limiting condition of $V_{DS} = (V_{GS} - V_T)$ to Equation (9.7), the drain current in the saturation region is obtained as

$$I_D = \frac{k}{2}(V_{GS} - V_T)^2 \quad (9.8)$$

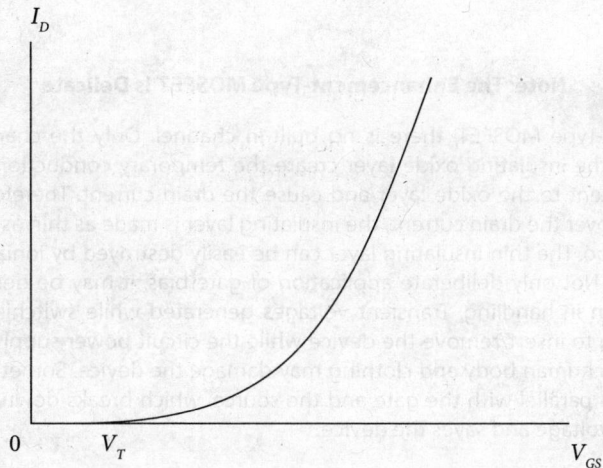

Figure 9.8 Transfer characteristic of n-channel enhancement-type MOSFET

Figure 9.8 sketches the transfer characteristic curve of the n-channel enhancement-type MOSFET. The whole parabolic curve resides at the positive side of the voltage axis and the current is zero for $V_{GS} \leq V_T$. The transfer characteristic for p-channel enhancement-type MOSFET would be the mirror image of that in Figure 9.8 about the current axis.

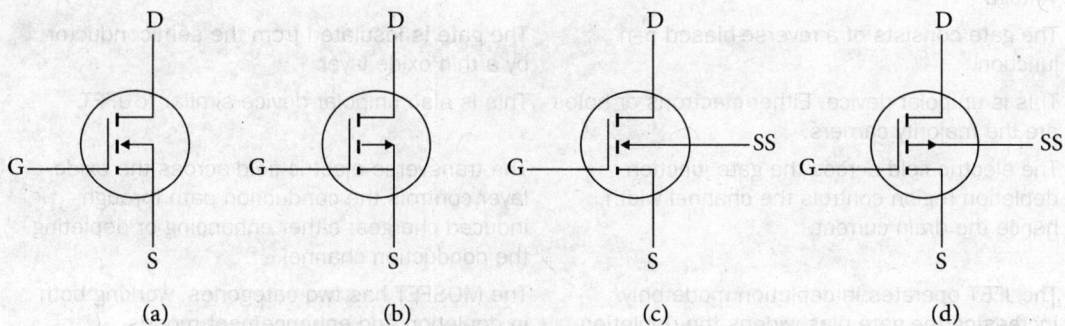

Figure 9.9 Symbols for enhancement-type MOSFET with internally connected source (S) and substrate (SS) for (a) n-channel and (b) p-channel devices and with separate external substrate (SS) connection for (c) n-channel, and (d) p-channel devices

Figure 9.9 introduces the symbols of n- and p-channel enhancement type MOSFETs. The dashed line between the drain and the source implies that no permanent channel exists between these two. By virtue of its threshold voltage, the enhancement-type MOSFET is quite suitable for use as a switching device. It is used widely in switching purpose, both for large currents in power electronics and small currents in digital circuits, particularly those of computers. Table 9.4 summarizes the main features of JFET and MOSFET discussed so far and compares the similarities and difference in construction and performance.

> **Note: The Enhancement-Type MOSFET Is Delicate**
>
> In the enhancement-type MOSFET, there is no built-in channel. Only the charges induced by the gate voltage across the insulating oxide layer create the temporary conduction channel within the semiconductor adjacent to the oxide layer and cause the drain current. Therefore, in order to have precise gate control over the drain current, the insulating layer is made as thin as possible. This makes the device delicate too. The thin insulating layer can be easily destroyed by ionization with excessive gate-source voltage. Not only deliberate application of gate bias, it may be destroyed easily unless precautions are taken in handling. Transient voltages generated while switching on-off the power supply or attempting to insert/remove the device while the circuit power supply is on or even static charges coming from human body and clothing may damage the device. Sometimes a built-in Zener diode is fabricated in parallel with the gate and the source, which breaks down before reaching the danger level of gate voltage and saves the device.

Table 9.4 Comparison of the construction and working of JFET and MOSFET

JFET	MOSFET
It has three terminals, namely source, drain and gate.	It also has the three terminals: source, drain and gate. Sometimes the substrate is used as a fourth terminal.
This is voltage-controlled device. The output (drain) current is controlled by the input (gate) voltage.	This is also voltage-controlled device, the drain current being controlled by the gate voltage.
The gate consists of a reverse-biased p–n junction.	The gate is insulated from the semiconductor by a thin oxide layer.
This is unipolar device. Either electrons or holes are the majority carriers.	This is also unipolar device similar to JFET.
The electric field across the gate junction depletion region controls the channel width, hence the drain current.	The transverse electric field across the oxide layer controls the conduction path through induced charges; either enhancing or depleting the conduction channel.
The JFET operates in depletion mode only. Increasing the gate bias widens the depletion width within the channel and causes reduction of the drain current.	The MOSFET has two categories, working both in depletion and enhancement modes.
It has very high input resistance, of the order of 10^8 Ω, due to the reverse-biased gate junction.	It has still higher input resistance, of the order of 10^{10} Ω or more, because of the insulated gate.

It is a *normally off* device because large drain current can flows with zero gate voltage also.	The depletion-type MOSFET is *normally on* similar to the JFET but the enhancement-type MOSFET is *normally off* because no drain current flows with zero gate voltage.
Simultaneous use of n-channel and p-channel is not possible.	Constructing both p-channel and n-channel MOSFETs on the same substrate (CMOS) is possible.
Suitable for low noise, small signal amplifiers and switches.	Suitable for digital circuits and also for power electronic circuits.

9.6 FET Model

The small signal, low frequency equivalent circuit of JFET or MOSFET can be represented by that of Figure 9.10. It is related to Equation (9.5) that expresses the drain current in terms of the input and output parameters. Only the ac components are involved. The input port comprised of the gate (G) and the source (S) is open circuited because the input resistance is supposed to be infinite. The ac input voltage (v_{gs}) appears across these two terminals. The output is a voltage dependent current source $g_m v_{gs}$ where the proportionality constant is the mutual conductance (g_m). The ac output voltage (v_{ds}) appears across the ac drain resistance (r_d), which exists in parallel with the current source. Table 9.5 compares this FET model with the hybrid equivalent model of a BJT discussed in Chapter 7.

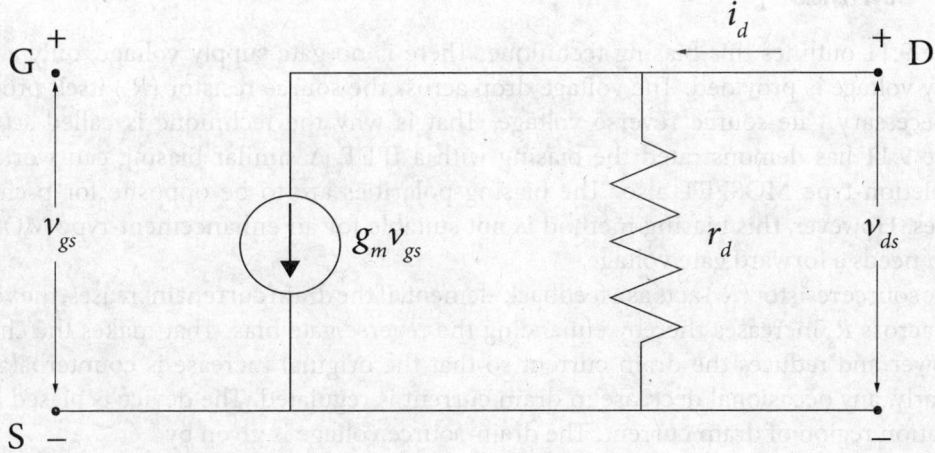

Figure 9.10 Small signal low frequency FET model

Table 9.5 Comparison of the small signal low frequency equivalent circuits of FET and BJT

FET Model	BJT Hybrid Model
It has Norton equivalent output circuit.	It also has Norton equivalent output circuit.
The output is voltage dependent current source.	This output is also current source but current dependent.
The input resistance is infinite.	The input resistance is finite.
The output resistance (r_d) is high.	The output resistance ($1/h_o$) is high.
There exists only output current. The input current is zero due to infinite resistance.	Both input and output currents exist.
Only the drain current is significant as output current.	Both the collector current and the emitter current are significant as output current.
It has no feedback from the output to the input.	It has a voltage feedback (though negligibly small) through reverse voltage transfer ratio (h_r).

9.7 FET Biasing

Similar to the BJT, the FET also requires biasing for stabilizing the operating point. The techniques are, however, different from those of BJT. Some popular biasing techniques for FETs are introduced here. Some are applicable to both JFET and MOSFETs bur some others are applicable to specific types of device only.

9.7.1 Self-Bias

Figure 9.11 outlines this biasing technique. There is no gate supply voltage; only a drain supply voltage is provided. The voltage drop across the source resistor (R_S) itself produces the necessary gate-source reverse voltage. That is why the technique is called *self-bias*. Figure 9.11 has demonstrated the biasing with a JFET. A similar biasing can work with a depletion-type MOSFET also. The biasing polarities are to be opposite for p-channel devices. However, this biasing method is not suitable for an enhancement-type MOSFET, which needs a forward gate voltage.

The source resistor (R_S) acts as a feedback element. If the drain current increases, the voltage drop across R_S increases thereby enhancing the reverse gate bias. That makes the channel narrower and reduces the drain current so that the original increase is counterbalanced. Similarly, any occasional decrease in drain current is regulated. The device is biased in the saturation region of drain current. The drain-source voltage is given by

$$V_{DS} = V_{DD} - I_D(R_D + R_S) \tag{9.9}$$

Since negligible current flows through the gate resistor (R_G), the gate voltage with respect to the ground is zero. Therefore, the gate-source voltage can be written as

$$V_{GS} = V_G - I_D R_S = 0 - I_D R_S \tag{9.10}$$

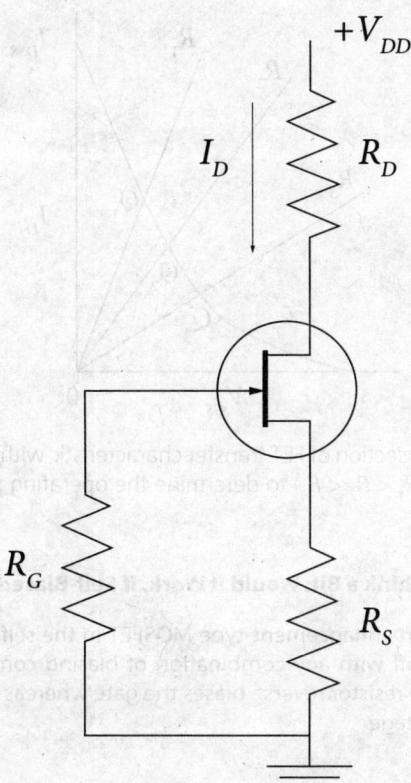

Figure 9.11 Self-bias circuit with n-channel JFET

Consequently,

$$I_D = -\frac{V_{GS}}{R_S} \tag{9.11}$$

Equation (9.11) is another justification for the term 'self-bias'. The gate-source voltage is controllable with the drain current itself. The graphical representation of Equation (9.11) for a fixed R_S is known as *self-bias line*. It has the following properties.

(i) It is a straight line with the slope of $-1/R_S$.
(ii) It determines the operating point of the FET by intersecting with its transfer characteristic, as interpreted in Figure 9.12.

The change of the slope of the bias line with the source resistor ($R_1 < R_2 < R_3$) and the corresponding change in the operating point (Q_1, Q_2 and Q_3) are indicated in Figure 9.12.

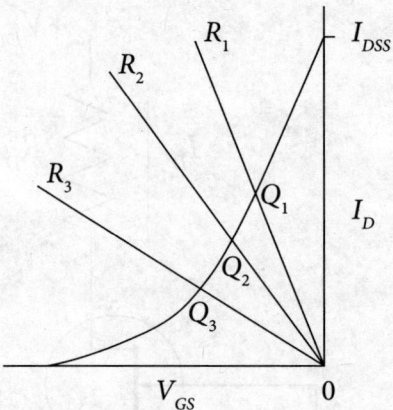

Figure 9.12 Intersection of FET transfer characteristic with self-bias line for different source resistors ($R_1 < R_2 < R_3$) to determine the operating points (Q_1, Q_2 and Q_3)

Think a Bit: Would it Work, if Self-Biased?

If the JFET were replaced by an enhancement-type MOSFET in the self-bias circuit of Figure 9.11, the circuit would remain turned off with any combination of biasing components. It is so because the voltage drop across the source resistor reverse biases the gate whereas a forward gate bias is required to overcome the threshold voltage.

9.7.2 Drain-Feedback Bias

This biasing technique, outlined in Figure 9.13 is suitable for enhancement-type MOSFETs. It contains a feedback resistor (R_f) connected between the drain and the gate, analogous to the collector-feedback bias of BJT. The resistor R_f serves the following two purposes.

(i) It provides with a forward gate bias, the required condition for the enhancement-type device.
(ii) It tends to compensate for any change in the drain current.

In this circuit, $V_{GS} \approx V_{DS}$, since almost no current flows through gate and R_f. If the drain current (I_D) has any instantaneous increase, the larger $I_D R_D$ drop decreases V_{DS}. This, in turn, reduces V_{GS}, hence V_{DS} and compensates for the increase in I_D. Similar interpretation holds for the momentary decrease in I_D. The proper value of the drain resistance required to maintain a specified value of I_D is determined by

$$R_D = \frac{V_{DD} - V_{DS}}{I_D} \qquad (9.12)$$

Assuming $V_{GS} = V_{DS}$, Equation (9.12) can be modified as

$$V_{GS} = V_{DD} - I_D R_D \qquad (9.13)$$

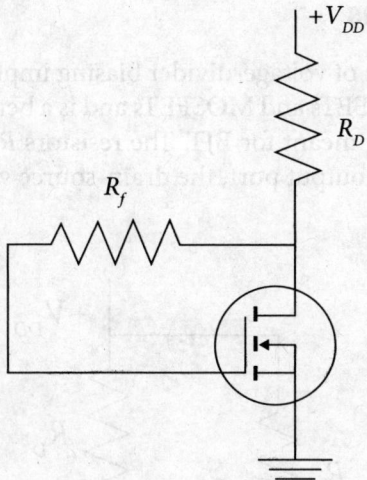

Figure 9.13 Drain-feedback bias circuit with n-channel enhancement-type MOSFET

9.7.3 Gate Bias

The gate biasing circuit diagram using a JFET is given in Figure 9.14. This is analogous to the base bias of BJT and can be applied to JFET and MOSFETs of both types. However, this is not an efficient biasing technique because the gate supplies a fixed bias voltage to the gate and the Q-point is highly sensitive to the particular device used in the circuit. The drain current varies widely from device to device of the same class.

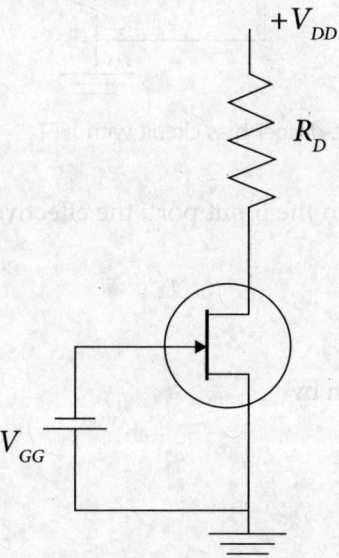

Figure 9.14 Gate bias circuit with JFET

9.7.4 Voltage-Divider Bias

Figure 9.15 shows an example of voltage-divider biasing implemented to JFET. This biasing technique is applicable to all JFETs and MOSFETs and is a better type of biasing. It is similar to the voltage-divider biasing meant for BJT. The resistors R_1 and R_2 establish the voltage-divider bias. Considering the output port, the drain-source voltage is given by

$$V_{DS} = V_{DD} - I_D(R_D + R_S) \tag{9.14}$$

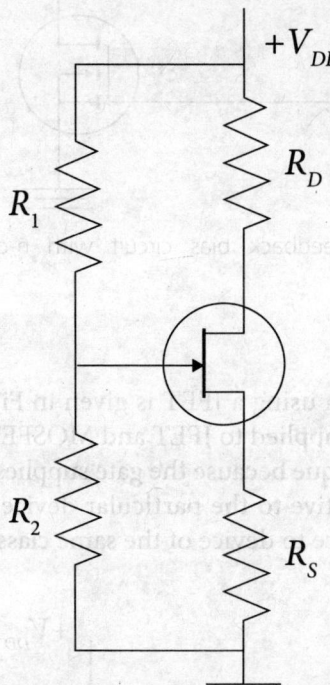

Figure 9.15 Voltage-divider bias circuit with JFET

Applying Thevenin's theorem to the input port, the effective gate supply voltage comes out to be

$$V_{GG} = \frac{R_2 V_{DD}}{R_1 + R_2} \tag{9.15}$$

The gate-source voltage is given by

$$V_{GS} = V_{GG} - I_D R_S \tag{9.16}$$

Therefore,

$$I_D = \frac{V_{GG} - V_{GS}}{R_S} \tag{9.17}$$

Field-Effect Transistor (FET)

> **Note: Benefits of Voltage-Divider Bias**
>
> Comparing Equation (9.17) with Equation (9.11), it is understood that similar to the self-bias, this voltage-divider bias also offers the facility of controlling V_{GS} with I_D for a fixed gate supply voltage (V_{GG}). Equation (9.15) suggests that such a fixed gate voltage, similar to that of gate bias (Figure 9.14) can be established easily with the drain supply voltage. Thus the voltage-divider bias, offering the beneficial features of both the gate bias and the self-bias techniques is a versatile, general-purpose biasing technique for all FETs.

Equation (9.17) represents the bias line for the FET with voltage-divider bias. Figure 9.16 shows its position and intersection is noted that the bias line in voltage-divider bias does not pass through the origin like that of self-bias. The voltage-axis intercept of the bias line represents the gate supply voltage (V_{GG}) and the intersection with the transfer curve represents the FET operating point (Q).

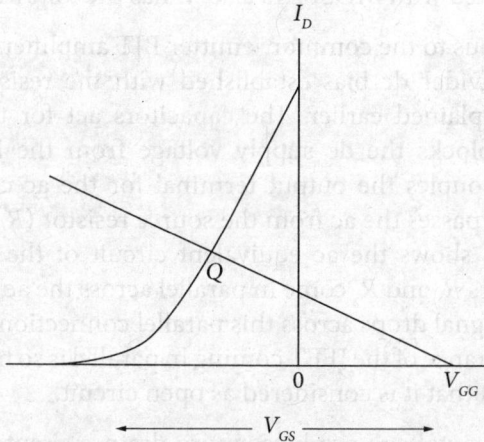

Figure 9.16 JFET transfer characteristic and bias line intersection for voltage-divider bias showing the Q-point

It can be inferred from the above illustrations that the FET biasing techniques have some advantages. For instance, the JFET can be biased in both the ohmic and the saturation region to make it equivalent to a resistor and a current source, respectively. The self-bias can correlate the gate-source voltage and the drain current for both JFET and depletion-type MOSFET. The drain-feedback Bias can provide with a forward gate bias and compensate for any change in drain current. The voltage-divider bias, offering the benefits of both the gate bias and the self-bias can be implemented to both JFET and MOSFET as a general biasing technique. Nevertheless, there are some disadvantages also. The self-bias technique cannot be applied to enhancement-type MOSFET. Also the inherent nonlinear variation of I_D with V_{GS} persists leading to movement of Q-point over a nonlinear path (Figure 9.12).

9.8 FET Amplifiers

Similar to the BJT, the JFET and the MOSFETs can be used in three configurations, taking one terminal in common with both input and output ports. Anyway the nomenclature is quite different from that of the BJT. The FET amplifiers can be common-source, common-drain and common-gate. The voltage-divider bias, suitable for different types of FETs, is adopted for the dc biasing. A small ac signal input superimposed on the dc bias causes fluctuation in the gate-source voltage, hence in the drain current. Because of the alternating voltage drop across the drain resistor (or external load resistor joined in addition), an amplified version of the ac input is obtained. A moderate frequency, of the order of kilohertz is assumed for the input signal.

9.8.1 Common-Source (CS) Amplifier

Figure 9.17(a) presents the circuit diagram for CS amplifier fabricated with JFET. A similar circuit could be constructed with MOSFETs also. It has the following properties.

(i)   This is analogous to the common-emitter BJT amplifier.
(ii) The voltage-divider dc bias established with the resistors works in the usual manner, as explained earlier. The capacitors act for the ac components. The capacitor C_B blocks the dc supply voltage from the input signal source, the capacitor C_C couples the output terminal for the ac component only and the capacitor C_S bypasses the ac from the source resistor (R_S).
(iii) Figure 9.17(b) shows the ac equivalent circuit of the CS amplifier where the biasing resistors R_1 and R_2 come in parallel across the ac input voltage source (v_i). The input ac signal drops across this parallel connection and is defined by v_{gs}.
(iv) The input resistance of the JFET coming in parallel is so high (that of the MOSFET is even higher) that it is considered as open circuit.

An increasing gate-source voltage produces more drain current and larger voltage drop across R_D so that the increasing input corresponds to decreasing drain voltage. The positive half-cycle of the input produces the negative half-cycle of the output. Thus a phase inversion of 180° is obtained in CS amplifier. Considering this property, the ac output voltage is given by

$$v_o = -g_m v_{gs} R_D \tag{9.18}$$

The negative sign accounts for the phase inversion. The CS amplifier has the following parameters.

Voltage Gain: Since $v_i = v_{gs}$, the voltage gain of the CS amplifier is expressed as

$$A_v = \frac{v_o}{v_{gs}} = -g_m R_D \tag{9.19}$$

If any additional load resistor (R_L) is added, as indicated by dotted line in Figure 9.17(a), then R_D in Equation (9.19) is replaced by $R_D // R_L$.

Input Resistance: The effective input resistance of the CS amplifier is $R_1//R_2$ because the input resistance of the FET is infinitely large and is ignored in parallel combination with $R_1//R_2$.

Output Resistance: The effective output resistance in absence of any external load is the drain biasing resistor (R_D), which is much smaller than the drain resistance (r_d) of the FET and the later can be ignored in the parallel combination $r_d // R_D$.

If r_d and R_D happen to be of the same order, the output voltage [Equation (9.18)] is modified as

$$v_o = -g_m v_{gs}(r_d // R_D) \tag{9.20}$$

whereas the input voltage remains the same (v_{gs}). Therefore, the voltage gain [Equation (9.19)] is modified as

$$A_{v1} = -g_m(r_d // R_D) \tag{9.21}$$

9.8.2 Common-Drain (CD) Amplifier

The circuit diagram is given in Figure 9.18(a). The ac input is applied to the gate and the output is obtained across the unbypassed source resistor. The output voltage is almost equal to and in phase with the input voltage. Therefore it is also called *source follower*. Indeed it is comparable to the common-collector amplifier or emitter-follower circuit obtained with BJT. The following parameters of the CD amplifier can be derived from Figure 9.18(b), which is the ac equivalent circuit of the CD amplifier of Figure 9.18(a).

Voltage Gain: The input voltage is

$$v_i = v_{gs} + g_m v_{gs} R_S \tag{9.22}$$

The output voltage is

$$v_o = g_m v_{gs} R_S \tag{9.23}$$

Therefore, the voltage gain is defined as

$$A_v = \frac{v_o}{v_i} = \frac{R_S}{R_S + 1/g_m} \tag{9.24}$$

If $R_S \gg 1/g_m$ in Equation (9.24), the voltage gain of the CD amplifier approaches unity.

Input Resistance: It is the gate-to-source resistance, which is infinitely high.

Output Resistance: Equation (9.24) implies that v_i is applied to R_S and $1/g_m$ in series and the output voltage (v_o) is obtained across R_S. So the output resistance is the Thevenin equivalent resistance, which is the parallel combination of R_S and $1/g_m$. If $R_S \gg 1/g_m$, the output resistance of the CD amplifier is $1/g_m$ approximately.

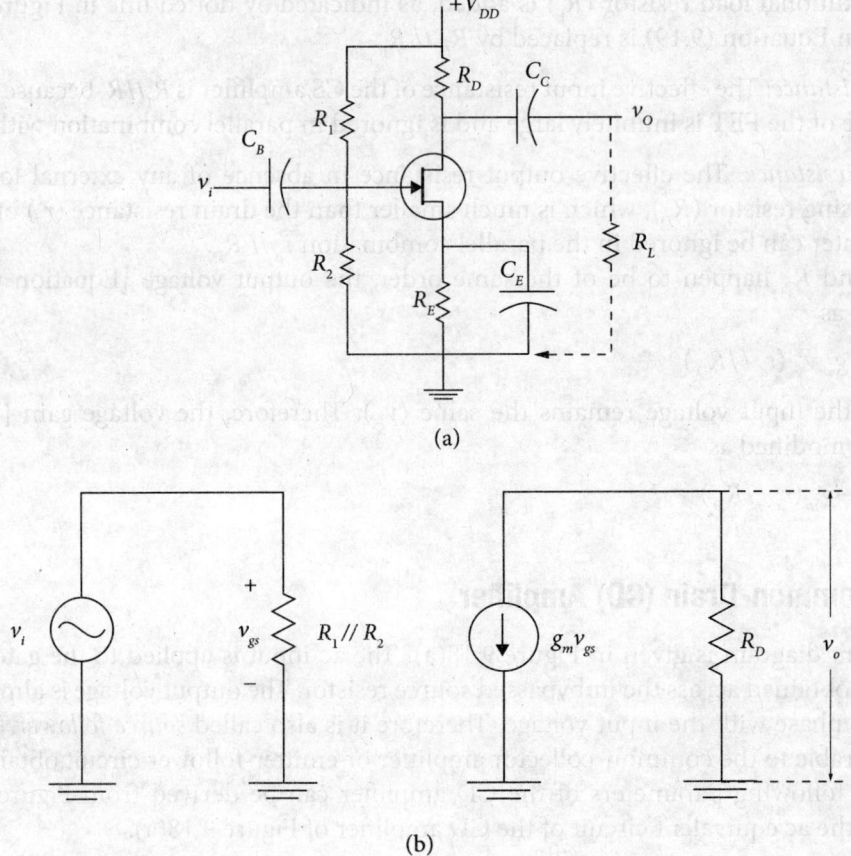

Figure 9.17 Common-source (CS) amplifier constructed with n-channel JFET: (a) circuit diagram and (b) ac equivalent circuit

9.8.3 Common-Gate (CG) Amplifier

This circuit is comparable to BJT common-base amplifier. Figure 9.19(a) shows the circuit diagram for CG amplifier and Figure 9.19(b) displays its AC equivalent circuit. The parameters are estimated as follows.

Voltage Gain: The input voltage (v_i) to the CG amplifier is v_{gs} and the output voltage (v_o) is $g_m v_{gs} R_D$ so that the voltage gain is given by

$$A_v = \frac{v_o}{v_i} = g_m R_D \tag{9.25}$$

Input Resistance: The ac input current (i_i) is the drain current ($g_m v_{gs}$) so that the input resistance is

$$\frac{v_{gs}}{i_i} = \frac{1}{g_m} \tag{9.26}$$

Field-Effect Transistor (FET)

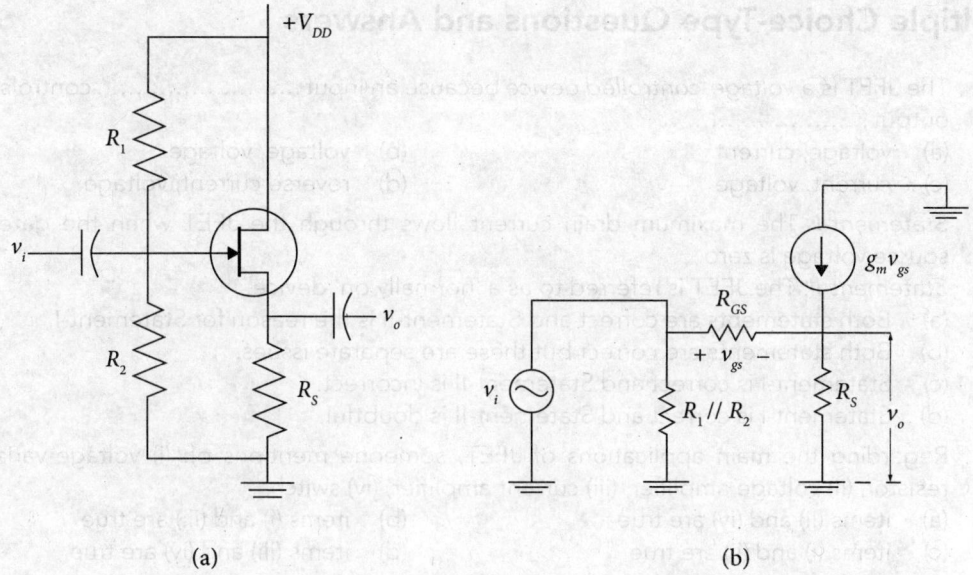

Figure 9.18 Common-drain (CD) amplifier constructed with n-channel JFET: (a) circuit diagram and (b) ac equivalent circuit

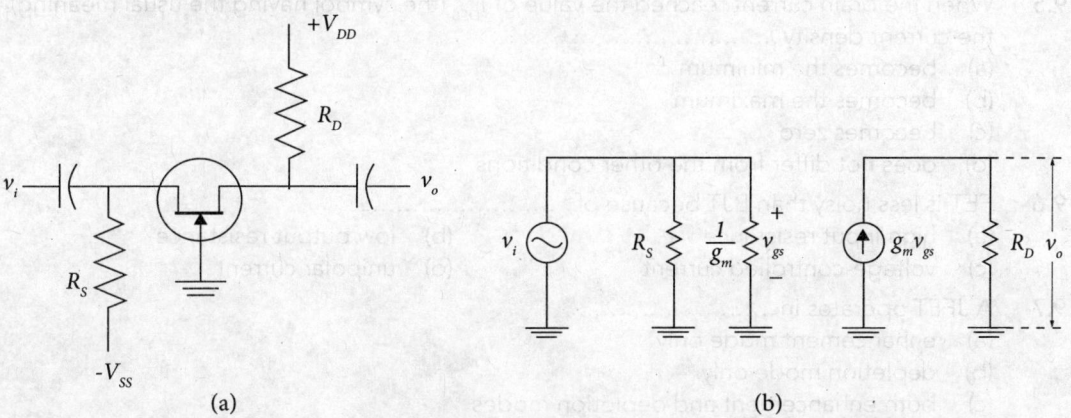

Figure 9.19 Common-gate (CG) amplifier constructed with n-channel JFET: (a) circuit diagram and (b) ac equivalent circuit

Equation (9.26) infers that the CD amplifier has low input resistance.

Output Resistance: The drain resistance of the FET is very high so that the external resistor (R_D) joined in the drain circuit acts as the effective output resistance, as denoted in Figure 9.19(b).

Multiple Choice-Type Questions and Answers

9.1 The JFET is a *voltage-controlled device* because an input .... controls the output ....
  (a) voltage, current
  (b) voltage, voltage
  (c) current, voltage
  (d) reverse current, voltage

9.2 *Statement-I*: The maximum drain current flows through the JFET when the gate-to-source voltage is zero.
  Statement-II: The JFET is referred to as a 'normally on' device.
  (a) Both statements are correct and Statement-II is the reason for Statement-I.
  (b) Both statements are correct but these are separate issues.
  (c) Statement-I is correct and Statement-II is incorrect.
  (d) Statement-I is correct and Statement-II is doubtful.

9.3 Regarding the main applications of JFET, someone mentions of: (i) voltage-variable resistor, (ii) voltage amplifier, (iii) current amplifier, (iv) switch
  (a) items (ii) and (iv) are true
  (b) items (i) and (iii) are true
  (c) items (i) and (ii) are true
  (d) items (iii) and (iv) are true

9.4 As the drain current of the FET shifts from ohmic region to saturation region for a certain gate voltage, the output resistance of the device ....
  (a) decreases
  (b) increases
  (c) remains constant
  (d) fluctuates

9.5 When the drain current reached the value of I_{DSS} (the symbol having the usual meaning), the current density ....
  (a) becomes the minimum
  (b) becomes the maximum
  (c) becomes zero
  (d) does not differ from the other conditions

9.6 FET is less noisy than BJT because of ....
  (a) high input resistance
  (b) low output resistance
  (c) voltage-controlled current
  (d) unipolar current

9.7 A JFET operates in ....
  (a) enhancement mode only
  (b) depletion mode only
  (c) both enhancement and depletion modes
  (d) saturation mode only

9.8 The gate-source p–n junction in a JFET is ....
  (a) reverse-biased for n-channel device and forward-biased for p-channel device
  (b) reverse-biased for p-channel device but forward-biased for n-channel device
  (c) reverse-biased for both n-channel and p-channel devices
  (d) reverse-biased for n-channel device and has no bias for p-channel device

9.9 I_{DSS} denotes the maximum drain current in the saturation region of operation of a JFET with the conditions of .... and ....
  (a) V_{GS} +ve, $V_{DS} \geq |V_P|$
  (b) V_{GS} –ve, $V_{DS} \geq |V_P|$
  (c) V_{GS} zero, $V_{DS} \geq |V_P|$
  (d) V_{GS} zero, $V_{DS} \leq |V_P|$

9.10 The JFET can act like a current source for a drain current .... I_{DSS}, the symbol having the usual meaning.
 (a) less than or equal to
 (b) only greater than
 (c) greater than or equal to
 (d) much less than

9.11 For which of the following levels of V_{GS} in a JFET, I_D becomes zero?
 (a) $V_{GS} < V_P$
 (b) $V_{GS} > V_P$
 (c) $V_{GS} = V_P$
 (d) $V_{GS} >> V_P$

9.12 In n-channel JFET, the drain current flows .... the device and in p-channel JFET, the drain current flows .... the device.
 (a) out of, into
 (b) into, into
 (c) out of, out of
 (d) into, out of

9.13 The biasing of an n-channel depletion-type MOSFET is changed in such a way that the drain current exceeds the prescribed I_{DSS} value. In such a case, the device ....
 (a) is destroyed
 (b) acts like short circuit
 (c) continues to work in depletion mode
 (d) continues to work in enhancement mode

9.14 The doping concentration of the source, drain and channel in an n-channel depletion-type MOSFET are always of .... type and the doping concentration of the .... is much lower than that of the ....
 (a) donor, channel, source
 (b) acceptor, channel, source
 (c) donor, source, channel
 (d) acceptor, source, channel

9.15 When $V_{GS} = 0$, the drain current of an n-channel enhancement-type and a p-channel depletion-type MOSFET are .... and ...., respectively. The symbols have the usual meaning.
 (a) I_{DSS}, zero
 (b) zero, I_{DSS}
 (c) I_{DSS}, $-I_{DSS}$
 (d) $-I_{DSS}$, I_{DSS}

9.16 An enhancement-type MOSFET working at either cutoff or Ohmic region resembles a ....
 (a) current source
 (b) voltage source
 (c) switching device
 (d) active load

9.17 CMOS devices use .... MOSFETs.
 (a) complementary depletion-type
 (b) complementary enhancement-type
 (c) depletion- and enhancement-type complementary connected
 (d) enhancement- and depletion-type complementary connected

9.18 The JFET acts as a voltage-controlled resistor in the .... region when the resistance .... with the increasing magnitude of the gate voltage.
 (a) active, increases
 (b) ohmic, increases
 (c) ohmic, decreases
 (d) saturation, decreases

9.19 There may be .... of V_{GS} between zero and pinchoff.
 (a) a single level
 (b) two levels
 (c) four to five levels
 (d) infinite number of levels

9.20 In the FET amplifier of Figure 9.17(a) the voltage gain will depend on ...., the symbol(s) having the usual meaning.
 (a) R_D, g_m and R_L
 (b) R_D and R_L
 (c) R_D and g_m
 (d) g_m and R_L

Answers

9.1	(a)	9.2	(a)	9.3	(c)	9.4	(b)
9.5	(b)	9.6	(d)	9.7	(b)	9.8	(c)
9.9	(c)	9.10	(a)	9.11	(c)	9.12	(d)
9.13	(d)	9.14	(a)	9.15	(c)	9.16	(c)
9.17	(b)	9.18	(b)	9.19	(d)	9.20	(a)

Reasoning-Type Questions and Answers

9.1 Is the high input resistance of FET a beneficial feature?

Ans. An ideal voltage amplifier should have infinite input resistance. Therefore the FET is quite appropriate in use as voltage amplifier. Sometimes the current gain feature of BJT is added to this benefit of FET by making a hybrid of FET and BJT in circuits like operational amplifiers.

9.2 Why does the FET have high input resistance?

Ans. In the case of JFET, the origin of high input resistance is reverse-biased gate junction and in the case of MOSFET, the insulated gate is the reason.

9.3 Compare the temperature dependence of BJT and FET actions.

Ans. In BJTs, the minority carrier current increases with temperature causing large fluctuation in current gain (β), hence in collector current. In FETs, the drain current depends on the transconductance (g_m), which decreases with increase in temperature because the carrier mobility decreases with increasing temperature. So the majority carrier current in FET decreases with increasing temperature and a negative temperature coefficient of drain current is obtained. Thus the temperature stability of FET is better and thermal runaway is avoided.

9.4 How should the input current–voltage characteristic of a FET look like?

Ans. Both the JFET and the MOSFET have very high input resistance and negligible gate current flows in any case, whatever may be the value of the gate voltage. Since no direct current is drawn through the gate, the gate current versus gate voltage curve would pass adjacent to the voltage axis showing zero current with voltage variation all the way.

9.5 Justify the statement, 'A BJT is a current-controlled device while a FET is a voltage-controlled device' with both diagrams and equations.

Ans. In BJT, the input current (I_B in CE mode and I_E in CB mode) controls the output (collector) current that can be represented pictorially by the output static characteristics, for example, by figures 5.4(b) and 5.5(b). In the corresponding equations: $I_C = \beta I_B$ and $I_C = \alpha I_E$ with β and α constants, the input currents (I_B and I_E) are the controlling variables for the output current (I_C). Also the hybrid model represents a current-controlled current source.

In the case of FET, the input voltage at the gate controls the drain current at the output in different extents. This can be shown pictorially for JFET (Figure 9.3) and MOSFETs [figures 9.4(b) and 9.7(b)]. Equation (9.2) approves the same fact where I_{DSS} and V_P are constants and V_{GS} is the controlling variable for I_D. Also the FET model represents a voltage-controlled current source.

9.6 What do the components 'metal', 'oxide' and 'semiconductor' in the MOSFET correspond to?

Ans. The metal stands for the gate and also the source and drain connections. The oxide represents the insulating layer of silicon dioxide in between the gate metal and the semiconductor channel. The semiconductor incorporates the structure with the substrate and the source, drain and channel regions embedded.

9.7 How does the phrase 'field-effect' originate in the MOSFET?

Ans. The transverse electric field generated from the immobile charges in the depletion region caused by both the gate voltage and the drain voltage affects the channel by either depleting or enhancing with induced charges.

9.8 The FET is called 'unipolar' and the BJT is called 'bipolar' whereas in both cases, only electron current flows through the external circuit. Then what is the justification of such naming?

Ans. It is known that holes have existence within the semiconductor device only and not in the external metallic wires. If we consider the interior of the two devices, namely the BJT and the FET, the nomenclature is found to be appropriate. In the BJT, both electrons and holes take part in conduction. As the majority carriers (electrons in n-side and holes in p-side) cross the junction, they become minority carriers. Thus both majority and minority carriers of two opposite polarities take part in the conduction. In the case of JFET or MOSFET, the carriers do not cross any junction and flow through a bulk semiconductor (electrons in n-channel and holes in p-channel device) between the source and the drain. Therefore, only majority carriers of single polarity is involved in the current transport.

Solved Numerical Problems

9.1 Calculate the drain current in saturation region for an n-channel JFET at gate-source voltage of –1.5 V, if the pinchoff voltage be –5.0 V and the saturation drain current at zero gate bias be 10 mA.

Soln. Using Equation (9.2), the required drain current is

$10 \times (1 - 1.5/5)^2 = 4.9$ mA

9.2 On changing the V_{GS} of a JFET from –1 V to –1.5 V keeping V_{DS} constant, the I_D changes from 7 to 6 mA. Calculate the mutual conductance of the JFET. If the drain resistance be 20 kΩ, calculate the amplification factor.

Soln. The mutual conductance [using Equation (9.6a)] is

$$\frac{(7-6) \times 10^{-3}}{-1-(-1.5)} = 2 \text{ mS (millisiemens)}$$

The amplification factor [using Equation (9.6c)] is

$2 \times 10^{-3} \times 20 \times 10^3 = 40$

9.3 A common source FET amplifier has R_L = 250 kΩ, the AC drain resistance is 200 kΩ and the transconductance is 0.1 mA/V. Calculate the voltage gain of the amplifier.

[C.U. 2011]

Soln. Since the drain resistance and the load resistance are comparable, the effective load resistance is

$(250 \times 200)/(250 + 200) = 111.11 \text{ k}\Omega$

The voltage gain is

$-0.1 \times 10^{-3} \times 111.11 \times 10^3 = -11.11$

9.4 Consider the CS amplifier circuit of Figure 9.17(a). The load resistor (R_L) shown with dotted line is now actually joined and it is of 10 kΩ. The value of R_D is 4.7 kΩ. The mutual conductance of the JFET is 2 mS. Determine the output voltage of the amplifier for an input voltage of 2 mV.

Soln. Following the ac equivalent circuit of Figure 9.17(b), the effective load resistance will be the parallel combination of R_L and R_D given by

$(4.7 \times 10)/(4.7 + 10) = 3.2 \text{ k}\Omega$ (approx.)

The voltage gain of the amplifier will be

$-2.0 \times 10^{-3} \times 3.2 \times 10^3 = -6.4$

Therefore, the input of 2 mV will produce the output of $2 \times (-6.4) = -12.8$ mV

9.5 Establish that the mutual conductance of a FET varies as the square root of its drain current.

Soln. Equation (9.6a) defines the mutual conductance (g_m) of a FET in terms of the rate of change of its drain current (I_D). Differentiating Equation (9.2) with respect to V_{GS} and rearranging, one may obtain

$$g_m = g_{mo}\left(1 - \frac{V_{GS}}{V_P}\right) = -\frac{2}{V_P}\sqrt{I_{DSS} I_D}$$

The symbol $g_{mo} = -2I_{DSS}/V_P$ represents the value of g_m for $V_{GS} = 0$. Since I_{DSS} and V_P are of opposite sign, g_{mo} is a positive quantity. The above expression shows that g_m varies as the square root of I_D.

9.6 A depletion-type MOSFET in self-bias has $V_{GS(off)} = -3$V and $I_{DSS} = 8$ mA, the symbols having the usual meaning. If the drain current at the operating point of saturation region is 1 mA, determine the transconductance of the device.

Soln. Using Equation (9.2), the gate-source voltage corresponding to the drain current of 1 mA can be calculated as

$-3 \times (1 - \sqrt{1/8}) = -1.94$ V approx.

Using the expression derived just above, in solved numerical problem 9.5, the required mutual conductance is determined as

$$\frac{-2 \times 8 \times 10^{-3}}{-3}\left(1 - \frac{-1.94}{-3}\right) = 1.88 \text{ mS}$$

Exercise

Subjective Questions

1. Make a simple sketch of the device structure of a JFET and draw its output characteristics showing the variation of drain current with gate voltage. Explain (i) the ohmic region, (ii) the saturation region and (iii) the breakdown region of drain current.

2. What is the pinchoff voltage of a JFET? Explain how the constant value of drain current is maintained after this voltage. How is the pinchoff voltage related to the gate voltage?
3. Outline the device structure of a depletion-type MOSFET and explain with diagram how it can operate both in depletion mode and enhancement mode.
4. Explain with diagram the working principle of an enhancement-type MOSFET. What is threshold voltage related to it?
5. What are normally on and normally off devices in connection with FETs?
6. What are the parameters mutual conductance, drain current and amplification factor of a FET. Are these relevant to both JFETs and MOSFETs?
7. Compare the device construction and electrical properties of the FET and the BJT.
8. Make a comparison between the structure and working principles of a JFET and a MOSFET.
9. Compare the structure and performance of depletion-type and enhancement-type MOSFETs.
10. Draw the circuit diagram for self-bias of a FET. Is it applicable to both JFETs and MOSFETs?
11. Suggest, with necessary diagram, a biasing technique suitable for enhancement-type MOSFETs.
12. Draw the circuit diagram of a FET common-source amplifier and also draw its ac equivalent circuit. Derive expressions for its voltage gain, input impedance and output impedance.

Conceptual Test

1. Can any FET have drain current greater than I_{DSS}?
2. Can one set the Q-point of a depletion-type MOSFET at $V_{GS} = 0$ V?
3. The BJT is often used as power amplifier where a high level of current is produced. The FET is also used in power amplification processes but mainly as a power switch turning large currents on and off. Justify.
4. The CS, CD and CG amplifiers discussed in this chapter have mentioned of the voltage gain, input resistance and output resistance. There is no mention of any 'current gain'. What is your comment on this point?
5. The JFET is sometimes referred to as a *depletion-mode device*. Can you justify the sentence?
6. The following paragraph contains some correct but jumbled information on the depletion-type MOSFET. Can you prepare a meaningful paragraph using these pieces of information?
   The depletion-type MOSFET has a relatively low voltage gain. It has limited use, mainly in radio frequency circuits. It has low levels of electrical noise and high input impedance values for both negative and positive V_{GS}. It is not suitable for use in Class-B or Class-AB operations. A large drain current flows for zero gate voltage.
7. Can the enhancement-type MOSFET be used in Class AB amplifier by biasing with a gate-source voltage slightly above the threshold voltage?
8. The enhancement-type MOSFET is used mainly as a switching device. In which current zone it is likely to be biased, ohmic region or saturation region?

Numerical Problems

1. The following data were recorded in an experiment on the dc characteristics of a FET. Calculate the ac drain resistance and the mutual conductance of the device.

V_{GS} (V)	V_{DS} (V)	I_D (mA)
0	7	10
0	14	10.25
0.5	14	9

2. A common-source amplifier constructed with a depletion-type MOSFET having transconductance of 2 mS has ac drain resistance of 100 kΩ and external load resistance of 10 kΩ. Calculate the voltage gain of the amplifier.

3. In the JFET gate bias circuit of Figure 9.14, V_{DD} = 12 V, R_D = 5.6 kΩ and V_{DS} = 5 V. Determine the drain current.

4. An n-channel enhancement-type MOSFET has a saturation drain current of 5 mA for a gate-source voltage of 6 V. If the threshold voltage is 2 V, calculate the drain current for a gate-source voltage of 6 V.

5. The voltage-divider bias circuit of Figure 9.15 has V_{DD} = 15 V, R_1 = 10 MΩ, R_2 = 5 MΩ, R_D = 1 kΩ, R_S = 1 kΩ, I_{DSS} = 12 mA and pinchoff voltage = –2 V. Determine the drain current (I_D) and also decide whether the JFET is operating in saturation region.

6. The CS amplifier circuit of Figure 9.17(a) has the following parameters. V_{DD} = 20 V, R_1 = 1 MΩ, R_2 = 2 MΩ, R_D = 4.7 kΩ, R_S = 1 kΩ and the externally joined load resistor (R_L) is of 10 kΩ. If the mutual conductance be 4 mS and the drain resistance be 50 kΩ, determine the voltage gain of the amplifier. If the signal source introduces a resistance of 5 kΩ in series, how is the voltage gain modified?

7. If, in the circuit of Figure 9.20, V_i = –2 V, C = 10 pF, V_{DD} = 16 V, R_D = 2 kΩ, I_{DSS} = 10 mA and V_P = –8 V, the voltage across D and S is
   (a) 11.125 V (b) 10.375 V (c) 5.75 V (d) 4.75 V

   [NET June 2017]

8. If the voltage across the source resistor in Figure 9.21 be 1 V, the drain voltage is ....

   [GATE 2015]

Project Work on Chapter 9

To draw the output current–voltage characteristics of junction field-effect transistor.

Theory:
The variation of drain current (I_D) of a junction field-effect transistor (JFET) with drain-source voltage (V_{DS}) for either zero or negative gate-source (–V_{GG}) plotted in a graph [similar to Figure 9.2(b)] refers to its output current–voltage characteristics. Two distinct types of current variation with voltage, namely Ohmic region and saturation region are noted on the I_D–V_{DS} curve.

Computer Analysis:
The experiment was simulated with LTspice open software with the following specifications. From the given choices of n-channel JFET (njft), 2N3819 was selected arbitrarily. The simulation command 'DC sweep' selected from the 'Simulate' menu specified the first source as V_{DD} (for drain bias) and the second as V_{GG} (for gate bias). The assembled circuit is shown in Figure P9.1. In the present case, V_{DS} is same as the drain bias (V_{DD}) and V_{GS} is same as the gate bias (V_{GG}).

Field-Effect Transistor (FET)   311

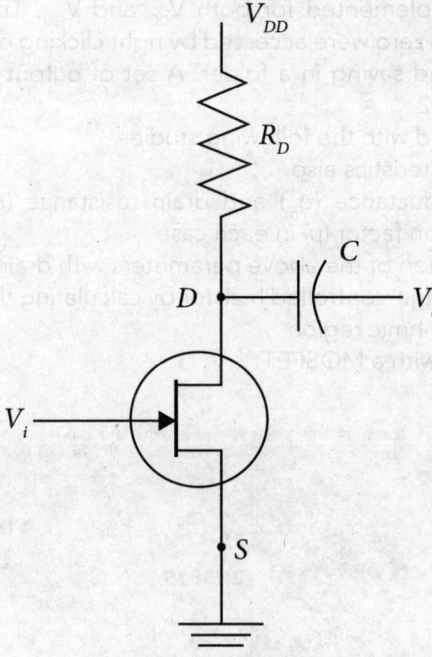

Figure 9.20

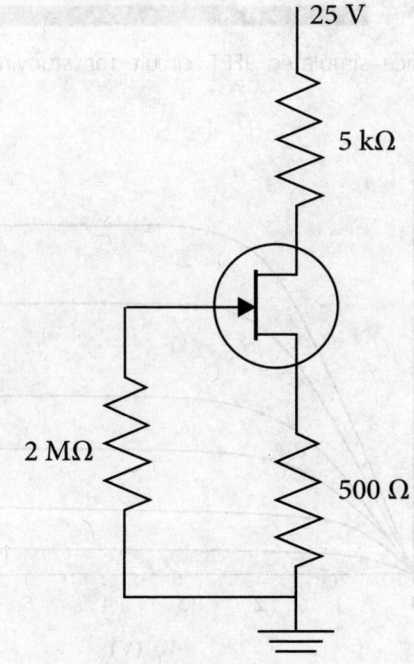

Figure 9.21

Linear sweep was implemented for both V_{DD} and V_{GG}. The current–voltage data for different gate bias, including zero were accessed by right clicking on the viewer, selecting 'File' and 'Export data as text' and saving in a folder. A set of output characteristics plotted with Origin is shown in Figure P9.2.

The work should be extended with the following studies.
- Draw the transfer characteristics also.
- Determine mutual conductance (g_m) and drain resistance (r_d) from the graphs, hence calculate the amplification factor (µ) in each case.
- Note the range of variation of the above parameters with drain current.
- Study the JFET as a voltage-controlled resistor by calculating the change of resistance with gate voltage within the ohmic region.
- Repeat the experiment with a MOSFET.

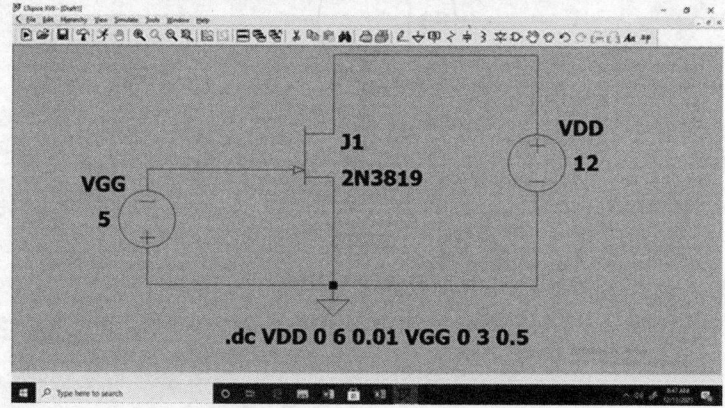

Figure P9.1 LTspice simulated JFET circuit for studying the current–voltage characteristics

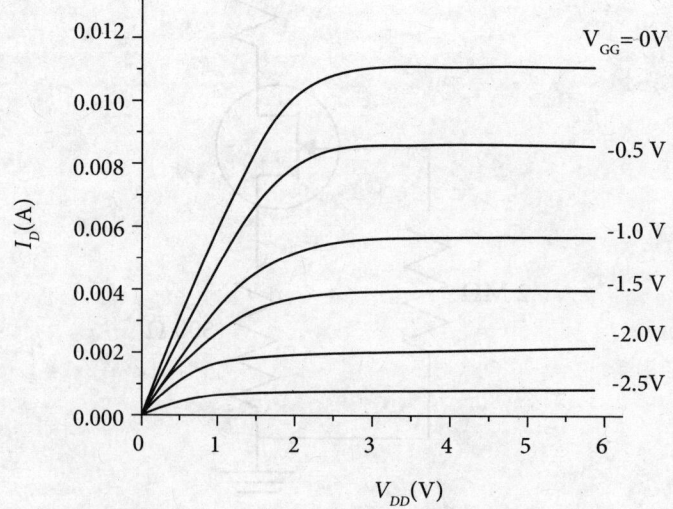

Figure P9.2 LTspice simulated output current–voltage characteristics of JFET

10

Feedback Amplifiers and Oscillators

This chapter presents the concept of feedback, which means looping back a portion of the output voltage or current of an amplifier to control the system within itself. Different types of feedback and the advantages of negative feedback are clarified. The relationship between positive feedback and oscillation is explained. Two different categories of oscillator circuits are elucidated, namely the oscillators based on positive feedback and resonant circuit, such as Hartley oscillator, Colpitts oscillator, Wien bridge oscillator, phase-shift oscillator and crystal oscillator, and the switching oscillators, popularly known as multivibrators.

10.1 Concept of Feedback

Incorporating feedback to an amplifier changes its gain and other properties. During the feedback process, a portion of either the output voltage or the output current is sampled and returned to the input either in series or in parallel with the original signal. The working principle of feedback implemented to an amplifier can be interpreted with the block diagram of Figure 10.1. It comprises the following elements.

Signal Source: It may be either a voltage source or a current source. The symbol X_s denotes the signal, either the voltage or the current obtained from it.

Mixing Unit: It combines the source signal (X_s) and the feedback signal (X_f) to produce the input signal (X_i) for the amplifier.

Amplifier: It performs the basic job of amplification of a signal. It may be any one of the BJT (Chapters 5 through 8) or the FET (Chapter 9) or any circuit made of those. It may also be an operational amplifier (Chapter 11) circuit. Whatever it may be, the amplifier has a transfer gain (A) and it amplifies the input signal to produce the output of X_o. The transfer gain of the basic amplifier is given by

$$A = \frac{X_o}{X_i} \tag{10.1}$$

Sampling Unit: It picks up a fraction of the output signal (X_o) and delivers to the feedback element. The mixing and sampling units are actually series and/or parallel connections of passive elements.

Feedback Element: It is a two-port network, generally made of passive elements, supplying the sampled fraction (X_f/X_o) of the output to the mixing unit. The fraction

$$\beta = \frac{X_f}{X_o} \qquad (10.2)$$

is called the *reverse transmission factor* of the feedback element and is popularly termed as the *feedback ratio*. Please note that this symbol is similar to that for BJT current gain in CE configuration but has completely different meaning.

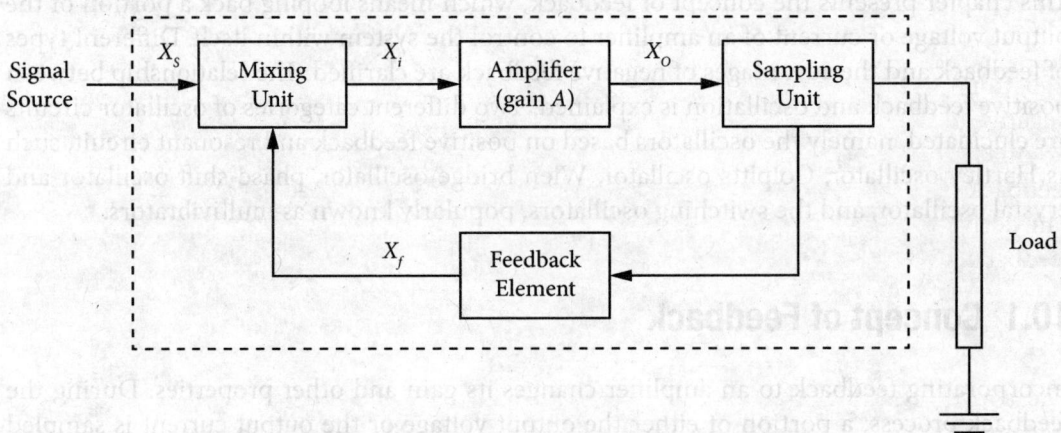

Figure 10.1 Schematic diagram of an amplifier with feedback

If the feedback be in the same phase with the original input from the source, it is called *positive feedback*. The feedback in opposite phase with respect to the original input is called *negative feedback*. The input (X_i) to the basic amplifier is given by

$$X_i = X_s + X_f \text{ (for positive feedback)} \qquad (10.3a)$$

and

$$X_i = X_s - X_f \text{ (for negative feedback)} \qquad (10.3b)$$

The amplifier with feedback as a whole is outlined with dashed lines in Figure 10.1. At present we are considering negative feedback. The transfer gain of the amplifier with negative feedback is expressed, using Equation (10.3b) as

$$A_f = \frac{X_o}{X_s} = \frac{X_o}{X_i + X_f} \qquad (10.4)$$

Rearranging Equation (10.4), one can write

$$A_f = \frac{X_o / X_i}{1 + \frac{X_f}{X_o}\frac{X_o}{X_i}} \qquad (10.4a)$$

Substituting for A from Equation (10.1) and for β from Equation (10.2), we obtain

$$A_f = \frac{A}{1 + A\beta} \qquad (10.5)$$

Equation (10.5) is the generalized expression for the transfer gain of an amplifier with negative feedback. It is called the *closed-loop gain* of the feedback amplifier in terms of the original gain (A) of the basic amplifier, known as the *open-loop gain* and the amount of feedback (β). The quantity ($-A\beta$) is called the *loop gain*. Proceeding in a similar way, one can derive the following formula for the closed-loop gain of the amplifier with positive feedback.

$$A_f = \frac{A}{1 - A\beta} \qquad (10.5a)$$

The above equations are based on the simplified assumption that the input signal (X_i) transmits through the amplifier alone and not through the feedback element. Similarly the feedback signal passes through the feedback element only; not through the amplifier and the feedback ratio (β) does not depend on the source and load impedances.

> **Note: Positive and Negative Feedback**
>
> Feedback in an electronic circuit is the process of cycling back a fraction of the output voltage or current of an amplifier to its input through passive network of resistor capacitor etc. and thereby modifying the gain of the amplifier. Equation (10.5) indicates that the closed-loop gain (A_f) becomes less than the open-loop gain on applying negative feedback. However, it will be discussed later that the negative feedback has many other advantageous features, such as stability in operation and improvement of impedance properties, at the cost of the gain. The gain with positive feedback, as apparent from Equation (10.5a), increases with respect to the open-loop gain. Under some special circumstances, known as the Barkhausen criterion, the denominator of Equation (10.5a) can even be zero, making the closed-loop gain infinitely large. This makes the amplifier an oscillator that generates an electrical signal of its own. These topics are to be discussed later in this chapter.

Table 10.1 Comparison of different types of feedback in amplifiers

Feedback type	Amplifier used	Closed-loop gain	Feedback ratio (β) and dimension	Effects on input and output impedance
Voltage–series	Voltage amplifier (voltage gain A_v)	$A_{Vf} = \dfrac{A_v}{1 + A_v \beta}$	$\dfrac{V_f}{V_o}$, dimensionless	Increases input impedance. Decreases output impedance
Voltage–shunt	Transresistance amplifier (open circuit mutual resistance R_M)	$R_{Mf} = \dfrac{R_M}{1 + R_M \beta}$	$\dfrac{I_f}{V_o}$, inverse resistance	Decreases input impedance. Decreases output impedance
Current–series	Transconductance amplifier (open circuit mutual conductance G_M)	$G_{Mf} = \dfrac{G_M}{1 + G_M \beta}$	$\dfrac{V_f}{I_o}$, resistance	Increases input impedance. Increases output impedance
Current–shunt	Current amplifier (current gain A_I)	$A_{If} = \dfrac{A_I}{1 + A_I \beta}$	$\dfrac{I_f}{I_o}$, dimensionless	Decreases input impedance. Increases output impedance

10.2 Types of Feedback

Depending on the sampling of voltage or current, application to the input and the category of the basic amplifier, several distinct types of feedback configurations can be implemented, as summarized in Table 10.1. It states that all the amplifiers are not suitable for all types of feedback. A specific type of feedback requires a particular type of amplifier whose properties are introduced in Table 10.2. This section illustrates the working principles of different types of feedback with block diagrams. Examples of actual electronic circuits are given somewhere in this book.

Table 10.2 Different categories of amplifiers used in feedback and their properties

Amplifier category	Properties
Voltage amplifier Examples: BJT emitter-follower (Chapter 6), op-amp non-inverting amplifier (Chapter 11)	• Amplifies the input voltage (V_i). • Transfer gain is the voltage gain (A_v). Input voltage is amplified to the output as $V_o = A_v V_i$. • It has very high (ideally infinite) input impedance and very low (ideally zero) output impedance.
Current amplifier Examples: Two BJTs cascaded [Figure 10.5(b)]. Op-amp circuits are also possible.	• Amplifies the input current (I_i). • Transfer gain is the current gain (A_I). Input current is amplified to the output as $I_o = A_I I_i$. • It has very low (ideally zero) input impedance and very high (ideally infinite) output impedance.

Transconductance Amplifier

Examples: CE amplifier with un-bypassed emitter resistance. Op-amp circuits are also possible.

- Converts the input voltage (V_i) to a proportional output current (I_o).
- Transfer gain is represented by the open-circuit mutual conductance (G_M) so that $I_o = G_M V_i$.
- It has very high (ideally infinite) input impedance and very high (ideally infinite) output impedance.

Transresistance Amplifier

Examples: BJT in collector-feedback bias (Chapter 6), op-amp inverting amplifier (Chapter 11).

- Converts the input current (I_i) to a proportional output voltage (V_o).
- Transfer gain is represented by the open-circuit mutual resistance (R_M) so that $V_o = R_M I_i$.
- It has very low (ideally zero) input impedance and very low (ideally zero) output impedance.

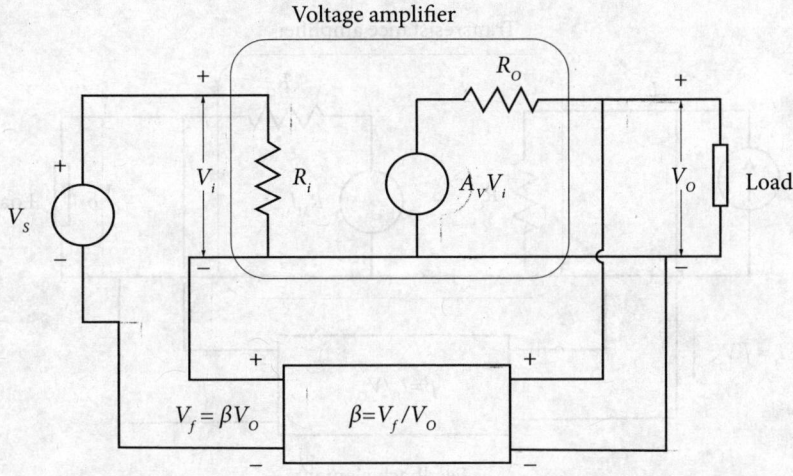

Figure 10.2 Voltage–series feedback: schematic diagram showing the voltage amplifier (outlined) having voltage gain A_v, input resistance R_i and output resistance R_o and the feedback element providing feedback voltage V_f in series with opposite phase to the source voltage V_s. The net input voltage to the voltage amplifier is V_i.

10.2.1 Voltage–series Feedback

The schematic layout of voltage–series feedback circuit is sketched in Figure 10.2. The sampled part (V_f) of the output voltage (V_o) developed across the load is supplied back in series and in opposite phase with the original source voltage (V_s) to the input of a voltage amplifier (outlined) having voltage gain of A_v. The feedback ratio is given by $\beta = V_f/V_o$. The actual input voltage to the voltage amplifier is

$$V_i = V_s - V_f \tag{10.6a}$$

Hence, the output voltage is

$$V_o = A_V V_i \tag{10.6b}$$

Using Equations (10.6a) and (10.6b) and considering $\beta = V_f/V_o$, the overall voltage gain (the closed-loop gain) in presence of voltage–series feedback is derived as

$$A_{Vf} = \frac{V_o}{V_s} = \frac{A_V}{1 + A_V \beta} \tag{10.7}$$

Equation (10.7) agrees with the general expression of Equation (10.5) for closed-loop gain with negative feedback. The voltage gain of the internal voltage amplifier without any feedback is $A_V = V_o/V_i$, which reduces to A_{Vf} on applying the voltage–series feedback.

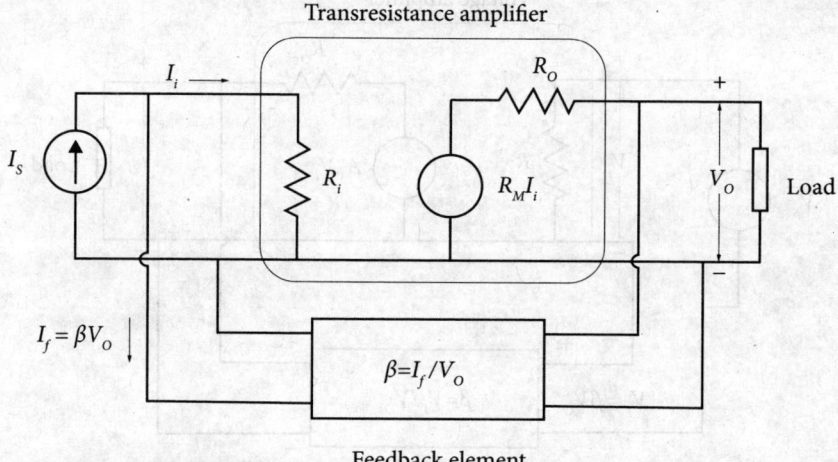

Figure 10.3 Voltage–shunt feedback: block diagram showing the transresistance amplifier with transfer gain R_M, input resistance R_i and output resistance R_o and the feedback element providing feedback current I_f. The net current to the transresistance amplifier is I_i.

10.2.2 Voltage–shunt Feedback

Figure 10.3 depicts the block diagram for this process. Here the basic amplifier is a transresistance amplifier (Table 10.2). The feedback element produces a current (I_f) proportional to the sampled portion of the output voltage (V_o), the feedback ratio being $\beta = I_f/V_o$. This feedback current is applied in parallel and in opposite phase with the source current (I_s) provided by the source voltage (V_s) so that the net input current is

$$I_i = I_s - I_f \tag{10.8a}$$

The transfer gain of the transresistance amplifier is its open circuit mutual resistance (R_M) so that the output voltage is

$$V_o = R_M I_i \tag{10.8b}$$

Considering Equations (10.8a) and (10.8b), the closed-loop transfer gain of the amplifier with voltage–shunt feedback can be represented as

$$R_{Mf} = \frac{V_o}{I_s} = \frac{R_M I_i}{I_i + I_f} \tag{10.8c}$$

Equation (10.8c) can be rearranged by putting $\beta = I_f/V_o$ to make it similar to the general expression of Equation (10.5) as

$$R_{Mf} = \frac{R_M}{1 + R_M \beta} \tag{10.9}$$

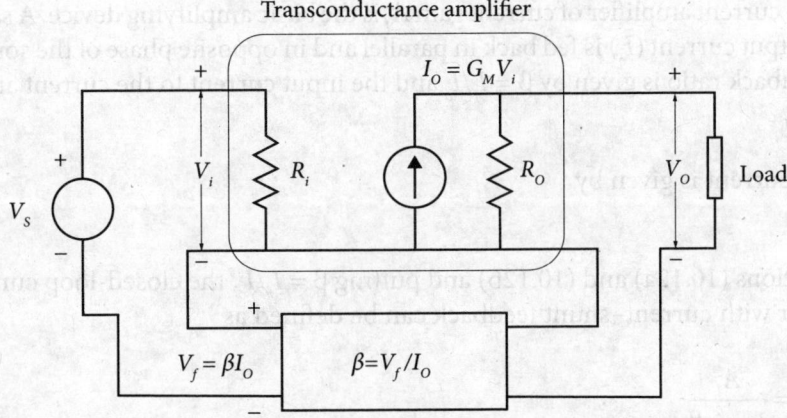

Figure 10.4 Current–series feedback: schematic sketch showing the transconductance amplifier with transfer gain G_M, input resistance R_i and output resistance R_o and the feedback element providing feedback voltage V_f. The net input voltage to the transconductance amplifier is V_i

10.2.3 Current–series Feedback

Figure 10.4 is the schematic representation of the current–series feedback process where the basic amplifying element is a transconductance amplifier (Table 10.2). The feedback element supplies a feedback voltage (V_f) proportional to the sampled output current, the feedback ratio being $\beta = V_f/I_o$. This feedback voltage is applied to the input of the transconductance amplifier in opposite phase to the original source voltage (V_s). Therefore, the actual input voltage is

$$V_i = V_s - V_f \tag{10.10a}$$

The transfer gain of the transconductance amplifier is its open circuit mutual conductance (R_M) so that the output voltage can be defined as

$$I_o = G_M V_i \tag{10.10b}$$

Using Equations (10.10a) and (10.10b) and using $\beta = V_f/I_o$, the closed-loop transfer gain of the amplifier with current–series feedback gain can be expressed as

$$G_{Mf} = \frac{I_o}{V_s} = \frac{G_M}{1 + G_M \beta} \tag{10.11}$$

Equation (10.11) is comparable with the general expression for closed-loop gain with negative feedback.

10.2.4 Current–shunt Feedback

The working principle of this type of feedback is explained with the block diagram of Figure 10.5 where a current amplifier of current gain A_I is the basic amplifying device. A sampled part (I_f) of the output current (I_o) is fed back in parallel and in opposite phase of the source current (I_s). The feedback ratio is given by $\beta = I_f/I_o$ and the input current to the current amplifier is

$$I_i = I_s - I_f \tag{10.12a}$$

The output current is given by

$$I_o = A_I I_i \tag{10.12b}$$

Using Equations (10.12a) and (10.12b) and putting $\beta = I_f/I_o$, the closed-loop current gain of the amplifier with current–shunt feedback can be defined as

$$A_{If} = \frac{I_o}{I_s} = \frac{A_I}{1 + A_I \beta} \tag{10.13}$$

Equation (10.13) is comparable with the general expression of (10.5).

> ### Example 10.1
>
> Negative feedback is applied to an amplifier of voltage gain 100. If the feedback ratio be 1%, calculate (i) the loop gain, (ii) the closed-loop gain, (iii) the output voltage of the feedback amplifier for an input voltage of 20 mV and (iv) the feedback voltage.
>
> **Solution**
>
> The feedback ratio (β) is 1% = 0.01. The original gain (A) is 100. Therefore,
> (i) The loop gain is: $-A\beta = -100 \times 0.01 = -1$.
> (ii) The closed-loop gain, using Equation (10.5) is: $100/(1+100 \times 0.01) = 50$
> (iii) The output voltage with feedback is $50 \times 20 \times 10^{-3} = 1$ V
> (iv) The feedback voltage (βV_o) is: $0.01 \times 1 = 0.01$ V

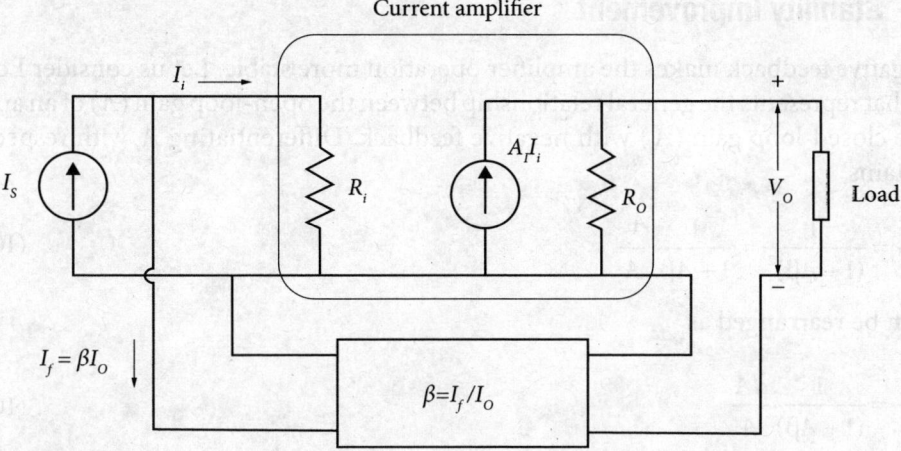

Figure 10.5 Current–shunt feedback: schematic representation showing current amplifier of current gain A_I, input resistance R_i and output resistance R_o and the feedback element supplying feedback current I_f. The net input current to the current amplifier is I_i

Example 10.2

The feedback (F) is sometimes expressed in decibel (dB) as: $F = 20\log_{10}(A_f/A)$, symbols having the usual meaning. Using this formula, calculate the extent of feedback in the case of Example 10.1 given above.

Solution

Here $A = 100$ and $A_f = 50$. Therefore the extent of feedback is
$F = 20\log_{10}(50/100) = -6.02$ dB

10.3 Advantages of Negative Feedback

It is obvious from the previous section that all sorts of negative feedback reduce the gain of the amplifier. It is, of course, a disadvantage. However, this section presents that there are a number of other advantages obtainable from the negative feedback at the cost of the gain.

10.3.1 Stability Improvement

The negative feedback makes the amplifier operation more stable. Let us consider Equation (10.5) that represents the general relationship between the open-loop gain (A) of an amplifier and the closed-loop gain (A_f) with negative feedback. Differentiating A_f with respect to A, one obtains

$$\frac{dA_f}{dA} = \frac{1}{(1+A\beta)^2} = \frac{1}{1+A\beta}\frac{A_f}{A} \tag{10.14}$$

This can be rearranged as

$$\frac{dA_f}{A_f} = \frac{1}{(1+A\beta)}\frac{dA}{A} \tag{10.15}$$

Equation (10.15) implies that the fractional change in gain without feedback (dA/A) is multiplied by a factor $1/(1+A\beta)$ to define the fractional change in gain with feedback (dA_f/A_f). Since the factor is much less than unity, (particularly for large A), any fluctuation in the gain without feedback is much reduced when negative feedback is applied. Hence the gain becomes more stable. The term $1/(1+A\beta)$ is called *sensitivity* and its inverse, $(1+A\beta)$ is called *desensitivity*.

The stability with negative feedback can also be understood implicitly. If the open-loop gain is very high, $A\beta \gg 1$ and $A_f \approx 1/\beta$. Since β is a constant term, it approximates a stable gain.

10.3.2 Impedance Improvement

'Impedance' is a general term that denotes the ohmic resistance plus the reactance: capacitive or inductive, in a circuit. The same concept is applicable to the input and output properties of an amplifier. Referring to Table 10.1 and Table 10.2, the diverse impedance properties of different types of amplifiers may be noted. It is a marvel that the application of negative feedback to an amplifier changes both of its input and output impedances to the desired direction. In particular:

- A voltage amplifier and a transconductance amplifier have very high input impedance and the negative feedback enhances both of these further.
- A voltage amplifier and a transresistance amplifier have very low output impedance and the negative feedback further lowers both of those.
- A current amplifier and a transresistance amplifier have very low input impedance and the negative feedback makes those further diminished.
- A current amplifier and a transconductance amplifier have very high output impedance and the negative feedback makes those higher.

Table 10.3 Effects of negative feedbacks on the input and output impedances of an amplifier

Feedback type	Input impedance	Output impedance
Voltage–series	increases	decreases
Voltage–shunt	decreases	decreases
Current–series	increases	increases
Current–shunt	decreases	increases

In brief, a series feedback tends to increase the input impedance and a shunt feedback tends to decrease it. A voltage feedback has the tendency to decrease the output impedance whereas a current feedback increases it. Table 10.2 present a summary of the above statements.

Two specific examples from Table 10.2, namely the input and output properties with voltage–series feedback are demonstrated here. Also we make some simplification by approximating the impedance with the resistance. It is a reasonable approximation for simple (one/two active elements) circuits at moderate (below MHz) frequencies when the reactance can be replaced by open/short circuits depending on the signal frequency. The schematic diagrams of amplifiers in Figures 10.2 through 10.5 are also simplified with resistances to make those consistent with the actual circuits of other chapters. Though the present demonstrations are made with resistance improvements, these properties are general and are applicable to the input and output impedances also.

A. Input resistance with voltage–series feedback
Referring to Figure 10.2, the input current to the amplifier is given by

$$I_i = \frac{V_i}{R_i} = \frac{V_s - V_f}{R_i} \tag{10.16}$$

Putting $V_f = \beta V_o = \beta A_V V_i$, Equation (10.16) can be rearranged to derive the source voltage as

$$V_s = I_i R_i + I_i \beta A_V R_i \tag{10.17}$$

Using Equation (10.17), the input resistance with feedback can be derived as

$$R_{if} = \frac{V_s}{I_i} = R_i(1 + A_V \beta) \tag{10.18}$$

Equation (10.18) implies that the original input resistance (R_i) of the voltage amplifier increases by a factor of $(1 + A_V \beta)$ on applying voltage–series negative feedback.

B. Output resistance with voltage–series feedback
In Figure 10.2, the output voltage (V_o) can be expressed in terms of the input voltage (V_i), the voltage gain (A_V), the output current (I_o) and the output resistance (R_o) of the voltage amplifier as

$$V_o = I_o R_o + A_V V_i \tag{10.19}$$

The output resistance with feedback (R_{of}) is determined by looking into the open circuited output terminals assuming the external signal source (V_s) short-circuited. Putting $V_s = 0$ in Equation (10.6a), the input voltage becomes $V_i = -V_f = \beta V_o$. Putting this condition in Equation (10.19), the output resistance with feedback is obtained as

$$R_{of} = \frac{V_o}{I_o} = \frac{R_o}{1 + A_V \beta} \tag{10.20}$$

Equation (10.20) indicates that the output resistance of a voltage amplifier decreases by a factor of $1/(1+A_v\beta)$ on applying voltage–series negative feedback.

Example 10.3

An amplifier has $A = -100$, $\beta = -0.1$, $R_i = 10$ kΩ and $R_o = 20$ kΩ, symbols having the usual meaning. Determine the input resistance and the output resistance for a voltage–series feedback.

Solution

Using Equation (10.18), the input resistance with feedback is calculated as
$[1 + (-100)(-0.1)] \times 10 \times 10^3 = 110$ kΩ
Using Equation (10.20), the output resistance with feedback is
$(20 \times 10^3)/[1 + (-100)(-0.1)] = 1.818$ kΩ

10.3.3 Bandwidth Enhancement

The frequency response of an amplifier discussed in Chapter 7 may be revisited. An amplifier can amplify the input signal over a specific range of its frequency, which is quantified as *bandwidth*. The negative feedback widens the bandwidth, as discussed below. Following Equation (7.47), the frequency (f) dependent gain at high frequency can be expressed in terms of the maximum gain (A_o) at mid-frequency and the upper cutoff frequency (f_{cu}) as follows.

$$A = \frac{A_o}{1 + j(f/f_{cu})} \tag{10.21}$$

Following the general concept of negative feedback [Equation (10.5)], the frequency dependent gain with negative feedback can be expressed as

$$A_f = \frac{\dfrac{A_o}{1 + j(f/f_{cu})}}{1 + \dfrac{\beta A_o}{1 + j(f/f_{cu})}} \tag{10.22}$$

With some algebraic simplifications, Equation (10.22) can be converted to

$$A_f = \frac{A_{of}}{1+j(f/f_{cuf})} \tag{10.23}$$

where

$$A_{of} = \frac{A_o}{1+A_o\beta} \tag{10.23a}$$

is the mid-band gain with feedback and

$$f_{cuf} = f_{cu}(1+A_o\beta) \tag{10.23b}$$

is the upper cutoff frequency with feedback. Equation (10.23b) states that the value of the upper cutoff frequency increases from f_{cu} by a factor of $(1+A_o\beta)$, thereby increasing the bandwidth on the higher frequency side. The mid-band gain, however, decreases, as apparent from Equation (10.23a).

Similarly the frequency dependent gain at low frequency, in terms of the lower cutoff frequency (f_{cl}) is given by [following Equation (7.36)]

$$A = \frac{A_o}{1-j(f_{cl}/f)} \tag{10.24}$$

The frequency dependent gain with negative feedback can be derived as

$$A_f = \frac{A_{of}}{1+j(f_{clf}/f)} \tag{10.25}$$

The parameter A_{of} is already defined by Equation (10.23a) and the lower cutoff frequency with feedback is

$$f_{clf} = \frac{f_{cl}}{1+A_o\beta} \tag{10.25a}$$

Equation (10.25 a) states that the value of the lower cutoff frequency decreases from f_{cl} by a factor of $1/(1+A_o\beta)$, which causes an increase of the bandwidth on the lower frequency side. At the same time, the mid-band gain also decreases.

Considering Equations (10.23a), (10.23b) and (10.25a), it is understood that the negative feedback decreases the lower cutoff frequency, increases the upper cutoff frequency and decreases the mid-band gain. As a combined effect, the capability of the amplifier gets enhanced over a larger range of frequency but the maximum achievable gain is diminished. Ideally the *gain-bandwidth product* remains constant.

> **Think a Bit: Does the Negative Feedback Keep the Gain-Bandwidth Product Constant?**
>
> Let originally the gain of the amplifier be A, the lower cutoff frequency f_{cl} and the upper cutoff frequency f_{cu}. Therefore, the bandwidth is defined as $f_{cu} - f_{cl} \approx f_{cu}$ because $f_{cu} \gg f_{cl}$. Thus the initial gain-bandwidth product is $f_{cu}A$.
>
> With negative feedback, the gain decreases to $A_f = A/(1+A\beta)$, the upper cutoff frequency increases to $f_{cuf} = f_{cu}(1+A\beta)$ and the lower cutoff frequency decreases to $f_{clf} = f_{cl}/(1+A\beta)$. The bandwidth is $f_{cuf} - f_{cul} \approx f_{cuf}$ because $f_{cuf} \gg f_{cul}$. Now the gain-bandwidth product is
>
> $$f_{cuf} A_f = f_{cu}(1+A\beta)\frac{A}{1+A\beta} = f_{cu}A$$
>
> Thus the gain-bandwidth product remains unchanged with negative feedback.

10.3.4 Noise Reduction

This is another beneficial feature of negative feedback that can be interpreted in the following way. If the gain of an amplifier be A, any noise voltage (V_n) appearing at the input would be amplified at its output as $V_{no} = AV_n$. In presence of negative feedback, the gain reduces to A_f and the noise output becomes $V_{nof} = A_f V_n$. Following Equation (10.5),

$$V_{nof} = \frac{AV_n}{1+A\beta} = \frac{V_{no}}{1+A\beta} \tag{10.26}$$

Equation (10.26) indicates that negative feedback reduces the effect of noise in amplification by the factor of $1/(1+A\beta)$.

10.3.5 Reduction of Nonlinear Distortion

As discussed in Chapter 8, the nonlinearity in the transfer curve of an amplifier causes distortion to its output, particularly at larger amplitudes of the input signal. Let a distortion D be generated at the output of the amplifier of gain A due to its nonlinearity. When a negative feedback is applied, the distortion at the output becomes D_f. It has two components:

- the original distortion (D) generated within the amplifier, and
- the component ($-\beta D_f$) fed back to the input.

Therefore, the distortion with feedback is: $D_f = D + A(-\beta D_f)$, which yields

$$D_f = \frac{D}{1+A\beta} \tag{10.27}$$

Equation (10.27) states that the nonlinear distortion reduces by the factor of $1/(1+A\beta)$ on applying negative feedback.

10.4 Oscillators

The oscillator is an electronic circuit consisting of active and passive elements that produces a periodic voltage waveform, sinusoidal or non-sinusoidal without the application of any external input signal. It converts the dc power derived from the battery or the power supply into ac power and delivers the same to the load. Oscillators of different working principles are available, as outlined below.

Oscillators with positive feedback: The positive feedback introduced in Equation (10.5a) can be adjusted for unit magnitude of the loop gain and no phase difference between the input and the output by the use of some resonant circuit or phase-shift circuit. That causes electrical oscillations in the circuit and a steady voltage waveform is generated. A number of popular oscillator circuits, such as Hartley oscillator, Colpitts oscillator, Wien bridge oscillator and phase-shift oscillator are constructed by this method. These all generate sinusoidal waveform.

Oscillators with switching circuit: Another approach of obtaining oscillation is to implement periodic on/off switching of the source of voltage through the charging and discharging of capacitors. Multivibrators are of this category. The output waveforms are generally square pulses.

Oscillators with negative resistance: Sometimes high frequency oscillations are obtained with the use of negative resistance, a unique feature of specific electronic devices, such as tunnel diodes. This category of oscillator is beyond the scope of this book.

Table 10.4 presents a summary of the different types of oscillators. The output voltage waveform generated by the oscillator is divided broadly into two categories, sinusoidal and non-sinusoidal. The oscillators generating non-sinusoidal waveforms, such as square wave, triangular wave and saw-tooth wave are termed as *relaxation oscillator*.

Table 10.4 Different types of oscillators based on construction and performance

Name	Frequency-selective element and mechanism of oscillation	Waveform generated	Frequency range
Hartley oscillator and Colpitts oscillator	LC resonant circuit with positive feedback amplifier. The output is in phase with the input only for a single frequency depending on the L & C values.	Sinusoidal	Radio frequency $\geq$ MHz, $<$ GHz
Phase-shift oscillator	RC network with BJT, FET or op-amp. Phase shift of 180° is made by each of the amplifier and the RC network.	Sinusoidal	Audio frequency $\geq$ kHz, $<$ MHz
Wien bridge oscillator	RC network in the form of Wien bridge with the amplifier. 360° or no phase change is caused by the amplifier and the Wien bridge.	Sinusoidal	Audio frequency, Hz to $\geq$ kHz
Multivibrator	One or more active device with RC network acting as on/off switch	Square	Audio frequency Hz to $\geq$ kHz
Negative resistance oscillator	Tunnel diode, Gunn diode and IMPATT diode with resonant cavity	Sinusoidal	Microwave frequency $\geq$ GHz

10.4.1 Positive Feedback and Oscillation

When an electronic circuit, especially an amplifier is just switched on, a transient noise voltage appears at the input because of the thermal agitation of the electrons within the resistors. This produces a transient output voltage. Now the following situations may occur.

(i) In the case of negative feedback, the output fraction returned to the input opposes the noise voltage to make it decay down spontaneously and the amplification becomes stable.

(ii) In the case of positive feedback, the noise transient gets supported so that it persists and blows up until it is limited by the nonlinearity of the circuit. This condition is quite unstable.

(iii) If the positive feedback element includes some frequency sensitive components, such as capacitors or inductors having reactance, then the feedback ratio (β) and the quantity $A\beta$ become frequency dependent.

(iv) The combined effect of the instability and the frequency dependence of the positive feedback gives rise to oscillation. We will see later that at a specific condition of such positive feedback, when the quantity $|A\beta|$ becomes unity, sustained oscillation takes place.

The quantity $A\beta$ can be represented mathematically by a complex quantity and is interpreted in a number of ways to represent mathematically the stability or the instability of a feedback amplifier. The following three are the most widely known methods.

Niquist Criterion: The quantity $A\beta$ is denoted as a point in the complex plane, the real component and the imaginary component being plotted along the x-axis and the y-axis, respectively. The locus of all the points representing the values of $A\beta$ in the complex plane is a closed curve. If the point corresponding to $A\beta$ be within the closed curve, that represents positive feedback and unstable condition. The point outside the curve indicates negative feedback and stable amplification.

Bode Plot: Instead of plotting $A\beta$ in the complex plane, either the magnitude $|A\beta|$ in decibel or the phase angle of $A\beta$ can be plotted as a function of frequency, which is called *Bode Plot*. The value of $|A\beta|$ at the frequency corresponding to phase angle of 180° is called the *gain margin*. A negative gain margin signifies negative feedback and a positive gain margin indicates instability of amplification.

Barkhausen Criterion: This is the most relevant one in the present context. It represents a special case of positive feedback when

(i) the magnitude of the loop gain is unity, i.e., $|A\beta| = 1$.
(ii) the phase angle of $A\beta$ is 360° or its integral multiples.

These two conditions are collectively termed as the *Barkhausen criterion*, the extreme condition of instability in positive feedback with infinite gain leading to sustained oscillation of electrical signal in an amplifier. Several of the oscillators mentioned in Table 10.4, such as Hartley, Colpitts, Wien bridge and phase-shift are based on the Barkhausen criterion, as discussed in the subsequent sections.

10.4.2 Resonant Circuit Oscillators

These oscillators generally work at high frequencies, typically above 1 MHz and less than 1 GHz. An essential component of the circuit is a combination of inductor and capacitor, known as tank circuit. A BJT or a FET is used as the basic amplifier. The general layout of a resonant circuit oscillator is displayed in Figure 10.6 (a). Let the basic amplifier has gain A_o and output resistance R_o. The load is comprised of three impedances, namely Z_1, Z_2 and Z_3, as organized in the figure. Clearly, the total load resistance is

$$Z_L = \frac{Z_2(Z_1 + Z_3)}{Z_1 + Z_2 + Z_3} \tag{10.28}$$

and the feedback ratio is

$$\beta = \frac{Z_1}{Z_1 + Z_3} \tag{10.29}$$

The gain without feedback in presence of this load is given by

$$A = \frac{A_o Z_L}{R_o + Z_L} \tag{10.30}$$

It is understood from Equation (10.29) that when R_o is negligibly small with respect to Z_L, $A \approx A_o$. The loop gain can be estimated, using Equations (10.29) and (10.30) as

$$-A\beta = -\frac{A_o Z_1 Z_L}{(Z_1 + Z_3)(Z_L + Z_o)} \tag{10.31}$$

Substituting for Z_L from Equation (10.28) and simplifying, one can arrive at

$$-A\beta = -\frac{A_o Z_1 Z_2}{R_o(Z_1 + Z_2 + Z_3) + Z_2(Z_1 + Z_3)} \tag{10.32}$$

It is now assumed that the three impedances are purely reactive. Indeed it is a necessary condition for the resonant circuit oscillator. Replacing impedance Z_1, Z_2 and Z_3 by reactance X_1, X_2 and X_3, respectively, Equation (10.32) changes to

$$-A\beta = \frac{A_o X_1 X_2}{jR_o(X_1 + X_2 + X_3) - X_2(X_1 + X_3)} \tag{10.33}$$

The loop gain should be a real quantity and zero phase shift is required. Therefore the imaginary part of Equation (10.33) becomes zero leading to

$$X_2 = -(X_1 + X_3) \tag{10.34}$$

Omitting the imaginary part of Equation (10.33) and substituting for X_2 from Equation (10.34), we have the expression for loop gain

$$-A\beta = \frac{A_o X_1}{X_2} \tag{10.35}$$

Since the loop gain should be unity for sustained oscillation, the right side of Equation (10.35) denotes that reactances X_1 and X_2 must be of the same sign X_3 is of opposite sign. This is a general condition for resonant circuit oscillator. Two very popular practical oscillator circuits based on the above conditions are: Colpitts oscillator [Figure 10.6(b)], having X_1 and X_2 capacitive and X_3 inductive and Hartley oscillator [Figure 10.6(c)], having X_1 and X_2 inductive and X_3 capacitive. The simplified ac equivalent circuit of Figure 10.6(d) are applicable to both of these two.

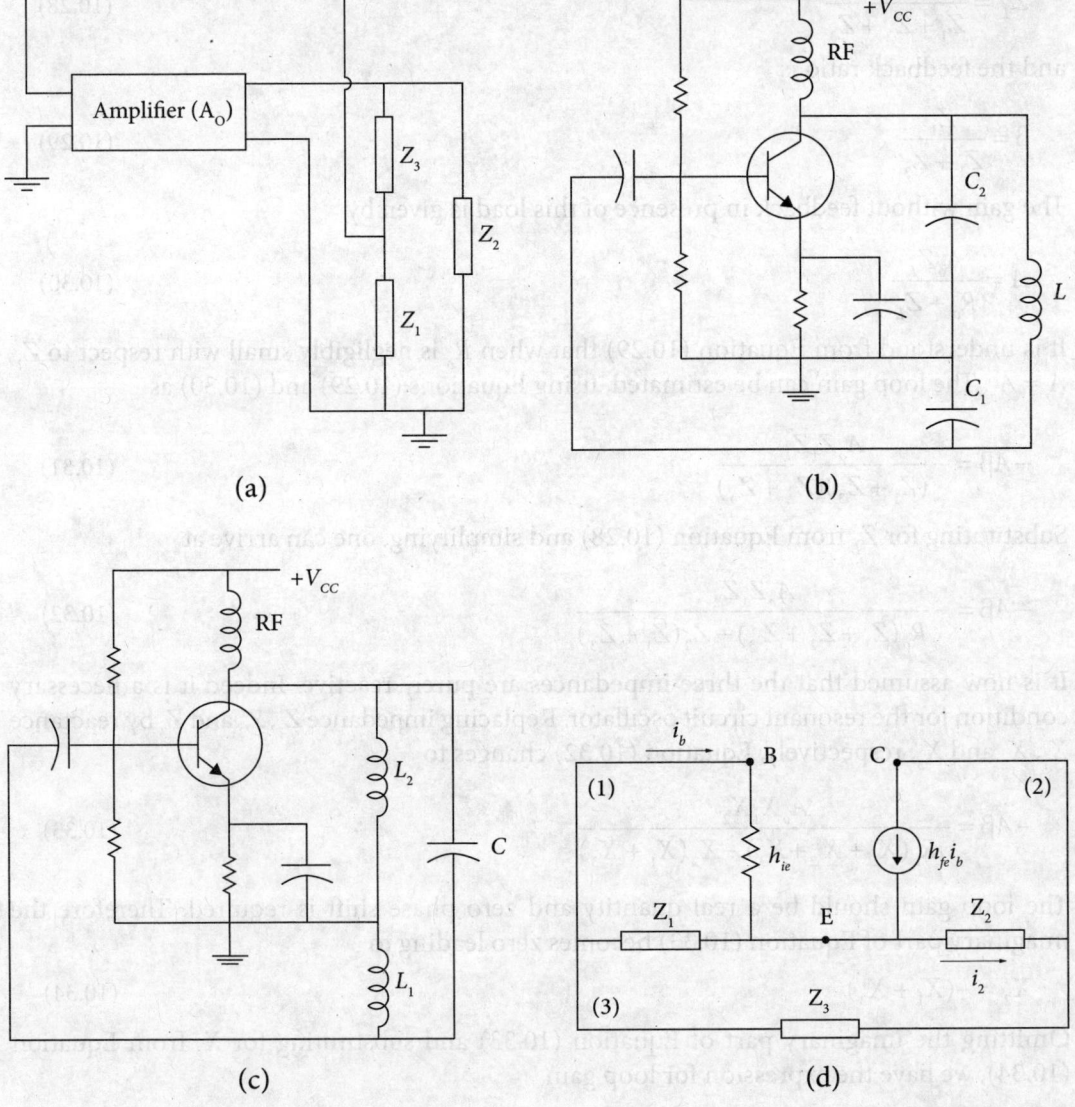

Figure 10.6 Resonant circuit oscillator: (a) general structure, practical realization as (b) Colpitts oscillator and (c) Hartley oscillator, and (d) ac equivalent circuit

> **Think a Bit: Why Sinusoidal Oscillation**
>
> We consider a hypothetical amplifier of gain A along with a matching feedback element. An external signal X_s applied to the input of the amplifier produces an output $X_o = AX_s$ and a portion $X_f = \beta X_o$ produces the feedback signal. Let the circuit parameters be adjusted in such a way that the feedback signal becomes just equal to X_s. In such a case, if the external source were removed but the feedback signal remained connected to the amplifier input, there would not be any difference to the amplifier and it would continue to operate and produce the same output. The above picture does not apparently impose any compulsion to (i) the nature of the input signal waveform, (ii) the linearity of the amplifier and (iii) the preservation of the wave shape in transmitting through the amplifier. However, the single requirement of $X_s = X_f$ encompasses indirectly all these stipulations because that needs their instantaneous values to be exactly same at all the instants of time and it is only the sinusoidal waveform that can keep the amplitude, frequency and phase of X_s and X_f identical. This explains why all the oscillators based on positive feedback and the Barkhausen criterion, generate sinusoidal waveform only. Moreover, the condition of the feedback signal being equal to the input signal calls for unity loop gain and the oscillator generates sine wave of a fixed frequency at which the total phase shift in traversing from the input to the amplifier, from the amplifier to the feedback and back to the input is either zero or integral multiples of 2π. This cannot be satisfied by more than one frequency at a time.

10.4.3 Colpitts Oscillator

The circuit diagram is given in Figure 10.6(b). The dc operating conditions are established with voltage-divider bias (Chapter 6). The blocking and bypass capacitors are also present for ac operation. However, there is no input ac signal. Instead, the combination of C_1, C_2 and L provide with the feedback satisfying the Barkhausen criterion. Another difference is the replacement of the collector resistor by a radio frequency choke coil (RF) to reduce the dc power loss and to prevent the high frequency signal from reaching the power supply.

The ac equivalent circuit of Figure 10.6(d) is applicable to Colpitts oscillator with $Z_1 = 1/j\omega C_1$, $Z_2 = 1/j\omega C_2$ and $Z_3 = j\omega L$. If the current through C_2 be i_2, the current through C_1 is $(h_{fe}i_b + i_b - i_2)$ and the current through L is $(h_{fe}i_b - i_2)$. Applying Kirchhoff's voltage law to loop-1 and loop-3, we have

$$h_{ie}i_b + (h_{fe}i_b + i_b - i_2)Z_1 = 0 \tag{10.36}$$

and

$$i_2 Z_2 - (h_{fe}i_b - i_2)Z_3 - (i_b + h_{fe}i_b - i_2)Z_1 = 0 \tag{10.37}$$

Substituting for i_2 from Equation (10.36) to Equation (10.37) and simplifying,

$$h_{ie}(Z_1 + Z_2 + Z_3) + (1 + h_e)Z_1 Z_2 + Z_1 Z_3 = 0 \tag{10.38}$$

Putting the values of Z_1, Z_2 and Z_3 for Colpitts oscillator in Equation (10.38) and equating the real and imaginary parts, we obtain

$$\frac{1 + h_{fe}}{\omega^2 C_2} = L \tag{10.39a}$$

and

$$\frac{1}{\omega C_1} + \frac{1}{\omega C_2} = \omega L \quad (10.39b)$$

Equation (10.39b) gives the frequency ($f = \omega/2\pi$) of the sinusoidal waveform generated by the Colpitts oscillator as

$$f = \frac{1}{2\pi\sqrt{LC_{eq}}} \quad (10.40)$$

The equivalent capacitance (C_{eq}) in Equation (10.40) is given by

$$C_{eq} = \frac{C_1 C_2}{C_1 + C_2} \quad (10.40a)$$

From Equations (10.39a) and (10.39b), one obtains the minimum value of the transistor current gain necessary to maintain the sustained oscillation as

$$h_{fe} = \frac{C_2}{C_1} \quad (10.41)$$

Example 10.4

A Colpitts oscillator has an inductor of 0.1 mH and two capacitors of 100 pF and 300 pF, respectively as part of the feedback network. Determine the frequency of the waveform generated by it.

Solution

The equivalent capacitance is [Equation (10.40a)]

$(100\times300)/(100 + 300) = 75$ pF

Using Equation (10.40), the frequency is

$$f = \frac{1}{2\pi\sqrt{0.1\times10^{-3}\times75\times10^{-12}}} = 1.838 \text{MHz}$$

10.4.4 Hartley Oscillator

The circuit diagram is shown in Figure 10.6(b) and the ac equivalent circuit of Figure 10.6(d) is applicable to this also. Consequently Equation (10.38) deduced above is valid for Hartley oscillator also. This time the impedance values are $Z_1 = j\omega L_1$, $Z_2 = j\omega L_2$ and $Z_3 = 1/j\omega C$. Putting these values in Equation (10.38) and equating the real and imaginary parts, we have

$$(1+h_{fe})\omega^2 L_2 = \frac{1}{C} \tag{10.42a}$$

and

$$\frac{1}{\omega C} = \omega(L_1 + L_2) \tag{10.42b}$$

Equation (10.42b) gives the frequency (f) of the sinusoidal waveform generated by the Hartley oscillator as

$$f = \frac{1}{2\pi\sqrt{LC}} \tag{10.43}$$

The equivalent inductance (L) is given by

$$L = L_1 + L_2 \tag{10.43a}$$

From Equations (10.42a) and (10.42b), the minimum necessary value of the transistor current gain for sustained oscillation can be deduced as

$$h_{fe} = \frac{L_1}{L_2} \tag{10.44}$$

The above two oscillators work well for generating high frequency signals. By changing the capacitor values, a range of change in frequency can be achieved. However, these are not suitable for low frequencies, such as audio frequencies. Ohmic resistance, instead of inductance is used in such cases with the capacitor. Two popular RC oscillator circuits, namely Wien bridge oscillator and phase-shift oscillator are illustrated in the subsequent sections.

10.4.5 Wien Bridge Oscillator

This oscillator circuit is suitable for generating sinusoidal voltage waveforms of frequencies ranging from several hertz to several kilohertz. It uses a Wien bridge, a sort of AC bridge, acting as both the feedback network and the frequency determining element. Figure 10.7(a) presents the oscillator circuit in a convenient form showing the position of the Wien bridge in it.

The circuit consists of an operational amplifier indicated by the triangle. Chapter 11 is fully devoted to the operational amplifier. Here it is adequate to state that it is a high voltage gain amplifier with high input impedance having two input terminals and a single output terminal. The input indicated by –ve sign is called *inverting input* because the output becomes 180° out of phase with respect to the input phase. The other input, indicated by +ve sign is called *noninverting input* because the output remains in the same phase as that of the input. The voltage difference (v_i) between these two input terminals appears, in amplified form, as the output voltage (v_o) and is fed back to the Wien bridge as its voltage supply. The following two types of feedback are implemented.

- A positive feedback is established through the series and parallel RC networks of the bridge. It is also known as lead-lag network. It helps the oscillation to build up.
- A negative feedback is obtained through the R_1–R_2 voltage divider of the bridge. It tends to reduce the loop gain.

The diode–resistor part encircled in Figure 10.7(a) is added to a practical circuit and will be discussed later. For the time being, it is excluded. Figure 10.7(b) shows the oscillator circuit in a modified form to indicate the feedback paths properly. The following features of the Wien bridge oscillator are illustrated referring to this figure.

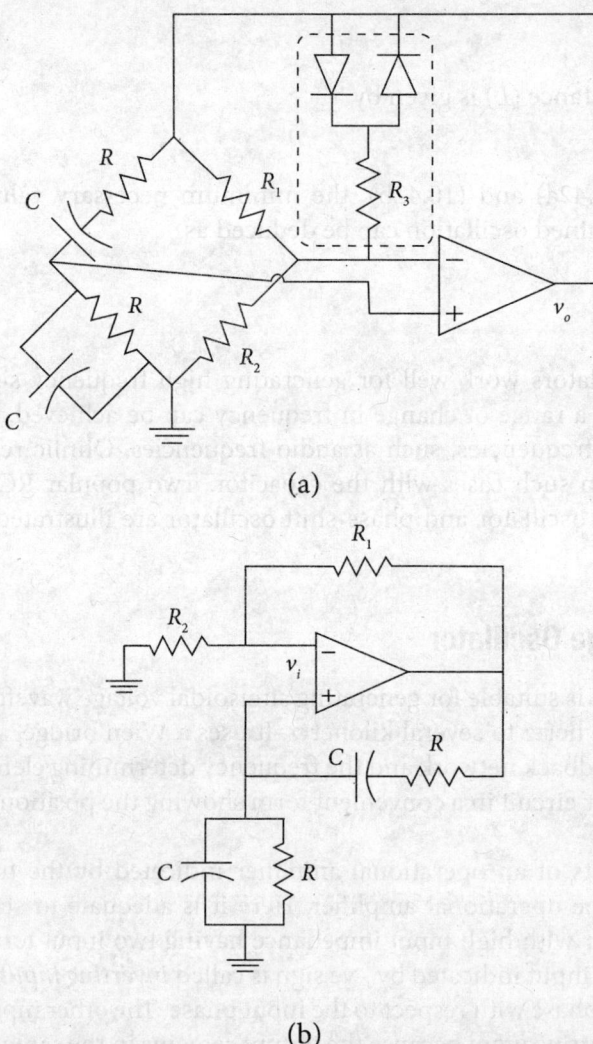

Figure 10.7 Wien bridge oscillator: (a) circuit diagram showing the bridge and (b) the circuit redrawn to show the feedback networks. The two diodes in (a) act as nonlinear elements that are omitted in (b)

> **Note: Wien Bridge Is an AC Bridge**
>
> The Wheatstone bridge consists of a rectangular connection of resistors, one in each arm with two bridge-like diagonal connections: one including a dc or ac power supply and the other joining a detector, e.g. galvanometer. A certain ratio of the resistances establishes a balanced condition of the bridge when the voltage difference across the detector becomes zero and no current passes through it. This is a dc bridge and is independent of the frequency of the input voltage. If, instead of resistances, some or all of the four arm contain some impedance (L, C or combinations of those with R), the circuit becomes frequency sensitive and is called ac bridge. It has two conditions of balance, one for the magnitude and the other for the phase. The ac bridge developed by German physicist Max Wien (1866–1938) is one such circuit. There are many other ac bridges proposed by scientists like Anderson, Maxwell and others, each having different connection of impedances and separate balance conditions.

A. Frequency of Oscillation

Since the Wien bridge itself does not introduce any phase shift between the input and the output at the frequency of oscillation, the amplifier also should not produce any phase shift in order to maintain the condition for positive feedback. Therefore, the present operational amplifier circuit is made to act as noninverting amplifier whose voltage gain (A) can be defined as (Chapter 11)

$$A = 1 + \frac{R_1}{R_2} \tag{10.45}$$

The feedback ratio, as apparent from Figure 10.7(b) is

$$\beta = \frac{Z_p}{Z_s + Z_p} \tag{10.46}$$

In Equation (10.46), the impedance (Z_s) of the RC series connection and the impedance (Z_p) of the RC parallel connection are given by

$$Z_s = R + 1/j\omega C \tag{10.46a}$$

and

$$Z_p = \frac{R/j\omega C}{R + 1/j\omega C} \tag{10.46b}$$

Respectively the expression for β in Equation (10.46) can be modified by algebraic rearrangement as

$$\beta = \frac{1}{3 - j(1/\omega CR - \omega CR)} \tag{10.47}$$

The loop gain ($-A\beta$) is a real quantity for oscillation. Therefore, making the phase part of Equation (10.47) equal to zero, we obtain

$$\omega = \frac{1}{CR}, \text{ hence } f = \frac{1}{2\pi CR} \tag{10.48}$$

Equation (10.48) represents the expression for the frequency (f) of the waveform generated by the Wien bridge oscillator.

B. Balanced Condition of Wien Bridge
The net input voltage appearing to the input of the operational amplifier, which is also the unbalanced output voltage of the Wien bridge, is given by

$$v_i = \frac{Z_p}{Z_s + Z_p} v_o - \frac{R_2}{R_1 + R_2} v_o \quad (10.49)$$

At the frequency of oscillation [Equation (10.48)], Z_s in Equation (10.46a) becomes $R(1-j)$ and Z_p in Equation (10.46b) becomes $(R/2)(1-j)$. Thus $Z_s = 2Z_p$ and

$$\frac{Z_p}{Z_s + Z_p} = \frac{1}{3} \quad (10.50)$$

It may be noted that the same result can be obtained by simply making the imaginary part of Equation (10.47) and comparing with Equation (10.46). If, at the same time, R_1 becomes $2R_2$,

$$\frac{R_2}{R_1 + R_2} = \frac{1}{3} \quad (10.51)$$

The conditions represented by Equation (10.50) and (10.51) make the Wien bridge fully balanced and the input voltage (v_i) to the amplifier [Equation (10.49)] becomes zero. So the Wien bridge oscillator does not work at perfectly balanced condition of the Wien bridge. It should be slightly unbalanced, as interpreted below.

C. Barkhausen Criterion
In order to initiate oscillation, the condition of Equation (10.51) is slightly modified. The resistance R_1 is made marginally larger than the resistance R_2 such that

$$\frac{R_2}{R_1 + R_2} = \frac{1}{3} - \frac{1}{\delta} \quad (10.52)$$

The parameter δ is a number slightly greater than 3 but not very large to make $1/\delta \to 0$. Imposing the condition of Equation (10.52) to Equation (10.49), the input to the amplifier becomes

$$v_i = \frac{v_o}{3} - \left(\frac{1}{3} - \frac{1}{\delta}\right) v_o \quad (10.53)$$

The required voltage gain of the amplifier is given by

$$\frac{v_o}{v_i} = \delta \quad (10.54)$$

The Barkhausen criterion requires $|A\beta| = 1$, hence $\beta = 1/\delta$. At the same time, no phase shift occurs because the operational amplifier works in noninverting mode. Thus both the conditions of the Barkhausen criterion are satisfied.

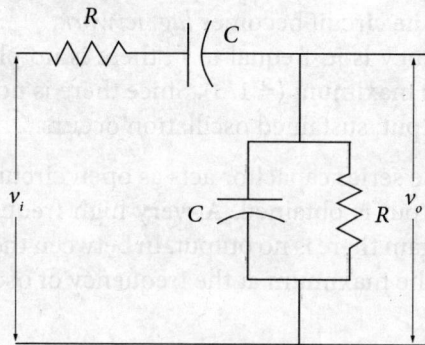

Figure 10.8 Lead-lag network in Wien bridge oscillator

D. Lead-Lag Network

The RC feedback circuit of the Wien bridge oscillator is called *lead-lag network* because depending on the frequency of the input signal, the output may lead or lag in phase. The oscillator has no separate input, the amplifier output fed back to the bridge serves as the input. However, the operation of the circuit is demonstrated by showing the input voltage (v_i) and the output voltage (v_o) separately in Figure 10.8. The output voltage (v_o) is defined in terms of the series RC impedance (Z_s) and the parallel RC impedance (Z_p) [Equations (10.46a) and (10.46b)] as

$$v_o = \frac{Z_p v_i}{Z_p + Z_s} \qquad (10.55)$$

Using Equations (10.47) and (10.55), the magnitude of the output–input ratio is obtained as

$$\left|\frac{v_o}{v_i}\right| = \frac{1}{\sqrt{3^2 + (1/\omega CR - \omega CR)^2}} \qquad (10.56)$$

It becomes maximum at $\omega = 1/RC$, the frequency of oscillation. Also v_o changes in phase with v_i for different frequencies and the phase angle is given by

$$\varphi = \tan^{-1}\left(\frac{1/\omega CR - \omega CR}{3}\right) \qquad (10.57)$$

Denoting the frequency of oscillation as $f_o = 1/(2\pi RC)$ and any other frequency by f, Equation (10.57) can be modified to

$$\varphi = \tan^{-1}\left(\frac{f_o/f - f/f_o}{3}\right) \qquad (10.58)$$

The following three cases may occur.

(i) At lower frequencies, when $f < f_o$, Equation (10.58) yields $\varphi > 0$ implying that the phase angle is positive and the circuit acts as a *lead network*.
(ii) At higher frequencies, when $f > f_o$, Equation (10.58) gives $\varphi < 0$ denoting negative phase angle and the circuit becomes *lag network*.
(iii) When the frequency is just equal to f_o, there is no phase shift and the feedback fraction becomes maximum ($\approx 1/3$). Since there is no phase change between the input and the output, sustained oscillation occurs.

Physically what happens, the series capacitor acts as open circuit to the input signal at very low frequencies and no output is obtained. At very high frequencies, the shunt capacitor works as short circuit and again there is no output. In between these two limiting conditions, the output voltage reaches the maximum at the frequency of oscillation.

E. Nonlinear Element

Since the amplifier gain becomes infinity at the condition of sustained oscillation, the output should go on increasing indefinitely until it was limited by the power supply. However, that would lead to distortion in the generated signal thereby clipping the sinusoidal waveform. In order to limit such increase and to stabilize the output, a nonlinear resistance is added to the Wien bridge oscillator circuit. The encircled portion of Figure 10.7(a) provides one such method. The two p–n junction diodes are in parallel and opposite connection in series with a resistor R_3. The whole combination is joined in parallel to the resistor R_1. When the power is just switched on, the diodes remain off. The feedback fraction is less than 1/3 because R_1/R_2 is more than 2. The output signal builds up and as it reaches a certain level, the diodes start conducting, one in each half-cycle. Then R_3 comes in parallel with R_1 and increases the feedback fraction to almost 1/3. Ultimately the loop gain becomes unity and the output becomes stabilized.

10.4.6 Phase-Shift Oscillator

In logical sense, all the previously mentioned oscillators are 'phase shift' because making the feedback signal phase a multiple of 2π with respect to the input signal is the key factor of oscillation. However, now we are talking about such an oscillator circuit that demonstrates explicitly how the output signal phase of an amplifier is gradually shifted further using several *RC* networks and ultimately reaches the condition of sustained oscillation.

Figure 10.9(a) displays the circuit diagram of a phase-shift oscillator constituted of a transistor CE amplifier and *RC* networks. The CE amplifier produces, on its own, 180° phase shift of the output signal. The remaining shift of 180° of the phase necessary for oscillation is accomplished by a feedback network of three cascaded combinations of identical capacitor (*C*) and resistor (*R*). The last resistor (R_1) is shown with a different value in the figure but ultimately it gets added to the transistor input resistance (h_{ie}) to produce the same value of *R*. Figure 10.9(b) represents the ac equivalent circuit of the phase-shift oscillator of Figure 10.9(a). Recalling the properties of hybrid parameters from Chapter 7, the following approximations can be made on it.

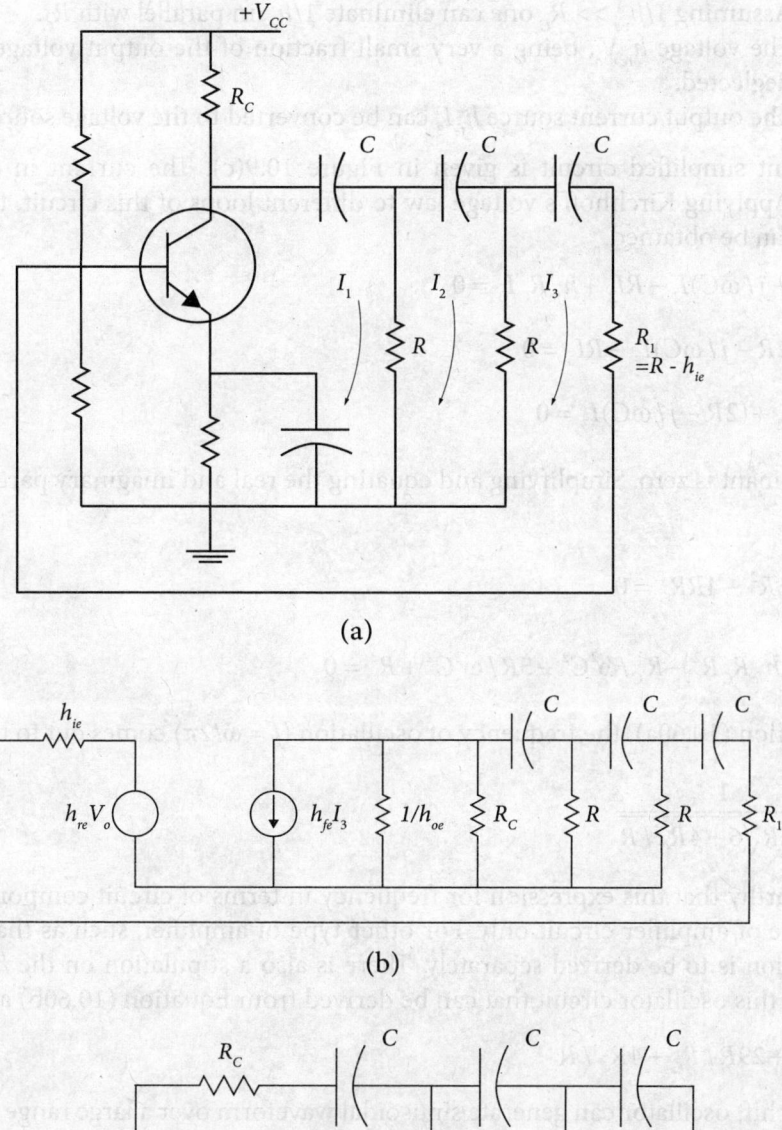

Figure 10.9 Phase-shift oscillator fabricated with BJT: (a) circuit diagram, (b) ac equivalent circuit, and (c) simplified ac equivalent circuit showing three symmetric *RC* combinations of feedback network

(i) Assuming $1/h_{oe} \gg R_C$ one can eliminate $1/h_{oe}$ in parallel with R_C.
(ii) The voltage $h_{re}V_o$, being a very small fraction of the output voltage (V_o) can be neglected.
(iii) The output current source $h_{fe}I_3$ can be converted to the voltage source $h_{fe}I_3R_C$.

The resultant simplified circuit is given in Figure 10.9(c). The current in each loop is indicated. Applying Kirchhoff's voltage law to different loops of this circuit, the following equations can be obtained.

$$(R_C + R - j/\omega C)I_1 - RI_2 + h_{fe}R_C I_3 = 0 \tag{10.59a}$$

$$-RI_1 + (2R - j/\omega C)I_2 - RI_3 = 0 \tag{10.59b}$$

$$0.I_1 - RI_2 + (2R - j/\omega C)I_3 = 0 \tag{10.59c}$$

The determinant is zero. Simplifying and equating the real and imaginary parts to zero one can obtain

$$\frac{1}{\omega^2 C^2} - 6R^2 - 4RR_C = 0 \tag{10.60a}$$

$$3R_C R^2 + h_{fe}R_C R^2 - R_C/\omega^2 C^2 - 5R/\omega^2 C^2 + R^3 = 0 \tag{10.60b}$$

From Equation (10.60a), the frequency of oscillation ($f = \omega/2\pi$) comes out to be

$$f = \frac{1}{2\pi CR\sqrt{6 + 4R_C/R}} \tag{10.61}$$

It is noteworthy that this expression for frequency in terms of circuit components is valid for this type of amplifier circuit only. For other type of amplifier, such as that with JFET, the expression is to be derived separately. There is also a stipulation on the h_{fe} of the BJT involved in this oscillator circuit that can be derived from Equation (10.60b) as

$$h_{fe} = 23 + 29R/R_C + 4R_C/R \tag{10.62}$$

The phase-shift oscillator can generate sinusoidal waveform over a large range of frequency from several hertz to several kilohertz. Ganged capacitors with three sections can be used to equally vary the three capacitances simultaneously.

> **Note: Minimum h_{fe} for BJT Phase-Shift Oscillator**
>
> Denoting R_C/R by k, Equation (10.62) can be re-written as: $h_{fe} = 23 + 29/k + 4k$. The minimum value of h_{fe} is obtained with $dh_{fe}/dk = 0$; or, $-29/k^2 + 4 = 0$ implying $k = 2.7$. Thus the transistor must have a minimum current gain of 44.5 in order to act as the active device for phase-shift oscillators.

10.4.7 Crystal Oscillator

It uses a suitably cut piece of piezoelectric crystal that behaves like an *LCR* circuit and generates a waveform of fixed but extremely stable frequency. The different aspects of a crystal oscillator are introduced here.

A. Relevance

The *LC* and *RC* oscillators mentioned so far have the advantageous feature of delivering voltage waveform of variable frequency by changing the *L*, *C* and *R* values. Nevertheless these passive components may undergo fluctuations in the values causing fluctuation in the frequency of oscillation. There are certain applications, such as computer hardware and telecommunication where perfectly stable frequency of the waveform is necessary. In such cases, the crystal oscillator is preferred.

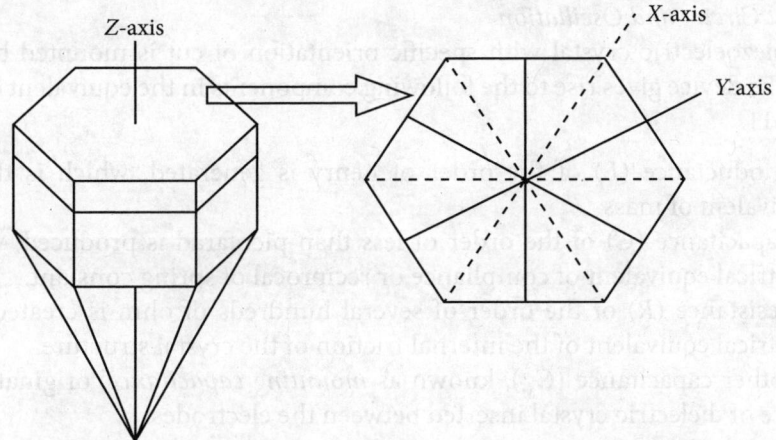

Figure 10.10 Piezoelectric crystal showing the hexagonal cross-section, the *Z*-axis, the three *X*-axes and the three *Y*-axes

B. Piezoelectric Effect

This is the basic physical phenomenon that the oscillator depends on. Some crystals, such as quartz and Rochelle salt have asymmetric unit cells in the lattice. The crystal as a whole is electrically neutral. When some mechanical stress is imposed, atoms get displaced and the charge balance condition is disturbed. The individual dipole moments no longer cancel one another and a net separation of positive and negative charges appear on two opposite faces of the crystal. It results in a voltage across the two faces. Conversely, if some potential difference is applied across these faces, the atoms tend to nullify the effect by displacing their positions and thus a mechanical stress, known as *electrostriction* is created. Such reversible electromechanical process of generating voltage under mechanical stress and causing mechanical deformation under the application of voltage in a crystal is termed as *piezoelectric effect*. The phenomenon can be understood with Figure 10.10 showing the hexagonal cross-section of quartz crystal. There are three specific directions, namely X-axis, Y-axis and Z-axis and there are three directions of each of X- and Y-axis. A piece of crystal having the flat sides perpendicular to X-axis is called 'X-cut' and a piece having the flat side perpendicular to Y-axis is called 'Y-cut'. Many other cuts are possible depending on the orientation with respect to X, Y and Z axes. For instance, the cut along the plane oriented almost 35°15′ with respect to Z-axis is called 'AT cut'.

The three X-axes, each separated by 120° with respect to the next one are called *electric axes* and the three Y-axes separated by 120° are called *mechanical axes*. Each X-axis corresponds to the Y-axis at right angle to it and vice-versa for establishing the piezoelectric effect. If voltage is applied in the direction of X-axis, a mechanical strain is produced in the direction of the Y-axis perpendicular to it. Conversely, if mechanical stress is applied along Y-axis, voltage is produced across the faces perpendicular to the X-axis at right angle to the Y-axis.

C. Equivalent Circuit and Oscillation

A piece of piezoelectric crystal with specific orientation of cut is mounted between two electrodes. The device gives rise to the following components in the equivalent circuit given in Figure 10.11.

- An inductance (L) of the order of henry is generated, which is the electrical equivalent of mass.
- A capacitance (C) of the order of less than picofarad is produced, which is the electrical equivalent of compliance or reciprocal of spring constant.
- A resistance (R) of the order of several hundreds of ohm is created that is the electrical equivalent of the internal friction of the crystal structure.
- Another capacitance (C_M), known as *mounting capacitance*, originates from the piece of dielectric crystal inserted between the electrodes.

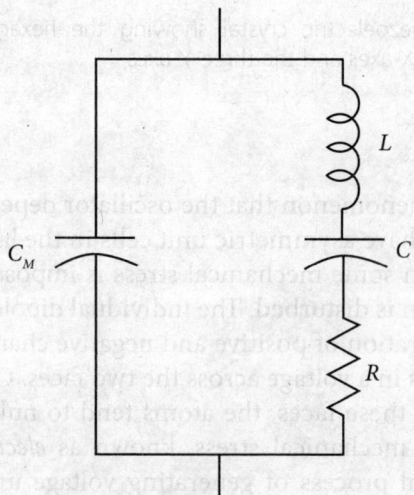

Figure 10.11 Equivalent circuit of piezoelectric crystal mounted between electrodes

The *quality factor* ($= \omega L/R$) of the equivalent circuit is very high because L is very large and R is very small. There is a natural frequency of the crystal depending on its size, orientation of cut and mounting. If an alternating voltage is applied to the piezoelectric crystal, alternating mechanical vibration takes place that, in turn, gives rise to an alternating voltage interacting with the voltage source. If the frequency of the generated voltage matches the natural resonant frequency, sustained oscillation occurs at that fixed frequency. The crystal exhibits two resonant frequencies as follows.

Series Resonant Frequency: This is also termed as zero impedance frequency and it corresponds to the condition when the reactance of the L and the C in the series arm (Figure 10.11) become equal and opposite at a certain condition of $\omega = \omega_s$ causing

$$\omega_s L - \frac{1}{\omega_s C} = 0 \tag{10.63}$$

Hence, the series resonance frequency is obtained as

$$f_s = \frac{1}{2\pi\sqrt{LC}} \tag{10.64}$$

At this frequency, the net impedance of the series LCR arm becomes purely resistive and very low (because R is low).

Parallel Resonant Frequency: This is also referred to as infinite impedance frequency that corresponds to the condition when the reactance of the LCR arm becomes equal to the reactance of C_M at a certain condition of $\omega = \omega_p$ causing

$$\omega_p L - \frac{1}{\omega_p C} = \frac{1}{\omega_p C_M} \tag{10.65}$$

Hence the parallel resonant frequency is derived as

$$f_p = \frac{1}{2\pi\sqrt{LC_E}} \tag{10.66}$$

The equivalent capacitance (C_E) is given by

$$C_E = \frac{CC_M}{C + C_M} \tag{10.66a}$$

The parallel resonance is actually the condition of anti-resonance when the impedance of the circuit becomes very large. In practice, $C_M \gg C$ so that $C_E \approx C$. Consequently f_s and f_p are almost the same, f_p being a bit large because two capacitances in series always produce a capacitance smaller than either. Crystal oscillators are fabricated by joining the piezoelectric crystal with the amplifier either in series resonant mode or in parallel resonant mode, depending on the requirement. Two such examples are given in Figures 10.12 (a) and (b).

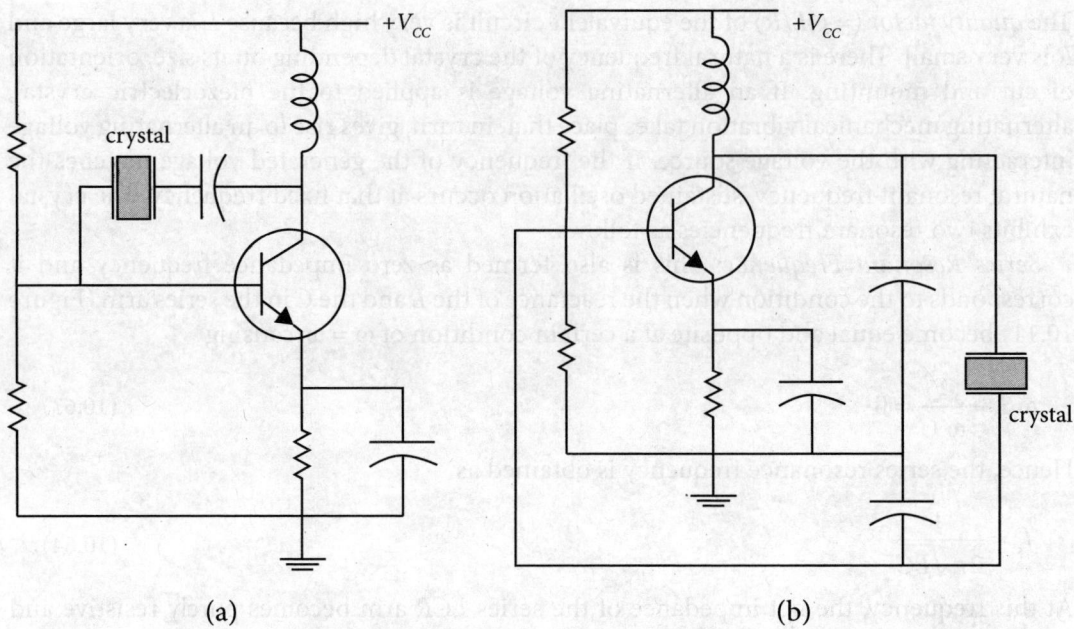

Figure 10.12 Crystal oscillator circuit operating (a) at series resonant frequency and (b) at parallel resonant frequency

D. Impedance Variation with Frequency

The resistance (R) in Figure 10.11 is a small quantity. Therefore, neglecting R, the equivalent reactance (jX) can be approximated as a parallel combination of $1/j\omega C_M$ and $j(\omega_L - 1/\omega_C)$ and can be expressed as

$$X = \frac{\omega L - 1/\omega C}{1 - \omega C_M(\omega L - 1/\omega C)} \tag{10.67}$$

This can be written in terms of the series and parallel resonant frequencies as

$$X = \frac{f^2 - f_s^2}{2\pi f C_M(f_p^2 - f^2)} \tag{10.68}$$

The following conclusions may be derived from Equation (10.68). These conditions are represented pictorially in Figure 10.13.

    (i)   At frequencies less than f_s, the reactance is capacitive that decreases with frequency.
    (ii)   At series resonant frequency (f_s), the reactance is zero (actually R).
    (iii) At frequencies above f_s, the series reactance becomes inductive and increases with frequency.
    (iv) At parallel resonant frequency (f_p), the reactance becomes infinitely large.
    (v)   At frequencies above f_p, the reactance is again capacitive and approaches zero, as the frequency becomes infinitely high.

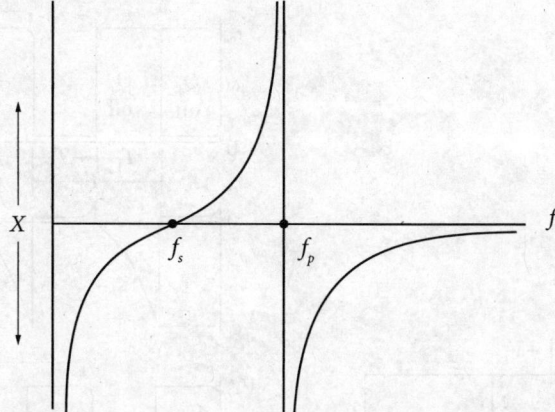

Figure 10.13 Variation of reactance of piezoelectric crystal with frequency

10.5 Multivibrators

Though multivibrator is an oscillator, it is of a different category. All the oscillators discussed so far generate sinusoidal waveforms whereas a multivibrator generates any of the followings.

- Square wave, symmetric or asymmetric, of varying time period (astable multivibrator)
- Square pulse of varying duration (monostable multivibrator)
- No wave at all, only two discrete stable states (bistable multivibrator)

Since all the above are somehow 'non-sinusoidal', the multivibrator is a relaxation oscillator. Actually the square wave output of the multivibrator consists of multiple harmonics. The non-sinusoidal output may be treated as a combination of harmonics of a fundamental sinusoidal wave. Depending on the nature of stability of the output, there are the following three types of multivibrators.

10.5.1 Astable Multivibrator

It consists of two identical transistors Q_1 and Q_2 [Figure 10.14(a)], each one switching between saturation and cutoff states, the duration being determined by the charging and discharging of two capacitors C_1 and C_2 through different values of resistors. The resistors R_{C1} and R_{C2} in the collector circuits and those in the base circuits (R_{B1} and R_{B2}) are shown in the figure. The output of the circuit, at the collector of each transistor, generates a square wave whose amplitude depends on the supply voltage (V_{CC}) and the frequency depends on the base resistor and capacitor combinations. The circuit, having no stable output state and vibrating between high and low quasi-stable states with many possible periods is called *astable multivibrator*.

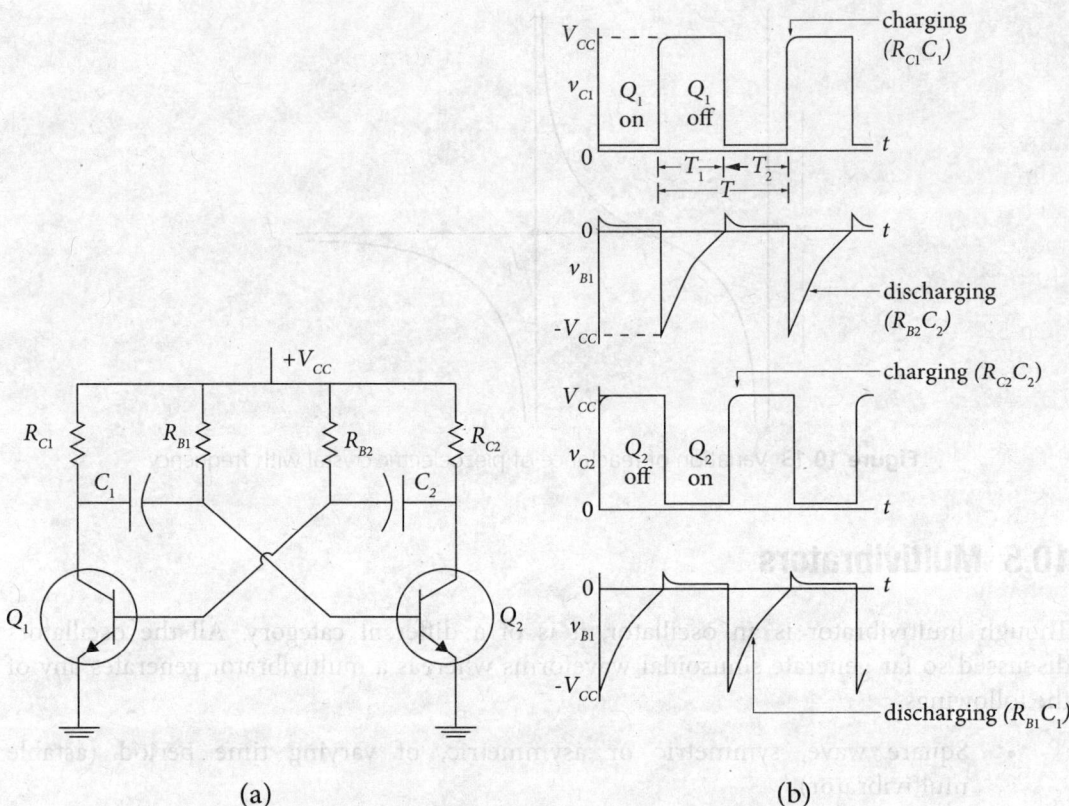

Figure 10.14 Astable multivibrator: (a) circuit diagram and (b) voltage waveforms obtained at the collector of Q_1 (v_{C1}), the base of Q_1 (v_{B1}), the collector of Q_2 (v_{C2}) and the base of Q_2 (v_{B2}).

The working principle of the circuit is based on the fact that no two transistors can be exactly same and one transistor may conduct more than the other because of the difference in thermal current and the fluctuation of resistance values. The circuit operation may be understood through the following steps.

Step-1: Let the supply voltage (V_{CC}) be switched on and Q_1 starts conducting more than Q_2. The collector voltage of Q_1 (v_{C1}) decreases because the increasing collector current causes larger voltage drop across R_{C1}.

Step-2: The decreasing v_{C1} coupled through C_1 decreases the forward bias of Q_2 tending it to cutoff. That increases its collector voltage (v_{C2}). The increasing v_{C2}, through C_2, increases the forward bias of Q_1 driving it to saturation. This condition of Q_1 saturated (hence $v_{C1} \approx 0$) and Q_2 cutoff (hence $v_{C2} \approx V_{CC}$) is *one of the two quasi-stable states* of the circuit.

The above feedback of voltage from one transistor to the other is referred to as *regenerative feedback*. It makes the conducting transistor to conduct more leading to saturation and the less conducting transistor to conduct less thereby driven to cutoff.

Step-3: When Q_1 is on and Q_2 is off, C_2 charges to V_{CC} through R_{C2} and the base–emitter junction of Q_1. The off time of Q_2 is determined by the time constant $R_{B1}C_1$.

Step-4: The sudden decrease of v_{C1} up to about zero causes a negative going change across C_1, almost $-V_{CC}$ that holds Q_2 in cutoff and C_1 starts discharging through R_{B1}. After discharging completely, C_1 attempt to charge again toward $+V_{CC}$ because the voltage source is connected all the way. However, as soon as C_1 gets charged up to 0.7 V, that positive going base voltage makes Q_2 turn on. Now C_2, that was charged during the previous cycle starts discharging through R_{B2}.

Step-5: The negative going voltage across the discharging C_2 turns off Q_2 and now C_1 starts charging through R_{C1} and the base-emitter junction of Q_2. This condition of Q_1 off and Q_2 on is *the other quasi-stable output state of the circuit* when $v_{C1} \approx V_{CC}$ and $v_{C2} \approx 0$.

Step-6: After total discharge C_2 starts charging toward $+V_{CC}$ through R_{B2} but on reaching 0.7 V it turns on Q_1 and C_1 starts discharging through R_{B1}.

The above steps are repeated periodically. Figure 10.14(b) shows the periodic waveforms produced at the collectors and the bases of the two transistors. The following features are noted in the graphs.

(i) Each of the transistors Q_1 and Q_2 generate voltage waveform of almost square shape with maximum of about $+V_{CC}$ and minimum of about zero (actually $V_{C(sat)} \approx 0.2$ V).

(ii) The rising edge of the square wave is a bit rounded due to the finite time constants of charging of C_1 through R_{C1} and C_2 through R_{C2}.

(iii) The falling edge of the square wave is very sharp because the charging capacitor causes a momentary increase in the forward bias of the transistor and aid the turning on.

(iv) The square waveform may be symmetric or asymmetric depending on the on/off time of the transistors that depends on the capacitor and base resistor values, as follows.

The time period of the voltage waveform is derived in the following way. The instantaneous base voltage of transistor Q_1 during the discharge of capacitor C_1 is given by

$$v_{B1} = V_{CC} - i_{C2} R_{B2} \tag{10.69}$$

where the initial collector current of transistor Q_2 is

$$i_{C2} = \frac{V_{CC} - (-V_{CC})}{R_{B2}} \tag{10.69a}$$

The current decays exponentially with a time constant $R_{B2}C_2$ so that Equations (10.69) and (10.69a) yield the base voltage of Q_1 as

$$v_{B1} = V_{CC} - 2V_{CC} \exp\left(-\frac{t}{R_{B2}C_2}\right) \tag{10.70}$$

The voltage required to switch on the transistor is around 0.7 V; much less than the supply voltage (V_{CC}). So one can assume that the transistor switches on at $v_{B1} = 0$ and write Equation (10.70) as

$$V_{CC}\left[1 - 2\exp\left(-\frac{t}{R_{B2}C_2}\right)\right] = 0 \tag{10.71}$$

Equation (10.71) gives the off time (T_1) for Q_1 (or the on time for Q_2), by putting $t = T_1$

$$T_1 = R_{B2}C_2 \ln 2 \tag{10.72}$$

Similarly the off time for Q_2 (or the on time for Q_1) can be obtained as

$$T_2 = R_{B1}C_1 \ln 2 \tag{10.73}$$

The total time period of the waveform, obtained from Equations (10.72) and (10.73) is given by

$$T = T_1 + T_2 = (R_{B1}C_1 + R_{B2}C_2)\ln 2 \tag{10.74}$$

When $R_{B1} = R_{B2} = R$ and $C_1 = C_2 = C$ in Equation (10.74), a symmetric square wave is obtained with the time period of

$$T = 1.38RC \tag{10.74a}$$

Some more features of the output waveform of the astable multivibrator and its time period are discussed below.

(i) Since the square wave is created out of the periodic saturation and cutoff conditions of the transistors, the square wave amplitude varies between V_{sat} (almost zero) and V_{CC}. Thus the amplitude of the square wave can be changed by changing the value of the supply voltage, the frequency remaining unchanged.

(ii) The ratio of the on time (T_1 or T_2) to the total time period (T) of the square wave expressed in percentage is termed as the *duty cycle* of the square wave. It can be varied by varying R_{B1}, R_{B2}, C_1 and C_2. For a symmetric wave, the duty cycle is 50%.

(iii) The circuit of Figure 10.14(a) can be modified by using a separate voltage supply for the base resistors R_{B1} and R_{B2}. By varying that voltage, the discharging time of the capacitors, hence the time period of oscillation can be varied. Such a circuit is referred to as *voltage-controlled oscillator* (VCO). It can be shown that for symmetric circuit, the time period of the VCO is $T \approx 2RC\ln(1 + V_{CC}/V_{BB})$.

(iv) The present discussion on the multivibrator circuit is carried out with n–p–n transistor. The same circuit performance can be achieved with p–n–p transistors also. The polarity of the supply voltage and the output waveform are just the opposite.

10.5.2 Monostable Multivibrator

This circuit has one stable state of the output. A triggering pulse causes transition to a quasi-stable state. After a certain time, the output returns to the stable state automatically. Figure 10.15(a) presents the circuit diagram of a monostable multivibrator.

The *stable state* of the circuit is with Q_1 off and Q_2 on, as can be understood with the following interpretation. When the supply voltage (V_{CC}) is switched on, a large collector current flows through Q_2 because its base–emitter junction is forward biased through R_B. The collector voltage of Q_2 becomes almost zero, which makes the base voltage (V_{B1}) of Q_1 almost zero. At the same time, there is a positive voltage drop (V_E) across the emitter resistor R_E, which keeps the transistor Q_1 cutoff. If Q_2 is turned on, resistors R_{C2} and R_B effectively come in parallel and assuming $R_{C2} \ll R_B$, one can obtain

$$V_E = \frac{R_E V_{CC}}{R_E + R_{C2}//R_B} \approx \frac{R_E}{R_E + R_{C2}} V_{CC} \tag{10.75}$$

The base voltage of Q_1 is obtained, assuming R_1 and R_2 much larger than R_{C2} as

$$V_{B1} = \frac{R_2 V_{CC}}{R_1 + R_2 + R_{C2}} \approx \frac{R_2 V_{CC}}{R_1 + R_2} \tag{10.76}$$

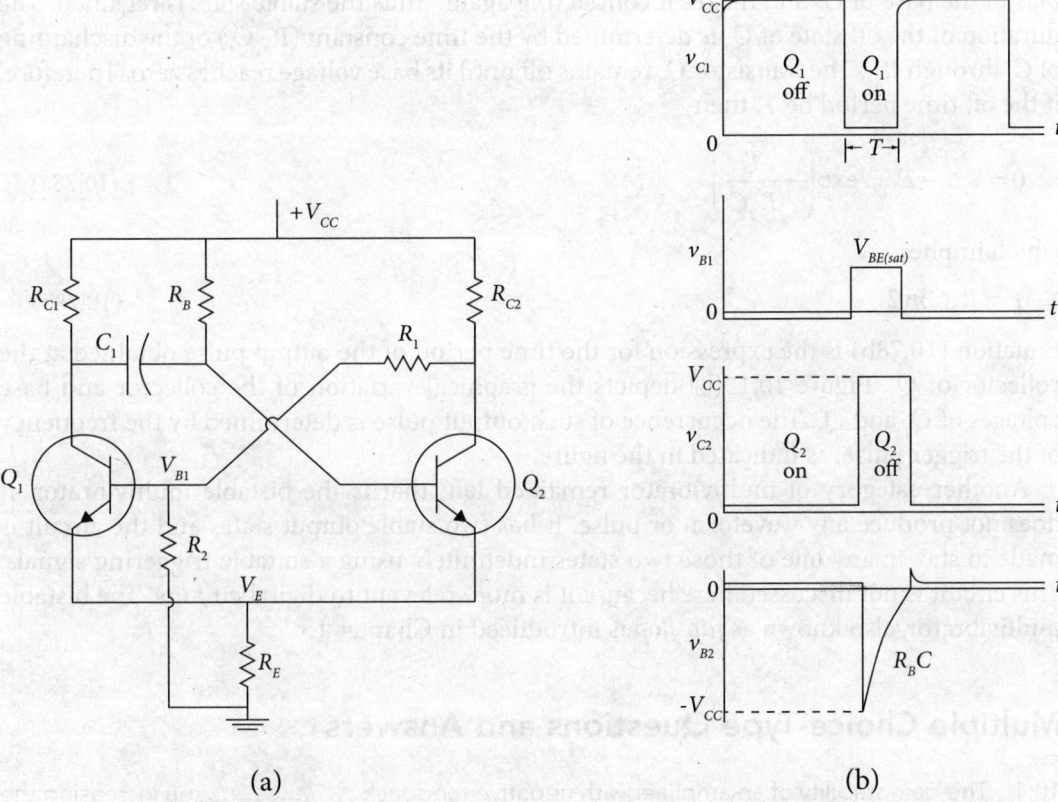

Figure 10.15 Monostable multivibrator: (a) circuit diagram and (b) voltage waveforms obtained at the collector of Q_1 (v_{C1}), the base of Q_1 (v_{B1}), the collector of Q_2 (v_{C2}) and the base of Q_2 (v_{B2}).

Thus the condition for the stable state, that is the condition of Q_1 remaining cutoff is $V_{B1} < V_E$, which implies from Equations (10.75) and (10.76) that

$$\frac{R_2}{R_1 + R_2} < \frac{R_E}{R_E + R_{C2}} \tag{10.77}$$

The *quasi-stable state* of the circuit is with Q_1 on and Q_2 off, which can be achieved by applying any of the following signals.

- A positive triggering pulse to the base of Q_1 or to the collector of Q_2
- A negative triggering pulse to the collector of Q_1 or to the base of Q_2.

The present discussion considers the first case. If the trigger voltage is greater than $(V_E - V_{B1})$, it drives Q_1 to saturation and its collector voltage decreases up to almost zero value. This negative going change appearing at the base of Q_2 makes it cutoff.

The capacitor C_1, which was charged through R_{C1} up to V_{CC} during the cutoff condition of Q_1, now starts discharging through R_B and Q_1. As C_1 fully discharges, it attempts to charge in the reverse direction toward V_{CC}. However, as it charges up to 0.7 V, it imposes a forward bias to the base of Q_2 and makes it conducting again. Thus the stable state is regained. The duration of the off state of Q_2 is determined by the time constant $(R_{B1}C_1)$ of the discharging of C_1 through R_{B1}. The transistor Q_2 remains off until its base voltage reaches zero. Therefore, if the off time period be T, then

$$0 = V_{CC} - 2V_{CC} \exp\left(-\frac{T}{R_B C}\right) \tag{10.78a}$$

which implies,

$$T = R_B C \ln 2 \tag{10.78b}$$

Equation (10.78b) is the expression for the time period of the output pulse obtained at the collector of Q_2. Figure 10.15(b) depicts the graphical variation of the collector and base voltages of Q_1 and Q_2. The occurrence of such output pulse is determined by the frequency of the trigger pulse, as indicated in the figure.

Another category of multivibrator remained left, that is the bistable multivibrator. It does not produce any waveform or pulse. It has two stable output states and the circuit is made to stay in any one of those two states indefinitely using a suitable triggering signals. This circuit is not discussed here because it is more relevant to digital circuits. The bistable multivibrator, also known as *flip-flop* is introduced in Chapter 15.

Multiple Choice-Type Questions and Answers

10.1 The gain stability of an amplifier with negative feedback .... on increasing the feedback ratio.
    (a) increases   (b) decreases
    (c) remains unchanged   (d) fluctuates

10.2 The negative feedback in an amplifier .... its lower cutoff frequency and .... the upper cutoff frequency.
   (a) increases, increases
   (b) increases, decreases
   (c) decreases, increases
   (d) decreases, decreases

10.3 The feedback ratio in voltage amplifier has .... whereas the feedback ratio in transresistance amplifier has ....
   (a) the dimension of voltage, the dimension of resistance
   (b) the dimension of current, the dimension of conductance
   (c) no dimension, the dimension of resistance
   (d) no dimension, the dimension of inverse resistance

10.4 An amplifier has open-loop gain of 100 and bandwidth of 50 kHz. On applying negative feedback, the closed-loop gain becomes 50 and the bandwidth becomes ....
   (a) 25 kHz
   (b) 100 kHz
   (c) 75 kHz
   (d) 500 kHz

10.5 In transconductance amplifier, ...
   (a) input is voltage, output is current
   (b) input is voltage, output is power
   (c) input is current, output is voltage
   (d) input is power, output is voltage

10.6 Feedback in an amplifier is always meant for ....
   (a) controlling the output
   (b) increasing the gain
   (c) decreasing the gain
   (d) stabilizing the gain

10.7 A transistor amplifier can act as a voltage amplifier having high input resistance and low output resistance by applying .... feedback.
   (a) voltage–shunt
   (b) current–shunt
   (c) voltage–series
   (d) current–series

10.8 The feedback ratio in the case of negative feedback amplifier is .... whereas the feedback ratio in the case of oscillator is ....
   (a) always real, always complex
   (b) real or complex, always complex
   (c) always complex, real or complex
   (d) real or complex, always real

10.9 The gain of the amplifier in a Wien bridge oscillator is ....
   (a) just equal to 3
   (b) slightly greater than 1/3
   (c) less than 3
   (d) greater than 3

10.10 The amplifier of a phase-shift oscillator operates in ....
   (a) class B
   (b) class A
   (c) class AB
   (d) class C

10.11 The frequency of oscillation of a Colpitts oscillator with capacitors of 250 pF and 375 pF and inductor of 37 µH is ....
   (a) 1.04 MHz
   (b) 0.152 MHz
   (c) 0.468 MHz
   (d) 2.1 MHz

10.12 A Hartley oscillator has RF choke of 0.5 mH and blocking capacitor of 0.005 pF. The magnitudes of resistance of the choke and the capacitor at 2.2 MHz are .... and ...., respectively.
   (a) 6.91 kΩ, 14.5 MΩ
   (b) 14.5 MΩ, 6.91 kΩ
   (c) 6.91 MΩ, 14.5 kΩ
   (d) 14.5 kΩ, 6.91 MΩ

10.13 An oscillator capable of generating a signal of 1Hz may be a ....
(a) Hartley oscillator
(b) Colpitts oscillator
(c) phase-shift oscillator
(d) crystal oscillator

10.14 An oscillator related to radio broadcasting may be ....
(a) Wien bridge oscillator
(b) Hartley oscillator
(c) phase-shift oscillator
(d) RC oscillator

10.15 The time period of a relaxation oscillator depends on ....
(a) negative resistance effect
(b) Barkhausen criterion
(c) resonance or phase-shift
(d) charging and discharging of capacitor

10.16 An emitter follower has .... feedback and the feedback resistance is equal to the .... resistance.
(a) negative, collector
(b) negative, emitter
(c) positive, input
(d) negative, input

Answers

10.1 (a)   10.2 (c)   10.3 (d)   10.4 (b)
10.5 (a)   10.6 (a)   10.7 (c)   10.8 (b)
10.9 (d)   10.10 (b) 10.11 (d) 10.12 (a)
10.13 (c) 10.14 (b) 10.15 (d) 10.16 (b)

Reasoning-Type Questions and Answers

10.1 Is feedback applicable to any amplifier?

Ans. Feedback is applicable to any voltage, current, transconductance and transresistance amplifier. The only stipulation is maintaining the linearity because the feedback signal combines with the input signal.

10.2 In what sense is the stability of gain with negative feedback useful?

Ans. The gain of an amplifier may change with the fluctuation in temperature, variation in device parameter, such as current gain and mutual conductance and even with the fluctuation of supply voltage. The usefulness of negative feedback is to reduce the effects of these all on the gain; though the gain is reduced.

10.3 Is there any other difference between positive and negative feedback except for the phase and gain?

Ans. The negative feedback has the purpose of reducing the gain of the amplifier being applied in opposite phase to the input. If the amplifier itself causes 180° phase shift, no additional phase change is required. In such a case, the negative feedback network may be constituted of resistors only. The positive feedback is mainly applied for establishing oscillation where the Barkhausen criterion relating both gain and phase is involved. So the feedback network must contain reactive elements like capacitor and inductor in order to participate in phase change. Thus there is a constructional difference.

10.4 What is the physical significance of the term 'loop gain' in a feedback amplifier?

Ans. The signal (X_i) traverses a closed loop starting from the input round the basic amplifier, the feedback element and then back to the input through the mixing unit. It acquires the following gains through the different units along the path.

(i) In passing through the amplifier, the signal is multiplied by gain A and becomes $X_o = AX_i$.

(ii) While passing through the feedback element, the signal is multiplied by β and becomes $X_f = \beta X_o$.

(iii) In course of being added to the original signal at the mixing unit, it is multiplied by −1 for negative feedback and by +1 for positive feedback so that

$X_i = X_s - A\beta X_i$ for negative feedback and

$X_i = X_s - (-A\beta)X_i$ for positive feedback.

In both cases, the quantity −Aβ represents the loop gain.

10.5 The oscillator generates a sustained waveform without any input. Is it an example of perpetual motion?

Ans. No, the oscillator converts the dc power of the power supply or the battery into ac power. No energy is created. The initial spurious input signal comes from noise, which is amplified and served back in phase with the input to develop the sustained waveform.

10.6 How does crystal oscillator maintain a stable frequency in the generated waveform?

Ans. The inductor, the capacitor and the resistor involved in oscillation originate from the same crystal accounting for less variation. Also the circuit has a very high Q value because of large inductance and small resistance.

10.7 Both the phase-shift oscillator and the Wien bridge oscillator are suitable for generating audio frequency sinusoidal signal over a wide range of frequency. Does one of these have any advantage over the other?

Ans. Yes, both have some advantageous features of their own. The continuous variation of frequency in a Wien bridge oscillator can be achieved by using a variable gang capacitor of only two sections whereas three capacitors are to be varied simultaneously in a phase-shift oscillator. In contrast, the phase-shift oscillator has a simple circuit and can be fabricated easily with BJT, FET and operational amplifier.

Solved Numerical Problems

10.1 An amplifier with negative feedback produces an output voltage of 2 V with an input voltage of 0.1 V. On removing the feedback, only 0.02 V of input voltage produces the same output. Determine the value of the feedback ratio.

Soln. The gain with feedback is 2/0.1 = 20

The gain without feedback is 2/0.02 = 100

Let the feedback ratio be β. Using Equation (10.5),

$$20 = \frac{100}{1+20\beta}$$

Hence $\beta = 0.04$

10.2 It is found that the loop gain of an amplifier gets changed by 15% due to changes in different parameters. However, the application requirement does not allow a change of more than 2% and a negative feedback amplifier is being designed to address this problem. If the open-loop gain of the amplifier be 100, estimate the necessary value of the feedback ratio.

Soln. Let the required value of feedback ratio be β. Using Equation (10.15),

$$2 = \frac{1}{1+100\beta} \times 15$$

Hence β is 0.065 or 6.5%.

10.3 Demonstrate that an emitter follower discussed in chapters 5 and 6 is a practical example of voltage–series feedback. Find out an expression for the closed-loop voltage gain.

Soln. Figure 10.16(a) shows the circuit diagram of the emitter follower comprising voltage–series feedback. Comparing with Figure 10.2, it is understood that that R_E is the feedback element and the feedback voltage (v_f) appears across it. The source voltage (v_s) and the net input voltage (v_i) appearing across the base and the emitter of the transistor are indicated there. The circuit is further simplified with its ac equivalent shown in Figure 10.16(b).

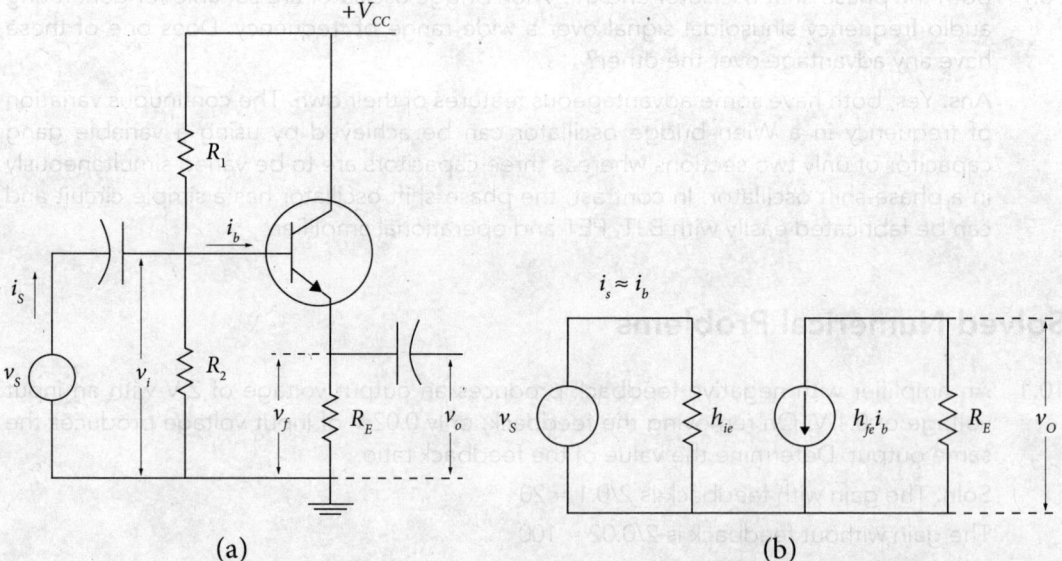

Figure 10.16 Emitter follower as voltage–series feedback amplifier: (a) circuit diagram showing the source, input and feedback voltages and (b) ac equivalent circuit

Since the capacitor acts as short circuit, $i_s \approx i_b$ and the voltage gain of the transistor amplifier without feedback (the ratio of the output voltage to the input voltage), i.e., the open-loop voltage gain is

$$A_V = \frac{h_{fe}i_b R_E}{h_{ie}i_b} = \frac{h_{fe}R_E}{h_{ie}}$$

Putting the above value of A_V to Equation (10.5) and simplifying, the closed-loop gain is obtained as

$$A_{Vf} = \frac{h_{fe}R_E}{h_{ie} + h_{fe}R_E \beta}$$

If β is considered to be unity (as in emitter follower) and $h_{ie} \ll h_{fe}R_E$, $A_{Vf} \approx 1$.

10.4 Establish that a transistor amplifier with collector-feedback bias discussed in Chapter 6 is a practical example of voltage–shunt feedback. Find out an expression for the closed-loop transfer gain.

Soln. Figure 10.17 presents the circuit diagram of the voltage–shunt feedback circuit. The output voltage (v_o) is much larger than the input voltage (v_i) and is 180° out of phase with respect to it. The feedback current shown in the figure can be expressed as

$$i_f = \frac{v_i - v_o}{R_f} \approx -\frac{v_o}{R_f}$$

The above expression implies that the feedback current is proportional to the output voltage thereby confirming voltage–shunt feedback. The feedback ratio is $-1/R_f$ having the dimension of inverse resistance. The transfer gain is given by

$$R_{Mf} = \frac{v_o}{i_s} \approx \frac{v_o}{i_f} = -R_f, \text{ a fixed quantity.}$$

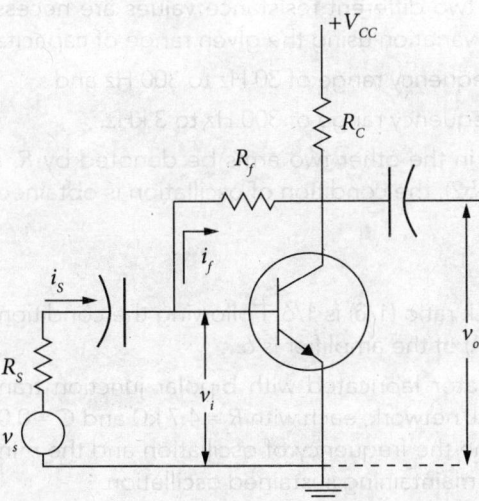

Figure 10.17 BJT voltage–shunt feedback amplifier

10.5 The frequency of a Hartley oscillator is to be varied between the range of 0.6 MHz and 1.2 MHz. The capacitor can be varied continuously over the range of 100 pF to 400 pF. Determine the required values of the inductance. The transistor used in the oscillator has current gain of 50.

Soln. Equation (10.43) is used here. Since the given capacitance varies over 1:4 ratio, the required range of frequency (1:2 ratio) can be achieved with a fixed pair of inductors.

Let the inductance be L. The minimum frequency corresponds to the maximum capacitance so that

$$0.6 \times 10^6 = \frac{1}{2\pi\sqrt{L \times 400 \times 10^{-12}}} \Rightarrow L = 1.76 \times 10^{-4} H$$

This L is the series combination of two inductances L_1 and L_2 so that $L_1 + L_2 = 1.76 \times 10^{-4}$ H. Also, using Equation (10.44), $L_1/L_2 = 50$. Solving these two, the inductances come out to be 0.1726 mH and 3.451 µH.

10.6 A Wien bridge oscillator is used for a frequency range of 30 Hz to 3 kHz. The variable capacitance has a range of 50 pF to 500 pF. Find out the resistance values required. If the resistance in the other arms are in the ratio of 5:1, find out the gain of the amplifier.

[C.U. 2001]

Soln. For a fixed resistance, if the capacitance is increased, the frequency of oscillation decreases, as apparent from Equation (10.48). Considering the capacitance (C) of 500 pF and the lower frequency limit of 30 Hz, the corresponding resistance (R) is calculated as

$$R = \frac{1}{2\pi \times 30 \times 500 \times 10^{-12}} = 10.61\,M\Omega$$

The given range of capacitance variation is in 10:1 ratio. Therefore keeping R fixed at 10.61 MΩ and varying C up to 50 pF would increase f up to 10 times, that is 300 Hz only. In order to achieve the range of 300 Hz to 3 kHz, R must be 1/10th of 10.61 MΩ or 1.061 MΩ. Thus the following two different resistance values are necessary to cover the required range of frequency variation using the given range of capacitance.

10.61 MΩ for the frequency range of 30 Hz to 300 Hz and

1.061 MΩ for the frequency range of 300 Hz to 3 kHz.

Let the resistances in the other two arms be denoted by R_1 and R_2 such that $R_1 = 5R_2$. Using Equation (10.52), the condition of oscillation is obtained as

$$\frac{R_2}{5R_2 + R_2} = \frac{1}{3} - \frac{1}{\delta}$$

Hence, the feedback ratio (1/δ) is 1/6. Following the condition of ($A\beta = 1$) for oscillation, the required gain (A) of the amplifier is 6.

10.7 A phase-shift oscillator fabricated with bipolar junction transistor (BJT) has three RC sections as feedback network, each with $R = 4.7$ kΩ and $C = 0.01$ µF. If the load resistance be 3.3 kΩ, determine the frequency of oscillation and the minimum allowed value of h_{fe} of the transistor for maintaining sustained oscillation.

Soln. It is understood from the circuit of BJT phase-shift oscillator, such as that of Figure 10.9(a) that the resistance of the collector resistor (R_c) is equivalent to the load resistance. Therefore, using Equation (10.61), the frequency of oscillation is

$$\frac{1}{2\pi \times 0.01 \times 10^{-6} \times 4.7 \times 10^3 \sqrt{6 + 4(3.3/4.7)}} = 1.14 \text{ kHz}$$

The minimum value of h_{fe} of the transistor is, using Equation (10.62),

$$23 + 29(4.7/3.3) + 4(3.3/4.7) = 67 \text{ (approx.)}$$

10.8 A crystal oscillator has inductance of 0.5 H, capacitance of 0.02 pF and resistance of 1 kΩ in the equivalent circuit. If the mounting capacitance be 20 pF, calculate the series and parallel resonant frequencies.

Soln. The series resonant frequency [Equation (10.64)] is

$$\frac{1}{2\pi\sqrt{0.5 \times 0.02 \times 10^{-12}}} = 1.5915 \text{ MHz}$$

The equivalent capacitance [Equation (10.66a)] is

$$\frac{0.02 \times 20}{0.02 + 20} = 0.01998 \text{ pF}$$

Therefore, the parallel resonant frequency [Equation (10.66)] is

$$\frac{1}{2\pi\sqrt{0.5 \times 0.01998 \times 10^{-12}}} = 1.5923 \text{ MHz}$$

Exercise

Subjective Questions

1. Explain the working principle of a feedback amplifier and deduce an expression for the gain with feedback in terms of the transfer gain of the basic amplifier and the feedback ratio.
2. Outline different types of feedback with block diagrams and compare the properties of the feedback ratios in these different cases.
3. Show that negative feedback improves the input and output resistances of a voltage amplifier.
4. State briefly about some important uses of oscillators. Mention the basic requirements for getting steady oscillation with positive feedback.
5. Show that all the reactance values associated with the feedback element in a resonant circuit oscillator cannot be of the same sign.
6. Draw the circuit diagram of a Hartley oscillator and derive an expression for its frequency of oscillation. Compare its feedback network with that of Colpitts oscillator.
7. What is piezoelectric effect and how it is utilized in a crystal oscillator? Draw the electrical equivalent circuit of a crystal oscillator.

8. Distinguish the fields of application of oscillators using LC and RC networks. Sketch the circuit diagram of an astable multivibrator and explain how it generates a square wave of minimum value almost zero and maximum value almost equal to the supply voltage.
9. Construct a monostable multivibrator circuit using two transistors and other passive components and obtain the conditions for (i) its stable state and (ii) the duration of its quasi-stable state.

Conceptual Test

1. The word 'feedback' is a general term applicable to any physical or logical system (even theoretical) when a component of the output is served back to the input along with the original contribution so as to control the whole system. Can you find out a few such examples?
2. The cistern and flush system in a toilet is a mechanical feedback system. Can you identify the input, the output and the feedback element and compare each of those with an electronic feedback amplifier?
3. How is the working of a multivibrator different from that of a resonant circuit oscillator?
4. How is the lead-lag network associated with the frequency of oscillation of a Wien bridge oscillator?
5. One of the two transistors in an astable multivibrator has the discharging time constant greater than that of the other. In that case, sketch the collector and base voltage waveforms (expected to be asymmetric) for the two transistors.
6. Modify the circuit of an astable multivibrator to achieve a voltage-controlled oscillator that can generate a square wave of frequency changeable with supply voltage.
7. Justify that the monostable multivibrator can be called a 'timer'.

Numerical Problems

1. An amplifier has 10 dB of negative feedback. If the open-loop gain of the amplifier is increased by 5%, what will be the corresponding change in the closed-loop gain?
2. The open-loop gain of an amplifier is 100 and it can fluctuate by 10%. A feedback amplifier constructed with it allows only 1% of fluctuation. Determine the closed-loop gain of the feedback amplifier.
3. A Hartley oscillator is to span a frequency range of 500 kHz to 1.5 MHz. The variable capacitor has a range of 40 pF to 400 pF. If one inductance is 3.3 μH, what is the other one?
4. A Wien bridge oscillator spans a frequency range of 20 Hz to 20 kHz. The variable capacitor has a value from 40 pF to 400 pF. Calculate the resistance values required to generate voltage waveform with frequencies over the specified range.
5. Determine the operating frequency of a BJT Colpitts oscillator, if the two capacitors be of 0.01 μF and 0.001 μF and the inductor be of 10 μH.
6. A Hartley oscillator is to span a frequency range of 500 kHz to 1.5 MHz. If the inductors are of 1 mH and 10 μH, calculate the required range of the variable capacitor.

Project Work on Chapter 10

Design an astable multivibrator capable of generating square voltage waveform of 1 kHz frequency using two identical bipolar junction transistors, each having $h_{fe} = 200$. Given that the supply voltage is 5 V and a minimum collector current of 5 mA should be maintained. Assume the output waveform to be symmetric.

Theory:
The minimum base current required for each transistor is given by
$$I_B = I_C/h_{fe}$$
Therefore, the maximum allowable base resistance is
$$R_B = (V_{CC} - V_{BE})/I_B$$
The collector resistance is
$$R_C = (V_{CC} - 0.2)/I_C$$
The time period for the symmetric astable multivibrator is [Equation (10.74a)]
$$T = 1.38 R_B C$$

Computer Analysis:
The circuit was simulated with LTspice (Figure P10.1) using the following values of components.
$$I_B = (5 \times 10^{-3})/200 = 25\ \mu A$$
$$R_B = (5 - 0.7)/(25 \times 10^{-6}) = 172\ k\Omega$$
$$R_C = (5 - 0.2)/(5 \times 10^{-3}) = 0.96\ k\Omega$$
Required frequency 1 kHz, hence time period 1 ms. Therefore,
$$C = (1 \times 10^{-3})/(1.38 \times 172 \times 10^3) = 4.2\ nF$$

Although simulation, the above calculated values were replaced by the nearest standard available values: $R_B = 150\ k\Omega$, $R_C = 1\ k\Omega$ and $C = 4\ nF$. The circuit assembled in LTspice is shown in Figure P10.1.

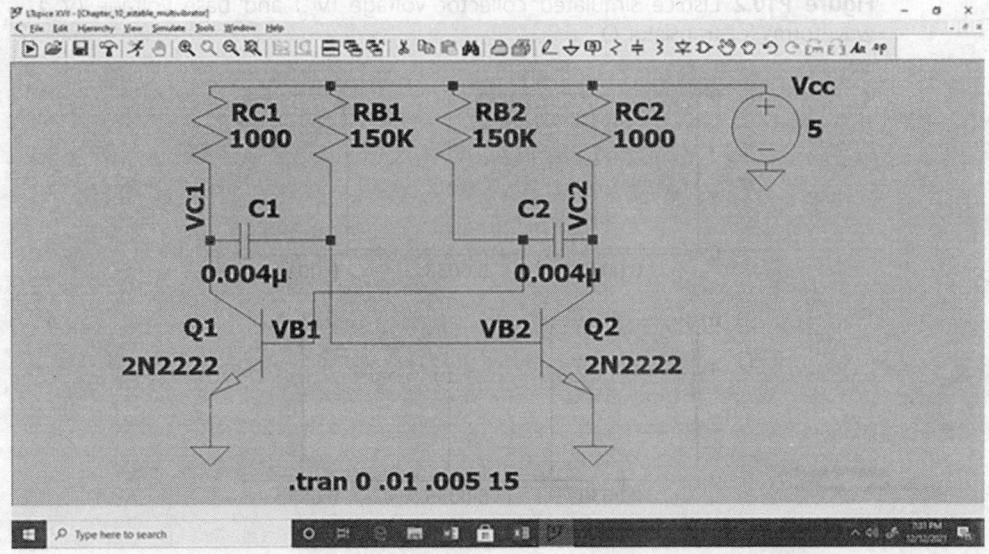

Figure P10.1 LTspice simulated astable multivibrator circuit diagram

The collector voltage waveform (V_{c1}) and the corresponding base voltage waveform (V_{b1}) of transistor Q_1 are plotted in Figure P10.2. Similar waveforms can be obtained with transistor Q_2. The following features of the waveforms may be noted from Figure P10.2.

(i) The periodic switching of the transistor between cutoff and saturation produces the square wave.
(ii) During cutoff, the square wave has amplitude of 5 V; equal to V_{CC}.
(iii) The minimum value is almost 0.2 V that occurs across the saturated transistor.
(iv) It takes some small but finite time to develop the sustained waveform. The initial premature waves are also shown in the graph.

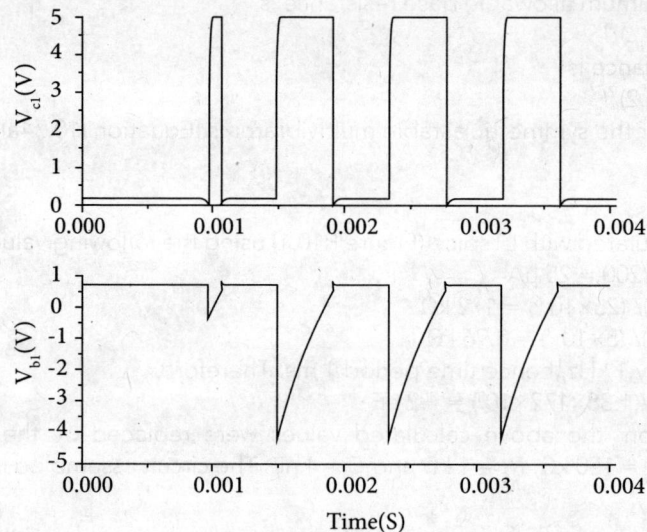

Figure P10.2 LTspice simulated collector voltage (V_{c1}) and base voltage (V_{b1}) waveforms for transistor Q_1

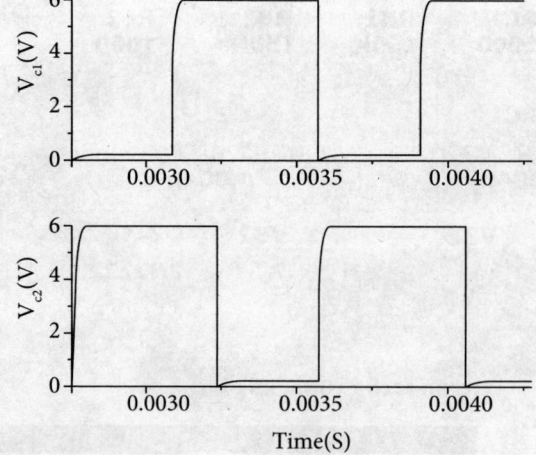

Figure P10.3 LTspice simulated collector voltage waveforms (V_{c1} and V_{c2}) of the two transistors Q_1 and Q_2

Figure P10.3 compares the collector voltage waveforms (V_{c1} and V_{c2}) of the two transistors Q_1 and Q_2. The complementary on/off operation of the two transistors is obvious from the graphs. Also the enlarged view reveals the rounded end of the waveform while charging through RC.

The study should be extended with the following investigations.

- Note the change in the waveform with the changes of R_B and C and compile the results in tabulated form.
- Make the output waveform asymmetric by putting two unequal base resistors [Equation (10.74)].
- Observe happens to the circuit operation on changing the supply voltage from 5 V to 10 V keeping all circuit parameters unchanged.
- Observe the change in output waveform, if R_C is increased.

11

Operational Amplifier

The operational amplifier, abbreviated as op-amp is an amplifier circuit comprised of active and passive components. It has high voltage gain, high input impedance, and wide range of frequency of input signal that it can handle. The most remarkable feature is that the op-amp can be used for the purpose of numerical computations, the numbers being represented by the input and output voltage values. This chapter brings together the basic properties of the op-amp and the most popular circuits constructed with it, such as inverting and noninverting amplifiers, adder, differentiator, integrator, active filter, comparator, Schmitt trigger, and logarithmic amplifier. Also some novel applications of op-amp, such as equation solving and waveform generating are introduced.

11.1 A Review on Amplifiers

In Chapters 5 through 10, we have come across different types of amplifiers made of bipolar junction transistors or field-effect transistors. Now we are quite familiar with the fact that the same device may not be suitable for voltage, current, and power amplification and at all frequencies of the input signal. In this chapter, we are about to start a new type of high gain voltage amplifier, known as operational amplifier, abbreviated as *op-amp*, which is made up of a number of transistors, either bipolar or field-effect or both, combined in a circuit. In this occasion, it may be relevant to have a quick recapitulation on the overall features of a typical amplifier.

Whatever may be the actual circuit configuration, an amplifier can be considered as an electronic circuit comprising one or more active device producing a magnified replica of the input voltage or current. In general, an amplifier has the following properties.

(i) The output signal has generally the same frequency and waveform as that of the input signal. If not, it is a special form of amplifier, known as oscillator or waveform generator.

(ii) The output amplitude is mostly different from the input amplitude and in maximum cases, the former is increased. Sometimes the output amplitude may decrease also when the circuit is called *attenuator*.
(iii) The phase of the output waveform may or may not change with respect to the input depending on the circuit operation and device property.
(iv) Each amplifier has a certain range of frequency for the input signal over which it can exhibit its properties.

The amplifier largely consists of the following components.

Input: The input signal applied across the two input terminals (one may be the ground) can be either alternating or direct voltage and/or current.

Load: The load across the two output terminals (one may be the ground) is any passive element: resistor, capacitor, inductor or network of these components. Sometimes active load is also used in op-amp and digital circuits.

Power Supply: A source of steady power establishes a steady voltage across the circuit and provides a steady current through the circuit. In most cases, it is a positive voltage with respect to the ground. As we will see, op-amps require both positive and negative voltages with respect to the ground.

Ground: Usually the ground is connected to one of each of the input and output terminals. However, it is not compulsory in the case of op-amp circuits. For instance, the differential amplifier discussed in this chapter contains two floating input terminals isolated from the ground.

In addition, some portion of the output may be returned to the input as feedback, as was illustrated in Chapter 10. The op-amp also contains the above components and the processes of feedback with op-amp circuits will be exemplified in this chapter.

11.2 Features of Op-Amp

The op-amp is actually a complicated circuit comprised of several amplifier stages including many active and passive components. We will avoid that complication and will get acquainted with the fundamental features of the building blocks of a typical op-amp outlined in Figure 11.1(a). Most modern op-amps are available in the form of integrated circuit (IC). The principles of IC will be discussed in Chapter 12. The op-amp can be represented by the symbol of Figure 11.1(b) and this symbol will be used throughout this book.

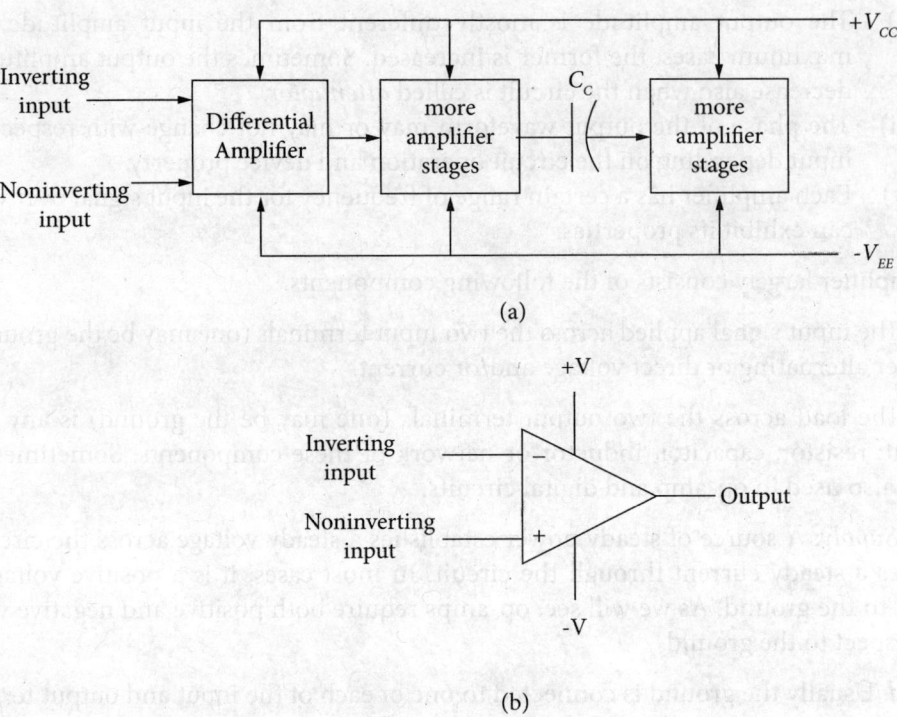

Figure 11.1 Operational amplifier (op-amp), (a) block diagram and (b) symbol

As sketched in Figure 11.1(a), the first stage of the op-amp is a differential amplifier, which is discussed in detail in Section 11.3. The output of this stage is further amplified by more amplifier stages including emitter–follower and push-pull amplifier. Typically the input voltage is differential (the voltage difference between the two input terminals) and the output is single-ended (with reference to the ground). Since the supply voltages are both positive and negative, the quiescent output voltage in absence of any input voltage should be zero.

Note: Compensating Capacitor

The op-amp is widely applied to provide negative feedback in conjunction with external network of passive components. The presence of any reactive element (frequency sensitive) in the feedback loop may cause delay of the signal. The cumulative effect of the delays may even result in a phase lag of 180° causing unwanted positive feedback and oscillation. To counteract this effect, a *frequency compensation* is established by incorporating some on-chip components. The inclusion of the capacitor (C_c), known as *compensating capacitor* across one of the internal gain stages is such a remedy, as indicated in Figure 11.1(a). Together with the output resistance of the differential amplifier stage, it builds up an RC network that counterbalances the phase lag effect caused by the reactance present in the feedback network. In IC, small capacitances are easier to fabricate. The actual value of C_c is small, of the order of pF. By virtue of Miller, it acts like a larger capacitance ($\approx$ nF).

The op-amp is not just an amplifier, it has many amazing features. As an amplifier circuit, has the following characteristic properties.

(i) The voltage gain of the op-amp is very high, ideally infinity. The input impedance is very high, ideally infinity and the output impedance is very small, ideally zero. These properties make it to act as an ideal voltage amplifier.

(ii) The op-amp can amplify both dc signals (zero frequency) and ac signals over a large frequency range. Ideally the bandwidth is infinity.

(iii) As denoted in Figure 11.1(b), it has two input terminals, namely inverting (–ve sign) and noninverting (+ve sign). The input signal applied to the inverting input terminal results in an amplified version with opposite phase at the output terminal. The input signal applied to the noninverting input terminal gets amplified in the same phase.

(iv) The op-amp amplifies the difference of the signals appearing at the two input terminals. Any signal common to both the input terminals has no effect on the output.

The followings are the most significant and useful characteristics of an op-amp from the user's point of view.

(i) The whole circuit can be operated like a single device with the input and output terminals and its performance can be controlled simply by joining a few passive components to the external terminals as feedback elements.

(ii) The amplifier can execute mathematical and logical operations, such as addition, integration, differentiation and comparison in terms of voltage. This process is termed as analog computation. Indeed the capability of doing mathematical and logical operations is the origin of the name 'operational amplifier'.

The above mentioned unique features have established the op-amp as useful tool for not only amplification but for diversified applications, such as oscillator, active filter, voltage regulator, analog-digital conversions and many others.

11.3 Differential Amplifier

The circuit is widely used as the input stage of an op-amp. Indeed it can be treated as the heart of the op-amp. Figure 11.2(a) introduces the general form of a differential amplifier. Two similar transistors Q_1 and Q_2 are connected face-to-face and two input voltages v_1 and v_2 are applied to their respective base terminals. The amplified voltage difference (v_o) is obtained between their collectors. Figure 11.2(b) is a more practical form of differential amplifier where the noninverting input (v_1) and the inverting input (v_2) terminals are made distinct without and with the use of the collector resistance (R_C). The operations are explained below.

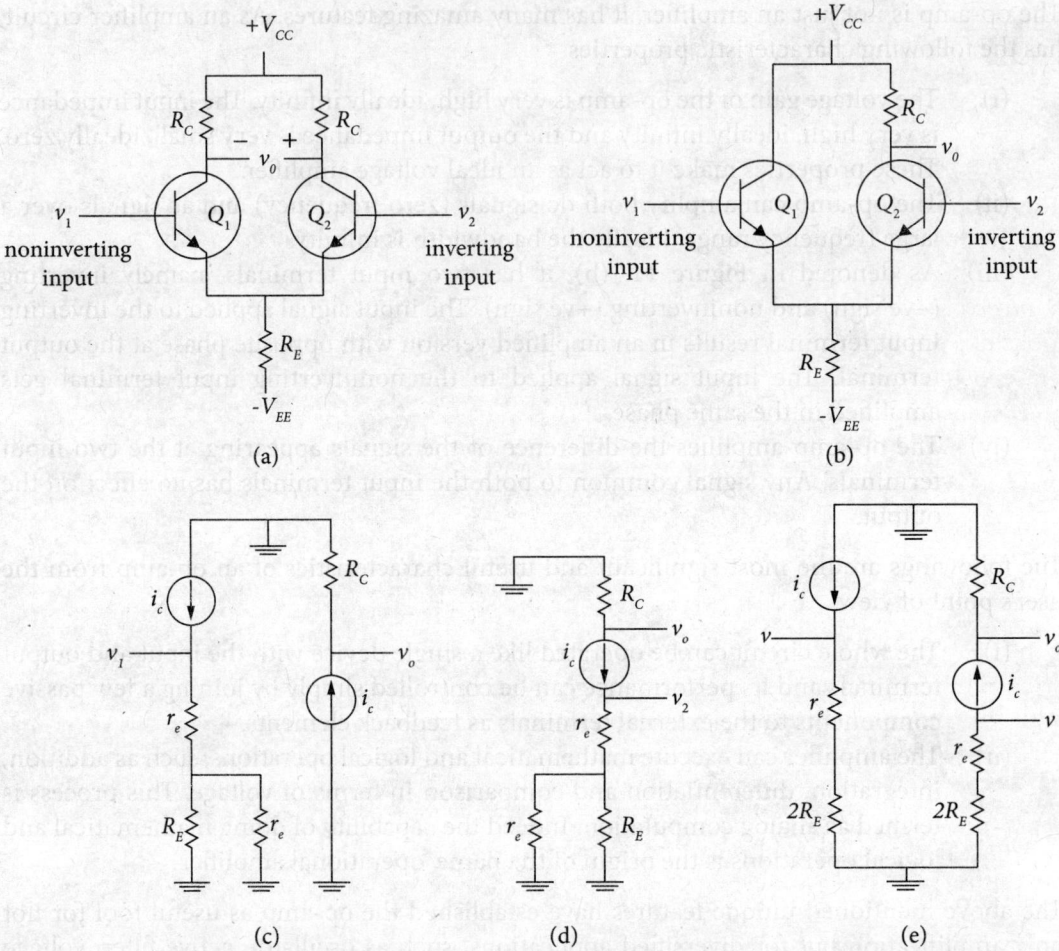

Figure 11.2 Differential amplifier, (a) basic construction with $v_1 > v_2$, (b) more popular form distinguishing noninverting (v_1) and inverting (v_2) input terminals, (c) ac equivalent when v_2 is grounded, (d) ac equivalent when v_1 is grounded and (e) ac equivalent in common mode

A. *Noninverting Input*: When v_2 is made zero by grounding this terminal, the ac equivalent circuit becomes similar to that of Figure 11.2(c). The transistor Q_1 acts like an emitter–follower. Its output drives the transistor Q_2, which acts like a common-base amplifier. Since Q_1 and Q_2 are identical transistors, both can be represented as current source i_c and both have ac emitter resistance r_e. Since $R_E \gg r_e$, the parallel combination of R_E and r_e is effectively r_e so that the emitter current, hence the collector current can be approximated as

$$i_c = \frac{v_1}{2r_e} \qquad (11.1)$$

Using Equation (11.1) and the output voltage (v_o) being equal to $i_c R_C$, the voltage gain is given by

$$\frac{v_o}{v_1} = \frac{R_C}{2r_e} \qquad (11.2)$$

There is no phase inversion and the final output (v_o) is in phase with the input (v_1). Therefore this input terminal is designated as *noninverting input*.

B. Inverting Input: When v_1 is made zero by grounding this terminal, the ac equivalent circuit looks like that of Figure 11.2(d). This time Q_2 drives Q_1 and the ac emitter current is

$$i_e = \frac{v_2}{2r_e} \qquad (11.3)$$

Expressing the collector current (i_c) as $-i_e$ and the output voltage (v_o) as $-i_e R_C$, the voltage gain can be stated as

$$\frac{v_o}{v_2} = -\frac{R_C}{2r_e} \qquad (11.4)$$

The negative sign in Equation (11.4) denotes that the final output (v_o) is in opposite phase with the input (v_2). Therefore this input terminal is termed as *inverting input*. Comparing Equation (11.4) with Equation (11.2) it is observed that the gain with the inverting input has the same magnitude as that with the noninverting input but the phase is opposite.

C. Differential Gain: If both the input voltages v_1 and v_2 become active simultaneously, the net output voltage becomes, using Equations (11.2) and (11.4),

$$v_d = \frac{R_C}{2r_e}(v_1 - v_2) \qquad (11.5)$$

The quantity $R_C/2r_e$ in Equation (11.5) is designated as the *differential mode gain* of the differential amplifier. Evidently it is a large quantity, much greater than unity.

D. Common Mode Gain: If v_1 becomes equal to v_2, it means the same voltage (v) is applied to both the inverting and the noninverting input, the ac equivalent circuit becomes similar to that of Figure 11.2(e). It contains the double of the large unbypassed emitter resistor in each arm. Therefore, the voltage gain is

$$\frac{v_o}{v} = -\frac{R_C}{r_e + 2R_E} \qquad (11.6)$$

Obviously the gain expressed by Equation (11.6) is a small quantity, often much less than unity. Thus in common mode, the amplifier does not amplify at all. This property of the differential amplifier is quantified in terms of common mode rejection ratio discussed in the next section.

11.4 Common Mode Rejection Ratio

The previous section presents the fact that a differential amplifier magnifies the difference of the two input voltages with an appreciable voltage gain of $R_C/2r_e$. Any signal common to both the inputs is not amplified and has no effect to the output. As if the amplifier 'rejects' it. Such property of a differential amplifier to discard the signal common to both of its input terminals is quantified in terms of a figure of merit known as *common mode rejection ratio* (CMRR) deduced below.

Let v_1 and v_2 be the voltage applied to the noninverting input and the inverting input, respectively, of the differential amplifier. The difference signal is given by

$$v_d = v_1 - v_2 \qquad (11.7)$$

The common mode signal is given by

$$v_c = \frac{v_1 + v_2}{2} \qquad (11.8)$$

Using Equations (11.7) and (11.8), the two input signals can be expressed as

$$v_1 = v_c + \frac{v_d}{2} \qquad (11.9a)$$

$$v_2 = v_c - \frac{v_d}{2} \qquad (11.9b)$$

The output voltage is given by

$$v_o = A_1 v_1 + A_2 v_2 \qquad (11.10)$$

A_1 is the gain when v_2 is grounded and A_2 is the gain when v_1 is grounded. Substituting for v_1 and v_2 from Equations (11.9a) and (11.9b) and rearranging,

$$v_o = A_d v_d + A_c v_c \qquad (11.11)$$

In Equation (11.11), $A_d = (A_1 - A_2)/2$ is the differential mode gain and $A_c = (A_1 + A_2)$ is the common mode gain. The common mode rejection ratio is defined in terms of these two gains in the following way.

$$\mathrm{CMRR} = \left| \frac{A_d}{A_c} \right| \qquad (11.12)$$

In Equation (11.12), A_d is reasonably large and A_c is very small, ideally zero. So the CMRR of an op-amp is a large quantity, ideally infinity. Two other associated features should be remembered.

- The CMRR is also expressed as $20\log(A_d/A_c)$ in decibel.
- The CMRR degrades at higher frequencies due to reactive effects in the circuit.

Operational Amplifier

> **Example 11.1**
>
> A differential amplifier has common mode rejection ratio of 80 dB and differential gain of 10^5. If the voltages at the inverting and the noninverting input terminal be -0.2 mV and -0.22 mV, respectively, what will be the output voltage?
>
> **Solution**
>
> Let the differential mode gain be A_d and the common mode gain be A_c.
>
> Given, 80 dB = $20 \log(A_d/A_c)$. Hence,
>
> $A_d/A_c = 10^4$
>
> Given, $A_d = 10^5$. Hence $A_c = 10$.
>
> The differential input is: $-0.22 - (-0.2) = -0.02$ mV
>
> The common mode input is: $(-0.22 - 0.2)/2 = -0.21$ mV
>
> Using Equation (11.11), the output voltage is
>
> $-0.02 \times 10^5 - 0.21 \times 10 = -2.0021$ V

> **Example 11.2**
>
> The CMRR of a differential amplifier using op-amp is 100 dB. The output voltage for a differential input of 200 µV is 2 V. Determine the common mode gain.
>
> [C.U. 2014]
>
> **Solution**
>
> The differential mode gain is calculated as
>
> $A_d = (2 \text{ V})/(200 \text{ µV}) = 10^4$.
>
> Let the common mode gain be A_c. Using the given CMRR,
>
> $20\log(A_d/A_c) = 100$, hence $A_d/A_c = 10^5$.
>
> Therefore, the common mode gain is
>
> $A_c = 10^4/10^5 = 0.1$

11.5 Diff-Amp to Op-Amp

It is the differential amplifier (diff-amp) itself that gives birth to the main advantageous features of an op-amp, namely high voltage gain, high input resistance and large bandwidth. Revising the previous section, the following properties of the diff-amp may be highlighted.

- The diff-amp has a voltage gain of magnitude $R_C/2r_e$, which is a considerably large quantity, since r_e is very small and R_C is of the order of kilo ohms.
- It has an input resistance of $2\beta r_e$ because each transistor contributes to βr_e. Thus the input resistance is a fairly large quantity.
- The two transistors are coupled directly and not through any reactive element so that the amplifier can be operated with both dc and ac inputs.

It is outlined in Figure 11.1 that a practical op-amp uses the diff-amp as its input stage along with other circuits. It enhances the above mentioned properties using other techniques. Some are mentioned below.

- The voltage gain is increased by cascading several amplifier stages. Some principles of multistage amplifiers are discussed in Chapter 8.
- The input resistance is enhanced by the use of emitter–follower (Chapter 6) or two transistors cascaded as Darlington pair (Chapter 8).
- A transistor is made to act as a resistor, the technique being known as *active load*. It eliminates the power loss cross a resistor and provides with current gain.
- Sometimes high input resistance devices, such as JFETs (Chapter 9) are used in combination with bipolar junction transistors.

The CMRR of the diff-amp is further increased by the increase of differential gain and the dc level of the final output is adjusted so as to ensure zero output for zero input. It provides symmetric output swing with reference to the ground.

> **Note: BIFET Technology**
>
> Modern general-purpose op-amps often combine the fabrication of JFETs and BJTs on the same chip, which is termed as BIFET Technology. This brings about the benefits of both the devices. The use of JFETs at the input stage results in higher input resistance and lower input bias and offset currents. Using BJTs in the later stages offer more voltage gain and bandwidth. Thus the performance characteristics get enhanced in terms of higher input impedance, lower bias current, higher slew rate, higher voltage gain and wider bandwidth.

11.6 Offset Parameters

An ideal op-amp should have zero output voltage when the input voltages become equal or both the input terminals are grounded. However, the two transistors of the differential amplifier at the input stage of the op-amp may not be perfectly matched. The following mismatches may creep in.

- The two transistors can have different values of current gain (β).
- The V_{BE} may be different.
- Also the two collector resistances may be slightly different.

In consequence, unequal bias currents flow through the two input terminals that results in an unwanted error voltage at the output even in absence of any input voltage. This is called the *output offset voltage* and is defined as the amount of direct voltage appearing at the output terminal when both the input terminals are grounded.

The output offset voltage can be nullified by using specific external network of resistors. The principle of such offset nullifying circuit is to implement a small voltage drop at the input so as to reduce the output to zero in absence of any input signal. The manufacturer specifies a parameter known as *input offset voltage* for a particular op-amp. It is numerically equal to the voltage difference that must be applied between the two input terminals in order to make the output voltage zero in absence of any input signal.

Figure 11.3 presents an example to demonstrate the effect of offset voltage in a typical op-amp. The symbol of Figure 11.1 is used. The ±V supply terminals are often discarded for convenience and we will do so in the subsequent op-amp circuits. A resistor (R_f), known as *feedback resistor* is regularly connected between the inverting input and the output terminals. Another resistor (R_1) is joined to the inverting input terminal. This is a very popular configuration of op-amp that we will be using many times. What we see in Figure 11.3, both the input terminals of the op-amp are grounded thereby having no input voltage. However, some output offset voltage $V_{O(off)}$ appears there. We can assume that this occurs due to an input offset voltage $V_{I(off)}$ appearing at the input. We will see later that the gain of this op-amp circuit is (R_f/R_1) so that the output offset voltage due to the input offset voltage is

$$V_{O(off)} = \frac{R_f}{R_1} V_{I(off)} \tag{11.13}$$

Equation (11.13) implies that the error voltage at the output can get amplified in different extents depending on the circuit configuration.

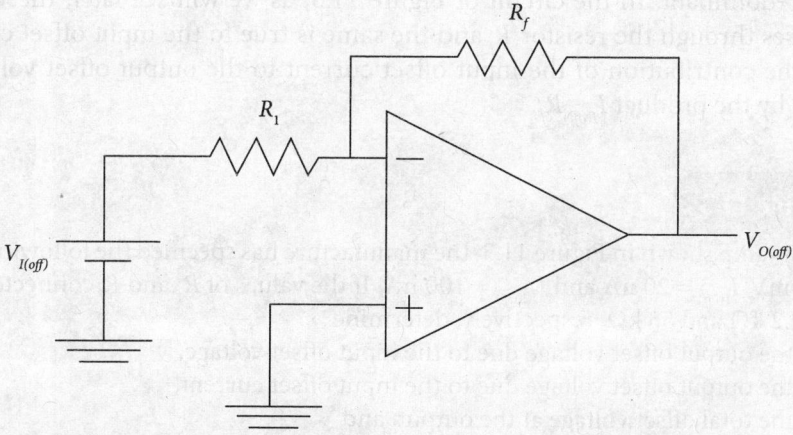

Figure 11.3 Offset voltages at the output and the input of an op-amp

Another parameter, named *input offset current* is useful in connection with the offset properties of the op-amp. It is the difference between the base current of the two transistors of the differential amplifier. The input offset current, in turn, is expressed in terms of the *input bias current*, which is the mean of the two input currents flowing through the bases of the two transistors of the differential amplifier.

If the noninverting and the inverting input terminals are denoted by the subscripts '1' and '2', respectively, the base currents at these two inputs are indicated by I_{B1} and I_{B2}, respectively. The input offset current $I_{I(off)}$, defined as the difference of these two currents is written as

$$I_{I(off)} = I_{B1} - I_{B2} \qquad (11.14)$$

The input bias current $I_{I(bias)}$ is given by

$$I_{I(bias)} = \frac{I_{B1} + I_{B2}}{2} \qquad (11.15)$$

The manufacture specifies the input offset current and the input bias current also, for a particular op-amp, along with the input offset voltage. Therefore, rearranging Equations (11.14) and (11.15), one can define the two input currents as

$$I_{B1} = I_{I(bias)} + \frac{I_{I(off)}}{2} \qquad (11.16a)$$

$$I_{B2} = I_{I(bias)} - \frac{I_{I(off)}}{2} \qquad (11.16b)$$

It is assumed $I_{B1} > I_{B2}$; if the opposite is true, the equations can be rearranged accordingly.

The total offset voltage appearing at the output is the combined effect of the input offset voltage and the input offset current. Equation (11.13) estimates the former one, which is generally predominant. In the circuit of Figure 11.3, as we will see later, the whole input current passes through the resistor R_f and the same is true to the input offset current also. Therefore, the contribution of the input offset current to the output offset voltage can be represented by the product $I_{I(off)} R_f$.

Example 11.3

For the op-amp shown in Figure 11.3, the manufacture has specified the following values: $V_{I(off)} = 2$ mV, $I_{I(off)} = 20$ nA and $I_{I(bias)} = 100$ nA. If the values of R_1 and R_f connected by the user be 2.2 kΩ and 56 kΩ, respectively, determine
    (i)   the output offset voltage due to the input offset voltage,
    (ii) the output offset voltage due to the input offset current,
    (iii) the total offset voltage at the output, and
    (iv) the base currents for the inverting and noninverting inputs.

Solution

(i) Using Equation (11.13), the output offset voltage due to the input offset voltage is
$$(56/2.2) \times 2 = 50.91 \text{ mV}$$

(ii) The output offset voltage due to the input offset current is
$$20 \times 10^{-9} \times 56 \times 10^3 = 1.12 \text{ mV}$$

(iii) The total offset voltage appearing the output is
$$50.91 + 1.12 = 52.03 \text{ mV}$$

(iv) Using Equations (11.16a) and (11.16b), the base currents are
$$100 + 20/2 = 110 \text{ nA}$$
and
$$100 - 20/2 = 90 \text{ nA}$$

11.7 Slew Rate

The capacitance (of the order of pF) of the compensating capacitor within the op-amp, as sketched in Figure 11.1(a) brings about a lot of things. Due to Miller effect it gets multiplied by the voltage gain of the internal amplifier stages to produce an equivalent capacitance of the order of nF. Together with the output resistance of the diff-amp, it produces a lag circuit that prevents unwanted oscillations from reactive feedback elements. This network has a very low cutoff frequency (≈ 10 Hz) that enables the op-amp to amplify both dc and ac input signals but the amplification goes on diminishing with the increase of frequency. That is why an op-amp cannot amplify signals of too high (> MHz) frequencies. The frequency when the voltage gain reduces to 1, is termed as the *unity-gain frequency* of the op-amp. The compensating capacitor also induces another type of constraint on the frequency response of the op-amp because of the finite time of its charging and discharging. It imposes a limit on how fast the op-amp output can change state. This limitation is quantified in terms of *slew rate*.

The slew rate represents how fast the op-amp output can 'slew' or 'slide' toward a new value in response to the input change. Generally the slew rate is expressed in *volt per microsecond*. For instance, the slew rate of IC 741 is about 0.5 V/μs. If a larger step of input voltage within this time span, such as 1 V/μs is imposed, the output remains unable to follow that and instead of a step, it produces a ramp of incorrect voltage. Similarly, a large amplitude, high frequency sinusoidal signal gets distorted like a vertically squeezed triangular wave. Thus the slew rate of the op-amp gives rise to a distortion for high frequency and large signal operations. The followings are the remarkable features associated with the slew rate of an op-amp.

- The slew rate is a parameter of the op-amp itself and it is to be taken into account while constructing any amplifier circuit with that op-amp.
- The slew rate of an op-amp limits its large-signal response and causes a signal distortion beyond a limit of both amplitude and frequency of the input signal.

- An op-amp with a high slew rate signifies that the op-amp can slew at a higher rate of voltage per unit time. The manufacture specifies the slew rate of a particular op-amp.

It is also worth mentioning that negative feedback (Chapter 10) can reduce nonlinear distortions but not the distortion due to slew rate. It is so because when the distortion due to slew rate takes place, the input voltage becomes high enough to saturate one transistor of the diff-amp and cut off the other. So the op amp no longer operates in the linear region and the negative feedback effect does not work.

> **Note: Power Bandwidth**
>
> The discussion on the slew rate makes it apparent that there is a maximum frequency (f_m) of an input signal with peak voltage (V_p) that can be amplified without distortion caused by slew rate. This frequency limit is also referred to as the power bandwidth of the op-amp and is given by $f_m = $ (slew rate)$/(2\pi V_p)$.

> **Example 11.4**
>
> A perfect square wave of 1 V amplitude and 1 MHz frequency is applied to the op-amp input but the output waveform visible in a cathode ray oscilloscope is a ramp that changes to 0.5 V in 1 μs. What is the slew rate of the op amp? Can we observe the actual square wave with this op-amp?
>
> **Answer**
>
> The slew rate is what we are seeing, that is 0.5 V/μs. This is the fastest response that the op-amp can execute. If we reduce the square wave amplitude below 0.5 V, the op-amp will be able to follow the time rate of change and the square waveform will be visible.

11.8 Linear Op-Amp Circuits

It is the wide variety of applications that has made op-amp so unique a device. The electronic circuits constructed with op-amps may be classified into two broad categories: linear and nonlinear. Some of the op-amp circuits preserve the shape and proportionality of the input signal at the output and these are called *linear op-amp circuits*. Some important linear op-amp circuits, namely inverting amplifier, noninverting amplifier, differential amplifier, instrumentation amplifier, and active filter are discussed here. Yet there are other circuits fabricated with the op-amp known as *nonlinear op-amp circuits* where the output appears to be something different from the input. Those will be introduced later in this chapter.

11.8.1 Inverting Amplifier

The name of the circuit implies that the input signal gets amplified with a phase inversion. The circuit diagram for an inverting amplifier is given in Figure 11.4(a). It has the following properties.

Voltage–Shunt Feedback: It is actually a negative feedback amplifier realized with op-amp. [Can you please have a quick look at Table 10.2 and solved numerical problem 10.4?] The input voltage (V_i) in Figure 11.4(a) is applied to the inverting input through an input resistor (R_1). The noninverting input terminal is grounded. The voltage–shunt feedback is obtained through feedback resistor (R_f) connected between the output and the inverting input terminal.

Virtual Ground: Let us consider two separate phenomena. First, due to the very high voltage gain of the op-amp, almost zero voltage difference (V_d) between the input terminals can produce a finite output. Therefore, if the other input, i.e., the noninverting input terminal is actually grounded, the point P (inverting input) is also effectively pulled down to zero potential. Secondly, because of the very high input resistance of the op-amp, almost no current enters the inverting input and practically the whole of the input current (I_i) bypasses through R_f, as indicated with arrow in the figure. Thus the point P is at a condition of about zero potential and almost zero current, which is referred to as *virtual ground*. This is different from an actual ground that may have zero voltage but not zero current.

Current-to-Voltage Conversion: The inverting input causes the output voltage (V_o) to be in opposite phase with respect to the input voltage. Therefore,

$$V_o = -(I_i R_f + V_d) \tag{11.17}$$

If the voltage gain (infinitely large) of the op-amp be A, the output voltage can be written as

$$V_o = A V_d \tag{11.18}$$

The input current is given by

$$I_i = \frac{V_i}{R_1} \tag{11.19}$$

Equations (11.17), (11.18) and (11.19) can be rearranged to define the output voltage as

$$V_o = -\frac{A}{1+A} I_i R_f \tag{11.20}$$

Since $A \gg 1$, Equation (11.20) effectively becomes

$$V_o = -R_f I_i \tag{11.21}$$

Equation (11.21) indicates that the output voltage of the inverting amplifier is proportional to the input current. Thus the inverting amplifier can be treated as a *current-to-voltage converter*.

Amplification: Using Equations (11.21) and (11.19), the following expression for the voltage gain (A_V) of the inverting amplifier can be obtained.

$$A_V = \frac{V_o}{V_i} = -\frac{R_f}{R_1} \qquad (11.22)$$

The following properties of the inverting amplifier can be inferred from (11.22).

(i) The gain is adjustable. It can be varied by changing the ratio R_f/R_1.
(ii) The negative sign denotes a phase inversion of the output with respect to the input.
(iii) The above formulation is derived using dc values but these are equally applicable to ac voltages also unless the high frequency limit is reached.

Transfer Characteristics: Figure 11.4(b) sketches the graphical variation of the output voltage (V_o) with respect to the input voltage (V_i) for different gains, hence different ratios of feedback and input resistances. The plots are linear for different voltage gains: A_{V1}, A_{V2} and A_{V3}, supporting the fact that the inverting amplifier is a linear application of op-amp for both positive and negative input voltages. The opposite phase of the output is also apparent from the plots. The linear slope increases with voltage gain and the linearity gets destroyed when the gain becomes so large that the output tends to exceed the supply voltage. This saturates the output at the supply voltage without any further increase with input voltage, as indicated with dotted lines in the figure.

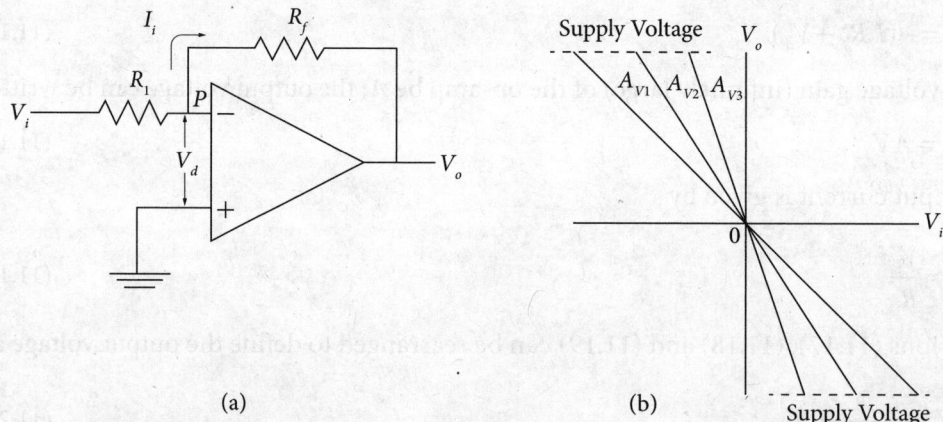

(a) (b)

Figure 11.4 Inverting amplifier constructed with op-amp, (a) circuit diagram and (b) transfer characteristics with different magnitudes of voltage gain: $A_{V1} < A_{V2} < A_{V3}$

Input Resistance: Since the input current (I_i) flows almost entirely through R_f and the inverting input is at virtual ground,

$$|I_i| = \frac{V_o}{R_f} = \frac{AV_d}{R_f} \qquad (11.23)$$

The input resistance can be derived from Equation (11.23) as

$$\frac{V_d}{I_i} = \frac{R_f}{A} \qquad (11.24)$$

Since $A \gg R_f$, Equation (11.24) indicates that the input resistance of the inverting amplifier is a very small quantity.

> **Example 11.5**
>
> The inverting amplifier of Figure 11.4 (a) has $R_1 = 1$ kΩ and $R_f = 2.2$ kΩ. Determine the output voltage, the input current and the input resistance for an input voltage of 1 V. Given: the open-loop voltage gain of the op-amp is 10^5.
>
> **Solution**
>
> Using Equation (11.22), the output voltage is $-(2.2 \text{ k}\Omega)/(1 \text{ k}\Omega) \times (1 \text{ V}) = -2.2$ V
>
> Using Equation (11.19), the input current is $(1 \text{ V})/(1 \text{ k}\Omega) = 1$ mA
>
> Using Equation (11.24), the input resistance is $(2.2 \text{ k}\Omega)/(10^5) = 0.022$ Ω

> **Think a Bit: Is the Gain Infinity?**
>
> We recall that the op-amp has very large voltage gain, if not infinity. However, the expression for voltage gain in Equation (11.22) indicates that the gain is adjustable to any value. How these two statements are matched? What Equation (11.22) shows is not the gain of any op-amp but the closed-loop gain of a negative feedback amplifier named 'inverting amplifier' made of op-amp. The infinitely large gain of the original opamp always exists in the background and the virtual ground is its evidence in the present circuit.

> **Note: Current-Controlled Voltage Source and Transresistance Amplifier**
>
> As mentioned earlier, the output voltage of the inverting amplifier is proportional to the input current. As if the input current [Equation (11.19)] controls the output voltage. Therefore the inverting amplifier can be considered as a *current-controlled voltage source*. The inverting amplifier is also an example of *transresistance amplifier* (Chapter 10) because in course of transferring from the input port to the output port, the value of the feedback resistor serves as a conversion factor that determines the output voltage for a given input current [Equation (11.21)] and the voltage gain for a given input voltage (11.22).

11.8.2 Noninverting Amplifier

It amplifies and produces output in the same phase as that of the input. The circuit diagram of a noninverting amplifier is given in Figure 11.5(a). For practical purpose, it is represented more simply in the form of Figure 11.5(b). However, the circuit operation is easily explained with Figure 11.5(a) revealing the following properties.

Voltage–Series Feedback: This is a negative feedback amplifier fabricated with op-amp where the input voltage (V_i) is applied to the noninverting input terminal. The voltage–series feedback achieved with the $R_f - R_1$ resistor chain is applied to the inverting input terminal. Thus none of the input is grounded and this circuit has no scope for establishing virtual ground. However, it has a *virtual short*, to be discussed later. The feedback voltage in Figure 11.5(a) is given by

$$V_f = \beta V_o \qquad (11.25)$$

the feedback ratio being

$$\beta = \frac{R_1}{R_1 + R_f} \qquad (11.25a)$$

The voltage difference between the two input terminals is

$$V_d = V_i - \beta V_o \qquad (11.26)$$

Amplification: Using Equation (11.26), the expression for the output voltage (V_o) becomes

$$V_o = AV_d = \frac{AV_i}{1 + A\beta} \qquad (11.27)$$

Since $A \gg 1$, Equation (11.27) effectively becomes

$$V_o = \frac{V_i}{\beta} \qquad (11.28)$$

From Equations (11.28) and (11.25a), the voltage gain of the noninverting amplifier is derived as

$$A_V = \frac{V_o}{V_i} = 1 + \frac{R_f}{R_1} \qquad (11.29)$$

Equation (11.29) represents the closed-loop voltage gain of a negative-feedback amplifier with voltage–series feedback constructed with op-amp, which is known as 'noninverting amplifier'. It has the following properties.

(i) The voltage gain can be adjusted by varying the ratio R_f/R_1. The minimum gain is unity.
(ii) There is no sign change, hence no phase change of the output with respect to the input.
(iii) The above formulation of the amplifier gain is applicable to both dc and ac input voltages except for very high frequencies.

Transfer Characteristics: Figure 11.5(c) drafts the output voltage (V_o) variation with respect to the input voltage (V_i) for different voltage gains: A_{V1}, A_{V2} and A_{V3}, which can be changed by changing the ratio of feedback and input resistances. The following features are indicated in the graph of Figure 11.5(c).

The output values are in phase with the input values. The plots are linear implying that the noninverting amplifier is a linear application of op-amp for both positive and negative input voltages. The linear slope increases with voltage gain ($A_{V3} > A_{V2} > A_{V1}$) and the linearity ends when the gain becomes comparable to the supply voltage. Then V_o gets saturated at the supply voltage without any more increase with increasing V_i, as indicated with dotted lines. In absence of feedback resistance, the infinitely large open-loop voltage gain makes V_o to saturate to ± supply voltage for almost zero plus/minus values of V_i.

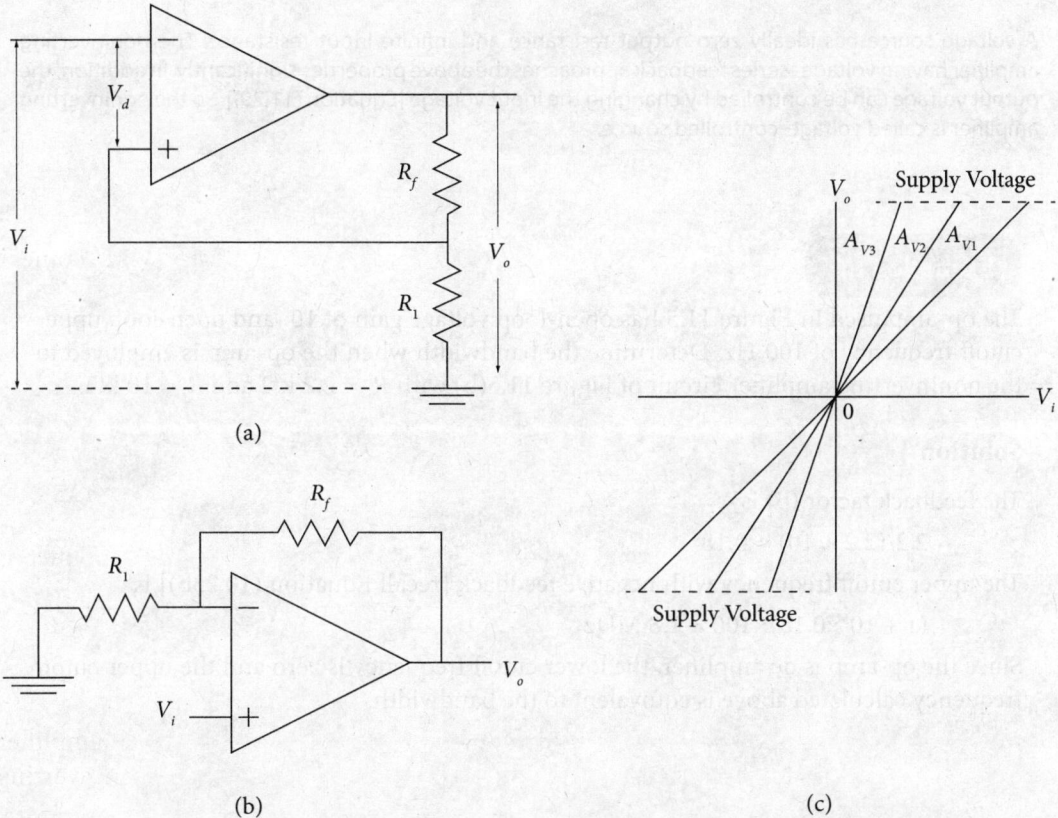

Figure 11.5 Noninverting amplifier constructed with op-amp, (a) circuit diagram, (b) more practical form of the circuit, and (c) transfer characteristics with different magnitudes of voltage gain: $A_{V1} < A_{V2} < A_{V3}$.

Input Resistance: The voltage difference (V_d) between the two inputs can be written as the product $R_i I_i$ where R_i is the input resistance between the two input terminals and I_i is the current flowing through the input terminals. Putting this value of V_d in Equation (11.27) and rearranging, the closed-loop input resistance of the overall amplifier can be expressed as

$$R_{if} = \frac{V_i}{R_i} = R_i(1 + A\beta) \qquad (11.30)$$

Equation (11.30) implies that the input resistance (R_i) of the op-amp is increased $(1 + A\beta)$ times with the voltage–series feedback in the noninverting amplifier.

The inverting and the noninverting amplifiers illustrated above are two fundamental applications of op-amp with negative feedback that are used in many applications. A comparative view of these two circuits is presented in Table 11.1.

> **Note: Voltage-Controlled Current Source**
>
> A voltage source has ideally zero output resistance and infinite input resistance. The noninverting amplifier having voltage–series feedback approaches the above properties significantly. In addition, the output voltage can be controlled by changing the input voltage [Equation (11.29)]. So the noninverting amplifier is called voltage-controlled source.

Example 11.6

The op-amp used in Figure 11.5 has open-loop voltage gain of 10^5 and open-loop upper cutoff frequency of 100 Hz. Determine the bandwidth when the op-amp is employed to the noninverting amplifier circuit of Figure 11.5(b) with $R_1 = 2.2$ kΩ and $R_f = 10$ kΩ.

Solution

The feedback factor (β) is

$$2.2/(2.2 + 10) = 0.18$$

The upper cutoff frequency with negative feedback [recall Equation (10.23b)] is

$$(1 + 10^5 \times 0.18) \times 100 \approx 1.8 \text{ MHz}$$

Since the op-amp is dc amplifier, the lower cutoff frequency is zero and the upper cutoff frequency calculated above is equivalent to the bandwidth.

> **Note: Frequency Response of Op-Amp**
>
> The op-amp, by virtue of its circuit construction, is dc amplifier. It can amplify both dc signals and ac signals over a wide range of frequency. The use of negative feedback, as is implemented to the inverting and the noninverting amplifier, enhances this property by increasing the bandwidth. However, the 'frequency response' of an amplifier refers to the continuous variation of the gain with input frequency, often represented graphically. For large range of frequency, this axis is plotted in logarithmic scale.

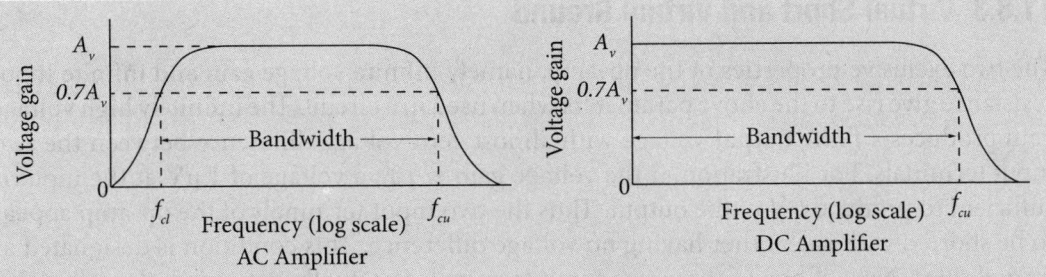

The frequency response of an ac amplifier constructed with discrete elements, such as BJT (left side) and the frequency response of an op-amp (right side) are compared above. For the ac amplifier (left side), the gain decreases at low frequencies because the coupling and bypass capacitors start acting as open circuits. Also the gain falls at high frequency because the junction capacitances and the stray wiring capacitances become prominent. The limits are quantified by the lower cutoff frequency (f_{cl}) and the upper cutoff frequency (f_{cu}), respectively. In the middle of these two limits, there is a large range of frequency where the gain remains constant at a highest value (A_v). Equations (7.36) and (7.47) may be revisited in this occasion. The phrase 'mid-band' refers to that range of input frequency (f) when $f_{cl}/f \approx 0$ as well as $f/f_{cu} \approx 0$. For the op-amp (right side), there exists only the upper cutoff frequency and the lower cutoff frequency reduces to zero. Input signals of any frequency down to zero hertz can be amplified. In this respect, the op-amp is preferred to the transistor for voltage amplifier designing because the op-amp has dc amplifying capability, high voltage gain, high input impedance and low output impedance.

Table 11.1 Comparison of op-amp inverting and noninverting amplifiers

Inverting Amplifier	Noninverting Amplifier
It is a negative feedback amplifier with the use of op-amp where voltage–shunt feedback is applied.	This is also negative feedback amplifier obtained with op-amp where voltage–series feedback is implemented.
The output is in opposite phase with respect to the input.	The output is in the same phase as that of the input.
There exists a virtual ground at the inverting input terminal.	There is no virtual ground.
This is a linear amplifier having gain adjustable with the ratio of the feedback resistance (R_f) and the external resistance (R_1) at the input terminal. The voltage gain (R_f/R_1) can be greater than, less than or equal to the unity depending on the ratio.	This is also linear amplifier with variable gain, depending on the feedback resistance (R_f) and the external input resistance (R_1). The voltage gain ($1 + R_f/R_1$) can be greater than or equal to unity but not less than unity.
The inverting amplifier can be modified to fabricate many useful circuits, such as adder and differential amplifier.	The noninverting amplifier also can be altered to achieve advantageous circuits, such as adder and voltage follower.

11.8.3 Virtual Short and Virtual Ground

The two exclusive properties of the op-amp, namely infinite voltage gain and infinite input resistance give rise to the above parameters when used in a circuit. The infinitely high voltage gain produces a finite output voltage with almost zero voltage difference between the two input terminals. For illustration, if the voltage gain is 10^6, a voltage of 1 µV at the input is sufficient to produce 1 V at the output. Thus the two input terminals of the op-amp appear to be shorted with each other having no voltage difference. This condition is designated as *virtual short*. Now, if one of these two input terminals is actually grounded, the other also comes down to almost zero potential in order to maintain the virtual short condition. This situation is called *virtual ground*.

However, the zero potential is not the sufficient condition for a virtual ground. The infinitely high input resistance of the op-amp allows negligible current to flow between the input terminals. Therefore, the point of virtual ground in the circuit is a point of zero current also. Considering the above features, the virtual ground may be distinguished from an actual ground in two ways.

(i) An actual ground in a circuit is a point of zero potential but not zero current whereas the virtual ground has both the voltage and current as zero.

(ii) The actual ground represents the resultant condition of an actual connection with zero potential (of the power supply) and the consequent current flow. The virtual ground is not any actual connection with zero potential and does not mean actually cutting off the current. These phenomena are virtually realized in a particular circuit.

The concept of virtual short may be revised with Figure 11.5 (a). It is redrawn in a modified manner in Figure 11.6. The op-amp position is indicated by dotted lines. Because of the virtual short between the two inputs (indicated by the thick arrow), the circuit of noninverting amplifier gets the input voltage of

$$V_i = \frac{R_i V_o}{R_i + R_f} \qquad (11.31)$$

Therefore, the voltage gain is

$$\frac{V_o}{V_i} = 1 + \frac{R_f}{R_i} \qquad (11.32)$$

This is same as the voltage gain expressed by Equation (11.29).

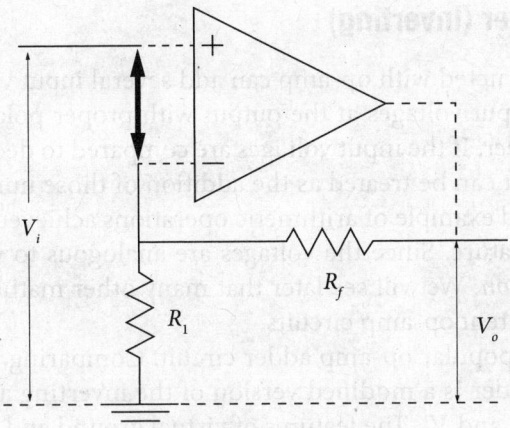

Figure 11.6 Virtual short in op-amp

11.8.4 Voltage Follower

This op-amp circuit is also known as *unity-follower*. It is a special configuration of the non-inverting amplifier where the feedback resistance (R_f) is made zero by shorting the inverting input and the output terminal. The resistance (R_1) connected to the input is opened. The resultant circuit becomes similar to that of Figure 11.7. The voltage gain of the voltage follower is unity, as may be verified by putting $R_f = 0$ and $R_1 = \infty$ in Equation (11.29). This circuit has the following properties.

(i) The unity voltage gain signifies that the input signal is reproduced at the output without any change in magnitude. Also there is no change in phase since the voltage follower is a particular case of noninverting amplifier.

(ii) The input resistance of the voltage follower is very high, as with the noninverting amplifier [Equation (11.30)].

(iii) Recall the use of CC amplifier as buffer (Chapter 6). The voltage follower also can act as a buffer or impedance matcher similar to the BJT emitter–follower. For instance, it can enhance the internal resistance of a voltmeter.

(iv) For both type of input signals, dc and ac, it can produce a replica at the output.

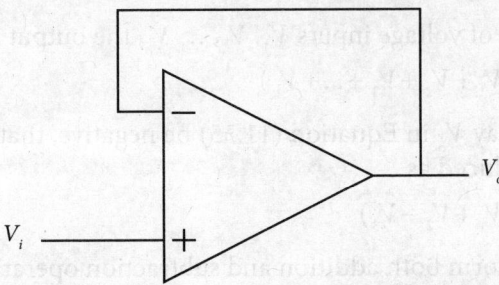

Figure 11.7 Voltage follower circuit fabricated with op-amp

11.8.5 Op-Amp Adder (Inverting)

The adder circuit constructed with op-amp can add several input voltages and produce the algebraic sum of the input voltages at the output with proper polarity. This circuit is also named *summing amplifier*. If the input voltages are compared to decimal numbers (integers or fractions), the output can be treated as the addition of those numbers with proper sign. Thus the adder is a good example of arithmetic operations achieved with op-amps, thereby signifying the nomenclature. Since the voltages are analogous to numbers, the process is called *analog computation*. We will see later that many other mathematical operations can be performed with different op-amp circuits.

Figure 11.8 shows a popular op-amp adder circuit. Comparing with Figure 11.4(a) it is understood that this adder is a modified version of the inverting amplifier with a number of inputs, such as V_1, V_2 and V_3. The features of virtual ground and voltage–shunt feedback persist there. The input currents, proportional to the respective input voltages are combined at the inverting input terminal and the output produces a voltage of inverted polarity and magnitude proportional to the algebraic sum of the input voltages. Point P is at virtual ground so that the total input current flowing through the feedback resistor (R_f) is given by

$$I_i = \frac{V_1}{R_1} + \frac{V_2}{R_2} + \frac{V_3}{R_3} \tag{11.33}$$

Substituting for I_i from Equation (11.21) and rearranging, the output voltage (V_o) can be derived as

$$V_o = -R_f \left(\frac{V_1}{R_1} + \frac{V_2}{R_2} + \frac{V_3}{R_3} \right) \tag{11.34}$$

At a particular condition, when $R_1 = R_2 = R_3 = R$, the output voltage of the adder defined by Equation (11.34) can be simplified as

$$V_o = -(V_1 + V_2 + V_3) \tag{11.35}$$

Equation (11.35) yields the actual sum of the input voltages. The polarity of the output voltage is, of course, inverted. From symmetry, the result of Equation (11.35) can be extended further for the following conditions.

(i) For a number of voltage inputs V_1, V_2, ... V_n, the output is given by

$$V_o = -(V_1 + V_2 + V_3 + ... + V_n) \tag{11.35a}$$

(ii) If any input, say V_3 in Equation (11.35) be negative, that gets subtracted and the output is produced as

$$V_o = -(V_1 + V_2 - V_3) \tag{11.35b}$$

Thus the circuit can perform both addition and subtraction operations in terms of voltage.

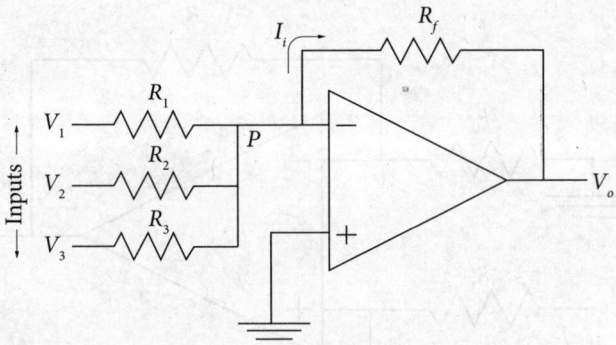

Figure 11.8 Op-amp adder circuit, with inverting output voltage

11.8.6 Op-Amp Adder (Noninverting)

One can also fabricate an op-amp adder using the noninverting amplifier having virtual short and voltage–series feedback. Figure 11.9 depicts such as op-amp adder that uses the noninverting input terminal for collecting the input signals and produces an output voltage of noninverting polarity and magnitude proportional to the algebraic sum of the input voltages.

Let V_1, V_2 and V_3 be the input voltages and V be the voltage at the noninverting input terminal. Since negligible current enters the op-amp,

$$\frac{V_1 - V}{R_1} + \frac{V_2 - V}{R_2} + \frac{V_3 - V}{R_3} = 0 \tag{11.36}$$

$$\Rightarrow V = \frac{V_1/R_1 + V_2/R_2 + V_3/R_3}{1/R_1 + 1/R_2 + 1/R_3} \tag{11.37}$$

The same voltage (V) as above appears at the inverting input by virtue of virtual short and using Equation (11.29), the output voltage can be deduced as

$$V_o = \left(1 + \frac{R_f}{R_i}\right) \frac{V_1/R_1 + V_2/R_2 + V_3/R_3}{1/R_1 + 1/R_2 + 1/R_3} \tag{11.38}$$

At a specific condition, when $R_1 = R_2 = R_3 = R_i = R$ and $R_f = 2R$, the output voltage given by Equation (11.38) is simplified as

$$V_o = (V_1 + V_2 + V_3) \tag{11.39}$$

It represents the actual sum of the input voltages and, comparing with Equation (11.35) it is found that the output does not invert phase with the present adder. In general, for a number of voltage inputs V_1, V_2, ... V_n, if the conditions: $R_1 = R_2 = ... = R_n = R_i = R$ and $R_f = (n - 1)R$ are satisfied, the output can represent the actual sum given by

$$V_o = (V_1 + V_2 + V_3 + ... + V_n) \tag{11.39a}$$

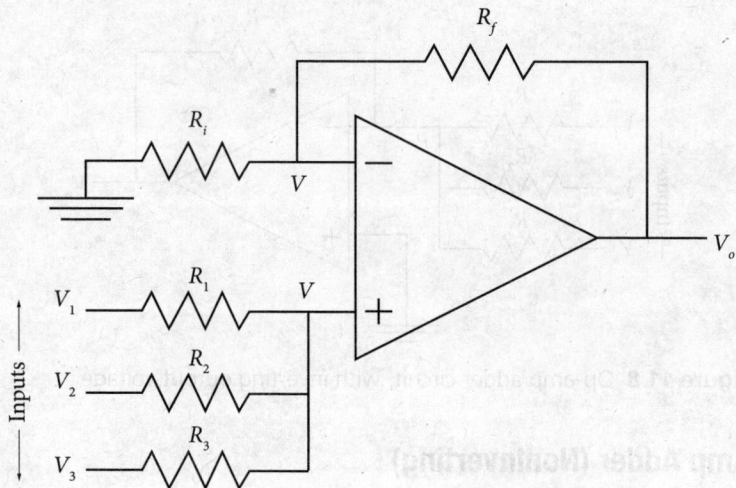

Figure 11.9 Op-amp adder circuit with noninverting output voltage

11.8.7 Differential Amplifier

This op-amp circuit, shown in Figure 11.10 can amplify the difference of two input voltages applied to the two input terminals. Sometimes it is also called the *voltage subtractor*. Let voltages V_1 and V_2 be applied to the noninverting input and the inverting input terminal, respectively. The noninverting input terminal is connected to the ground through resistor R_2. The resistors connecting the input voltages to the input terminals are R_1 and R_3, respectively. The resistor R_4 acts as the feedback element. The voltage appearing at the noninverting and the inverting terminal are denoted by V_3 and V_4, respectively. The circuit operates on the following two assumptions.

(i) Because of virtual short between the inputs, $V_3 \approx V_4$.
(ii) Because of very high input resistance, negligible current enters the noninverting and inverting inputs and almost all the input currents I_1 and I_2 get bypassed through R_2 and R_4, respectively.

Therefore, one can arrive at the following expressions for voltage and current.

$$V_3 = \frac{R_2 V_1}{R_1 + R_2} \tag{11.40a}$$

$$\frac{V_2 - V_4}{R_3} = \frac{V_4 - V_o}{R_4} \tag{11.40b}$$

The algebraic simplification of Equations (11.40a) and (11.40b) with the assumption of $V_3 = V_4$ leads to the following expression for the output voltage.

$$V_o = \frac{R_2}{R_1+R_2}\left(1+\frac{R_4}{R_3}\right)V_1 - \frac{R_4}{R_3}V_2 \qquad (11.41)$$

If the circuit of Figure 11.10 is made symmetric by making $R_4 = R_2$ and $R_3 = R_1$, the expression for output voltage given by Equation (11.41) is simplified as

$$V_o = \frac{R_2}{R_1}(V_1 - V_2) \qquad (11.42)$$

Equation (11.42) defines the output of the differential amplifier explicitly as the difference of the two input voltages. If the circuit is further simplified by making $R_1 = R_2$ then the output simply becomes $(V_1 - V_2)$.

The advantage of the differential amplifier is that the difference between any two voltages can be accepted as the input and the same can be amplified with an adjustable gain depending on the ratio of R_1 and R_2. No ground terminal is required.

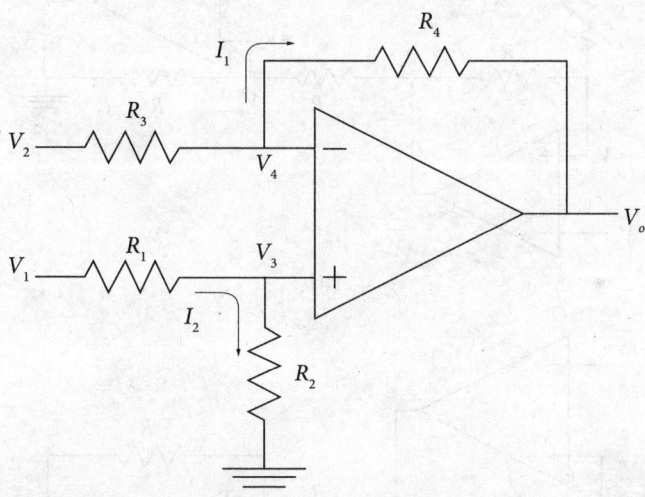

Figure 11.10 Op-amp differential amplifier circuit

Note: CMRR of a Differential Amplifier

A critical parameter of the differential amplifier is the CMRR of the circuit because the input signal is not necessarily a differential voltage alone. A common-mode voltage, such as a noise may also appear. Unless the circuit has a high CMRR, the common-mode input signal may create a distortion at the output. The overall CMRR of the circuit has two components: one due to the op-amp itself and the other is the common-mode error due to the difference in tolerance of the resistors. When $R_1 = R_3$ and $R_2 = R_4$ perfectly, the net common-mode input voltage across the op-amp input terminals becomes zero. However, if there is any mismatch of the resistances values, a differential input voltage is produced. A good remedy to nullify the resistance mismatch effect is to increase the input impedance of the circuit, for instance, by using a voltage follower at the input. That principle is used in an op-amp instrumentation amplifier.

11.8.8 Instrumentation Amplifier

This circuit is actually an advanced version of the differential amplifier described above. An important field of op-amp applications is instrumentation and the instrumentation amplifier is of that category. It has high input resistance and high CMRR. It is generally used for amplifying small voltage difference in presence of noise.

The circuit diagram of the instrumentation amplifier is given in Figure 11.11(a). It is comprised of three op-amps. The input voltages are V_1 and V_2 and R_1 is an adjustable resistance. Assuming virtual short and high input resistance of op-amp one can write

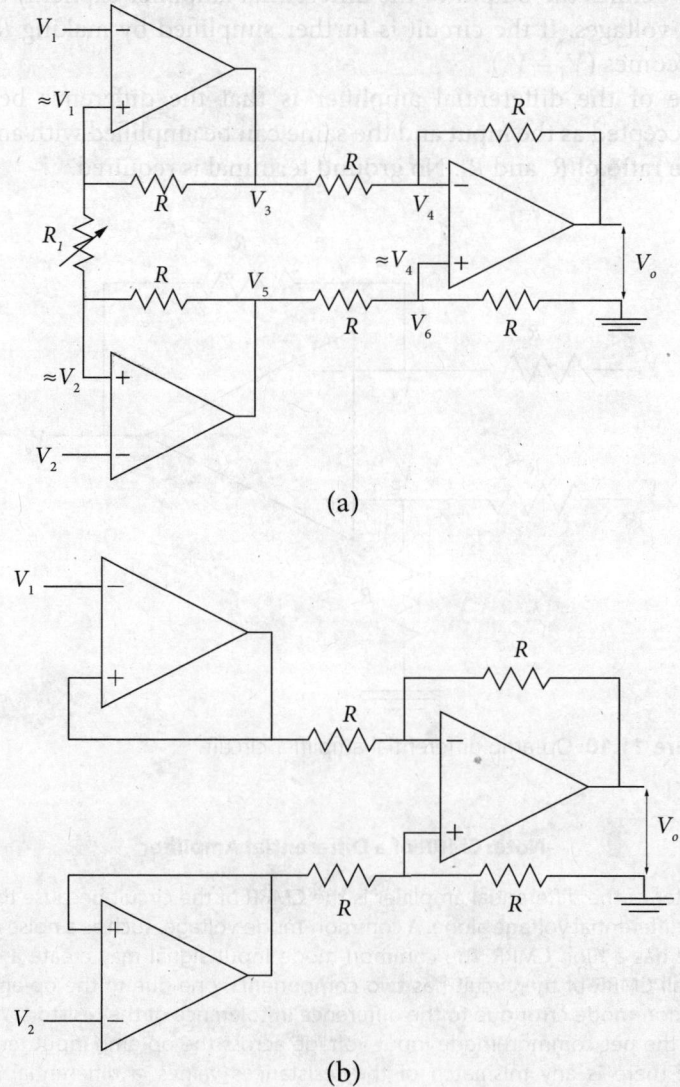

Figure 11.11 Instrumentation amplifier, (a) actual circuit and (b) simplified circuit diagram

$$\frac{V_3 - V_1}{R} = \frac{V_1 - V_2}{R_1} = \frac{V_2 - V_5}{R} \tag{11.43}$$

Equation (11.43) can be solved to obtain

$$V_3 = \frac{V_1(R + R_1)}{R_1} - \frac{RV_2}{R_1} \tag{11.44a}$$

$$V_5 = \frac{V_2(R + R_1)}{R_1} - \frac{RV_1}{R_1} \tag{11.44b}$$

Similarly, one can write

$$\frac{V_4 - V_3}{R} = \frac{V_0 - V_4}{R} \Rightarrow V_o = 2V_4 - V_3 \tag{11.45a}$$

Assuming $V_6 \approx V_4$,

$$\frac{V_5 - V_4}{R} = \frac{V_4 - 0}{R} \Rightarrow V_4 = \frac{V_5}{2} \tag{11.45b}$$

Using Equations (11.45a) and (11.45b) and substituting for V_3 and V_5 from Equations (11.44a) and (11.44b), respectively, the output voltage can be obtained as

$$V_o = (V_2 - V_1)\left(1 + \frac{2R}{R_1}\right) \tag{11.46}$$

Equation (11.46) implies that the output (V_o) is an amplified version of the input voltage difference ($V_2 - V_1$) and the extent of amplification can be adjusted by changing the variable resistor (R_1).

The instrumentation amplifier circuit can be further simplified, as shown in Figure 11.11(b), by using the two input op-amps as voltage follower and opening R_1. In that case, Equation (11.46) is reduced to

$$V_o = (V_2 - V_1) \tag{11.47}$$

11.8.9 Passive and Active Filters

A filter is a frequency selective circuit that allows electrical signals of some pre-determined range of frequency to pass through it and attenuates the signals of other frequencies. The range depends on the filter components. Filters have many important applications in telecommunication, instrumentation and signal processing. An essential component of a filter is capacitor or inductor that has frequency dependent reactance. If the filter is constructed of passive components only, such as resistors, capacitors and inductors, it is called *passive filter*. The filters discussed in Chapter 4 in connection with rectifiers are passive filters. If active components, such as op-amps are involved in the construction of the filter, it is called *active filter*.

Figure 11.12 gives the idea of constructing passive filters with capacitance (C) and (R). The left side network acts as a low-pass filter. For dc (zero frequency) or low frequency of the input voltage (v_i), the capacitor acts as open circuit and the same voltage appears as the output voltage (v_o). At high frequencies, when $1/\omega C$ becomes very small, the capacitor acts as short circuit and v_o gets attenuated. On interchanging the positions of C and R, as shown in the right side of Figure 11.12, the circuit becomes a high-pass filter. At dc or low frequency of v_i, the capacitor works like open circuit and no signal reaches the output. Above a certain frequency, the capacitor starts working like a short circuit and the input voltage is transmitted to the output terminal. Now you try on your own to justify that an LR circuit can also act as a filter.

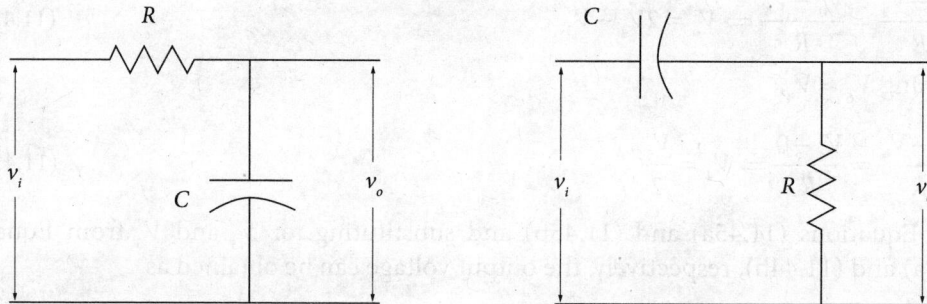

Figure 11.12 CR circuits used as low-pass (left) and high-pass (right) filters

The above-mentioned filtering property of CR circuit is further enriched with gain, buffering and other parameters in active filters using op-amps, as will be discussed in the next sections. Both the passive and active filters may be of the following types.

Low-Pass Filter: It allows dc and ac signals up to a certain cutoff frequency and attenuates all signals above that frequency.

High-Pass Filter: It transmits all signals above a cutoff frequency and attenuates all signals below that frequency.

Band-Pass Filter: These filters transmit the signals between a range (band) of frequency and attenuate all signals below and above that frequency range.

Band-Stop Filter: These attenuate the signals within a range (band) of frequency and transmit all signals beyond that band. This is also called *band-reject* filter.

Figure 11.13 schematically outlines the frequency response of these filters. In all the cases, the frequency thresholds for pass or stop is determined by the reactance of the filter components. However, a practical filter circuit, either passive or active, cannot produce such sharply demarcated pass and stop bands for the input signal. Only some approximations can be made by defining a cutoff frequency where the output becomes $1/\sqrt{2}$ times the maximum value. As indicated in Figure 11.13(e), active filters attempt to approach the ideal condition with the use of higher 'orders', which is relate to the number of reactive components in the filter. There are many categories of filter design named after Butterworth, Chebyshev and others, each one having its own qualities and restrictions. This chapter deals with simple *first order active filters* which contain only one CR network.

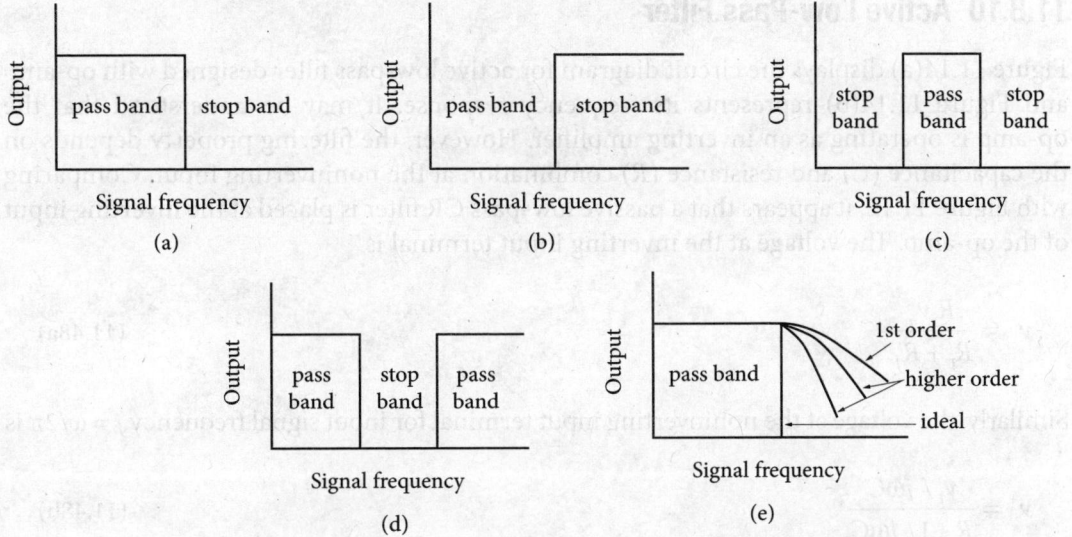

Figure 11.13 Frequency response of ideal filters, (a) low-pass, (b) high-pass, (c) band pass and (d) band stop filters, and (e) active low-pass filter response for different orders

The active filter has the following advantageous features over the passive filter.

Amplification: The output level is adjustable by changing the gain of the amplifier involved in the filter.

Signal Isolation: The high input impedance and low output impedance of the active filter enables it to act as a buffer and establish good isolation between different stages.

Wide Frequency Range: Filter design from almost dc to the order of megahertz of input frequency is possible.

Economic: The active filter is compatible with IC technology. It is rugged and inexpensive.

Compatibility: Active filters can be implemented on the same IC with other networks including digital circuits.

In spite of all the above benefits, active filters have several limitations also. For instance,

- the active filter needs a separate power supply for its own operation.
- it is not suitable for high frequencies and
- the bandwidth is less than that of passive filters because the limitations of both the passive elements and the active elements are to be taken into account.

11.8.10 Active Low-Pass Filter

Figure 11.14(a) displays the circuit diagram for active low-pass filter designed with op-amp and Figure 11.14(b) represents its frequency response. It may be understood that the op-amp is operating as an inverting amplifier. However, the filtering property depends on the capacitance (C) and resistance (R) combination at the noninverting input. Comparing with Figure 11.12, it appears that a passive low-pass CR filter is placed at the inverting input of the op-amp. The voltage at the inverting input terminal is

$$v^- = \frac{R_1 v_o}{R_1 + R_f} \tag{11.48a}$$

Similarly, the voltage at the noninverting input terminal for input signal frequency $f = \omega/2\pi$ is

$$v^+ = \frac{v_i / j\omega C}{R + 1/j\omega C} \tag{11.48b}$$

Assuming virtual short at the op-amp input, Equations (11.48a) and (11.48b) are equated to obtain the transfer function (gain) of the filter as

$$\frac{v_o}{v_i} = \frac{1 + R_f / R_1}{1 + j\omega CR} \tag{11.49}$$

The magnitude of the gain can be expressed by

$$\left|\frac{v_o}{v_i}\right| = \frac{1 + R_f / R_1}{\sqrt{1 + (\omega/\omega_{cl})^2}} = \frac{1 + R_f / R_1}{\sqrt{1 + (f/f_{cl})^2}} \tag{11.50}$$

The term f_{cl} in Equation (11.50) is called the *lower cutoff frequency* of the active low-pass filter and is given by

$$f_{cl} = \frac{1}{2\pi CR} \tag{11.50a}$$

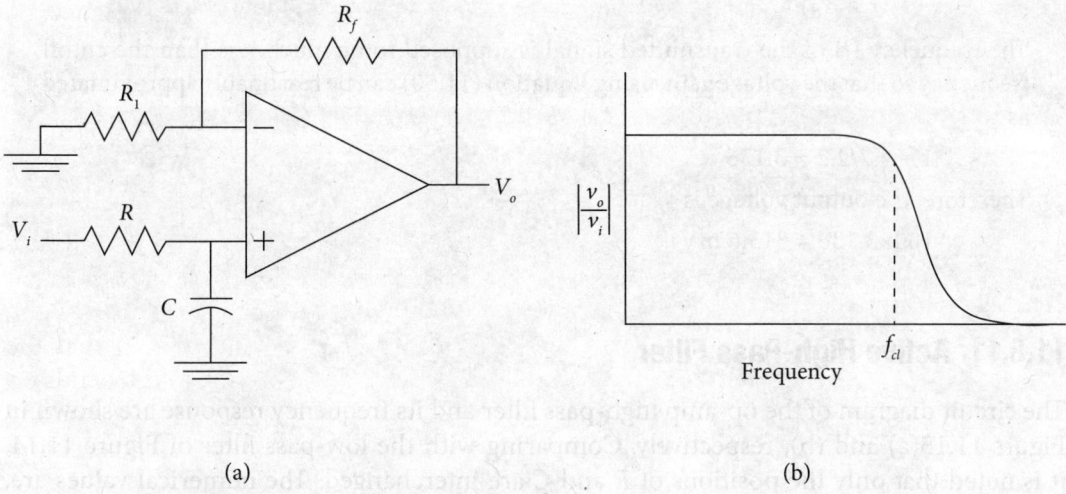

Figure 11.14 Op-amp active low-pass filter, (a) circuit diagram and (b) frequency response curve showing lower cutoff frequency (f_{cl})

The following relationships of the cutoff frequency with the input signal frequency (f) and the filter transfer function can be inferred from Equation (11.50).

(i) When $f = 0$ (dc input) or $f \ll f_{cl}$, the term f/f_{cl} becomes very small and the gain assumes a constant value of $(1 + R_f/R_1)$. Thus the low frequency signals can pass through the filter circuit without attenuation and the gain can be adjusted by varying the ratio of R_f/R_1.

(ii) When the frequency (f) of the input signal becomes equal to f_{cl}, the gain reduces to $1/\sqrt{2}$ times the maximum value of $(1 + R_f/R_1)$.

(iii) When $f \gg f_{cl}$, the term f/f_{cl} in the denominator blows up and the gain gets diminished, which indicates that the high frequency signals cannot transmit through this filter and the lower cutoff frequency (f_{cl}) represents the threshold value.

Example 11.5

A sinusoidal voltage of amplitude 100 mV and variable frequency is applied to the low-pass filter of Figure 11.14(a). If $R = 1$ kΩ, $C = 0.022$ µF, $R_1 = 2.2$ kΩ and $R_f = 4.7$ kΩ, determine the cutoff frequency for the input voltage and the amplitude of the output voltage transmitted through the filter.

Solution

Using Equation (11.50a), the cutoff frequency is

$$1/(2\pi \times 1 \times 10^3 \times 0.022 \times 10^{-6}) = 7.23 \text{ kHz}$$

> The frequency (f) of the transmitted signal is supposed to be much less than the cutoff frequency so that the voltage gain, using Equation (11.50) can be reasonably approximated as
>
> $$1 + 4.7/2.2 = 3.136$$
>
> Therefore, the output voltage is
>
> $$100 \times 3.136 = 313.6 \text{ mV}$$

11.8.11 Active High-Pass Filter

The circuit diagram of the op-amp high-pass filter and its frequency response are shown in Figure 11.15(a) and (b), respectively. Comparing with the low-pass filter of Figure 11.14, it is noted that only the positions of R and C are interchanged. The numerical values are, of course, different. The voltage (v^-) at the inverting input terminal of the high-pass filter is, obviously the same as that of Equation (11.48a). The voltage at the noninverting input terminal is different, given by

$$v^+ = \frac{Rv_i}{R + 1/j\omega C} \qquad (11.51)$$

Assuming virtual short at the op-amp input, the gain or transfer function of the high-pass filter can be obtained as

$$\frac{v_o}{v_i} = \frac{1 + R_f/R_1}{1 + 1/j\omega CR} \qquad (11.52)$$

The magnitude of the gain is given by

$$\left|\frac{v_o}{v_i}\right| = \frac{1 + R_f/R_1}{\sqrt{1 + (f_{cu}/f)^2}} \qquad (11.53)$$

The term f_{cu} in Equation (11.53) is called the *upper cutoff frequency* of the active low-pass filter and is given by

$$f_{cu} = \frac{1}{2\pi CR} \qquad (11.53a)$$

Comparing Equations (11.53a) and (11.50a) it is realized that the lower and upper cutoff frequencies of the active filter have similar expressions but the values of R, C and ω are different. Equation (11.53) leads to the following conclusions about the performance of the active high-pass filter.

(i) When the input signal frequency (f) is zero (dc input) or very low ($<< f_{cu}$), the denominator becomes very large and the gain is reduced. So the low frequency signals get attenuated.

(ii) When $f >> f_{cu}$, the f_{cu}/f term becomes negligibly small and the gain attains a constant value of $(1 + R_f/R_1)$. Thus the signals of higher frequencies can transmit through the high-pass filter, f_{cu} being a threshold value. The gain at high frequencies can be adjusted by changing the R_f/R_1 ratio.

(iii) When $f = f_{cu}$, the gain becomes $1/\sqrt{2}$ times the maximum value of $(1 + R_f/R_1)$.

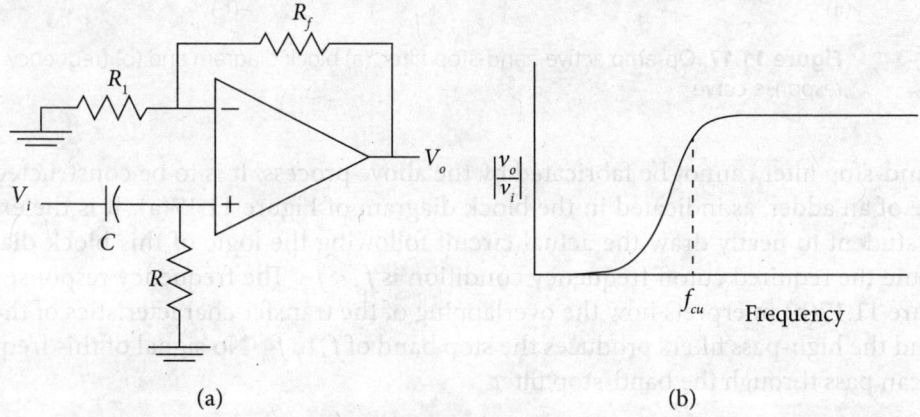

Figure 11.15 Op-amp active high-pass filter, (a) circuit diagram and (b) frequency response curve showing the upper cutoff frequency (f_{cu})

11.8.12 Active Band-Pass and Band-Stop Filters

A low-pass and a high-pass filter can be cascaded, as shown in Figure 11.16(a), to constitute a band-pass filter. The stipulation is $f_{cl} > f_{cu}$. Because of this condition, the transfer characteristics of the low-pass and the high-pass filter get overlapped to produce the pass band of f_{cu} to f_{cl}, as sketched in Figure 11.16(b).

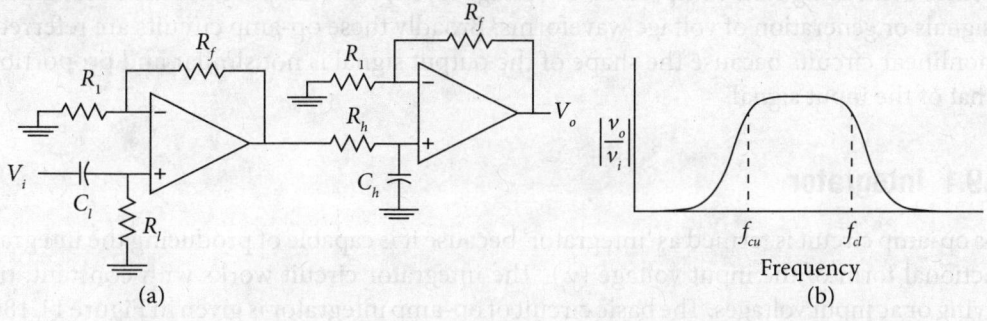

Figure 11.16 Op-amp active band-pass filter, (a) circuit diagram and (b) frequency response curve

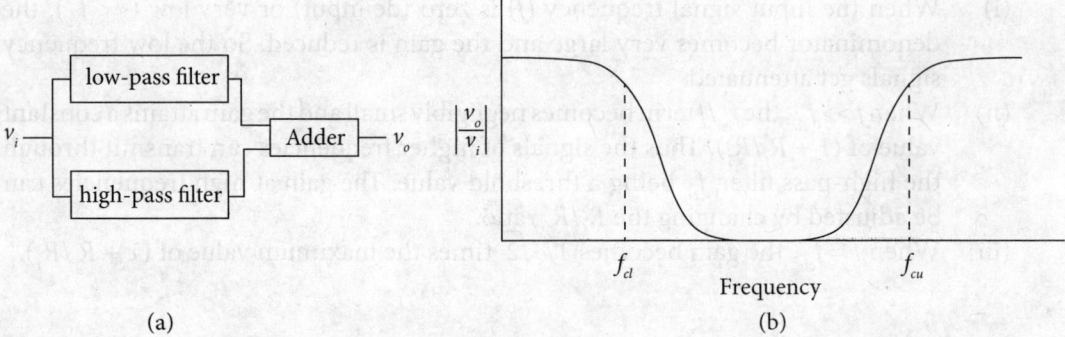

Figure 11.17 Op-amp active band-stop filter, (a) block diagram and (b) frequency response curve

The band-stop filter cannot be fabricated by the above process. It is to be constructed with the use of an adder, as indicated in the block diagram of Figure 11.17(a). It is the exercise of the student to neatly draw the actual circuit following the logic of this block diagram. This time the required cutoff frequency condition is $f_{cl} < f_{cu}$. The frequency response curve of Figure 11.17(b) interprets how the overlapping of the transfer characteristics of the low-pass and the high-pass filters produces the stop band of f_{cl} to f_{cu}. No signal of this frequency range can pass through the band-stop filter.

11.9 Nonlinear Op-Amp Circuits

The linear applications of op-amp mentioned in the previous sections were aimed mainly at the amplification of the input signal in a controlled manner. There were frequency-selective output or limiting of amplitude at the output but the input waveform was preserved at the output. Now some other applications of op-amp, namely integrator, differentiator, comparator, Schmitt trigger, logarithmic amplifier and precision rectifier are being introduced where the amplification remains at the background but the prime objective becomes something else: comparison of voltage levels, wave shaping, rectification of small ac signals or generation of voltage waveforms. Broadly these op-amp circuits are referred to as nonlinear circuits because the shape of the output signal is not similar and proportional to that of the input signal.

11.9.1 Integrator

This op-amp circuit is termed as 'integrator' because it is capable of producing the integrated functional form of the input voltage (v_i). The integrator circuit works with constant, time varying or ac input voltages. The basic circuit of op-amp integrator is given in Figure 11.18(a). It resembles an inverting amplifier with the feedback resistor replaced by a capacitor (C). The virtual ground at the inverting input terminal exists in this circuit also. Therefore, the input current (i_i) flowing through the resistor (R_1) connected to the inverting input terminal flows through the feedback capacitor also. The output voltage (v_o) for a time interval 0 to t can be expressed in terms of the charge (q) across the capacitor as

$$v_o = -\frac{q}{C} = -\frac{1}{C}\int_0^t i\,dt \qquad (11.54)$$

Putting $i_i = v_i/R_1$ in Equation (11.54), the output voltage is obtained as

$$v_o = -\frac{1}{CR}\int_0^t v_i\,dt \qquad (11.55)$$

Equation (11.55) denotes that the output voltage of the integrator is proportional to the time integral of the input voltage. This is another example of mathematical operation executed by the op-amp in terms of voltage. A few specific examples will clarify the circuit performance.

- If the input voltage be a fixed dc quantity, the output would be varying linearly over the given time interval, thereby producing a ramp voltage.
- A square wave input voltage produces triangular voltage waveform.
- A sine wave input yields a cosine wave.

In all the above cases, there exists a voltage inversion and the quantity 1/CR serves as a scale multiplier. For proper functioning of the integrator, the CR time constant should be much larger than the time variation of the input signal.

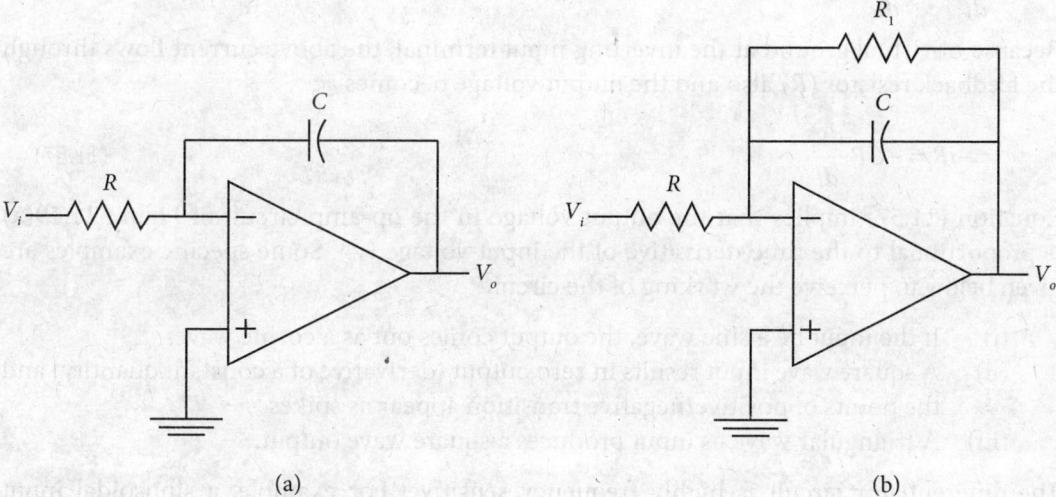

Figure 11.18 Integrator constructed with op-amp, (a) basic circuit and (b) practical circuit

The integrator circuit of Figure 11.18(a) is correct in principle but suffers from a practical drawback. For dc inputs, the capacitor acts as open circuit and there exists no negative feedback. Therefore, the input offset voltage, which is a very common property of the op-amp acts like the input voltage and the charging of the capacitor makes the output voltage saturated up to the supply voltage. To get rid of this problem, the integrator circuit of Figure 11.18(b) is used where a resistor R_1 ($\gg R$) is joined in parallel with the capacitor in the feedback circuit. Since $R_1 \gg R$ (say 10 times), the closed-loop voltage gain keeps the output offset voltage at a finite level. At the same time, this additional resistor has almost no effect on the charging of the capacitor. Therefore, when the actual alternating input voltage is present, the 'integration' process through the capacitor is not disturbed. The practical method of choosing R_1 is to make $R_1 \gg 1/\omega C$ where ω is the angular frequency of the input signal. This makes the time constant CR_1 much larger than the time period of the input signal. Sometimes a resistor $R_2 = R_1 // R$ is connected between the noninverting input and the ground in order to minimize the effect of offset.

11.9.2 Differentiator

This op-amp circuit can differentiate an input voltage signal by producing its time derivative. Figure 11.19(a) displays the basic op-amp differentiator circuit. Comparing with the integrator of Figure 11.18(a) it may be noticed that the positions of the resistor (R) and the capacitor (C) have been interchanged in the differentiator. At any instant of time t, the charge on the capacitor is $q = Cv_i$ so that the input current is

$$i = \frac{dq}{dt} = C\frac{dv_i}{dt} \tag{11.56}$$

Because of virtual ground at the inverting input terminal, the above current flows through the feedback resistor (R) also and the output voltage becomes

$$v_o = -iR = -CR\frac{dv_i}{dt} \tag{11.57}$$

Equation (11.57) implies that the output voltage in the op-amp circuit of Figure 11.19(a) is proportional to the time derivative of the input voltage (v_i). Some specific examples are given below to perceive the working of the circuit.

(i) If the input be a sine wave, the output comes out as a cosine wave.
(ii) A square wave input results in zero output (derivative of a constant quantity) and the points of positive/negative transition appear as spikes.
(iii) A triangular wave as input produces a square wave output.

The differentiator circuit is highly frequency sensitive. For example, a sinusoidal input of $V_o\sin\omega t$ results in the output of $-\omega CRV_o\cos\omega t$, which increases with frequency. The differentiator circuit amplifies high frequency noise and switching spikes. This situation can be counterbalanced by joining a resistor R_1 ($\ll R$) in series with C, as shown in Figure 11.19(b). It limits the high frequency gain and makes the circuit operating properly below the limiting frequency of $1/(2\pi CR_1)$.

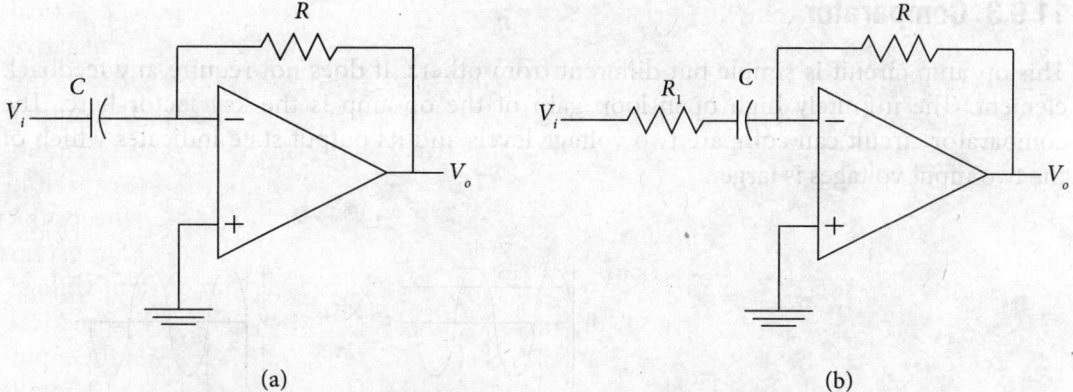

Figure 11.19 Differentiator constructed with op-amp, (a) basic circuit and (b) practical circuit

Example 11.8

An input signal (mV) given by $v_i = 10\sin1000\pi t$ is applied to the integrator of Figure 11.18(a) having $R = 100$ kΩ and $C = 0.1$ μF. Determine the output voltage.

Solution

Using Equation (11.55), the output voltage is

$$-\frac{1}{100\times10^3 \times 0.1\times10^{-6}} \int_0^t 10\sin1000\pi t\, dt$$

$$= -\frac{10^3}{1000\pi}\left[-\cos1000\pi t\right]_0^t$$

$$= \frac{1}{\pi}(\cos1000\pi t - 1)$$

Example 11.9

The input voltage of the differentiator circuit of Figure 11.19(a) increases linearly at a rate of 1 volt per millisecond. If $R = 1$ kΩ and $C = 0.01$ μF, determine the nature of the output voltage.

Solution

Using Equation (11.57), the output voltage is

$$-0.01 \times 10^{-6} \times 1 \times 10^3 \times (1\text{V}/1\text{ms}) = -0.01 \text{ V}$$

Thus a square pulse of -10 mV amplitude and 1 ms duration is obtained.

11.9.3 Comparator

This op-amp circuit is simple but different from others. It does not require any feedback element. The infinitely high open-loop gain of the op-amp is the key factor here. The comparator circuit can compare two voltage levels and its output state indicates which of the two input voltages is larger.

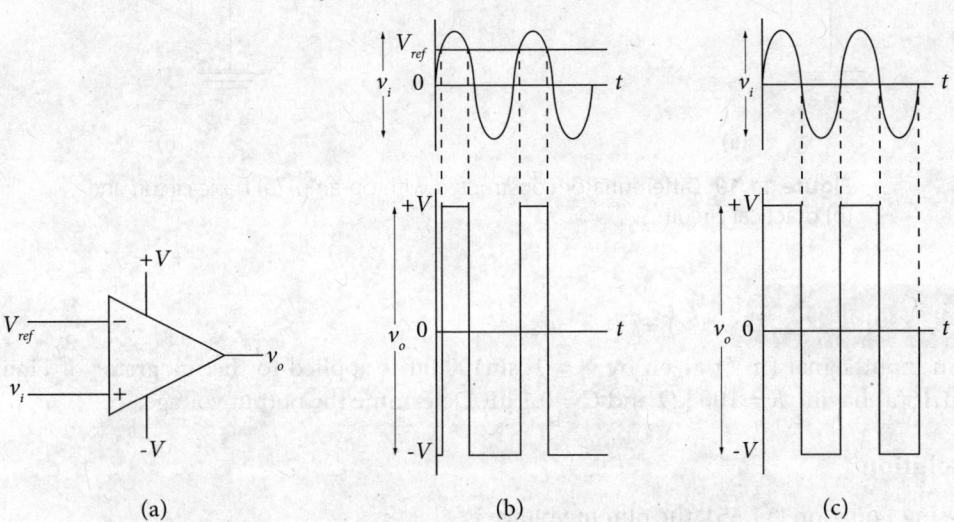

Figure 11.20 Op-amp used as comparator, (a) circuit diagram, (b) input-output variation with respect to reference voltage (V_{ref}), and (c) zero-crossing detector

Figure 11.20(a) introduces the op-amp comparator. The op-amp has supply voltages of $+V$ and $-V$ and a fixed reference voltage (V_{ref}) is connected to the inverting input terminal. The input voltage (v_i) is applied to the noninverting input terminal. The following two cases may occur when v_i is varied.

(i) So long as v_i is less than V_{ref}, the output voltage (v_o) gets fixed at $-V$ because of the very high open-loop voltage gain of the op-amp.
(ii) As soon as v_i exceeds V_{ref}, the output turns to $+V$ because of the same reason.

The mode of operating of the comparator is explained in Figure 11.20(b) with a sinusoidal input voltage. The input (v_i) varies continuously but the output (v_o) switches state only on crossing the reference voltage level thereby causing an asymmetric square wave of $\pm V$.

A special case of comparator operation, shown in Figure 11.20(c) is called the *zero-crossing detector* when the reference voltage is made zero. The inverting input is grounded. When v_i just crosses the zero level and tends to become slightly positive, v_o switches to $+V$ and when v_i crosses again the zero level and becomes slightly negative, v_o becomes $-V$. A symmetric square wave is thus obtained.

11.9.4 Schmitt Trigger

This is actually an improved version of the op-amp comparator named after American biomedical engineer O. H. Schmitt (1913–1998). The conventional comparator illustrated above has the shortcoming that any possible noise present at the input may cause unwanted crossing of the reference level and cause false switching of the output. The Schmitt trigger, as will be explained, can eliminate the effect of noise.

Figure 11.21(a) depicts the circuit diagram of Schmitt trigger. The original comparator did not contain any feedback but this circuit has a feedback at the noninverting input through the R_1–R_2 voltage divider network. The input voltage (v_i) is applied to the inverting input terminal. A fixed reference voltage V_R is connected with the voltage-divider resistor network. Using superposition principle, the feedback voltage is given by

$$v_f = \frac{R_2 v_o}{R_1 + R_2} + \frac{R_1 V_R}{R_1 + R_2} \tag{11.58}$$

The feedback factor is $R_2/(R_1+R_2)$. Let v_i starts from zero voltage and starts increasing but remains much less than v_f. The output gets saturated at $+V_{sat}$, almost equal to the positive supply voltage. Putting $v_o = +V_{sat}$ in Equation (11.58), the feedback voltage becomes

$$v_f = \frac{R_2 V_{sat}}{R_1 + R_2} + \frac{R_1 V_R}{R_1 + R_2} = V_1 \text{ (say)} \tag{11.59}$$

The following two cases of output saturation may occur, which are interpreted graphically in Figure 11.21(b).

Case-I: When v_i is being increased, so long as $v_i < V_1$, the output remains fixed at $+V_{sat}$. As soon as v_i exceeds V_1, the output switches to $-V_{sat}$ and remains fixed there on increasing v_i further. Now the feedback voltage is [Equation (11.58)]

$$v_f = \frac{-R_2 V_{sat}}{R_1 + R_2} + \frac{R_1 V_R}{R_1 + R_2} = V_2 \text{ (say)} \tag{11.60}$$

Case-II: When v_i is being decreased, so long as $v_i > V_2$, the output remains constant at $-V_{sat}$. When v_i just crosses V_2 and tends to decrease further, the output switches again to $+V_{sat}$ and remains constant there on decreasing v_i further. The situation is already described by Equation (11.59).

Thus the output switching takes place at two different input voltage values: at V_1 while increasing and at V_2 when decreasing. These two are separated by [Eqs (11.59) and (11.60)]

$$V_1 - V_2 = \frac{2 R_2 V_{sat}}{R_1 + R_2} \tag{11.61}$$

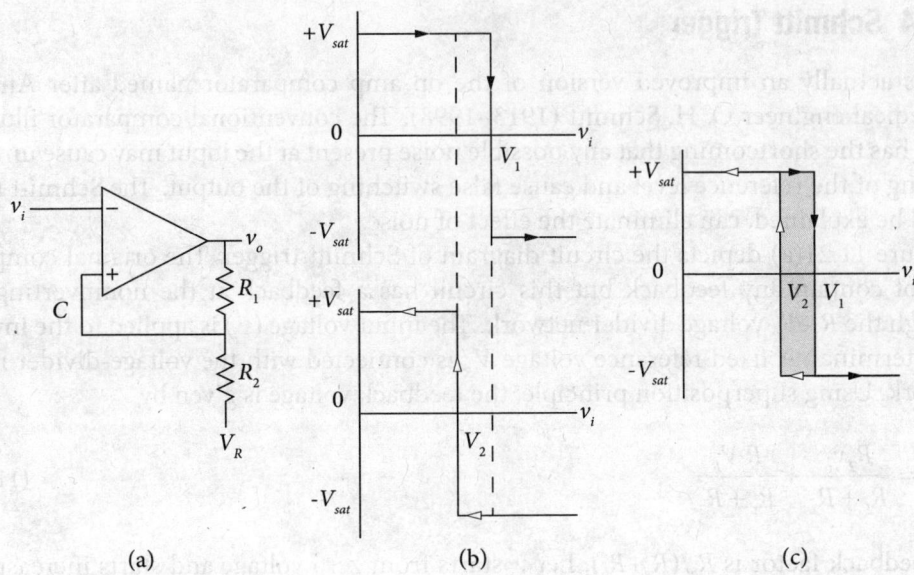

Figure 11.21 Schmitt trigger: (a) circuit diagram, (b) transfer characteristics for increasing and decreasing input voltage, and (c) hysteresis (shaded) shown by superimposing the two transfer characteristic curves

Figure 11.21(c) illustrates the phenomenon graphically by superimposing the two curves of Figure 11.21(b). A hysteresis-like (shaded) closed region is obtained, which indicates the prevention of noise from causing false switching. To affect the circuit, any noise voltage must have a magnitude larger than $(V_1 - V_2)$, which is seldom possible.

11.9.5 Logarithmic Amplifier

This op-amp circuit is different from all the previous applications in the sense that the feedback element is an active device: a diode. The circuit diagram of the logarithmic amplifier is given in Figure 11.22. The input voltage (V_i) is applied to the inverting terminal through the resistor R. The circuit is basically a sort of inverting amplifier where the virtual ground exists and the input current $I = V_i/R$ flows through the diode also. Therefore,

$$I = I_o \exp\left(\frac{qV_f}{kT}\right) \tag{11.62}$$

In Equation (11.62), V_f is the voltage across the forward biased diode. This is also the feedback voltage. I_o is the usual reverse saturation current. The unity term is neglected in the diode current expression of Equation (11.62) because qV_f/kT is likely to be much greater than unity. From Equation (11.62), the feedback voltage can be obtained as

$$V_f = \frac{kT}{q} \ln\left(\frac{I}{I_o}\right) = \frac{kT}{q} \ln\left(\frac{V_i}{I_o R}\right) \tag{11.63}$$

Using Equation (11.63), the output voltage can be expressed as

$$V_o = -V_f = -\frac{kT}{q}\ln\left(\frac{V_i}{I_o R}\right) \qquad (11.64)$$

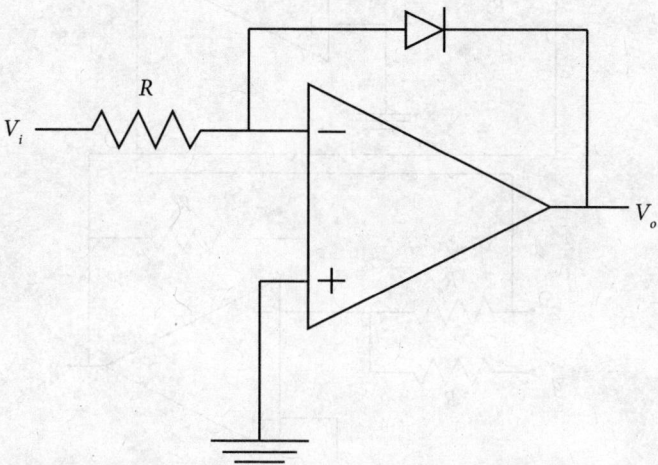

Figure 11.22 Logarithmic amplifier constructed with op-amp

Since I_o is a constant quantity at a fixed temperature, the output voltage in Equation (11.64) is proportional to the logarithm of the input voltage. This is another mathematical operation realized with the op-amp.

11.9.6 Solving Algebraic Equation

Op-amps being used as adders can execute the algebraic operation of solving simultaneous equations. An example is presented in Figure 11.23. Let the given pair of simultaneous equations be

$$ax + by = c$$
$$px + qy = r$$

Solving these, one can obtain

$$x = \frac{c}{a} - \frac{b}{a}y$$
$$y = \frac{r}{q} - \frac{p}{q}x$$

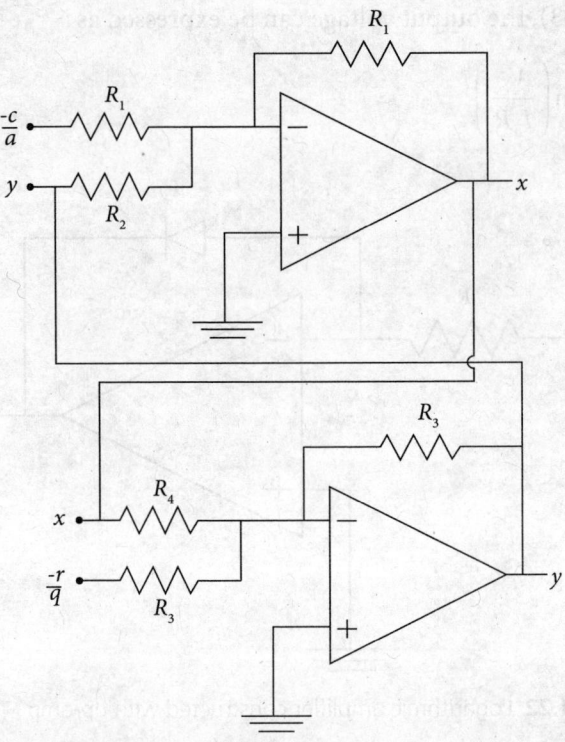

Figure 11.23 Solution of simultaneous equations using op-amps. In the present case, $R_1/R_2 = b/a$ and $R_3/R_4 = p/q$

These variables and constant terms are implemented in terms of voltages in Figure 11.23 where the two op-amps are functioning as adders and the output of one is fed to the input of the other. An external input voltage of $-c/a$ is applied to the first adder and R_1 and R_2 are selected in such a way that $R_1/R_2 = b/a$. As the result, the adder produces x as the output voltage, which is applied as an input to the second adder. This second adder has external input voltage of $-r/q$ and $R_3/R_4 = p/q$ so that it produces y as the output voltage, which is fed back to the first adder. Thus the op-amp circuit solves for x and y simultaneously in terms of voltage. A practical circuit uses precision resistors and potentiometers to adjust the input voltage and the resistance ratio.

11.9.7 Solving Differential Equation

Op-amp used as integrator can execute the algebraic operation for solving a differential equation. Figure 11.24 gives such an example. Let us consider a simple first order linear differential equation

$$\frac{dv}{dt} + Pv = 0$$ where P is a constant.

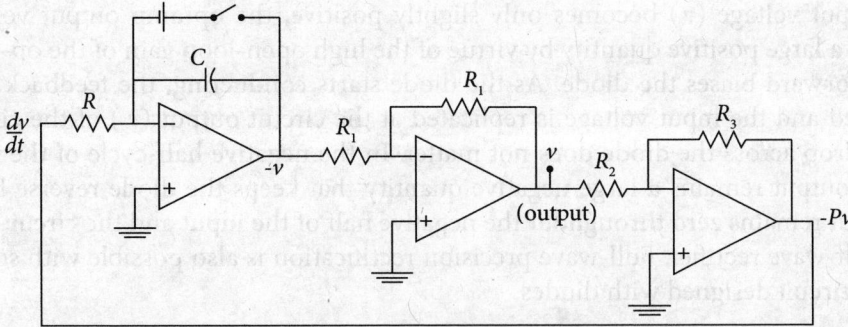

Figure 11.24 Solving differential equation using op-amp

Let us assume that dv/dt is somehow available in terms of voltage at the input of an integrator, as shown in Figure 11.24. Actually it is obtained from feedback, as will explained shortly. The capacitance (C) and the resistance (R) in the integrator are so selected that the time constant CR is 1 second and consequently the output of the integrator becomes $-v$. An inverting amplifier with equal input and feedback resistors (both R_1) inverts the phase and produces v as the output. The second inverting amplifier with $R_3/R_2 = P$ produces $-Pv$ as the output, which is fed back to the integrator input. Since, according to the given equation, $dv/dt = -Pv$, the integrator receives the proper input voltage and the output (v) is produced at the point indicated in Figure 11.24.

The initial condition of v at $t = 0$ is introduced by applying proper dc value of voltage across the capacitor by closing the switch. The output is detected keeping the switch open. In a practical circuit, the switch is a relay and the detector is a cathode ray oscilloscope.

11.9.8 Precision Rectifier

An ordinary diode rectifier (Chapter 4) cannot rectify alternating voltages of very small amplitude, comparable to the diode forward cut-in voltage. However, a diode in combination with an op-amp can build up such a rectifier that can rectify an ac signal of amplitude much smaller than the diode cut-in voltage. Such op-amp circuit is called precision rectifier. A simple example of such precision rectifier is given in Figure 11.25. The op-amp is used as a voltage follower accommodating a diode in the feedback loop.

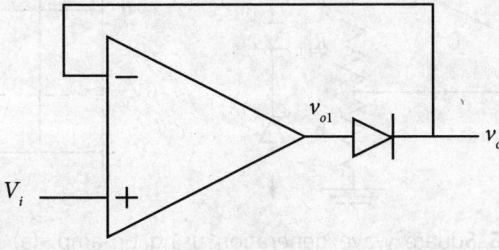

Figure 11.25 Precision rectifier fabricated with op-amp

If the input voltage (v_i) becomes only slightly positive, the op-amp output voltage (v_{o1}) becomes a large positive quantity by virtue of the high open-loop gain of the op-amp. This voltage forward biases the diode. As the diode starts conducting, the feedback path gets completed and the input voltage is replicated at the circuit output (v_o) of the circuit. The voltage drop across the diode does not matter. In the negative half-cycle of the input, the op-amp output remains a large negative quantity that keeps the diode reverse biased. So the output remains zero throughout the negative half of the input and the circuit performs like a half-wave rectifier. Full-wave precision rectification is also possible with some other op-amp circuit designed with diodes.

11.10 Op-Amp Waveform Generators

All the previously mentioned linear or nonlinear applications of op-amp are related to amplification and some input signal is needed. The op-amp can also be used for the construction of oscillators that generate a steady waveform, sinusoidal or non-sinusoidal, without any regular input signal. The properties of oscillators can be recapitulated from Chapter 10. This section illustrates some examples of op-amp oscillators.

11.10.1 Square Wave Generator

The op-amp circuit of Figure 11.26(a) produces a square wave [Figure 11.26(b)] whose frequency is determined by the capacitance (C) and the resistance (R) and the amplitude is equal to the saturation voltage of the op-amp. However, the amplitude can be adjusted by joining two Zener diodes in opposite connection, as indicated in the diagram. The amplitude is then gets limited by the Zener breakdown voltage. This circuit is a relaxation oscillator because the waveform produced is non-sinusoidal. The operating principle of the circuit is outlined below.

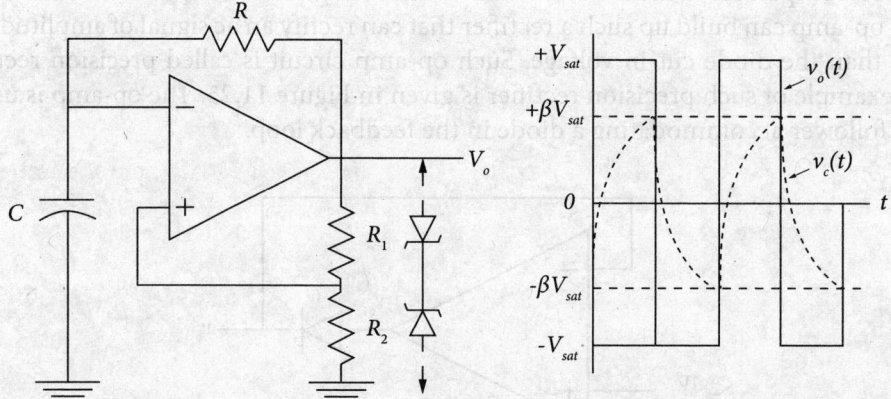

Figure 11.26 Square wave generation using op-amp: (a) circuit diagram and (b) graphical representation of the output voltage $v_o(t)$ (solid line) and the capacitor voltage $v_c(t)$ (dotted line)

When the power supply is just turned on, any stray voltage caused by noise or residual charge in the capacitor appears as a random input voltage. Depending on the polarity of this voltage, the high voltage gain of the op-amp causes the output to saturate at either $+V_{sat}$ or $-V_{sat}$, both close to the supply voltage. The output voltage finds the following two feedback paths.

(i) The capacitor starts charging through the resistor with the time constant CR thereby providing a time-varying voltage $v_c(t)$ at the inverting input terminal.

(ii) A fraction $\beta = R_2/(R_1 + R_2)$ of the output voltage is fed back to the noninverting input terminal.

Let at $t = 0$, the output voltage $v_o(t)$ be $+V_{sat}$. The voltage at the noninverting input is $+\beta V_{sat}$. When the capacitor voltage $v_c(t)$ exceeds this value while charging, the output voltage switches to $-V_{sat}$ and the feedback voltage to the noninverting input terminal becomes $-\beta V_{sat}$. The capacitor now discharges until $v_c(t)$ reaches $-\beta V_{sat}$ and just crosses it. Then the output voltage switches again to $+V_{sat}$ and the capacitor starts charging again.

The above sequence continues and a square voltage waveform is obtained at the output, as sketched in Figure 11.26(b). The exponential variation of the capacitor voltage is also drawn there. If the two Zener diodes of equal breakdown voltage (V_z) were joined, the output remains confined to $\pm V_z$ instead of $\pm V_{sat}$. To determine the time period (T) of the square wave, the instantaneous voltage across the capacitor can be derived as

$$v_c(t) = V_{sat}[1 - (1+\beta)e^{-t/CR}] \quad (11.65)$$

At time $t = T/2$, $v_c = +\beta V_{sat}$ and after another $T/2$ time it comes back to $-\beta V_{sat}$. Putting $t = T/2$ and $v_c = \beta V_{sat}$ in Equation (11.65) and simplifying,

$$T = 2CR \ln \frac{1+\beta}{1-\beta} \quad (11.66)$$

Equation (11.66) is the formula for the time period of the op-amp square wave generator. This is actually an astable multivibrator realized with op-amp. The waveform may be compared with that of the transistor multivibrator of Chapter 10 and the 555 timer of Chapter 15.

11.10.2 Triangular Wave Generator

This op-amp circuit integrates a square wave voltage and thus produces the triangular voltage waveform. The circuit diagram is given in Figure 11.27(a). The left-side op-amp circuit is a comparator that generates a square wave with amplitude of $\pm V_{sat}$. The right-side op-amp circuit is an integrator that converts the square voltage to triangular voltage waveform. This triangular voltage, in turn drives the comparator. Figure 11.27(b) is a schematic representation of both the waveforms.

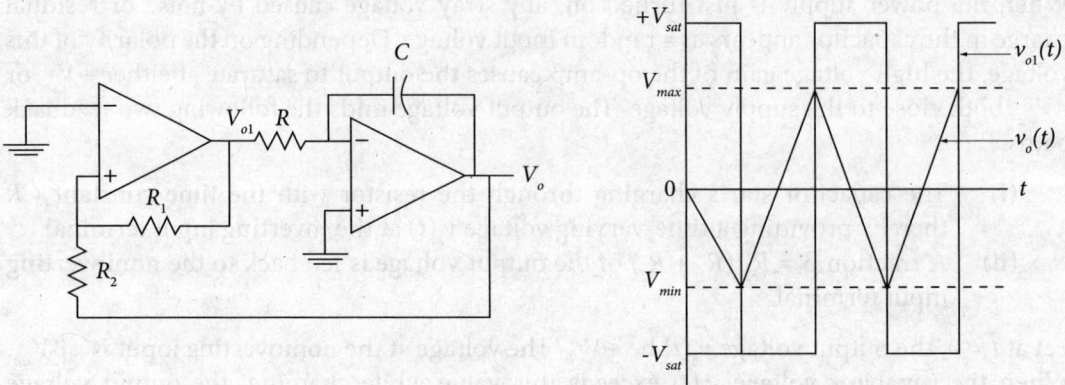

Figure 11.27 Triangular wave generation with op-amps: (a) circuit diagram and (b) output waveforms

Let v_{o1} and v_o be the instantaneous value of the comparator output and the integrator output, respectively. The voltage at the noninverting input terminal of the comparator can be expressed by the use of superposition principle as

$$v_1 = \frac{R_2 v_{o1}}{R_1 + R_2} + \frac{R_1 v_o}{R_1 + R_2} \tag{11.67}$$

The following two cases may take place.

Case-I: When v_{o1} is $+V_{sat}$, the integrator output (v_o) decreases linearly until the voltage v_1 becomes zero. From Equation (11.67), the minimum value of v_o is

$$v_{min} = -\frac{R_2}{R_1} V_{sat} \tag{11.68}$$

Case-II: When v_o crosses the above value, the comparator output (v_{o1}) switches to $-V_{sat}$. Now the integrator output increases linearly until v_1 becomes zero again. From Equation (11.67), the maximum value of v_o is

$$v_{max} = \frac{R_2}{R_1} V_{sat} \tag{11.69}$$

Thus the integrator output voltage varies linearly between v_{max} and v_{min} and a triangular wave is obtained, as sketched in Figure 11.27(b). The simultaneous variation of the comparator output (square wave) is also shown there. The time period of the triangular wave can be deduced as

$$T = \frac{4CRR_2}{R_1} \tag{11.70}$$

11.10.3 Sine Wave Generator

The circuit diagram of the op-amp sine wave generator is shown in Figure 11.28. This is actually a phase-shift oscillator fabricated with op-amp and may be compared with the BJT phase-shift oscillator of Chapter 10.

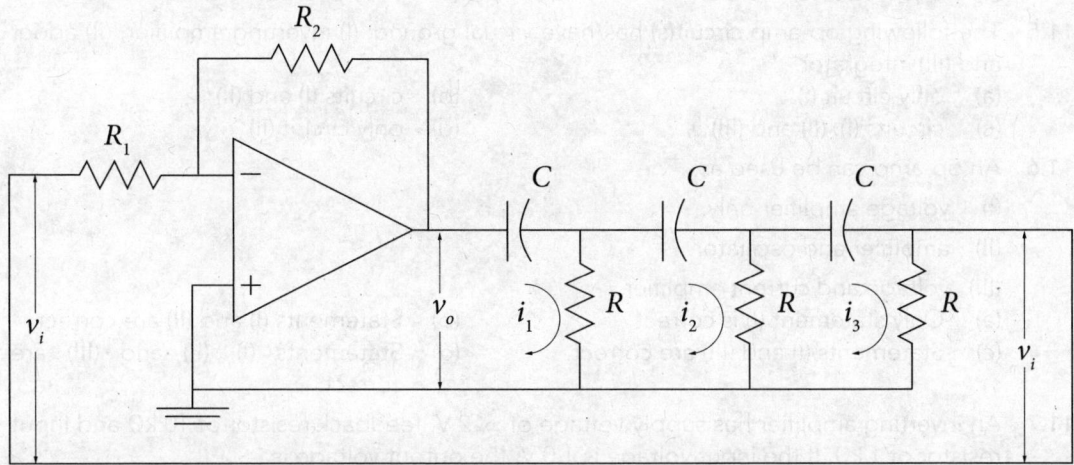

Figure 11.28 Op-amp phase-shift oscillator to generate sine wave

The op-amp inverting amplifier causes 180° phase shift of the output with respect to the input. Another 180° phase shift is caused by the feedback network comprising three identical sets of capacitor (C) resistor (R). The voltage (v_i) across the resistor R acts as the input, as denoted in the figure. Since there is 360° or no phase shift, the Barkhausen criterion is satisfied. Applying Kirchhoff's voltage law to the three CR loops, simplifying the algebraic equations and taking the real parts one can derive the frequency of the sine wave generated by the phase-shift oscillator as

$$f = \frac{1}{2\pi\sqrt{6}CR} \tag{11.71}$$

Multiple Choice-Type Questions and Answers

11.1 An inverting amplifier constructed with op-amp has .... input resistance and .... output resistance.
   (a) high, low   (b) low, high   (c) low, low   (d) high, high

11.2 A noninverting amplifier constructed with op-amp has .... input resistance and .... output resistance.
   (a) high, low   (b) low, high   (c) low, low   (d) high, high

11.3 A device has very high input resistance and very low output resistance. It may be ....
(a) FET and BJT (b) FET and op-amp
(c) BJT and op-amp (d) op-amp only

11.4 The voltage gain of an inverting amplifier is ....
(a) greater than unity (b) less than unity
(c) equal to unity (d) of any value

11.5 The following op-amp circuit(s) has/have virtual ground: (I) inverting amplifier, (II) adder and (III) integrator.
(a) only circuit (I) (b) circuits (I) and (II)
(c) circuits (I), (II) and (III) (d) only circuit (II)

11.6 An op-amp can be used as:
(I) voltage amplifier only.
(II) amplifier and oscillator.
(III) voltage and current amplifier.
(a) Only statement (I) is correct. (b) Statements (I) and (II) are correct.
(c) Statements (I) and (III) are correct. (d) Statements (I), (II) and (III) are correct.

11.7 An inverting amplifier has supply voltage of ±12 V, feedback resistor of 10 kΩ and input resistor of 1 kΩ. If the input voltage is 1.0 V, the output voltage is ....
(a) 12 V (b) −12 V (c) 10 V (d) −10 V

11.8 If, in the above example, the supply voltage is made ±15 V, the output voltage will be ....
(a) 15 V (b) −15 V (c) 10 V (d) −10 V

11.9 An op-amp circuit produces output voltage of −1 V from the input current of 1 mA. It may be ....
(a) inverting amplifier with feedback resistance of 1 kΩ
(b) inverting amplifier with voltage gain of 1000
(c) noninverting amplifier with voltage gain of 1000
(d) inverting amplifier with current gain of 1000

11.10 A voltage-controlled voltage source has negative feedback involving .... and ....
(a) input voltage and output current
(b) input voltage and output voltage
(c) input current and output voltage
(d) input current and output current

11.11 The output voltage waveforms of both integrator and differentiator look the same for .... wave input.
(a) sine (b) square
(c) triangular (d) rectangular

11.12 Negative feedback has no effect on .... distortion.
(a) nonlinear (b) harmonic (c) slew rate (d) noise

11.13 An open-loop op-amp circuit is .... for linear applications.
  (a) suitable (b) unsuitable (c) efficient (d) uncertain

11.14 The transresistance of an amplifier is the ratio of its .... to ....
  (a) input voltage, output current (b) input voltage, input current
  (c) output voltage, input current (d) output current, input voltage

11.15 The feedback resistor of an inverting amplifier has got open because of defective circuit construction. The output will be ....
  (a) zero (b) saturated
  (c) equal to the input (d) less than the input

11.16 A noninverting amplifier circuit has some constructional defect. The output is always zero. What may be the possible reason?
  (a) The feedback resistor is open. (b) The feedback resistor is shorted.
  (c) The load resistor is open. (d) The load resistor is shorted.

11.17 The following statements are given about a linear op-amp circuit.
  (I) The input waveform is preserved at the output.
  (II) The output is proportional to the input.
  (III) The op-amp does not go into saturation.
  (a) Statements (I) and (II) are correct. (b) Only statement (I) is correct.
  (c) Only statement (II) is correct. (d) Statements (I), (II) and (III) are correct.

11.18 The following statements are given about a nonlinear op-amp circuit.
  (I) The output waveform does not resemble the input waveform.
  (II) The output is not proportional to the input.
  (III) The op-amp may saturate.
  (a) Statements (I), (II) and (III) are correct. (b) Only statement (I) is correct.
  (c) Only statement (I) is correct. (d) Statements (I) and (II) are correct.

11.19 If the input to a zero-crossing detector be a triangular wave, the output becomes a ....
  (a) triangular wave (b) series of voltage spikes
  (c) rectangular wave
  (d) sine wave

11.20 The Schmitt trigger makes use of .... feedback.
  (a) positive (b) negative
  (c) regenerative (d) no

11.21 If the parameters y and x are related by $y = \log(x)$, then the circuit that can be used to produce an output voltage V_o varying linearly with x is

[NET Dec. 2015]

Hints. In the logarithmic amplifier of Figure 11.22, if the positions of the C and the R are interchanged, the circuit acts as an exponential amplifier.

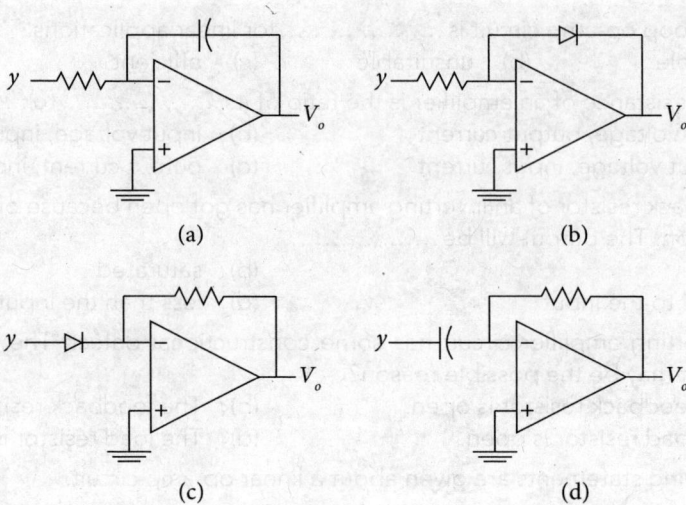

Figure 11.29

11.22 If a constant voltage +V is applied to the input of the op-amp circuit of Figure 11.30 for a time t, the output voltage will approach ....
(a) +V exponentially
(b) −V exponentially
(c) +V linearly
(d) −V linearly

[JAM 2016]

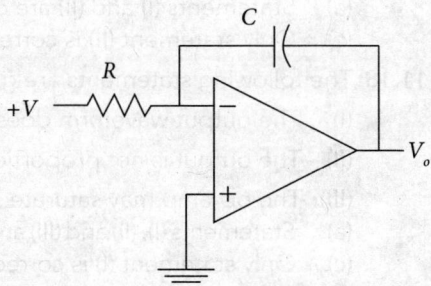

Figure 11.30

11.23 If the op-amp of Figure 11.31 is ideal, the potential at point A is ....
(a) 1 V (b) 0 V (c) 5 V (d) 25 V

[JAM 2014]

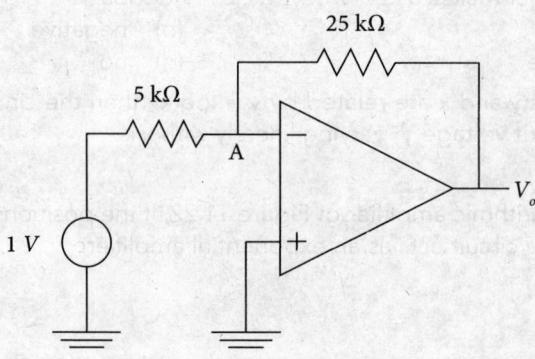

Figure 11.31

11.24 If the op-amp open-loop gain is $A = 10^5$, the feedback configuration of Figure 11.32 and the closed-loop gain (A_f) are
  (a) series-shunt, $A_f = 9$
  (b) series-series, $A_f = 10$
  (c) series-shunt, $A_f = 10$
  (d) shunt-shunt, $A_f = 10$

[GATE 2015]

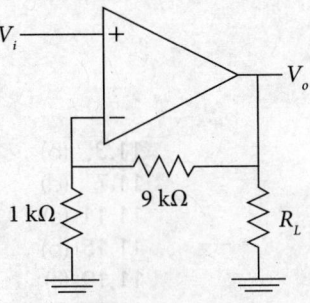

Figure 11.32

Hints. This is voltage–series feedback with gain = 1 + 9/1 = 10. The word 'series' means joining the feedback signal in series with the input signal voltage and the word 'shunt' refers to connecting the feedback signal in parallel (shunt) with the input current source. Here the feedback voltage at the input is in series and the sampling at the output for feedback is done in parallel.

11.25 In the ideal op-amp circuit of Figure 11.33, $R_1 = 3$ kΩ, $R_2 = 1$ kΩ and $v_i = 0.5\sin\omega t$ (in volt). Which of the following statements are true?
  (a) The current through R_1 is equal to the current through R_2.
  (b) The potential at P is $R_2 v_o / R_1$.
  (c) The amplitude of v_o is 2V.
  (d) The output voltage v_o is in phase with v_i.

[JAM 2015]

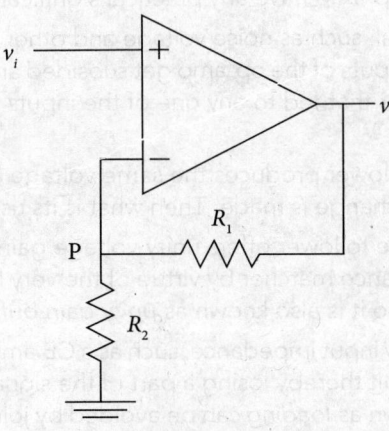

Figure 11.33

Hints. Noninverting amplifier, voltage gain = 1+3/1 = 4.

11.26 For optimal performance of an op-amp based current-to-voltage converter circuit, the input and output impedance should be
(a) low input impedance and high output impedance
(b) low input impedance and low output impedance
(c) high input impedance and high output impedance
(d) high input impedance and low output impedance

[NET Jun. 2019]

Answers

11.1 (c)	11.2 (a)	11.3 (b)	11.4 (d)
11.5 (c)	11.6 (b)	11.7 (d)	11.8 (d)
11.9 (a)	11.10 (b)	11.11 (a)	11.12 (c)
11.13 (b)	11.14 (c)	11.15 (b)	11.16 (d)
11.17 (d)	11.18 (a)	11.19 (c)	11.20 (a)
11.21 (c)	11.22 (d)	11.23 (b)	11.24 (c)
11.25 (c) and (d)	11.26 (b)		

Reasoning-Type Questions and Answers

11.1 Can all op-amp circuits be made with the same op-amp?

Ans. No, similar to the different transistors meant for voltage or power amplification and low and high frequency operations, op-amps are also manufactured in different categories. Those have wide difference in power rating and bandwidth and variance in optimization for input offsets, noise, and other parameters. There are commercially available op-amps for almost all analog applications. However, there are some general purpose op-amps, such as IC 741, which can be used in the fabrication of almost any op-amp circuit up to a moderate range (several kHz) of frequencies of the input signal.

11.2 Does the CMRR of an op-amp have any practical significance?

Ans. Yes, any stray signal, such as noise voltage and other electromagnetic interferences picked up by both the inputs of the op-amp get subsided and are not amplified along with the original signal, either that fed to any one of the inputs or the difference between the inputs.

11.3 The op-amp voltage follower produces the same voltage at the output as the input. No amplification or phase change is made. Then what is its use?

Ans. Though the voltage follower offers unity voltage gain and no phase shift, it can act as a buffer or an impedance matcher by virtue of the very high input resistance and very low output resistance. So it is also known as *unity gain buffer*.

An amplifier of low input impedance, such as a CB amplifier joined to a signal source may act like a short circuit thereby losing a part of the signal at the input of the amplifier. This phenomenon, known as *loading* can be avoided by joining a buffer (device with high input resistance and low output resistance) between the signal source and the amplifier.

11.4 In the inverting amplifier of Figure 11.4(a), if the resistances of R_1 and R_f are made equal, the magnitude of the voltage gain becomes unity [Equation (11.22)]. In such a case, can the circuit be treated as a voltage follower?

Ans. No, the unity gain is not the sufficient condition for being a buffer. The input resistance of the present circuit is very low. Moreover the output is in opposite phase with respect to the input.

11.5 Using op-amps, one can fabricate circuits capable of executing addition (figures 11.8 and 11.9) and subtraction (Figure 11.10) processes in terms of voltage. Can multiplication be achieved with the op-amp?

Ans. Yes, the inverting amplifier [Figure 11.4(a)] provides with such an application and the process is known as *scale changer*. The resistance ratio (R_f/R_1) serves as the multiplication factor. Using $R_f = nR_1$, the input voltage can be multiplied n-times at the output.

The above process introduces a phase change of 180° at the output. That can be eliminated by using another inverting amplifier of unity (R_f/R_1) ratio. Such an op-amp circuit that changes the phase without disturbing the amplitude is known as *phase shifter*.

11.6 There is a category of filter termed as 'all pass' that passes signals of all frequencies; ideally zero to infinity. Why is such a filter employed instead of receiving the signal directly?

Ans. By using the filter circuit, especially the active filter, the amplitude and phase of the output can be adjusted. In addition, the filter isolates the signal source and the load and establishes a suitable impedance matching between these two. Such facilities are not available with direct receiving of the signal.

Solved Numerical Problems

11.1 Explain the mode of working of the circuit of Figure 11.34 when the connection is made with point A and point B, respectively.

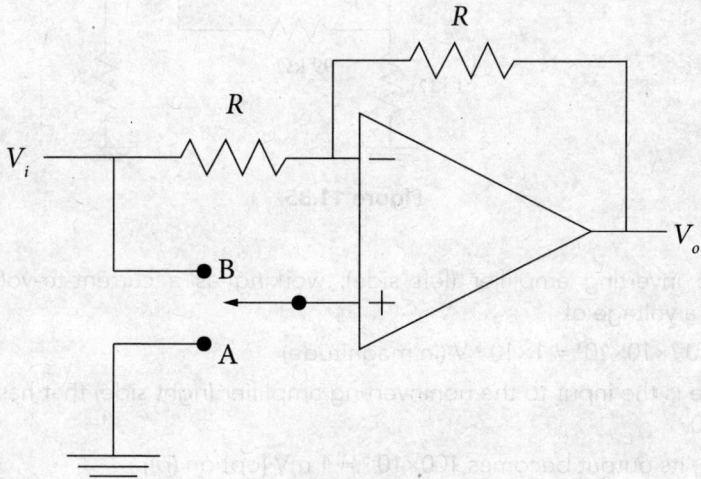

Figure 11.34

Soln. When the connection is made with A (ground), the circuit becomes an inverting amplifier with voltage gain

$$-R/R = -1$$

When the connection is made with B, the input signal gets connected to both the inverting and noninverting inputs.

For the inverting path, the gain is still

$$-R/R = -1$$

For the noninverting path, the voltage gain is

$$1 + R/R = 2$$

The net voltage gain is

$$-1 + 2 = 1$$

Thus the circuit produces an output voltage of magnitude equal to that of the input voltage in both cases but the phase can be either the same or opposite by 180°.

11.2 What is the voltage at the output (V_o) of the amplifier circuit of Figure 11.35 out of the following options?
(a) 1 V
(b) 1 mV
(c) 1 µV
(d) 1 nV

[JEST 2015]

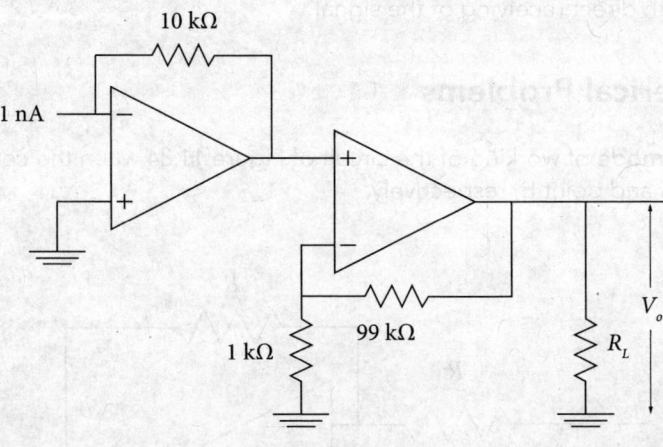

Figure 11.35

Soln. The inverting amplifier (left side), working as a current-to-voltage converter, produces a voltage of

$$1 \times 10^{-9} \times 10 \times 10^3 = 1 \times 10^{-5} \text{ V (in magnitude)}$$

The above is the input to the noninverting amplifier (right side) that has the gain of 1 + 99/1 = 100.

Therefore, its output becomes $100 \times 10^{-5} = 1$ mV [option (b)]

11.3 Determine the output voltage (V_o) of the op-amp circuit of Figure 11.36.

[JAM 2016]

Soln. Clearly, it is a noninverting adder, similar to that of Figure 11.9. Therefore, using Equation (11.39), the output voltage (V_o) is

$$1 + 2 + 3 = 6 \text{ V}$$

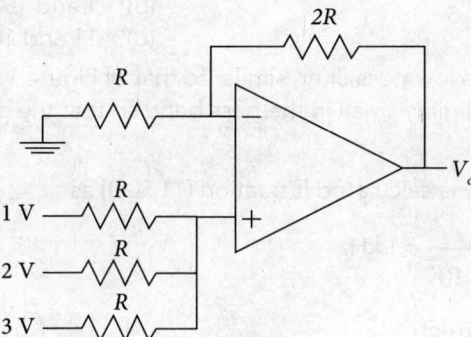

Figure 11.36

11.4 For the given circuit and the given input voltage (v_i) of Figure 11.37, which of the following waveforms correctly represents the output voltage (v_o)?

[NET Jun. 2016]

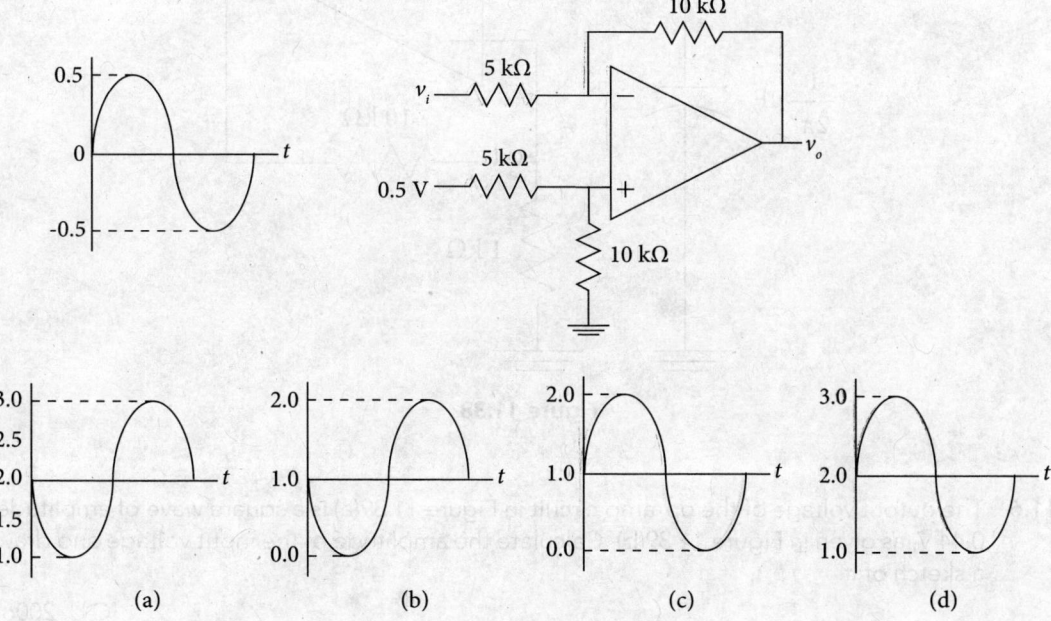

Figure 11.37

Soln. The circuit is a differential amplifier with output [Equation (11.42)]

$(10/5)(0.5 - v_i)$
$= 1 - 2 \times 0.5 \sin\theta$
$= 1 - \sin\theta$ [option (b)]

11.5 In the op-amp circuit of Figure 11.38, the dc gain and the cutoff frequency, respectively, are ....
(a) 1 and 1 kHz
(b) 1 and 100 kHz
(c) 11 and 1 kHz
(d) 11 and 100 Hz

Soln. This is an active low-pass filter, similar to that of Figure 11.14(a). In Equation (11.50), the (f/f_{cl}) term is negligibly small in the pass band so that the gain is

$1 + 10/1 = 11$

The cutoff frequency is calculated [Equation (11.50a)] as

$$\frac{1}{2\pi \times \frac{1}{2\pi} \times 10^{-6} \times 1 \times 10^{3}} = 1 \text{ kHz}$$

Thus option (c) is correct.

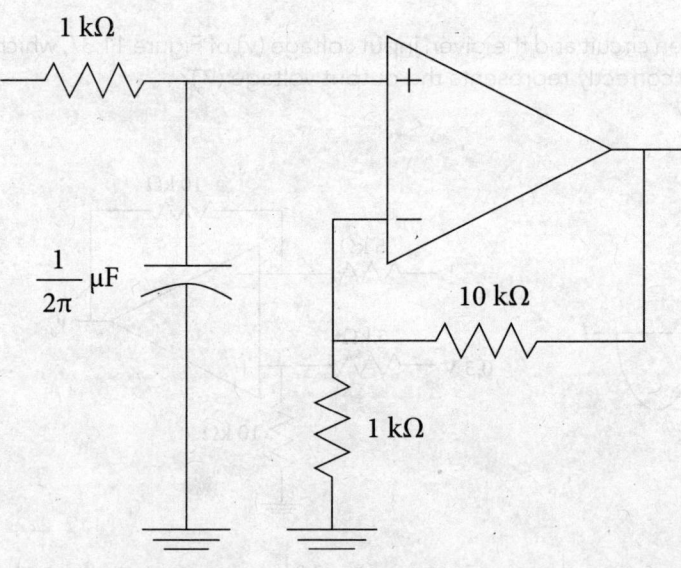

Figure 11.38

11.6 The output voltage of the op-amp circuit in Figure 11.39(a) is a square wave of amplitude 0.44 V, as given in Figure 11.39(b). Calculate the amplitude of the input voltage and draw a sketch of it.

[C.U. 2006]

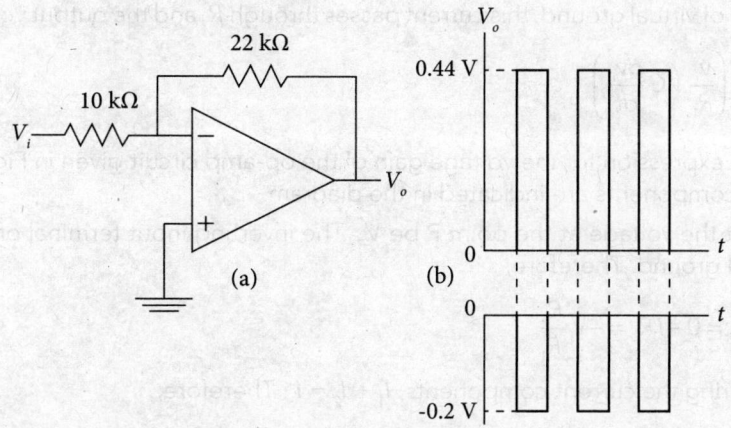

Figure 11.39

Soln. The output voltage amplitude can be related to the input and feedback resistances by

$$0.44 = -(22/10)V_i$$

Therefore, $V_i = -0.2$ V.

When the output voltage is zero, V_i is also zero.

Thus the input voltage is a negative square pulse of amplitude 0.2 V. This is sketched below Figure 11.39(b) comparing with the output waveform.

11.7 Show that the output voltage in the circuit of Figure 11.40 is: $v_o = -\dfrac{R_2}{R_1}v_i - CR_2\dfrac{dv_i}{dt}$.

[C.U. 2010]

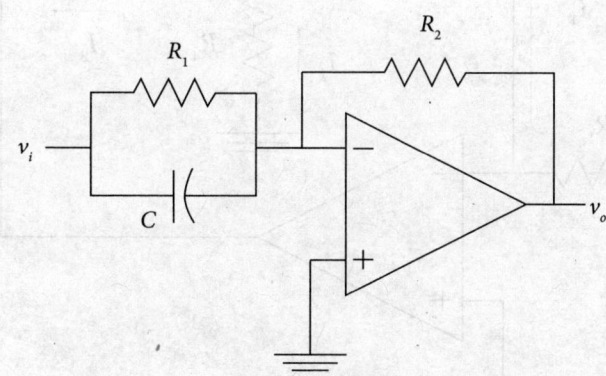

Figure 11.40

Soln. The current through R_1 is $\dfrac{v_i}{R_1}$ and the current through C is $C\dfrac{dv_i}{dt}$. Therefore, the total input current is $\dfrac{v_i}{R_1} + C\dfrac{dv_i}{dt}$.

Because of virtual ground, this current passes through R_2 and the output voltage becomes

$$v_o = -R_2\left(\frac{v_i}{R_1} + C\frac{dv_i}{dt}\right)$$

11.8 Find the expression for the voltage gain of the op-amp circuit given in Figure 11.41. The current components are indicated in the diagram.

Soln. Let the voltage at the point P be V_P. The inverting input terminal of the op-amp is at virtual ground. Therefore,

$$V_P = 0 - I_1 R_2 = -V_i \frac{R_2}{R_1} \tag{1}$$

Considering the current components, $I_1 + I_2 = I_3$. Therefore,

$$\frac{V_P - 0}{R_2} + \frac{V_P}{R_4} = \frac{V_o - V_P}{R_3}$$

$$\Rightarrow V_P\left(\frac{1}{R_1} + \frac{1}{R_2} + \frac{1}{R_3}\right) = \frac{V_o}{R_3} \tag{2}$$

From (1) and (2), the voltage gain is obtained as

$$\frac{V_o}{V_i} = -\frac{R_2}{R_1}\left(1 + \frac{R_3}{R_2} + \frac{R_3}{R_4}\right) \tag{3}$$

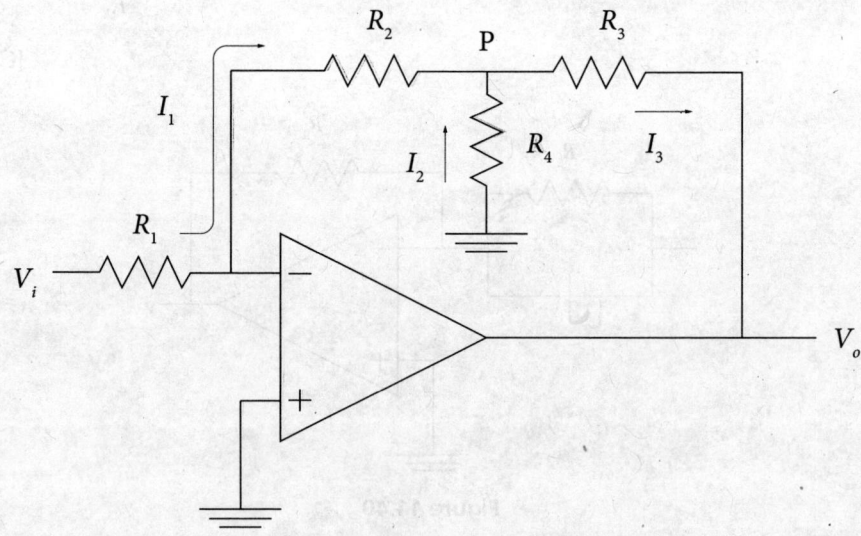

Figure 11.41

> **Note: Significance of the Circuit of Figure 11.41**
>
> The circuit of Figure 11.41 is useful when high voltage gain is required with high input resistance. A conventional inverting amplifier in such a case needs extremely high feedback resistance. For instance, if $R_1 = 10$ kΩ and the required voltage gain is 100, that calls for a large value of 1 MΩ for the feedback resistance. For such a large value of resistance, even a slight tolerance may cause a fluctuation in gain. The present circuit can adjust the ratio of R_3 to R_4 and provide with high voltage gain using feedback resistance values comparable with R_1. It may be noted that when the resistor R_4 in the circuit of Figure 11.41 is opened, $R_4 \approx \infty$ in Equation (3) above and the expression for voltage gain of a conventional inverting amplifier is obtained with feedback resistance of $(R_2 + R_3)$.

11.9 A sinusoidal signal of peak-to-peak amplitude 1 V and unknown time period is input to the circuit of Figure 11.42 for 5 seconds duration. If the counter measures a value (3E8)H then the time period of the input signal is ....
(a) 2.5 ms   (b) 4 ms   (c) 10 ms   (d) 5 ms

[NET Dec. 2015]

Soln. [Some relevant information may be given here. The counter, a digital circuit to be discussed in Chapter 15 can count the number of voltage pulses. The counting is given here in hexadecimal number system, which is explained in Chapter 13. The given number converted to decimal is 1000.]

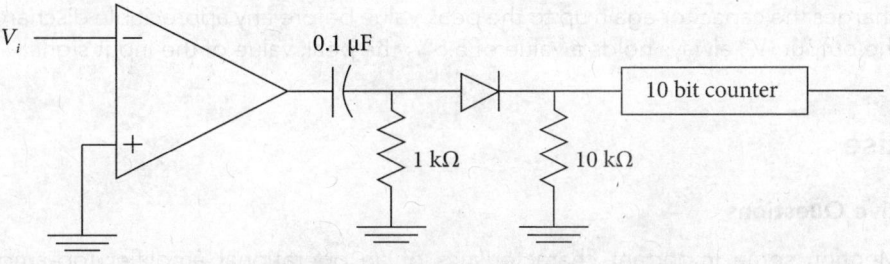

Figure 11.42

The op-amp comparator produces a square wave. The CR circuit converts it into pulses and the diode eliminates the negative pulses. The number of pulses is equal to the number of time periods.

The number of pulses in 5 seconds is 1000 so that the time period is 5/1000 = 5 ms [option (d)].

11.10 The capacitor in the op-amp circuit of Figure 11.43 has the value of 100 µF and the load resistor in the same circuit has the value of 10 kΩ. The numerical values related to the sinusoidal voltage (in volts) applied to the noninverting input terminal are given in the figure. Explain the circuit operation.

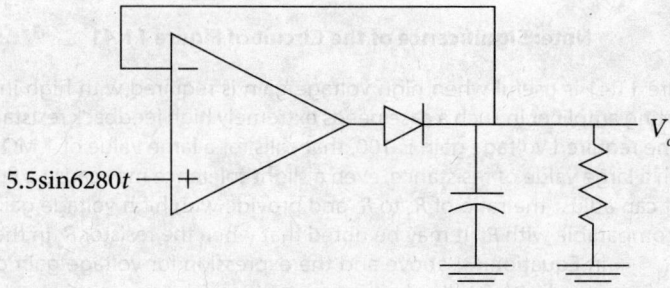

Figure 11.43

Soln. From the given data, the peak value of the sinusoidal input is 5.5 V and the frequency is 1 kHz, hence the time period is 1 ms.

The time constant of the *CR* combination is: $100\times10^{-6}\times10\times10^{3} = 1$ s, much larger than the time period of the input signal. Indeed it is a necessary condition for proper functioning of a circuit like the present one.

The given circuit is known as *active peak detector*. It can detect the peak value of low amplitude signals. When the input voltage is positive, the diode is turned on and the negative feedback produces an output resistance of almost zero value so that the capacitor charges quickly up to 5.5 V, the positive peak value of the input voltage. When the input voltage is negative, the diode is turned off and the feedback path is open. The capacitor starts discharging through the load resistor. Since the time constant is much longer than the time period of the input signal, the subsequent positive input voltage charges the capacitor again up to the peak value before any appreciable discharge. Thus the output (V_o) always holds a value of 5.5 V, the peak value of the input signal.

Exercise

Subjective Questions

1. Mention some important characteristics of an operational amplifier (op-amp). What specific operation does it perform so that it is termed as 'operational amplifier'?
2. Explain how an op-amp can be used to perform the arithmetic operation of addition. What type of feedback is used in op-amp adder?
3. What is virtual ground related to an op-amp circuit? How it is different from an actual ground?
4. Explain, with circuit diagram, the working principle of op-amp voltage follower. Mention some of its applications.
5. What is CMRR of an op-amp? How it is related to the CMRR of a differential amplifier circuit made of op-amp?
6. Draw the circuit diagram of an op-amp integrator and explain what it integrates, with respect to what and how.
7. Justify the statement, 'Noise voltages of higher frequency have more pronounced effect on op-amp differentiator circuit.'

8. Discuss the similarities and distinction of the Schmitt trigger in comparison with a conventional comparator.
9. What is the use of precision rectifier and how it can be constructed with op-amp?
10. Derive Equation (11.65) that describes the instantaneous voltage across the capacitor in an op-amp square wave generator.
11. Deduce Equation (11.70) denoting the time period of the op-amp triangular wave generator.
12. Deduce Equation (11.71) for the frequency of op-amp sine wave generator.

Conceptual Test

1. A magnifying glass or a microscope can produce an enlarged image of an object. Is it an amplification?
2. Do all op-amp circuits have (i) virtual ground and (ii) virtual short? Explain.
3. Can any op-amp circuit have more than one virtual ground?
4. What will be the output, if the feedback resistor of the inverting amplifier is made smaller than the resistor connected to the inverting input terminal?
5. Can you use the differential amplifier for testing the balanced condition of a Wheatstone bridge?
6. Can you modify the adder circuit of Figure 11.8 in such a way that it can produce the average of the three input voltages?
7. A simple CR circuit can be used as an integrator or a differentiator depending on its time constant. How the op-amp integrator or differentiator is different from that?
8. Does the logarithmic amplifier function with a negative input voltage?

Numerical Problems

1. An op-amp has a slew rate of 1.5 V/μs. What maximum closed-loop voltage gain that can be used while constructing an amplifier circuit with this op-amp when the input signal varies by 0.5 V in 10 μs?
2. In a differential amplifier, when the inputs are 105 μV and 95 μV, the output is 1.01 mV and when the inputs are 5 μV and −5 μV, the output is 1.0 mV. Calculate the CMRR of the differential amplifier.
3. Consider the op-amp integrator circuit of Figure 11.18(a) where the resistance and capacitance are of 100 kΩ and 0.01 μF, respectively. A square voltage waveform of amplitude ± 5 V and frequency 250 Hz is applied to the circuit as input. Sketch the nature of the output voltage with proper values.
4. Derive a suitable expression for the output voltage of the op-amp circuit given in Figure 11.44. Also suggest how the expression will be modified, if $R_f \gg R_1$.
5. The input voltage in the circuit of Figure 11.45 is V_i = 2 V. If V_{CC} = 16 V, R_2 = 2 kΩ and R_L = 10 kΩ, the value of R_1 required to deliver 10 mW of power across R_L is ....
   (a) 12 kΩ　　　(b) 4 kΩ　　　(c) 8 kΩ　　　(d) 14 kΩ
   [NET Dec. 2016]
6. Determine the output voltage (v_o) in the circuit of Figure 11.46 in terms of input voltages v_1 and v_2.

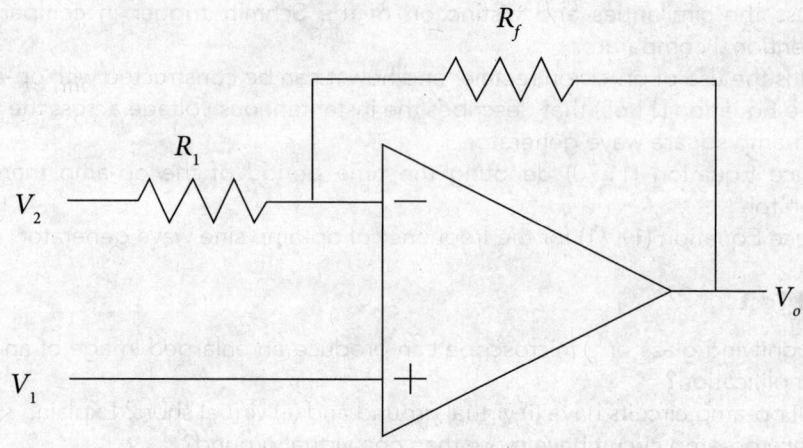

Figure 11.44

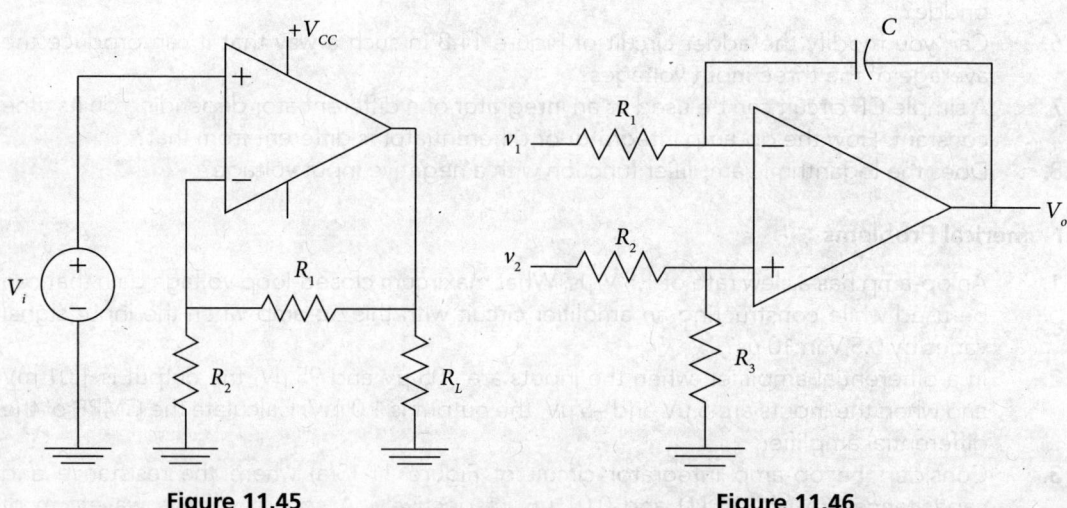

Figure 11.45   **Figure 11.46**

7. Construct a circuit using one or more op-amp that has two input signals v_1 and v_2 and the output is given by $v_o = 4v_1 + 6v_2$.

8. Determine and trace the output waveform in the circuit of Figure 11.47, if the input voltage be a square wave of time period of 1 ms, maximum value of 10 mV and minimum value of zero.

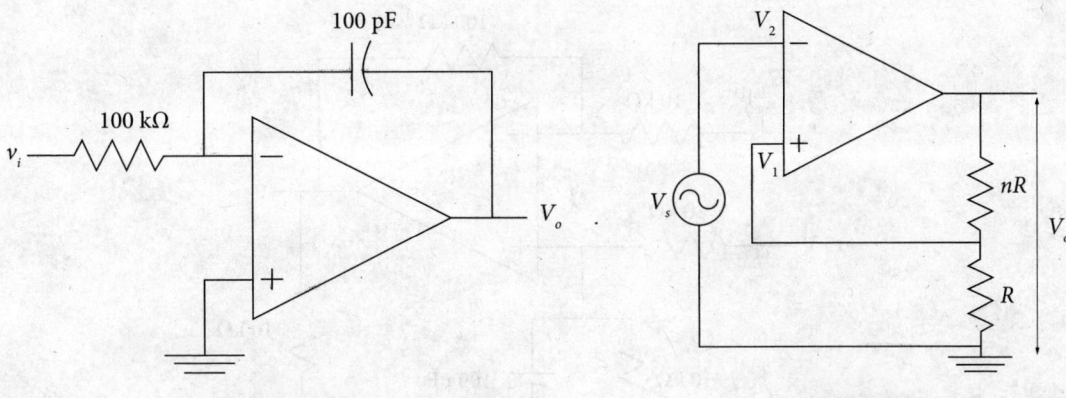

Figure 11.47   Figure 11.48

9. For the differential amplifier shown in Figure 11.48, assume infinite input resistance, zero output resistance and finite differential gain: $A_v = V_o/(V_2 - V_1)$. Show that in the limit $A_v \to \infty$, the gain $A_{vf} = n+1$.

[C.U. 2008]

10. In the circuit of Figure 11.49, $C_i = 10$ nF, $R_i = 20$ kΩ, $R_F = 200$ kΩ and $C_F = 100$ pF. The magnitude of the gain at an input signal frequency of 16 kHz is ....
   (a) 67   (b) 0.15   (c) 0.3   (d) 3.5

[NET Jun. 2017]

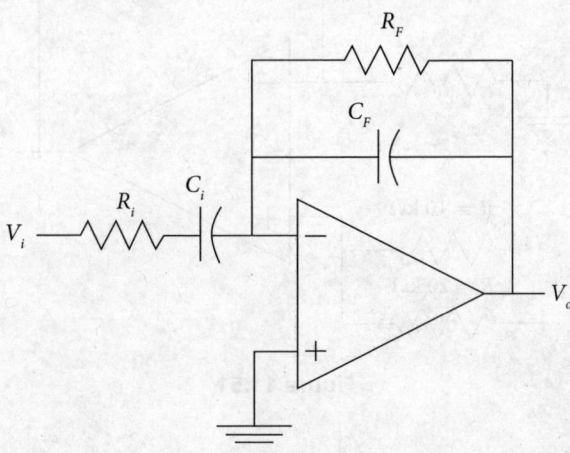

Figure 11.49

11. The input in the circuit of Figure 11.50 is v_i = sinusoidal signal of 1 mV peak-to-peak.

   (a) What is the lower cutoff frequency at which the gain is down by 3 dB as compared to the gain at midband? If the bandwidth of the amplifier is 1 MHz for unity gain, what will be the bandwidth of the given circuit?

   (b) What is the output voltage (v_o) at 15 kHz?

[JAM 2011]

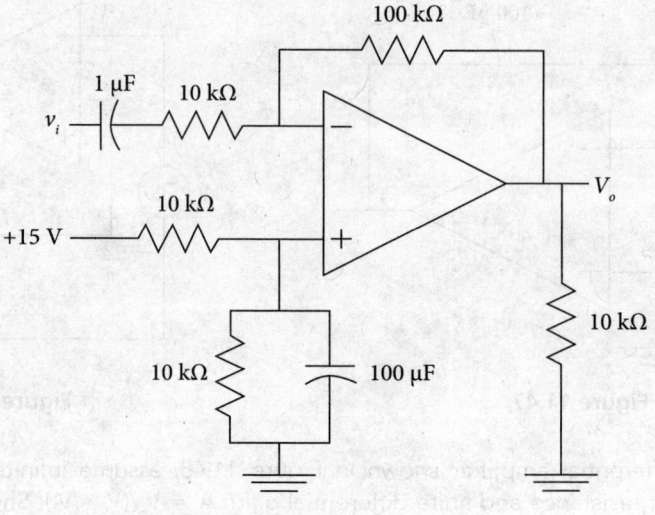

Figure 11.50

12. Determine the output voltage of the op-amp circuit of Figure 11.51.

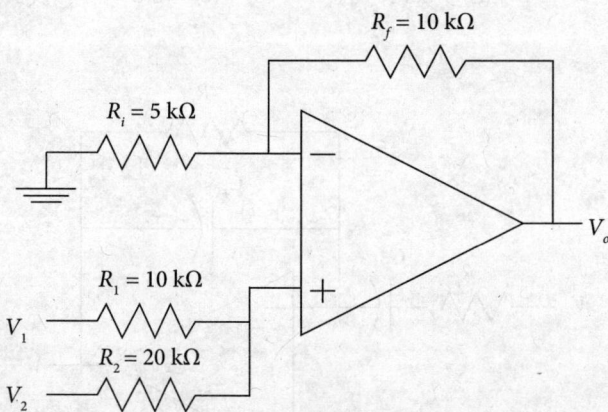

Figure 11.51

13. Using the op-amp circuit of Figure 11.51, fix the resistance values R_i, R_f, R_1 and R_2 in such a way that the output voltage (V_o) becomes $4V_1 + 6V_2$ where V_1 and V_2 are the two input voltages.

14. The input voltage (V_{in}) to the circuit shown in Figure 11.52 is $2\cos(100t)$ V. The output voltage (V_{out}) is $2\cos(100t - \pi/2)$ V. If $R = 1$ kΩ, the value of C (in μF) is
    (a) 0.1 (b) 1 (c) 10 (d) 100

[GATE 2020]

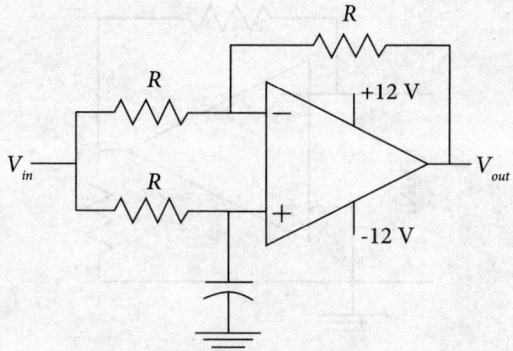

Figure 11.52

15. The gain of the circuit given in Figure 11.53 is $-1/\omega RC$. The modification of the circuit required to introduce a dc feedback is to add a resistor
    (a) between a and b
    (b) between positive terminal of the op-amp and ground
    (c) in series with C
    (d) parallel to C

    [NET June 2017]

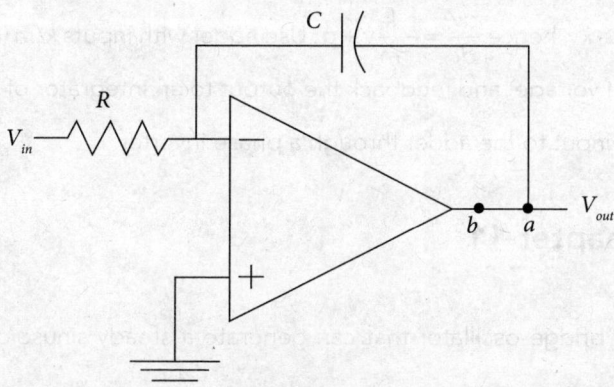

Figure 11.53

[Give a brief explanation for the choice of option in problem 15 above.]

16. Consider the circuit of Figure 11.54, which involves an op-amp and two resistors with an input voltage marked INPUT. Which of the following circuit components, when connected across the input terminals, is most likely to create a problem in the normal operation of the circuit?
    (a) A voltage source with a very high Thevenin resistance
    (b) A current source with a very high Norton resistance
    (c) A voltage source with a very low Thevenin resistance
    (d) A current source with a very low Norton resistance

    [TIFR 2018]

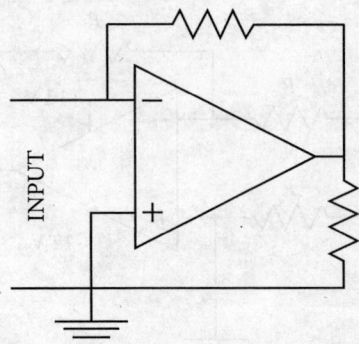

Figure 11.54

[Give a brief explanation for the choice of option in problem 16 above.]

17. Design an analog computational circuit for determining the velocity of an object falling freely under gravity.

    Hints. $F = mg$, hence velocity is $v = g\int dt$. The velocity is given to an integrator circuit in terms of $v_i = 1$ volt. The time constant $RC = 1/g$.

18. Construct an analog computational system for determining the velocity of an object falling through a viscous liquid.

    Hints. $F = mg - kv$, hence $\dfrac{dv}{dt} = -\dfrac{k}{m}v + g$. Use adder with inputs k/m (resistance ratio) and $-g$ (in terms of voltage) and feedback the output to an integrator of time constant unity.

    The output is input to the adder through a phase inverter.

Projects on Chapter 11

Project 1:
To fabricate a Wien bridge oscillator that can generate a steady sinusoidal voltage of 2 kHz frequency.

[It is good to revise the section on Wien bridge oscillator from Chapter 10 and the properties of op-amp from this chapter.]

Theory:
The circuit of Figure 10.7(a) is used with proper values of resistors and capacitors and including the diode-R_3 combination. The frequency (f) of the waveform generated by the oscillator is given by [Equation (10.48)]

$$f = \dfrac{1}{2\pi CR} \tag{P11.1}$$

In Equation (P11.1), R and C represent the resistor and capacitor, respectively, used in both the series and parallel arm of the Wien bridge.

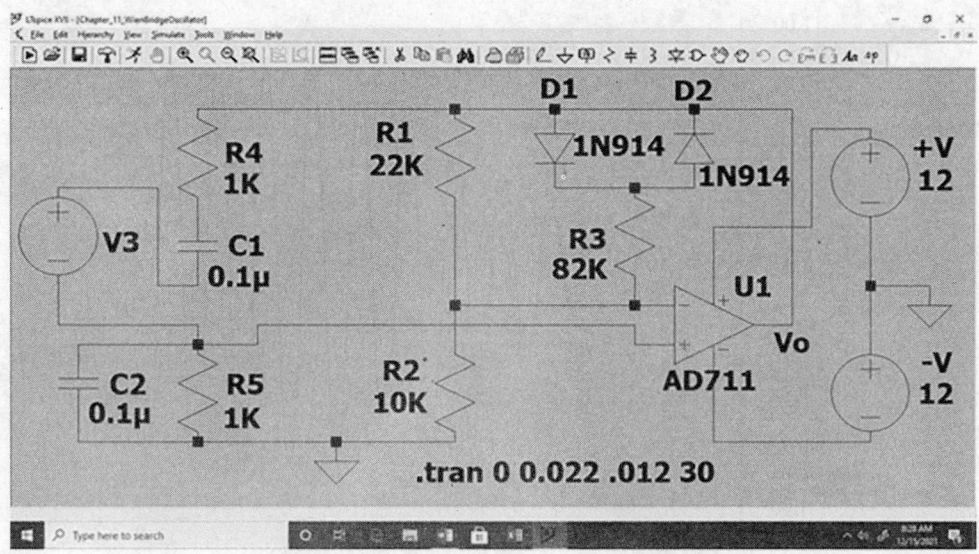

Figure P11.1 LTspice simulated Wien bridge oscillator circuit

Computer Analysis:
The circuit of Figure 10.7(a) was simulated with LTspice, as shown in Figure P11.1 with the following specifications.

(i) The diodes and the op-amp were chosen arbitrarily.
(ii) The op-amp power supply with +ve and −ve terminals was constructed by joining two equal voltage sources.
(iii) The resistors R_1, R_2 and R_3 were denoted by symbols similar to those of the theoretical circuit of Figure 10.7(a). However, the simulated circuit cannot have two circuit parameters with the same name. So the two equal resistors of the bridge were designated as R_4 and R_5, respectively. The two corresponding capacitors were denoted by C_1 and C_2, respectively.
(iv) The voltage source V_3 is not required in a practical circuit where the noise voltage initiates the voltage triggering for oscillation. Since the present one is a simulation having no noise, a small pulse generator of 1 mV amplitude and 1 ms duration was added to the circuit for the initial triggering.
(v) The given frequency is 2 kHz. Putting this value in Equation (P11.1) and using $R = 1\,k\Omega$, the value of C comes out to be approximately 0.08 μF. The nearest available value of 0.1 μF was used in the circuit. Thus the simulated circuit has $R_4 = R_5 = 1\,k\Omega$ and $C_1 = C_2 = 0.1\,\mu F$. A large value of R_3 (= 82 kΩ) is used.
(vi) The initial choice of R_1 and R_2 were 22 kΩ and 10 kΩ, respectively to comply with both the condition of $R_1 \geq 2R_2$ and all the resistors in the bridge being of the same order.

From the menu: 'Simulate', then 'Edit Simulation Command' and 'Transient' option was selected. Since the formation of sustained oscillation takes some small but finite time, the time to start saving data was specified not zero but as a few milliseconds. The maximum time step was entered as 30.

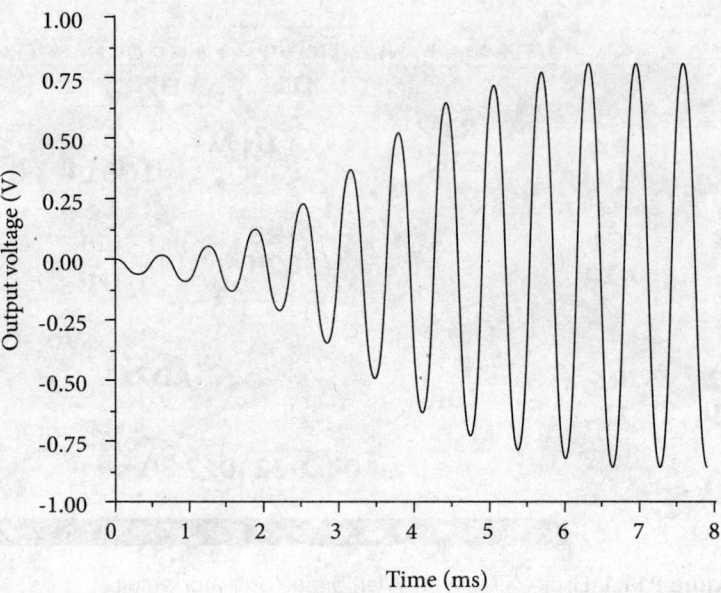

Figure P11.2 Output voltage waveform of the LTspice simulated Wien bridge oscillator

The time variation of the output voltage V_o is plotted in Figure P11.2. It is obvious that the oscillation builds up gradually and takes the form of a sinusoidal wave. The study should be extended with the following investigations.

- Note the effect of changing R_3 on the frequency, amplitude and shape of the waveform. What happens when it is simply open circuited (infinity)?
- What happens, if the small pulse generator V_3 is eliminated from the simulation?
- After getting the steady waveform, compare the calculated frequency of oscillation with the frequency of the simulated waveform.
- Keeping other parameters constant, note the change in output frequency with C and R of the bridge. Do not forget to change simultaneously in the series and parallel arm.
- Read 'lead-lag network' of Wien bridge from Chapter 10. Using a sinusoidal input voltage to the lead-lag network, compare the phases of the input and output waves at different frequencies including the frequency of oscillation.
- Repeat the simulation with another set of op-amp and diodes.

A general suggestion is that always try to design the simulated circuit with standard available values of components.

Project 2:

The circuit given in Figure P11.3 has input current I_i and output voltage V_o. Both the op-amps have supply voltage of ±12 V. Execute the following jobs with it.

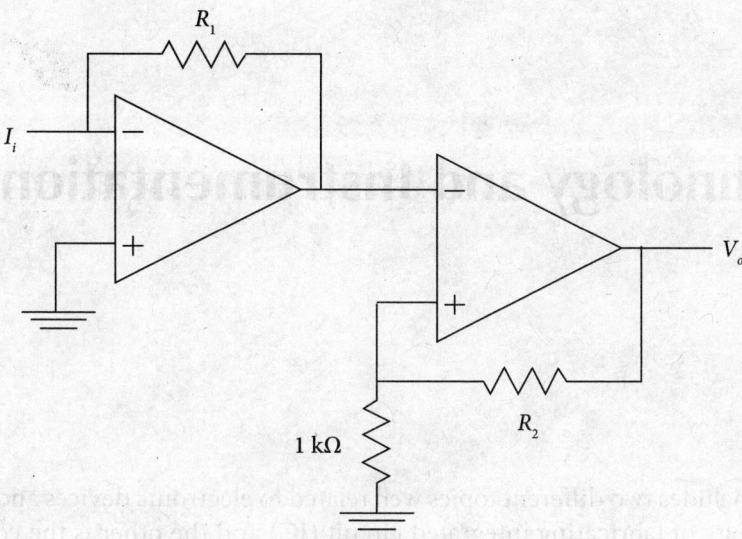

Figure P11.3

1. Work out the formula for V_o in terms of I_i and R_1.

   Hints. The left op-amp acts as inverting amplifier and the right op-amp acts as non-inverting amplifier.

2. (i) Change I_i from 1 to 100 nanoampere taking at least 10 values. Considering $R_1 = 10\,k\Omega$ and $R_2 = 100\,k\Omega$, calculate the corresponding values of V_o.

   (ii) Repeat the above for $R_2 = 200\,k\Omega$.

3. Draw V_o versus I_i graphs for the above. Decide, which one should be on the x-axis. The graphs may be either hand-drawn or generated by application software. However, that should present a complete view.

4. What properties of the circuit are reflected from the graphs?

5. Explain the operation of the circuit, if I_i varies from 1 μA to 100 μA.

6. Is there any virtual ground in the circuit? If yes, then where and how and if not, then why?

12

IC Technology and Instrumentation

This chapter includes two different topics well related to electronic devices and circuits. One is the technology of fabricating integrated circuit (IC) and the other is the construction of measuring instruments, mainly concentrated on cathode ray oscilloscope (CRO).

12.1 Integrated Circuit (IC)

An integrated circuit (IC) is a complete miniature circuit comprising active elements, such as diodes and transistors and passive elements, such as resistors and capacitors, all fabricated on a single chip of semiconductor, mostly silicon. American engineer Jack Kilby (1923–2005) and American physicist Robert Noyce (1927–1990) independently invented the principle of IC in 1958 and 1959, respectively. ICs are indispensable components of almost all modern electronic appliances ranging from computer, television, smartphone and other household gadgets to the critical advancing on missiles and artificial satellites.

> **Note: Noyce Is Nicknamed "The Mayor of Silicon Valley"**
>
> Silicon Valley is a region of California (US) famous for global technological companies, especially manufacturers of silicon-based IC, computer, information technology and related applications. Noyce is credited with the realization of the first monolithic IC made of silicon that identified silicon as a suitable material for IC and paved the way for modern personal computers and other digital systems. It may be mentioned that Kilby's IC was made of germanium with external wire connections. Noyce was the co-founder of world famous Fairchild Semiconductor and Intel corporations and he can be counted among the few physicist entrepreneurs, such as Guglielmo Marconi (1874–1937), inventor of long range wireless communication and Chester Carlson (1906–1968), inventor of photocopiers.

Use of IC in electronic circuits has a number of advantageous features over the bulky circuits made of interconnected discrete components. The most significant benefits are the followings.

Miniaturization: Fabrication of small-size devices is the most noted advantage of IC. Indeed the word 'microelectronics' became popular after the arrival of IC. Accommodating more than 10^8 transistors within a chip of only 1 cm^2 is a very common feature in today's IC technology.

Batch Processing: Millions of ICs can be fabricated at a time from a single piece of semiconductor. The process reduces the cost and provides with good matching among the devices.

Improved Performance: The reliability of the working of ICs is very good because the possible faults due to interconnection of discrete components is eliminated. Small size and light weight make the circuit suitable for airborne, space-based and other vital equipment.

Maintenance: It consumes less power and the entire circuit can be replaced easily at less cost.

In spite of the above benefits, ICs have some limitations, as follows.

- Transistors capable of handling high power cannot be fabricated on IC.
- Inductors and transformers cannot be fabricated on IC.
- IC technology is not suitable for high value capacitors, such as 100–1000 µF.
- Precision resistors having tolerance less than 1% are not possible with IC.

As a whole, the range of IC-fabricated passive components is quite restricted. The power handling capability of IC components is low and the components may be sensitive to voltage and static charge. Nevertheless, the advantages of ICs are much more than the limitations. By virtue of the high packing density of components at low cost, high stability and least maintenance, ICs have made

- complex electronic circuits easily available,
- computer, mobile phone and other consumer electronic products so handy, and
- scientific and technical instruments upgraded to a great extent.

12.2 IC Classification

Different types of IC are available for various purposes. Some broad classifications can be made on the basis of construction and objective, as outlined in Table 12.1 through Table 12.4.

Table 12.1 IC classification on the basis of fabrication

Monolithic IC	This is the most popular and widely used category of IC. All the active and passive components with their interconnections are fabricated on a single silicon chip. 'Monos' means 'single' and 'lithos' means 'stone'. The word *monolithic* is coined like that.
Thick-film and thin-film IC	The passive components like resistors and capacitors are fabricated as films of varying thickness and the active components like diodes and transistors are connected separately with wire bonds.
Hybrid IC	Two or more individual chips are connected by metallized patterns. Such ICs are suitable for high power operations.

Table 12.2 IC classification based on the number of components per chip

Small-scale integration (SSI)	The number of circuit components is not so large; generally less than 100. The ICs made for basic logic gates are such examples.
Medium-scale integration (MSI)	The number of circuit components per chip is a bit more but not more than 1000. ICs for 8-bit register and 4-bit counter are such examples.
Large-scale integration (LSI)	It has more than 1000 but less than 10000 components. Example: ROM IC.
Very large-scale integration (VLSI)	As the qualitative nomenclature suggests, the number of components is larger, exceeding 10000. The microprocessor meant for instrumentation, such as 8085 is one such IC.
Ultra large-scale integration (ULSI)	ICs containing more than 10^6 components are categorized this way. Example: Pentium microprocessors of today's home computers.

Table 12.3 IC classes on the basis of the mode of operation

Linear IC	These are analog IC. The output is similar to the input in nature and varies continuously over a range of values depending on the input. Operational amplifiers (Chapter 11) are available in this form of IC. Voltage regulator ICs are other such examples.
Digital IC	These involve two-state voltage operations only. The input and the output can assume either of two discrete states. Digital ICs cover a wide range extended from basic gates to microprocessors.
Mixed IC	These contain both analog and digital circuits on a single chip. The 555 timer IC (Chapter 15) and analog-to-digital converters (Chapter 16) are such examples.

Table 12.4 IC classification depending on the type of package

Dual-in-line (DIP) pack	Rectangular, plastic-coated package with two parallel rows of connecting pins. Some examples are given in Table 12.5.
Flat pack	Square, flat, plastic-coated package containing pin-connections on one side (e.g. 78XX), two or four sides.
Transistor pack	Transistor-like metallic package with connecting pins. These are used for high power applications.

Each IC, along with the manufacturer's symbol bears a distinct number code that signifies its category and purpose. For examples,

Texas Instruments:	SN
Motorola:	MC, MFC
National Semiconductor:	LM, LH, LF
Fairchild:	µA
Signetics:	NE/SE

Table 12.5 Introduction to some common ICs

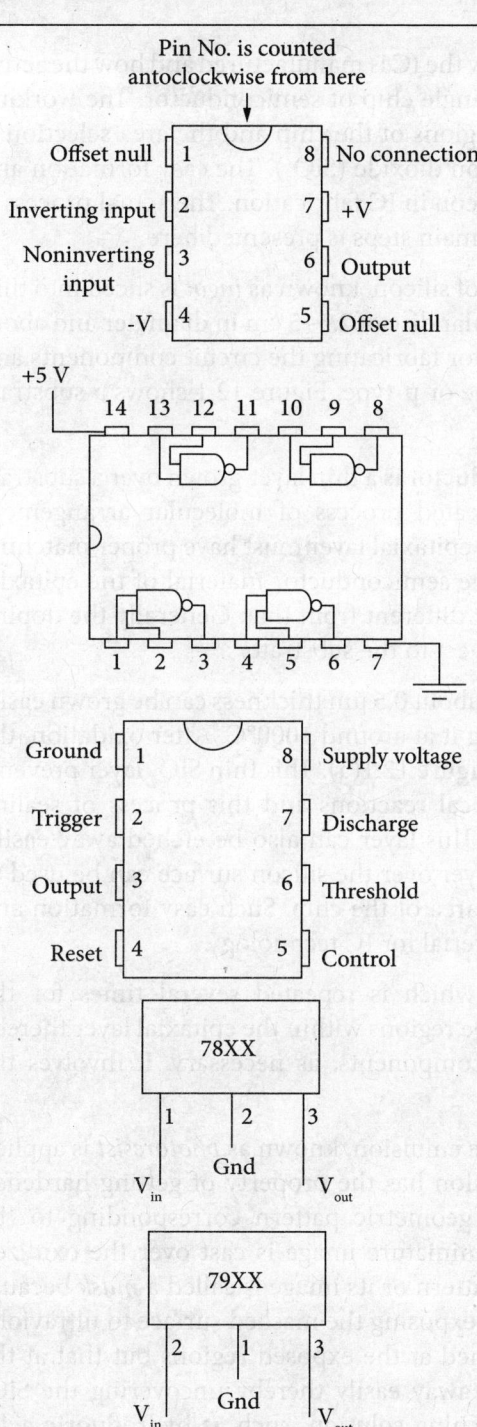

IC 741 (8-pin DIP): Common, all-purpose operational amplifier.
The circuit symbol given in Chapter 11 can be compared with this pin configuration to identify the input and output terminals and the connecting points for the supply voltage and the offset voltage.

IC 7400 (14-pin DIP): Quad 2-input NAND gate
The IC contains four NAND gates whose positions are drawn in the diagram. The supply voltage should be fixed at 5 V.
Some other digital ICs with exactly the same pin-configuration are
IC 7408: Quad 2-input AND gate
IC 7432: Quad 2-input OR gate
IC 7486: Quad 2-input Ex-OR gate
These gates are discussed in Chapter 14.

IC 555 (8-pin DIP): Timer IC
It generates rectangular voltage pulses with or without trigger voltage at pin-2. The time period is determined by external resistors and capacitors. The use of this IC is elaborated in Chapter 15.

78XX: Series of fixed positive voltage regulators of different values where XX denotes 05, 06, 08, 09, 10, 12, 15, 18 or 24 volt of the regulated output voltage.

79XX: Series of fixed negative voltage regulators of different values where XX denotes −5, −8, −12 or −15 volt of the regulated output voltage.

12.3 IC Fabrication

The most relevant topic is yet to discuss that is how the IC is manufactured and how the active and passive components are embedded to the single chip of semiconductor. The working principle involves selective doping of specific regions of the chip and the area selection is accomplished by using a protective layer of silicon dioxide (SiO_2). The easy formation and selective removal of SiO_2 facilitates the use of silicon in IC fabrication. The actual process is a complicated technology. A brief outline of the main steps is presented here.

Substrate Preparation: A bulk single crystal bar of silicon, known as *ingot* is sliced into thin plates, termed as *wafers*. A typical wafer is a circular disc of 2.5–5 cm in diameter and about 150 μm in thickness. The wafer acts as the base for fabricating the circuit components and is referred to as *substrate*. It may be either n-type or p-type. Figure 12.1 shows a substrate of p-type.

Epitaxial Growth: An epitaxial layer of semiconductor is a thin layer grown over a substrate either by chemical reactions or by a sophisticated process of molecular arrangement through evaporation of the solid in vacuum. The epitaxial layer must have proper matching of lattice structure with that of the substrate. The semiconductor material of the epitaxial layer may be the same as that of the substrate or different from that. Generally the doping of the epitaxial layer is of opposite type with respect to the substrate.

Oxidation: A thin silicon dioxide (SiO_2) layer of about 0.5 μm thickness can be grown easily on the silicon epitaxial surface layer by oxidizing it at around 1000°C. After oxidation, the cross-section of the structure looks like that of Figure 12.1(a). This thin SiO_2 layer prevents the semiconductor surface from further chemical reactions and this process of sealing off the silicon surface is known as *passivation*. This layer can also be etched away easily. Therefore, as mentioned previously, the oxide layer over the silicon surface can be used as a convenient covering while doping a selective area of the chip. Such easy formation and removal of oxide has made silicon a suitable material for IC technology.

Photolithography: This is the main process, which is repeated several times for the fabrication of different layers of n-type and p-type regions within the epitaxial layer thereby giving rise to diodes, transistors and passive components, as necessary. It involves the following steps.

    (i) *Masking and Etching*: A photosensitive emulsion, known as *photoresist* is applied to the oxide-coated wafer. The emulsion has the property of getting hardened on exposure to ultraviolet light. A geometric pattern corresponding to the required circuit is prepared and its miniature image is cast over the oxidized wafer coated with photoresist. This pattern or its image is called a *mask* because it partly shades the oxide surface. On exposing the masked surface to ultraviolet radiation, the photoresist gets hardened at the exposed regions but that at the unexposed regions can be dissolved away easily thereby uncovering the SiO_2 layer. Immersing that wafer in an etching solution, such as hydrofluoric acid, the uncovered oxide regions are etched away leaving the bare semiconductor surface. The hardened photoresist is also removed by mechanical abrasion with hot sulphuric acid. After these processes, the structure with selectively removed SiO_2 layer looks like that of Figure 12.1(b).

(ii) *Isolation Diffusion*: The wafer with selective exposure of silicon surface and selective covering with SiO_2 is now doped by diffusion. The doping type is opposite to and the doping concentration is much higher than that of the substrate. After this doping process, the continuous epitaxial layer get divided into isolated n-type island-like regions and this process is termed as *isolation diffusion*. The depletion regions occurring between the substrate and these islands keep the substrate isolated from the integrated components. This mechanism is known as *depletion-layer isolation*. The isolation is further enhanced by high doping of opposite type, as indicated by p^+ in Figure 12.1(c).

(iii) *Base Diffusion*: The remaining oxide covers are now etched away and a new oxide layer is formed again over the surface. The masking, etching and doping processes mentioned above are repeated with other selective openings. Now the opposite type of doping is implemented to construct p–n junctions, as sketched in Figure 12.1(d). These junctions can be either left as diodes or treated as the base-emitter junctions of transistors, hence this step is termed as *base diffusion*.

(iv) *Emitter Diffusion*: A fresh SiO_2 layer is formed again over the entire silicon surface and the above processes are repeated so that openings can be made over the base regions. This time n-type doping is implemented to create the emitter regions of the transistors, as shown in Figure 12.1(e). The components to be preserved as p–n junction diode are masked from the emitter diffusion step.

(v) *Metallization*: Again a new SiO_2 layer is formed over the entire silicon surface and the masking and etching processes are repeated with suitable openings. Vacuum deposition of a thin (≈ 1 μm) layer of metal (generally aluminum) is made all over the surface. Then applying the photoresist technique, the undesired areas of metal coating are etched away. Thus the various components are interconnected.

Packaging: The above mentioned process is also referred to as *batch processing* because a large number of complete circuits are fabricated on a single wafer simultaneously. Such concurrent production of a huge number of circuits on a single wafer is the key reason for the low cost of ICs. The wafer is cut into many small rectangular pieces, known as chips. Each chip contains a complete, independent circuit. Each such circuit is wire-bonded for external connection and encapsulated in an inert atmosphere so as to avoid any reaction with the wafer.

12.4 IC Components: Active and Passive

The marvel of IC fabrication technique is that howsoever complicated the circuit may be, the fabrication process is basically the accurate repetition of several routine processes, such as repeated oxidation and etching windows, forming p- and n-type islands as necessary and interconnecting the components. The photolithography technique mentioned above has outlined the procedure of constructing p–n junction diode and bipolar transistor. Resistors and capacitors also can be fabricated by the same technique in the following ways.

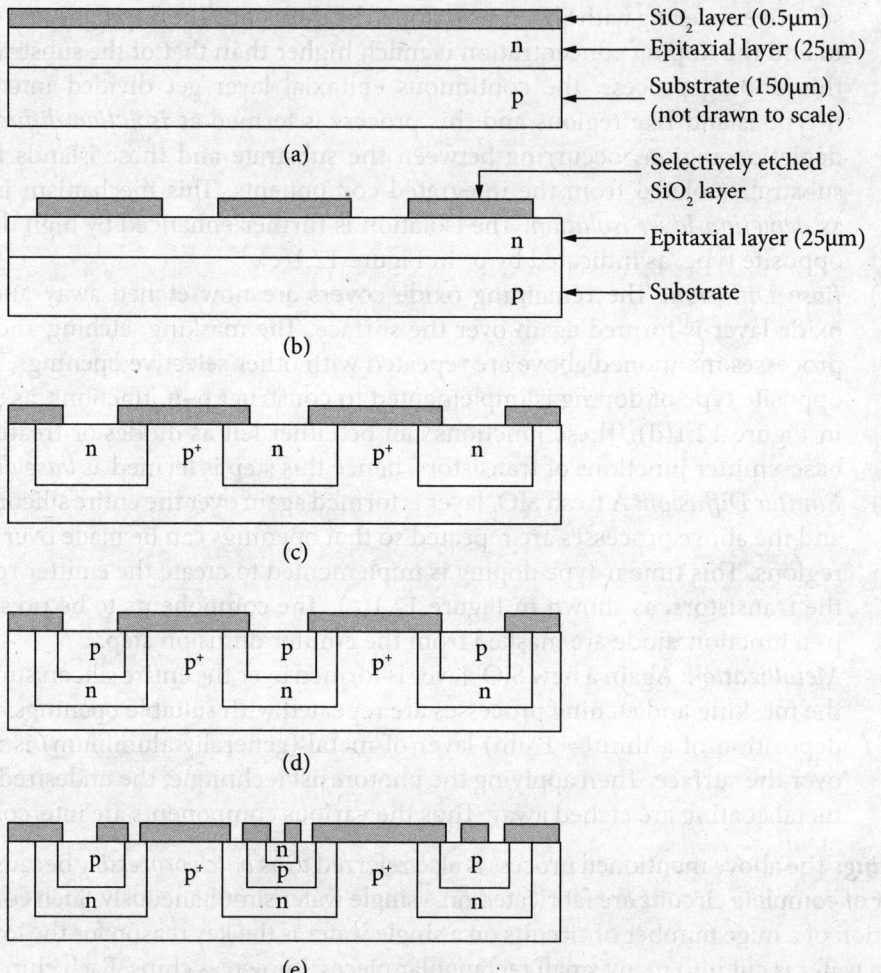

Figure 12.1 Different conditions of photolithography: (a) ready for masking after oxidation, (b) after selective removal of silicon dioxide (SiO$_2$) layer, (c) after isolation diffusion, (d) after base diffusion, and (e) after emitter diffusion, containing both p–n junction and n–p–n transistor

Resistor: A sheet of semiconductor is used as resistor. The resistance can be changed by varying the doping concentration and the depth of diffusion. The range of resistor values obtained this way varies from ohms to hundreds of kilo ohms. A useful term in this connection is

$$\text{Sheet resistance} = \text{resistivity} \div \text{thickness} \tag{12.1}$$

The resistance of a semiconductor sheet of given length and width is calculated as

$$\text{Resistance} = (\text{sheet resistance}) \times \text{length} \div \text{width} \tag{12.2}$$

Capacitor: The depletion region between the p- and n-side of a p–n junction can be made to act as the dielectric of a capacitor and low values of capacitance, of the order of up to around 100 pF can be obtained by this way. Also higher values of capacitance can be achieved with MOS structure. A layer of silicon dioxide is used as dielectric. One plate of the capacitor is formed by diffusing a heavily doped n-region and the other plate is formed by depositing a film of aluminum on the silicon dioxide dielectric on the wafer surface.

Figure 12.2 shows a hypothetical electronic circuit and the cross-sectional view of its monolithic IC equivalent. The diode and the transistor are obvious. The resistor is formed by the bulk p-type material. The capacitor is formed by the MOS structure containing the aluminum top coating, the SiO_2 layer as dielectric and a heavily doped n^+ region with the metal coating.

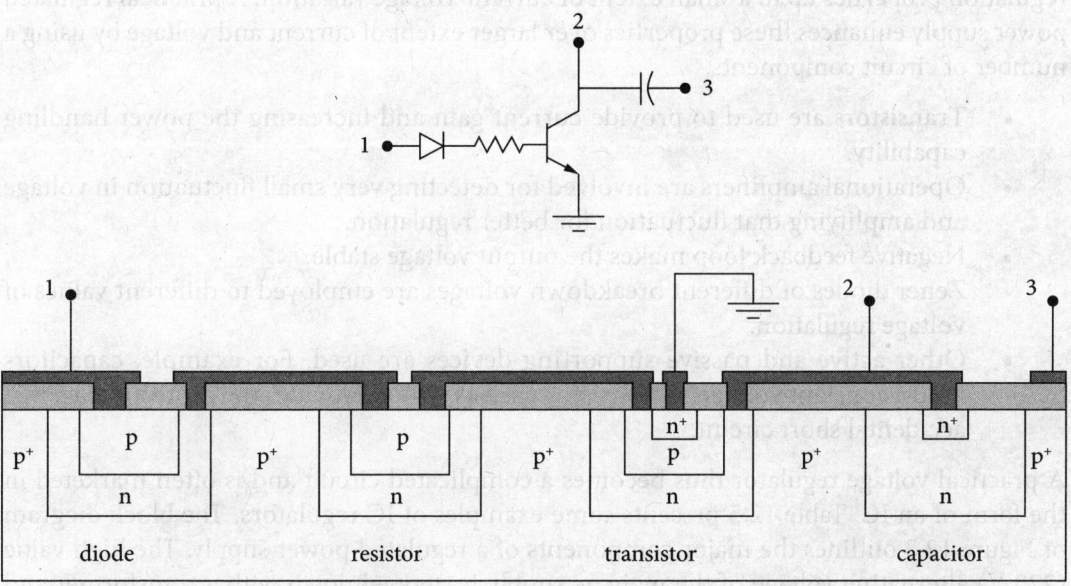

Figure 12.2 Example of an electronic circuit (top) and its equivalent monolithic integrated circuit layout (bottom)

12.5 Regulated Power Supply

An instrument indispensable to all the electronic circuits given in the previous chapters is a dc power supply for providing the required fixed voltage. In our real life also, all electronic gadgets need a power supply. That may be a chemical battery, such as that of laptop computer, wrist watch or mobile phone. However, an electronic power supply is used in most cases, as in television, desktop computer and digital music systems. A common feature of all such electronic power supplies is to convert the main ac supply (220 V RMS) to the required level of dc supply. Not only that, it is compulsory that the power supply is to be *regulated*, which means it should be capable of providing with a constant voltage to the load in spite of fluctuation in the load and in the main ac input. A regulated power supply must have the following two qualities.

(i) The capability of holding a constant voltage across the load despite the variation in load resistance (hence the load current), which is termed as *load regulation* and is quantified as the percentage change of the output voltage for a given change of output (load) current.

(ii) The competency of producing a fixed voltage across the load against the fluctuation of the input voltage of ac main supply, which is referred to as *line regulation* and is defined as the percentage change in the output voltage for a given change in input voltage.

The Zener diode (Chapter 3) is the basic voltage regulator possessing both the load and line regulation properties up to a small extent of current/voltage variation. A practical regulated power supply enhances these properties over larger extent of current and voltage by using a number of circuit components.

- Transistors are used to provide current gain and increasing the power handling capability.
- Operational amplifiers are involved for detecting very small fluctuation in voltage and amplifying that fluctuation for better regulation.
- Negative feedback loop makes the output voltage stable.
- Zener diodes of different breakdown voltages are employed to different values of voltage regulation.
- Other active and passive supporting devices are used. For example, capacitors hold a constant voltage. A transistor used as switch provides the protection against accidental short circuit.

A practical voltage regulator thus becomes a complicated circuit and is often marketed in the form of an IC. Table 12.5 presents some examples of IC regulators. The block diagram of Figure 12.3 outlines the major components of a regulated power supply. The high value (220 V) alternating voltage of the main ac supply is stepped down with a transformer and is rectified with a suitable rectifier (Chapter 3). The filter (Chapter 3) smoothens the wavy nature of the rectified output voltage. Then the regulator unit is added to obtain a stabilized output voltage unaffected by the fluctuations of the ac input voltage and the dc load current up to a specified extent. For instance, a regulated power supply specified by 12V, 1A signifies its capability of supplying a constant voltage of 12V up to the load current of 1A.

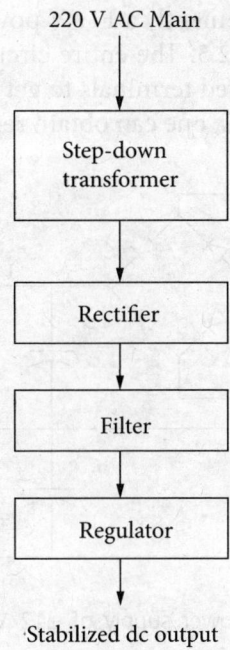

Figure 12.3 Block diagram of a regulated power supply

Figure 12.4 presents a simple but practical example of regulated power supply where the Zener diode holds a fixed reference voltage at the base of a transistor, which works as emitter follower (Chapter 6) thereby providing large current gain. The output voltage is almost 0.7 V less than the Zener voltage.

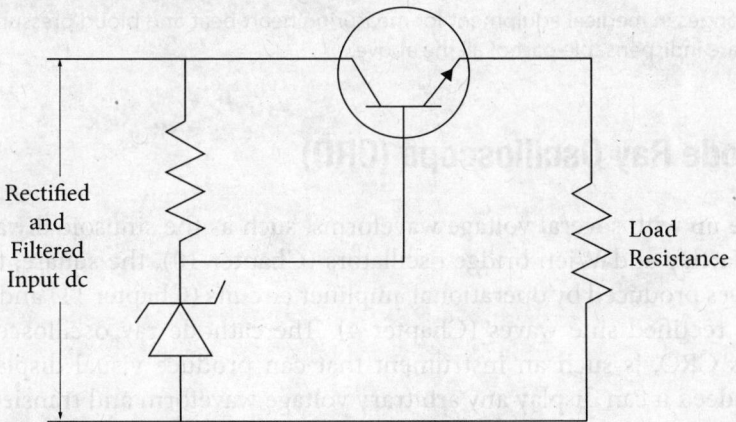

Figure 12.4 A simple regulated power supply constructed with Zener diode and transistor

Figure 12.5 demonstrates how conveniently a ±12 V power supply can be built up using the regulator ICs mentioned in Table 12.5. The entire circuits are built-in and simply the in/out pins are connected to the required terminals to get the regulated output voltage of the desired level. Just by changing the IC, one can obtain regulated voltage of different values.

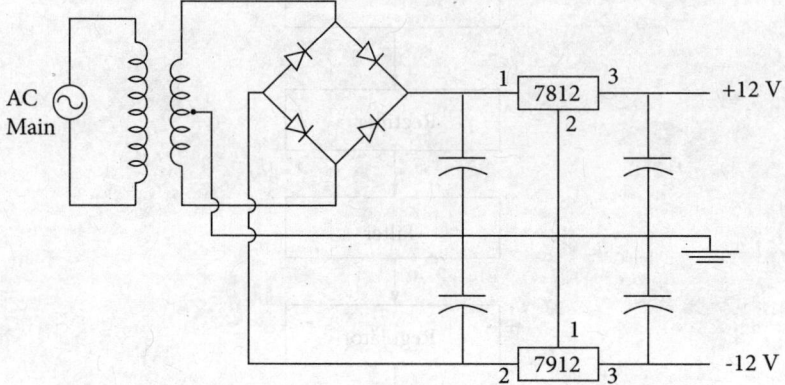

Figure 12.5 Regulated power supply of ±12 V constructed with regulator IC 7812 and 7912

> **Note: Electronics and Instrumentation**
>
> Measurement and instrumentation are inseparable components of experimental physics. Innumerable systems of instruments are available for measuring physical quantities ranging from the thickness of a wire to the luminosity of a star and from the dimension of a molecule to the age of a fossil. Electronics has added (Chapter 1) new concepts in the field of instrumentation encompassing virtually everything. Replacement of photographic film by charge-coupled device in digital camera, substitution of traditional acoustic musical instruments by synthesizer generating sound from electronic oscillators and radical changes in medical equipment for measuring heart beat and blood pressure are few such examples. ICs are indispensable part of all the above.

12.6 Cathode Ray Oscilloscope (CRO)

We have come up with several voltage waveforms, such as the sinusoidal waves generated by Colpitts, Hartley and Wien bridge oscillators (Chapter 10), the square, triangular and sinusoidal waves produced by operational amplifier circuits (Chapter 11) and the half-wave and full-wave rectified sine waves (Chapter 4). The cathode ray oscilloscope, popularly abbreviated as CRO, is such an instrument that can produce visual display of all these waveforms. Indeed it can display any arbitrary voltage waveform and transient over a wide range of amplitude and frequency. The CRO is a very popular and useful instrument for studying electronic circuits. It is not just an *electronic graph-drawing machine*, it has the following salient features.

(i) The CRO can quantify an electrical signal by measuring its amplitude, time period and phase.

(ii) This instrument can be utilized for visual interpretation of a circuit and device, such as
- analyzing the Fourier components in a wave and
- tracing the characteristic curves of a transistor.

(iii) It is a successful combination of several fundamental principles of physics, such as
- thermionic emission of electron,
- application of equipotential surface for converging electron beam, and
- the use of fluorescence.

(iv) The concept of building up the oscilloscope with deflection of electron beam realized other analytic instruments such as
- waveform monitor in television broadcasting system,
- spectrum analyzer for direct observation of signal amplitudes at different frequencies,
- electrocardiogram for medical diagnosis, and
- digital storage oscilloscope for permanent recording of the waveform in the form of database.

> **Note: Development of Cathode Ray Oscilloscope**
>
> The idea of creating the graphical representation of a waveform was conceived prior to the development of electronics. During the early nineteenth century, hand-drawn oscillograms were established to represent the graphical variation of wave by recording the voltage of a spinning rotor at different points around its axis. After the invention of cathode ray tube in the late 19th century for demonstrating the properties of electrons (then referred to as cathode ray), the earlier electromechanical systems were upgraded with the use of cathode ray tube. The waveform was now represented by the fluctuation of the electron beam caused by the electrical signal under test. German physicist Ferdinand Braun (1850–1918) developed the first cathode ray oscilloscope in 1897, the same year when English physicist Joseph John Thomson (1856–1940) discovered electron by propounding that cathode rays are actually particles of definite charge and mass.

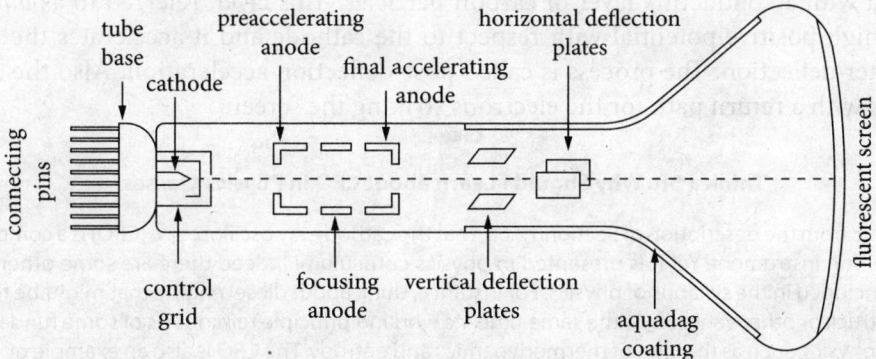

Figure 12.6 Construction of cathode ray oscilloscope comprising the provisions for generating, focusing and deflecting electron beam within an evacuated tube

12.6.1 Construction of CRO

As sketched in Figure 12.6, the oscilloscope contains an evacuated cathode ray tube with several metallic electrodes for producing an electron beam, deflecting it in accordance with the variation of the input electrical signal and tracing the waveform on the flattened front end of the tube acting as a screen. The main components of the CRO are the followings.

Electron Gun: It produces a focused beam of electrons using the following units.

- *Cathode*: This emits the electrons by thermionic emission.
- *Control Grid*: This tubular structure with a small aperture allows a narrow beam of electrons through it. The brightness of the trace of the waveform can be adjusted by varying the control grid voltage.
- *Preaccelerating Anode*: This is at positive potential with respect to the cathode and speeds up the electrons toward the screen.
- *Focusing Anode*: It converges the electrons to a sharp beam establishing electric fields and equipotential surfaces. The details are given in a subsequent section.
- *Final Accelerating Anode*: It accelerates the focused electron beam toward the screen.

Deflection System: The electron beam produced by the electron gun, as mentioned above is allowed to pass through two pair of parallel plates where the beam can be deflected electrostatically in horizontal and vertical directions by applying proper voltages to the plates. The extent of deflection on the screen is quantified in terms of two measuring terms, namely *deflection sensitivity* and *deflection factor*, to be discussed in a subsequent section.

Fluorescent Screen: The front end of the cathode ray tube is flattened and its inner surface is coated with a fluorescent material known as *phosphor*. It has the property of absorbing the kinetic energy of the electrons and re-emitting a portion of the absorbed energy as visible light. This makes the waveform visible on the screen as the electron beam gets deflected in proportion to the amplitude and frequency of the input signal. The screen also contains graphical grids drawn over it for measuring the amplitude and time period of the wave.

Aquadag: The inner portion of the tube from the screen to the final accelerating anode is coated with a conducting layer of carbon particles. This layer, referred to as *aquadag* is kept at high positive potential with respect to the cathode and it accelerates the electron beam after deflection. The process is called post-deflection acceleration. Also the aquadag provides with a return path for the electrons striking the screen.

> **Think a Bit: Why Should I Learn about CRO in Physics Classes?**
>
> It appears from the description of Section 12.6.1 that the cathode ray oscilloscope (CRO) is a complicated engineering instrument. Yet it is presented in physics curriculum! Indeed there are some other similar topics included in the syllabus of physics. For instance, think about diesel engine that might be relevant to bus, truck or other vehicles. At the same time, its working principle reminds us of some fundamental laws of physics, such as the laws of thermodynamics and entropy. The CRO is also an example of such an ingenious appliance that has assembled a number of principles of basic physics. Thermionic emission, fluorescence, equipotential surface, all these are apparently unrelated topics of physics that have been utilized to a single gadget.

12.6.2 Working Principle

The production and deflection of the electron beam and the tracing of the waveform on the fluorescent screen discussed with Figure 12.6 are the main activities of a CRO but that is not the complete picture. Figure 12.7 explains the mode of working of different units of the CRO with a simplified block diagram. The following components are involved in the process.

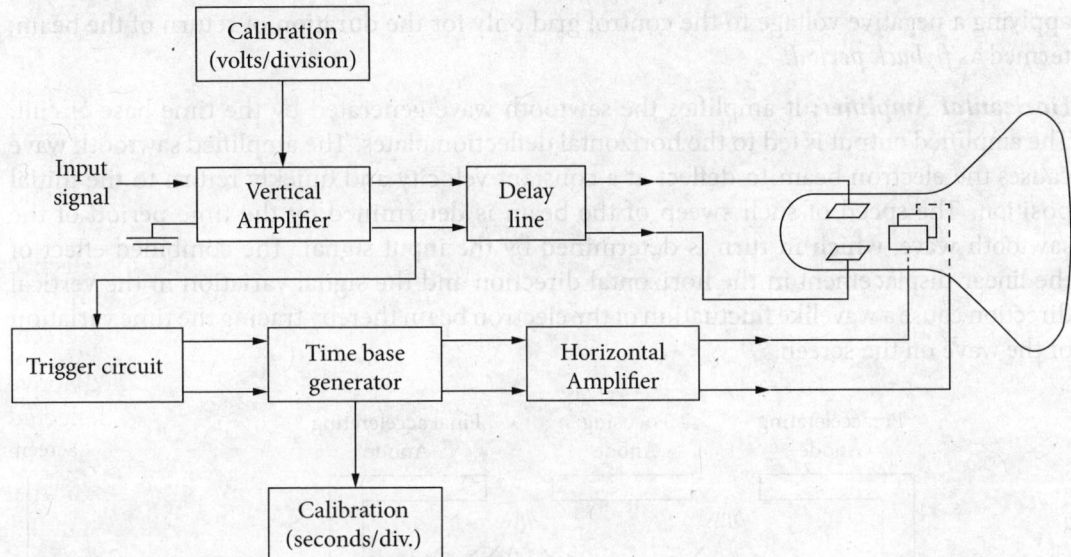

Figure 12.7 Block diagram for different components of cathode ray oscilloscope. The position of the cathode ray tube of Figure 12.6 is indicated

Vertical Amplifier: This amplifies the input electrical signal. The amplified output is connected to both the vertical deflection plates through a delay line and to a trigger circuit activating the time base generator. The gain of the vertical amplifier is variable and it has a provision for calibration so that the user can determine the output value (volt per division) on the graphical screen. The vertical amplifier is of large bandwidth, several hertz to several megahertz so as to accommodate for input signals of wide frequency range.

Delay Line: The vertical amplifier output causes the vertical deflection of the electron beam. The deflection should be harmonized with the horizontal deflection, which goes through a series of processes and takes some time. To synchronize with this delay, the vertical amplifier output is delayed deliberately by the same amount before reaching the vertical deflection plates. The delay line unit is meant for this purpose. It is made of LC networks, which hold the signal for some time.

Trigger Circuit: In order to obtain a stable wave pattern on the screen, the horizontal deflection rate is synchronized with the input signal frequency. For this purpose, the leading edge of the vertical amplifier output is applied for a trigger pulse. That initiates the time base generator, which produces a sawtooth wave having linear increase of voltage with time followed by sudden fall to zero.

Time Base Generator: The trigger pulse causes the time base circuit to produce a sawtooth wave of proportional frequency. The circuit involves the charging of a capacitor through constant current source and then sudden discharge through a short circuit. The circuit output is calibrated so that the user can estimate time per division for the wave pattern at the graphical screen.

The sawtooth wave causes linear displacement of the electron beam and retracting to the initial position. The electron spot on the screen is blanked out during the retrace by applying a negative voltage to the control grid only for the duration of return of the beam, termed as *flyback period*.

Horizontal Amplifier: It amplifies the sawtooth wave generated by the time base circuit. The amplified output is fed to the horizontal deflection plates. The amplified sawtooth wave causes the electron beam to deflect at a constant velocity and quickly return to the initial position. The speed of such sweep of the beam is determined by the time period of the sawtooth wave, which in turn is determined by the input signal. The combined effect of the linear displacement in the horizontal direction and the signal variation in the vertical direction cause a wavelike fluctuation of the electron beam thereby tracing the time variation of the wave on the screen.

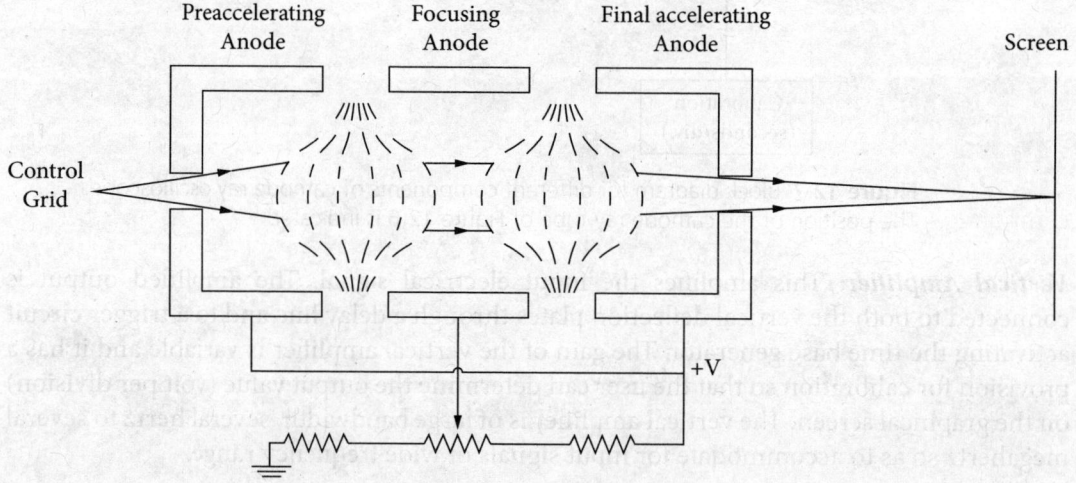

Figure 12.8 Enlarged view of the anodes to indicate electrostatic focusing of electron beam

12.6.3 Electrostatic Focusing

It was mentioned in Section 12.6.1 that the anodes of the cathode ray tube make the electrons convergent to form a sharp beam. Now the process is explained in detail with the help of Figure 12.8, which represents an enlarged view of the three anodes mentioned earlier. The three anodes are coaxial cylinders, the focusing anode being at lower potential with respect to the accelerating anodes. Consequently a non-uniform electric field is generated between the electrodes and the equipotential surfaces indicated by dashed lines work like convex lenses. The electrons emerging from the control grid has a tendency to diverge because of mutual repulsion and variation of energy. When the electron beam is passed through the three coaxial anodes, as shown in Figure 12.8, the electrons passing through the equipotential surfaces are deflected in a direction perpendicular to the surface and thus tend to get focused.

12.6.4 Electrostatic Deflection

As mentioned previously, the electron beam is deflected vertically or horizontally by applying a voltage across the plates. It is actually the electric field caused by the voltage that gives rise to the deflection. This section interprets the physical principle of this phenomenon.

Figure 12.9 shows a pair of deflection plates, each of length l, separated by a distance d. The screen is at a distance L from the mid-point of the plates. The deflecting voltage is V_d and the accelerating voltage is V_a. Let the electron moves along x-direction and the deflecting field E_d ($= V_d/d$) acts along y-direction. The velocity component of the electron along y-axis is given by

$$v_d = u_y + f_y t \qquad (12.3)$$

where

$$f_y = -\frac{qE_y}{m} \qquad (12.3a)$$

is the acceleration due to the field, m is the mass of the electron and u_y is its initial velocity component along y-axis. Since there is no initial velocity along y-axis, putting $u_y = 0$ in Equation (12.3) one obtains

$$v_y = -\frac{qE_y t}{m} \qquad (12.4)$$

The displacement of the electron with zero initial velocity along the y-axis is

$$Y = \frac{1}{2} f_y t^2 = -\frac{qE_y t^2}{2m} \qquad (12.5)$$

The displacement of the electron along the x-axis moving with constant velocity u_x is

$$X = u_x t + \frac{1}{2} f_x t^2 = u_x t \qquad (12.6)$$

It is so because there is no acceleration (f_x) along the x-axis. Substituting for t from Equation (12.6) to Equation (12.5) and simplifying,

$$Y = -\frac{qE_y}{2u_x^2 m} X^2 \qquad (12.7)$$

Equation (12.7) represents a parabola, which indicates that the path of the electron under the action of the deflecting voltage is parabolic, as sketched in Figure 12.9.

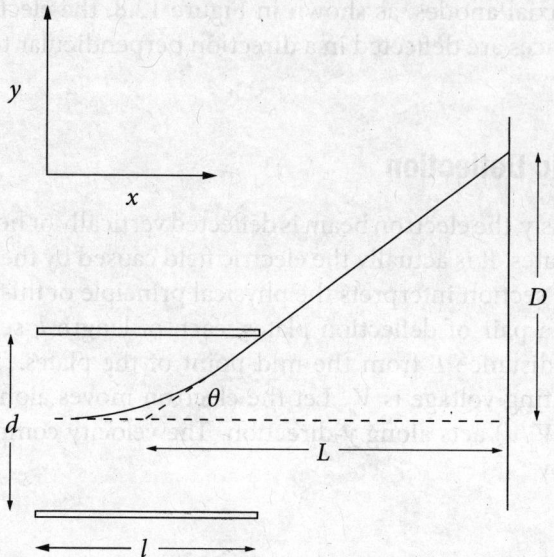

Figure 12.9 Deflection of electron beam in parabolic path under the action of deflecting field between the plates of length *l* separated by distance *d*. The extent of deflection on the screen at distance *L* from the mid-point of the plates is *D*

If the angle of deflection be θ, one can obtain from Equation (12.7)

$$\tan\theta = \frac{dY}{dX} = -\frac{qE_y X}{u_x^2 m} \tag{12.8}$$

Putting $X = l$ (the effective length of the deflection plates) in Equation (12.8) and considering only the magnitude, the deflection from the mean position on the screen can be obtained as

$$D = L\tan\theta = \frac{qE_y lL}{u_x^2 m} \tag{12.9}$$

The kinetic energy of the electron entering the space between the deflection plates with an initial velocity u_x can be equated to the accelerating voltage as follows.

$$\frac{1}{2}mu_x^2 = qV_a \tag{12.10}$$

Using Equations (12.9) and (12.10) and putting $E_y = V_d/d$, the deflection can be expressed as

$$D = \frac{lLV_d}{2dV_a} \tag{12.11}$$

Equation (12.11) expresses the deflection in terms of the geometrical parameters: l (length of each deflecting plate), L (distance of the screen from the center of the plates), d (separation between the plates) and the two applied voltages: V_a (final accelerating anode voltage) and V_d (deflecting voltage between the plates). The sensitiveness of the deflection is quantified by the following two terms.

- Deflection sensitivity $(S) = \dfrac{D}{V_d} = \dfrac{lL}{2dV_a}$ \hfill (12.12)

- Deflection factor $(G) = \dfrac{1}{S}$ \hfill (12.12a)

12.6.5 Waveform Display

The combined effect of the horizontal linear displacement of the electron beam and its vertical fluctuation according to the signal variation produces a visual display of the time variation of the wave on the fluorescent screen. Let the input signal applied to the vertical deflection plates be a sine wave. Simultaneously a sawtooth wave of the same time period is applied to the horizontal deflection plates. The shape of the sawtooth wave is sketched in Figure 12.10. It increases linearly with time over the period t_s, referred to as *sweep time* and then falls rapidly to zero within a very short (ideally zero) time t_f, referred to as *flyback time*.

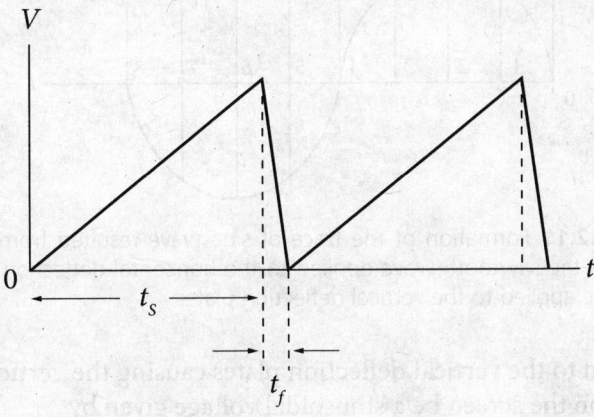

Figure 12.10 Sawtooth voltage waveform applied to the horizontal deflection plates

The electron beam spot on the fluorescent screen undergoes a horizontal deflection proportional to time and a vertical deflection proportional to the magnitude and polarity of the input signal. The resultant movement of the beam traces the input signal waveform on the screen. Figure 12.11 explains the mechanism of the wave formation. The input signal and the sawtooth wave have the same time period. The linear increase of voltage causes a gradual horizontal shift and the sudden fall of the voltage retracts the beam. During this time, the vertical deflection is proportional to the instantaneous value of the input voltage. The resultant is the time variation of the wave. Figure 12.11 sketches the wave for eight discrete points of time. Actually there are infinite number of such points for the continuous tracing of the waveform. The figure shows a single sine wave for a single sawtooth wave. This is the case of both time periods being equal. If the time period of the input signal becomes $1/n$ times that of the sawtooth wave, then n cycles of the signal is displayed on the screen. This adjustment is done by the time/division calibration of the time base generator indicated in Figure 12.7.

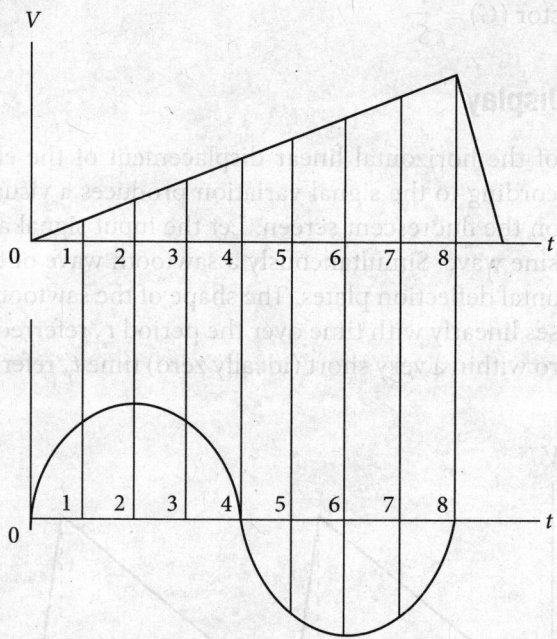

Figure 12.11 Formation of the trace of sine wave resulted from the combined action of the sawtooth wave applied to the horizontal deflection plates and the sine wave applied to the vertical deflection plates

Let the signal applied to the vertical deflection plates causing the vertical movement of the electron beam spot on the screen be a sinusoidal voltage given by

$$y = V_y \sin \omega t = V_y \sin \frac{2\pi}{T} t \tag{12.13}$$

The amplitude and the time period of the signal are given by V_y and T, respectively. The sawtooth voltage applied to the horizontal deflection plates causing the horizontal movement of the beam spot is given by

$$x = \frac{V_x}{T} t \tag{12.14}$$

The amplitude and the time period of the signal are given by V_x and T, respectively. Actually it is the sweep time and is approximately equal to the time period because the retrace time is very small. Assuming the sawtooth wave having the same frequency as that of the sine wave, one can obtain from Equations (12.13) and (12.14)

$$y = V_y \sin \frac{2\pi x}{V_x} \qquad (12.15)$$

Equation (12.15) indicates that the resultant motion of the electron beam spot traces a stable waveform on the screen. It has the following properties.

(i) The magnification of the wave amplitude can be adjusted by the vertical gain knob calibrated in volts per division.
(ii) The expansion of the wave can be adjusted by the horizontal gain knob calibrated in time per division.
(iii) If the sawtooth time period is n times that of the signal waveform, then n number of complete waves are seen on the screen.
(iv) The trigger circuit establishes a phase synchronization each time the time-base sweep starts with the signal.

12.6.6 Applications of CRO

The CRO not only displays the waveform of an electrical signal, it allows to measure the amplitude and time period (hence the frequency) of the wave with reasonable accuracy. The horizontal input is a built-in time base signal and its time period can be adjusted by a 'time/div.' knob. The vertical input is the signal under measurement and its amplitude can be adjusted by the 'volts/div.' knob. Generally two such channels are provided for simultaneous observation of two waveforms. The measurement procedures are demonstrated with the following examples.

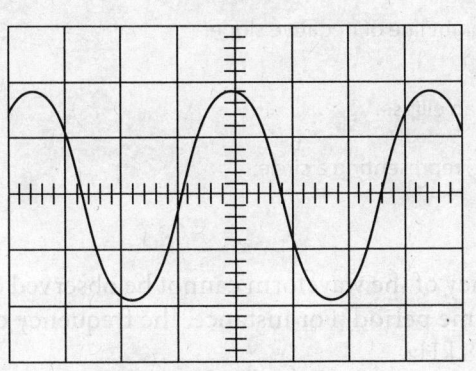
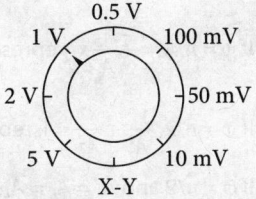
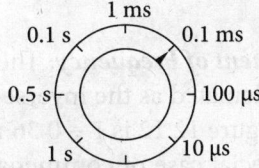

Figure 12.12 Schematic representation of the CRO front panel with graduated screen and adjustment knobs

Measurement of Voltage: Figure 12.12 is a schematic representation of the CRO front panel. A sinusoidal voltage waveform is visually cast on the graduated screen. Each division of the screen is divided into five equal parts. The peak-to-peak voltage spans almost 3.8 divisions of the vertical axis. The volt/div knob is fixed at the scale of 1V/div, as indicated by the arrowhead. Therefore, the peak-to-peak voltage is 3.8×1 = 3.8 V, the voltage amplitude is 3.8/2 = 1.9 V and the rms voltage is $1.9/\sqrt{2}$ = 1.34 V.

Measurement of Time Period: The time/div scale is to be adjusted in such a way that at least a full waveform becomes visible. In Figure 12.12, the full wave spans almost 3.6 divisions. Such non-coinciding with any graduation mark is a common occurrence and the more inclined mark is to be considered. The time/div knob is fixed at 0.1 ms/div, as indicated by the arrowhead. Therefore, the time period of the given waveform is 3.6×0.1 = 0.36 ms.

> **Note: Lissajous Figures**
>
> French physicist J. A. Lissajous (1822–1880) experimentally established that superposition of two simple harmonic motions acting at right angles to each other results in different geometrical figures depending on the frequency ratio and phase difference. His apparatus comprised two tuning forks with mirrors and reflection of light tracing curves on the wall. The mathematical background of the above was known earlier. American mathematician N. Bowditch (1773–1838) propounded that the superposition of parametric equations $x(t) = A\cos(\omega_x t - \varphi_x)$ and $y(t) = B\cos(\omega_y t - \varphi_y)$ can generate straight line, circle, ellipse and complicated curves.
>
> If A_1 and A_2 be the amplitudes of two sine waves of the same frequency and phase difference of φ acting along x-axis and y-axis, respectively. The resultant motion has the form of
>
> $$\frac{y^2}{A_2^2} - \frac{2xy}{A_1 A_2}\cos\varphi + \frac{x^2}{A_1^2} = \sin^2\varphi$$
>
> The following are some special cases.
>
> (i) If no phase difference ($\varphi = 0$), $y = \frac{A_2}{A_1}x$ representing a straight line.
>
> (ii) If $\varphi = \pi$, $y = -\frac{A_2}{A_1}x$ representing straight line of negative slope.
>
> (iii) If $\varphi = \pi/2$, $\frac{y^2}{A_2^2} + \frac{x^2}{A_1^2} = 1$, representing an ellipse.
>
> (iv) If $\varphi = \pi/2$ and $A_1 = A_2 = A$, $x_2 + y_2 = A_2$, representing a circle.

Measurement of Frequency: The frequency of the waveform cannot be observed directly. It is to be calculated as the inverse of the time period. For instance, the frequency of the sine wave in Figure 12.12 is 1 ÷ 0.36 ms = 2.78 kHz.

In a special case of commensuration frequencies with one known frequency, Lissajous figures can be used for estimating the unknown frequency based on the following rules of thumb.

- If there be n loops side-by-side, the ratio of the vertical signal frequency (f_v) to the horizontal signal frequency (f_h) is n:1.
- If there be n loops top-down, the ratio of the vertical signal frequency (f_v) to the horizontal signal frequency (f_h) is 1:n.

The CRO has a provision, termed as *X-Y* mode, for superposing the two input signal voltages. The two signals are applied simultaneously to the vertical and horizontal deflection plates. The resultant formation of loops for different frequency ratio are drawn in Figure 12.13.

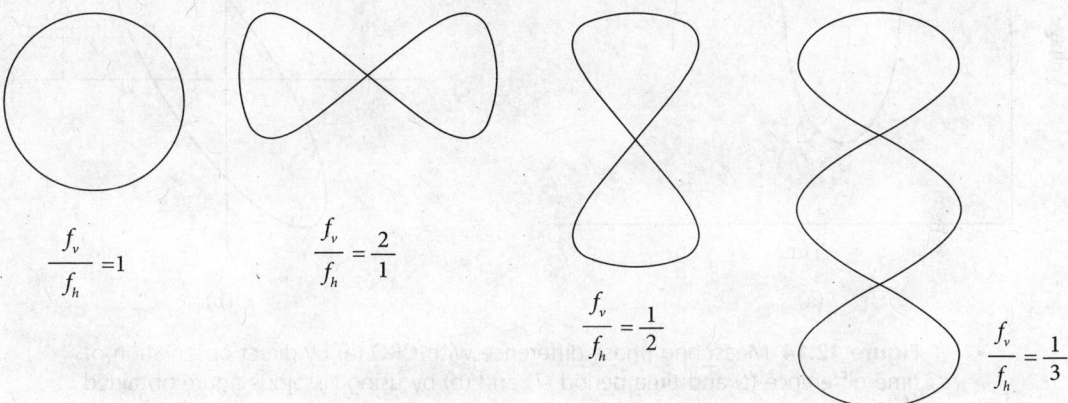

Figure 12.13 Examples of Lissajous figures obtained with different ratios of vertical signal frequency (f_v) to the horizontal signal frequency (f_h)

Measurement of Phase: The phase difference of a certain voltage signal with respect to another signal of similar kind can be estimated by two different methods.

Method-1: The two voltage signals are applied to the two input channels of a dual trace CRO. The phase difference can be determined from direct observation, as shown in Figure 12.14(a). Since the time period (T) corresponds to 2π phase change, the time difference (t) observed on the screen stands for the phase difference of

$$\varphi = \frac{2\pi t}{T} \quad (12.16)$$

Method-2: The phase difference can also be estimated by the use of Lissajous figures, as explained with Figure 12.14(b). The given note on Lissajous figures may be followed. The two signals are applied to the two channels and the CRO knob is kept at *X–Y* mode to obtain the ellipse. The point of intersection (y) of the ellipse with y-axis and the maximum displacement (A) are shown in the figure. Both y and A can be measured directly on the screen and the phase difference can be calculated as

$$\varphi = \sin^{-1}\frac{y}{A} \quad (12.17)$$

This is the case when $x = 0$, $A_1 = A_2 = A$ and $0 \le \varphi \le 90°$. For $90° \le \varphi \le 180°$, the ellipse occurs between the 2nd and 4th quadrants. In the case of $\varphi = 90°$ or $180°$, the ellipse converts to a straight line.

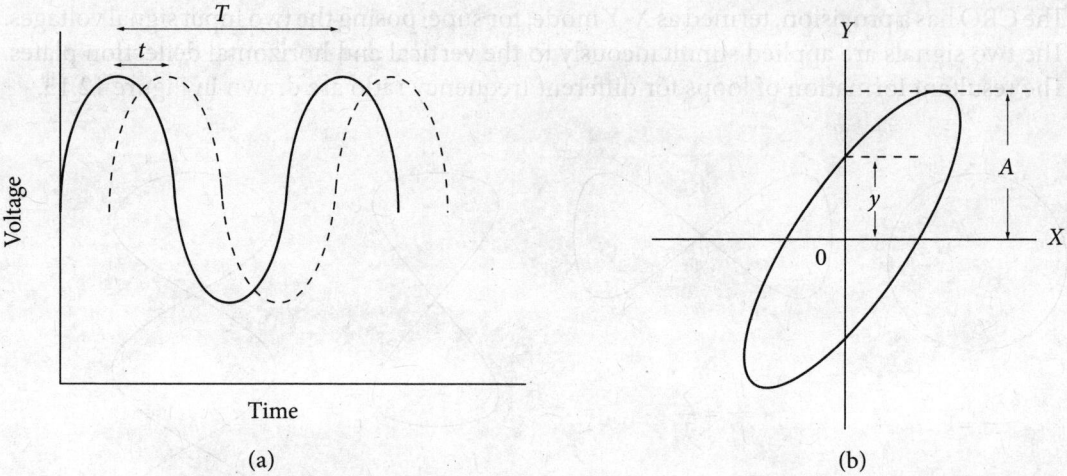

Figure 12.14 Measuring phase difference with CRO (a) by direct observation of time difference (t) and time period (T) and (b) by using Lissajous figure obtained in X-Y mode

12.7 Digital Storage Oscilloscope

The next chapter will detail the difference between 'analog' and 'digital'. The CRO illustrated above is of 'analog' nature. It can display the continuous changes of the input signal voltage. The movement of the electron beam (the output) is continuously influenced by the input variable (input voltage).

A digital oscilloscope, however, samples the given input signal condition at regular interval of time by using an analog-to-digital (A/D) converter. This process is called *digitizing*. The original signal is reconstructed form the sampled data points and is displayed on the cathode ray tube. Also the facility of liquid crystal display (LCD) can be availed of. In addition to the visual display of the waveform, the instrument can also store the data in computer-readable binary format.

The block diagram of a digital storage oscilloscope is given in Figure 12.15. It differs from the analog oscilloscope in terms of triggering, storage, and measurement. The analog input signal is amplified up to the specified level. An internal clock divides the input signals into separate time slots for the purpose of digitizing the instantaneous amplitudes. This signal is then fed to an analog to digital (A/D) converter and a trigger detector. Generally it uses flash-type A/D converter (Chapter 16). The digital data are stored in the memory. (The concept of memory is illustrated in Chapter 16.) The waveform is regenerated at a predetermined clock rate. For perfection in reconstruction, intermediate numerical values are generated by using interpolation technique. The maximum frequency of the input signal that can be measured by the oscilloscope depends on both the sampling rate and the A/D converter properties. The sampled and converted data are stored in shift register (Chapter 15) that can hold 1s and 0s and can shift their positions. The storage in shift register can compensate for high speed requirement.

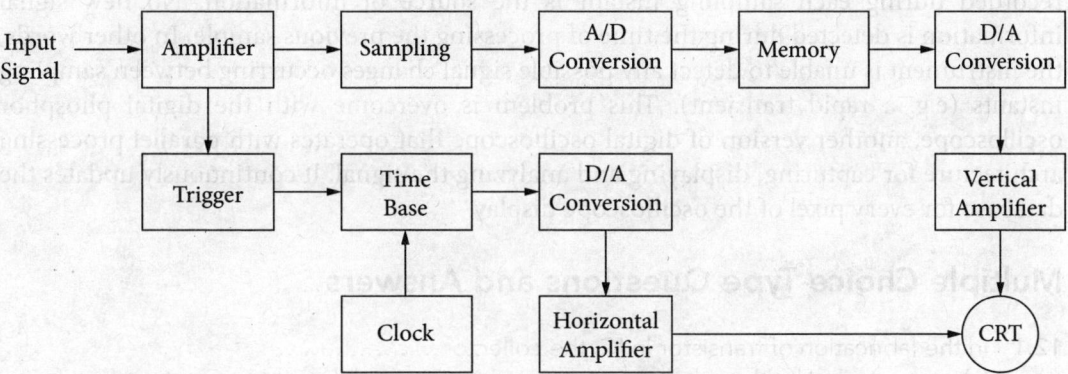

Figure 12.15 Block diagram of digital storage oscilloscope

The acquisition, storage, and display of signals can be operated in three modes: roll mode, store mode, and save/hold mode.

- In *roll mode*, the input signals are displayed on the screen without triggering. It is a basic mode of operation similar to that of a conventional analog CRO.
- In the *store mode*, the digitized signals are stored into the memory.
- In *hold mode*, the data are saved before getting stored in the digital memory. In this mode, the signal traces can be analyzed by the user's choice.

The triggering of the oscilloscope allows to stabilize the signal and display the waveform repeatedly. Any change in the input re-triggers the oscilloscope causing the memory to reset for the new data.

The digital storage oscilloscope has the following advantageous features.

- *Storage*: It can hold the data for infinite time. The provision for storage of data provides with easy debugging of circuits, particularly transient studies.
- Holding the data in digital format averts signal degradation.
- It is appropriate for the testing of a circuit during manufacturing.
- *Versatility*: In addition to the straight measurement of time period, frequency and voltages, device properties like capacitance, inductance and transistor output characteristics can be easily estimated.
- *Many Channels*: The salient feature of the digital storage oscilloscope is that a number of different input channels can be used. Nevertheless, the available memory to each channel goes on reducing. A common storage place can be used by all the channel by the technique of multiplexing (Chapter 14).

- *Computer Interfacing*: Because of easy computer connectivity, the digital storage oscilloscope is suitable for diversified applications, such as telecommunications, medical science and miscellaneous analytical studies.
- Computation and signal processing within the instrument is possible.

The major limitation of a digital storage oscilloscope is that only the signal information recorded during each sampling instant is the source of information. No new signal information is detected during the time of processing the previous sample. In other words, the instrument is unable to detect any possible signal changes occurring between sampling instants (e.g. a rapid transient). This problem is overcome with the digital phosphor oscilloscope, another version of digital oscilloscope that operates with parallel processing architecture for capturing, displaying and analyzing the signal. It continuously updates the database for every pixel of the oscilloscope display.

Multiple Choice-Type Questions and Answers

12.1 In the fabrication of transistor in IC, the collector ....
  (a) is provided by the substrate
  (b) is diffused separately at first
  (c) is diffused separately at the end
  (d) need not be fabricated separately

12.2 In the fabrication of diode in IC, some transistors are converted to diodes by employing the .... junction as diode and the .... being short circuited.
  (a) emitter–base, emitter and collector
  (b) collector–base, emitter and collector
  (c) emitter–base, collector and base
  (d) collector–base, emitter and base

12.3 The output voltage of an ideal voltage regulator is independent of ....
  (a) output current
  (b) load impedance
  (c) output current and load impedance
  (d) input voltage alone

12.4 The brightness of the trace of the waveform visible at CRO screen can be changed by adjusting the .... voltage.
  (a) cathode
  (b) anode
  (c) filament
  (d) control grid

12.5 The CRO may be considered as a *linear voltage-sensitive device* because its ....
  (a) deflection is proportional to the deflecting voltage
  (b) deflection factor is proportional to the deflecting voltage
  (c) deflection sensitivity is proportional to the accelerating voltage
  (d) deflection sensitivity is inversely proportional to the accelerating voltage

12.6 The deflection sensitivity of CRO can be increased by decreasing the accelerating voltage. The brightness of the spot on the screen .... in this process.
  (a) increases
  (b) decreases
  (c) fluctuates
  (d) remains unchanged

12.7 Two sinusoidal signals of frequencies ω_x and ω_y having same amplitude are applied to X- and Y-channels of a cathode ray oscilloscope, respectively. The stationary figure of Figure 12.16 will be observed when ....
  (a) $\omega_y = \omega_x$
  (b) phase difference is 0
  (c) $\omega_y = 2\omega_x$
  (d) phase difference is $\pi/2$

[JAM 2016]

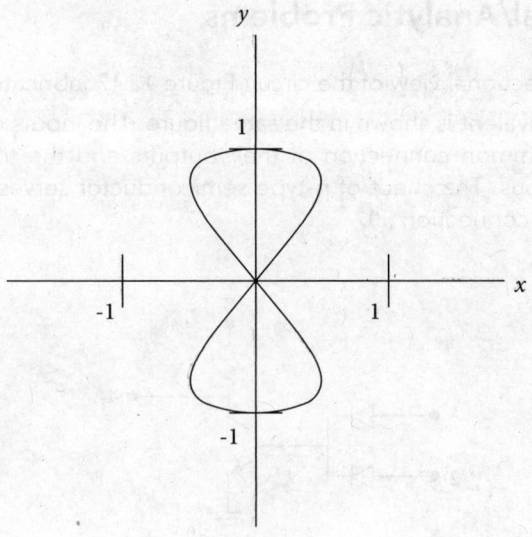

Figure 12.16

Answers

12.1 (b)	12.2 (c)	12.3 (c)	12.4 (d)
12.5 (a)	12.6 (b)	12.7 (b)	

Reasoning-Type Questions and Answers

12.1 How are the different components of an integrated circuit isolated electrically?

Ans. During the isolation diffusion, the discrete regions of doping opposite to that of the substrate act as back-to-back p–n junctions and get electrically isolated. The subsequent components, such as diodes or transistors fabricated within these regions also remain isolated.

12.2 The regulated power supply contains a voltage regulator. Can there be any current analogy too?

Ans. Yes, there are current regulators, which can supply constant current through a load irrespective of its change, hence the change of the voltage across the load.

12.3 Where go the electrons after hitting the CRO screen?

Ans. The electrons in the focused beam of the cathode ray tube trace the waveform by impinging on the fluorescent screen. Then the electrons must come back to the anode so as to complete the circuit. Otherwise the negative charge accumulated at the screen would repel the electron beam. The return path for the electrons is provided by *aquadag*; the sidewalls coated with carbon particles kept at high positive potential.

Solved Numerical/Analytic Problems

12.1 Draw the cross-sectional view of the circuit Figure 12.17 fabricated in monolithic IC.

Soln. The IC equivalent is shown in the same figure. The inputs of the two diodes (point 1 and 2), the common connection of their outputs and the transistor with grounded emitter are obvious. The sheet of p-type semiconductor serves the role of the resistor with the external connection (3).

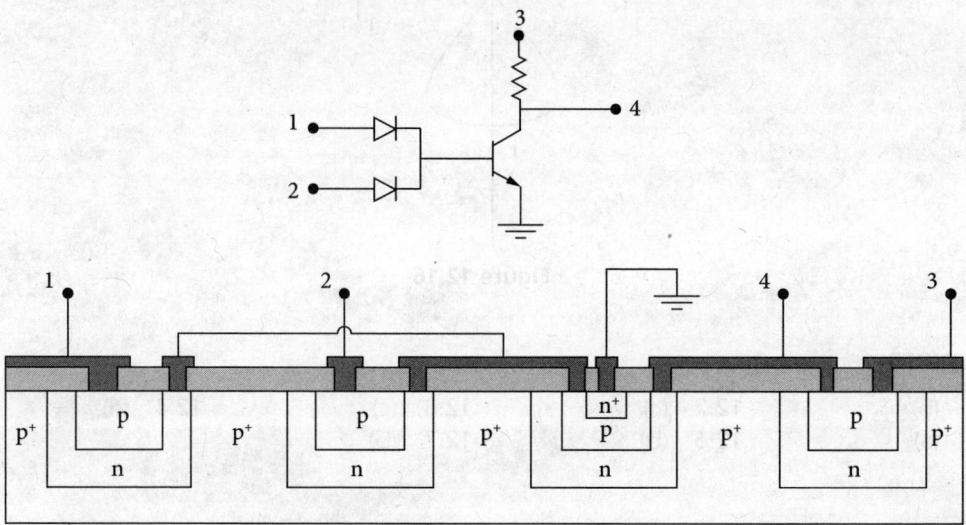

Figure 12.17

12.2 A regulated power supply produces output voltage of 12 V for a load current of 10 mA. The voltage reduces to 11.75 V for load current of 150 mA. Determine the output resistance of the regulator.

Soln. Considering only the magnitudes of change in voltage and current, the output resistance can be calculated as

$$(12 - 11.75)/[(150 - 10) \times 10^{-3}] = 1.78 \, \Omega$$

12.3 The deflection plates of a CRO have length of 3 cm and separation of 5 mm. The distance of the screen from the centre of the plates is 20 cm. If the required deflection sensitivity is 0.5 mm/V, determine the necessary voltage to be applied to the accelerating anode.

Soln. The given deflection sensitivity may be written as 0.05 cm/V, the plate length in 3 cm, the plate separation is 0.5 cm and the screen distance is 20 cm. Let the necessary voltage be V. Hence, using Equation (12.12),

$$0.05 = \frac{3 \times 20}{2 \times 0.5 \times V} \quad \text{so that } V = 1200 \text{ volt.}$$

Exercise

Subjective Questions

1. Why is silicon a convenient material for IC? Briefly introduce the basic steps involved in fabricating a monolithic IC using necessary diagrams.
2. Explain how resistors and capacitors are fabricated on a monolithic IC. What is the significance of the term *sheet resistance*?
3. Explain, how the electron beam is produced and focused in a cathode ray oscilloscope. What are the functions of (i) aquadag and (ii) delay line?

Conceptual Test

1. Is the silicon dioxide layer grown on silicon as an epitaxial layer? Justify your inference.
2. Draw the circuit diagram of a +6 V regulated power supply using a step down transformer and any regulator IC chosen with the help of Table 12.5.
3. The regulated power supply circuit of Figure 12.3(b) is called *series regulator* because the regulator acts as a variable resistance in series with the load. The voltage across the variable resistance is adjusted in such a way that the voltage across the load remains fixed. Now explain what this 'variable resistance' is.
4. It is fact that a regulated power supply provides with a constant voltage across the load. Would the term 'regulated voltage supply' be a more appropriate nomenclature?

Numerical Problems

1. A rectangular semiconductor sheet has resistivity of 0.5 Ω cm and thickness of 5 μm. If the length be twice the width, what would be the resistance of the sheet along the length?
2. The deflection sensitivity of a CRO tube is 0.5 mm/V and an applied sinusoidal voltage can trace a straight line of 5 cm length on the screen. Calculate the rms value of the sinusoidal voltage.

13

Digital Principles and Boolean Algebra

The working principles of computers and other digital systems are based on binary and hexadecimal number systems. This chapter introduces different number systems, such as binary, octal, and hexadecimal, and their relationships with our familiar decimal system. Some popular digital codes are also mentioned briefly. Arithmetic operations with binary numbers, development of algebraic techniques with Boolean variables and simplifications of Boolean expressions are illustrated. Indeed this chapter deals with the mathematical and theoretical portion of the digital system. All practical implementations with electronic circuits are included in the subsequent chapters.

13.1 The Digital System

The word 'digital' originates from digit, meaning separate organs or components like fingers of human hand and each of the keys of a keyboard. In electronic operations, the digit represents two discrete electrical states, such as on/off conditions of a switch or presence/absence of voltage at a terminal. In the present context, digital refers to those electronic circuits which realize binary numbers in terms of two discrete voltage levels and perform mathematical and logical operations based on the principles of Boolean algebra.

13.1.1 Analog and Digital

The word 'analog' is derived from the Greek word *analogos*, which means '*according to a certain ratio*'. The computation with the operational amplifier, as explicated in Chapter 11, is analog in nature because the voltage values are analogous to numbers. The increment and decrement of numbers is in proportion to the voltage. The following characteristics are the major demarcations between analog and digital systems.

- *Continuous and Discrete*: An analog device consists of continuously changing quantity. For example, a continuous curve in XY-plane is an analog graph. The voltage change in operational amplifier is continuous. In contrary, a bar chart can be treated as a digital graph. The digital electronic circuits handle only two discrete levels of voltage.

- *Size-Dependent Accuracy*: The precision of an analog instrument depends on its size, marking of scale and other physical parameters. The least count of a slide calipers and that of a travelling microscope are different. A small size transistor has less power handling capability than that of a larger one. In contrast, the accuracy of a digital instrument does not depend on its size. For instance, the ancient abacus is a digital equipment having precision of calculation independent of the wire length or the bead size. The accuracy of a digital circuit fabricated with large size discrete electronic components and that of a miniature similar circuit fabricated on IC may be of the same accuracy.

- *Task-Specific or Not*: An analog system is dedicated to a specific job only. The integrator circuit made with the operational amplifier cannot differentiate a voltage waveform. A screw gauge or a spherometer is a sort of mechanical analog calculator made for some specific purpose. However, a digital calculator or a digital computer can perform varieties of mathematical operations.

13.1.2 Pros and Cons of Digital System

The word 'digital' is now becoming of growing importance in various fields of modern science and technology as well as in daily life. In addition to computer, telecommunication, and Internet, we often talk about digital camera, digital photocopy, digital x-ray, digital sound recording, digital thermometer, and many others of such category. Whatever can be processed with computer is designated as 'digital'. Non-electrical quantities can be made digital first by converting to proportional electrical quantities using a suitable transducer and then transforming the analog electrical signal to equivalent digital quantity using suitable electronic circuits. The digital system has got a lot of advantageous features.

- The digital system is less affected by noise because it can accommodate a range of variation for a discrete state. A digital picture copied several times retains the original accuracy.
- The precision and accuracy of a digital system can be increased easily by adding as many digits as required.
- The full advantage of IC technology can be availed of and the use of bulky devices, such as high value capacitors, precision resistors and inductors can be minimized.
- Storage of information is easier in terms of 0 and 1 that has built up the huge regime of cyber world ranging from a tiny pen drive to the world-wide cloud for computation.

However, the digital system is not similar to the world we are familiar with. The binary and hexadecimal numbers are not as much familiar to us as the decimal system. The physical quantities around us, namely temperature, pressure, position, and velocity are all analog in nature. These are to be converted to digital machine-readable form and this stipulation imposes a limitation on the accuracy. There are certain requirements, such as amplifying an electrical signal, where the analog system is easier.

13.2 Number Systems and Conversions

We are so conversant with decimal numbers that their construction is seldom thought about. Anyway there is a general rule followed by the number systems and in the case of binary and other systems we have to take care of that consciously. Let us start with our familiar decimal system. It has the following features.

(i) Ten different digits: 0, 1, 2, 3, 4, 5, 6, 7, 8 and 9 are used for constructing a decimal number. Each digit represents certain quantity.

(ii) Any decimal number has a *base* of 10. An integer and a fraction can be expressed as a polynomial of increasing and decreasing powers of 10, respectively. For example: $2022.56 = 2 \times 10^3 + 0 \times 10^2 + 2 \times 10^1 + 2 \times 10^0 + 5 \times 10^{-1} + 6 \times 10^{-2}$.

(iii) The position of the individual digits indicate different quantities. For instance, the decimal numbers 2021, 1202 and 2201 use the same digits at different positions and represent different quantities. Hence the decimal system is a *positional number system*. Each position of a digit is assigned a *weight*, which is the power of 10.

Generalizing the above concept, any number N in a positional number system of base b can be expressed as

$$N = a_{n-1}b^{n-1} + a_{n-2}b^{n-2} + \ldots + a_1b^1 + a_0b^0 + a_{-1}b^{-1} + a_{-2}b^{-2} + \ldots + a_{-k}b^{-k} \qquad (13.1)$$

In more compact form it is written as

$$N = \sum_{i=-k}^{n-1} a_i b^i \qquad (13.2)$$

Equation (13.2) has the following specifications. The base (b) is also called the *radix* of the number system. The coefficients $0 \leq a_i \leq b - 1$ represent different powers of b, a_{n-1} is called the *most significant digit* and a_{-k} is called the *least significant digit* of the number.

Table 13.1 Comparison of decimal and binary number systems

Decimal	Binary
It is base-10 system comprising ten different digits 0, 1, 2, ..., 9.	It is base-2 system comprising only two different digits 0 and 1. The binary digit is abbreviated as *bit*.
It is positional system. Each position of the digit indicates its weight as the power of 10. The decimal weights for integers are $10^0 = 1$, $10^1 = 10$, $10^3 = 1000$ etc. and the decimal weights for fractions are $10^{-1} = 0.1$, $10^{-2} = 0.01$, $10^{-3} = 0.001$ etc.	It is also positional system. Each position of a bit indicates its weight as the power of 2. For integers, the binary weights are $2^0 = 1 = 1$ (binary) $2^1 = 2 = 10$ (binary) $2^2 = 4 = 100$ (binary) and so on. For fractions, the binary weights are $2^{-1} = 0.5 = 0.100$ (binary) $2^{-2} = 0.25 = 0.010$ (binary) $2^{-3} = 0.125 = 0.001$ (binary) etc.
To count decimal numbers, we start at 0 and count up to 9. Then the count is reset to 0 and restarts from the next higher position. Examples:	Binary numbers start from 0 and count up to 1 only. Then the position is reset to 0 and starts again at the next higher position. Examples of binary and equivalent decimal:

0	11	99	Binary	Decimal	Binary	Decimal	
1	12	100	0	0	1001	9	
2	–	–	1	1	1010	10	
–	19	999	10	2	1011	11	
9	20	1000	11	3	1100	12	
10	–		100	4	1101	13	
			101	5	1110	14	
			110	6	1111	15	
			111	7	10000	16	
			1000	8			

13.2.1 Binary Numbers

This number system has base of 2 and only two digits, 0 and 1. All other numbers are represented by combinations of 0 and 1. The weight of each position is a power of 2. For example, the binary equivalent of decimal 5.75 is 101.11, which is composed as $1 \times 2^2 + 0 \times 2^1 + 1 \times 2^0 + 1 \times 2^{-1} + 1 \times 2^{-2}$.

The binary system is useful to computer and other digital systems because only two discrete numbers 0 and 1 can be indicated easily by the presence and absence of voltage. Table 13.1 makes a comparison of the properties of decimal and binary numbers. It is understood that the binary weights for integers are 2, 4, 8, 16, 32 etc. and that for fractions are 0.5, 0.25, 0.125 etc. Sometimes the binary number resembles a decimal number but pronounced in a different way. For example binary 4 is 100, pronounced as 'one zero zero'. Binary 9 (1001) is not 'one thousand one' but 'one zero zero one'. The left most bit of a binary number is called its *most significant bit* (MSB) and the right most bit is called its *least significant bit* (LSB).

> **Note: Bit, Byte, KB and MB**
>
> A binary digit, either 0 or 1 is called a bit. A group of eight bits is known as a byte. It is considered as a unit of memory size in digital systems. There is another unit named *nibble*, which is a 4-bit aggregation. However, the use of byte is more prevalent. A collection of 1024 bytes (close to 10^3) is called *a kilobyte* and is denoted by KB. Similarly, a group of 1048576 bits, close to 10^6, is termed as a megabyte (MB). A gigabyte (GB) is close to 10^9. The origin of such naming is the power of 2 in binary system. For instance, $1024 = 2^{10}$. Since the prefixes 'kilo', 'mega' etc. of the International System of Units (SI) are applicable to powers of 10, the International Electrochemical Commission (IEC) prescribed another system of nomenclature, such as *kibibyte* (KiB), *mebibyte* (MiB) and *gibibyte* (GiB). However, the binary representation is still of widespread use and the present book follows the same.

13.2.2 Binary to Decimal Conversion

Following the general rule [Equation 13.1(a) and (b)], any number can be expressed as a sum of digits multiplied by powers of the base. Therefore, the given binary number can be converted to decimal by multiplying each bit (0 or 1) by the corresponding power of 2 and adding the results. Few examples of such binary to decimal conversion are given below.

Binary	Decimal
111	$1 \times 2^2 + 1 \times 2^1 + 1 \times 2^0 = 7$
10101	$1 \times 2^4 + 0 \times 2^3 + 1 \times 2^2 + 0 \times 2^1 + 1 \times 2^0 = 21$
1101.10	$1 \times 2^3 + 1 \times 2^2 + 0 \times 2^1 + 1 \times 2^0 + 1 \times 2^{-1} = 13.5$
11001.101	$1 \times 2^4 + 1 \times 2^3 + 0 \times 2^2 + 0 \times 2^1 + 1 \times 2^0 + 1 \times 2^{-1} + 0 \times 2^{-2} + 1 \times 2^{-3} = 25.625$

Sometimes the base is indicated as subscript against the number. Following that concept, the decimal and binary equivalents can be written as, for example,

$(1011)_2 = (11)_{10}$
$(110011.101)_2 = (51.625)_{10}$
$(1111.1111)_2 = (15.9375)_{10}$

13.2.3 Decimal to Binary Conversion

There are two popular methods of converting a given decimal number to its binary equivalent. The first one, known as *sum of weight method* expresses the decimal number as a sum of binary weights. Each available weight position is indicated by 1 and each absent weight position is indicated by 0. Examples:

$(11)_{10} = 8 + 2 + 1$
$\qquad = 2^3 + 0 + 2^1 + 2^0$
$\qquad = (1011)_2$

$(27)_{10} = 16 + 8 + 2 + 1$
$= 2^4 + 2^3 + 0 + 2^1 + 2^0$
$= (11011)_2$

$(0.625)_{10} = 0.5 + 0 + 0.125$
$= 2^{-1} + 0 + 2^{-3}$
$= (0.101)_2$

$(17.3125)_{10} = 16 + 1 + 0.25 + 0.0625$
$= 2^4 + 0 + 0 + 0 + 2^0 + 0 + 2^{-2} + 0 + 2^{-4}$
$= (10001.0101)_2$

The second method for decimal to binary conversion, known as *double-dabble method* is based on repeated division by 2 for integers and repeated multiplication by 2 for fractions. For integers, the given decimal number is repeatedly divided by 2 until zero quotient is reached. The remainders (of course 0 or 1) for each division collected in the reverse sequence represents the resultant binary number. The following example of converting decimal 14 to binary equivalent with line-by-line interpretation clarifies the process.

```
2|14  0 (LSB)    ↑    (14 is divided by 2 producing quotient of 7 and remainder of 0)
2|7   1               (7 is divided by 2 producing quotient of 3 and remainder of 1)
2|3   1               (3 is divided by 2 producing quotient of 1 and remainder of 1)
2|1   1 (MSB)         (1 cannot be divided by 2, quotient is 0 and remainder is 1)
  0
```

Thus $(1110)_2$ is the equivalent of $(14)_{10}$. Two more examples for integer conversion are given below.

```
2|25  1 (LSB)  ↑      2|35  1 (LSB)  ↑
2|12  0               2|17  1
2|6   0               2|8   0
2|3   1               2|4   0
2|1   1 (MSB)         2|2   0
  0                   2|1   1 (MSB)
                        0
```

Thus $(25)_{10} = (11001)_2$ $(35)_{10} = (100011)_2$

For fractions, the given decimal number is multiplied by 2 and the integral part of the product (0 or 1) is collected. The resultant fraction of the product is again multiplied by 2 and the integral part is collected. This process is continued until the fractional part becomes zero. The collection of 1 and 0 in forward sequence constitutes the binary equivalent. The following example of converting decimal fraction 0.3125 to the binary equivalent demonstrates the method. The arrows are given only for the sake of easy understanding.

$0.3125 \times 2 \quad = 0.625$
$\quad \quad \quad \quad \quad \quad \quad \rightarrow 0$ (MSB)

$0.625 \times 2 \quad = 1.25$
$\quad \quad \quad \quad \quad \quad \quad \rightarrow 1$

$0.25 \times 2 \quad = 0.5$
$\quad \quad \quad \quad \quad \quad \quad \rightarrow 0$

$0.5 \times 2 \quad = 1.0$
$\quad \quad \quad \quad \quad \quad \quad \rightarrow 1$ (LSB)

The above conversion results in $(0.3125)_{10} = (0.0101)_2$.

Two more examples are given below to clarify the technique. Hope the arrow signs are no more needed.

$0.8125 \times 2 \quad = 1.625$
$\quad \quad \quad \quad \quad \quad \quad 1$ (MSB)

$0.625 \times 2 \quad = 1.25$
$\quad \quad \quad \quad \quad \quad \quad 1$

$0.25 \times 2 \quad = 0.5$
$\quad \quad \quad \quad \quad \quad \quad 0$

$0.5 \times 2 \quad = 1.0$
$\quad \quad \quad \quad \quad \quad \quad 1$ (LSB)

$0.6875 \times 2 \quad = 1.375$
$\quad \quad \quad \quad \quad \quad \quad 1$ (MSB)

$0.375 \times 2 \quad = 0.75$
$\quad \quad \quad \quad \quad \quad \quad 0$

$0.75 \times 2 \quad = 1.5$
$\quad \quad \quad \quad \quad \quad \quad 1$

$0.5 \times 2 \quad = 1.0$
$\quad \quad \quad \quad \quad \quad \quad 1$ (LSB)

Therefore, $(0.8125)_{10} = (0.1101)_2 \quad \quad (0.6875)_{10} = (0.1011)_2$

It may so happen that the fractional part does not become zero even after a number of steps. In that case, the process is terminated after achieving a predetermined accuracy. Example: decimal fraction 0.98 is being converted to binary.

$0.98 \times 2 \quad = 1.96$
$\quad \quad \quad \quad \quad \quad \quad 1$

$0.96 \times 2 \quad = 1.92$
$\quad \quad \quad \quad \quad \quad \quad 1$

$0.92 \times 2 \quad = 1.84$
$\quad \quad \quad \quad \quad \quad \quad 1$

$0.84 \times 2 \quad = 1.68$
$\quad \quad \quad \quad \quad \quad \quad 1$

$0.68 \times 2 \quad = 1.36$
$\quad \quad \quad \quad \quad \quad \quad 1$

$0.36 \times 2 \quad = 0.72$
$\quad \quad \quad \quad \quad \quad \quad 0$

$0.72 \times 2 = 1.44$
    1
$0.44 \times 2 = 0.88$
    0

The zero condition of the fractional part has not reached as yet. If we terminated after 4 bits, the result would be $(0.1111)_2 = (0.9375)_{10}$, a bit different ($\approx 4\%$) from the original number. If we converted up to 8 bits, as given above, the result would be $(0.11111010)_2 = (0.976)_{10}$, much closer (< 1%) to the original value. Thus the extent of conversion depends on the specific requirement of accuracy. It is worth mentioning that increasing the number of bits may not always lead to accuracy. For instance, if we continued up to 9 bits in the above case, the result would be $(0.9804)_{10}$, exceeding the original value.

If a decimal number contains both integral and fractional parts, the integer and the fraction are converted separately and joined together. For example, if the given decimal number be 15.3125, first $(15)_{10}$ is converted to $(1111)_2$, then $(0.3125)_{10}$ is converted to $(0.0101)_2$ and then the complete binary equivalent of the given decimal number is written as 1111.0101. The above rule for inter-conversion is followed for the other number systems also.

13.2.4 Octal Numbers

This number system has the base of 8 and it uses eight digits: 0, 1, 2, 3, 4, 5, 6 and 7. The next octal number is 10 (equivalent to decimal 8). Some decimal and equivalent octal numbers are given below for easy understanding. Total counts with n octal digits is 8^n.

Decimal	Octal	Decimal	Octal	Decimal	Octal
0	0	9	11	17	21
1	1	10	12	18	22
2	2	11	13	19	23
3	3	12	14	20	24
4	4	13	15	21	25
5	5	14	16	22	26
6	6	15	17	23	27
7	7	16	20	24	30
8	10				

The octal number system is a compromise between the machine-readable binary system and the decimal system that human beings are familiar with. The octal, binary and decimal numbers can be interconverted easily in the following ways.

A. *Octal to Binary Conversion*: Each octal digit is converted to its binary equivalent and is written in 3-bit form. Zero is added to the left, if necessary. The resultant chain of 1s and 0s represents the binary equivalent. Examples:

$(35)_8 = (011101)_2$

$(273)_8 = (010111011)_2$

$(24.76)_8 = (010100.111110)_2$

B. *Binary to Octal Conversion*: This is just the opposite process. The given binary number is arranged in groups of three bits. If necessary, zeros are added to the left. Then each group of three bit is converted to the corresponding octal digit. Examples:

$(1101)_2 = (001\ 101)_2 = (15)_8$
$(10110)_2 = (010\ 110)_2 = (26)_8$
$(1011.011)_2 = (001\ 011.011)_2 = (13.3)_8$
$(110.01101)_2 = (110.011\ 010)_2 = (6.32)_8$

C. *Octal to Decimal Conversion*: The conversion of octal to decimal and vice versa is similar to binary–decimal conversions; just the base being replaced by 8. The weights are ascending powers of 8 for integers and descending powers of 8 for fractions. Examples:

$(24)_8 = 2 \times 8^1 + 4 \times 8^0 = (20)_{10}$
$(255.14)_8 = 2 \times 8^2 + 5 \times 8^1 + 5 \times 8^0 + 1 \times 8^{-1} + 4 \times 8^{-2} = (173.19)_{10}$

The decimal fraction is approximated in the second example.

D. *Decimal to Octal Conversion*: This can be done by repeated division by 8 for integers and repeated multiplication by 8 for fractions in a similar way as decimal–binary conversion. Examples for integers:

```
8|38   6 (LSB)          8|273   1 (LSB)
8|4    4 (MSB)          8|34    2
  0                     8|4     4 (MSB)
                          0
```

Thus

$(38)_{10} = (46)_8$ $(273)_{10} = (421)_8$

For fractions, the given decimal fraction is multiplied by 8 and the integral part of the product is collected. The remaining fraction is multiplied again by 8 and the integral part is collected. Examples:

$0.111 \times 8 = 0.888$ → 0 (MSB)
$0.888 \times 8 = 7.104$ → 7
$0.104 \times 8 = 0.832$ → 0
$0.832 \times 8 = 6.656$ → 6 (LSB)

$0.1425 \times 8 = 1.1400$ → 1 (MSB)
$0.14 \times 8 = 1.12$ → 1
$0.12 \times 8 = 0.96$ → 0
$0.96 \times 8 = 7.68$ → 7 (LSB)

The results are:

$(0.111)_{10} = (0.0706)_8$ $(0.1425)_{10} = (0.1107)_8$

In both the cases, the process has been terminated after several digits. Reconverting the octal fraction does not return the exact value of the original decimal fraction. However, the error is neglected.

13.2.5 Hexadecimal Numbers

This number system has the base of 16 and it uses sixteen digits. In addition to decimal digits 0 through 9, letters A, B, C, D, E and F are used as digits assigning numerical values of 10, 11, 12, 13, 14 and 15, respectively. The sequence of hexadecimal numbers is given below.

Decimal	Hexadecimal	Decimal	Hexadecimal	Decimal	Hexadecimal	Decimal	Hexadecimal
0	0	9	9	17	11	25	19
1	1	10	A	18	12	26	1A
2	2	11	B	19	13	27	1B
3	3	12	C	20	14	28	1C
4	4	13	D	21	15	29	1D
5	5	14	E	22	16	30	1E
6	6	15	F	23	17	31	1F
7	7	16	10	24	18	32	20
8	8						

The hexadecimal number is a better compromise between the machine-readable binary and the human-friendly decimal system. The hexadecimal, binary and decimal numbers can be interconverted in the following ways.

A. Hexadecimal to Binary Conversion: Each hexadecimal digit is converted to its binary equivalent and is written in 4-bit form. Zeros are added to the left, if necessary. Examples:

$(2A8)_{16} = (0010\ 1010\ 1000)_2$

$(C05E)_{16} = (1100\ 0000\ 0101\ 1110)_2$

$(3F.71)_{16} = (0011\ 1111.0111\ 0001)_2$

B. Binary to Hexadecimal Conversion: The given binary number is arranged into groups of 4 bits and each group is converted to hexadecimal equivalent. Examples:

$(11011010)_2 = (1101\ 1010)_2 = (DA)_{16}$

$(1011001)_2 = (0101\ 1001)_2 = (59)_{16}$

$(1111.0101001)_2 = (1111.0101\ 0010)_2 = (F.52)_{16}$

C. Hexadecimal to Decimal Conversion: Each hexadecimal digit is multiplied by the corresponding power of 16 and the results are added. Examples:

$(FF)_{16} = F \times 16^1 + F \times 16^0 = 15 \times 16 + 15 \times 1 = (255)_{10}$

$(F8A)_{16} = F \times 16^2 + 8 \times 16^1 + A \times 16^0 = (3978)_{10}$

$(1E.4B)_{16} = 1 \times 16^1 + E \times 16^0 + 4 \times 16^{-1} + B \times 16^{-2} = (31.293)_{10}$ approx.

D. Decimal to Hexadecimal Conversion: Now we are supposed to get familiar with the general rule of conversion and attempt both integer and fraction at a time. Suppose the decimal number 3978.323 is to be converted to hexadecimal. The integer part is converted as

16|3978   10 = A (LSB)
16|248   8
16|15   15 = F (MSB)

Thus $(3978)_{10} = (F8A)_{16}$. Now the fractional part is converted as follows.

$0.323 \times 16 \quad = 5.168$
$\qquad\qquad\qquad 5$ (MSB)
$0.168 \times 16 \quad = 2.688$
$\qquad\qquad\qquad 2$
$0.688 \times 16 \quad = 11.008$
$\qquad\qquad\qquad 11 = B$ (LSB)
$0.008 \times 16 \quad = 0.128$
$\qquad\qquad\qquad 0$

We obtain $(0.323)_{10} = (52B)_{16}$. Combining the converted integral and fractional components, the actual hexadecimal equivalent of the given decimal 3978.323 is obtained as F8A.52B.

13.3 Digital Codes

The arithmetic and logical operations within computers, calculators and other digital instruments are executed in terms of binary numbers. The electronic circuits within these instruments can understand only two discrete states of voltage representing 0 and 1, respectively. Similarly, the alphabetic and symbolic instructions are also specified in terms of 0s and 1s. Such specifications of information are called *codes*. Both the binary number system and digital coding systems are based on the two-state logic of 0 and 1. However, there are the following differences.

- The number system has a certain base and all the numbers are represented as weighted sums of the digits involving the powers of the base. The coding system need not be bound to such regulation.
- The number system is positional. Each position of a digit has certain weight, in terms of the power of base, determining its numerical value. The code may be non-positional and not weighted. Gray code is such an example.

> **Note: Why Octal and Hexadecimal**
>
> The binary system fits well to the operation of digital equipment with two discrete states. It is very difficult to devise ten different indicators to implement the decimal system to the machine. However, the human user is more familiar with the decimal system. The octal and the hexadecimal numbers give the user the comfort of handling binary numbers in a similar way as that of decimal numbers. The octal system was widely used in earlier computing systems working with 12-bit, 24-bit and 36-bit binary words. For instance, any 12-bit word could be represented by four octal digits and no extra symbol, other than the decimal digits were needed. Modern computers work with the group of 16, 32 and 64 bits. In such a case, the hexadecimal system is more convenient to system designers and programmers to describe memory locations and to make binary codes suitable for computer coding. However, it requires six extra symbols, A through F, used as numerals in addition to the ten decimal digits.

Many diversified types of codes are used in the scientific and commercial applications of digital electronics. Only a few are introduced very briefly.

A. ASCII Code: The alphabetic characters, decimal numbers and other symbols, such as punctuation marks on the keyboard should be somehow coded in terms of 0 and 1 because the central processing unit of the computer does not understand anything else. Such coding of alphabetic characters, decimal numbers and other symbols of the keyboard is called *alphanumeric coding*. A popular alphanumeric code is *American Standard Code for Information Interchange* (ASCII). It is a 7-bit code of 0 and 1 for standardizing the computer keyboard. Another coding system of this category is 8-bit *Extended Binary Coded Decimal Interchange Code* (EBCDIC).

B. Universal Product Code: Consumer products like stationeries, books, electric bills etc. contain a machine-readable label comprising a number of parallel black and white bars of variable width. The digits 0 through 9 are expressed in terms of the bar width variation. The code also contains information on the manufacturer and the product identity. It is easily read by laser-aided optical reader.

C. Error Detection Code: The 0 and the 1 in a digital system is simulated by two discrete states of voltage. If somehow some false 0 or 1 gets inserted due to noise or circuit malfunctioning, especially in the case of long range digital data transmission, the receiving end gets erroneous information. There are several coding systems for detecting such errors. One such technique is *parity checker*, discussed in Chapter 14. A code named *Gray code* after American physicist Frank Gray (1887–1969) is used to facilitate error correction in digital communications. Two successive values in this system differ in only one bit. The use of Gray code in distinguishing the grid cells of Karnaugh map is introduced in section 13.8.

D. Binary Coded Decimal (BCD): This coding system is applied to the display of decimal digits in digital instruments like calculator, digital watch and laboratory instruments. To convert a decimal number to BCD, each decimal digit is converted to its 4-bit binary equivalent, called *nibble*. Examples:

$(527)_{10} = (0101\ 0010\ 0111)_{BCD}$

$(36.8)_{10} = (0011\ 0110.1000)_{BCD}$

Similarly a BCD to decimal conversion involves grouping of bits into nibbles with 0 added, if necessary. Examples:

$(1001001)_{BCD} = (0100\ 1001)_{BCD} = (49)_{10}$

$(1100011.01)_{BCD} = (0110\ 0011.0100)_{BCD} = (63.4)_{10}$

The BCD is sometimes called *8421 code* because the weight of the digit positions from MSB to LSB are 8, 4, 2 and 1, respectively. The range of a BCD digit can be 0000 through 1001 (0 through 9). The higher 4-bit binary numbers cannot be incorporated because the highest decimal digit is 9.

13.4 Binary Arithmetic

The arithmetic operations with binary numbers can be carried out in a similar way as that with decimal numbers. Few examples are given below.

```
Addition: 1101 + 1010              Subtraction: 1101 – 1001
1101 (= 13)     augend             1101 (minuend)
+1010 (= 10)    addend             –1001 (subtrahend)
 10111 (= 23)   sum                 0100 (difference)

Multiplication: 1101 × 11          Division: 10101 ÷ 111
     1101                          111)10101(11
       11                              111
     1101                              00111
    1101×                                111
   100111                                  ×
```

First of all, the young reader must not make a habit of cross-checking binary operations by converting into decimal, as is done in one of the above examples. Always try to identify binary numbers in binary form. The next point to be noted is that the above examples of mathematical operations with binary numbers have been executed in a human-like way. However, the arithmetic processes with binary numbers in electronic circuits are performed in an absolutely different way.

- Subtraction is done as negative addition. The negative of a number is generated by specific conversion techniques, known as *1's complement* and *2's complement* discussed later.
- Multiplication is done as repeated addition.
- Division is done as repeated subtraction (negative addition).

The addition of two bits at any position of binary numbers may have the following situations.

$0 + 0 = 0$ (sum = 0, carry = 0)

$0 + 1 = 1$ (sum = 1, carry = 0)

$1 + 1 = 10$ (sum = 0, carry = 1)

Also the carry bit from the previous position may add to produce $1 + 1 + 1 = 11$, when sum = 1 and carry = 1.

Another very significant point to note is that the number of bits in the binary numbers involved in the arithmetic process should be specified and maintained. If the result exceeds the number of digits allotted, the condition is referred to as *overflow*.

> **Note: Overflow May Be Dangerous**
>
> A rocket named Ariane 5, developed and operated by European Space Agency exploded and lost immediately after the launch in 1996 due to loss of computer guidance originated from the malfunctioning of the controlling software because of data overflow.

13.4.1 1's Complement and 2's Complement

The *1's complement* of a binary number is the result of complementing its each bit: changing each 0 to 1 and each 1 to 0. When a 1 is added to this 1's complement, it results in the *2's complement* of the binary number. The followings are some examples.

Binary number	1's complement	2's complement
1100	0011	0100 (0011 + 1)
11000101	00111010	00111011
00110110	11001001	11001010
1101.10	0010.01	0010.10

Making the 2's complement of a binary number gives back the original one. The 2's complement is a useful tool for binary subtraction. The principle of the subtraction process is to make the negative of the number by converting to its 2's complement. When that negative is added to another number, the effect of subtraction is obtained, as illustrated with the following examples.

> **Example 13.1**
>
> Execute 15 − 7 in binary form using 2's complement method.
>
> **Solution**
>
> The job is equivalent to 15 plus 2's complement of 7 in binary. $(15)_{10} = (1111)_2$ and $(7)_{10} = (111)_2$ but this must be written as $(0111)_2$ in the present case to make the number of bits the same for both the numbers. This is an important point to remember in 2's complement technique' the number of bits should be equal for all the numbers.
>
> Now 7 = 0111, 1's complement of 7 = 1000 and 2's complement of 7 = 1001. So 15 − 7 is executed as
>
> ```
>     1111
>    +1001
>    11000
> ```
>
> The left-most carry bit (1) is to be discarded because the operation is done with 4 bits. Deleting the left-most 1, the remaining 4 bits represent 8, which is the correct answer.

> **Example 13.2**
>
> Repeat the above subtraction of 15 − 7 with 8 bits.
>
> **Solution**
>
> Now 15 = 0000 1111, 7 = 0000 0111 and the 2's complement of 7 is 11111001. Therefore 15 − 7 in 8 bits is executed as
>
> ```
>      0000 1111
>     +1111 1001
>     10000 1000
> ```
>
> Again discarding the left-most carry 1, the remaining 8 bits represent 0000 1000 = 8.
>
> The result becomes interesting, if we attempt to perform subtraction of larger numbers, as given in the following example.

Example 13.3

Execute 8 − 17 in binary using 2's complement method.

Solution

Now and 8 = 01000, 17 = 10001 and the 2's complement of 17 = 01111.

```
  01000
 +01111
  10111
```

Is the result −9, as expected? Yes, it is 9 in 2's complement form. If it is made 2's complement again, we get 01001, i.e., 9.

Example 13.4

Add −9 and −4 in binary.

Solution

This can be done by adding their 2's complements. Making the number of bits equal, 9 = 1001 and 4 = 0100. Making 2's complements and adding,

```
 −9 = 0111
 −4 = 1100
      10011
```

Ignoring the extra carry at the left, 0011 represents −13. If you don't believe, make the 2's complement again!

Example 13.5

Add 4 and −9 by 2's complement method.

Solution

4 = 0100 and −9 = 0111.

```
  0100
  0111
  1011        (The result is −5)
```

Using the 1's complement of a binary number can also be a useful technique for binary addition and subtraction operations. As will be discussed later, the positive and the negative sign of a binary number is denoted by putting 0 and 1, respectively, to the left. This convention is necessary to the 1's complement method.

- In the case of addition by 1's complement technique, the augend and the addend are written in signed binary form for positive numbers and in 1's complement form for negative numbers.
- In the case of subtraction, the 1's complement of the subtrahend is added to the minuend in signed binary form.

Any extra carry appearing in the sum is removed and added to the sum as *end-around carry*. The following examples will clarify the process.

Example 13.6

Add 7 and 5 by 1's complement method.

$$7 = 0111 \text{ (signed binary for 7, 0 for +ve)}$$
$$\underline{5 = 0101} \text{ (signed binary for +5)}$$
$$12 = 1100 \text{ (no extra carry in the result)}$$

Example 13.7

Subtract 5 from 9 using 1's complement method.

$$9 = 01001 \text{ (signed binary for +9)}$$
$$\underline{-5 = 11010} \text{ (1's complement of 5, equating the number of bits)}$$
$$100011 \text{ (sum with extra carry)}$$
$$\underline{\qquad\qquad 1} \text{ (extra carry added)}$$
$$00100 \text{ (= 4, the result)}$$

Example 13.8

Add −7 and −3 using 1's complement method.

$$-7 = 1000 \text{ (1's complement of 7)}$$
$$\underline{+(-3) = 1100} \text{ (1's complement of 3)}$$
$$10100 \text{ (sum with extra carry)}$$
$$\underline{\qquad\qquad 1} \text{ (carry added)}$$
$$0101 \text{ (result, 1's complement of 10)}$$

13.4.2 Radix Complements

Now we talk about the general definition of complementing a digit for any number system of base (radix) b.

- the *radix-minus-1 complement* of the number is obtained by subtracting each digit from $(b - 1)$, the largest digit of the system and
- the *radix complement* or *true complement* of the number is that obtained by adding 1 to the radix-minus-1 complement.

For decimal system, the base is 10 and the radix-minus-1 complement is called the *9's complement* of the number. It is obtained by subtracting each digit from 9 (= 10 − 1). For example, the 9's complement of 2021 is 7978. The true complement in decimal system is called 10's complement. For example, the 10's complement of 2021 is 7978+1 = 7979.

For a binary system, the base is 2 and so that the radix-minus-1 complement is called the 1's complement (each bit subtracted from 1) and the true complement is called 2's complement. Some examples are already given earlier.

We are so familiar with decimal arithmetic that it may seem boring to repeat it. Anyway we should complete our exercise on radix complementing. Similar to binary subtraction by 2's complement method, a decimal subtraction can be made by 10's complement method, as shown in the following example. Making the number of digits equal for both numbers by adding zeros, as necessary, is followed always.

Example 13.9

Subtract decimal 199 from decimal 273 by 10's complement method.

Solution

The 9's complement of 199 is 800 and the 10's complement of 199 is 801. The work of 273 − 199 is performed as

```
   273
  +801
  1074
```

Discarding the left-most 1 of carry, the remaining three digits give the result of 74.

Subtraction of decimal numbers by 9's complement method is possible by adding the 9's complement of the subtrahend to the minuend. The carry is not discarded but added to the sum. The following example clarifies the process.

> **Example 13.10**
>
> Subtract 23 from 57 by 9's complement method.
>
> ```
>     57
>   +76 (9's complement of 23)
>    133
>     +1 (carry 1 is added)
>     34 (result)
> ```

Subtraction of binary numbers by 1's complement method can be executed in a similar way. The 1's complement of the subtrahend is added to the minuend and the carry, if any, is added to the sum.

> **Example 13.11**
>
> Perform 1100 − 1001 by 1's complement method.
>
> ```
>    1100
>  +0110 (1's complement of 1001)
>   10010
>      +1
>    0011 (result)
> ```

13.4.3 Signed Binary Numbers

It is a very common requirement to indicate both the sign and the magnitude of a number, such as +7 and −7, +255 and −255 and so on. Similar necessity is faced with binary numbers also. In general, the negative sign for a base-b number system is indicated by $b - 1$. For binary, it is $2 - 1 = 1$. The distinction of positive and negative binary numbers can be made in several ways, as follows.

2's Complement Method: This is the most common way to denote signed binary numbers. The positive number is expressed by the magnitude with a sign bit of 0 at the left, similar to 1's complement method. The negative number is represented by the 2's complement of the signed magnitude (1 added to the 1's complement). The following examples explain that the same signed binary number stated in different number of bits have different forms but the left-most bit is always 0 for positive and always 1 for negative numbers.

Signed number	binary equivalent (5 bits)	binary equivalent (6 bits)
+12	01100	001100
−12	10100	110100
+7	00111	000111
−7	11001	111001

With n bits, the 2's complement method can accommodate for 2^n numbers ranging from -2^{n-1} to $(2^{n-1} - 1)$ including zero. Example: using 4 bits, $2^4 = 16$ numbers ranging from $-2^3 = -8$ (= 1000) to $2^3 - 1 = 7$ (= 0111) including 0000 can be represented. Note that proper distinction of +8 (= 01000) and −8 (= 11000) needs 5 bits.

1's Complement Method: To express positive numbers, the magnitude of the number is written with a 0 at the extreme left indicating positive sign. To express the negative number, each bit of the positive number including the sign bit is complemented. Examples:

$$+9 = 01001 \qquad -9 = 10110$$
$$+15 = 01111 \qquad -15 = 10000$$

Some spot values are given above. All of these contain five bits. The combination of 5 bits can express all the integers from +15 through −15 including 0. The combination of 8 bits can represent +127 to −127, total 255 numbers including 0. In general, the span of n bits can accommodate for $2^n - 1$ signed binary numbers ranging from $-(2^{n-1} - 1)$ to $(2^{n-1} - 1)$ including zero.

Sign-Magnitude Method: In this system, the magnitude of the number is written in binary form, as usual. Then the sign bit, 0 for positive and 1 for negative, is put to the left-most position. Thus the sign bit plus the magnitude bit constitute the full signed number. Since the computer works with a fixed length of bits, such as 8-bit, 16-bit, 32-bit etc., the sign-magnitude notation is formatted accordingly. For example, in 8-bit format,

$$+7 = 0000\ 0111$$
$$-7 = 1000\ 0111$$
$$+15 = 0000\ 1111$$
$$-15 = 1000\ 1111$$
$$+127 = 0111\ 1111$$
$$-127 = 1111\ 1111$$

The extent of n bits can include $2^n - 1$ signed binary numbers ranging from $-(2^{n-1} - 1)$ to $(2^{n-1} - 1)$ including zero. The sign-magnitude method is straight forward in principle but the circuit implementation is not convenient.

13.5 Boolean Algebra

In daily life, we often come across propositions that come true when some specific conditions are satisfied. For example,

"The door is open, if the latch is unlocked."

"The office closes when the time exceeds 6 PM."

Such statements are called propositional logic. These are based on two-state conditions like Yes/No and True/False. Irish mathematician George Boole had propounded (1854) some symbolic logics involving only two conditions. Subsequent researches of Shannon, De Morgan, and Karnaugh et al. gave birth to the subject as a powerful mathematical tool capable of formulating two-state logical conditions. In the honor of Boole, it is called *Boolean algebra* and it has the following features.

- The rules of Boolean algebra are similar to those of conventional algebra but there are certain differences. The *Boolean variables*, i.e., the variables involved in Boolean algebra can assume only two possible values, 0 or 1.
- The well-known laws of common algebra, such as commutative, associative and distributive laws are applicable to Boolean variables also.
- In addition, Boolean variables support some special mathematical operation, such as OR, AND, NOT, NAND and NOR that can be practically implemented with digital electronic circuits named gates illustrated in Chapter 14.
- A Boolean variable in its original form or in complemented form is called *literal*. For instance, A and $\bar{A}$ are two literals of a Boolean variable. However, $A+B$ and $A + \bar{B}$ are distinct variables. These are the literals that do not have the same variable. The AND of one or more literals is termed as *product* or *implicant*. An implicant that cannot be combined further with others to produce fewer literals is called a *prime implicant*.

The rules of Boolean algebra are based on a several *axioms* or *definitions*, as given in Table 13.2. Each axiom has a *dual*, which means the statement remains correct, if the values of 0 and 1 and the OR (+) and AND (.) operators are interchanged.

Table 13.2 Axioms and the duals for any Boolean variable (A)

Axiom	Dual	Meaning
If $A \neq 0$ then $A = 1$	If $A \neq 1$ then $A = 0$	Variables are binary numbers
$\bar{1} = 0$	$\bar{0} = 1$	NOT operation
$1 + 1 = 1$	$0.0 = 0$	OR/ AND operations
$0 + 0 = 0$	$1.1 = 1$	
$1+0 = 0+1 = 1$	$0.1 = 1.0 = 0$	

Based on the above axioms and their duality property, some theorems on Boolean variables are established, as introduced in Table 13.3 and Table 13.4. The theorems also have duality properties.

Table 13.3 Theorems on any single Boolean variable (B)

Theorem	Dual	Meaning
$B + 0 = B$	$B.1 = B$	*Identity*: Any Boolean variable OR 0 equals that variable. Any Boolean variable AND 1 equals that variable.

$B + 1 = 1$	$B.0 = 0$	**Null element**: In OR operation, 1 is the null element because it nullifies the effect of any other variable. In AND operation, 0 is the null element because it nullifies the effect of any other variable.
$B + B = B$	$B.B = B$	**Idempotency**: Any Boolean variable OR itself is equal to just the variable itself. Likewise, a variable AND itself is same as itself.
	$\bar{\bar{B}} = B$	**Involution**: Complementing a Boolean variable twice gives back the original variable.
$B + \bar{B} = 1$	$B.\bar{B} = 0$	**Complements**: A Boolean variable AND its complement is 0. A Boolean variable OR its complement is 1.

Table 13.4 Theorems on several Boolean variables (A, B and C)

Theorem	Dual	Meaning
$A + B = B + A$	$A.B = B.A$	*Commutative Law*: The order of input variables for OR/ AND operation does not affect the output
$(A + B) + C = A + (B + C)$	$(A.B).C = A.(B.C)$	*Associative Law*: Specific groupings of input variables for OR/ AND operation do not affect the output
$A.(B + C) = A.B + B.C$	$A + (B.C) = (A+B).(A+C)$	*Distributive Law*: AND distributes over OR and also OR distributes over AND. (In traditional algebra, addition does not distribute over multiplication.)
$A + (B.C) = A$	$A.(B + C) = A$	*Covering Theorem*, it permits to eliminate redundant variables.
$(A + B).(A + \bar{B}) = A$	$A.B + A\bar{B} = A$	*Combining Theorem*, this also allows to remove redundant variables.
$(A + B).(\bar{A} + C).(B + C)$ $= (A + B).(\bar{A} + C)$	$A.B + \bar{A}.C + B.C$ $= A.B + \bar{A}.C$	*Consensus Theorem*, lets discarding redundant variables.

Though the above axioms and theorems represent the most fundamental properties of Boolean variables, many other useful theorems, such as De Margan's theorems introduced in a subsequent section have enriched the practical implementation of those variables. The Boolean algebraic expressions defined in terms of binary numbers and implemented in terms of voltage in digital electronic circuits constitute the basic building blocks of computer and other digital instruments.

13.5.1 OR Operation

Consider the logical statement: "The college is closed on Sundays or holidays." The output is two-state (open/closed), depending on the occurrence of any one event: Sunday *or* holiday *or* both because the holiday too may be a Sunday. Statements like this can be formulated in Boolean algebra in the following way.

Let Y be the Boolean variable representing the output condition that depends on two other Boolean variables A and B. When either or both of these become true, the output becomes true. The statement is written as

$$Y = A + B \tag{13.3}$$

Equation (13.3) is the expression for OR operation. This is pronounced as 'Y equals A or B'. The '+' symbol is not 'plus' but OR operation. If the occurrence of any condition is represented by 1 and the non-occurrence by 0, all possible combinations of input (A and B) and output (Y) variables can be compiled as that given in Table 13.5. It is called the *truth table* for OR operation. The word 'truth' stands for occurrence of the conditions. The conditions of Table 13.5 can be implemented in terms of presence and absence of voltage with a digital electronic circuit named OR gate discussed in Chapter 14.

Table 13.5 Truth table for 2-input OR operation

A	B	Y
0	0	0
0	1	1
1	0	1
1	1	1

For a number of Boolean variables $A, B, \ldots, N$, Equation (13.3) can be extended as

$$Y = A + B + \ldots + N \tag{13.3a}$$

The output (Y) becomes true when any one or more inputs become true. Also the following properties of any Boolean variable (A), introduced earlier in Table 13.3 can be inferred from Table 13.5 also.

$$A + 0 = A$$
$$A + 1 = 1$$
$$A + A = A$$

These are the unique properties of Boolean variables. In common arithmetic, $1 + 1 = 1$ has no meaning. A common algebraic variable x cannot satisfy the condition $x + x = x$ for non-zero values. In the present case, the symbol '+' denotes OR operation and not addition operation.

13.5.2 AND Operation

Consider the statement, "The light glows, if the switch is turned on and the bulb is not burnt out." The two-state output condition of glowing or not glowing depends on simultaneous occurrence of the two events: turned on switch *and* the bulb at good condition. Statements like this can be formulated with Boolean variables as follows.

Let the output condition of an event be represented by the Boolean variable Y that becomes true when two other Boolean variables A and B are true simultaneously. It is stated as

$$Y = AB \tag{13.4}$$

Equation (13.4) is the logical expression for AND operation pronounced as 'Y equals A and B'. This is not multiplication of A and B. Sometimes the dot (.) indicating the AND operation is ignored. This book also follows the same convention. Table 13.6 represents the truth table all possible combinations of input (A and B) and output (Y) variables for AND operation. These can be implemented with electronic circuit named AND gate illustrated in Chapter 14.

For a number of Boolean variables, $A, B, \ldots, N$, Equation (13.4) can be extended as

$$Y = AB\ldots N \tag{13.4a}$$

The output becomes true when all the input variables become true simultaneously.

Table 13.6 Truth table for 2-input AND operation

A	B	Y
0	0	0
0	1	0
1	0	0
1	1	1

Table 13.6 reveals the following properties of any Boolean variable A, which were introduced in Table 13.3.

$A.0 = 0$
$A.1 = A$
$A.A = A$

13.5.3 NOT Operation

Consider the statement, "If expenditure increases, pocket money decreases." Such statements relating two opposite conditions of input and output can be treated with Boolean variables, as follows.

Let Y be a Boolean variable that occurs only when some other Boolean variable A does not occur. This is stated as

$$Y = \overline{A} \tag{13.5}$$

Equation (13.5) is the logical statement for NOT operation. The notation $\overline{A}$ is known as 'not-A' or 'A-bar' or 'A-complemented'. It is the condition of variable A not being true. Table 13.7 is the truth table for the NOT operation. The electronic implementation of this is called NOT gate, discussed in Chapter 14.

Table 13.7 Truth table of NOT gate

A	Y
0	1
1	0

The following properties of any Boolean variable A can be arrived at by combining the NOT, OR and AND operations. These were introduced in Table 13.3.

$$A + \overline{A} = 1$$
$$A\overline{A} = 0$$
$$\overline{\overline{A}} = A$$

13.5.4 De Morgan's Theorems

Based on the above mentioned OR, AND and NOT operations, De Morgan propounded the following two useful theorems.

For any two Boolean variables A and B,

$$\overline{A + B} = \overline{A}\,\overline{B} \tag{13.6a}$$

and

$$\overline{AB} = \overline{A} + \overline{B} \tag{13.6b}$$

For a number of Boolean variables $A, B, \ldots, N$, the above two equations can be generalized as

$$\overline{A + B + \ldots + N} = \overline{A}\,\overline{B}\ldots\overline{N} \tag{13.7a}$$

and

$$\overline{AB\ldots N} = \overline{A} + \overline{B} + \ldots + \overline{N} \tag{13.7b}$$

Two interesting features come out of the above statements.

NOR Operation: The expressions $\overline{A + B}$ and $\overline{A + B + \ldots + N}$ are OR operation followed by NOT operation and this combined operation is termed as NOR operation.

NAND Operation: The expressions $\overline{AB}$ and $\overline{AB\ldots N}$ are AND operation followed by NOT and are termed as NAND operations.

The practical implementation of the above operations with digital circuits, known as NOR gate and NAND gate, respectively, are discussed in Chapter 14. It is also explained there that these two are called 'universal gate' because all other gates like AND, OR and NOT can be realized with these gates.

13.6 Boolean Simplification

Logical conditions expressed with Boolean variables get complicated when many variables are involved. Realizing the logical expressions with digital circuits becomes difficult. A designer, therefore, attempts for simplifying the logical expression for the sake of minimizing the circuit components. The simplification is also important for theoretical purpose in order to make the expression shorter and easily understandable. Simplification of Boolean expressions is similar to conventional algebraic simplifications up to a certain extent. Careful use of the laws and theorems given in Tables 13.2, 13.3 and 13.4 and manipulation of Boolean variables can achieve the simplification analytically. Several examples are given below.

Example 13.12

Prove that $AB + \bar{A}B + A\bar{B} = A + B$ for any two Boolean variables A and B.

Solution

$$AB + \bar{A}B + A\bar{B}$$
$$= AB + AB + \bar{A}B + A\bar{B} \quad [\text{since } AB + AB = AB]$$
$$= A(B + \bar{B}) + B(A + \bar{A}) \quad [\text{distributive law}]$$
$$= A + B \quad [\text{since } A + \bar{A} = 1,\ B + \bar{B} = 1]$$

Example 13.13

Prove that $A + \bar{A}B = A + B$

Solution

$$A + B = A + B(A + \bar{A})$$
$$= A + AB + \bar{A}B$$
$$= A(1 + B) + \bar{A}B$$
$$= A + \bar{A}B \quad [\text{since } 1 + B = 1]$$

[The reason for each step, as written in the above two examples need not be mentioned always.]

Example 13.14

Simplify: $AB + AC + B\overline{C}$

Solution

The given expression can be modified as

$$AB(C + \overline{C}) + AC + B\overline{C}$$
$$= ABC + AB\overline{C} + AC + B\overline{C}$$
$$= AC(B + 1) + B\overline{C}(A + 1)$$
$$= AC + B\overline{C}$$

Example 13.15

Prove the Boolean identity: $(A + B)(B + C)(C + A) = AB + BC + CA$

Solution

The given expression can be written as

$$(AB + AC + B + BC)(C + A)$$
$$= ABC + AB + AC + BC$$
$$= AB(C + 1) + BC + AC$$
$$= AB + BC + CA$$

Example 13.16

Simplify $xy + \overline{x}z + yz$ for Boolean variables.

Solution

The given expression is modified as

$$xy + \overline{x}z + yz(x + \overline{x})$$
$$= xy + \overline{x}z + xyz + \overline{x}yz$$
$$= xy(1 + z) + \overline{x}z(1 + y)$$
$$= xy + \overline{x}z$$

Digital Principles and Boolean Algebra

> **Note: Duality Theorem**
>
> It may be identified that Example 13.16 is Consensus theorem. Example 13.18 given later is its dual. As mentioned in tables 13.2, 13.3 and 13.4, the duality is a useful relationship for Boolean expressions. It states about the interchange of OR and AND signs and 0 and 1 in a given expression. Using the duality theorem, the result of Example 13.16 could be readily converted to the result of Example 13.18.

Example 13.17

Verify: $(A+B)(A+C) = A + BC$

Solution

$$(A+B)(A+C)$$
$$= A + AC + AB + BC$$
$$= A(1+C) + AB + BC$$
$$= A + AB + BC$$
$$= A(1+B) + BC$$
$$= A + BC$$

Example 13.18

Simplify: $(x+y)(\bar{x}+z)(y+z)$

Solution

The given expression can be modified as

$$(x+y)(\bar{x}+z)(y+z+x\bar{x})$$
$$= (x+y)(\bar{x}+z)(y+z+x)(y+z+\bar{x})$$
$$= (x+y)(x+y+z)(\bar{x}+z)(\bar{x}+z+y)$$
$$= (x+y)(1+z)(\bar{x}+z)(1+y)$$
$$= (x+y)(\bar{x}+z)$$

Example 13.19

Prove that $\overline{A}B\overline{C} + AB\overline{C} + ABC = AB + B\overline{C}$

Solution

The given expression can be written as

$$\overline{A}B\overline{C} + AB\overline{C} + AB\overline{C} + ABC$$
$$= (\overline{A} + A)B\overline{C} + AB(\overline{C} + C)$$
$$= AB + B\overline{C}$$

Think a Bit: Why Boolean Simplification?

The motivation of simplification is to combine the implicants so as to reduce the number of literals. The Boolean simplification leads to a minimal expression (may not be unique) comprising prime implicants. The simplified expression, when implemented to digital electronic circuit, reduces the number of components thereby making it simple and economic and, of course, faster. In the subsequent chapters, we will come across a technical term 'propagation delay' that refers to the small but finite duration of a digital circuit to respond to a change in the input state. The minimal logical expression implemented to the circuitry also minimizes the effect of propagation delay.

13.7 Sum-of-Products and Product-of-Sums

The Boolean expressions presented so far either contained or did not contain all the variables in each term. A Boolean expression can be presented as standard functions involving $2n$ combinations of n variables in the following two forms.

- Sum-of-Products form
- Product-of-Sums form

Each term of any of the above two forms contains all the variables involved in either normal or complemented form.

13.7.1 Sum-of-Products (SOP)

In this format, several Boolean algebraic terms are combined with OR operation. Each term, called *fundamental product* or *minterm* contains all the variables in complemented or non-complemented state connected by AND operation. Any arbitrary Boolean expression can be converted to SOP form by the following steps.

(i) If there is already any mintrm, it is kept unchanged.
(ii) The variables absent in each term are identified.
(iii) For any absent variable x, the product term is multiplied by $(x+\bar{x})$. Rearranging makes each term a minterm.
(iv) The non-complemented variables of the minterm are assigned with 1 and the complemented variables are allotted 0.
(v) The algebraic SOP form is condensed as a shorthand canonical form where the binary numbers represented by the minterms are expressed as decimal equivalents.

Let us have some practice on the above steps by converting the Boolean expression $AC + BC + \bar{B}\bar{C}$ to canonical SOP form. There is no existing minterm. Following steps (ii) and (iii), we modify the given expression as

$$AC(B+\bar{B}) + BC(A+\bar{A}) + \bar{B}\bar{C}(A+\bar{A})$$
$$= ABC + A\bar{B}C + \bar{A}BC + A\bar{B}\bar{C} + \bar{A}\bar{B}\bar{C}$$

The above is a valid SOP form for Boolean variables A, B and C. Each minterm contains all the three variables in complemented or non-complemented form. Now following steps (iv) and (v), it can be written as a Boolean function

$$f(A,B,C) = 111 + 101 + 011 + 100 + 000$$
$$= 000 + 011 + 100 + 101 + 111$$
$$= \sum m(0,3,4,5,7)$$

The symbol Σ is used for SOP form and the letter m indicates the minterms.

Example 13.20

Express $\bar{A}B + \bar{C} + ABC$ in sum-of-products form.

Solution

$\bar{A}B + \bar{C} + ABC$

$= \bar{A}B(C+\bar{C}) + \bar{C}(A+\bar{A})(B+\bar{B}) + ABC$

$= \bar{A}BC + \bar{A}B\bar{C} + AB\bar{C} + \bar{A}B\bar{C} + A\bar{B}\bar{C} + ABC$

$= \bar{A}\bar{B}\bar{C} + \bar{A}B\bar{C} + \bar{A}BC + A\bar{B}\bar{C} + AB\bar{C} + ABC$

It can be also written as: $f(A,B,C) = \sum m(0,2,3,4,6,7)$

> **Example 13.21**
>
> Determine the SOP form for the following Boolean conditions, which are the outcomes of some unknown Boolean function.
>
> | P | Q | R | Y |
> |---|---|---|---|
> | 0 | 0 | 0 | 0 |
> | 0 | 0 | 1 | 1 |
> | 0 | 1 | 0 | 0 |
> | 0 | 1 | 1 | 0 |
> | 1 | 0 | 0 | 1 |
> | 1 | 0 | 1 | 1 |
> | 1 | 1 | 0 | 1 |
> | 1 | 1 | 1 | 1 |
>
> **Solution**
>
> Collecting the '1' conditions of the output (Y) from the table,
> $$Y = \bar{P}\bar{Q}R + P\bar{Q}\bar{R} + P\bar{Q}R + PQ\bar{R} + PQR$$
>
> Assigning 1 to complemented and 0 to non-complemented variables, the binary equivalents are: 001 + 100 + 101 + 110 + 111
>
> Therefore, the canonical SOP form, in terms of decimal equivalents is
> $$Y(P,Q,R) = \sum m(1,4,5,6,7)$$
>
> [N.B. Try to simplify the above expression analytically to obtain $Y = P + \bar{Q}R$. A regular technique for any Boolean simplification, known as Karnaugh map will be introduced in a subsequent section.]

13.7.2 Product-of-Sums (POS)

This format consists of several Boolean algebraic terms combined with AND operations. Each term, called *fundamental sum* or *maxterm* contains all the variable in complemented or non-complemented state connected by OR operation. Any arbitrary Boolean expression can be converted to POS form by the following steps.

(i) If there is already any maxterm, it is kept unchanged.
(ii) The variables absent in each term are identified.
(iii) For any absent variable x, a term of $x\bar{x}$ is added to the sum term. Rearrangement makes each term a maxterm.
(iv) The non-complemented variables of the maxterms are assigned with 0 and the complemented variables are assigned with 1.
(v) The algebraic POS form can be condensed as a shorthand canonical form where the binary numbers represented by the maxterms are expressed as decimal equivalents.

We are now trying to convert the Boolean expression $\overline{A}(\overline{B}+C)$ to canonical POS form using the above steps. There is no existing maxterm. Following steps (ii) and (iii), the given expression is modified as

$$(\overline{A}+B\overline{B}+C\overline{C})(A\overline{A}+\overline{B}+C)$$
$$=[(\overline{A}+B+C)(\overline{A}+\overline{B}+C)][(\overline{A}+B+C)(\overline{A}+\overline{B}+C)](A\overline{A}+\overline{B}+C)$$
$$=(\overline{A}+B+C)(\overline{A}+\overline{B}+C)(\overline{A}+B+\overline{C})(\overline{A}+\overline{B}+\overline{C})(A+\overline{B}+C)$$

The above is a valid POS expression for the given Boolean variables A, B and C. Each maxterm comprises all the three variables in complemented or non-complemented form. Following steps (iv) and (v), this is converted to Boolean function as

$$f(A,B,C) = (100)(110)(101)(111)(010)$$
$$= (010)(100)(101)(110)(111)$$
$$\prod M(2,4,5,6,7)$$

The binary numbers are expressed as decimal equivalents. The symbol $\prod$ is used for POS form and the letter M indicates the maxterms.

Example 13.22

Express $Y(x,y,z) = \prod M(3,4,6)$ in algebraic POS form.

Solution

Converting the given maxterms to binary equivalent, $Y = (011)(100)(110)$.
Assigning 0 to the non-complemented variables and 1 to the complemented variables,

$$Y = (x+\overline{y}+\overline{z})(\overline{x}+y+z)(\overline{x}+\overline{y}+z)$$

Example 13.23

Determine the POS form for the truth table given with Example 13.21.

Solution

This time the '0' conditions of the output (Y) are collected, the non-complemented variables are assigned 0 and the complemented variables are assigned 1 so that

$$Y = (P+Q+R)(P+\overline{Q}+R)(P+\overline{Q}+\overline{R})$$
$$= (000)(010)(011)$$

Therefore, $Y(P,Q,R) = \prod M(0,2,3)$

[N.B. An interesting feature comes out on comparing the above result of POS with the SOP obtained in Example 13.21. The possible range of numbers with three variables is 000 through 111 (0 through 7). The numbers missing in the POS appear in the SOP and vice versa.]

Example 13.24

Determine (i) the SOP form and (ii) the POS form for the following Boolean conditions.

A	B	C	Y
0	0	0	0
0	0	1	0
0	1	0	0
0	1	1	1
1	0	0	0
1	0	1	1
1	1	0	1
1	1	1	1

Solution

(i) Collecting the '1' conditions of the output (Y), the SOP form is obtained as

$$Y(A,B,C) = \overline{A}BC + A\overline{B}C + AB\overline{C} + ABC$$
$$= 011 + 101 + 110 + 111$$
$$= \sum m(3,5,6,7)$$

[N.B. Based on the finding of the previous example, we could conclude at a glance that the POS contained the missing numbers 0, 1, 2 and 4 and could write immediately the POS as $\prod M(0,1,2,4)$. However, the matter is treated independently.]

(ii) Considering the '0' conditions of the output, assigning 1 and 0 for complemented and non-complemented states, respectively,

$$Y(A,B,C) = (A+B+C)(A+B+\overline{C})(A+\overline{B}+C)(\overline{A}+B+C)$$
$$= (000)(001)(010)(100)$$
$$= \prod M(0,1,2,4)$$

Example 13.25

Show that the POS result in the above example could be derived from the SOP result.

Solution

The SOP in the above example stated that Y was true for minterms corresponding to 3, 5, 6 and 7. Hence Y would be false for the remaining minterms, corresponding to 0, 1, 2 and 4.

Therefore, $\overline{Y} = \overline{A}\,\overline{B}\,\overline{C} + \overline{A}\,B\,C + A\overline{B}\,\overline{C} + AB\,\overline{C}$

$Y = \overline{\overline{Y}} = \overline{\overline{A}\,\overline{B}\,\overline{C} + \overline{A}\,B\,C + A\overline{B}\,\overline{C} + AB\,\overline{C}}$

Following De Morgan's theorems,

$Y = (\overline{\overline{A}\,\overline{B}\,\overline{C}})(\overline{\overline{A}\,BC})(\overline{A\overline{B}\,\overline{C}})(\overline{AB\overline{C}})$

$= (\overline{\overline{A}} + \overline{\overline{B}} + \overline{\overline{C}})(\overline{\overline{A}} + \overline{B} + \overline{C})(\overline{A} + \overline{\overline{B}} + \overline{\overline{C}})(\overline{A} + \overline{B} + \overline{\overline{C}})$

$= (A+B+C)(A+\overline{B}+\overline{C})(\overline{A}+B+C)(\overline{A}+\overline{B}+C)$

The above function of Y is the required POS form.

Note: SOP and POS Equivalence

The sum-of-products (SOP) and the product-of-sums (POS) are two alternative methods of representing Boolean functions. These two are equivalent and subject to mutual interconverting. For a Boolean function containing n variables, there may be 2^n minterms or maxterms indicated by decimal numbers 0 through $2^n - 1$. Each minterm is the complement of the corresponding maxterm and vice versa. The missing numbers in one form constitute the other. For example, a Boolean function involving three variables A, B and C has the possible span of decimal numbers is 0 through 7. In such a case,

(i) $\sum m(3,5,6,7)$ and $\prod M(0,1,2,4)$ are equivalent. Therefore,

$\overline{A}BC + A\overline{B}C + AB\overline{C} + ABC = (A+B+C)(A+B+\overline{C})(A+\overline{B}+C)(\overline{A}+B+C)$

(ii) $\sum m(1,4,5,6,7)$ and $\prod M(0,2,3)$ are equivalent. Therefore,

$\overline{A}\,\overline{B}C + A\overline{B}\,\overline{C} + A\overline{B}C + AB\overline{C} + ABC = (A+B+C)(A+\overline{B}+C)(A+\overline{B}+\overline{C})$

13.8 Karnaugh Map

The Boolean simplifications presented in the previous examples did not follow any fixed methodology. Clever use of the formula and situation-specific algebraic manipulations led to the solution. Such analytic simplification may not be convenient to complicated expressions involving many variables. Also that is not suitable for computer programming. Thus the necessity of some systematic technique for Boolean simplification is realized. The *Karnaugh map* technique invented by American physicist Maurice Karnaugh (1924 –) is such a standard and popular method for simplifying Boolean expressions. This is a pictorial representation of a Boolean function in the form of a tableau and involves the aforementioned sum-of-products (SOP) and product-of-sums (POS) techniques. The algebraic simplification is, after all a trial-and-error process. The Karnaugh map technique is more orderly process with well-defined steps and it ensures the result of a minimal expression. The Karnaugh mapping technique for simplifying a Boolean expression has the following steps.

Step-1: The given Boolean expression is first converted to either SOP or POS form.

Step-2: A matrix-like tableau is constructed that contains 2^n cells for n variables. The cells of the tableau are assigned according to the possible combinations of the variables. The possible input states are specified at the top row and the left column. The order is same as that of Gray code. Each cell corresponds to a row in the truth table and contains the output value (0 or 1) for that row.

Step-3: Any two adjacent cells can differ only in one variable complemented or non-complemented.

Step-4: For SOP form, the true conditions of the minterms are indicated by putting '1' in the corresponding cells. For POS form, the true conditions of the maxterms are denoted by putting '0' in the respective cells.

Step-5: Simplification is done by grouping the cells into pairs, quads and octets, as available. These technical terms are introduced below.

Pair: When two 1s in SOP or two 0s in POS appear in two vertically or horizontally (but not diagonally) adjacent cells, the variable complemented or non-complemented between these two cells is eliminated.

Quad: When four 1s in SOP or four 0s in POS appear in horizontal, vertical or tiled cells, two complemented and non-complemented variables can be eliminated among the cells.

Octet: When eight 1s in SOP or 0s in POS appear in two consecutive rows or columns, three variables and their complements can be eliminated.

Attempt is made always to enclose the largest possible grouping for 2, 4,..., 2^n cells. The following exercise clarifies the applications of the above steps. Let us simplify the Boolean expression $A + \overline{A}B$ by the Karnaugh map.

Step-1: The given expression is converted to SOP form as: $AB + A\overline{B} + \overline{A}B$.

Step-2 & 3: Since only two variables are involved, a tableau of $2^2 = 4$ cells is constructed [Figure 13.1(a)]. Any two adjacent cells differ in only one variable complemented or non-complemented.

Step-4: The true conditions represented by the SOP are indicated by 1, as shown in Figure 13.1(b).

Step-5: Two pairings are possible [Figure 13.1(c)]. One pair results in A eliminating B and $\overline{B}$ and the other pair results in B eliminating A and $\overline{A}$. Thus the simplified expression is $A + B$.

The above simplification could be achieved also with other constructions of tableau, as given in Figures 13.1(d) and (e). These demonstrate that there is no restriction of the order of variables in the tableau and the sequence of mapping and grouping may be different but the condition of changing only one variable state between any two adjacent cells must be maintained.

Digital Principles and Boolean Algebra

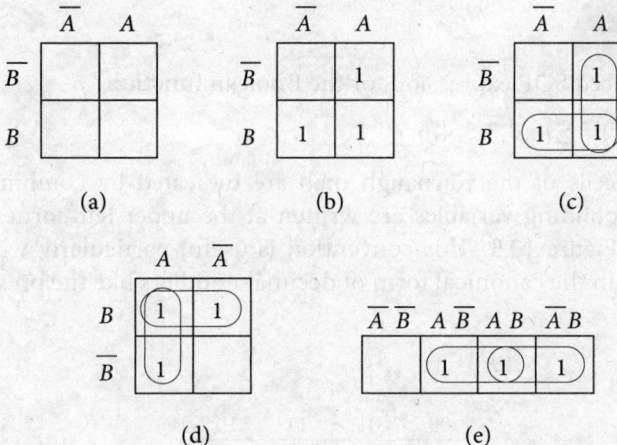

Figure 13.1 Karnaugh map simplification for $A + \overline{A}B$: (a) construction of tableau, (b) indication of true conditions, (c) simplification by grouping, (d) an alternative simplification, and (e) another alternative simplification

Example 13.26

Simplify by the Karnaugh map: $AB + AC + B\overline{C}$

Solution

The required SOP form of the given expression is: $ABC + AB\overline{C} + A\overline{B}C + \overline{A}B\overline{C}$

Since three variables are involved, a tableau of $2^3 = 8$ cells is constructed, as shown in Figure 13.2(a). The mapping of 1s and the simplification by pairing are also completed there. The simplified expression is $AC + B\overline{C}$.

[N.B. The 1s already grouped should not be grouped again, as shown in Figure 13.2(b). Such a grouping becomes superfluous and is called *redundant group*.]

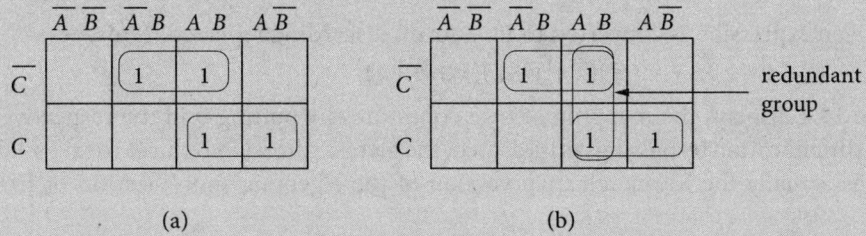

Figure 13.2 Karnaugh map for $AB + AC + B\overline{C}$: (a) mapping and simplification, (b) pointing out redundant group

Example 13.27

Find out the reduced SOP expression for the Boolean function

$$f(A,B,C) = \sum m(0,2,3,4,6,7)$$

[Sometimes the cells of the Karnaugh map are indicated by combinations of 0 and 1 and the corresponding variables are written at the upper left corner of the tableau, as elucidated in Figure 13.3. This convention is useful particularly when the Boolean function is given in the canonical form of decimal numbers like the present one.]

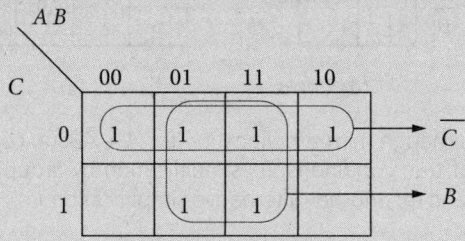

Figure 13.3

Solution

The binary equivalents of the given decimal numbers representing the minterms are mapped, as given in Figure 13.3. The largest possible grouping is with the two quads, as encircled. The simplified expression is $B + \overline{C}$.

Example 13.28

Simplify $(x+y)(\overline{x}+z)(y+z)$.

Solution

The given expression is converted to POS form, discarding repeated terms, as

$$(x+y+z)(x+y+\overline{z})(\overline{x}+y+z)(\overline{x}+\overline{y}+z)$$

Figure 13.4 displays the mapping of true conditions by putting 0 in the respective cells. After simplification by pairing, as indicated, the given expression reduces to $(x+y)(\overline{x}+z)$. [This is actually the Karnaugh map version of the algebraic simplification of Example 13.18.]

Digital Principles and Boolean Algebra

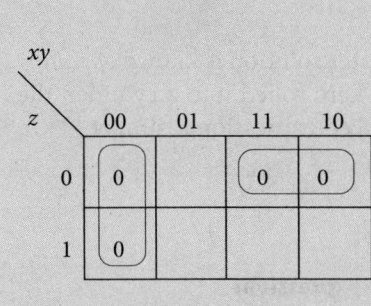

Figure 13.4

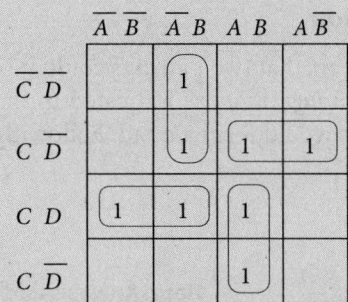

Figure 13.5

Example 13.29

The Karnaugh mapping for a certain Boolean function involving four variables is given in Figure 13.5 with the grouping indicated. What will be the simplified expression?

Solution

Eliminating the complemented/non-complemented variable from each pair, the simplified expression is $\overline{AB}C + ACD + \overline{A}CD + AB\overline{C}$.

Example 13.30

Figure 13.6 represents the Karnaugh mapping for a certain Boolean function involving four variables. What will be the simplified expression?

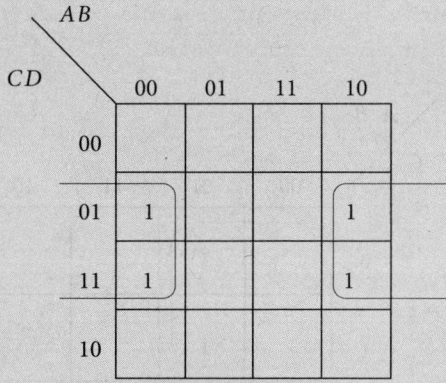

Figure 13.6

Solution

It appears that two pairings could be done. However, a better solution is known as *rolling of Karnaugh map*, as indicated in the figure. If the map were rolled into a cylinder, the four 1s would form a quad. Following that quad, the reduced expression is simply $\overline{B}D$.

Note: Analysis and Synthesis of Boolean Expressions

Constructing a truth table from a logical expression is called its *analysis* or *evaluation*. For n number of input variables, there may be 2^n possible binary variables in the truth table. The opposite process, deriving a logical expression from a given truth table is termed as *synthesis* of a Boolean expression. The *sum-of-products* and *product-of-sums* discussed above are two standard forms of such synthesizing that are useful to simplify Boolean expressions systematically. The simplifications can also be achieved algebraically, as exemplified earlier. Also sometimes the analysis with truth table becomes useful to prove Boolean identities. The method is termed as *proof by perfect induction*. All possible combinations of the input variables are tabulated for both sides of the given logical expression by providing 0 and 1, as necessary and it is verified whether both sides yield the same values.

13.8.1 Don't Care Conditions

In a digital circuit, it may so happen that certain input conditions never occur and the corresponding output states never appear. Such a condition is designated as *don't care condition* and is represented by '×' in a truth table. One may let it equal either 0 or 1, whichever leads to the simpler logical expression. Figure 13.7 interprets the use of don't care condition. The don't care condition (×) is treated as 1 and included to the quad along with the other three 1s, as shown in the figure. This results in a simplified logical expression of $\overline{A}$.

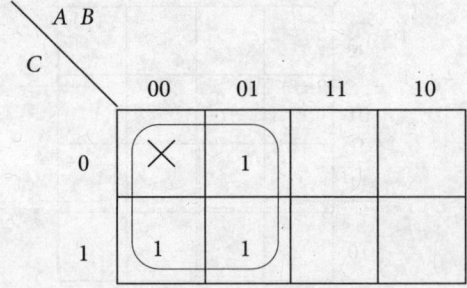

Figure 13.7 Example of don't care condition

13.8.2 Simplification is not Unique

The Karnaugh map simplification need not always result in the same logical expression. Depending on the choice of grouping, more than one minimal logical expressions may be obtained. Figure 13.8 demonstrates such an example. On simplification, the grouping of Figure 13.8 results in the implicant of $\overline{A}BC + \overline{A}CD + ABD + BCD$ whereas that of Figure 13.8(b) yields $\overline{A}\overline{B}D + ABC + \overline{A}C\overline{D} + BCD$, both are acceptable.

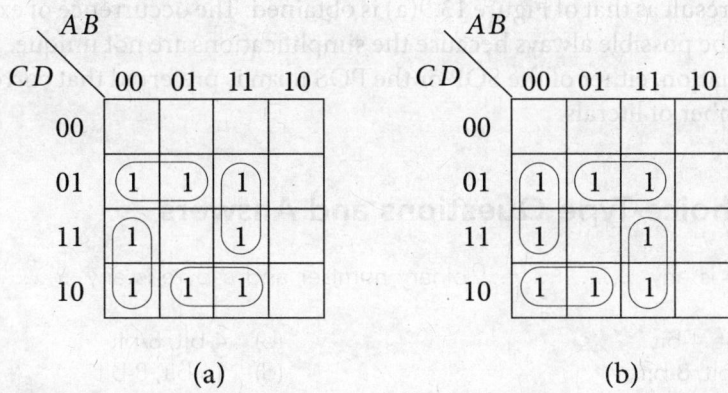

(a)   (b)

Figure 13.8 Example of more than one simplification with Karnaugh map

13.8.3 SOP and POS are Equivalent

The Karnaugh map simplification of a logical expression with either SOP form or POS form provides with equivalent results. This is interpreted with an example in Figure 13.9.

The mapping of a certain logical expression is given in Figure 13.9(a). There are two quads available for grouping and the usual simplification through these groupings produces the simplified implicant: $Y = AB + \overline{A}D$.

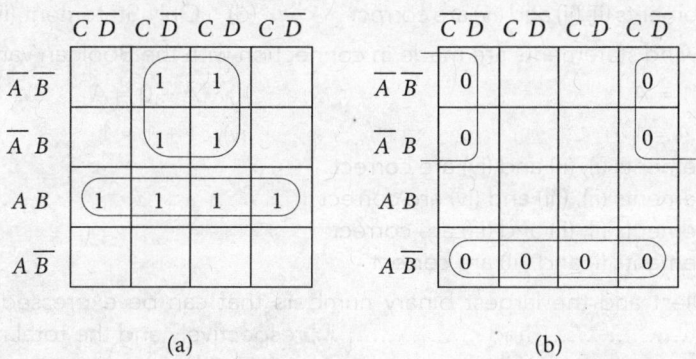

(a)   (b)

Figure 13.9 Example of SOP and POS equivalence in Karnaugh map simplification

Figure 13.9(b) displays the complementary map considering the 0 states. The simplification (including a rolling of map) yields: $\overline{Y} = \overline{A}D + AB$. Let us complement it again. That yields

$$\overline{\overline{Y}} = Y = \overline{\overline{A}D + AB}$$
$$= AB + \overline{AD + BD} \qquad \text{[a few steps using De Morgan]}$$
$$= AB + \overline{A}D \qquad \text{[which theorem?]}$$

Thus the same result as that of Figure 13.9(a) is obtained. The occurrence of exactly the same result may not be possible always because the simplifications are not unique. Depending on the specific situation, either of the SOP or the POS form is preferred that corresponds to the minimum number of literals.

Multiple Choice-Type Questions and Answers

13.1 A *nibble* is any .... binary number and a *byte* is any .... binary number.
  (a) 8-bit, 4-bit
  (b) 4-bit, 8-bit
  (c) 12-bit, 8-bit
  (d) 16-bit, 8-bit

13.2 A kilobyte is equal to .... bytes.
  (a) 1000
  (b) 10^3
  (c) 2^{10}
  (d) 2^{20}

13.3 The largest decimal number that can be represented by eight binary digits is ....
  (a) 128
  (b) 64
  (c) 256
  (d) 255

13.4 The minimum number of bits necessary to express decimal 64 is ....
  (a) 8
  (b) 7
  (c) 6
  (d) 16

13.5 The following statements are made in connection with Boolean variables X, Y and Z.
  (i) X + XY + XYZ = X
  (ii) X(X + Y) = X
  (iii) X(X + Y) = Y
  (iv) X.0 = 0
  (a) Only Statement (iv) is correct
  (b) Statements (ii) and (iv) are correct
  (c) Statements (i), (ii) and (iv) are correct
  (d) Only Statement (iii) is correct

13.6 The following statements are made in connection with the Boolean variable A.
  (i) A + 1 = A
  (ii) A + 0 = A
  (iii) A + A = 1
  (iv) 1 + 1 = 1
  (a) Statements (i), (ii) and (iv) are correct
  (b) Statements (ii), (iii) and (iv) are correct
  (c) Statements (i), (ii) and (iii) are correct
  (d) Statements (i) and (iii) are correct

13.7 The smallest and the largest binary numbers that can be expressed using 8 bits are decimal .... and ...., respectively and the total number of counts is ....
  (a) 0, 255, 256
  (b) 1, 256, 256
  (c) 0, 256, 256
  (d) 1, 255, 255

13.8 The total number of counts in octal using two octal digits is equal to the total counts in 2^n binary numbers. What is n?
(a) 2   (b) 4   (c) 6   (d) 8

13.9 If an x-digit octal number be equal to y-digit hexadecimal number, the minimum values of x and y are .... and ...., respectively.
(a) 2, 1   (b) 4, 3   (c) 3, 2   (d) 8, 6

Hint. $8^x = 16^y$

13.10 The largest negative and the largest positive decimals that can be represented with 12 bits are .... and ...., respectively.
(a) −2048, 2048   (b) −2047, 2047
(c) −2047, 2048   (d) −2048, 2047

Hints. -2^{11} to $2^{11} - 1$ including zero, one bit being used for the sign.

13.11 The number of bits required to represent decimal − 33 through 33 in 2's complement system is ....
(a) 4   (b) 5   (c) 6   (d) 7

13.12 In the output column of the truth table for n-input OR gate, the number of zeros is ....
(a) always 1
(b) always n − 1
(c) equal to the number of any input being 0
(d) equal to the number of any one input being 1

13.13 The Boolean expression $ABC + A\bar{B} + AB\bar{C} + \bar{A}$ can be simplified to ....
(a) A   (b) B   (c) C   (d) 1

Hints. $AB(C + \bar{C}) + A\bar{B} + \bar{A} = A(B + \bar{B}) + \bar{A} = A + \bar{A}$

13.14 The expression $\overline{AB} + \bar{A} + AB$ can be simplified as ....
(a) 1   (b) A   (c) B   (d) $\bar{A}$

Hints. For any Boolean variable, $\overline{AB} + AB = 1; \bar{A} + 1 = 1$

13.15 The Boolean expression $(A+B)(A+\bar{B})(\bar{A}+B)$ is equivalent to ....
(a) A + B   (b) AB   (c) $\overline{AB}$ (d) $\bar{A}B$

13.16 The Boolean expression AC + ABC can be written as:
(i) AC   (ii) AB   (iii) $\overline{\bar{A}+\bar{C}}$ (iv) BC
(a) Only expression (i) is true   (b) Expressions (i) and (ii) are true
(c) Expressions (i) and (iii) are true   (d) Expressions (ii) and (iv) are true

13.17 If the Boolean function $Z = PQ + PQR + PQRS + PQRST + PQRSTU$ then $\bar{Z}$ is
(a) $\bar{P}\bar{Q} + \bar{R}(\bar{S}+\bar{T}+U)$ (b) $\overline{PQ}$
(c) $\bar{P}+\bar{Q}$ (d) $\bar{P}+\bar{Q}+\bar{R}+\bar{S}+\bar{T}+\bar{U}$

[JAM 2017]

Hints. One can simplify $Z = PQ$, hence $\bar{Z} = \overline{PQ} = \bar{P}+\bar{Q}$

13.18 Which of the followings is an incorrect Boolean expression?
(a) $\bar{P}Q + PQ = Q$ (b) $(P+\bar{Q})(P+Q) = P$
(c) $P(P+Q) = Q$ (d) $(PQR + \bar{P}QR + P\bar{Q}R + P\bar{Q}\bar{R}) = \bar{Q}$

[JAM 2009]

13.19 The Boolean algebraic expression $\overline{A}B + AB\overline{C} + ABC$ is equal to ....
(a) AB (b) B (c) $A + B$ (d) A

13.20 Consider the following truth table. The logical expression for F is ....

A	B	C	F
0	0	0	1
0	0	1	0
0	1	0	0
0	1	1	0
1	0	0	1
1	0	1	1
1	1	0	1
1	1	1	0

(a) $A\overline{B} + BC + CA$ (b) $A\overline{B} + A\overline{C} + \overline{B}\,\overline{C}$
(c) $C\overline{A}\,\overline{B} + A\overline{B}$ (d) $\overline{C}(A + \overline{B}) + A\overline{B}$

[JAM 2010]

Hints. $F = \overline{A}\,\overline{B}\,\overline{C} + A\overline{B}\,\overline{C} + A\overline{B}C + AB\overline{C}$. Karnaugh map simplification yields $F = A\overline{B} + A\overline{C} + \overline{B}\,\overline{C}$.

13.21 Each cell in a Karnaugh map represents a .... that differs from an adjacent cell by .... literal.
(a) maxterm, the change of a single
(b) minterm, no change of
(c) maxterm, no change of
(d) minterm, the change of a single

13.22 One hexadecimal digit stores .... and two hexadecimal digits store ....
(a) two nibbles, one byte
(b) one nibble, one byte
(c) one nibble, one byte
(d) two nibbles, two bytes

13.23 An n-digit decimal number represents one of .... possibilities ranging from 0, 1, 2, 3, ... up to ....
(a) 10^n, $10^n - 1$
(b) 10^{n-1}, $10^n - 1$
(c) 10^n, 10^{n-1}
(d) $10^n - 1$, 10^n

13.24 Which one is the correct binary equivalent of hexadecimal F6C?
(a) 0110 1111 1100
(b) 1111 0110 1100
(c) 1100 0110 1111
(d) 011011000111

[GATE 2020]

13.25 The Boolean expression $Y = AB\overline{C}\,\overline{D} + \overline{A}B\overline{C}\,\overline{D} + \overline{A}\,\overline{B}\,\overline{C}D + \overline{A}BCD + \overline{A}BCD + AB\overline{C}D$ can be simplified to
(a) $\overline{A}\,\overline{B}C + A\overline{D}$
(b) $\overline{A}B\overline{C} + A\overline{D}$
(c) $AB\overline{C} + \overline{A}D$
(d) $AB\overline{C} + \overline{A}D$

[GATE 2011]

Answers

10.1 (b)	10.2 (c)	10.3 (d)	10.4 (b)
10.5 (c)	10.6 (a)	10.7 (a)	10.8 (c)
10.9 (b)	10.10 (d)	10.11 (d)	10.12 (a)
10.13 (d)	10.14 (a)	10.15 (b)	10.16 (c)
10.17 (c)	10.18 (c)	10.19 (b)	10.20 (d)
10.21 (d)	10.22 (c)	10.23 (a)	10.24 (b)
10.25 (c)			

Reasoning-Type Questions and Answers

13.1 What is the significance of a truth table?

Ans. The truth table of a Boolean expression analyzes all the possible combinations of input variables and the resulting outputs. It depicts the true conditions of a Boolean expression so that the expression can be synthesized from the truth table.

13.2 What is the range of (i) unsigned and (ii) signed binary numbers that can be accommodated to 8 bits?

Ans. In the case of unsigned numbers, the range is 0 through 255 (0000 0000 through 1111 1111). In the case of signed numbers, one bit is used up for the sign. The remaining 7 bits can accommodate for −127 to +127 in sign-magnitude system and −128 to +127 in 2's complement system, zero being included in both cases.

13.3 What is the minimum number of bits required to represent decimal numbers −52 to +52?

Ans. One bit is required for the sign. The maximum decimal number represented by 5 bits is 31 and that by 6 bits is 63. Since 52 in in between these two, the minimum number of bits required is 7; one for the sign and the remaining six for the magnitude of the number.

13.4 Is the rolling of Karnaugh map a logically valid process?

Ans. Yes, because the assigned states, 1 for SOP and 0 for POS at one edge of the map differs in only a single variable state from the assigned states on the other edge.

Solved Numerical and Logical Problems

13.1 Convert the binary number 1111 1111 to decimal.

Soln. It can be converted as: $1\times2^7 + 1\times2^6 + 1\times2^5 + 1\times2^4 + 1\times2^3 + 1\times2^2 + 1\times2^1 + 1\times2^0 = 255$.

[N.B. There is another efficient way when all the bits are 1 in the n-bit binary number like that in the present case. In such a case, the decimal equivalent is $2^n - 1$. Therefore, we could convert the given 8-bit binary number as $2^8 - 1 = 255$.

13.2 Make the following conversions: (i) $(C5E2)_{16}$ to binary, (ii) $(2478)_{10}$ to hexadecimal,

Soln.
(i) $(C5E2)_{16} = (1100\ 0101\ 1110\ 0010)_2$

(ii)
```
16 | 2478     14 = E
16 |  154     10 = A
16 |    9      9
        0
```
Thus $(2478)_{10} = (9AE)_{16}$.

13.3 Make the following conversions: (i) $(23.8125)_{10}$ to binary, (ii) $(13.65625)10$ to binary.

Soln.
(i) The integral and the fractional parts are converted separately, as follows.
```
2 | 23  1
2 | 11  1
2 |  5  1
2 |  2  0
2 |  1  1
     0
```
Thus $(23)_{10} = (10111)_2$

$0.8125 \times 2 = 1.6250 \rightarrow 1$

$0.625 \times 2 = 1.250 \rightarrow 1$

$0.25 \times 2 = 0.50 \rightarrow 0$

$0.5 \times 2 = 1.0 \rightarrow 1$

Thus $(0.8125)_{10} = (0.1101)_2$

Combining these two parts, $(23.8125)_{10} = (10111.1101)_2$

(ii) In a similar way as above, the integral part is converted as $(13)_{10} = (1101)_2$. The fractional part is converted as

$0.65625 \times 2 = 1.31250 \rightarrow 1$

$0.3125 \times 2 = 0.6250 \rightarrow 0$

$0.625 \times 2 = 1.25 \rightarrow 1$

$0.25 \times 2 = 0.50 \rightarrow 0$

$0.5 \times 2 = 1.0 \rightarrow 1$

$(0.65625)_{10} = (0.10101)_2$

Combining the two parts, $(13.65625)_{10} = (1101.10101)_2$.

13.4 Convert the decimal numbers 49 and 25 to binary by sum of weight method, express both in 8-bit form and execute the subtraction 49 − 25 in binary by 2's complement method.

Soln.

$(49)_{10} = 32 + 16 + 1 = (110001)_2 = (0011\ 0001)_2$ in 8-bit form.

$(25)_{10} = 16 + 9 = (11001)_2 = (0001\ 1001)_2$ in 8-bit form.

The 2's complement of 25 in 8-bit form is 1110 0111. Therefore, 49 − 25 is executed as

```
  0011 0001
+ 1110 0111
  10001 1000
```

Ignoring the left-most carry bit, the result is 11000, which can be identified as decimal 24.

13.5 Perform the binary subtraction 11011 − 01101 by both 1's complement and 2's complement methods and compare the results.

Soln. 1's complement of 01101 is 10010, hence the subtraction by 1's complement method is performed as:

```
 11011
+10010
101101
    ↘1
  1110
```

The 2's complement of 01101 is 10011, hence the subtraction by 2's complement method is executed as:

```
 11011
+10011
101110
```

Ignoring the carry, the result is 1110. Thus both the methods lead to the same result.

13.6 Prove the Boolean identity: $(A+B)(\overline{A}+C) = AC + \overline{A}B$

Soln. The given expression $(A+B)(\overline{A}+C)$ can be modified as

$A\overline{A} + AC + \overline{A}B + BC(A+\overline{A})$

$= AC + \overline{A}B(1+C) + ABC$

$= AC + \overline{A}B + ABC$

$= AC(1+B) + \overline{A}B$

$= AC + \overline{A}B$

13.7 Prove algebraically: $\overline{A}D + \overline{C}D + A\overline{B} = \overline{A}\overline{C}D + A\overline{C}D + A\overline{B}\overline{C} + A\overline{B}C + \overline{A}CD$

Soln. The expression at the right side is rearranged as

$\overline{A}D(C+\overline{C}) + \overline{C}D(A+\overline{A}) + A\overline{B}(C+\overline{C})$

$= \overline{A}D + \overline{C}D + A\overline{B}$

13.8 Prove that $(A+B)(A+\bar{B})(\bar{A}+C) = AC$

Soln. The left side expression is modified as

$(A+A\bar{B}+AB+B\bar{B})(\bar{A}+C)$
$= A\bar{A}+AC+A\bar{A}\bar{B}+A\bar{A}B+\bar{A}B\bar{B}+B\bar{B}C$ (All other terms contain variables like $x\bar{x}$ and get vanished.)
$= AC$

13.9 If $Y = \bar{A}B+\bar{B}A$, find $\bar{Y}$.

Soln.

$\bar{Y} = \overline{\bar{A}B+\bar{B}A}$
$= \overline{(\bar{A}B)} \cdot \overline{(\bar{B}A)}$
$= (A+\bar{B})(B+\bar{A})$
$= AB+A\bar{A}+\bar{B}B+\bar{B}\bar{A}$

Thus $\bar{Y} = AB+\bar{A}\bar{B}$

[N.B. Chapter 14 deals with these two expressions in more detail. The given expression for Y is called 'exclusive-OR' and its complement is 'exclusive-NOR', which are used in determining the equality condition of two bits.]

13.10 Simplify the Boolean function: $AB+B\bar{C}+\bar{A}(B+C)+\bar{B}C+\bar{A}\bar{B}\bar{C}+ABC$

Soln. The given expression is converted to sum-of-products form as

$AB(C+\bar{C})+B\bar{C}(A+\bar{A})+\bar{A}B(C+\bar{C})+\bar{A}C(B+\bar{B})+\bar{B}C(A+\bar{A})+\bar{A}\bar{B}\bar{C}+ABC$
$= ABC+AB\bar{C}+\bar{A}B\bar{C}+\bar{A}BC+\bar{A}\bar{B}C+A\bar{B}C+\bar{A}\bar{B}\bar{C}$ (discarding repetitions)

This is put into Karnaugh map (Figure 13.10) and simplified as: $\bar{A}+B+C$. Three overlapping quads and the outcome of each one are indicated in the figure.

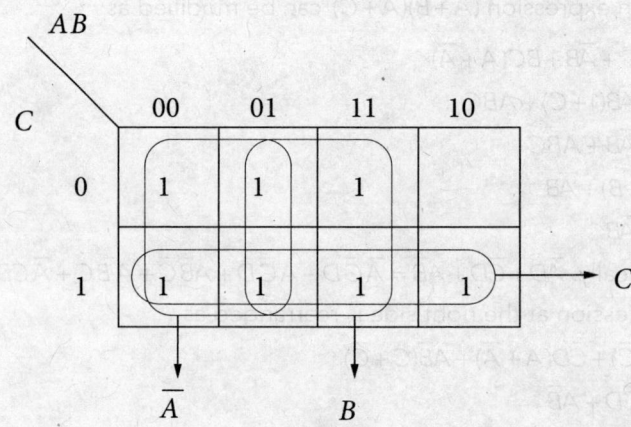

Figure 13.10

Exercise

Subjective Questions

1. What is meant by the word 'digital' in modern science and technology? Mention some advantageous features of a digital systems and some of its limitations.
2. What is the *base* or *radix* of a number system? Justify that both decimal and binary systems are *positional* and *weighted*.
3. What are meant by *byte* and *nibble*? Explain what BCD is and mention some of its significant uses.
4. Explain, with examples, how signed binary numbers can be expressed by (i) sign-magnitude, (ii) 1's complement and (iii) 2's complement methods. Mention the allowable range in each case.
5. Explain the sum-of-products and product-of-sums formats for Boolean expressions.

Conceptual Test

1. Demonstrate with two Boolean variables that AND operation can be performed using OR and NOT operations.
2. Show, using two Boolean variables that OR operation can be realize with AND and NOT operations.
3. Justify the statement: AND operation for positive logic functions as OR operation for negative logic.

    Hints. If $Y = AB...N$, $\overline{Y} = \overline{AB...N} = \overline{A} + \overline{B} + ... + \overline{N}$. If the output and all the inputs are complemented, 1 becomes 0 and vice versa.
4. There are 256 identical packets of food products and only one of those contains slightly less amount of material than the others. You are given a sensitive balance that can accommodate any number of packets on either side. Can you sort out the lighter packet using this balance? Suggest the most efficient method.
5. The *sexagesimal system* is a number system with 60 as its base. Astonishingly, it originated with the ancient Babylonian civilization about 4000 years ago. Can you state, how many bits of information is conveyed with one sexagesimal digit? (ii) Can you write decimal 4000 in sexagesimal?

Numerical and Logical Problems

1. Convert to binary: (i) decimal 18.50, (ii) decimal −19.25

    Answer (i) 10010.10, (ii) 01101.01
2. Perform the following binary operations where the subtraction is to be done by 1's complement method. Repeat the work by 2's complement method.

    (i) 11111 + 1101 − 111   (ii) 1011 + 0001 − 1101
3. Convert hexadecimal E3B to equivalent (i) binary, (ii) octal and decimal.

    Answer (i) 1110 0011 1011, (ii) 7073, (iii) 3643
4. (i) Given $(1123)_x = (A3)_{16}$, find x. (ii) $(3245.25)_{10} = (X)_2 = (Y)_8 = (Z)_{16}$. Find X, Y and Z.

    [C.U 2007]

    Answer (i) 5, (ii) X = 110010101101.01, Y = 6255.2, Z = CAD.4

5. Simplify the following Boolean expressions.
   (i) $\overline{A}+B+\overline{A}+\overline{B}$ Answer A
   (ii) $A+AB+ABC+ABCD$ Answer A
   (iii) $A+ABC+AB\overline{C}$ Answer A
6. Prove the following Boolean identities algebraically.
   (i) $(A+B)(\overline{A}+\overline{B}) = A\overline{B}+\overline{A}B$
   (ii) $AB+BC+\overline{A}C = AB+\overline{A}C$ Hints. LHS $= AB+BC(A+\overline{A})+\overline{A}C$
   (iii) $\overline{A}B+\overline{C}+ABC = B+\overline{C}$
   Hints. $\overline{A}B(C+\overline{C})+\overline{C}(B+\overline{B})+ABC \rightarrow BC+B\overline{C}+\overline{B}\overline{C} \rightarrow B(C+\overline{C})+\overline{C}(B+\overline{B})$
7. Find out the minimum expression for $f(A,B,C) = \Sigma m(0,1,2,3,6)$
        Answer $\overline{A}+B\overline{C}$
8. Derive the minimum SOP and POS expressions for the function $f(a,b,c) = \Sigma m(0,2,4,6)$
        Answer $\overline{c}$
9. Determine all possible simplifications obtainable with the Karnaugh map of Figure 13.11.
10. The increasing speed of a car is indicated by five discrete levels: v_1, v_2, v_3, v_4 and v_5. These are converted to proportional electrical signals with proper transducers and the analog electrical signal is converted to 3-bit digital number in terms of voltage. These voltage levels drive five LEDs, each glowing for the corresponding speed level or higher than that. Considering the glowing condition of an LED as the 1 state, determine the minimal condition for the LED indicating that speed level v_3 has reached or exceeded.

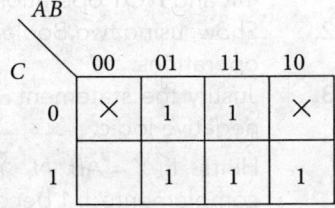

Figure 13.11

Projects on Chapter 13

Project 1:
Tom, Dick and Harry are the attendants of your laboratory who often like to abstain from work. An important experiment is to be conducted tomorrow and you are suspecting that some of the attendants may not be present at your help. After a prolonged interview, you have been able to retrieve the following possibilities from their honest confessions.

(i) None of them will turn up.
(ii) Only Tom will show up.
(iii) Only Harry will appear.
(iv) Both Tom and Harry will attend but not Dick.
(v) Both Dick and Harry will arrive but not Tom.

Buzzling or puzzling? Can you decipher a simplified condition for tomorrow?

Hints:
Denoting Tom as T, Dick as D and Harry as H, the above statements can be coded in logical form as: $\overline{T}\overline{D}\overline{H} + T\overline{D}\overline{H} + \overline{T}\overline{D}H + T\overline{D}H + \overline{T}DH$. Now simplify by Karnaugh mapping.

Project 2:
First understand the working principle of counters from Chapter 15. Then address the following quests.
  (i) What is the decimal counting range for a 6-bit counter?
  (ii) What is its octal equivalent?
  (iii) Can you arrange the 6-bit counter outputs in such a way that those represent the octal numbers in binary?

Hints:
The six output bits of the counter are, in sequence,
    (MSB) Q6, Q5, Q4, Q3, Q2 and Q1 (LSB).
If those are arranged in two groups as Q6Q5Q4 and Q3Q2Q1 then each one will represent the octal number and will vary from 000 through 111.

It is recommended to write down all the counter outputs from 000000 through 111111. That will automatically verify the answers to the first two questions.

14

Combinational Logic Circuits

The fundamental theories and principles related to digital systems, such as the number systems, codes and the Boolean algebra are discussed in Chapter 13. This chapter presents the practical enactments of those concepts with digital electronic circuits. The basic logic gates, namely AND, OR and NOT and universal gates (NAND and NOR) are illustrated and the working principles of the following types of digital circuits based on the combinations of these gates are explained.

- Circuits executing arithmetic operations: *adders* and *subtractors*.
- Circuits capable of comparing the magnitudes of two binary numbers, known as *comparators*.
- Data processing circuits: multiplexer, demultiplexer, decoder, and encoder.

The different categories of combinations of electronic components for realizing logic gates, known as logic families are also introduced.

14.1 Boolean Algebra and Digital Electronics

The algebraic expressions involving Boolean variables (Chapter 13) and their practical realization in terms of voltage levels associated with electronic circuits are two different but interrelated subjects. Boolean algebra is a mathematical theory involving two-state variables and their combinations whereas digital electronics realizes those variables and their algebraic operations in terms of electrical signals of two distinct levels.

English mathematician George Boole (1815–1864) propounded (1854) several symbolic logical expressions involving two states only. The possibility of its practical application was first realized when American engineer Claude E. Shannon (1916–2001) applied (1938) such concept to the analysis of telephone switching circuits, the 'closed' and 'open' conditions of relay switches being indicated by the two states of Boolean logic. This was the pioneering work on the application of Boolean algebra to digital electronics. Then it got further maturity with the contributions of Karnaugh, De Morgan, and other notable scientists.

A digital electronic circuit is constructed with our known elements, such as diodes, transistors, and FETs along with resistors and capacitors. The difference is that the digital circuits operate with only two discrete levels of voltage for inputs and outputs representing the two possible states of Boolean variables. Digital electronics deals with circuits performing mainly the following operations.

Arithmetic and Logic Operations: An adder circuit can execute binary addition, binary subtraction as negative addition, binary multiplication as repeated addition, and binary division as repeated negative addition. A comparator circuit can decide whether a given binary number is greater than, equal to or less than another.

Data processing: Multiplexing and demultiplexing circuits can transfer binary data from several lines into a single path or vice versa obeying a specified order.

Encoding and Decoding: These circuits convert a decimal number to binary or other code and vice versa in terms of two discrete levels of voltage.

Counting: These circuits can count binary numbers in terms of voltage pulse and can generate a particular sequence of numbers.

Data storage: Circuits named register and memory incorporate a memory property in instrumentation by storing, shifting, changing and accessing data to and from specific locations.

In addition, a regular voltage pulse generating circuit, known as clock is widely used in digital systems.

14.1.1 Combinational and Sequential Logic

Digital circuits can be classified broadly into two categories. In *combinational logic* circuits, the output depends only on the instantaneous values of the inputs and their specific combinations. No sequence is required for applying the inputs. All the digital circuits of this chapter are combinational in nature. In *sequential logic* circuits, the output depends on the sequence of the input states given at different time. Not only the instantaneous conditions but the previous conditions of the inputs are also to be considered. Such circuits are illustrated in Chapter 15. Table 14.1 presents a comparison between combinational and sequential digital systems.

Table 14.1 Comparison of combinational and sequential logic systems

Combinational Logic	Sequential Logic
The inputs and the outputs of the circuits are all two-state variables following the rules of Boolean algebra.	These are also circuits following Boolean algebra and comprising two-state inputs and outputs.
The output depends on specific combinations of the inputs. Only the instantaneous combinations of the inputs are necessary.	The output depends on specific combinations of the inputs at an instant of time but the instantaneous combinations art not enough. The previous and the next steps of operation are also to be considered.

It does not require any sequence of applying the inputs. For example, a two-input OR gate with any one input at high state can yield high output state.	The sequence of the inputs is most important. For instance, the high output state of a flip-flop (Chapter 15) is obtained with only a specific state of the two inputs.
The previous or the forthcoming states of the inputs need not be considered.	Both the instant conditions and the conditions at a previous instant of time are needed to decide the output state.
The output changes always with the change in the input. No data storage is involved.	The output may not change always with the input change. The circuit may have some 'memory' or capability of storing bits.
Examples: Arithmetic and logic circuits, such as adder; Data processing circuits, such as multiplexer	Examples: Data storing circuits, such as flip-flop and register; Counting circuits, such as counters

14.1.2 Positive and Negative Logic

Digital electronic circuits are also termed as *logic circuits* because the input and output conditions follow Boolean logic and the performance of a digital circuit can be analyzed with the rules of Boolean algebra. The following two types of logic conventions are incorporated to a digital circuit.

Positive Logic: The more positive voltage indicates the logical '1' and the negative or less positive voltage represents the logical '0'.

Negative Logic: The more negative voltage denotes the logical '1' and the positive or less negative voltage indicates the logical '0'. Table 14.2 exemplifies some combinations of such logic levels.

Table 14.2 Some specimens of positive and negative logic levels

Logic	Logical 1	Logical 0
Positive	2.5 V	− 2.5 V
	5.0 V	0 V
	− 0.5 V	− 5.0 V
Negative	− 2.5 V	2.5 V
	− 5.0 V	− 0.5 V
	0 V	5.0 V

The basic building blocks, which constitute the complex logical systems are called *gate*. A gate may operate on positive or negative logic depending on the electronic device constituting the gate.

14.2 Logic Gates

The gate is an electronic circuit based on Boolean logic having one or more input terminals but only one output terminal. All the input and out states assume either of two distinct states represented by the presence or absence of voltage. Gates provide with the basic circuit configuration in digital electronics. It is called 'gate' because the output state depends on specific combinations of inputs analogous to the open or closed conditions of an actual gate depending on the locks and the bolts. It is called 'logic gate' because the operation of the gate is based on a certain Boolean logic that rules the input–out conditions. Several types of gates, each with individual characteristic functions exist that act as the basic elements for constructing complex logical systems.

14.2.1 OR Gate

The OR operation with Boolean variables has been introduced in the previous chapter (Section 13.5.1). The OR gate is a digital circuit that implements the properties of OR operation electronically, in terms of voltage. The circuit of Figure 14.1(a) is a popular prototype to demonstrate the operation of 2-input OR gate. It contains two diodes and a resistor. The input voltages applied to the input terminals A and B are either 0 volt representing logical '0' or 5 volt representing logical '1'. The terminal Y stands for the output state in terms of the voltage across the resistor, depending on the diode conditions. The following input–output combinations may take place.

(i) When $A = 0$ and $B = 0$, both diodes are cut off and $Y = 0$
(ii) When $A = 0$ and $B = 1$, the diode with B gets forward biased and almost 5 V is obtained at the output and $Y = 1$.
(iii) Similarly, when $A = 1$ and $B = 0$, the diode with A gets forward biased and $Y = 1$.
(iv) When $A = 1$ and $B = 1$, both the diodes are forward biased and $Y = 1$.

The truth table of Figure 14.1(b) compiles all the above states for the 2-input OR gate. This is a practical realization of the conditions of Table 13.4 in terms of voltage levels. Figure 14.1(c) introduces the standard symbol for denoting a 2-input OR gate. Whatever may be the internal circuit, an OR gate is represented by this geometric shape. For a number of inputs, each of the N inputs are included in the symbol, as indicated in Figure 14.1(d).

Several important points may be noted. The above circuit is obeying positive logic. Instead of 5 V, it could be any other positive voltage well above the diode cut-in voltage. The truth table contains 4 rows. In general, the truth table of n-input OR gate contains 2^n rows and the output becomes high when any one or more of the inputs is high.

14.2.2 AND Gate

The AND operation with Boolean variables introduced in Chapter 13 can be implemented with the AND gate of Figure 14.2(a) and a truth table similar to Table 13.5 can be obtained in terms of voltage levels.

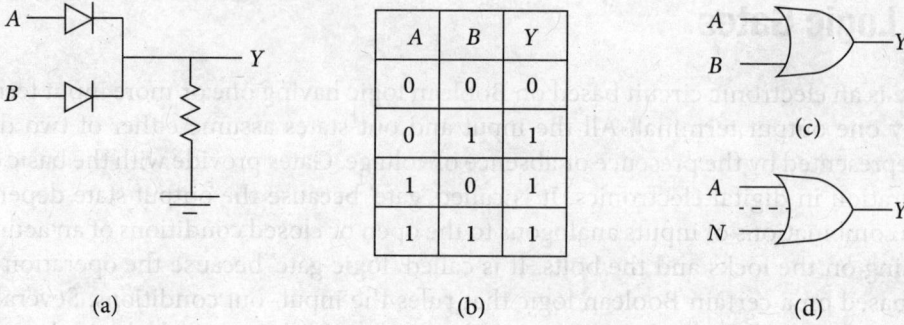

Figure 14.1 OR gate (a) circuit for two inputs, (b) truth table, (c) symbol of 2-input OR gate, and (d) indication for N-input OR gate

The diode–resistor circuit has two input terminals A and B where 5 V (or any other positive voltage much more than the diode cut-in voltage) and 0 V are applied as logical 1 and logical 0, respectively. The following operations may be executed.

(i) If A = 0 and B = 0, both the diodes get forward biased, the whole of 5 V drops across the resistor and the output (Y) is pulled down to almost ground potential so that Y = 0.
(ii) If A = 0 and B = 1, the diode with B remains cutoff due to reverse bias but the diode with A turns on and makes Y = 0.
(iii) Similarly, A = 1 and B = 0 makes Y = 0 because the diode with B is turned on.
(iv) When A = B = 1, both the diodes get reverse biased and the output becomes almost 5 V, hence Y = 1.

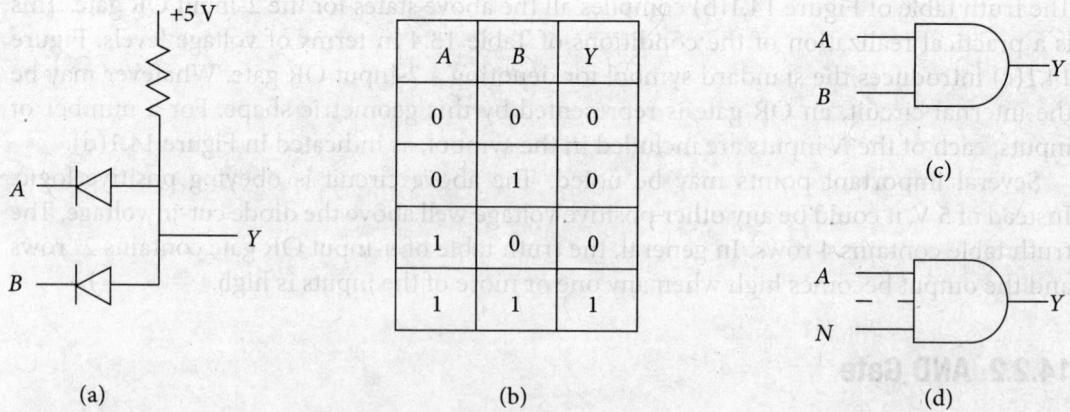

Figure 14.2 AND gate (a) circuit for two inputs, (b) truth table, (c) symbol of 2-input AND gate, and (d) indication for N-input AND gate

The above voltage states are collected in Table 14.2(b), which is same as the truth table (Table 13.5) of 2-input AND gate. Figure 14.1(c) introduces the symbol for the 2-input AND gate. The number of inputs can be increased, as indicated in Figure 14.2(d) but the same geometric shape is used to denote AND gate. For n-input AND gate, the number of rows in the truth table is 2^n and the output becomes high only when all the inputs are high.

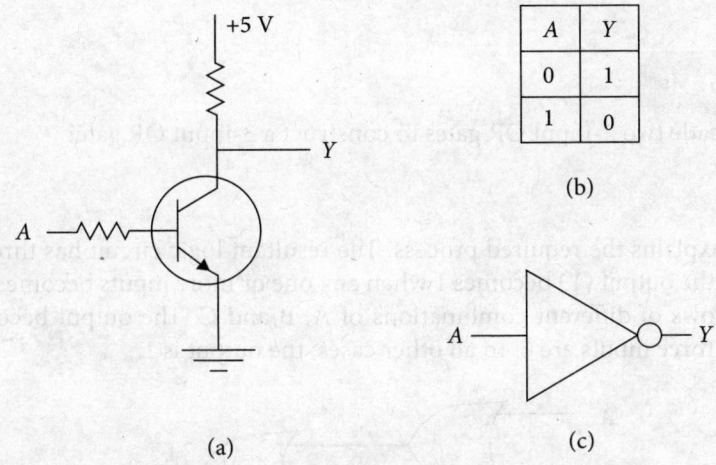

Figure 14.3 NOT gate (a) circuit diagram, (b) truth table, and (c) symbol

14.2.3 NOT Gate

The NOT operation (Section 13.5.3) and its truth table (Table 13.6) can be executed with the NOT gate of Figure 14.3(a). It consists of a transistor having its base as the input and the collector as output. The transistor is biased in such a way that it acts as switch with the following two conditions.

(i) When the input (A) is made high by connecting with V_{CC} (5 V in the present case), the transistor is driven into saturation and the output (Y) becomes almost grounded. Thus $A = 1$ leads to $Y = 0$.

(ii) When the input (A) is made low by grounding, the transistor becomes cutoff and the output shows almost the supply voltage. Thus $A = 0$ results in $Y = 1$.

The NOT gate has only one input and a single output. Compiling the above two conditions, the truth table for the NOT gate is obtained, as shown in Figure 14.3(b). The symbolic representation of the NOT gate is given in Figure 14.3(c). It is also called *inverter* because the output state is always the opposite of the input state.

> **Note: Buffer Is Another 1-Bit Gate**
>
> In addition to the NOT gate, there is another one-input logic gate, called *buffer*. It simply replicates the input to the output. Yet it is different from a simple connecting wire in terms of the output current level and input/output impedance difference. For instance, if the BJT emitter follower (Chapter 6) or the op-amp source follower (Chapter 11) works with only two discrete voltage levels, that may be designated as a digital buffer.

> **Example 14.1**
>
> Can you cascade two 2-input OR gates to construct a 3-input OR gate?
>
> **Solution**
>
> Figure 14.4 explains the required process. The resultant logic circuit has three inputs A, B and C and the output (Y) becomes 1 when any one or more inputs become 1. The truth table has 8 rows of different combinations of A, B and C. The output becomes 0 only when all the three inputs are 0. In all other cases, the output is 1.
>
>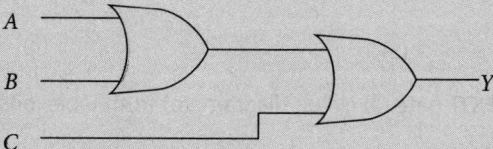
>
> **Figure 14.4** Constructing 3-input OR gate using two 2-input OR gates
>
> [N.B. One can construct 3-input AND gate in a similar way. The truth table has 8 rows and the output becomes 1 only when all the three inputs are 1. All other cases, when one or more input is 0, the output is 0.]

14.2.4 NOR Gate (Universal Gate)

'An OR gate followed by a NOT gate' is merged to a single gate, known as NOR gate. The symbol of a 2-input NOR gate is shown in Figure 14.5(a). The output (Y) of this gate is obviously represented by $\overline{A+B}$ and the input–output conditions are given in the truth table of Figure 14.5(b). The number of inputs can be increased similar to that of the OR gate, the logic remaining the same. The following properties of the NOR gate are noted from the truth table.

- The NOR gate gives a high output only when all inputs are low.
- If any input goes high, the output becomes low.
- The output conditions are the complements of those of the OR gate.

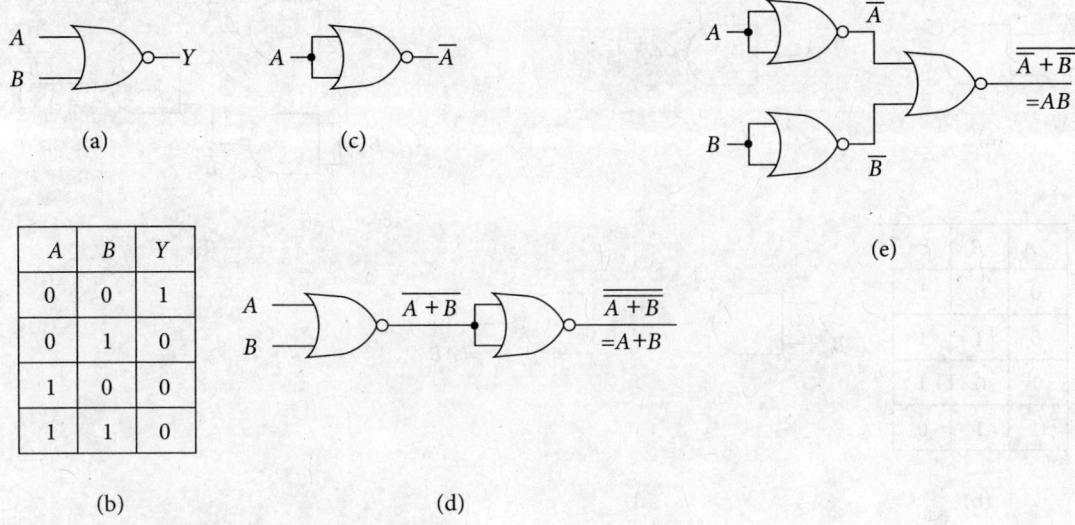

Figure 14.5 NOR gate (a) symbol, (b) truth table and fabrication of (c) NOT gate, (d) OR gate, and (e) AND gate with the NOR gate

Another very important property of the NOR gate is its universality. Any other gate and any sort of logic function can be generated by interconnecting NOR gates alone. Therefore, the NOR gate is termed as *universal gate*. The fabrication of all basic gates with NOR gates is demonstrated in Figure 14.5 (c) through (e). The intermediate logical expressions mentioned in the diagrams can be simplified with the use of De Morgan's theorems.

14.2.5 NAND Gate (Universal Gate)

This is a combined form of 'AND gate followed by NOT gate'. Figure 14.6(a) shows the symbol of 2-input NAND gate. The output (Y) is $\overline{AB}$ that yields the truth table of Figure 14.6(b). The number of inputs can be increased keeping the logic same. The NAND gate has the following properties.

- The NAND gate yields a low output, if all the inputs are high.
- If any input becomes low, the output goes high.
- The output conditions are the complements of those of the AND gate.

Similar to the NOR gate, the NAND gate also is a universal gate and any logical expression can be realized with proper combinations of this gate only. The formation of the basic gates with the use of NAND gates is explained in Figure 14.6(c) through (e).

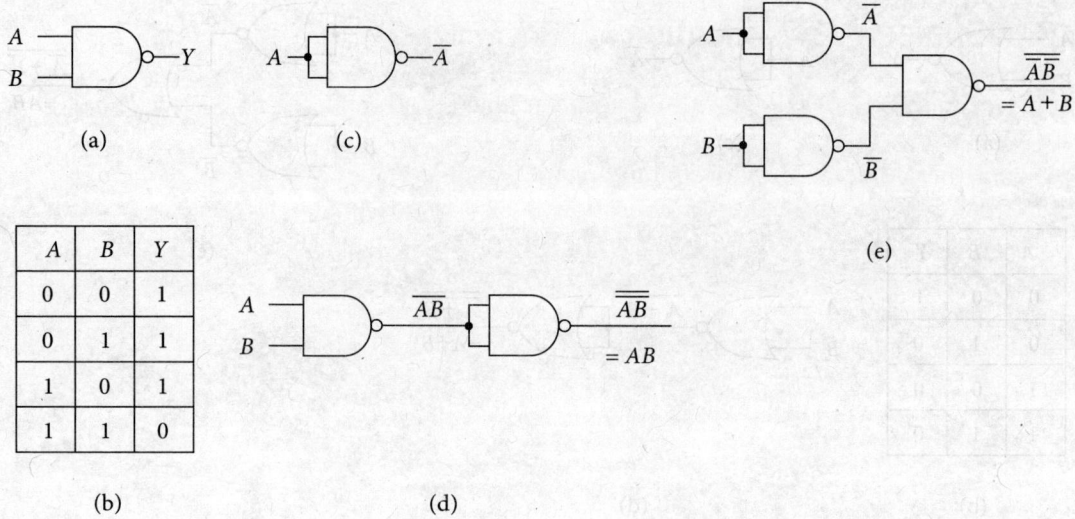

Figure 14.6 NAND gate (a) symbol, (b) truth table and fabrication of (c) NOT gate, (d) AND gate, and (e) OR gate with the NAND gate

14.2.6 Bubbled Gates

When a basic gate accepts its inputs in inverted form, it is called a *bubbled gate*. The bubble symbol represents inversion. This is not any separate type of gate. It is a concept applied to the basic gates that provides with more compactness of the circuit diagram and easy analysis techniques, as follows.

The output condition of the 2-input NOR gate of Figure 14.5(a), with the use of De Morgan's theorem can be expressed as

$$\overline{A+B} = \overline{A}\,\overline{B} \tag{14.1}$$

The right side of Equation (14.1) can be represented with NOT and AND gates, as drawn in Figure 14.7(a). However, it can be expressed in more abbreviated form by the *bubbled AND gate* of Figure 14.7(b) where the bubbles at the inputs denote the inversion of the inputs before the AND operation. Thus the bubbled AND gate is equivalent to the NOR gate.

Likewise, following De Morgan's theorem, the output condition of the 2-input NAND gate of Figure 14.6(a) can be stated as

$$\overline{AB} = \overline{A} + \overline{B} \tag{14.2}$$

The right side of Equation (14.2) could be realized with NOT and OR gates, as given in Figure 14.7(c) but the more condensed form is the *bubbled OR gate* of Figure 14.7(d). The bubbled OR gate is equivalent to the NAND gate.

Combinational Logic Circuits

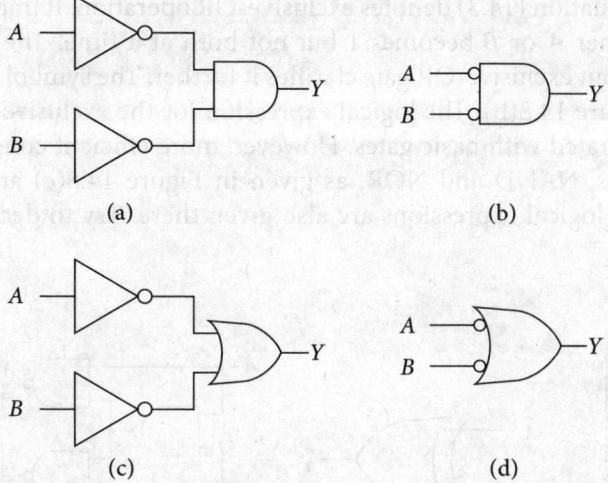

Figure 14.7 Bubbled gates: (a) NOR operation implemented with AND gate having inverted inputs, (b) the same abbreviated with bubbled AND gate, (c) NAND operation executed with OR gate having inverted inputs, and (d) the same abbreviated as bubbled OR gate

---

Note: Why Universal Gate

The NAND gate and the NOR gate is designated as universal gate because using only this single type of gate one can fabricate any other gate and any complicated logic circuit. This feature is quite beneficial to IC technology (Chapter 12) where a single type of gate is preferable while constructing logic circuits because it is easier to fabricate the same type of gate on a single chip instead of different types, the number of gates being immaterial.

Another significant benefit of the universal gate is direct implementation of any sum-of-products or product-of-sums expressions, hence any logical expression. Two basic examples are given below.

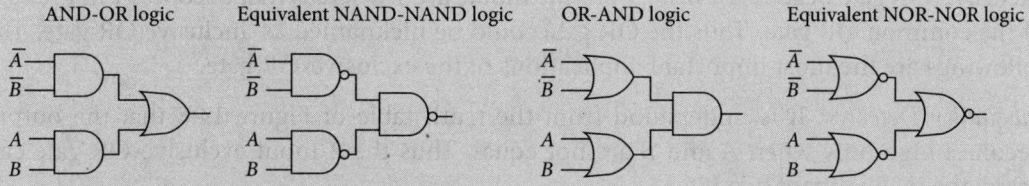

Now it is your job to verify the equivalence with both truth table and Boolean simplification.

14.2.7 Exclusive-OR (XOR) Gate

This gate produces a high output only when an odd number of inputs become high. In the simplest case of 2-input exclusive-OR gate having inputs A and B, the output (Y) is represented as

$$Y = A\bar{B} + \bar{A}B = A \oplus B \tag{14.3}$$

The ⊕ symbol in Equation (14.3) denotes exclusive-OR operation. It implies that Y becomes 1, if exclusively either A or B becomes 1 but not both at a time. The truth table [Figure 14.8(a)] of the 2-input exclusive-OR gate clarifies it further. The symbol of the exclusive-OR gate is given in Figure 14.8(b). The logical expression for the exclusive-OR gate [Equation (14.3)] can be generated with basic gates. However, more efficient constructions are those with universal gates, NAND and NOR, as given in Figure 14.8(c) and (d), respectively. Some intermediate logical expressions are also given there. Try to derive in full using De Morgan's theorems.

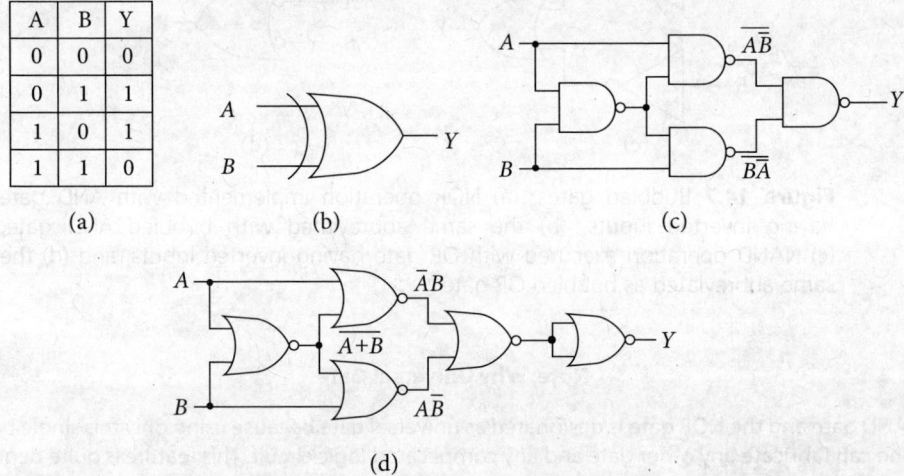

Figure 14.8 Exclusive-OR gate with two inputs: (a) truth table, (b) symbol and construction with (c) NAND gates, and (d) NOR gates

Comparing with the truth table of ordinary OR gate, it is noticed that the output of the exclusive-OR gate becomes 0 when both the inputs are 1 whereas that becomes 1 in the case of the common OR gate. Thus the OR gate could be nicknamed as 'inclusive OR gate'. The followings are the most important applications of the exclusive-OR gate.

Inequality Detector: It is understood from the truth table of Figure 14.8 that the output becomes high only when A and B are not equal. Thus the 2-input exclusive-OR gate can detect the inequality of two bits.

Equality Detector: If a NOT gate is added to the output of the exclusive-OR gate, it becomes an exclusive-NOR gate. Its output becomes high only when both A and B are equal; either both 0 or together 1. Thus it detects the equality of two bits. The logical condition for the equality can be obtained by inverting the right side of Equation (14.3), as given by

$$Y = AB + \overline{A}\,\overline{B} \qquad (14.4)$$

Controlled Inversion: This is discussed in Section 14.4.4.

Parity Checker and Generator: These are deliberated in Section 14.5.6.

Example 14.2

Construct a 3-input exclusive-OR gate using 2-input exclusive-OR gates and work out its truth table.

Solution

Figure 14.9 gives the required construction of 3-input exclusive-OR gate. It can accept three inputs A, B and C. The truth table is worked out by the side.

[N.B. It is noted that the output (Y) becomes 1 only when odd number of inputs (one or three in the present case) become 1. This is the general rule for the exclusive-OR gate.]

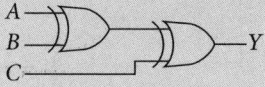

A	B	C	Y
0	0	0	0
0	0	1	1
0	1	0	1
0	1	1	0
1	0	0	1
1	0	1	0
1	1	0	0
1	1	1	1

Figure 14.9 Logical diagram and truth table for 3-input exclusive-OR gate constructed with 2-input exclusive-OR gates

Example 14.3

Derive the condition for the equality of two bits A and B.

Solution

Since the output of a 2-input exclusive-OR gate becomes high for the inequality of the two input bits, the equality would be determined by the inversion of that given by

$$A \oplus B = \overline{A}B + A\overline{B}$$

Using De Morgan theorems, the above expression can be modified as

$$\overline{\overline{A}B}\;\overline{A\overline{B}} = (A + \overline{B})(\overline{A} + B) = AB + \overline{A}\,\overline{B}$$

[The right-most expression is given in Equation (14.4).]

14.2.8 Timing Diagram

For all the gates mentioned above and also for any logic circuit, it is often convenient to represent the time-dependent input–output variations graphically. Such diagram containing the input and output waveforms of a logic circuit is called the *timing diagram* for the circuit.

Figure 14.10 displays two examples of timing diagram for arbitrary input states A and B of (a) 2-input OR gate and (b) 2-input AND gate. The results are self-explanatory. The output (Y) of the OR gate becomes high whenever any one or both of A and B become high. In the case of the AND gate, Y becomes high only when both A and B are high.

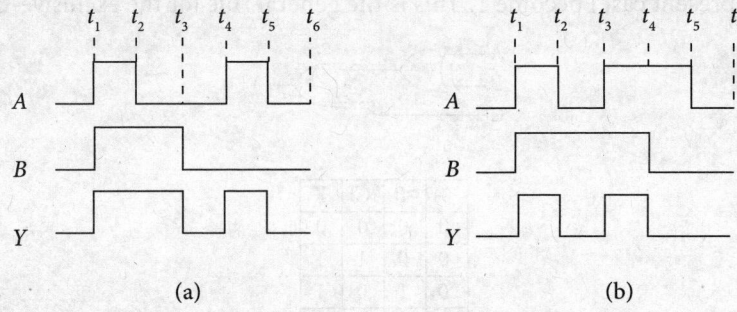

Figure 14.10 Timing diagrams for arbitrary inputs A and B of (a) 2-input OR gate and (b) 2-input AND gate

14.3 Logic Families

In most cases of digital circuit diagrams, the gates are drawn directly. However, it is justified to have a brief idea about the actual electronic circuit within a gate. The diode–resistor circuits for the OR and the AND gates, as given in Figures 14.1 and 14.2 are only models for understanding the concept. A practical gate circuit contains something else, bipolar junction transistors or field-effect transistors, as explained in this section.

First let us understand why the mere establishment of Boolean operations in terms of voltage does not qualify a circuit for being a logic gate. A digital circuit suitable for gate must fulfil the following requirements.

(i) The gate fabrication process should comply with the methods used in IC technology. The use of capacitors and resistors is minimized and transistors are used whenever possible.
(ii) The use of universal gates (NAND and NOR) is preferred to basic gates (AND, OR, NOT).
(iii) *Fan-in* and *Fan-out* are two important properties of a logic gate that must be taken into account. The fan-out of a gate is the number of other gate inputs it can drive and the fan-in is the number of inputs the gate can handle. To increase the fan-out of a stage (i.e., fan-in to the subsequent stage), the voltage and current should be amplified.

Considering the above aspects, active elements like bipolar and MOS transistors are used as the ingredients of logic gates. The gate circuits fabricated with one type of active device is termed as a *family*. Two popular families of logic gates are introduced here.

14.3.1 Transistor–Transistor Logic (TTL)

The basic constituents of all digital ICs of this family are bipolar junction transistors. Both the universal gates, namely NAND and NOR can be fabricated with TTL logic.

A. TTL NAND Gate

Figure 14.11 shows the circuit diagram for a 2-input TTL NAND gate. The TTL ICs fabricate special multiple-emitter transistors. The transistor Q_1 is of this category. It has two emitter terminals corresponding to two inputs A and B. The following two circuit conditions may occur thereby representing the NAND operation similar to that of the truth table in Figure 14.6(b).

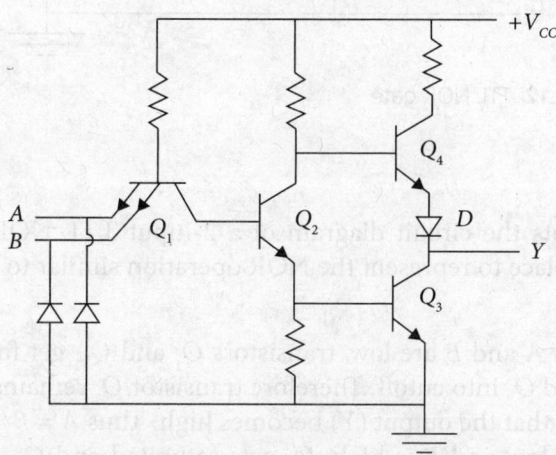

Figure 14.11 TTL NAND gate

Case-I: When either A or B is grounded, transistor Q_1 gets saturated because of its base–emitter junction being forward biased. Negligible current is supplied to the base of transistor Q_2 and it remains cutoff resulting in low voltage at the base of transistor Q_3 and high voltage at the base of transistor Q_4. Consequently Q_3 remains off and Q_4 turns on and a high voltage, almost equal to $+V_{CC}$ appears at the output (Y). Thus when either A = 0 or B = 0 or both are 0, Y = 1.

Case-II: When both A and B become high, Q_1 gets cutoff because its emitter–base junction becomes reverse biased. However, its base-collector junction gets forward biased that drives Q_2 into saturation. That causes a high voltage at the base of Q_3 driving this too into saturation. At the same time, the low collector voltage of saturated Q_2 keeps Q_4 cutoff. The diode (D) in series ensures the cutoff. As a result, low voltage appears at the output. Thus A = B = 1 makes Y = 0.

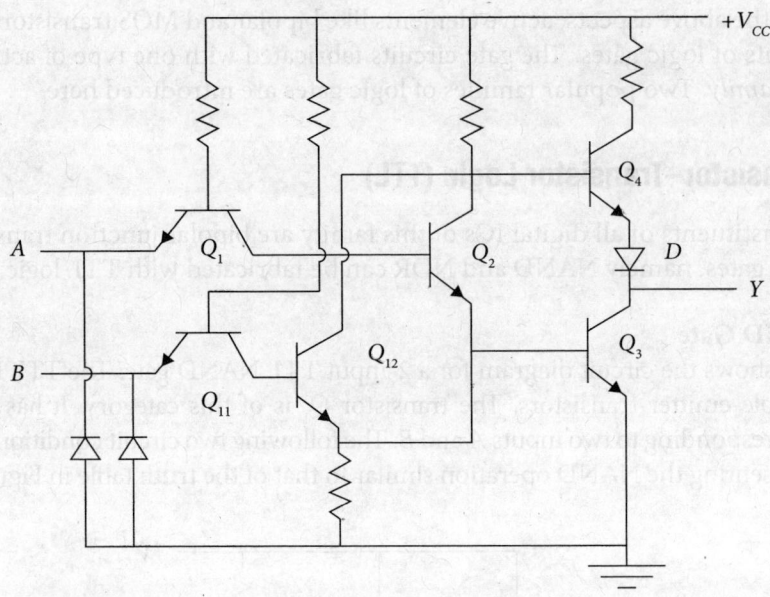

Figure 14.12 TTL NOR gate

B. TTL NOR Gate

Figure 14.12 represents the circuit diagram of a 2-input TTL NOR gate. The following conditions may take place to represent the NOR operation similar to that of the truth table in Figure 14.5(b).

(i) When both A and B are low, transistors Q_1 and Q_{11} get forward biased driving both Q_{12} and Q_2 into cutoff. Therefore transistor Q_3 remains off and transistor Q_4 turns on so that the output (Y) becomes high. Thus $A = B = 0$ makes $Y = 1$.

(ii) When A is low and B is high, Q_1 gets saturated and Q_{11} becomes cutoff. That makes Q_2 cutoff and Q_{12} turned on, which turns on Q_3 and turns off Q_4 making the output low. Thus $A = 0, B = 1$ makes $Y = 0$.

(iii) Similarly, when A is high and B is low, Q_2 is on and Q_{11} is off so that Q_2 turns on Q_3 and turns off Q_4 thereby making $Y = 0$. Thus $A = 1, B = 0$ makes $Y = 0$.

(iv) When both A and B are high, both Q_1 and Q_{11} are off, hence both Q_2 and Q_{12} are on. Thus Q_3 is on and Q_4 is off making the output low. Thus $A = B = 1$ means $Y = 0$.

> **Think a Bit: What Happens if Any Input in TTL Gate Is Kept Floating?**
>
> It is understood from Figures 14.11 and 14.12 that when A or B is kept floating without any connection, that serves the same purpose (turning off the corresponding transistor) as that of making high. So any unconnected input terminal in TTL IC acts as logical 1. If logical 0 is required at any input terminal, that must be grounded physically and should not be left unconnected.

> **Note: Totem–Pole Arrangement**
>
> The transistors Q_3 and Q_4 together in TTL NAND (Figure 14.11) and TTL NOR (Figure 14.12) gates create an iconic representation of low resistance. When Q_3 conducts, it acts as a grounded emitter follower having low output resistance. When Q_4 conducts, that too offers a low output resistance. The low output resistance of the circuit serves the following purposes.
>
> - It enhances the fan-out by increasing the output current.
> - It helps high speed operation because any stray output capacitance is quickly charged/discharged through the low resistance.
> - If there were an actual small ohmic resistance, there would be heating loss due to high current. The 'totem' or iconic resistance produced by the transistors avoids the power loss.
> - The totem-pole output is also known as active pull up. This is a salient feature of TTL logic family.

14.3.2 MOS Logic

The active device in this logic family is the MOSFET. The enhancement-type MOSFET is preferred because biasing the gate and the drain with the same polarity is a required phenomenon. Both n-channel and p-channel MOSFETs can be used. The p-channel device (PMOS) is easy to fabricate but the n-channel device (NMOS) is faster in operation and occupies smaller space. Therefore NMOS logic circuits are used widely. The working principles of the two universal gates using MOS logic are explained below.

A. MOS NAND Gate

Figure 14.13(a) presents the circuit diagram for 2-input NAND gate realized with MOS logic. Two MOSFETs Q_1 and Q_2 connected in series are the input devices and the MOSFET Q_3 acting as a resistor is termed *active load*. Both of its gate and drain terminals are connected with the supply voltage $(+V_{DD})$ so that Q_3 conducts all the way. The output terminal is indicated by Y.

If either input A or input B becomes low, the corresponding MOSFET turns off and the connection with the ground is broken. Consequently a high voltage, almost $+V_{DD}$ appears at the output. If both A and B are high, both Q_1 and Q_2 are turned on and act as closed switch so that the output becomes low. The output becomes 0 only when all the inputs become 1. Thus NAND operation is executed.

B. MOS NOR Gate

Figure 14.13(b) depicts the circuit diagram for a 2-input NOR gate. MOSFETs Q_1 and Q_2 are the input devices. When both the inputs A and B become low, both Q_1 and Q_2 get cutoff so that the output (Y) goes high. If any one of A or B becomes high, the corresponding MOSFET turns on and the output, being shorted to the ground becomes low. Thus NOR operation is performed. If any or both the inputs become 1, the output becomes 0.

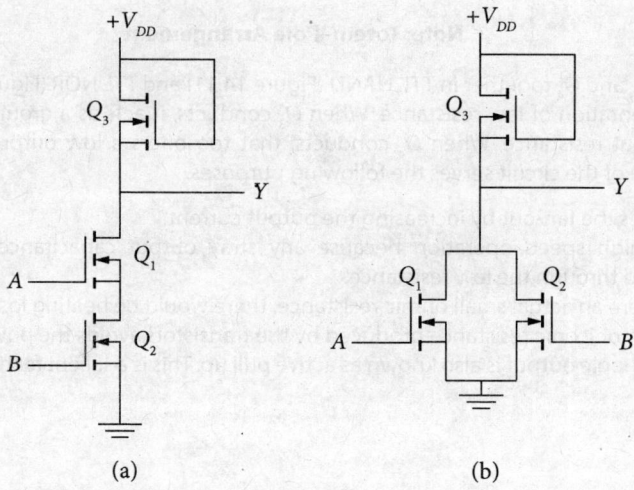

Figure 14.13 MOS logic for (a) 2-input NAND gate and (b) 2-input NOR gate

C. Complementary MOS (CMOS)

This family of logic circuits employs both n-channel and p-channel MOSFETs (complementary to each other) combined in push–pull arrangement. Since the two types of devices require opposite type of bias, one of the MOSFET remains off in each logic level resulting in negligible drain current, hence very low power consumption. Therefore complementary MOS (CMOS) circuits are used in battery-operated equipment, such as calculator, mobile phone and digital watch where limited consumption of power is a prime factor.

Figure 14.14 introduces the circuit diagram for executing NOT operation with CMOS logic. When the input is low, MOSFET Q_1 remains off because it is n-channel device. However, MOSFET Q_2 (p-channel) turns on and the output becomes high. When the input is high, Q_1 turns on but Q_2 remains off so that the output becomes low. Thus NOT operation is executed for either case of the input.

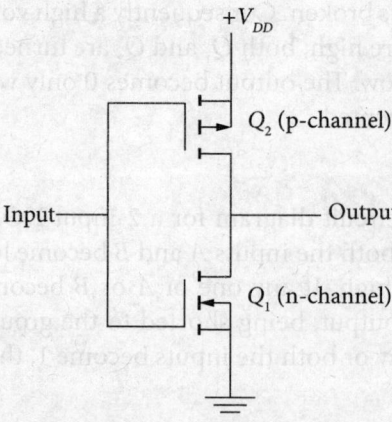

Figure 14.14 CMOS NOT gate

14.4 Arithmetic and Logic Circuits

The arithmetic logic unit of the computer contains digital electronic circuits that can execute mathematical and logical operations, such as addition, subtraction, multiplication, division, and comparison of two numbers, all in binary system in terms of voltage. The preliminary idea about such digital circuit is given in this section. The adder circuit is the most fundamental one because the same circuit can perform other arithmetic operations: subtraction as negative addition, multiplication as repetitive addition and division as repeated negative addition. The following categories of arithmetic circuits are generally available.

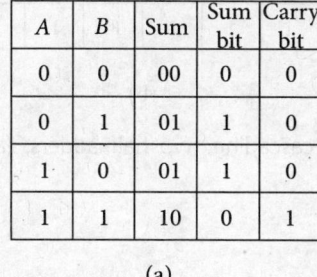
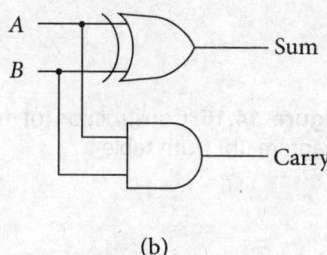

Figure 14.15 Half adder representing the addition of two bits A and B, (a) logical conditions and (b) circuit implementation

14.4.1 Half Adder

The simplest case of addition of two 1-bit numbers A and B can assume the possible combinations given in Figure 14.15(a). Looking separately to the sum and the carry bits, it is comprehended that the sum bit matches the logic states of exclusive-OR operation and the carry bit corresponds to that of AND operation. Therefore the combination of these two gates, as shown in Figure 14.15(b), establishes a 1-bit adder circuit. As noted in the figure, the digital circuit has inputs for only two bits A and B coming from any place of two binary numbers and it can produce sum bit and possible carry bit for that place. However, the circuit has no provision for accepting any carry bit from the previous place and thus the addition process remains incomplete. Because of this limitation, the circuit is named as 'half adder'.

For two input bits A_n and B_n of any n-th place of a binary number, the output conditions of a half adder are given by

$$S_n = A_n \oplus B_n = A_n \overline{B_n} + \overline{A_n} B_n \qquad (14.5a)$$

and

$$C_n = A_n B_n \qquad (14.5b)$$

Here S_n is the sum bit and C_n is the carry bit produced at the n-th place. The carry bit can be inserted to the next higher place using a full adder.

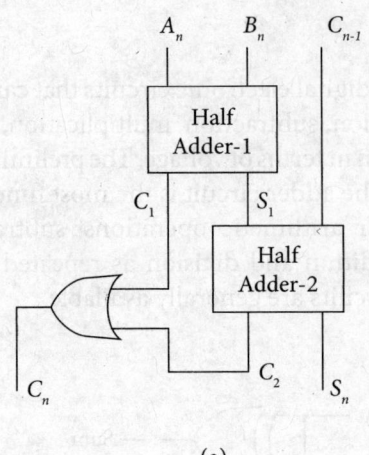

A_n	B_n	C_{n-1}	Sum	S_n	C_n
0	0	0	00	0	0
0	0	1	01	1	0
0	1	0	01	1	0
0	1	1	10	0	1
1	0	0	01	1	0
1	0	1	10	0	1
1	1	0	10	0	1
1	1	1	11	1	1

(a)   (b)

Figure 14.16 Construction of full adder cascading two half adders; (a) block diagram, (b) truth table

14.4.2 Full Adder

It is seen above that a half adder can add two binary digits (the addend and the augend) and can produce the sum and the carry. However, the summing process is said to be 'full', if it includes the provision for a carry input too from the next lower significant place. Figure 14.16(a) demonstrates how such a full adder can be constructed by cascading two half adders.

Let A_n and B_n be two bits of the n-th place of any n-bit binary number and C_{n-1} be the carry bit from the next lower place of that binary number. The first half adder adds A_n and B_n and produces a sum bit (S_1) and a carry bit (C_1). The second half adder adds S_1 and C_{n-1} to complete the addition process for the n-th place and produces the actual sum bit (S_n) for this place. The second half adder also can produce a carry bit (C_2) of its own. Either of the two carry bits C_1 and C_2 will produce the actual carry from the n-th place. Therefore these two are passed through an OR gate whose output is treated as the carry bit (C_n) from the n-th place. Obviously both C_1 and C_2 cannot be 1 at a time. Either both can be 0 or any one can be 1.

The truth table of the full adder is given in Figure 14.16(b). Collecting the true conditions for the sum and the carry, one can write

$$S_n = \bar{A}_n \bar{B}_n C_{n-1} + \bar{A}_n B_n \bar{C}_{n-1} + A_n \bar{B}_n \bar{C}_{n-1} + A_n B_n C_{n-1} \tag{14.6a}$$

which can be simplified as

$$S_n = (A_n \oplus B_n) \oplus C_{n-1} \tag{14.6b}$$

Similarly,

$$C_n = \overline{A}_n B_n C_{n-1} + A_n \overline{B}_n C_{n-1} + A_n B_n \overline{C}_{n-1} + A_n B_n C_{n-1} \tag{14.7a}$$

which can be simplified as

$$C_n = A_n B_n + (A_n \oplus B_n) C_{n-1} \tag{14.7b}$$

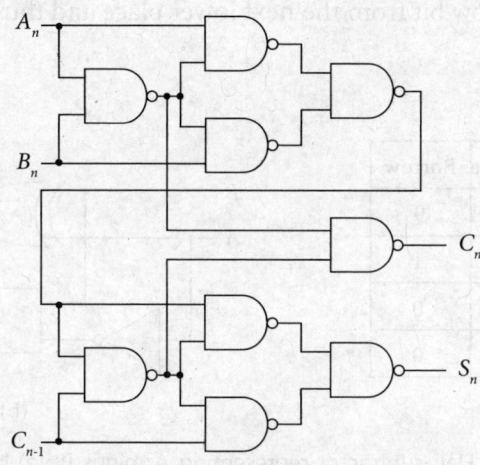

Figure 14.17 Full adder constructed with NAND gates

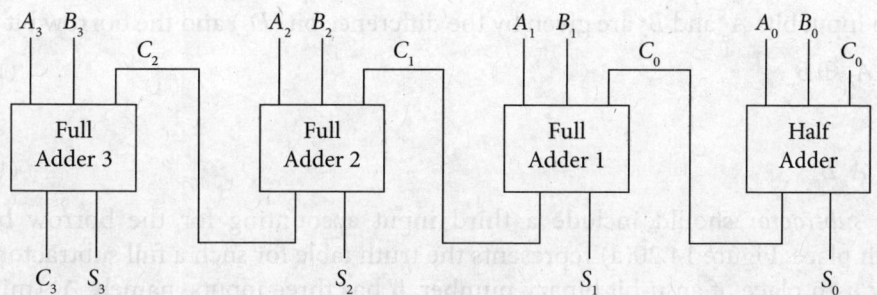

Figure 14.18 Construction of 4-bit adder with three full adders and a half adder

Figure 14.17 presents an actual logic circuit for the full adder constructed with NAND gates (universal) only. It has three inputs A_n, B_n and C_{n-1} and two outputs S_n and C_n, all five for the n-th place. Such a full adder can perform complete addition of two bits with possible carry. If it is required to construct an adder circuit for adding two 4-bit binary numbers $A_3 A_2 A_1 A_0$ and $B_3 B_2 B_1 B_0$, at least three full adders and one half adder are to be cascaded, as sketched in Figure 14.18. In general, addition of two n-bit numbers requires one half adder and $(n-1)$ full adders.

14.4.3 Half and Full Subtractors

The simplest case of 1-bit subtraction is considered. The table of Figure 14.19(a) compiles the possible conditions of the subtraction of bit B (subtrahend) from bit A (minuend). It is apparent that the difference bit (A – B) matches exclusive-OR operation and the borrow bit follows the logic $\overline{A}B$. The combination of the gate and the logical expression establishes a 1-bit subtractor, as shown in Figure 14.19(b). This is called a *half subtractor* because there is no provision for the borrow bit from the next lower place and thus the subtraction process is left incomplete.

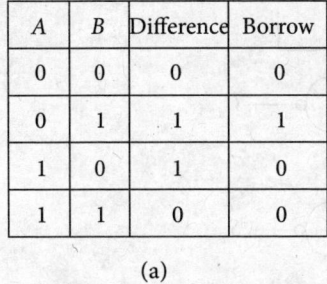
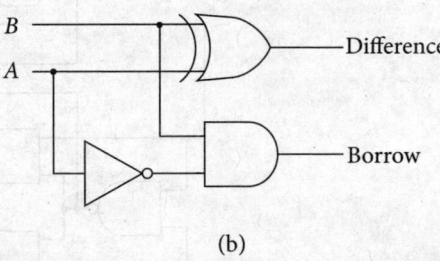

Figure 14.19 Half subtractor representing A minus B; (a) logical conditions for difference and borrow, (b) circuit implementation

In general, for any n-th place of a binary number, the output conditions of a half subtractor with two input bits A_n and B_n are given by the difference bit (D_n) and the borrow bit (Br_n) as

$$D_n = A_n \oplus B_n \tag{14.8a}$$

and

$$Br_n = \overline{A_n} B_n \tag{14.8b}$$

The *full subtractor* should include a third input accounting for the borrow bit from $(n - 1)$-th place. Figure 14.20(a) represents the truth table for such a full subtractor. For an arbitrary n-th place of an n-bit binary number, it has three inputs, namely A_n (minuend), B_n (subtrahend) and Br_{n-1} (borrow from the previous place) and two outputs namely D_n (difference) and borrow (Br_n) to be sent to the next higher place. The logical expressions for the difference and the borrow bit can be derived from the table as

$$D_n = A_n \oplus B_n \oplus C_{n-1} \tag{14.9a}$$

and

$$C_n = \overline{A_n} B_n + (\overline{A_n \oplus B_n}) C_{n-1} \tag{14.9b}$$

Figure 14.20(b) outlines the construction of the full subtractor by cascading two half subtractors.

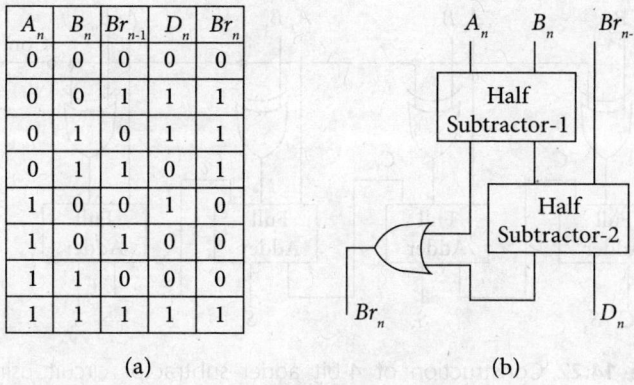

Figure 14.20 Full subtractor; (a) logical conditions and (b) block diagram showing the construction by cascading two half subtractors

14.4.4 Adder–Subtractor

This is to explain that instead of separate subtractor circuits mentioned above, a single circuit can act as both adder and subtractor at the user's disposal. It is achieved by 2's complement method (Chapter 13) in collaboration with another useful technique known as *controlled inversion* explained with Figure 14.21. It is observed that the 2-input exclusive-OR gate has a single input A and the other input terminal is used as *Control*. It follows from exclusive-OR logic that when Control = 0, the output (Y) equals A and when Control = 1, $Y = \overline{A}$. Thus making the control terminal 0 or 1, the input or its inversion can be communicated to the output.

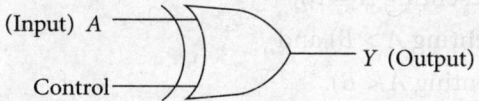

Figure 14.21 Controlled inversion of a bit

The above concept of controlled inversion is implemented to the 4-bit adder cum subtractor circuit of Figure 14.22. The bits of the first number $A_3A_2A_1A_0$ are applied directly to the adder to the adder inputs and the bits of the second number $B_3B_2B_1B_0$, are applied each through an exclusive-OR gate. The other inputs of the exclusive-OR gates are tied together and that connection is used as the control terminal. The carry input (C_{01}) of the first full adder is also joined to the control terminal. The following two arithmetic operations can be executed separately.

Addition: When Control = 0, $C_{01} = 0$ and the bits B_3, B_2, B_1 and B_0 get inserted directly to the respective adders. Thus the addition of two numbers $A_3A_2A_1A_0$ and $B_3B_2B_1B_0$ is performed.

Subtraction: When Control = 1, controlled inversion takes place and $\overline{B}_3, \overline{B}_2, \overline{B}_1$ and $\overline{B}_0$ are inserted to the adder thereby producing 1's complement of $B_3B_2B_1B_0$. At the same time, $C_{01} = 1$ so that 1 is added to this 1's complement producing the 2's complement of $B_3B_2B_1B_0$. Addition of this 2's complement with $A_3A_2A_1A_0$ performs the subtraction operation.

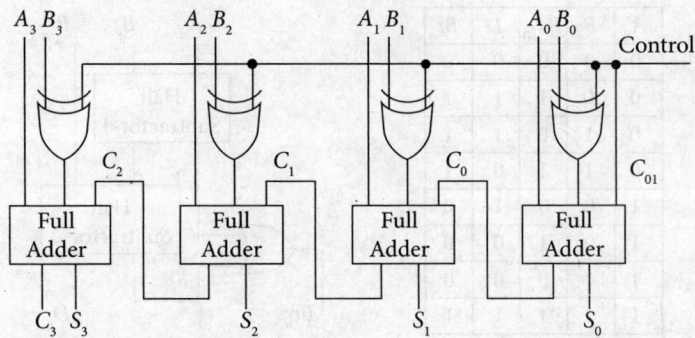

Figure 14.22 Construction of 4-bit adder–subtractor circuit using controlled inversion

14.4.5 Digital Comparators

Such circuits can detect the equality or inequality of single bit or multi-bit numbers. The exclusive-OR gate of Section 14.2.7 is a prototype comparator for detecting the equality or inequality of two single bit numbers. However, the comparator should have more provisions, as illustrated in this section.

Figure 14.23(a) shows the logical expressions and states for the 'equal to', 'greater than' and 'less than' conditions of two bits A and B. Assembling the logic gate implementations of these three conditions in a single circuit, we obtain a 1-bit comparator, as given in Figure 14.23(b). It has two inputs for the bits A and B and the following three outputs:

$Y_1 = \overline{A \oplus B}$ (representing $A = B$),
$Y_2 = A\overline{B}$ (representing $A > B$) and
$Y_3 = \overline{A}B$ (representing $A < B$).

Any one of these three outputs can go high depending on the condition of the compared bits.

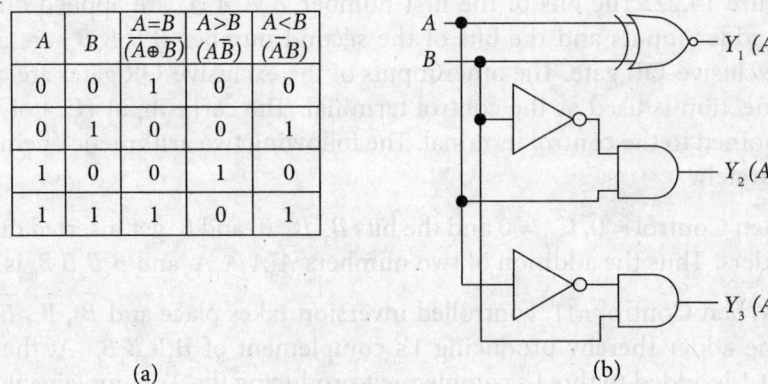

(a)   (b)

Figure 14.23 1-bit comparator (a) logical conditions (b) circuit implementation

In the case of multiple bit numbers, the comparison process becomes more complicated as several conditions are to be considered at a time. For instance, if there be two 2-bit numbers $A = A_1 A_0$ and $B = B_1 B_0$, the number A can be greater than the number B, if $A_1 > B_1$ or $A_1 = B_1$ but $A_0 > B_0$. Both 1 and 0 states have to be considered. For larger number of bits, the conditions are more elaborated. A general method is to simplify the conditions using Karnaugh map. Figures 14.24(a), (b) and (c) depict the Karnaugh map simplification for the condition $A > B$, $A < B$ and $A = B$, respectively, for 2-bit numbers. The results come out as

$$A > B \rightarrow A_1 \overline{B_1} + A_0 \overline{B_1} \overline{B_0} + A_1 A_0 \overline{B_0}$$

$$A < B \rightarrow \overline{A_1} \overline{A_0} B_0 + \overline{A_1} B_1 + \overline{A_0} B_1 B_0$$

$$A = B \rightarrow \overline{A_1 \oplus B_1} \cdot \overline{A_0 \oplus B_0}$$

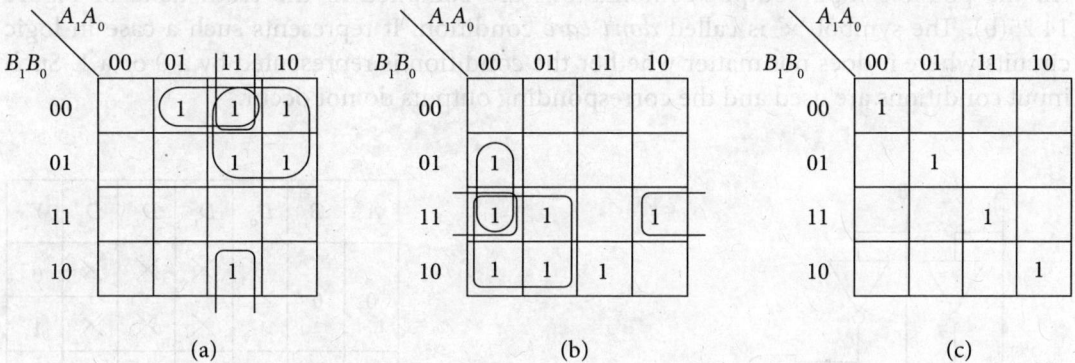

Figure 14.24 Karnaugh map representation and possible simplification for (a) $A > B$, (b) $A < B$ and (c) $A = B$

14.5 Data Processing Circuits

Digital systems need to execute innumerable exchanges of binary data whereas the circuit paths are limited. These limited tracks are to be made into use for unlimited data transfer without disturbing one another. Also the numbers should be decoded to and coded from the binary system. The specific circuits available for serving such purposes are collectively known as data processing circuits. The multiplexer, demultiplexer, decoder, encoder, etc. discussed in this section are of this category.

14.5.1 Multiplexer

'Multiplexing' means simultaneous transmission of multiple signals on the same line. The multiplexer circuit contains a number of data inputs and a single data output. Any one of the data inputs can be selected and transmitted to the output by suitable combination of several controlling terminals referred to as *select*. The multiplexer is also called *data selector*.

Figure 14.25(a) presents the logic circuit for a simple 4-to-1 multiplexer that has four data inputs D_0, D_1, D_2 and D_3 and a single output Y. The terminals A and B are called 'select' because changing their states one can select any specific input for the output.

(i) If $A = B = 0$, only D_0 reaches the output (Y), as explained below. Except for G_0, all the other AND gates have one or more inputs at 0 state so that these get fixed at 0 output state. The inputs D_1, D_2 and D_3 have no effect on these all. Only G_0, having the other input at 1 state, depends on D_0. If $D_0 = 0$, G_0 also gives 0 and if $D_0 = 1$, G_0 yields 1. This change of D_0 is echoed at Y through the OR gate.

(ii) If $A = 0$, $B = 1$, following the above reasoning, only the AND gate G_1 becomes active and only D_1 is communicated to Y.

(iii) Similarly, $A = 1$, $B = 0$ connects only the data input D_2 to Y and

(iv) $A = B = 1$ makes only D_3 communicated to Y.

All the possible input–output combinations are compiled in the truth table of Figure 14.25(b). The symbol 'x' is called *don't care* condition. It represents such a case in logic circuits where it does not matter whether the condition is represented by a 0 or a 1. Such input conditions are used and the corresponding outputs do not occur.

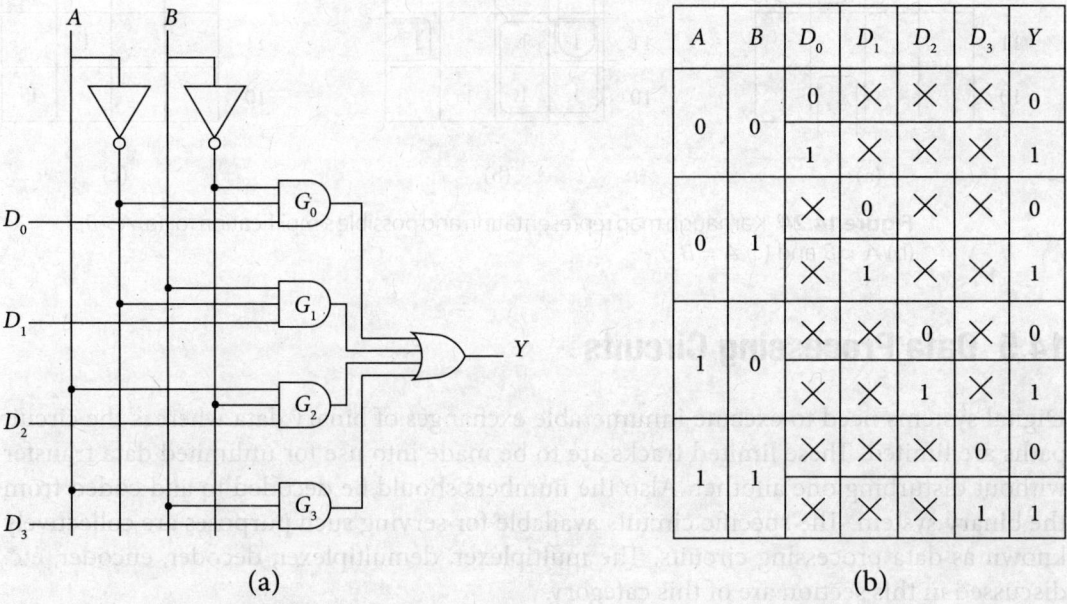

Figure 14.25 4-to-1 multiplexer: (a) circuit diagram, (b) truth table

A	B	D_0	D_1	D_2	D_3	Y
0	0	0	x	x	x	0
0	0	1	x	x	x	1
0	1	x	0	x	x	0
0	1	x	1	x	x	1
1	0	x	x	0	x	0
1	0	x	x	1	x	1
1	1	x	x	x	0	0
1	1	x	x	x	1	1

The multiplexer circuit can be more expanded, such as 8-to-1 with 3 select terminals and 16-to-1 with 4 select terminals. In general, a 2^n-input multiplexer has n number of selects. The circuit operation is similar to that discussed above. A multiplexer has the following applications.

- *Data Selection*: The main application of multiplexer is to select any one bit out of a number of inputs.
- *Parallel-to-Serial Conversion*: Let any *n*-bit binary number be applied parallel to the n inputs of a multiplexer such that D_0 and D_n represent the LSB and the MSB, respectively. Changing the select combination, the bits can be made to appear successively at the output one by one.
- *Sequential Data Selection*: Using a specific combination of select one or more data input can be selected repeatedly in a pre-determined sequence.
- *Building Logic Circuits*: The multiplexer can be utilized for generating logic circuits. Two standard methods of implementing a truth table are sum-of-products and product-of-sums, as discussed in Section 13.7. A third method is the multiplexer logic. This is universal logic because 2*n*-to-1 multiplexer can achieve any truth table for *n* variables.

Example 14.4

Construct 4-to-1 multiplexer using only 2-to-1 multiplexers.

Solution

For 2-to-1 multiplexer, the output (Y) condition is

$$Y = \overline{A}D_0 + AD_1$$

D_0 and D_1 being the two inputs and A being the single select terminal. For 4-to-1 multiplexer similar to that of Figure 14.25(a), the output condition is expressed as

$$Y = \overline{A}\,\overline{B}D_0 + \overline{A}BD_1 + A\overline{B}D_2 + ABD_3$$
$$= \overline{A}(\overline{B}D_0 + BD_1) + A(\overline{B}D_2 + BD_3)$$

The above expression can be realized with two 2-to-1 multiplexers using B as the select input. The outputs of these two multiplexers are joined as data inputs to a third 2-to-1 multiplexer where A acts as the select. The block diagram of the circuit is sketched in Figure 14.26.

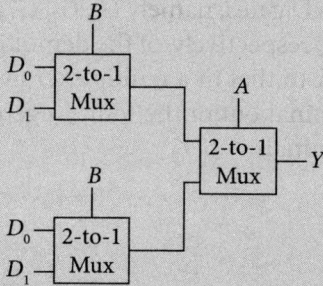

Figure 14.26 Fabricating 4-to-1 multiplexer with 2-to-1 multiplexer (2-to-1 Mux)

> **Example 14.5**
>
> Construct a 1-bit full adder using 4-to-1 multiplexer.
>
> **Solution**
>
> Comparing the truth table of full adder [Figure 14.16(b)] with that of 4-to-1 multiplexer [Figure 14.25(b)], the following equivalences are detected.
>
> | Full Adder | 4-to-1 Multiplexer |
> |---|---|
> | Inputs A_n and B_n | Selects A and B |
> | Carry C_{n-1} | Any input D_0, D_1, D_2, D_3 |
> | Sum bit (S_n) | Output (Y) |
> | Carry bit (C_n) | Output (Y) |
>
> The following conclusions are made on the basis of the above comparisons.
>
> (i) S_n and Y have the same state for D_0 and D_3 and opposite state for D_1 and D_2. Therefore, applying D_1 and D_2 through NOT gates can match the logic states of S_n and Y.
>
> (ii) C_n and Y states are the same for D_1 and D_2. For equalizing the other two, D_0 and D_3 are fixed at 0 and 1, respectively.
>
> Thus two 4-to-1 multiplexers are needed. Connecting the inputs D_0, D_1, D_2, D_3 and using select terminals as mentioned above, one multiplexer can produce the sum (S_n) and another multiplexer can produce the carry (C_n).

14.5.2 Demultiplexer

This is just the opposite of a multiplexer. The demultiplexer has a single data input and a number of outputs. Changing the select terminals, the input data can be sent to any selected one of the output terminals. The demultiplexer is also referred to as data distributor.

Figure 14.27 shows the logic diagram for a 1-to-4 demultiplexer. A single data input (D) is connected with all the four AND gates, namely G_0, G_1, G_2 and G_3 whose outputs represent the four outputs Y_0, Y_1, Y_2 and Y_3, respectively, of the demultiplexer. The two select terminals A and B operate in a way similar to that in a multiplexer and make the data D to appear at any pre-determined output terminal out of the four. In general, a 2^n-output demultiplexer contains n number of select terminals.

14.5.3 Decoders

'Coding' in a digital system refers to conveying some instruction in terms of specific combinations of bits. The process of identifying the presence of such a specific bit combination is called 'decoding' and the corresponding digital circuit is termed as *decoder*.

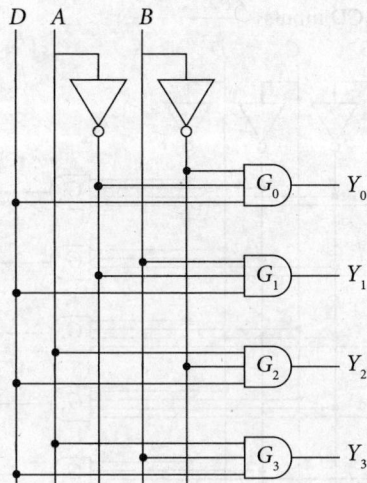

Figure 14.27 1-to-4 demultiplexer

In general a decoder has n inputs for n bits of the code and 2^n outputs, each one capable of responding to a specific input code. For example, a 4-bit input code can assume $2^4 = 16$ different values (0000 through 1111) so that the decoder should contain 16 different output terminals, each one responding to a specific input value. It is referred to as 1-of-16 decoder.

A more popular decoder is the BCD-to-decimal decoder shown in Figure 14.28, which converts each BCD (0000 through 1001) applied to the inputs A, B, C and D to one of the 10 possible decimal digit indications (0 through 9). Only one of the 10 output gates (G_0, ..., G_9) goes high depending on the input BCD value. When the input BCD is 0000, G_0 output is high, when the input is 0001, G_1 becomes high and so on. The BCD-to-decimal decoder is actually a truncated form of 1-of-16 decoder where the remaining six of the sixteen gates are eliminated. The working principle of the circuit is similar to that of demultiplexer. The data input is absent and the BCD inputs take the place of the select terminals.

14.5.4 Seven-Segment Display

The above principle of decoding can be implemented to other decoding systems also. A widely applied system is *BCD-to-Seven-Segment Decoder* that accepts BCD codes (0 through 9) as inputs and converts in such a way that an assembly of seven light-emitting diodes (LEDs) arranged in a geometrical shape turn on to display visually the equivalent decimal numeral. This decoding system is very useful in making the digital instruments directly readable.

Figure 14.29 explains how the arrangement of seven LEDs of elongated shape, labelled a through g in a specific geometric orientation can generate a numeral. The method of making the common electrical connection is also sketched. The table in the bottom explains how glowing selective LEDs out of the seven can make a visual simulation of the decimal digits 0 through 9. The BCD-to-Seven-Segment Decoder is a complicated digital circuit that causes the proper LED segments to get activated in accordance with the input BCD value.

538 Electronics: Analog and Digital

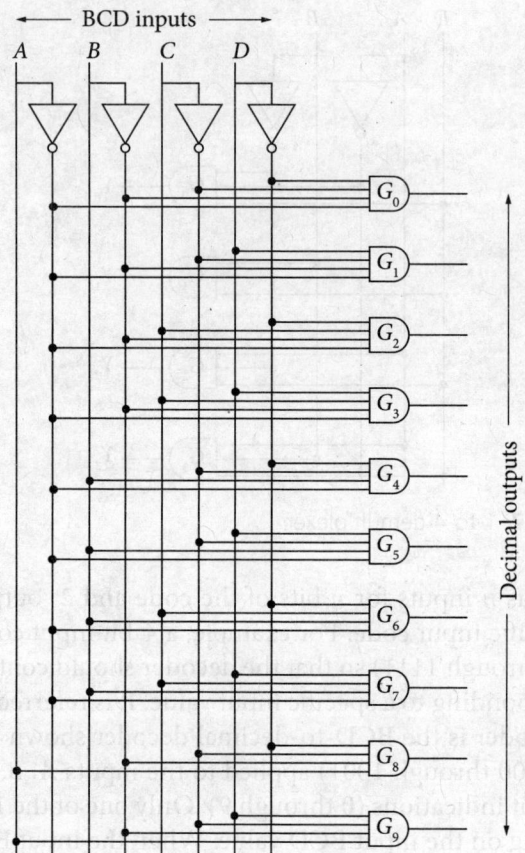

Figure 14.28 BCD-to-decimal decoder

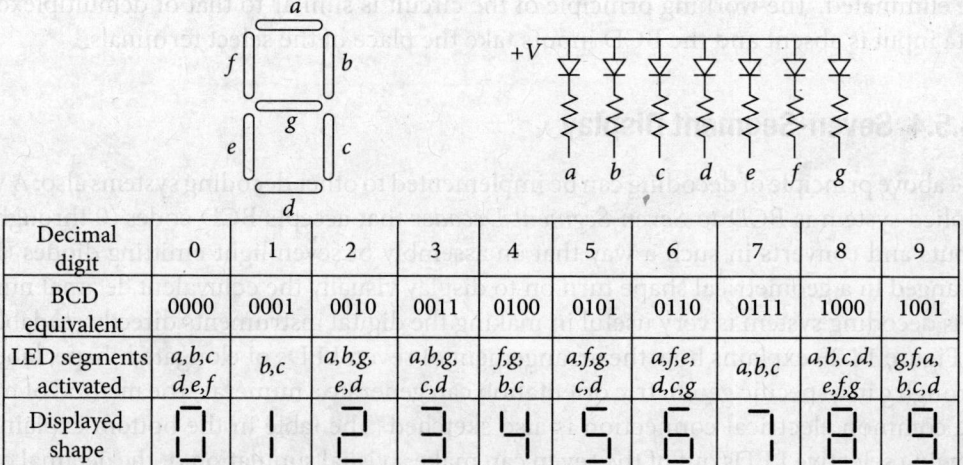

Decimal digit	0	1	2	3	4	5	6	7	8	9
BCD equivalent	0000	0001	0010	0011	0100	0101	0110	0111	1000	1001
LED segments activated	a,b,c d,e,f	b,c	a,b,g e,d	a,b,g c,d	f,g, b,c	a,f,g, c,d	a,f,e, d,c,g	a,b,c	a,b,c,d, e,f,g	g,f,a, b,c,d
Displayed shape	0	1	2	3	4	5	6	7	8	9

Figure 14.29 Seven-segment display: LED arrangement (left up), electrical connection (right up) and response on decoding (bottom)

Note: Word Multiplexer

Instead of a single data bit, a full *n*-bit word can be multiplexed, as outlined below. A 4-bit word, either $A_3A_2A_1A_0$ or $B_3B_2B_1B_0$ can be selected as the output $Y_3Y_2Y_1Y_0$ by making Select = 1 or 0, respectively.

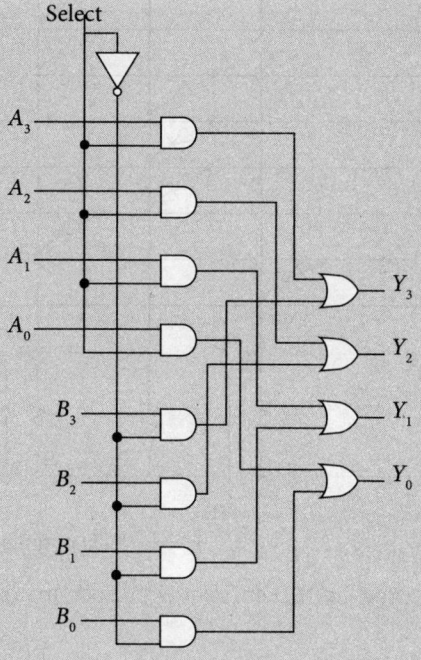

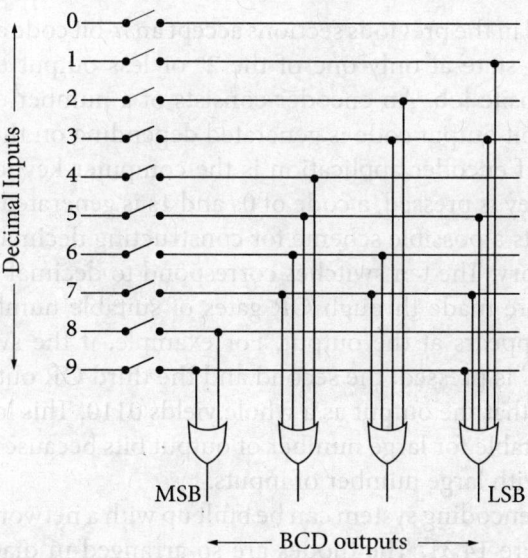

Figure 14.30 Decimal-to-BCD encoder

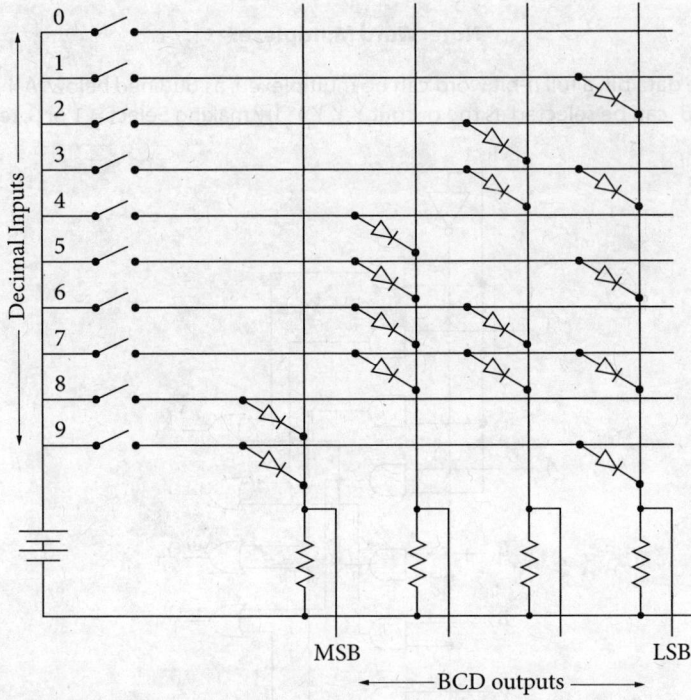

Figure 14.31 Decimal-to-BCD encoder with diode matrix

14.5.5 Encoders

The decoders discussed in the previous sections accept an n-bit code and identify a particular code by establishing 1 state at only one of the 2^n or less output terminals. The encoder performs just the opposite job. An encoder consists of a number of inputs, only of those being at 1 state. An n-bit output code is generated depending on the input specification. A well-known example of encoder application is the computer keyboard. When any one of the letter or numeral key is pressed, a code of 0s and 1s is generated that enters the CPU.

Figure 14.30 presents a possible scheme for constructing decimal-to-BCD encoder. The circuit is self-explanatory. The ten switches correspond to decimal digits 0 through 9. The internal connections are made through OR gates of suitable number of inputs such that the BCD equivalent appears at the output. For example, if the switch corresponding to the decimal number '6' is pressed, the second and the third OR outputs become 1 and the other two remain 0 so that the output as a whole yields 0110. This logic circuit is correct in principle but is not suitable for large number of output bits because it is quite inconvenient to fabricate OR gates with large number of inputs.

A more generalized encoding system can be built up with a network of diode connections, as explained with Figure 14.31. The diodes are so arranged in diagonal connections that turning on any decimal indicator switch makes some selected diodes turn on and the BCD equivalent is generated at the output. For instance, if the switch corresponding to the decimal number '7' is pressed, the voltage source forward biases only the three diodes connected diagonally in this line and all other diodes remain off. The voltage across the output resistors corresponding to the forward biased diodes generate the output of 0111. Such an encoder with diode matrix is also useful for designing an encoder with diode matrix is also useful for designing other complicated digital systems, such as *programmable array logic*.

14.5.6 Parity Checker and Generator

The encoders and decoders discussed above can work with a specific type of code, either the code is generated or that is detected. Sometimes logic circuits also require to convert one type of code to another. Such circuits are called *code converters*. Another category of digital circuit can verify whether the code is being generated properly. These circuits can detect the malfunctioning of a digital system with the help of a certain type of code known as *error detection code*. A brief idea of such code conversion and error detection is presented here with popular digital circuits termed as *parity checker* and *parity generator*.

A. Parity Checker

A binary number having an even number of 1s is stated to be of *even parity* and a binary number having odd number of 1s is itemized as of *odd parity*. For example, numbers 0111, 0001, 01111111 etc. have odd parity and 1100, 0011, 10011001 etc. are of even parity. A digital circuit capable of detecting the even or odd parity of a binary number is called *parity checker*. The exclusive-OR gate can act as a parity checker. It yields 1 state at the output for odd number of 1s at the input and gives 0 output state for even number of 1s at the input.

B. Parity Generator

Sometimes false 1 or 0 is introduced in a digital circuit due to noise or malfunctioning. Such situation is more likely to occur in digital data transmission over long distance and intercommunications of different units in a computer system. Such bit errors can be detected by attaching parity bits to all binary words such that the total number of 1s remains always even or always odd. The digital circuit generating such parity bits is referred to as *parity generator*. At the receiving end, a parity checker verifies whether the number of the 1s remains the same, either even or odd. If there is no error, the received data has the same parity. If any one of the transmitted bits gets changed from 0 to 1 or vice versa, the received data converts to the opposite parity and the error is detected.

Figure 14.32(a) demonstrates the construction of a 5-bit parity generator with exclusive-OR gate. If the four input bits A_0, A_1, A_2 and A_3 contain even number of 1s, the output Y' of the exclusive-OR gate becomes 0 and the output Y of the NOT gate becomes 1. The total 5-bit output for parity generation including Y and $A_3A_2A_1A_0$ becomes of odd parity. The true conditions of Y' and Y are mentioned in Figure 14.32(b). If, in contrast, the inputs contain odd number of 1s, $Y' = 1$ and $Y = 0$ so that the 5-bit output becomes again of odd parity. Thus the circuit acts as an *odd parity generator*. It produces a 5-bit output of odd parity whatever may be the input. If Y' is considered instead of Y, then Y' and $A_3A_2A_1A_0$ altogether produce an *even parity generator*.

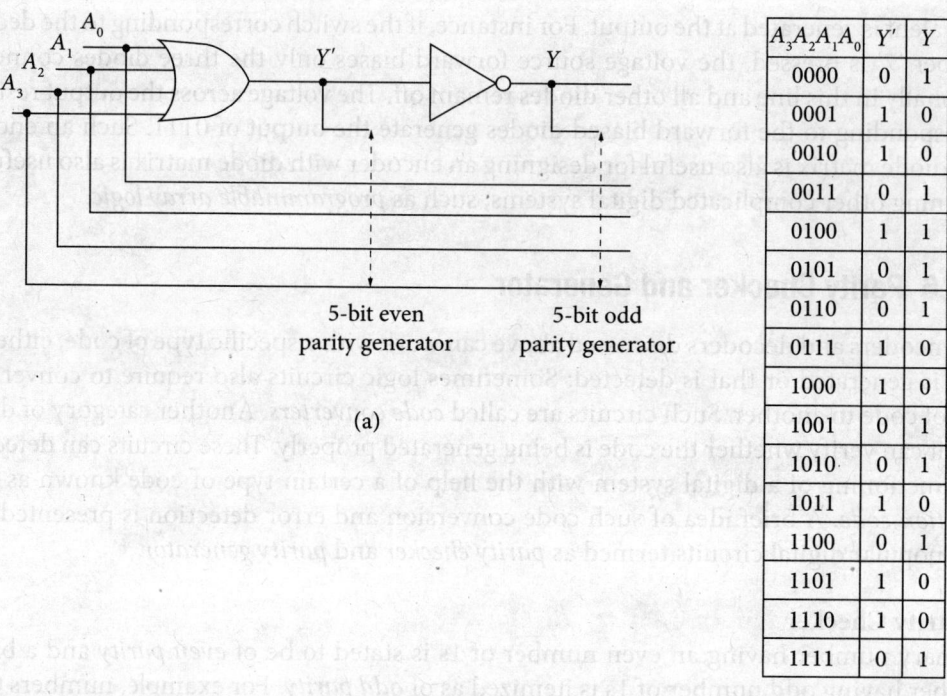

Figure 14.32 Parity generator (a) circuit diagram and (b) truth table

Multiple Choice-Type Questions and Answers

14.1 It is required to monitor two electrical lines and to indicate a high voltage level at the output whenever either or both lines are low. Which of the followings is suitable?
 (a) 2-input NAND gate
 (b) 2-input bubbled-OR gate
 (c) both of the above
 (d) none of the above

14.2 It is required to monitor the simultaneous occurrence of low states in two separate electrical lines by indicating a high output voltage level. Which of the following devices may be used?
 (a) 2-input NOR gate
 (b) 2-input bubbled-AND gate
 (c) both of the above
 (d) none of the above

14.3 True condition in positive logic produces a .... voltage and true condition in negative logic produces a .... voltage.
 (a) low, high
 (b) negative, positive
 (c) high, positive
 (d) high, low

14.4 The following four combinations of voltage levels are available for representing logical 0 and logical 1: (I) $-5\,V \equiv 0$, $5\,V \equiv 1$; (II) $-0.5\,V \equiv 0$, $4.5\,V \equiv 1$; (III) $-0.5\,V \equiv 1$, $-5\,V \equiv 0$; (IV) $-2.5\,V \equiv 0$, $2.5\,V \equiv 1$. Positive logic is represented by ....
 (a) only (I)
 (b) only (IV)
 (c) both (I) and (IV)
 (d) all the four

14.5 Which of the following circuits is represented by the Boolean expression $P + \bar{P}Q$ where P and Q are the inputs?
 (a) AND   (b) OR   (c) NOT   (d) NAND
 Hints. $P + \bar{P}Q = (P + \bar{P})(P + Q) = P + Q$

14.6 A 2-input NOR gate can be constructed with
 (a) four 2-input NAND gates
 (b) three 2-input NAND gates
 (c) one 2-input AND gate and one NOT gate
 (d) two NOT gates and one 2-input OR gate

14.7 Which one does not represent an exclusive-OR operation for inputs A and B?
 (a) $(A+B)\overline{AB}$
 (b) $A\bar{B} + B\bar{A}$
 (c) $(A+B)(\bar{A}+\bar{B})$
 (d) $(A+B)AB$

[GATE 2015]

14.8 A 2-input gate has inputs A and B and output Y. The input and the output waveforms are given in Figure 14.33(a). The gate may be ....
 (a) OR   (b) NOR   (c) AND   (d) NAND

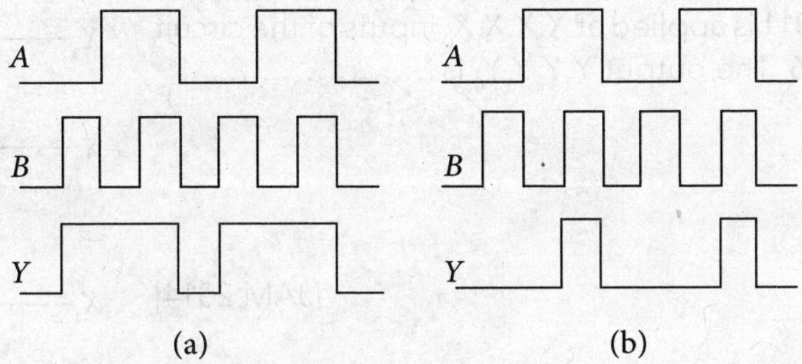

Figure 14.33

14.9 The input and the output waveforms of a 2-input gate are shown in Figure 14.33(b). The gate may be ....
 (a) NAND   (b) AND   (c) OR   (d) NOR

14.10 The required input condition (A, B, C) to make the output $(Y) = 1$ in the circuit of Figure 14.34 is ....
 (a) 1, 0, 1   (b) 0, 0, 1   (c) 1, 1, 1   (d) 0, 1, 1

[JEST 2015]

Hints. $Y = (A\bar{B} + \bar{A}B)(\bar{B}\bar{C} + BC)C = \bar{A}BC$

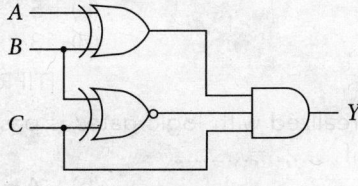

Figure 14.34

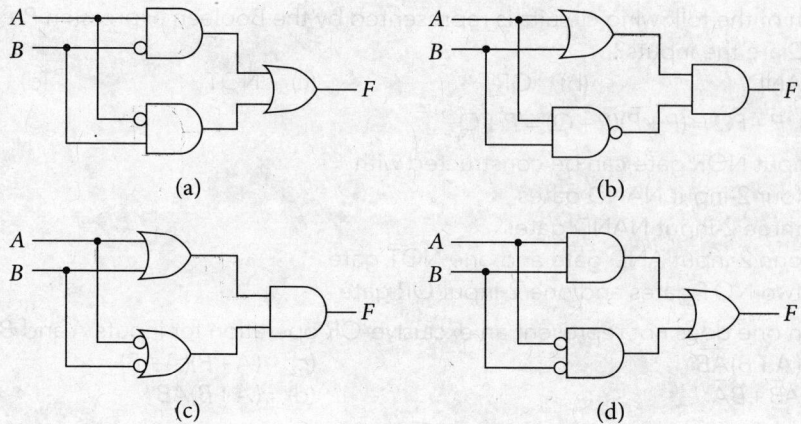

Figure 14.35

14.11 Which of the circuits of Figure 14.35 does not satisfy the Boolean expression $A\bar{B} + \bar{A}B = F$?

[JAM 2011]

14.12 Binary input 1011 is applied at $X_3X_2X_1X_0$ inputs of the circuit of Figure 14.36. The output $Y_3Y_2Y_1Y_0$ is ....
 (a) 1101
 (b) 1010
 (c) 1111
 (d) 0101

[JAM 2014]

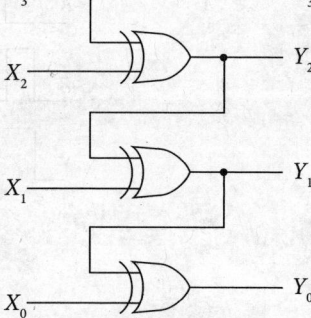

Figure 14.36

Hints.
$Y_3 = X_3$
$Y_2 = X_3\bar{X}_2 + \bar{X}_3 X_2$
$Y_1 = Y_2\bar{X}_1 + \bar{Y}_2 X_1$
$Y_0 = Y_1\bar{X}_0 + \bar{Y}_1 X_0$

14.13 The Boolean expression $P + \bar{P}Q$, where P and Q are the inputs to a circuit, represents the following gate.
 (a) AND (b) NAND (c) NOT (d) OR

[JAM 2012]

14.14 Consider the circuit of Figure 14.37. The minimum number of NAND gates required to design this circuit is
 (a) 6 (b) 5
 (c) 4 (d) 3

[TIFR 2012]

Figure 14.37

14.15 The expression $A(\bar{B} + \bar{C})$ realized with logic gates is passed through a NOT gate. The output produces 1 state, if ....
 (a) $A = 0, B = 0, C = 0$
 (b) $A = 1, B = 0, C = 1$
 (c) $A = 1, B = 1, C = 0$
 (d) $A = 1, B = 0, C = 0$

14.16 The truth table for the circuit of Figure 14.38 is ....

(a)			(b)			(c)			(d)		
J	K	Q	J	K	Q	J	K	Q	J	K	Q
0	0	1	0	0	1	0	0	0	0	0	0
0	1	0	0	1	0	0	1	1	0	1	1
1	0	1	1	0	0	1	0	0	1	0	1
1	1	0	1	1	1	1	1	1	1	1	0

[JAM 2006]

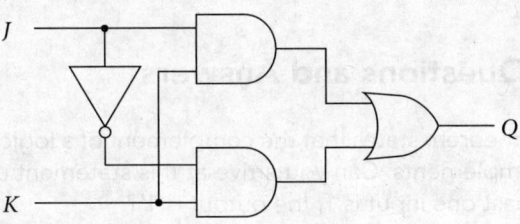

Figure 14.38

14.17 Consider the digital circuit shown in Figure 14.39(a), in which the input C is always high (1). The truth table for the circuit can be written as Figure 14.39(b). The entries in the Z column vertically are
(a) 1010 (b) 0100 (c) 1111 (d) 1011

[NET 2011]

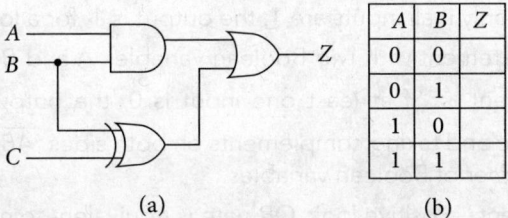

Figure 14.39

14.18 Four members of an administrative committee are electing a fifth member for the committee. If any of them vote against a member, the member gets disqualified. If an electronic machine were constructed to implement the required rule, which one of the following would be the most appropriate?
(a) XOR gate (b) XNOR gate (c) OR gate (d) AND gate

14.19 Consider the Boolean expression: $(\overline{A}+\overline{B})[\overline{A(B+C)}]+A(\overline{B}+\overline{C})$. It can be represented by a single three-input logic gate. Identify the gate.
(a) AND (b) OR (c) XOR (d) NAND

[GATE 2019]

Hints. Using De Morgan's theorem, the given expression can be simplified to: $\overline{A}+\overline{B}+\overline{C}+\overline{BC}$, hence $\overline{A}+\overline{B}+\overline{C}$, which is equivalent to $\overline{ABC}$.

14.20 The 2's compliment of 1111 1111 is ....
  (a). 00000001   (b) 00000000   (c) 1111 1111   (d) 1000 0000

  [JEST 2020]

Answers

14.1 (c)	14.2 (c)	14.3 (d)	14.4 (d)
14.5 (b)	14.6 (a)	14.7 (d)	14.8 (a)
14.9 (b)	14.10 (d)	14.11 (d)	14.12 (a)
14.13 (d)	14.14 (c)	14.15 (a)	14.16 (c)
14.17 (d)	14.18 (d)	14.19 (d)	14.20 (a)

Reasoning-Type Questions and Answers

14.1 De Morgan's first theorem states that the complement of a logical sum equals the logical product of the complements. Can you arrive at this statement using the statement on a logic gate: 'If at least one input is 1, the output is 1'?

Ans. For two Boolean variables A and B, the given statement can be expressed as $A+B$. The equivalent logical statement is: 'If all the inputs are 0, the output is 0', which can be expressed as $\bar{A}\bar{B}$. Equating these two expressions and taking complements on both sides, we get $\overline{A+B} = \bar{A}\bar{B}$. The above logic can be extended up to any number of variables.

14.2 De Morgan's second theorem states that the complement of a logical product equals the logical sum of the complements. Can you come up with the same statement with the statement: 'If and only if all inputs are 1, the output is 1' for a logic gate?

Ans. The given statement with two Boolean variables A and B is expressed as AB. The equivalent statement is: 'If at least one input is 0, the output is 0', written as $\bar{A}+\bar{B}$. Equating these two and taking complements on both sides, $\overline{AB} = \bar{A}+\bar{B}$. The above logic is true for any number of Boolean variables.

14.3 Justify the statement – Positive logic OR gate is equivalent to negative logic AND gate.

Ans. Let us consider the following input–output conditions of a 2-input gate.

Input-1	Input-2	Output
Low	Low	Low
Low	High	High
High	Low	High
High	High	High

If we consider positive logic, the High state represents 1 and the gate acts as OR gate. In negative logic, the High state represents 0 and the gate acts as AND gate.

[N.B. Similarly positive logic AND gate is equivalent to negative logic OR gate. Similar relationships exist between NAND and NOR gates.]

14.4 Can the output of a positive logic AND gate and that of a positive logic OR gate assume the same state for the same input conditions?

Ans. Yes, when all the inputs are either low or high.

[N.B. Similar condition is true for NAND and NOR gates and all the above are true for negative logic also.]

14.5 Can the exclusive-OR gate act as equality detector?

Ans. Yes, with negative logic. The output becomes zero when both the inputs become equal.

Solved Logical Problems

14.1 Construct a 2-bit equality detector with two 2-input exclusive-OR gates.

Soln. Since a 2-input exclusive-NOR gate acts as 1-bit equality detector [Equation (14.4)], applying the outputs of two such gates to an AND gate, the equality of two 2-bit inputs $A = A_1 A_0$ and $B = B_1 B_0$ can be detected, as depicts in Figure 14.40.

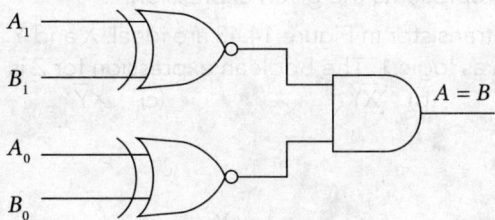

Figure 14.40

14.2 Explain that the logic circuit of Figure 14.41 is a 2-to-1 multiplexer.

Soln. It is obvious that the output is given by $Y = \overline{S}D_0 + SD_1$ that means when $S = 0$, $Y = D_0$ and when $S = 1$, $Y = D_1$. Thus S is the Select terminal that sends either of the inputs D_0 and D_1.

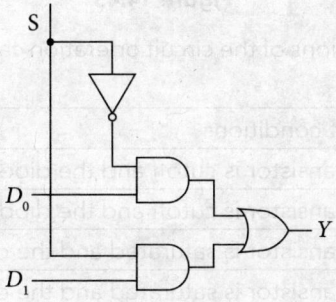

Figure 14.41

14.3 Which of the following circuits of Figure 14.42 represents the Boolean expression $S = \overline{\overline{P+QR} + \overline{\overline{Q}P}}$.

[JAM 2015]

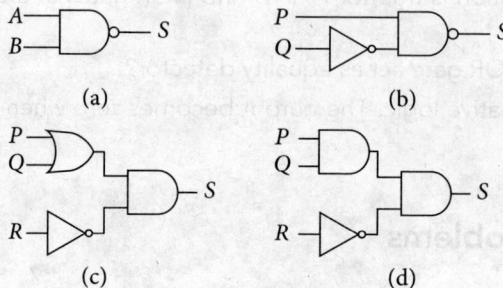

Figure 14.42

Soln. Applying De Morgan's theorems, the given expression is modified and simplified as

$$S = \overline{\overline{P\,QR} + Q + \overline{P}} = (\overline{\overline{P\,QR}})(\overline{Q+\overline{P}}) = (P+QR)(\overline{Q}P) = P\overline{Q}$$

Thus the circuit (b) represents the given expression.

14.4 The diode and the transistor in Figure 14.43 are ideal. X and Y are digital signals with 0 V as logic 0 and $+V_{CC}$ as logic 1. The Boolean expression for Z is
(a) XY (b) $\overline{X}Y$ (c) $X\overline{Y}$ (d) $\overline{XY}$

[GATE 2013]

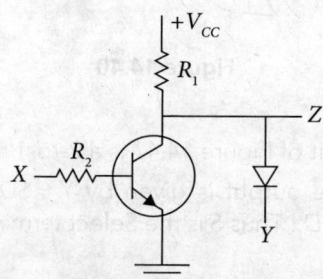

Figure 14.43

Soln. The different conditions of the circuit operation can be summarized in the following tabular form.

X	Y	Z	Circuit conditions
0	0	0	The transistor is cutoff and the diode is forward biased
0	1	1	The transistor is cutoff and the diode is reverse biased
1	0	0	The transistor is saturated and the diode is forward biased
1	1	0	The transistor is saturated and the diode is reverse biased

The above outcomes indicate that the output condition is $\overline{X}Y$, hence option (b) is correct.

14.5 Four digital outputs V, P, T and H monitor the speed (v), tire pressure (p), temperature (t) and relative humidity (h) of the car. These outputs switch from 0 to 1 when the values of the parameters exceed 85 km/h, 2 bar, 40°C and 50%, respectively. A logic circuit that is used to switch on a lamp at the output E is shown in Figure 14.44. Which of the following conditions will switch the lamp ON?

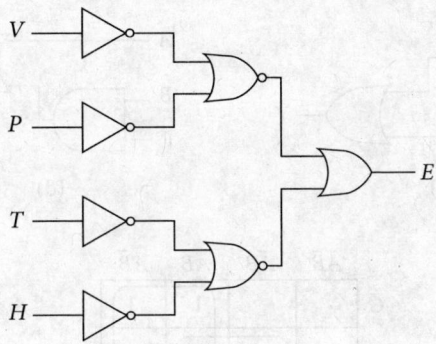

Figure 14.44

(a) $v < 85$ km/h, $p < 2$ bar, $t > 40°C$, $h > 50\%$
(b) $v < 85$ km/h, $p < 2$ bar, $t < 40°C$, $h < 50\%$
(c) $v > 85$ km/h, $p < 2$ bar, $t > 40°C$, $h < 50\%$
(d) $v > 85$ km/h, $p < 2$ bar, $t > 40°C$, $h > 50\%$

[NET 2013]

Soln. The output is given by $E = \overline{\overline{V}+\overline{P}} + \overline{\overline{T}+\overline{H}} = VP + TH$. The different states of E corresponding to the given conditions can be expressed in terms of logical conditions as follows.

Option	V	P	T	H	E
(a)	0	0	1	1	1
(b)	0	0	1	0	0
(c)	1	0	1	0	0
(d)	1	0	0	1	0

It is inferred that option (a) will switch on the lamp.

14.6 In a digital circuit for three input signals (A, B and C), the final output should be such that for inputs

A   B   C
0   0   0
0   0   1
0   1   0

The output (Y) should be low and for all other cases it should be high. Which of the digital circuits in Figure 14.45 will give such output?

[TIFR 2016]

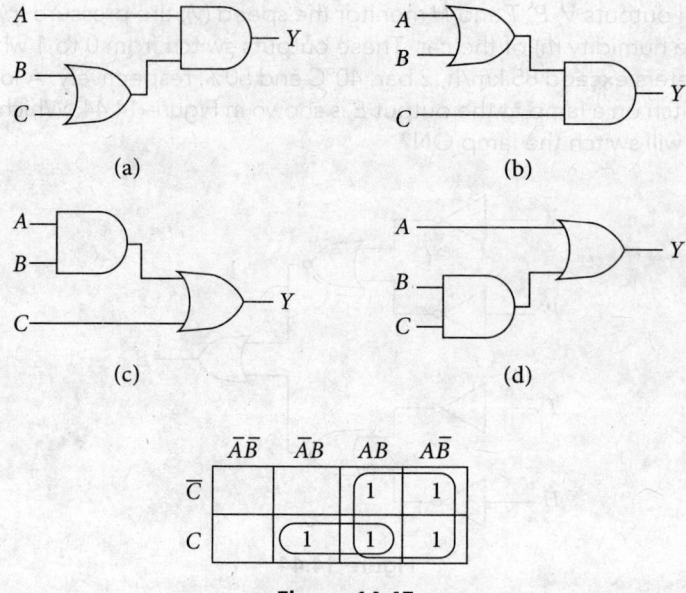

Figure 14.45

Soln. The high output condition can be expressed as $Y = \bar{A}BC + A\bar{B}\bar{C} + A\bar{B}C + AB\bar{C} + ABC$ and the Karnaugh map simplification given in Figure 14.43 yields $Y = A + BC$. Thus the circuit (d) corresponds to this output condition.

14.7 In Figure 14.46, X and Y are one-bit inputs. Which of the circuits corresponds to a one-bit comparator?

[NET June 2017]

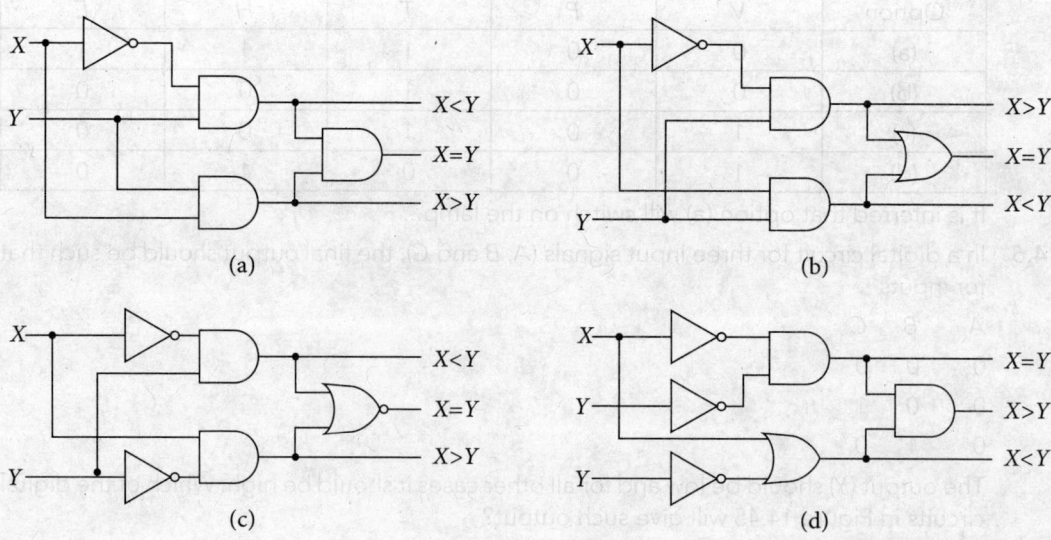

Figure 14.46

Soln. One can infer the followings at the first glance.

Circuit (a) is not applicable because the lower AND gate output XY cannot represent $X > Y$.
Circuit (b) is not applicable because the upper AND gate output $\overline{XY}$ cannot represent $X > Y$.
Circuit (d) is not true because the OR gate output $X + \overline{Y}$ cannot represent $X < Y$.
So only Circuit (c) is left but one must cross-check if this is the appropriate circuit. Circuit (c) has the following outputs.
The upper AND gate output $\overline{X}Y$ represents $X < Y$.
The lower AND gate output $X\overline{Y}$ represents $X > Y$.
The NOR gate output represents neither $X < Y$ nor $X > Y$, hence $X = Y$.
So this circuit actually operates as a one-bit comparator.
[N.B. The circuit of Figure 14.23(b) as another version of one-bit comparator. Many independent circuit realizations are possible for the same logical expression.

14.8 Which of the circuits in Figure 14.47 implements the Boolean function:
$F(A,B,C) = \Sigma(1,2,4,6)$?
These are 4-to-1 multiplexers.

[NET Dec. 2016]

Soln. The output for the given Boolean expression is $\overline{A}\overline{B}C + \overline{A}B\overline{C} + AB\overline{C} + AB\overline{C}$.
The output condition for the 4-to-1 multiplexer, in terms of the four inputs I_0, I_1, I_2 and I_3 and the two selects A and B can be written as $I_0\overline{A}\overline{B} + I_1\overline{A}B + I_2A\overline{B} + I_3AB$
The given four circuits give the output of:
(a) $\overline{C}\overline{A}B + C\overline{A}B + CA\overline{B} + CAB$,
(b) $\dot{C}A\overline{B} + \overline{C}\overline{A}B + \overline{C}AB + \overline{C}AB$,
(c) $C\overline{A}\overline{B} + \overline{A}B + A\overline{B}$ and
(d) $\overline{C}\overline{A}B + \overline{C}AB + CAB$, respectively.
It is understood that Circuit (b) matches the multiplexer output condition.

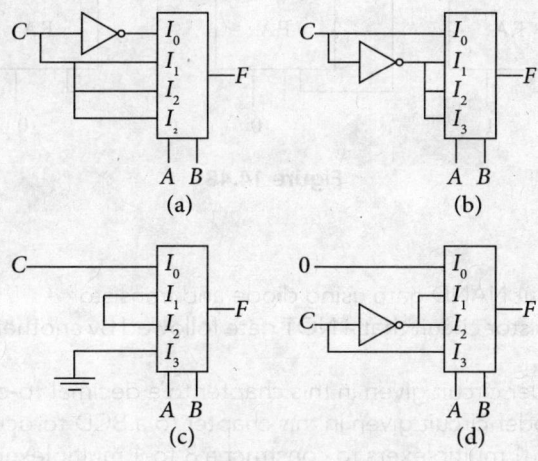

Figure 14.47

Exercise

Subjective Questions

1. Distinguish between combinational and sequential logic circuits. Mention some important operations performed by digital circuits.
2. What do you mean by positive and negative logic? Construct 3-input positive logic (i) OR gate and (ii) AND gate with diode and resistor and write down the corresponding truth tables.
3. What are the advantages of a universal gate? Design a 2-input exclusive-OR gate using (i) NOR gates and (ii) NAND gates exclusively.
4. Construct the logic circuit for exclusive-NOR gate using (i) basic gates and (ii) NAND gates only and compose the truth table.
5. What is a logic family? Draw the circuit diagram for a 2-input NAND gate using MOS logic and explain its working principle. Repeat the same using TTL logic.
6. Introduce CMOS inverter and its working principle. What is the benefit of using this circuit in digital electronics?
7. Design the logic circuit for 8-to-1 multiplexer using basic gates.
8. Distinguish between half adder and full adder. Draw the logic circuit of a full adder, specify its three inputs and two outputs and compose the truth table indicating all possible combinations of sum and carry.
9. Explain the process of controlled inversion. Using this principle, build up the block diagram of such a logic circuit that can execute either binary addition or binary subtraction operation at the user's disposal.
10. Draw the block diagram for the logic circuit that can add decimal 18 and 7, obviously both in binary form. Indicate the augend, addend, sum and carry bits.

    Hints. Take the help of Figure 14.48.

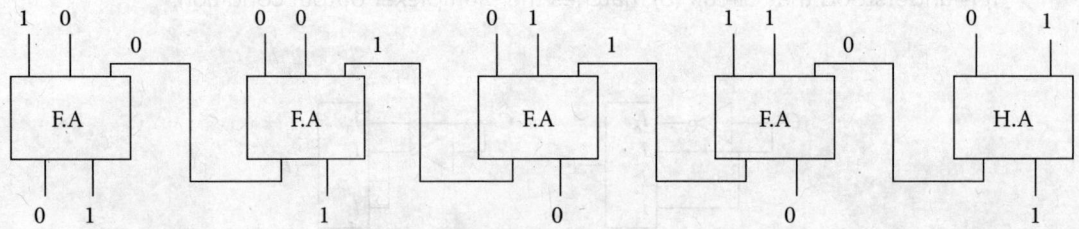

Figure 14.48

Conceptual Test

1. Construct a 2-input NAND gate using diode and transistor.
2. Explain with transistor circuit that a NOT gate followed by another NOT gate reproduces the original input.
3. Modify the encoder circuit given in this chapter to a decimal-to-octal encoder.
4. Convert the decoder circuit given in this chapter to a BCD-to-octal decoder.
5. Cascade two 4-to-1 multiplexers to construct a 8-to-1 multiplexer.

6. Develop a logic circuit on your own for checking the parity of a 3-bit number. That should yield high output only when odd number of inputs were at 1 state. Do you find any similar circuit in this chapter?
7. A digital adder circuit is to be fabricated for adding two 2-bit binary numbers. No full adder is available but a lot of half adders are obtainable. Can you execute the required job with half adders? Illustrate your logic diagram.
8. Build up a two-way switching system for turning on an LED with two different switches placed at two different positions, such as the two ends of a passage.

   Hints. Use exclusive-OR gate.

Logical Problems

1. Can you implement AND operation using OR and NOT gates?
2. Construct a NAND gate using NOR gates only.
3. Construct a NOR gate using NAND gates only.
4. Construct the logic circuit using NOR gate only for implementing the following Boolean expressions. (i) $AB+B\overline{C}$ (ii) $AB+\overline{A}C$
5. Repeat the above construction with NAND gates only.
6. Implement the circuit for $(A \oplus B) \oplus C$ using AND and OR gates only.
7. Two 2-bit binary numbers are $A = A_1 A_0$ and $B = B_1 B_0$. Solve, by Karnaugh map method for the condition of $A \geq B$ and fabricate the logic circuit for the same.
8. The output O of the circuit given in Figure 14.49 in cases I and II are, respectively
   (a) 1, 0   (b) 0, 1   (c) 0, 0   (d) 1, 1

   Case-I: $A, B = 1, C, D = 0, E, F = 1$ and $G = 0$
   Case-II: $A, B = 0, C, D = 0, E, F = 0$ and $G = 1$

   [NET 2012]

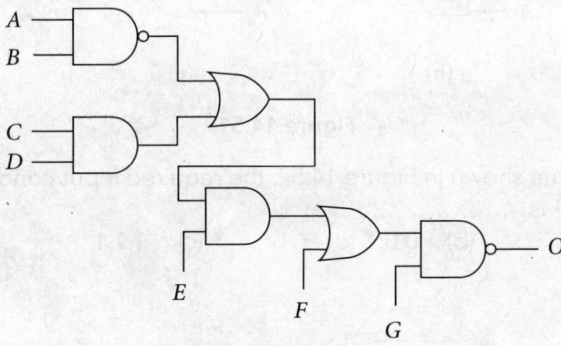

Figure 14.49

9. The logic circuit shown in Figure 14.50 implements the Boolean expression
   (a) $y = \overline{A.B}$ (b) $y = \overline{A}.B$ (c) $y = A.B$ (d) $y = A+B$

   [NET Dec. 2012]

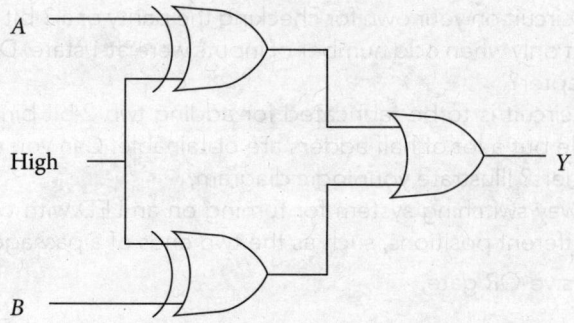

Figure 14.50

10. In a 2-to-1 multiplexer as shown in Figure 14.51, the output $X = A_0$, if $C = 0$, and $X = A_1$ if $C = 1$. Which one of the following is the correct implementation of this multiplexer?

[Gate 2018]

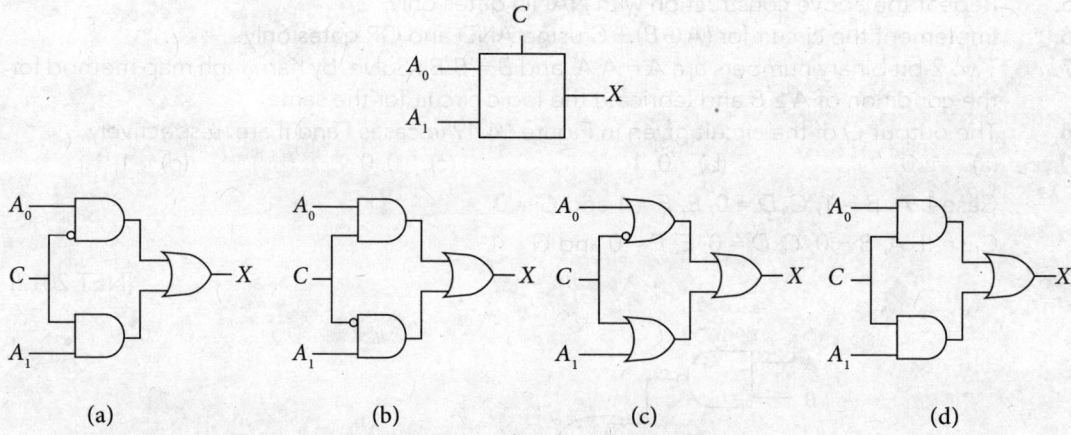

Figure 14.51

11. For the logic circuit shown in Figure 14.52, the required input condition (A, B, C) to make the output (X) =1 is ....
    (a) 1,0,1   (b) 0,0,1   (c) 1,1,1   (d) 0,1,1

[JEST 2015]

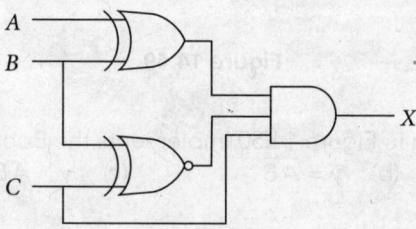

Figure 14.52

12. What is Y for the circuit shown in Figure 14.53?
   (a) $Y = \overline{(A+\bar{B})(\bar{B}+C)}$
   (b) $Y = \overline{(A+\bar{B})}(B+C)$
   (c) $Y = \overline{(\bar{A}+B)(\bar{B}+C)}$
   (d) $Y = \overline{(A+B)(\bar{B}+C)}$

[JEST 2017]

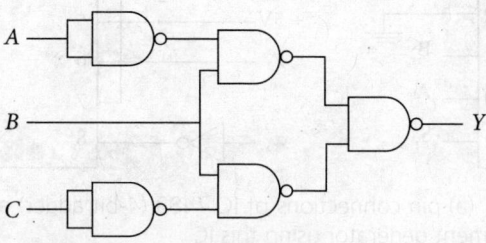

Figure 14.53

Project Work on Chapter 14

To construct a digital circuit that can generate 9's complement of a decimal number in binary form.

[It is suggested to revise the topics 'radix complements' from Chapter 13 and 'full adder', particularly 4-bit adder (Figure 14.18) from this chapter.]

Theory:
If r be the radix of a number system, the *radix-minus-1 complement* of the number is obtained by subtracting each digit of the number from $(r-1)$. In the present case, $r = 10$ so that the 9's complement of a decimal digit x is given by

$$y = 10 - 1 - x \qquad (P14.1)$$

For a decimal number x expressed in binary form, $15 - x$ is equivalent to its 1's complement. [For example, $9 \equiv 1001$ and $15 - 9 \equiv 0110$]. Therefore, Equation (P14.1) is modified as

$$y = 10 + 15 - x - 16 \qquad (P14.2)$$

Subtraction of 16 ($\equiv 10000$) is very easy, only the MSB is to be discarded and the remaining zeroes may be ignored.

Experiment:
The circuit uses a 4-bit adder IC 7483 whose pin diagram is given in Figure P14.1(a). There are input provisions for two 4-bit binary numbers $A_4A_3A_2A_1$ and $B_4B_3B_2B_1$. The respective sum bits are obtained at pins S_4, S_3, S_2 and S_1. The carry at MSB, if any, is obtained at pin C_4 and a carry at LSB, if need, can be applied to pin C_0.

Figure P14.1(b) explains how Equation (P14.2) can be realized with this adder IC. The number 10 ($\equiv 1010$) is given permanently to one set of input terminals ($B_4B_3B_2B_1$) by making the $B_4 = B_2 = 1$ and $B_3 = B_1 = 0$.

Let $x = x_4x_3x_2x_1$ be the given decimal number in binary form with its four bits x_4, x_3, x_2 and x_1 and $y = y_4y_3y_2y_1$ be the 9's complement in binary form with the four bits y_4, y_3, y_2 and y_1. The four bits of x are applied to the input set $A_4A_3A_2A_1$, as shown in Figure P14.1(b), each through a NOT gate so that the 1's complement of the number is inserted to the IC. The four bits of the 9's complement are obtained at the pins S_4, S_3, S_2 and S_1. The MSB carry at pin 14 is simply ignored. All the input–output combinations given below should be verified with the actual circuit.

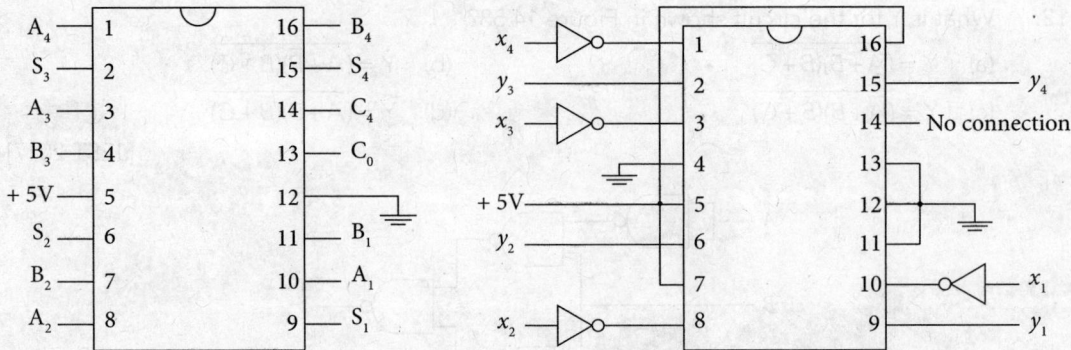

Figure P14.1 (a) pin connections of IC 7483 (4-bit adder) and (b) logic diagram of 9's complement generator using this IC

Decimal number	Input bits applied to the NOT gates connected to A_4, A_3, A_2 and A_1	9's complement	Output bits obtained at S_4, S_3, S_2 and S_1
0	0000	9 = (10 − 1) − 0	1001
1	0001	8 = (10 − 1) − 1	1000
2	0010	7 = (10 − 1) − 2	0111
3	0011	6 = (10 − 1) − 3	0110
4	0100	5 = (10 − 1) − 4	0101
5	0101	4 = (10 − 1) − 5	0100
6	0110	3 = (10 − 1) − 6	0011
7	0111	2 = (10 − 1) − 7	0010
8	1000	1 = (10 − 1) − 8	0001
9	1001	0 = (10 − 1) − 9	0000

15

Sequential Logic Circuits

The previous chapter has dealt with digital circuits having output states dependent only on the instantaneous combination of the inputs. There exist yet another set of digital circuits, known as *sequential logic circuits*, which are presented in this chapter. A sequential digital circuit is almost always associated with a clock or a timer. This chapter illustrates the properties of a clock and explains several popular sequential digital circuits, namely flip-flop, register and counter that are the building blocks for many complicated digital systems including computer hardware.

15.1 Clock and Timer

First let us understand the meaning of the word 'sequential' sticking to some digital circuits. It is named so because:

- the output is affected by the positional and temporal sequence of the input states,
- the sequence of the change of output states is different from that of the inputs, and
- a memory property is sometimes exhibited by holding one or more bits in terms of voltage levels.

Table 14.1 may be revisited in this occasion. Now the question arises, who makes this sequence to occur? That is the job of the clock and the timer. As explained in the previous chapters, digital circuits need two fixed and discrete voltage states as inputs. It is a very common phenomenon in a digital system, such as a computer that a number of circuits have to change the logic states in synchronism. That synchronizing is initiated by a timer and a clock.

The *timer* is an electronic circuit that generates a voltage waveform, generally square or rectangular, of well-defined amplitude and precise frequency. That voltage waveform of fixed amplitude and frequency generated by the timer is referred to as *clock signal*, *clock pulse* or simply *clock*. It defines a timing interval during which the circuit operations must be performed. Thus the timer is the 'heart' and the clock is the 'heartbeat' of a digital system. The clock has the following properties.

- It may be symmetric or asymmetric voltage waveform but should have fixed and pre-determined amplitude and frequency.
- The prime requirement for a clock is the perfection of its frequency and its stability.
- The high and low voltage levels of the clock should be distinct enough for denoting logical 1 and 0.
- The absolute voltage values should comply with the circuit requirements.
- The clock is a unidirectional pulse. It varies between zero and positive or zero and negative values and does not cross the zero.

15.1.1 Clock Parameters

The following characteristics of a clock pulse are important from the point of view of circuit operation.

Duty Cycle: It is defined as the ratio of the time duration of the pulse remaining high to the time of its period. Instead of ratio, it is often expressed in percentage. The duty cycle of a voltage pulse is a measure of the symmetry or asymmetry of the waveform. A symmetric voltage wave has equal on and off time, duty cycle of 50%. The use of duty cycle is meaningful for rectangular pulses having low and high states of fixed durations and fixed amplitudes.

Rise and Fall Time: Ideally the time required for the clock to change its state from low to high and vice versa should be zero. However, a practical timer circuit takes finite rise time (the time required for low to high transition) and fall time (the time for high to low transition). Conventionally these two are measured between the points of acquiring 10% and 90% of the maximum amplitude. For example, the rise time of a 5 V clock is the time taken for reaching 4.5 V from 0.5 V and the fall time is that for dropping down to 0.5 V from 4.5 V.

Propagation Delay: It is the small but finite time elapsed between the input clock transition and the resulting output change. The propagation delay limits the maximum clock frequency that the circuit can respond to.

> **Example 15.1**
>
> The propagation delay through a digital circuit is 25 ns. What is the upper limit of clock frequency that the circuit can afford?
>
> **Solution**
>
> The circuit must complete the change of state within one time period of the clock. The clock period can be greater than or equal to the propagation delay. In the limiting case, the clock period is
>
> $1/(25 \times 10^{-9}) = 40$ MHz
>
> The circuit cannot respond to a clock signal of greater than this frequency.

We have already come across several circuits made of transistors that can be used as timer or clock generator. The astable multivibrator (Section 10.5.1) and the monostable multivibrator (Section 10.5.2) are such examples. However, it is more convenient to fabricate similar circuits with timer IC 555 and that is more popular in use. The astable and monostable multivibrators realized with IC 555 are illustrated in the subsequent sections. It may also be mentioned that the crystal oscillator (Section 10.4.7) is used as a clock generator in digital circuits because of its excellent frequency stability.

15.1.2 Working Principles of IC 555

It is a unique combination of analog and digital circuits fabricated in such a way that the performance can be controlled simply by adding some resistors and capacitors from outside (similar to the case of an op-amp). Figure 15.1 outlines the internal circuitry of IC 555. The pin positions may be identified by comparing with the pin-diagram given in the inset. The circuit has the following components.

Resistor Chain: The IC contains a voltage divider comprised of three equal resistors, each of 5 kΩ in series, connected between the supply voltage terminal $(+V_{cc})$ (Pin 8) and the ground (Pin 1). The resistor chain produces two reference voltages, $2V_{cc}/3$ and $V_{cc}/3$, respectively.

Op-amp Comparators: There are two op-amps, each one acting as a comparator. The upper one is called *threshold comparator*. Its output (Op-1) becomes high when the threshold voltage (Pin 6) exceeds $2V_{cc}/3$, the reference voltage at the Control pin (Pin 5). The lower op-amp is termed as *trigger comparator* whose output (Op-2) becomes high when the trigger voltage (Pin 2) goes below $V_{cc}/3$.

Flip-Flop: This is a sequential digital circuit to be discussed immediately after the present topic. Right now suffice it to state that the flip-flop has two inputs: R and S and two outputs: Q and $\bar{Q}$, one being the complement to the other. Here only the $\bar{Q}$ output is made into use. The flip-flop acts like a switch depending on the two inputs as follows.

- When R is high and S is low, the $\bar{Q}$ output becomes high.
- When R is low and S is high, the $\bar{Q}$ output becomes low.
- When both and S are low, the output remains unchanged.

Reset and Control: The Reset (Pin 4) enters the flip-flop with inverted connection. When Pin 4 is made low by connecting to the ground, the $\bar{Q}$ output of the flip-flop becomes high and when Pin 4 is connected to the positive terminal of the power supply, it does not affect the flip-flop. The present circuits with IC 555 do not use the Reset option. The Control (Pin 5) can afford some external signal to modulate the threshold of $2V_{cc}/3$. The present circuits with IC 555 do not use the Control and it remains grounded through a capacitor in order to avoid stray noise.

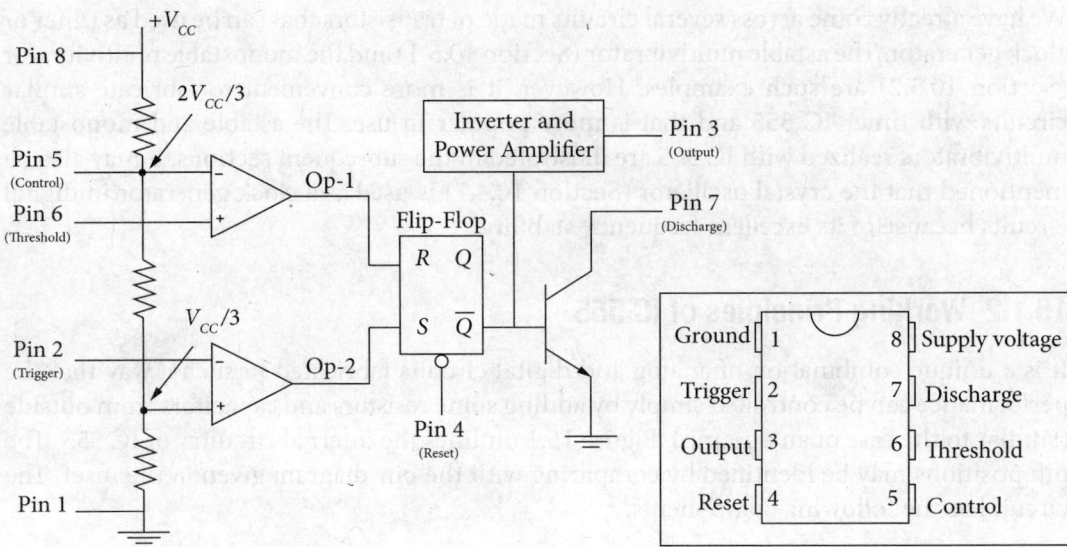

Figure 15.1 Internal circuit diagram of 555 timer IC; the pin configuration is given in the inset

Discharge Transistor: The transistor within the 555 circuit switches between saturation and cut off conditions depending on the $\bar{Q}$ output of the flip-flop. When saturated, the collector of the transistor makes the voltage of Pin 7 (Discharge) almost zero thereby allowing any capacitor connected to it to discharge.

Output: The $\bar{Q}$ output of the flip-flop gets inverted and amplified before appearing as the final output (Pin 3) of the 555 circuit. Thus when $\bar{Q}$ is high, the output at Pin 3 becomes low and vice versa.

15.1.3 Astable Multivibrator with IC 555

This circuit does not have any stable output state and it varies periodically between two quasi-stable states thereby producing a rectangular voltage waveform. A similar circuit was fabricated with two transistors in Chapter 10. However, the working principle of the present circuit is quite different. It depends on the mechanism of IC 555 introduced in the previous section.

Circuit Configuration: The astable multivibrator circuit fabricated with IC 555 is shown in Figure 15.2(a). It contains a common connection of the trigger and the threshold and employs two resistors R_A and R_B with a capacitor C. The time period of the output square wave depends on the charging-discharging of the capacitor through the two resistors.

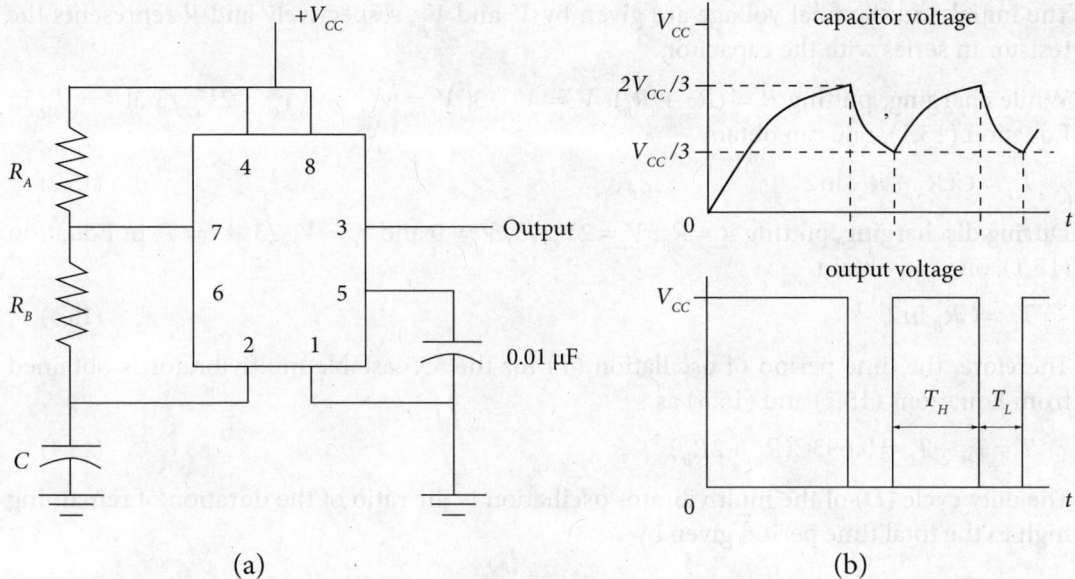

Figure 15.2 Astable multivibrator using IC 555, (a) circuit diagram, (b) output voltage and capacitor voltage waveforms

Working Principle: Initially, when the power supply is switched on, the capacitor (C) is not charged so that the output of trigger comparator (Op-2) is high, which makes the $\bar{Q}$ output of the flip-flop low. The transistor remains cutoff and the output (Pin 3) is high. Then the capacitor starts charging through the series combination ($R_A + R_B$) and as the capacitor voltage exceeds $V_{CC}/3$, Op-2 becomes low. The Op-1 output is already low. Both S and R thus being low, no change occurs at the flip-flop output and the charging of the capacitor continues. When the capacitor voltage exceeds $2V_{CC}/3$, Op-1 goes high. It makes $\bar{Q}$ high, which drives the transistor into saturation causing the capacitor to discharge through R_B only. When the capacitor voltage goes below $V_{CC}/3$, Op-2 becomes high again making $\bar{Q}$ low. This causes the transistor to cutoff and the capacitor starts charging again. The above sequences are repeated periodically. The capacitor charges up to $2V_{CC}/3$ with time constant $C(R_A + R_B)$ and discharges up to $V_{CC}/3$ with time constant CR_B. Figure 15.2(b) shows how the capacitor voltage and the output voltage, starting from the initial point ($t = 0$), acquire steady periodicity with fixed durations of remaining high (T_H) and remaining low (T_L) thereby producing a steady square voltage waveform.

Time Period: The time period of oscillation for the astable multivibrator is the sum of the durations of being high and low ($T_H + T_L$) and can be estimated in the following way.

The general expression for the voltage across the capacitor (C) at any instant of time t is given by

$$v_C = V_f + (V_i - V_f)e^{-t/RC} \qquad (15.1)$$

The initial and the final voltage are given by V_i and V_f, respectively and R represents the resistor in series with the capacitor.

While charging, putting $R = (R_A + R_B)$, $V_i = V_{CC}/3$, $V_f = V_{CC}$ and $v_C = 2V_{CC}/3$ at $t = T_H$ in Equation (15.1), one can obtain

$$T_H = C(R_A + R_B)\ln 2 \tag{15.2}$$

During discharging, putting $R = R_B$, $V_i = 2V_{CC}/3$, $V_f = 0$ and $v_C = V_{CC}/3$ at $t = T_L$ in Equation (15.1), one can obtain

$$T_L = CR_B \ln 2 \tag{15.3}$$

Therefore, the time period of oscillation (T) for the 555 astable multivibrator is obtained from Equations (15.2) and (15.3) as

$$T = T_H + T_L = 0.693 C(R_A + 2R_B) \tag{15.4}$$

The duty cycle (D) of the multivibrator oscillation is the ratio of the duration of remaining high to the total time period given by

$$D = \frac{T_H}{T}\% = \frac{R_A + R_B}{R_A + 2R_B}\% \tag{15.5}$$

Equations (15.4) and (15.5) state that the time period and the duty cycle of the astable multivibrator can be determined by the external components R_A, R_B and C.

15.1.4 Monostable Multivibrator with IC 555

Figure 15.3(a) displays the monostable multivibrator circuit fabricated with IC 555. This circuit has one stable and one quasi-stable output state and it produces a rectangular voltage pulse whose width can be varied by the external resistor (R) and capacitor (C). The pulse repetition frequency can be changed by the frequency of the trigger pulse applied to Pin 2 of the IC.

Stable State of the Circuit: When the supply voltage (V_{CC}) is switched on, the capacitor C charges through the resistor R and the threshold voltage increases. When it exceeds $2V_{CC}/3$, the threshold comparator output (Op-1) becomes high, which makes $\bar{Q}$ of the flip-flop high and drives the transistor into saturation. The capacitor fully discharges through the saturated transistor rapidly. The output (Pin 3) remains low because it is inverted $\bar{Q}$. This is the stable state of the circuit output in absence of any trigger input.

Quasi-Stable State of the Circuit: When a negative voltage pulse is applied to the trigger input (Pin 2), the voltage at this point becomes less than $V_{CC}/3$ so that the trigger comparator output (Op-2) becomes $+V_{CC}$, which makes $\bar{Q}$ low. The transistor gets cutoff and the discharge pin (Pin 7) gets isolated from the ground. The capacitor (C) charges again through R. When the capacitor voltage just exceeds $2V_{CC}/3$, the output of the threshold comparator (Op-1) goes high causing $\bar{Q}$ high. The transistor saturates again and the stable state is regained.

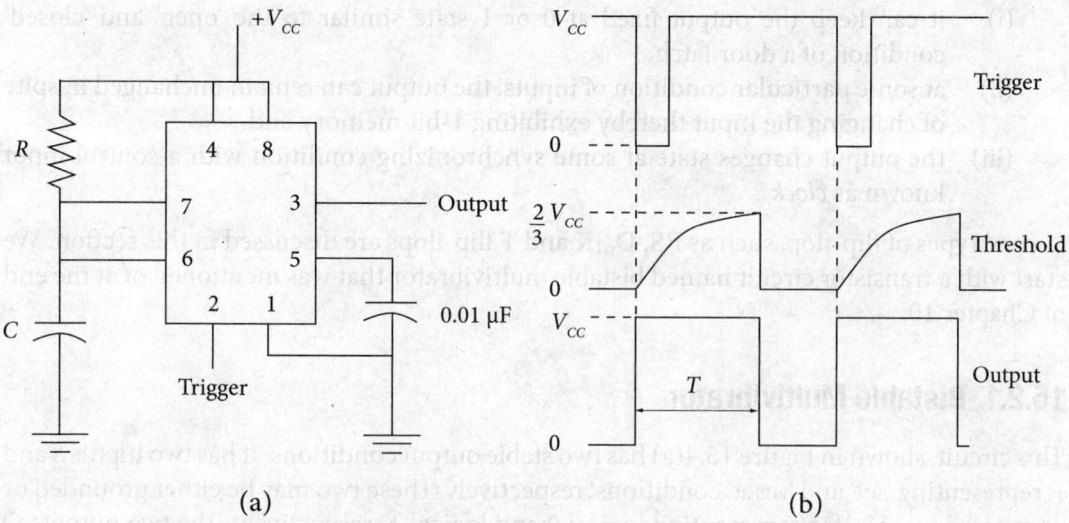

Figure 15.3 Monostable multivibrator using IC 555, (a) circuit diagram, (b) voltage waveforms for trigger pulse, threshold voltage and output pulse

The above cycles are repeated for each trigger pulse. Figure 15.3(b) sketches the trigger pulse and the resultant threshold and output voltages. The *pulse width* (T), meaning the time span of the pulse generated by the monostable multivibrator can be determined from the generalised expression of Equation (15.1).

Initially the capacitor is uncharged and its target voltage is V_{CC}. Therefore, at $t = 0$, putting $V_i = 0$ and $V_f = V_{CC}$ in Equation (15.1),

$$v_C = V_{CC}(1 - e^{-t/RC}) \tag{15.6}$$

At $t = T$, $v_C = 2V_{CC}/3$. Putting these values in Equation (15.6) and rearranging,

$$T = RC \ln 3 = 1.1 RC \tag{15.7}$$

Equation (15.7) states that the pulse width (T) of the 555 monostable multivibrator can be determined simply by two external components R and C. The square pulses generated by the multivibrators can be used as clock pulses in sequential digital circuits, such as flip-flops, as will be discussed in the subsequent sections.

15.2 Latch and Flip-Flop

The flip-flop is the most fundamental unit of sequential digital system. Generally it has two inputs and in some specific cases, it has a single input. The output has its complement too. The characteristics of the flip-flop circuit are such that:

(i) it can keep the output fixed at 0 or 1 state similar to the 'open' and 'closed' condition of a door latch,
(ii) at some particular condition of inputs, the output can remain unchanged in spite of changing the input thereby exhibiting 1-bit memory and
(iii) the output changes state at some synchronizing condition with a control input known as *clock*.

Several types of flip-flop, such as RS, D, JK and T flip-flops are discussed in this section. We start with a transistor circuit named bistable multivibrator that was mentioned of at the end of Chapter 10.

15.2.1 Bistable Multivibrator

This circuit, shown in Figure 15.4(a) has two stable output conditions. It has two inputs S and R representing 'set' and 'reset' conditions, respectively. These two may be either grounded or connected to $+V_{CC}$ for representing logical 0 and logical 1, respectively. The two outputs Q and $\bar{Q}$ are obtained at the collectors of two identical transistors T_1 and T_2, respectively. The following input–output conditions can take place.

(i) When $S = 1$ and $R = 0$, T_2 gets saturated and T_1 is cutoff so that Q becomes almost $+V_{CC}$, representing 1 and $\bar{Q}$ comes at almost ground potential representing 0. The two outputs are thus complements to each other and the output states are *latched* at $Q = 1, \bar{Q} = 0$ and remains fixed indefinitely unless some specific change is made in the input conditions acting like the key of the latch.
(ii) If S is disconnected keeping the other conditions unchanged, the input state becomes $S = 0, R = 0$, which has no effect on the output states because the high collector voltage of T_1 keeps T_2 at saturation condition. So the output states $Q = 1$, $\bar{Q} = 0$ are retained; as if the circuit has a *memory* of retaining a bit.
(iii) When $S = 0$ and $R = 1$, T_1 is saturated and T_2 is cut off so that $Q = 0$ and $\bar{Q} = 1$. This is another *latched* condition of the output states.
(iv) If R is detached keeping the other conditions unaltered, the output remains at the previous state showing again the *1-bit memory* property.
(v) If both $S = R = 1$, both transistors go into saturation. Though the transistors are identical, all the aspects, such as the current gain and the base resistance cannot be exactly the same and any one may conduct more. So it cannot be predicted, which transistor would saturate earlier. As if there would be a *racing* between them. If both S and R are now made zero, both the transistors tend to come out of saturation and again there would be a racing between them to be active first. If T_1 was faster, Q would be 0 and if T_2 was faster, Q would be 1. It is apparent that the whole process is uncertain and should be avoided. So this condition is designated as *forbidden state*.

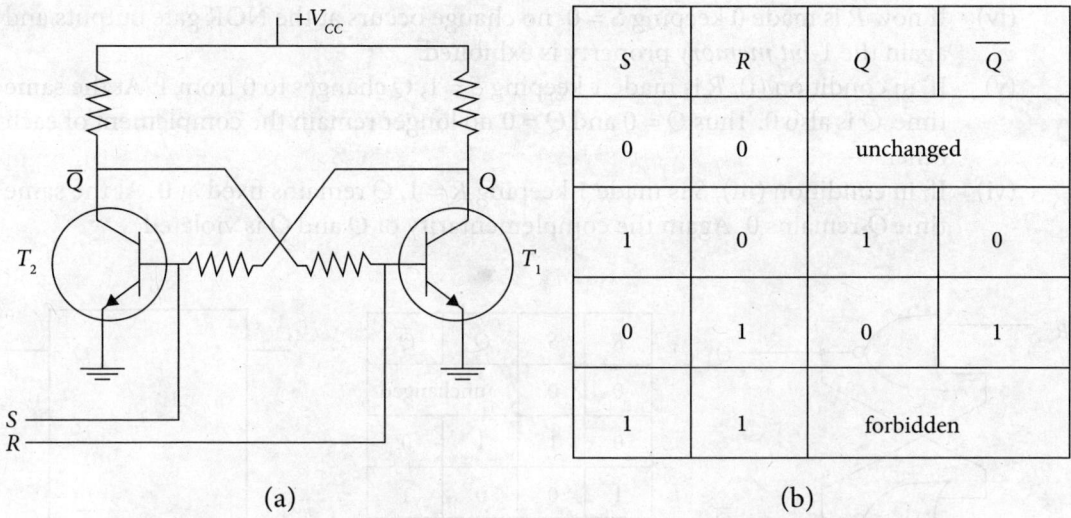

Figure 15.4 Transistor bistable multivibrator: (a) circuit diagram, (b) truth table

All the above input–output combinations are summarized in the truth table of Figure 15.4(b). This is the generalized truth table of a popular digital circuit named flip-flop introduced in the subsequent sections.

> **Note: Bistable Multivibrator and Flip-Flop**
>
> The bistable multivibrator possesses both the properties of a flip-flop: (i) getting latched at one of the two stable states and (ii) storing a bit as memory. It is a good model for latch and flip-flop. However, the actual flip-flops in digital circuits are constructed with logic gates in order to get compliance with IC technology and to maintain a specific logic family.

15.2.2 RS Flip-Flop with NOR Gates

Figure 15.5(a) shows the circuit of RS flip-flop made of NOR gates. The inputs are S (set) and R (reset) and the outputs are Q and $\bar{Q}$, which are complements to each other. The following input–output combinations may occur.

(i) When $R = 0$ and $S = 1$, $\bar{Q}$ becomes 0 because any input of 1 makes the output of a NOR gate 0. Thus both the inputs of the other NOR gate become zero making $Q = 1$. This condition of $Q = 1$, $\bar{Q} = 0$ is called the *Set* condition of the flip-flop.

(ii) If S is now made 0 keeping $R = 0$, there is no effect on the NOR gates and the previous output state of $Q = 1$, $\bar{Q} = 0$ remains unchanged. This represents the capability of the flip-flop to retain *one bit of memory*.

(iii) When $R = 1$ and $S = 0$, following the same reason as that of (i), Q becomes 0 and $\bar{Q}$ becomes 1. The condition of $Q = 0$, $\bar{Q} = 1$ is called the *Reset* condition of the flip-flop.

(iv) If now R is made 0 keeping S = 0, no change occurs at the NOR gate outputs and again the 1-*bit memory* property is exhibited.

(v) If, in condition (i), R is made 1 keeping S = 1, Q changes to 0 from 1. At the same time, $\bar{Q}$ is also 0. Thus Q = 0 and $\bar{Q}$ = 0 no longer remain the complement of each other.

(vi) If, in condition (iii), S is made 1 keeping R = 1, Q remains fixed at 0. At the same time $\bar{Q}$ remains 0. Again the complementarity of Q and $\bar{Q}$ is violated.

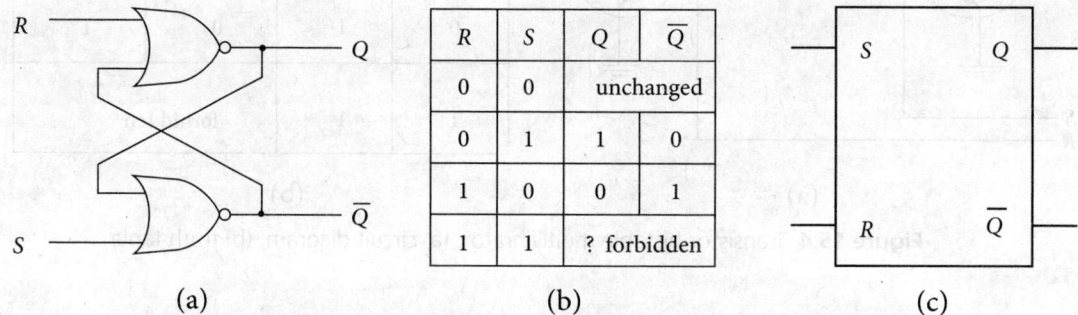

Figure 15.5 RS flip-flop with NOR gate: (a) circuit diagram, (b) truth table, and (c) logic symbol

Comparing conditions (v) and (vi) it is understood that in both cases, S = R = 1 but Q may or may not change. So it is an *uncertain* condition and is indicated by a question mark (?) in the truth table. Furthermore, Q = $\bar{Q}$ = 0 violates the complementarity of the output conditions. Because of these two reasons, the condition of S = R = 1 is considered *uncertain* and *forbidden* to RS flip-flop. All the above conditions are compiled in the truth table of Figure 15.5(b). This is same as the truth table of Figure 15.4(b). The S and R columns are changed only to show that the logical input–output combinations remain the same. Figure 15.5(c) represents the logic symbol of an RS flip-flop. This symbol is used directly while sketching the diagrams of logic systems comprising flip-flops. The presence of the power supply is understood.

Thank a Bit: Is it Sequential?

Comparing the truth tables of figures 15.4(b) and 15.5(b) it is found that changing the S and R columns do not matter the output conditions of the flip-flop. Then should it be treated as a sequential circuit? Yes, of course! We have merely rearranged the table keeping the circuit untouched. If we interchanged the R and S in the circuit of Figure 15.5(a), the positions of the outputs Q = $\bar{Q}$ would change absolutely. Try it on your own and realize that the inputs must be given in a certain sequence of position and time in order to get a certain output state.

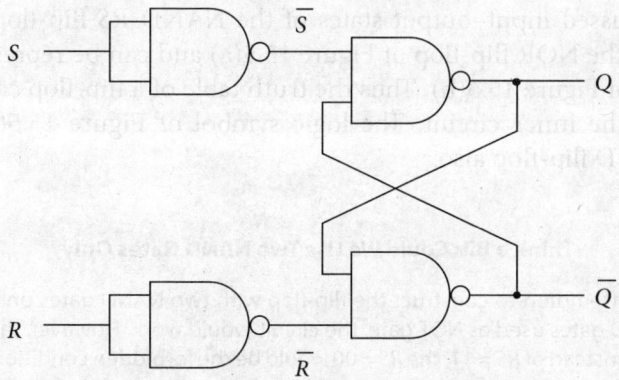

Figure 15.6 RS flip-flop with NAND gate. The truth table and the logic symbol are the same as that of figures 15.5(b) and 15.5(c), respectively

15.2.3 RS Flip-Flop with NAND Gates

The RS flip-flop can be constructed with NAND gates also, as demonstrated in Figure 15.6. Two NAND gates construct the basic latch circuit and two additional NAND gates, used as NOT gates invert the S and R inputs so that actually $\bar{S}$ and $\bar{R}$ enter the latch. The following input–output conditions are possible.

(i) When $S = 1$ and $R = 0$, $\bar{S} = 0$ and $\bar{R} = 1$. It makes $Q = 1$ because any input of 0 to a NAND gate makes its output 1. Since Q is fed back to the other NAND gate, that has both the inputs 1 making $\bar{Q} = 0$. This is the *Set* condition of the flip-flop.

(ii) If S is made 0 keeping $R = 0$, that makes $\bar{S} = 1$ and $\bar{R} = 1$. Since an input of 1 has no effect on the NAND output, the previous state of $Q = 1$, $\bar{Q} = 0$ is retained indicating the *memory* of one bit.

(iii) When $S = 0$ and $R = 1$, $\bar{S} = 1$ and $\bar{R} = 0$. Consequently $\bar{Q} = 1$. The other NAND gate, having both the inputs at 1 state makes $Q = 0$. This is the *Reset* condition of the flip-flop.

(iv) If R is made 0 keeping $S = 0$, that results in $\bar{S} = 1$ and $\bar{R} = 1$ having no effect on the previous output states of $Q = 0$, $\bar{Q} = 1$. The 1-bit storage property is exhibited again.

(v) In condition (i), if R is made 1 keeping $S = 1$, both $\bar{R}$ and $\bar{S}$ become 0 making $Q = \bar{Q} = 1$, which violates the basic definition of Q and $\bar{Q}$ in a flip-flop being the complement to each other. Comparing with condition (i), we note that $\bar{Q}$ changes state here.

(vi) In condition (iii), if S is made 1 keeping $R = 1$, both $\bar{S}$ and $\bar{R}$ become 0 making $Q = \bar{Q} = 1$. This also breaks the basic rule of Q and $\bar{Q}$ being mutually complements. Also, comparing with condition (iii) we find that this time $\bar{Q}$ does not change state. Therefore, conditions (v) and (vi) where $S = R = 1$ are *uncertain* and *forbidden* to this RS flip-flop and should be avoided.

All the above discussed input–output states of the NAND *RS* flip-flop (Figure 15.6) are similar to those of the NOR flip-flop of Figure 15.5(a) and can be represented by the same truth table as that of Figure 15.5(b). Thus the truth table of a flip-flop can be standardized, whatever may be the inner circuit. The logic symbol of Figure 15.5(c) can be used to represent this NAND flip-flop also.

> **Think a Bit: Could We Use Two NAND Gates Only?**
>
> If, in Figure 5.6, we attempted to construct the flip-flop with two NAND gates only and discarded the two additional NAND gates used as NOT gate, the circuit would work. However, the output conditions would get changed. Instead of $RS = 11$, the $RS = 00$ would be the forbidden condition thereby producing a different truth table. Now fill up the other input-output conditions of that truth table on your own. Your circuit would be referred to as 'active low' flip-flop because a low state of the input caused change in the output. The circuit of Figure 5.5(a) is 'active high' because a high input state changes the output. Thus we understand that a flip-flop can be constructed in a number of ways. Anyway the change of the truth table with each internal circuit is not a matter of convenience. Figures 15.5 and 15.6 explain a method to achieve an unaltered truth table with different circuits.

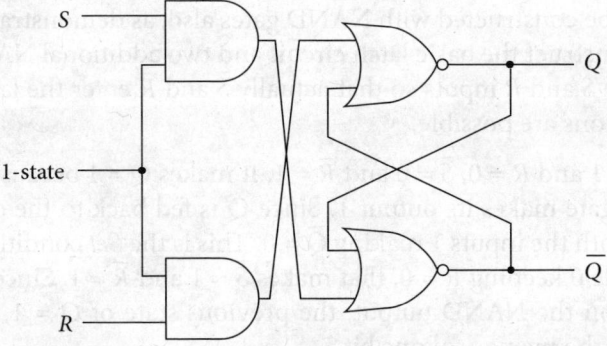

Figure 15.7

> **Example 15.2**
>
> Show that the logic diagram of Figure 15.7 acts as a flip-flop. What happens, if the 1-state becomes 0-state?
>
> **Solution**
>
> Since the two AND gates have one input always 1, the other inputs (*S* or *R*) will be transmitted to the NOR gates and the circuit will follow the output conditions of Figure 15.5(b). If one input of each AND gates is made 0, the AND gate outputs will always impose 0 to the NOR gates and no change of *S* and *R* inputs will be reflected anywhere.

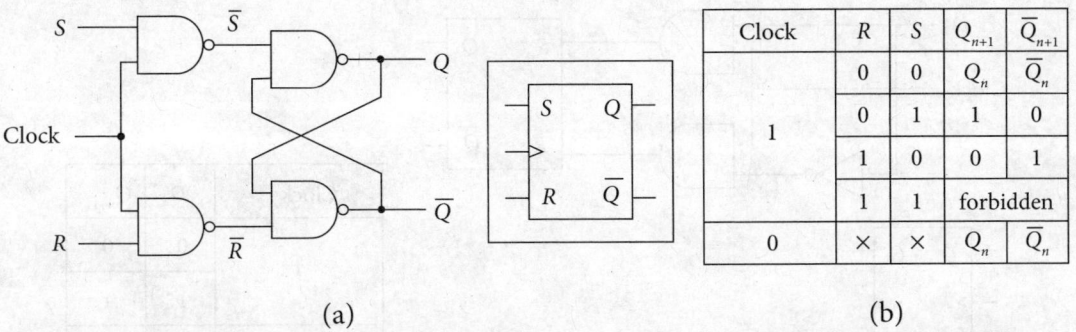

Figure 15.8 Clocked RS flip-flop (a) circuit diagram and symbol (inset) and (b) truth table

15.2.4 Clocked RS Flip-Flop

The previous discussions on the flip-flop have explained its two important properties: bistable state and 1-bit memory. Another salient feature of the flip-flop, hence of all sequential logic circuits is to control the input–output states and to synchronize with other logic circuits with the use of a clock.

Figure 15.8 explains the use of clock in a flip-flop circuit to enable or disable it, as per requirement. The symbol of a clocked RS flip-flop is similar to that without a clock, except for the arrow head symbol representing the clock, as sketch in the figure. Unless otherwise stated, it is assumed that the clock is effective at high-to-low transition. The truth table of the clocked RS flip-flop given in Figure 15.8 is similar to that of an ordinary flip-flop discussed earlier. However, it consists of the following two conditions.

Enabled Condition: When the clock is 1, inputs S and R are transmitted to the NAND latch as $\bar{S}$ and $\bar{R}$, respectively and the circuit acts as a common NAND flip-flop, as is indicated in the truth table of Figure 15.8. The symbols Q_n and $\bar{Q}_n$ represent the output states at the time before the clock transition; Q_{n+1} and $\bar{Q}_{n+1}$ are those after the clock transition. The uncertain (?), forbidden state is also present there. The circuit will continue to act in a similar way as long as the clock is high.

Disabled Condition: When the clock becomes 0, the output remains fixed retaining the input states received just before the moment of the clock transition. Since the changes of the inputs S and R are now immaterial, those are represented by a '×' symbol, known as *don't care* condition.

It is now clear that the clocked RS flip-flop can perform the set-reset and bit storage operations and also can hold the stored information for any desired period of time decided by the clock. The circuit of Figure 15.7 was indeed a clocked NOR flip-flop, if the 1 and 0 states were replaced by a clock.

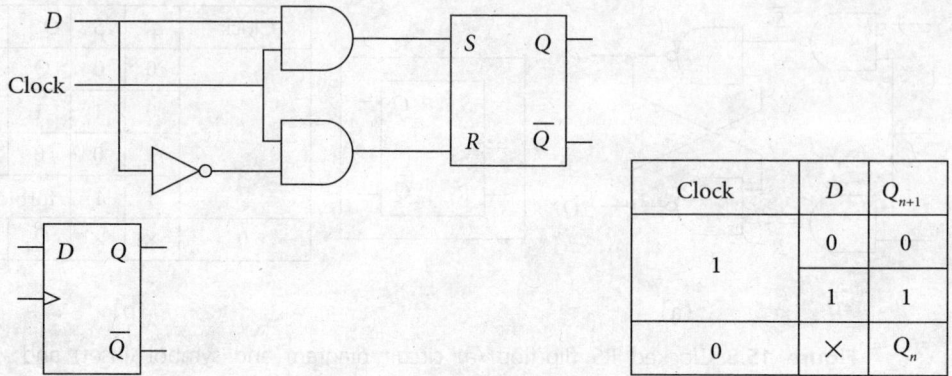

Figure 15.9 D flip-flop: circuit diagram (top), symbol (bottom left) and truth table (bottom right)

15.2.5 D Flip-Flop

This flip-flop has a single data input (D) and the use of clock is important. Figure 15.9 is a simple way to construct D flip-flop. The RS flip-flop is converted to D flip-flop by inserting a NOT gate between S and R so that the forbidden state of S = R = 1 can be avoided. The symbol of the D flip-flop and its truth table are given in Figure 15.9. As usual, Q_n represents the output states at the instant of time before the clock transition and Q_{n+1} is that after the instant of clock transition. The following two conditions are compiled there.

(i) When the clock is 0, a certain data bit (D) is withheld and is not transmitted to the output. The outputs Q and $\bar{Q}$ get locked to the previous state even if the data input is changed.

(ii) When the clock becomes 1, the D input and its complement are applied to S and R, respectively thereby changing the Q and $\bar{Q}$ states. Generally Q is used as the output because it reflects D in presence of clock.

> **Example 15.3**
>
> Can you construct a system having n-bit data input ($D_n...D_0$) and n-bit data output ($Q_n...Q_0$) with D flip-flops such that the input may be delayed for a desired time span?
>
> **Solution**
>
> Figure 15.10 introduces the required circuit. When the clock is 0, the input ($D_n...D_0$) of n-bits gets withheld and the n-bit output ($Q_n...Q_0$) remains fixed at the previous state. When the clock is 1, the current values of the input bits get transmitted to the corresponding output terminals.

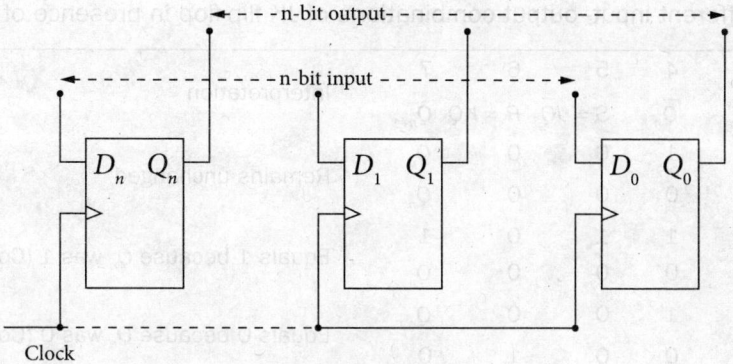

Figure 15.10

15.2.6 JK Flip-Flop

This is a modification of RS flip-flop that accounts for the forbidden state of $R = S = 1$. Similar to the RS flip-flop, the JK flip-flop has two inputs, namely J and K and two outputs Q and $\overline{Q}$. Similar to the RS flip-flop, the JK flip-flop also executes the bistable state and 1-bit storage action. In addition, the $J = K = 1$ state is made use, as discussed later.

Figure 15.11 shows a simple method of constructing JK flip-flop. The outputs Q and $\overline{Q}$ of an RS flip-flop are fed back to two AND gates along with the inputs K and J, respectively. Thus $S = J\overline{Q}$ and $R = KQ$. Table 15.1 illustrates all possible input–output combinations in presence of clock. Only the Q-output states are considered for convenience.

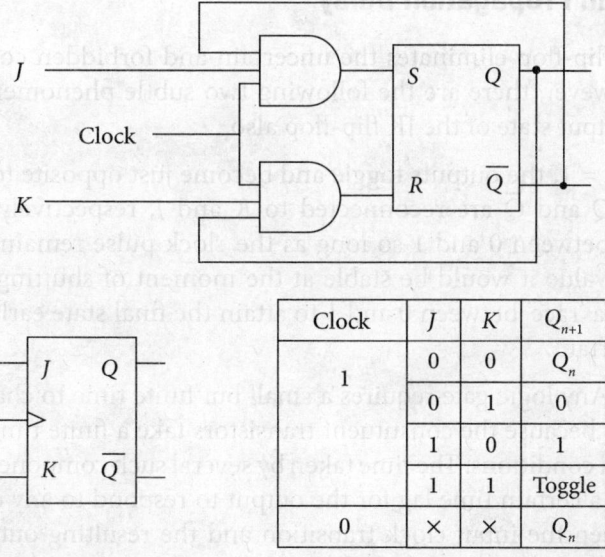

Figure 15.11 JK flip-flop: circuit diagram (top), symbol (bottom left) and truth table (bottom right)

Table 15.1 Different input–output combinations of JK flip-flop in presence of clock

1	2	3	4	5	6	7	Interpretation
J	K	Q_n	$\bar{Q}_n$	$S = J\bar{Q}$	$R = KQ$	Q_{n+1}	
0	0	0	1	0	0	Q_n	Remains unchanged
0	0	1	0	0	0	Q_n	
1	0	0	1	1	0	1	Equals 1 because Q_n was 1 (Column 3)
1	0	1	0	0	0	Q_n	
0	1	0	1	0	0	Q_n	Equals 0 because Q_n was 0 (Column 3)
0	1	1	0	0	1	0	
1	1	0	1	1	0	1	Equals $\bar{Q}_n$ (comparing columns 3 and 4)
1	1	1	0	0	1	0	

The different conditions of J and K are collected and entered in the truth table of Figure 15.11. The symbol Q_n represents the output state before the clock transition and Q_{n+1} is that after the clock transition. The following conclusions can be made from it.

(i) When $J = K = 1$, the output of the JK flip-flop becomes just the opposite of the previous state: 0 to 1 and vice versa. This phenomenon is known as *toggling*. This state is quite different from that of RS flip-flop.

(ii) For the other combinations of J and K, the JK flip-flop behaves like an RS flip-flop. The inputs J and K may be compared with S and R, respectively.

(iii) Similar to the RS flip-flop, the JK flip-flop also executes bistable and 1-bit memory.

15.2.7 Racing and Propagation Delay

Apparently the JK flip-flop eliminates the uncertain and forbidden condition of both the inputs being 1. However, there are the following two subtle phenomena that may lead to ambiguity in the output state of the JK flip-flop also.

Racing: When $J = K = 1$, the outputs toggle and become just opposite to the previous state. Since the outputs Q and $\bar{Q}$ are reconnected to K and J, respectively, the output would continue to toggle between 0 and 1 so long as the clock pulse remains high. One cannot predict, with what value it would be stable at the moment of shutting down of the clock pulse. As if there is a 'race' between 0 and 1 to attain the final state earlier. Thus the output state becomes uncertain.

Propagation Delay: Any logic gate requires a small but finite time to change the 0 or 1 state of its output. It is so because the constituent transistors take a finite time to switch between saturation and cutoff conditions. The time taken by several such components are cumulatively added and result in a certain time lag for the output to respond to any change in the input. This time lag between the input clock transition and the resulting output change and the resulting change at the output state is referred to as the *propagation delay* of the digital circuit. The TTL gates have a shorter propagation delay ($\approx$ 10 ns) and that of CMOS gates is a bit longer ($\approx$ 25 ns). The significance of propagation delay can be put up in two different ways.

(i) The propagation delay is one remedy to prevent the JK flip-flop from racing. If the duration of the clock pulse is made smaller than the propagation delay time, toggling more than once during a clock pulse can be avoided.

(ii) The propagation delay limits the maximum clock frequency that the circuit can respond to.

If such high frequency clock pulse and the matching logic circuits are not available, the uncertainty in toggling can be eliminated by an alternative method of cascading two JK flip-flops. That construction is known as *master–slave* combination, discussed in a subsequent section.

15.2.8 Edge- and Pulse-Triggering

In order to ensure that all the inputs change states at a unique point of time, the clock is converted to narrow pulses where either the negative-to-positive or the positive-to-negative transition edge is extracted and applied to the digital circuit. Such a process is called *edge-triggering*.

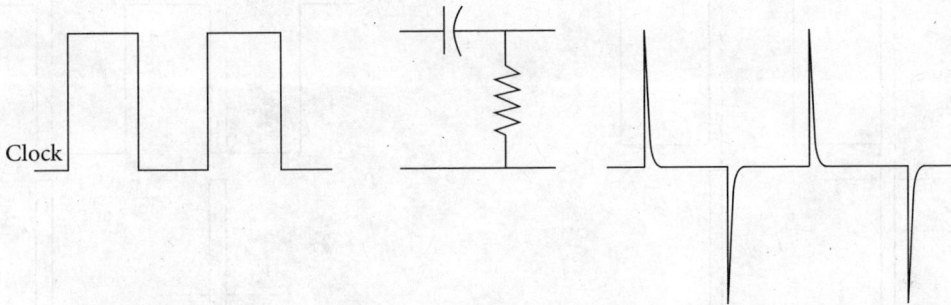

Figure 15.12 Converting a square clock pulse (left) to edge-spikes (right) using CR circuit (middle)

Figure 15.12 presents a simple technique to achieve edge-triggering with analog circuit. The square clock pulse (left) is applied to a CR circuit (middle) of small time constant so that the output is differentiated as spikes of positive and negative voltages (right). The voltage spikes take place corresponding to the positive and negative edges of the original square waveform. However, better triggering can be achieved with digital circuits presented below.

Positive Edge-Triggering: In the circuit of Figure 15.13(a), the square voltage of the clock is applied to a NAND gate (used as NOT) and an AND gate simultaneously. Both the gates are assumed to have the same propagation delay (t_p). The output (Y) of the NAND gate is delayed by t_p with respect to the clock. The output (P) of the AND gate becomes high for this small duration only when both the inputs become high and comes out as a narrow pulse of width t_p. The digital circuits associated with it change state in synchronism with the positive transition of the clock and are referred to as *positive edge-triggered* circuits.

Negative Edge-triggering: If, instead of AND gate, a NOR gate is used [Figure 15.13(b)], the pulse (P) becomes high corresponding to negative clock transition because the output of the NOR gate becomes high only when both the inputs become low. The digital circuits driven by such negative clock transitions are said to be *negative edge-triggered* circuits.

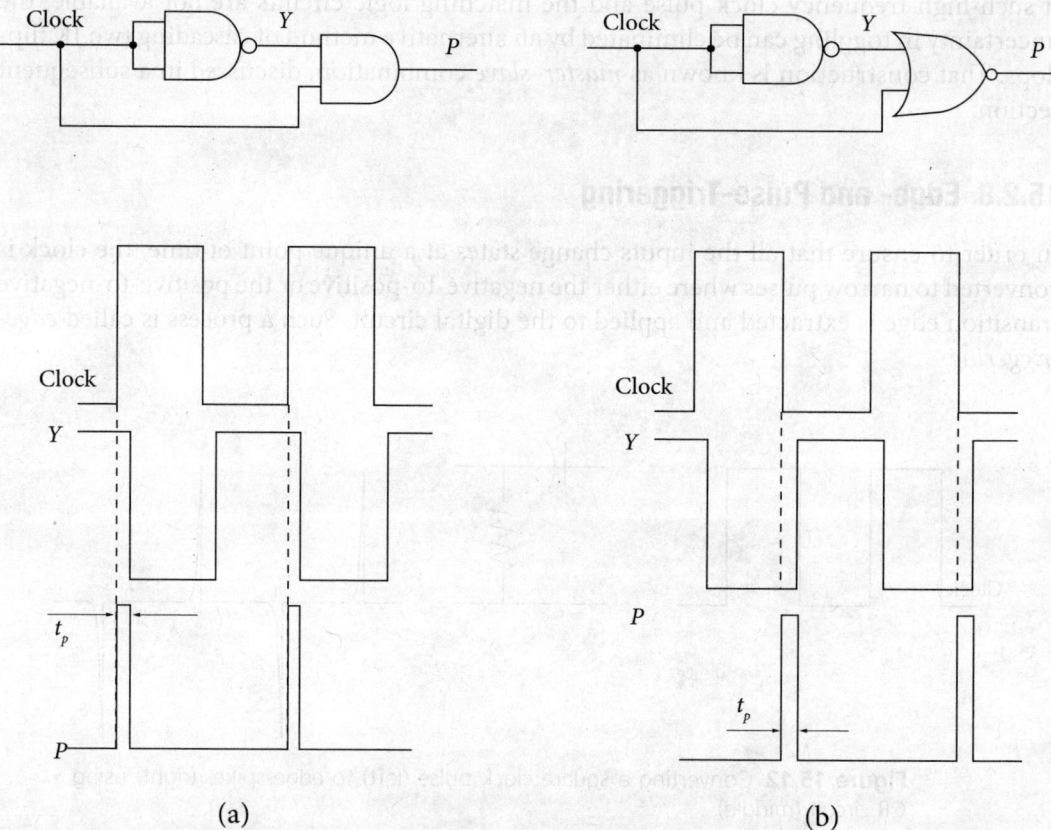

Figure 15.13 Edge-triggering circuits (top) and the corresponding clock, output (Y) and pulse (P) waveforms (bottom) for (a) positive and (b) negative edges

It may be noted from both Figures 15.13 (a) and (b) that in both cases the output of edge transition shifts further by t_p with respect to the clock because of the propagation delay of the AND or the NOR gate. However, the objective is to synchronize the time of state change for all the circuits triggered by the narrow pulse and that purpose is served. There is another category of circuit, such as the master–slave flip-flop that consumes the whole of the clock pulse to complete the operation. The output does not change state unless the clock pulse goes high and again makes the negative transition thereby completing the full clock cycle. This type of digital circuit is called *pulse-triggered* circuit.

15.2.9 JK Master–Slave Flip-Flop

This nickname, though sometimes not favoured in literature, indicates that the circuit involves two JK flip-flops, one controlling the other. The circuit as a whole is a pulse-triggered circuit because the output toggles only once during a full clock pulse irrespective of its frequency. It provides an alternative way to avoid multiple toggling and the uncertainty due to racing.

Figure 15.14 represents a simplified circuit diagram of JK master–slave flip-flop. The flip-flop named Master gets enabled at positive clock states. Its inputs J and K are the actual inputs of the circuit. The outputs Q_M and $\bar{Q}_M$ of the Master drive the inputs J_S and K_S, respectively, of the other flip-flop named Slave, which is enabled at negative clock states. The same clock as that of the Master is applied to the Slave through a NOT gate. The outputs Q and $\bar{Q}$ are the final outputs of the circuit. The whole circuit, represented by the dotted line is considered as a single unit. The logic symbol of the JK master–slave flip-flop is given in the inset of Figure 15.14. The bubble sign at the clock indicates that this flip-flop changes its final output state at the negative transition of the clock. The circuit performance is considered for the following input combinations.

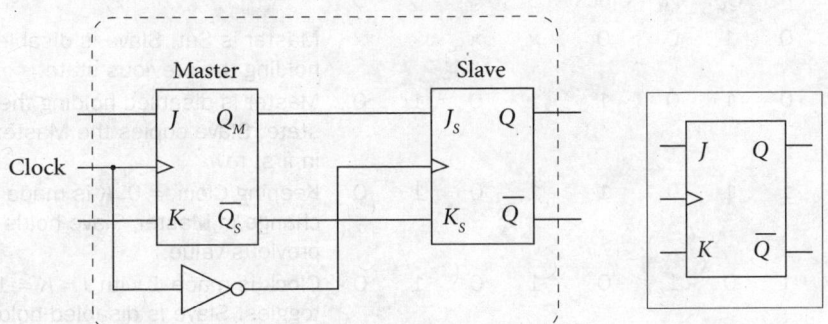

Figure 15.14 JK master–slave flip-flop, construction and symbol (inset)

Case-I: $J = 1$ and $K = 0$
When the clock is positive, the Master is enabled, $Q_M = 1$ and $\bar{Q}_M = 0$. The Slave is disabled and its outputs do not change.
When the clock becomes negative, the Master gets disabled but the Slave is enabled. The previous values of $Q_M = 1$ and $\bar{Q}_M = 0$ make $J_S = 1$ and $K_S = 0$ so that the final outputs become $Q = 1$ and $\bar{Q} = 0$. Thus the Slave copies the action of the Master irrespective of the clock speed.

Case-II: $J = 0$ and $K = 1$
When the clock is positive, the Master is enabled, $Q_M = 0$ and $\bar{Q}_M = 1$. The Slave is disabled. When the clock is negative, the Slave is enabled. The Master goes disabled holding the previous Q_M and $\bar{Q}_M$ values, which now make $J_S = 0$ and $K_S = 1$ thereby making $Q = 0$ and $\bar{Q} = 1$. Again the Slave copies, during the negative part of the clock cycle, whatever the Master does during the positive part of the clock cycle.

Case-III: $J = 0$ and $K = 0$

The Master remains unchanged in the positive clock state and the Slave remains unchanged in the negative clock state and the combination keeps the final output unchanged throughout the full clock cycle.

Case-IV: $J = 1$ and $K = 1$

The outputs Q_M and $\bar{Q}_M$ of the Master toggle during the positive clock state. The toggling message is transmitted to the inputs J_S and K_S of the Slave but it remains disabled. When the clock becomes negative, the Slave outputs toggle but the Master remains inactive. Thus a full clock pulse can cause a single toggle.

The above conditions are compiled in Table 15.2. The summary is that whatever the Master does during the positive clock state is repeated by the Slave during the negative clock state and the whole circuit changes the output states only once during the full clock period.

Table 15.2 Different input–output states of the JK master–slave flip-flop

	Master Unit				Slave Unit					Remark
Clock	J	K	Q_M	$\bar{Q}_M$	Clock	S	R	Q	$\bar{Q}$	
1	1	0	1	0	0	×	×	×	×	Master is Set. Slave is disabled holding the previous state.
0	1	0	1	0	1	1	0	1	0	Master is disabled holding the previous state. Slave copies the Master action in first row.
0	1	1	1	0	1	1	0	1	0	Keeping Clock = 0, K is made 1. No change in Master, Slave holds the previous value.
1	1	1	0	1	0	1	0	1	0	Clock is made 1 with J = K = 1. Master toggles. Slave is disabled holding the previous state.
0	1	1	0	1	1	0	1	0	1	No change in Master. Slave toggles and copies the Master action in fourth row.
1	1	1	1	0	0	0	1	0	1	Master toggles again. Slave is disabled.
0	1	1	1	0	1	1	0	1	0	Slave toggles again. It copies the Master action of sixth row.

Example 15.4

Can you support the statement, 'The NAND gate is a universal gate' by fabricating a JK master–slave flip-flop using NAND gates only?

Solution

Figure 15.15 shows the required circuit. The inputs J and K are accepted by the Master unit that has outputs Q_M and $\bar{Q}_M$. These are also the inputs S and R, respectively, to the Slave unit whose outputs Q and $\bar{Q}$ are the actual inputs of the circuit.

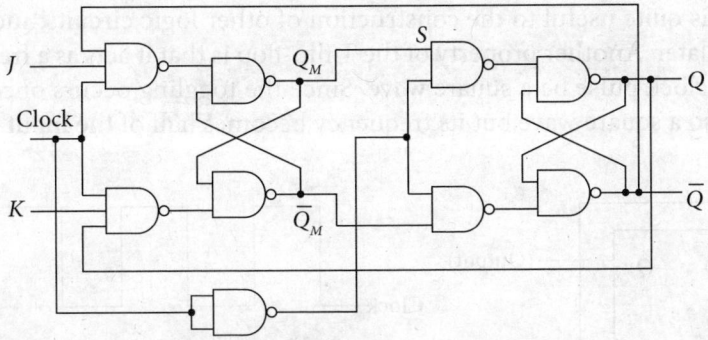

Figure 15.15

> ### Think a Bit: Truth Table of JK Master–Slave Flip-Flop?
>
> It need not be given separately. The JK master–slave flip-flop produces the same truth table as that of an ordinary JK flip-flop (Figure 15.11). This is so because the problem of racing or multiple toggling in a clock is not a regular circuit condition but a limitation of the circuit. That can be overcome in two different ways. If the clock pulse can be made sufficiently narrow, less than the propagation delay (as with edge-triggering), the JK flip-flop would get rid of the racing problem. This method is efficient for circuit fabrication but it imposes a condition of clock frequency being less than the propagation delay of the circuit. An alternative method, where the clock time need not be less than the propagation delay is the internal combination of two flip-flops (master–slave), which makes the external output change only once in a single clock pulse. Both the techniques help the circuit yielding the logic conditions perfectly.

15.2.10 T Flip-Flop

The toggle property of the JK flip-flop is utilized as a separate configuration known as T flip-flop where 'T' stands for 'toggle'. This circuit has only a clock input and the output simply toggles each time the clock changes state. Such a logic circuit can be realised by various methods. Figure 15.16 presents such a method where a T flip-flop has been constructed with a JK master–slave flip-flop. Both the J and K inputs are tied to the positive terminal of the supply voltage so that $J = K = 1$. The clock pulse is the single input to the circuit. Since both J and K are at 1 state, toggling occurs at each clock pulse and the output (Q) changes state on each negative transition of the clock.

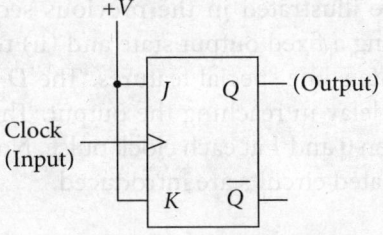

Figure 15.16 T flip-flop constructed with JK master–slave flip-flop

The T flip-flop is quite useful to the construction of other logic circuits, such as a counter, to be discussed later. Another property of the T flip-flop is that it acts as a frequency divider circuit. Let the clock pulse be a square wave. Since the toggling occurs once in each cycle, the output is also a square wave but its frequency becomes half of the input frequency.

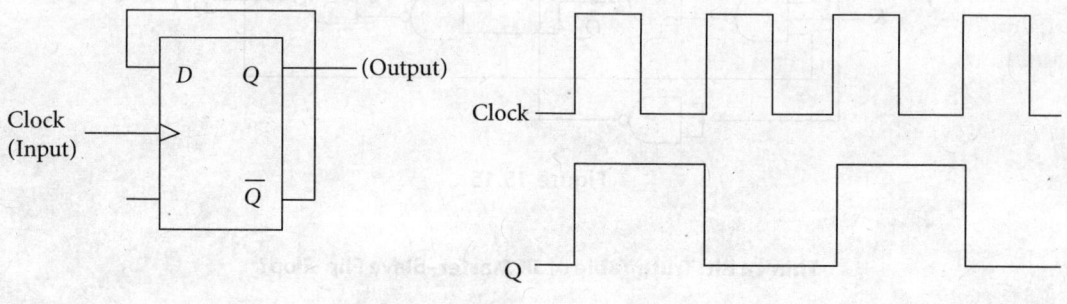

Figure 15.17

Example 15.5

Can you explain the timing diagram for the circuit constructed with D flip flop given in Figure 15.17? What is your inference about the circuit performance?

Solution

The circuit comprises a D flip-flop, the $\overline{Q}$ output being connected to D. Initially $Q = 0$ and $\overline{Q} = 1$, hence $D = 1$. When the clock becomes 1 for the first time, Q becomes 1 (updated with $D = 1$) and $\overline{Q}$ becomes 0. This 0 is fed back to D before the arrival of the next positive edge of the clock. The next positive clock makes $Q = 0$ and $\overline{Q} = 1$ again. Thus a toggling is performed. This sequence continues and Q (and also $\overline{Q}$) toggles every time a positive clock edge appears. It is inferred that the circuit is another construction of T flip-flop.

15.3 Flip-Flop Characterization

Several types of flip-flops are illustrated in the previous sections. All of these have two common features of (i) holding a fixed output state and (ii) retaining a bit as 'memory'. In addition, flip-flops can execute some special features. The D flip-flop can withhold a data bit at the input and make it delay in reaching the output. The JK flip-flop can toggle, i.e., switch the output state between 0 and 1 at each clock pulse. Now some standard methods to characterize flip-flops and related circuits are introduced.

15.3.1 Characteristic Equation

The characteristic equation of a flip-flop expresses its forthcoming output state (Q_{n+1}) as a function of the present output state (Q_n) and the input states. The equation is useful to explain the operation of more complicated digital circuits made of flip-flops. The characteristic equations of different flip-flops are derived from the Karnaugh maps for the truth tables of the flip-flops (Figure 15.18). The optimizations of the maps give rise to the following equations.

- RS flip-flop: $Q_{n+1} = S + \overline{R}Q_n$
- JK flip-flop: $Q_{n+1} = J\overline{Q}_n + \overline{K}Q_n$
- D flip-flop: $Q_{n+1} = D$
- T flip-flop: $Q_{n+1} = T\overline{Q}_n + \overline{T}Q_n$

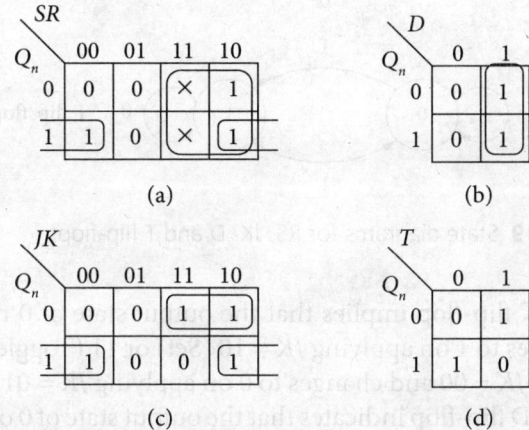

Figure. 15.18 Karnaugh map simplification of input–output states for (a) RS flip-flop, (b) D flip-flop, (c) JK flip-flop, and (d) T flip-flop

15.3.2 State Diagram

For sequential logic circuits, it is an alternative to truth table. The state diagram is a pictorial representation of the relationships among the present and the next input and output states. A state 0 or 1 is represented by a circle and the transition between the states is indicated by arrowed arcs. Figure 15.19 explains the process.

In the state diagram for RS flip-flop, if the Q-output stores 0 (left circle), the application of $SR = 00$ or 01 does not change this value, which is indicated by the arrowed arc. Similarly, if the Q-output stores 1 (right circle), the application of $SR = 00$ or 10 does not change it. Application of $SR = 10$ (Set) changes the output from 0 to 1, indicated by the left-to-right arrow. Applying $SR = 01$ (Reset) changes the output from 1 to 0 (right-to-left arrow). The $\overline{Q}$ output states also could be interpreted in a similar way but it is customary to deal with the Q-states.

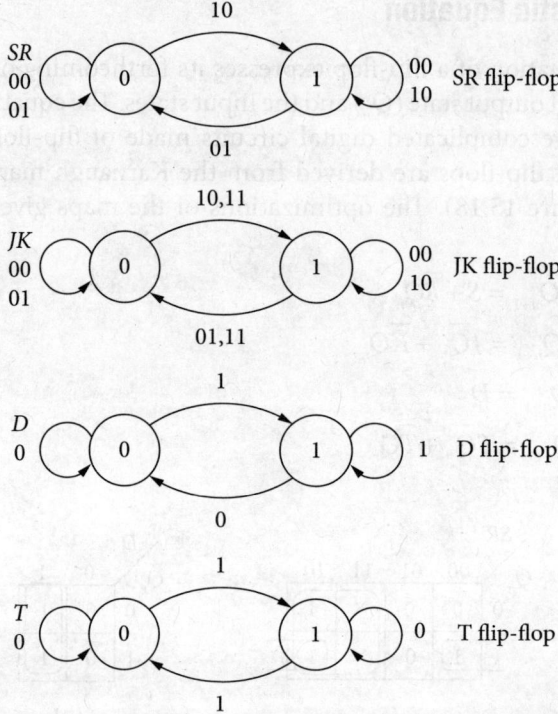

Figure 15.19 State diagrams for RS, JK, D and T flip-flops

The state diagram of JK flip-flop implies that the output state of 0 remains unchanged for JK = 00 or 01 but changes to 1 on applying JK = 10 (Set) or 11 (Toggle). The output state of 1 remains unchanged for JK = 00 and changes to 0 on applying JK = 01 (Reset) or 11 (Toggle).

The state diagram of D flip-flop indicates that the output state of 0 or 1 remains unchanged with the same input but changes state with the opposite input. Actually, the next output is the present input. The characteristic equation $Q_{n+1} = D$ may be recalled.

The output state of T flip-flop is not affected by 0 input state. Application of 1 changes the output from 0 to 1 and vice versa.

15.3.3 Preset and Clear

The inputs of a flip-flop, namely S, R, J, K or D may be termed as *synchronous inputs*, because these make the circuit operate in harmony with the clock. This inputs control the operating of the flip-flop in combination with the clock signal for a unique instant of time. However, the sequential logic circuits may possess *asynchronous inputs* also, which activate the sequential circuit independently of the clock and the synchronous inputs. These can override the other inputs and set the flip-flop to 0 or 1 state at any time, irrespective of the other input conditions. The preset and clear are such asynchronous inputs. These allow the digital system to be fixed at a pre-determined state.

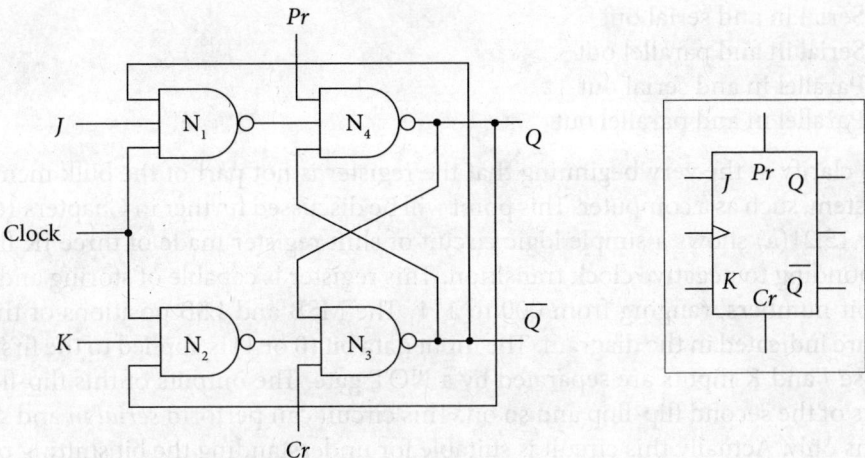

Figure 15.20 JK flip-flop with preset and clear provisions

Figure 15.20 demonstrates the effect of preset (Pr) and clear (Cr) with a simple model of NAND JK flip-flop. The output may be set to any one of the following two conditions.

Clear Condition: It fixes the output at $Q = 0$ and $\overline{Q} = 1$ by making $Cr = 0$ and $Pr = 1$. If the clock is at 0 state, the output of gate N_1 becomes 1 so that the gate N_3 gets all its three inputs at 1 state and Q becomes 0. Since $Cr = 0$, $\overline{Q} = 1$.

Preset Condition: It settles the output states at $Q = 1$ and $\overline{Q} = 0$ by making $Pr = 0$ and $Cr = 1$. If the clock is 0, the output of gate N_2 becomes 1 and gate N_4, having all the inputs at 1 state makes $\overline{Q} = 0$. Since $Pr = 0$, $Q = 1$.

15.4 Register

The register is an important sequential logic system made of flip-flops that can store binary information. Since a flip-flop can store a bit (0 or 1) of digital data, a collection of n flip-flops is expected to store n-bit data. The register serves that purpose by using flip-flops as the building blocks. Not only that, the flip-flops within the register are arranged in such a way that the bits may be shifted into the register or shifted out of it one by one or all at a time. Therefore, the circuit is also referred to as *shift register*. When the bits are shifted one at a time in a serial fashion, beginning with either the MSB or the LSB, the process is called *serial shifting* and when the bits are shifted all at a time, the procedure is known as *parallel shifting*.

The flip-flops for constructing the register may be of JK, RS or D type and the bit handling capability can be enhanced simply by increasing the number of flip-flops. In general, a register composed of n flip-flops can store the numbers 0 through $2^n - 1$. The storing and shifting of numbers in a register can be made in the following four ways.

- Serial in and serial out
- Serial in and parallel out
- Parallel in and serial out
- Parallel in and parallel out

This is to clarify at the very beginning that the register is not part of the bulk memory of a digital system, such as a computer. This point will be discussed further in Chapters 16 and 17.

Figure 15.21(a) shows a simple logic circuit of shift register made of three JK flip-flops, each responding to negative clock transition. This register is capable of storing and shifting up to 3-bit numbers, ranging from 000 to 111. The MSB and LSB positions of the stored number are indicated in the diagram. The input data bit (0 or 1) is applied to the first JK flip-flop whose J and K inputs are separated by a NOT gate. The outputs of this flip-flop drive the inputs of the second flip-flop and so on. This circuit can perform *serial in* and *serial out* operations only. Actually this circuit is suitable for understanding the bit shifting process.

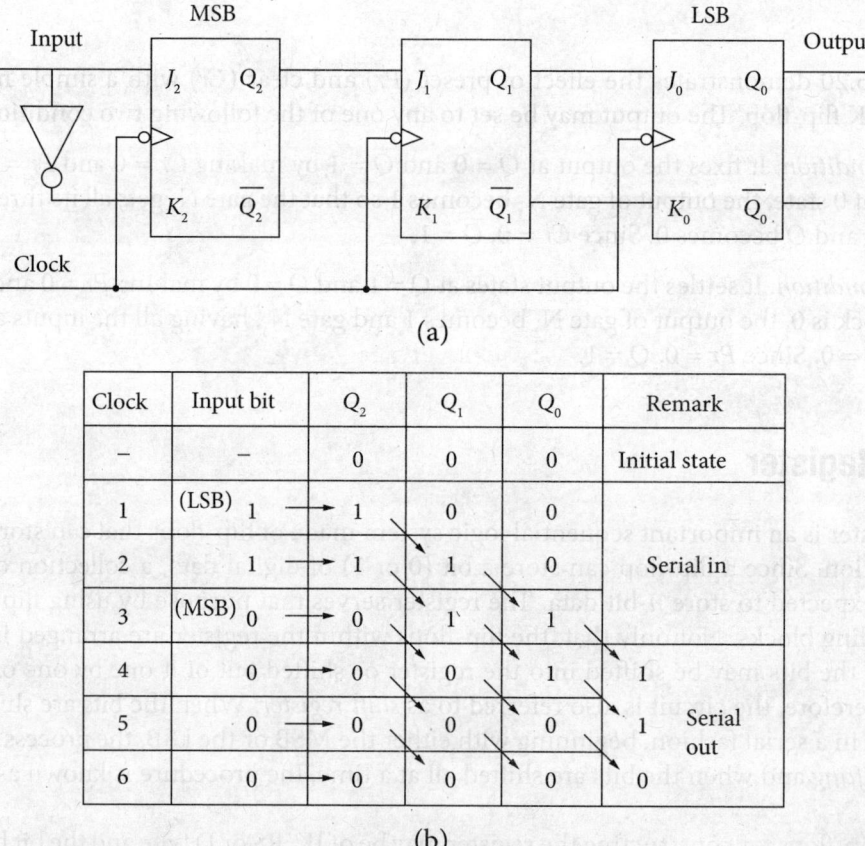

Figure 15.21 3-bit register for serial in and serial out operations constructed with JK flip-flops responding to negative clock transitions, (a) circuit diagram and (b) pictorial representation of shifting in and out the number 011 bit-by-bit serially

Let the three flip-flops of the register be at 0 state initially and the 3-bit number 011 is to be shifted into the register. The mechanism is represented pictorially in Figure 15.21(b). The shifting process starts with the LSB of the given number, 1 in the present case. It is applied to the input terminal. After the first clock, this 1 shifts into Q_2. The second clock transition makes $Q_1 = 1$, as if the 1 is shifted from Q_2 to Q_1. At the same time, the next higher input bit (another 1) applied to the input terminal enters Q_2. The third clock causes the LSB (0) to enter Q_2, at the same instant the 1 shifts from Q_2 to Q_1 and another 1 shifts from Q_1 to Q_0. Thus the input number 011 gets inserted to the register serially, the MSB–LSB positions being matched properly. The bits of the stored number can also be shifted out of the register one by one serially, as indicated in Figure 15.21(b), by applying 0 to the input at each clock pulse. The bits are obtained one by one, starting from the LSB, at the output Q_0.

The logic circuit of Figure 15.21(a) can interpret the basic working principle of a shift register. However, if one intends to achieve all the four serial and parallel shifting with the same circuit, this is not sufficient. More features must be incorporated to the circuit. The *parallel out* provision can be added to the circuit of Figure 15.21(a) very easily by connecting an output terminal to each of Q_2, Q_1 and Q_0. Nonetheless, Figure 15.22 depicts an independent register circuit capable of performing the serial in, serial out and parallel out operation. The main difference is instead of JK flip-flops, it employs three D flip-flops whose $\overline{Q}$ outputs simply remain unused. The action is self-explanatory. Have you noticed that the JK flip-flops in Figure 15.21(a) also were acting like D flip-flops?

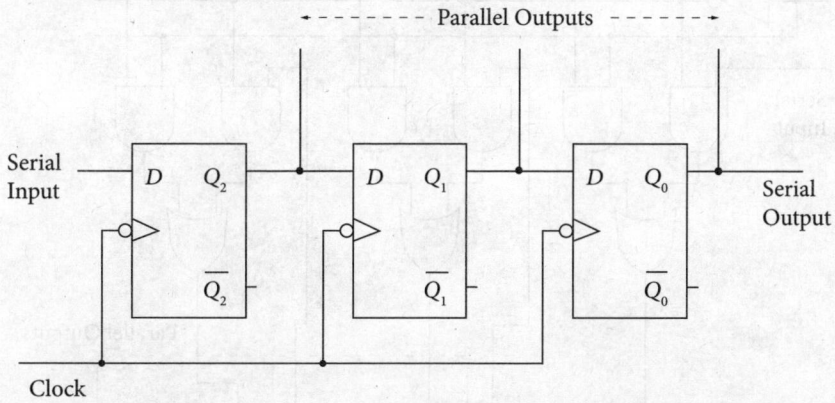

Figure 15.22 3-bit register for serial in, serial out and parallel out operations constructed with JK flip-flops responding to negative clock transitions

15.4.1 Register with Series and Parallel Shifting

Now we may be more ambitious and look for such a register that has the provisions for serial and parallel inputs and outputs and all possible combinations of those. Figure 15.23 depicts such an all-in-one register. It may look complicated at the first glance but is not actually so. Understanding the function of the 'two AND one OR' combination is important.

Serial In and Serial Out: When the Select terminal becomes 0, all the right-side AND gates get disabled and the associated parallel input bits are blocked. The NOT output being 1, the left-side AND gates are enabled. Therefore, the serial input can enter D_2 flip-flop through the OR gate, the output Q_2 can enter D_1 through OR gate and so on. The serial output can supply the output bits one by one. The clock operation is usual, similar to those in figures 15.21 and 15.22.

Serial In and Parallel Out: In the above operation, all the three outputs can be obtained simultaneously, as indicated in the diagram.

Parallel In and Series Out: If the Select terminal becomes 1, the NOT output turns 0 and all the left-side AND gates get disabled. Consequently, the serial operation is blocked. The right-side flip-flops are enabled and the associated parallel input bits can enter the flip-flops through the corresponding OR gates. Thus all the inputs can be inserted simultaneously in parallel and the serial output can supply the output bits one by one.

Parallel and Parallel Out: The serial output and the parallel outputs are independent of one another. So instead of getting one by one at the serial output, all the three outputs can be obtained simultaneously at the parallel output terminals.

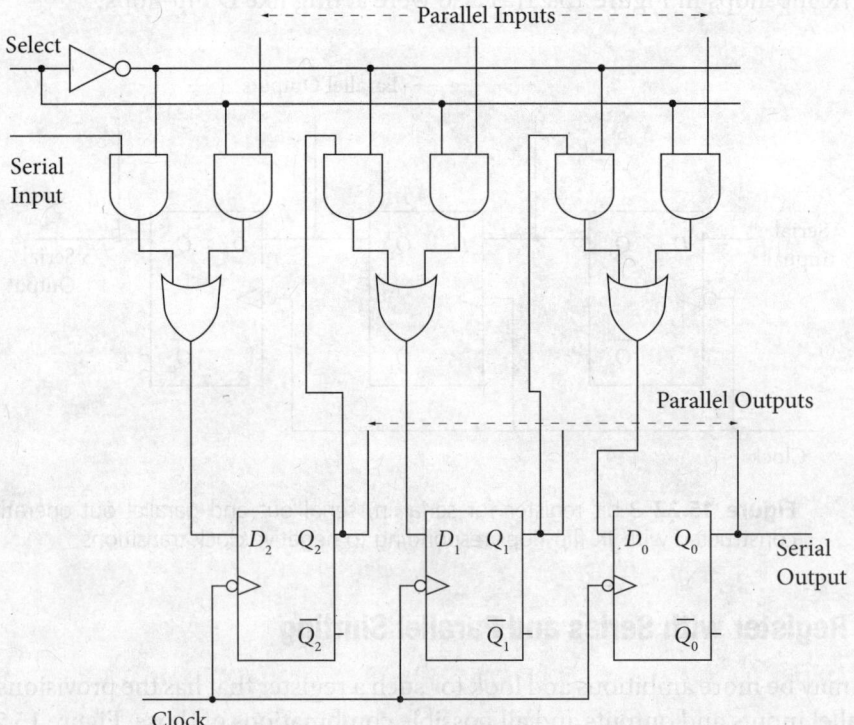

Figure 15.23 3-bit register with provisions for serial in, serial out, parallel in and parallel out operations

Example 15.6

The logic circuit of Figure 15.24 introduces another method of constructing register with all serial and parallel shifting provisions. It makes use of the preset and clear inputs mentioned earlier. Can you explain the circuit operation, if the number 1011 is shifted in and out?

Solution

The present circuit has four flip-flops, hence it can store and shift up to 4-bit numbers. Initially all the four Q-outputs are made 0 by using the clear terminals. Then any one of the following procedures can be executed.

Serial In Serial Out: The operation is similar to that of the circuit in Figure 15.21(a). Let the four Q-outputs in the present circuit be designated as Q_3, Q_2, Q_1 and Q_0, respectively, from MSB to LSB.

Serial In Parallel Out: The input is shifted in by the above process but all the Q-states are recorded simultaneously across the parallel output terminals.

The bit-by-bit shifting in and shifting out of the number 1011 in the above two cases is given in Table 15.3.

Table 15.3 Output states of the 4-bit register of Figure 15.24 for shifting in and out the number 1011 serially

Clock	Input bit	Q_3	Q_2	Q_1	Q_0		Remark
–	–	0	0	0	0		Initial states
1	1 (LSB)	1	0	0	0		
2	1	1	1	0	0		
3	0	0	1	1	0		Serial In
4	1 (MSB)	1	0	1	1		
5	0	0	1	0	1	1	
6	0	0	0	1	0	1	Serial Out
7	0	0	0	0	1	0	
8	0	0	0	0	0	1	

Parallel In Serial Out: Using the preset inputs, the bits of the given number are inserted to the corresponding flip-flops all at a time. The output is obtained bit-by-bit at the serial output terminal.

Parallel In Parallel Out: The four input bits are entered at a time through the preset inputs and the results are also observed simultaneously at the four parallel output terminals.

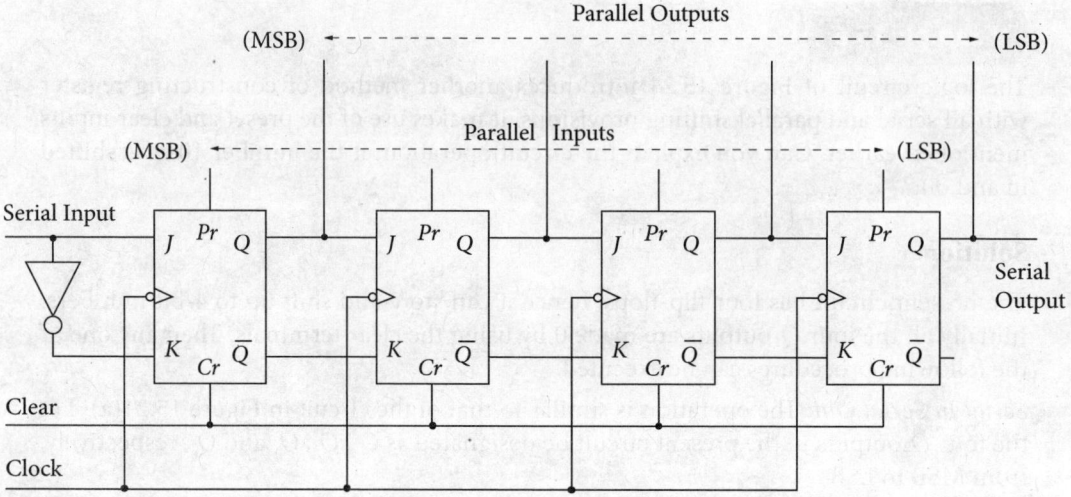

Figure 15.24 4-bit register using preset and clear for providing serial and parallel operations

15.4.2 Ring Counter

This is a modified register circuit that circulates a voltage pulse, representing logical 1, through a closed path in a sequential manner. The logic diagram of a 4-bit ring counter is sketched in Figure 15.25(a). This is similar to the 4-bit register of Figure 15.24 but the parallel inputs are absent and the serial output is fed back to the serial input. The output terminals of the four flip-flops are denoted by Q_3, Q_2, Q_1 and Q_0, respectively, to distinguish from one another. The circuit operates in the following steps.

(i) Initially only Q_3 is made 1 using the corresponding preset and all the other outputs are initialized as 0.

(ii) The timing diagrams for the clock pulse and the output states are sketched in Figure 15.25(b). The first negative clock transition makes $Q_3 = 0$ and $Q_2 = 1$ because before the clock, Q_0 fed back to the left-most flip-flop was 0 and Q_3 connected to the next flip-flop was 1. As if the 1 is shifted from Q_3 to Q_2.

(iii) Similarly, the second negative clock transition shifts the 1 from Q_2 to Q_1 and again a 0 is fed back to the left-most flip-flop keeping $Q_3 = 0$. This, in turn, keeps $Q_2 = 0$.

(iv) The third negative clock transition shifts the 1 from Q_1 to Q_0 and the fourth transition brings the 1 back to Q_3.

Thus a single 1 is retained in the circuit and is shifted around the closed ring path Q_3 to Q_0 and back, advancing one step at each clock pulse.

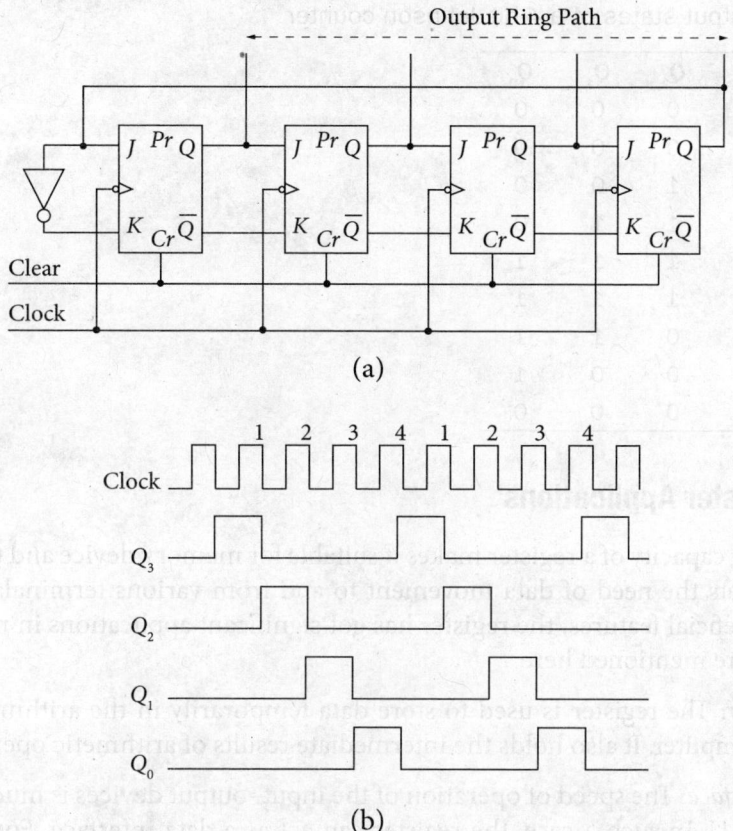

Figure 15.25 4-bit ring counter: (a) construction and (b) timing diagram

The ring counter circulates a voltage pulse in a definite sequence among specified terminals. Such a circuit is used for controlling events in computation and data transmission in a pre-determined sequence. There are other variations of logic circuits like the ring counter where the output is fed back to the input. For example, a fixed binary code may be stored in the register with feedback connection. On applying clock pulses, a fixed number code is generated sequentially. Such a circuit is called *sequence generator*.

15.4.3 Johnson Counter

This is similar to the ring counter but instead of the output of the last flip-flop in the ring counter, its complement ($\bar{Q}_0$) is returned to the input of the first flip-flop. This circuit produces a special sequence of bits. Assuming all the flip-flops are cleared initially, $\bar{Q}_0$ is left at 1 state and it turns the first flip-flop output to 1 state at the next clock. The subsequent clock pulses fill up all the flip-flops with 1s from left to right one by one and then fill up again with 0s one by one. Table 15.4 displays the states of a 4-bit Johnson counter at successive clock pulses. In general, a Johnson counter comprising n flip-flops produces $2n$ states.

Table 15.4 Output states of a 4-bit Johnson counter

Clock	Q_3	Q_2	Q_1	Q_0
0	0	0	0	0
1	1	0	0	0
2	1	1	0	0
3	1	1	1	0
4	1	1	1	1
5	0	1	1	1
6	0	0	1	1
7	0	0	0	1
8	0	0	0	0

15.4.4 Register Applications

The bit storing capacity of a register makes it suitable for memory device and the bit shifting capability meets the need of data movement to and from various terminals. By virtue of these two beneficial features, the register has got significant applications in many practical fields. A few are mentioned here.

In a Computer: The register is used to store data temporarily in the arithmetic logic unit (ALU) of a computer. It also holds the intermediate results of arithmetic operations there.

Acting as Interface: The speed of operation of the input–output devices is much slower than that of the ALU. In such a case, the register can act as a data interface. For instance, the register can accept the input data from the keyboard and transfer it to the ALU in proper time sequence. Similarly it can accept the output data from the ALU and disburse it to the monitor or the printer.

Data Conversion: The register can convert a set of parallel data input to serial sequence and vice versa.

Sequence Generator: A prescribed sequence of data can be generated or detected by the use of register. The ring counter is such an example.

15.5 Counters

The ring counter and similar circuits discussed above are termed as 'counter' because those can keep account of a specified sequence of bits. However, these do not count numbers in true sense. There is another family of digital circuits, known as counters that can truly count the number of bits.

The counter is a sequential logic circuit made of flip-flops that can keep account of the number of clock pulses applied to it. Since clock pulses occur at fixed time intervals, the counter can be used for measuring time and frequency also. The number of counts that the counter circuit is capable of is referred to as the *modulus* of the counter. It is the total number of output states through which the counter can progress. In general, a counter comprised of n flip-flops has 2^n output states and it can count 0 through $2^n - 1$. This is called a mod-2^n counter. For instance, a counter with four flip-flops has natural count of $2^4 = 16$, it can count from 0000 through 1111 and is called mod-16 counter. The modulus can also be modified to some intermediate number, as will be discussed later.

Counter circuits can be classified broadly in two categories. One of those is known as *asynchronous counter* or *ripple counter* where each flip-flop output drives the clock input of the next flip-flop. The other one is called *synchronous counter* or *parallel counter* where all the flip-flops are triggered simultaneously by the clock pulses.

15.5.1 Asynchronous Counter

As stated above, each flip-flop is triggered by the previous one, hence the flip-flops do not change states in exact synchronism with the applied clock pulses. This type of counter is also referred to as ripple counter because the response of the flip-flops ripples through one after another.

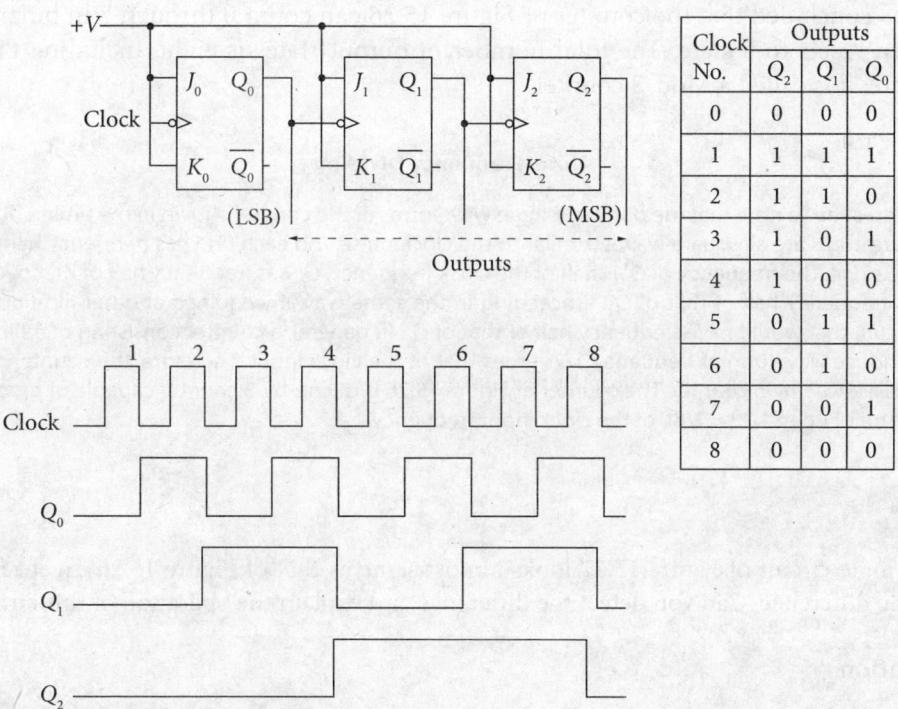

Figure 15.26 Mod-8 asynchronous counter logic diagram, truth table and timing diagram (clockwise)

Figure 15.26 presents an example of asynchronous counter. The JK flip-flops are made to act as T flip-flops by connecting the *J* and *K* inputs to the positive terminal (+V) of the power supply. The external clock pulse is applied only to the first flip-flop representing the LSB. The output (Q_0) of this flip-flop is used as the clock input to the next flip-flop and this sequence is continued up to the last flip-flop representing the MSB. The present counter circuit containing three flip-flops can count 000 through 111. The working of the circuit is outlined in the following steps.

(i) Before starting the clock pulse, $Q_2 = Q_1 = Q_0 = 0$ so that the total output is 000.
(ii) After the first clock pulse (negative transition), Q_0 becomes 1 because the first flip-flop toggles. The total output becomes 001.
(iii) After the second clock pulse, Q_0 becomes 0 because the first flip-flop toggles again. At the same time, Q_1 becomes 1 because Q_0, the clock to the second flip-flop undergoes negative transition from 1 to 0. The total output is now 010.
(iv) Proceeding in the same way up to the seventh pulse, the output represents 011 through 111, as compiled in the truth table and indicated by the timing diagram of Figure 15.26.
(v) After the seventh clock, all the output states are at 1 state. At the eighth clock transition from positive to negative, Q_0 turns 0, which makes $Q_1 = 0$, which in turn makes $Q_2 = 0$. The output returns to the initial state of 000.

It is thus concluded that the counter of Figure 15.26 can count 0 through 7 in binary form and then resets to 0 state. The total number of output states is eight, including the zero. Therefore, it is called a Mod-8 counter.

Note: Frequency Dividing

It is interesting to note that the output voltage waveforms of the counter shown in the timing diagram of Figure 15.26 are all square waves, similar to the clock pulse and each one has frequency half of the previous one. The frequency of Q_0 is half of the clock frequency, Q_1 has frequency half of that of Q_0 and Q_2 has frequency half of that of Q_1. Proceeding in the same way, if we joined another flip-flop Q_3 to the circuit, that would have frequency half of that of Q_2. In general, a counter consisting of *n* flip-flops can produce waveform of frequency $1/2^n$ times that of the clock input. Therefore, the counter can be used as a *divide-by-n* counter. The counter of Figure 15.26 is divide-by-3 counter capable of producing waveform of up to $1/2^3 = 1/8^{th}$ of the clock input frequency.

Example 15.7

The logic circuit of Figure 15.27 looks almost same as that of Figure 15.26 except for a subtle difference. Can you detect the difference and explain the operation of this circuit?

Solution

The only difference is that instead of the Q-outputs, the complements ($\overline{Q}$) of the flip-flops are used as the clock inputs to the subsequent flip-flops. This little change makes a big difference: the resultant circuit starts counting downward, 7 through 0 in descending order.

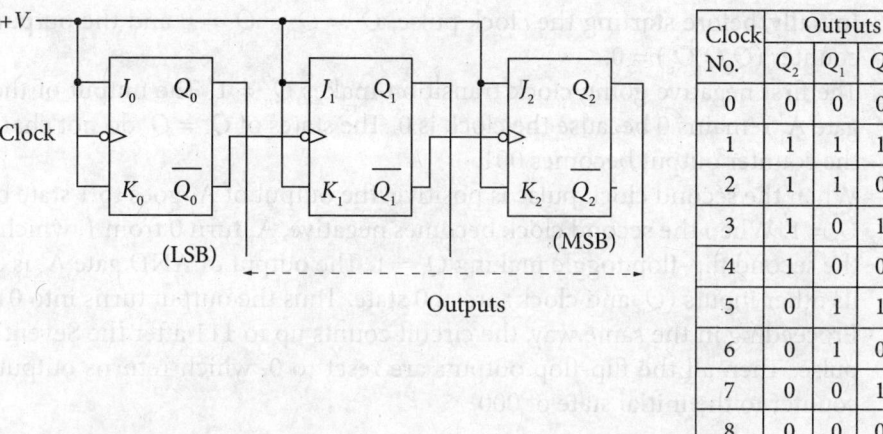

Figure 15.27 Mod-8 downward asynchronous counter logic diagram (left) and truth table (right)

Think a Bit: Does the asynchronous counter have any limitation?

We talked about asynchronous and synchronous counters. The two counters discussed so far in Figures 15.26 and 15.27 are obviously asynchronous counters. These may also be named *ripple counters* because the change of output states from one flip-flop to the subsequent one is transmitted like ripples. These two have performed well the job of counting, either in normal mode or in reverse mode, as necessary. Then why should we go for any other category of counter? One thing we have ignored so far, that is the propagation delay. Each flip-flop takes a finite time to change the output state in response to the input clock change. Since the output of a certain flip-flop drives the next one, the total time taken for changing the states of the whole counter circuit is the cumulative addition of the individual propagation delays of all the constituent flip-flops. This limits the speed of the asynchronous counter. If all the flip-flops can be triggered simultaneously, the time taken in transmitting the message of state change can be reduced to the propagation delay of a single flip-flop and the speed of operation can be enhanced. That is done with synchronous flip-flops. However, the circuit becomes more complicated, particularly for large number of flip-flops.

15.5.2 Synchronous Counter

This type of counter has all the flip-flops clocked at a time. Figure 15.28 demonstrates the procedure with a simple mod-8 synchronous composed of three JK flip-flops. Each one acts in toggle mode, similar to those in Figure 15.26. The truth table and the output timing diagrams are also the same as those of Figure 15.26 but the circuit construction and the operating mechanism are quite different. The input clock pulse is applied directly to the first flip-flop whose output (Q_0) represents the LSB. To the other flip-flops, the clock is applied through AND gates, along with the outputs of all the previous flip-flops. The circuit works in the following way.

(i) Initially, before starting the clock pulse, $Q_2 = Q_1 = Q_0 = 0$ and the output of the counter $(Q_2Q_1Q_0) = 0$.

(ii) The first negative going clock transition makes $Q_0 = 1$. The output of the AND gate A_1 remains 0 because the clock is 0. The states of $Q_2 = Q_1$ do not change and the counter output becomes 001.

(iii) When the second clock pulse is positive, the output of A_1 goes to 1 state because $Q_0 = 1$. When the second clock becomes negative, A_1 turn 0 from 1, which makes the second flip-flop toggle making $Q_1 = 1$. The output of AND gate A_2 is 0, since its other inputs (Q_0 and clock) are at 0 state. Thus the output turns into 010.

(iv) Proceeding in the same way, the circuit counts up to 111 after the Seventh clock pulse. Then all the flip-flop outputs are reset to 0, which returns output of the counter to the initial state of 000.

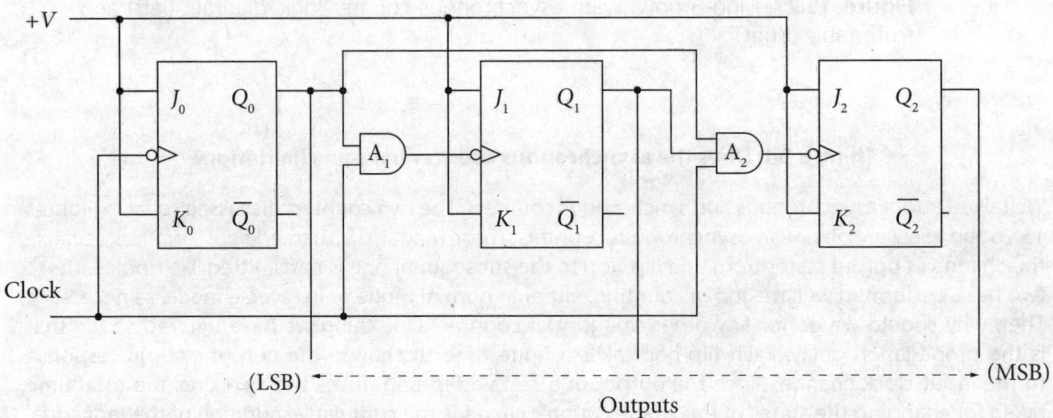

Figure 15.28 Mod-8 synchronous counter

15.6 Changing Counter Modulus

It is stated earlier that the natural count of a counter with n flip-flops is 0 through $2^n - 1$. The counter may be asynchronous or synchronous. The modulus of a counter can be modified so that it can count up to an intermediate number (N) such that $0 < N < 2^n - 1$. This is designated as *mod-N counter* or *divide-by-N counter*. The counter modulus can be modified by two different ways.

- A smaller modulus is obtained by skipping several states of a larger modulus counter through suitable feedback connections.
- A larger modulus is achieved by cascading two or more counters of smaller modulus.

Examples of both the above types of modulus change are presented here.

15.6.1 Mod-3 Counter

The circuit can count 00, 01 and 10 and then resets to 00. Figure 15.29 depicts the logic diagram of a mod-3 counter comprised of two JK flip-flops. The K-inputs of both flip-flops are made 1 permanently by connecting to the positive terminal $(+V)$ of the power supply. The different steps of counting are summarized below. The word 'clock' refers to negative clock transitions in the present cases.

(i) Initially, $Q_1 = 0$ and $Q_0 = 0$ so that the counter output $Q_1 Q_0 = 00$.
(ii) At this time, $J_0 (= \bar{Q}_1) = 1$ and $K_0 = 1$.
(iii) At the first clock, Q_1 remains unchanged but Q_0 toggles to 1, since $J_0 = K_0 = 1$. The counter output turns into 01.
(iv) This time, $J_0 (= \bar{Q}_1) = 1$ and $K_0 = 1$; $J_1 (= Q_0) = 1$ and $K_1 = 1$.
(v) The second clock makes both Q_1 and Q_0 to toggle, since both flip-flops now have $J = K = 1$. The toggling make $Q_1 = 1$ and $Q_0 = 0$ so that the output turns out to be 10.
(vi) Now $J_0 (= \bar{Q}_1) = 0$ and $K_0 = 1$; $J_1 (= Q_0) = 0$ and $K_1 = 1$. Both flip-flops thus have $J = 0, K = 1$.
(vii) The third clock resets Q_1 to 0 and Q_0 to 0 (already it was so), hence the output becomes 00, the initial state.

The mod-3 counter discussed above can be considered as a divide-by-3 counter because the output waveform at Q_1 is of frequency one-third of the input (clock) frequency.

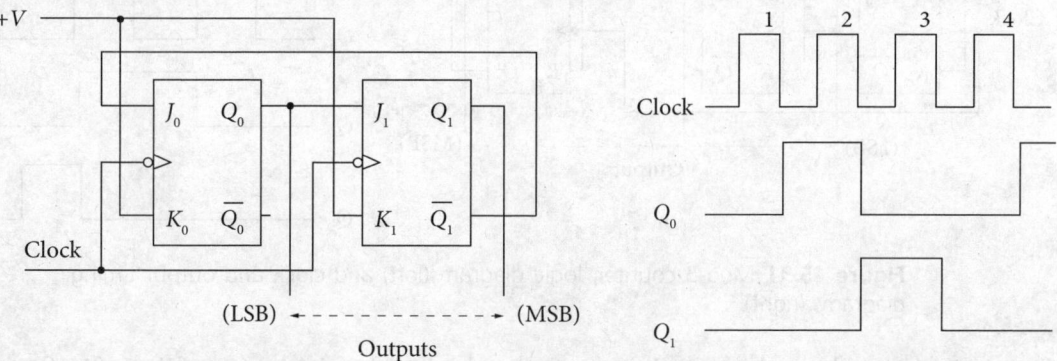

Figure 15.29 Mod-3 counter, logic diagram (left) and clock and output timing diagrams (right)

15.6.2 Mod-6 Counter

Figure 15.30 illustrates the process of constructing a mod-6 counter as $3 \times 2 = 6$ using the mod-3 counter of Figure 15.29 as a building block (shaded) and adding another flip-flop to it. The detailed circuit within the shaded block is repeated for the sake of easy understanding. The mod-3 counter is built. The output Q_1 of the mod-3 counter is used as the clock to the added flip-flop. The mod-6 flip-flop counts 000 through 110.

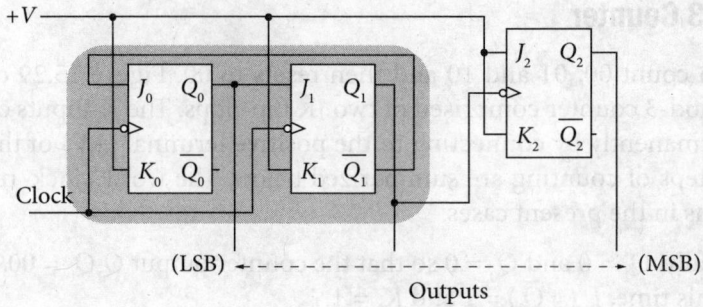

Figure 15.30 Construction of mod-6 counter using mod-3 counter as a building block

15.6.3 Mod-5 Counter

Binary five (101) is a 3-bit number, hence a mod-5 counter requires three flip-flops that has a natural count of 8. Additional connections are made in such a way that three counts are skipped over and it counts from 000 up to 100 through a binary sequence.

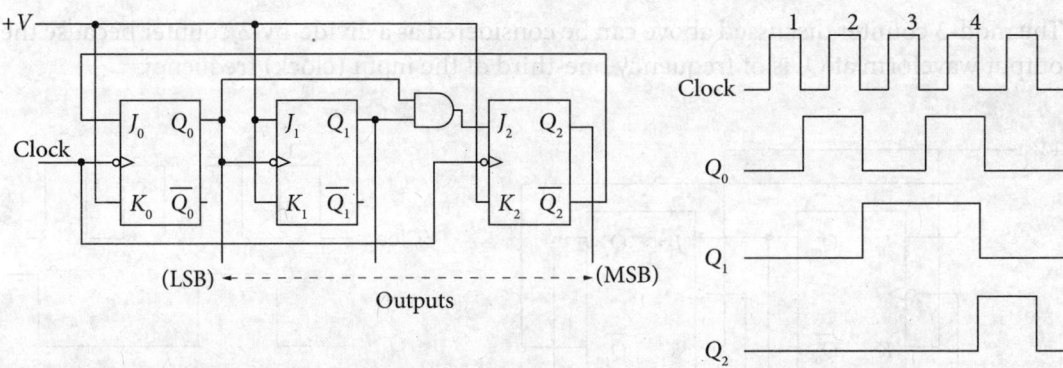

Figure 15.31 Mod-5 counter, logic diagram (left) and clock and output timing diagrams (right)

Figure 15.31 introduces the logic diagram for mod-5 flip-flop. It consists of three flip-flops, each having the K-input at 1 state permanently and each one changing state at negative clock transition. The complement of the MSB ($\overline{Q}_2$) is connected to J_0. The following counting sequence is followed.

Initial State: $Q_2 = Q_1 = Q_0 = 0$ and the counter output ($Q_2Q_1Q_0$) = 0. Since $\overline{Q}_2 = 1$, $J_0 = 1$.

First Clock: Q_0 having both J and K inputs at 1 state toggles to 1. Q_1 remains unchanged because its clock is inactive. Q_2 remains unchanged at 0 state because the AND output is 0, hence $J_2 = 0$. The counter output is 001.

Second Clock: Q_0 toggles again to 0. This makes Q_1 to toggle to 1. Q_2 remains at 0 state because the AND output is 0, hence $J_2 = 0$. The counter output is 010.

Third Clock: Q_0 toggles to 1 and Q_1 is unchanged (clock inactive) at 1 state. The AND output, hence J_2 is 1 and Q_2 waits at previous 0 state for the next clock. The counter output is 011.

Fourth Clock: Q_0 toggles from 1 to 0, which makes Q_1 also toggle to 0. Q_2 toggles to 1 for the first time. The counter output becomes 100.

After the fourth, when the next (fifth) negative clock transition takes place, Q_0 is reset to 0 (already it was so), Q_1 remains unchanged at 0 and Q_2 is reset to 0. As a whole, the output of the counter turns out to be 000. Thus it returns to the initial state after counting the five states sequentially.

15.6.4 Decade (Mod-10) Counter

Figure 15.32(a) illustrates the logic circuit for a decade counter that counts from 0000 up to 1001, i.e. from decimal 0 through 9 and then resets to the initial state. Following the method of cascading two counters, the decade counter has been constructed by connecting a flip-flop to a mod-5 counter in 2×5 configuration. The output state of the counter represented by $Q_3Q_2Q_1Q_0$ is zero (0000) initially and the circuit executes the following operations with negative clock transitions.

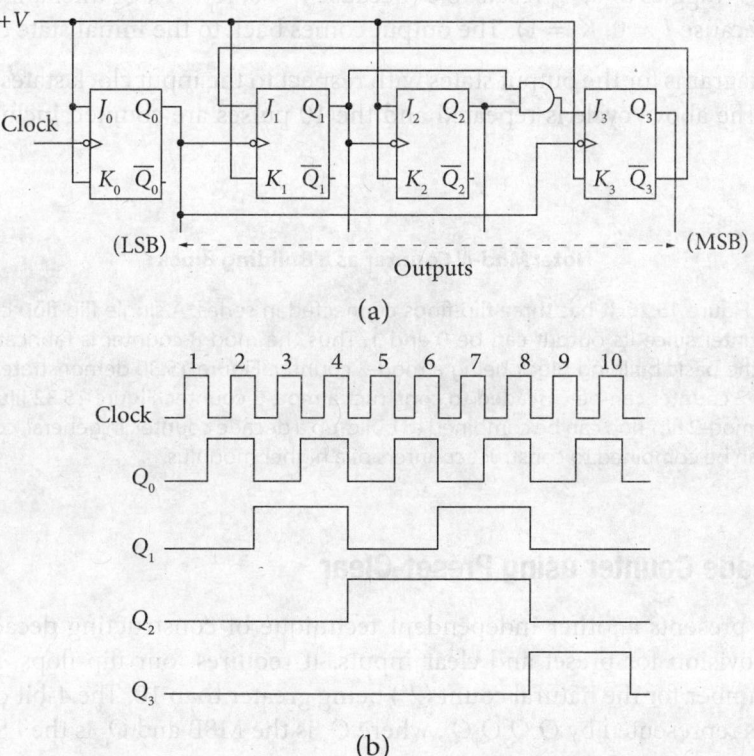

Figure 15.32 Decade or mod-10 counter, (a) logic diagram and (b) clock and output timing diagrams

First Clock: Q_0 toggles to 1. No change of the other flip-flops. The counter output becomes 0001.

Second Clock: Q_0 toggles to 0, which makes Q_1 to toggle to 1 and the others remain unchanged. The output is 0010.

Third Clock: Q_0 toggles to 1, others remain unchanged and the output is 0011.

Fourth Clock: Q_0 toggles to 0, which makes Q_1 to toggle to 0. This, in turn toggles Q_2 so that the output becomes 0100.

Fifth Clock: Q_0 toggles to 1, others remain unchanged and the output is 0101.

Sixth Clock: Q_0 toggles to 0, which makes Q_1 to toggle to 0. The output becomes 0110.

Seventh Clock: Q_0 toggles to 1, others remain unchanged and the output is 0111.

Eighth Clock: Q_0 toggles to 0, hence Q_1 to toggles to 0, hence Q_2 toggles to 0, which makes Q_3 to toggle to 1 because the AND output (hence J_3) was at 1 state 1 and K_3 was already 1. Now the output turns out to be 1000.

Ninth Clock: Only Q_0 toggles to 1, the output is 1001.

Tenth Clock: Q_0 toggles to 0, Q_1 resets to 0 (because $J_1 = 0$, $K_1 = 1$), Q_2 unchanged at 0 and Q_3 resets to 0 (because $J_3 = 0$, $K_3 = 1$). The output comes back to the initial state of 0000.

The timing diagrams for the output states with respect to the input clock states are displayed in 15.32(b). The above cycle is repeated and the 10 pulses are counted including the zero state.

> **Note: Mod-*N* Counter as a Building Block**
>
> Let us revisit Figure 15.26. It has three flip-flops connected in series. A single flip-flop can be treated as mod-2 counter since its output can be 0 and 1. Thus the mod-8 counter is fabricated by 2×2×2 connection, the basic building block being a mod-2 counter. Figure 15.30 demonstrates that a mod-3 and a mod-2 counter can be cascaded to construct a mod-6 counter. Figure 15.32 illustrates that a mod-5 and a mod-2 flip-flop can be combined to build-up a decade counter. In general, counters of any modulus *N* can be combined to construct counters of a higher modulus.

15.6.5 Decade Counter using Preset-Clear

Figure 15.33 presents another independent technique of constructing decade counter. It needs the provision for preset and clear inputs. It requires four flip-flops, since 4 is the minimum number for the natural count (2^4) being greater than 10. The 4-bit output state of the counter is represented by $Q_3Q_2Q_1Q_0$, where Q_3 is the MSB and Q_0 is the LSB.

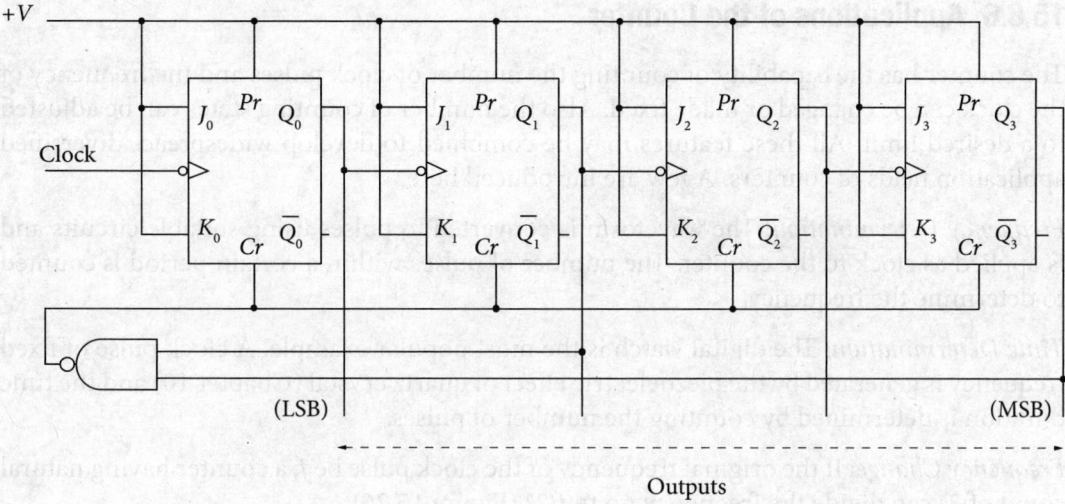

Figure 15.33 Decade (mod-10) counter constructed with JK flip-flops having preset-clear arrangements

The target is to stop and reset after counting 0 through 9. The trick lies in the connection of the preset and clear inputs. All the preset inputs are kept fixed at 1 state. All the clear inputs are the connected to the output of a NAND gate whose inputs are Q_3 and Q_1. It is interesting to verify that from 0000 up to 1001, these two outputs never become both 1 at a time. It is only for the first time when the state 1010 (= decimal 10) appears, $Q_3 = Q_1 = 1$. At this moment, the NAND output becomes 0, hence all clear inputs become 0 so that the next clock pulse reset all four flip-flop outputs to 0 state. Before this, the NAND output had been 1 all the way because one or more inputs used to be at 0 state. Thus the system starts counting from 0000, proceeds up to 1001 (= decimal 9) and then returns to the initial state. The timing diagrams are similar to those of the decade counter in Figure 15.32.

> **Think a Bit: Can any modulus be achieved with preset-clear?**
>
> Yes, not only the decade counter, any modulus of counter can be accomplished by using the preset-clear provision. The general technique for fabricating a mod-N counter is to assemble the counter with n flip-flops (each with preset and clear) such that $2^n - 1 > N$. Only those flip-flop output states are selected for the feedback through NAND gate, which assume 1 state all at a time just after counting the required modulus. For instance, the mod-s counter can be constructed with three flip-flops, the output being $Q_2 Q_1 Q_0$. Both Q_2 and Q_0 become 1 concurrently for the first time when the state 101 takes place. So Q_2 and Q_0 are used as the inputs of a NAND gate whose outputs are connected to all the clear inputs. If, in the circuit of Figure 15.33, we used a 3-input NAND gate with Q_3, Q_1 and Q_0 as the inputs, the counter would be able to count up to 1010 and reset with the next state 1011.

15.6.6 Applications of the Counter

The counter has the capability of counting the number of clock pulses and the frequency of the clock can be changed or made fixed. Also the number of counting states can be adjusted to a desired limit. All these features may be combined to develop widespread, diversified application fields of counters. A few are introduced here.

Frequency Determination: The waveform is converted to pulses using suitable circuits and is applied as clock to the counter. The number of pulses within a certain period is counted to determine the frequency.

Time Determination: The digital watch is the most popular example. A clock pulse of fixed frequency is generated by the piezoelectric effect of quartz crystal (Chapter 10) and the time duration is determined by counting the number of pulses.

Frequency Change: If the original frequency of the clock pulse be f, a counter having natural count of 2^n can divide the frequency up to $f/2^n$ (Figure 15.26).

Distance Measurement: Radar works on this principle. An electromagnetic signal of fixed frequency is transmitted in air, which gets reflected from a distant object. Counting the number of pulses, the time elapsed is determined and knowing the speed of electromagnetic wave the distance is determined.

Object Counting: The number of finished packets in industries, the number of pills put into a vial, the number of currency notes or similar objects can be counted very rapidly and accurately. The working principle is to focus a light source, generally laser, on a photosensitive device and to move the objects one by one across the light path. Each time the light is obstructed, the photosensitive device generates a voltage pulse that is counted.

Analog to Digital Conversion: The converter circuits use counters, as will be discussed in the next chapter.

Program Counter: The computer instructions in its central processing unit (CPU) are executed in a proper sequence. A counter circuit keeps account of the number of instructions loaded from the memory and executed.

Multiple Choice-Type Questions and Answers

15.1 The astable multivibrator of Figure 15.2(a) has $R_A = 2$ kΩ and $R_B = 1$ kΩ. The duration of high output state is .... the duration of low output state.
    (a) three times   (b) one-third of
    (c) equal to   (d) less than

15.2 The pulse frequency of a 555 monostable multivibrator depends on the .... of the negative trigger pulse.
    (a) frequency   (b) width
    (c) width and frequency   (d) frequency and amplitude

15.3 The pulse width of a 555 monostable multivibrator depends on the ....
 (a) the duration of the trigger pulse
 (b) the external C and R values
 (c) the frequency of the trigger pulse
 (d) the supply voltage value

15.4 The forbidden state of $R_n = 1$ and $S_n = 1$ is replaced by .... for $J_n = K_n = 1$ in the JK flip-flop.
 (a) $Q_n = \overline{Q}_{n+1}$
 (b) $Q_{n+1} = \overline{Q}_n$
 (c) $\overline{Q}_n = \overline{Q}_{n+1}$
 (d) $Q_{n+1} = Q_n$

15.5 The D flip-flop can be made of .... whereas the T flip-flop can be made of ....
 (a) JK flip-flop only, both RS and JK flip-flops
 (b) both RS and JK flip-flops, RS flip-flop only
 (c) both RS and JK flip-flops, JK flip-flop only
 (d) RS flip-flop only, both RS and JK flip-flops

15.6 In an edge-triggered RS flip-flop, the R and S inputs must be unchanged during the ....
 (a) short period of transition of the clock
 (b) high state of the clock
 (c) low state of the clock
 (d) full cycle of the clock

15.7 In the above case, R and S are allowed to change state at ....
 (a) the time when the clock is high
 (b) the time when the clock is low
 (c) the time when the clock transition is made
 (d) any time when the clock is either high or low

15.8 In a T flip-flop, the output is .... for $J = K = 0$ and the output is .... for $J = K = 1$.
 (a) $Q_{n+1} = Q_n$, $Q_{n+1} = \overline{Q}_n$
 (b) $Q_{n+1} = \overline{Q}_n$, $Q_{n+1} = Q_n$
 (c) $\overline{Q}_{n+1} = Q_n$, $Q_{n+1} = \overline{Q}_n$
 (d) $Q_{n+1} = \overline{Q}_n$, $\overline{Q}_{n+1} = Q_n$

15.9 A shift register made of RS flip-flops has the noninverting output (Q) of one flip-flop connected to the .... of the next flip-flop and the inverting output ($\overline{Q}$) connected to the .... of the next flip-flop.
 (a) S input, clock
 (b) S input, R input
 (c) clock, R input
 (d) R input, S input

15.10 The flip-flops in the above circuit effectively act like a .... flip-flop.
 (a) JK
 (b) RS
 (c) D
 (d) T

15.11 A counter capable of counting 0 through 127 in binary form requires at least .... flip-flops.
 (a) 8
 (b) 12
 (c) 5
 (d) 7

15.12 The MSB of a mod-16 counter produces a square wave of frequency 1 kHz and the clock pulse is a symmetric square wave. The clock frequency is ....
 (a) 4 kHz
 (b) 8 kHz
 (c) 16 kH
 (d) 32 kHz

15.13 At least .... flip-flops are needed to construct a mod-9 counter.
 (a) 10
 (b) 8
 (c) 9
 (d) (4)

15.14 The statement, 'A single flip-flop can provide with a mod-2 counter' is ....
(a) true (b) false
(c) ambiguous (d) dependent on the clock

15.15 Using four flip-flops and other accessories, one can construct a counter of any modulus between ...., both inclusive.
(a) 15 and 1 (b) 16 and 2 (c) 15 and 2 (d) 16 and 1

15.16 If the clock frequency be 1 MHz, the time taken to shift an 8-bit number in to an 8-bit register would be ....
(a) 8 μs (b) 1 μs (c) 16 μs (d) 0.5 μs

15.17 The total number of clock pulses required to serially enter a byte of data into an 8-bit register and again shift it out serially is written in hexadecimal as ....
(a) 08 (b) 0C (c) 10 (d) 11

15.18 The binary number 10110101 is serially shifted into an 8-bit parallel-out register that has an initial content of 11100100. The output condition after four clock pulses will be .... in binary and .... in hexadecimal.
(a) 01011110, 5E (b) 10110101, B5
(c) 01011110, 5F (d) 01001110, 4E

15.19 Each flip-flop of a mod-16 ripple counter has a propagation delay of 10 ns. The counter will take .... to change state from 1111 to 0000. If the above counter was made synchronous, the time taken would be ....
(a) 10 ns, 1 ns (b) 40 ns, 10 ns (c) 16 ns, 8 ns (d) 10 ns, 10 ns

15.20 A counter with 4 flip-flops is executing its natural counts. If it is initially zero, the count it will show after 70 clock pulses is ....
(a) 16 (b) 15 (c) 6 (d) uncertain

Answers

10.1 (a)   10.2 (d)   10.3 (b)   10.4 (b)
10.5 (c)   10.6 (a)   10.7 (d)   10.8 (a)
10.9 (b)   10.10 (c) 10.11 (d) 10.12 (c)
10.13 (d) 10.14 (a) 10.15 (b) 10.16 (a)
10.17 (c) 10.18 (a) 10.19 (b) 10.20 (c)

Reasoning-Type Questions and Answers

15.1 Is there any difference between a timer and a clock generator?

Ans. Both the circuits serve as the timing unit of a digital system and produce square or rectangular voltage pulses. However, the timer is a more generalized device. Conventionally the clock generator is a relaxation oscillator that generates a steady rectangular waveform (symmetric or asymmetric) and does not have any stable output state. The astable multivibrator is of this category. The timer includes those circuits also (such as, the monostable multivibrator), which have one stable and one quasi-stable output state and can produce voltage pulses (generally asymmetric).

15.2 What is the physical significance of duty cycle?

Ans. The duty cycle is defined as the percentage of the on-time of a rectangular wave with respect to its total time period. The power delivered to a load can be regulated by varying the duty cycle of a voltage waveform as input. Higher duty cycle corresponds to greater average power delivered to the load.

15.3 Justify the statement, "A flip-flop is a latch plus something".

Ans. A latch is a bistable electronic circuit, either analog or digital, that remains stable at any one of two possible states. It has analogy with the open or closed condition of a door latch. A bistable multivibrator made of transistors is an analog latch that can hold the high or low state of the output. A flip-flop, though ultimately made of transistors, is realized as a digital circuit. This is a digital latch that can hold logical 0 or 1 state. In addition, the flip-flop can execute three important operations.

(i) It can retain a 0 or 1, thus having a 1-bit memory.

(ii) The state change can be suitably synchronized by applying a clock pulse.

(iii) The output can be made fixed at a pre-determined state using preset and clear provisions.

15.4 Can two 3-input NOR gates be utilized to fabricate an RS flip-flop?

Ans. Yes, two inputs of each flip-flop can be tied together so that it can act as 2-input NOR gate. Then the circuit of Figure 15.5(a) can be used.

15.5 Does the D flip-flop possess any memory?

Ans. The output of a D flip-flop can be withheld for a desired period of time with the use of clock. So long as the clock is zero, the output does not change on changing the input. In this sense, the D flip-flop has some property similar to memory. However, the memory capability exhibited by RS and JK flip-flops is different from the above. Even in absence of any external clock, the flip-flop can retain the previous output state for $R = S = 0$ or $J = K = 0$ condition. Thus the memory is an inherent property of RS and JK flip-flops while that in D flip-flop is to be incorporated by using external clock pulse.

15.6 Can one state that a latch is a pulse-triggered flip-flop?

Ans. The flip-flop has several properties, as discussed in connection with Q. 15.3 above and the latch is only a part of these activities. A latch can be considered as a limited form of flip-flop that can stay at either of the two stable states. The latch is sensitive to the level of the input. Therefore, under the action of a clock, it can change state only when the clock attains a high or a low level. The state is retained until the clock changes the state. This is unlike an edge-triggered flip-flop that can change state with the positive or negative transition of the clock pulse. The latch requires the full change of clock pulse.

15.7 What is the condition of the largest time delay to change state in a mod-8 asynchronous counter?

Ans. The largest time is taken when all the three flip-flops change state. Two such cases occur, 011 to 100 and 111 to 000. The total delay is the summation of the propagation delay of the individual flip-flops.

15.8 Is the register a sort of memory?

Ans. The register, of course, stores binary data temporarily in limited scale. The memory refers to a long-term storage of large amount of digital information. The use of these two devices are different.

Registers are mainly used as parts of the CPU of a computer where the intermediate results of arithmetic and logic operations, the accounts of the instructions executed and the related information are stored. Registers are also used as buffers between the CPU and the input–output devices. Register contents change continuously with system operations. The memory is built as a separate unit meant exclusive for storage of large amount of binary information and there exists a provision for methodical data exchange with the CPU.

The materials are also different. The register is essentially made of semiconductor devices whereas the secondary memory or the auxiliary storage can be made of magnetic material or optical devices.

Solved Numerical Problems

15.1 The 555 astable multivibrator of Figure 15.2(a) has oscillation frequency of 10.93 kHz. If the capacitor (C) be of 0.022 µF and the duty cycle be 75%, determine the values of the resistors R_A and R_B.

Soln. The given duty cycle is D = 0.75

Using Equation (15.5),

$$\frac{R_A}{R_B} = \frac{2D-1}{1-D} = \frac{2 \times 0.75 - 1}{1 - 0.75} = 2$$
$$\Rightarrow R_A = 2R_B$$

The time period of oscillation is $1/(10.93 \times 10^{-3})$ = 0.0915 ms

Using Equation (15.4),

$$0.0915 \times 10^{-3} = 0.693 \times 0.022 \times 10^{-6} \times 4R_B$$
$$\Rightarrow R_B = 1.5 \text{ k}\Omega \text{ and } R_A = 3 \text{ k}\Omega$$

15.2 A digital system combining 10 identical flip-flops is operated by +5 V source. If each flip-flop draws 50 mA of current, what minimum current the source should provide?

Soln. The total power requirement is (5 V) × (50 mA) × 10 = 2.5 W

Therefore, the minimum current requirement is (2.5 W)/(5 V) = 0.5 A

15.3 The available devices are decade counters, mod-5 counters and single flip-flops; each category in abundance. And there is a source of 10 MHz clock pulse. How can a designer derive voltage pulses of the following frequencies: (i) 2.5 MHz, (ii) 62.5 kHz and (iii) 40 kHz?

Soln. Two flip-flops cascaded will produce (10/2) ÷ 2 = 2.5 MHz

A decade counter and four flip-flops cascaded make the following change in output frequency.

$$[(10 \text{ MHz}) \div 10] \div 1/2^4 = 62.5 \text{ kHz}$$

Two mod-5 counters and one decade counter cascaded make the following change in frequency.

$$[\{(10 \text{ MHz}) \div 5\} \div 5] \div 10 = 40 \text{ kHz}$$

15.4 Four T flip-flops are cascaded. Determine (i) the number of input pulses required to cause one output pulse and (ii) the final output frequency when the input clock frequency is 512 Hz.

Soln. Since each T flip-flop divides the signal frequency by 2, four T flip-flops in cascaded form divides the input frequency by $2^4 = 16$.

Therefore, (i) 16 input pulses are required to produce one output pulse and the final output frequency will be 512/16 = 32 Hz.

15.5 If each JK flip-flop has a propagation delay of 10 ns, what is the largest possible modulus of a counter built with such JK flip-flops operating at a frequency of 10 MHz?

Soln. Let n be the maximum number of flip-flops allowed.

$$1/[n \times (10 \text{ ns})] = 10 \text{ MHz}$$

Hence, $n = 10$ and the modulus is $2^{10} = 1024$

15.6 Write down the truth table and draw the state diagram for the logic circuit given in Figure. 15.34.

Soln. The D input is transferred to the output (Y) at each positive-going clock transition. The state diagram is also given in Figure 15.34 and the truth table is given below.

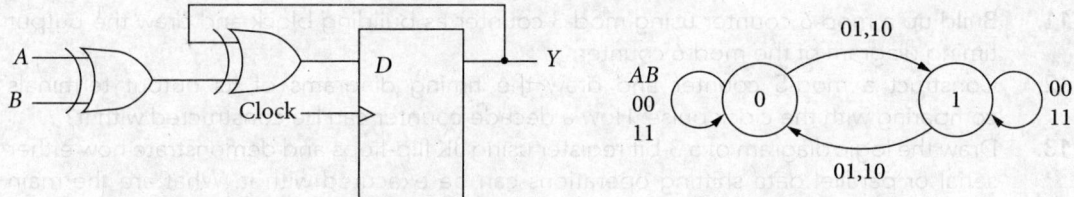

Figure 15.34

Inputs		Output (Y) states	
A	B	Present	Next
0	0	0	0
0	1	0	1
1	0	0	1
1	1	0	0
0	0	1	1
0	1	1	0
1	0	1	0
1	1	1	1

Exercise

Subjective Questions

1. Distinguish between clock and timer. What are the major clock parameters?
2. Draw the circuit diagram of an astable multivibrator using IC 555 and derive an expression for the time period of the waveform generated by this multivibrator. Represent the waveform graphically.
3. Draw the circuit diagram of a monostable multivibrator constructed with IC 555 and sketch the voltage waveform generated by it. Deduce the expression for the time period of this waveform.
4. Make a distinction between a latch and a flip-flop. Explain how the flip-flop exhibits a 1-bit memory capacity.
5. Convert a given RS flip-flop to JK flip-flop and explain how the JK flip-flop avoids the forbidden state of the RS flip-flop.
6. What is propagation delay? How does it impose a limitation to a digital circuit?
7. Explain the 'edge triggering' process related to a clock. Briefly outline how one can produce positive and negative edge-triggered circuits.
8. Derive the characteristic equations for (i) RS and (ii) JK flip-flop and draw the state diagrams.
9. Construct the logic diagram of a ripple counter circuit using JK flip-flops capable of counting 0 through 15.
10. Explain how a counter circuit can reduce the frequency of a square pulse and up to what extent. What are the major applications of the counter?
11. Build up a mod-6 counter using mod-3 counter as building block and draw the output timing diagram of the mod-6 counter.
12. Construct a mod-5 counter and draw the timing diagrams of its output terminals comparing with the clock pulse. How a decade counter can be constructed with it?
13. Draw the logic diagram of a 3-bit register using JK flip-flops and demonstrate how either serial or parallel data shifting operations can be executed with it. What are the main applications of a shift register?
14. Draw the logic diagram of (i) a Johnson counter and (ii) a ring counter and state about the applications of these circuits.

Conceptual Test

1. A sinusoidal input signal is applied to a diode half-wave rectifier. Does the duty cycle of the rectifier output have any significance?
2. The flip-flops described in this chapter as composed of either NAND or NOR gates; both being universal. Is the use of universal gate compulsory to the fabrication of flip-flops?
3. May the flip-flop be stated is a mod-2 counter? Justify your decision.
4. How long will it take to shift an 8-bit number serially, if the clock period is 1 μs?
5. How many flip-flops are needed to construct a shift register capable of storing up to decimal 33?
6. What can be the maximum count of a counter constructed of six flip-flops expressed (i) in decimal, (ii) in hexadecimal?
7. An 8-bit counter starts with all zero condition. What will be the output states after 520 clock pulses?

Numerical/Logical Problems

1. Figure 15.35 shows the $\bar{S}$ and $\bar{R}$ input states of an RS flip-flop at different instants of time. Write down the corresponding Q output states.

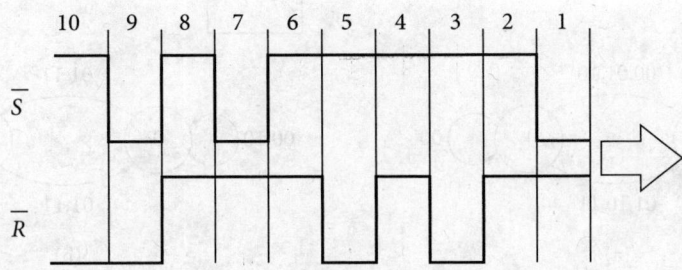

Figure 15.35

2. An 8-bit register has initial content of $E4_H$. The number $B5_H$ is being inserted serially, starting from the LSB. What will be the register content, expressed in hexadecimal after four clock pulses?

3. The initial contents of the 4-bit serial in parallel out, right shift register shown in Figure 15.36 is 0110. What will be the contents of the shift register after applying three clock pulses?

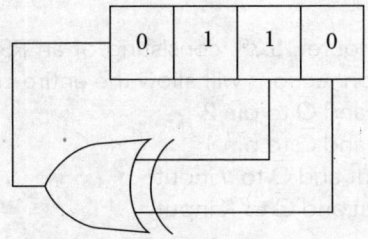

Figure 15.36

4. A counter can count from 0 through 1023. If the input clock frequency be 2 MHz, what will be the frequency of the output of the last flip-flop of the counter?

5. Determine the largest possible modulus of asynchronous counter constructed with JK flip-flops, each having the propagation delay of 10 ns such that the counter can respond to clock pulses up to 10 MHz.

6. Suggest a method of fabricating a divide-by-60 block for a digital clock with an input of 60-Hz square wave.

7. What is the maximum count of a 5-bit counter expressed in decimal number and how many JK flip-flops are required to fabricate the circuit? Can you convert it to a mod-12 counter? If yes, just mention the process.

8. Follow the logic circuit given at the top of Figure 15.37. Which one, out of the given four options at the bottom of the figure is the state diagram corresponding to the circuit?

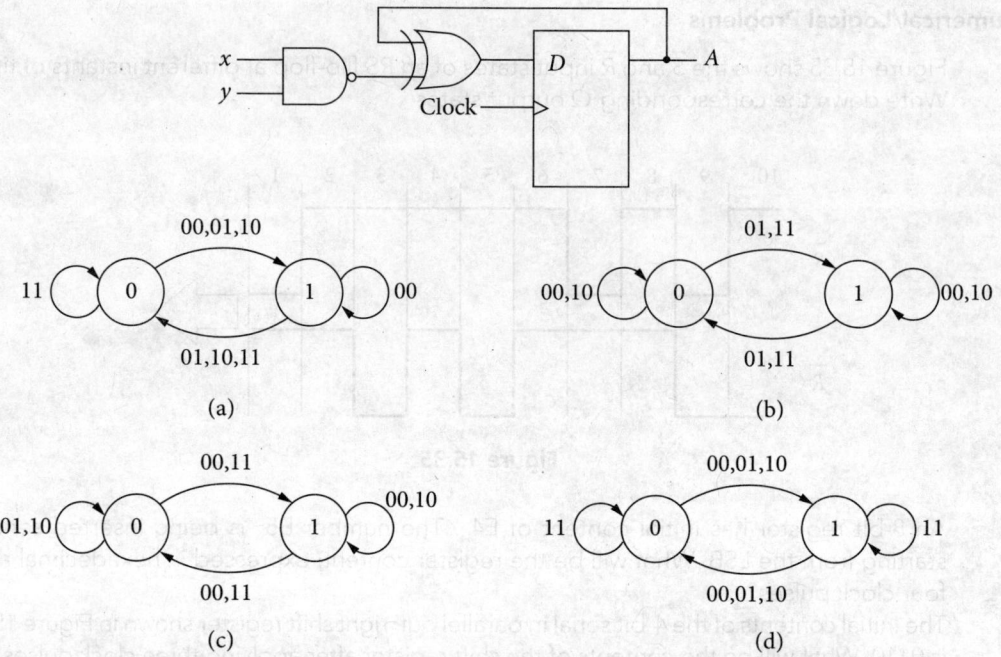

Figure 15.37

9. Consider the circuit of Figure 15.38, consisting of an RS flip-flop and two AND gates. Which of the following connections will allow the entire circuit to act as a JK flip-flop?
   (a) Connect Q to pin 1 and $\overline{Q}$ to pin 2
   (b) Connect Q to pin 2 and $\overline{Q}$ to pin 1
   (c) Connect Q to K input and $\overline{Q}$ to J input
   (d) Connect Q to J input and $\overline{Q}$ to K input

[NET Dec. 2018]

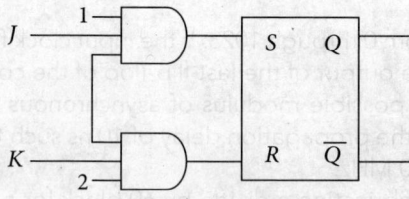

Figure 15.38

10. In the schematic diagram of Figure 15.39, the initial values of 4 bit shift registers A and B are 1011 and 0010, respectively. The values at SO_A and SO_B after the pulse t_2 are, respectively,
    (a) 1110 and 1001
    (b) 1101 and 1001
    (c) 1101 and 1100
    (d) 1110 and 1100

[NET Dec. 2015]

Sequential Logic Circuits

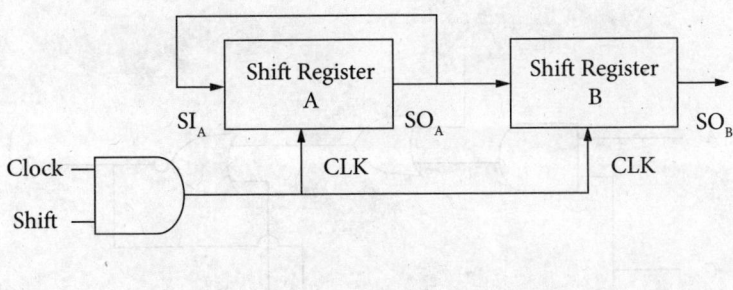

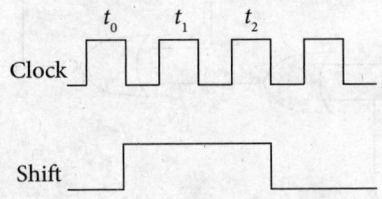

Figure 15.39

11. In Figure 15.40, each logic gate has the same propagation delay. The propagation delay of the circuit will be maximum when the logic inputs (A, B) make the transition
    (a) 0,1 to 1,1 (b) 1,1 to 0,1 (c) 0,0 to 1,1 (d) 0,0 to 0,1
    [NET Jun. 2016]

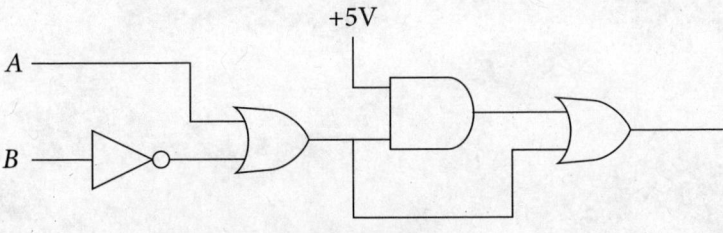

Figure 15.40

12. In Figure 15.41, the correct logic values for the entries X_2 and Y_2 in the truth table are ------------.
    (a) 1 and 0 (b) 0 and 0 (c) 0 and 1 (d) 1 and 1
    [GATE 2019]

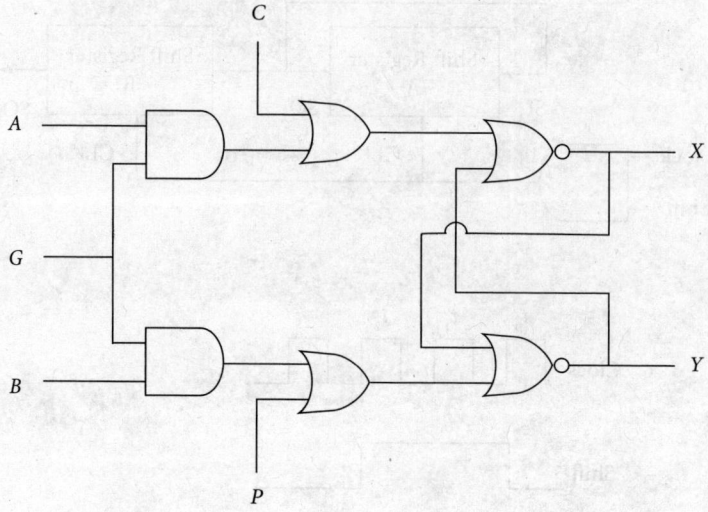

Figure 15.41

16

Analog–Digital Conversion and Memory

The digital circuits introduced so far, such as arithmetic, logical, data-processing, and sequential circuits are somehow associated with the core of a digital system, such as, the central processing unit of a computer. Yet there are two other very significant topics, namely analog and digital conversion circuits and different types of memory circuits for storage of digital information. This chapter deliberates upon several digital-to-analog (D/A) and analog-to-digital (A/D) conversion circuits and introduces the classification and fabrication of memory devices in digital systems.

16.1 Why D/A and A/D Conversions

A digital system is a special man-made technique whereas most of the natural entities, such as voltage, temperature and light intensity are analog in nature undergoing continuous variation. Indeed the basic unit of a digital circuit, namely the logic gate is made of bipolar transistors or MOSFETs that are analog of their own. These devices are made to operate in two discrete output states so as to comply with the binary system. Hence, for the purpose of interacting with a digital system, it is not sufficient to just convert the nonelectrical entity to a proportional electrical signal. The continuous variation of voltage or current must somehow be converted to a proportional digital change of two discrete states. The two-state response of the digital system also should be translated to a proportional continuous variation. There lies the significance of analog–digital interconversion systems. A few specific examples may be cited.

Printing Computer Outputs: The digital results processed by a computer must be communicated to the analog servomechanism through some equivalent analog signals.

Measuring Nonelectrical Quantities: The digital instruments for measuring nonelectrical quantities, such as temperature, speed or blood pressure first converts the variation of the quantity to a proportional electrical signal using an appropriate transducer. Then the analog electrical signal is converted to equivalent digital code.

Internet Communication: Both way analog–digital conversions are necessary for connecting telephone cables to computers for Internet communications, such as emailing, uploading and downloading data.

There are many household, industrial, scientific, and technical appliances where the need of analog and digital interconversion is realized. We start the discussion with D/A conversion because it is sometimes an integral part of A/D conversion. The basic objective of converting a digital signal into an equivalent analog signal is to convert a number of separate voltage levels to a single equivalent analog voltage, which is defined by an equivalent binary weighted voltage explained in the subsequent section.

16.2 Binary Equivalent Weight

Each successive bit in a binary system has a value twice that of the preceding bit. Therefore, for an *n*-bit binary number, the weight (w_m) of the bit for any *m*-th position is

$$w_m = \frac{2^m}{2^n - 1} \tag{16.1}$$

The sum of all such weights is equal to unity given by

$$\frac{1}{2^n - 1}(2^{n-1} + 2^{n-2} + \ldots + 2^1 + 2^0) = 1 \tag{16.2}$$

Following the above concept, a circuit can be built that converts an *n*-bit digital number given in terms of discrete voltage levels to an equivalent analog voltage given by

$$V_o = K(2^{n-1} S_{n-1} + 2^{n-2} S_{n-2} + \ldots + 2^1 S_1 + 2^0 S_0) \tag{16.3}$$

The constant of proportionality (K) depends on the circuit parameters and $S_0, S_1, \ldots, S_{n-1}$ denote the switching conditions for the input bits, either 0 or 1. The MSB and the LSB of the given digital input correspond to S_{n-1} and S_0, respectively.

> **Note: The Proportionality Constant in D/A Conversion**
>
> The analog output voltage of the D/A converter can be less than, exactly equal to or greater than the input digital number depending on the proportionality factor (*K*) offered by the circuit. When the analog output is in voltage, *K* is in voltage units and when the output is current, *K* is in current units.

> **Example 16.1**
>
> Determine the weight of each bit in a 4-bit binary number. Also determine the sum of the weights.

> **Solution**
>
> Each weight is determined using Equation (16.1).
>
> The LSB (2^0 position) has the weight of $\dfrac{2^0}{2^4-1} = \dfrac{1}{15}$
>
> The next higher power bit (2^1 position) has weight of $\dfrac{2^1}{2^4-1} = \dfrac{2}{15}$
>
> The next higher power bit (2^2 position) has weight of $\dfrac{2^2}{2^4-1} = \dfrac{4}{15}$
>
> The MSB (2^3 position) has the weight of $\dfrac{2^3}{2^4-1} = \dfrac{8}{15}$
>
> The sum of the weight is $\dfrac{1}{15} + \dfrac{2}{15} + \dfrac{4}{15} + \dfrac{8}{15} = 1$

16.3 Digital-to-Analog (D/A) Conversion

The aforementioned principle of digital-to-analog conversion, henceforth denoted by D/A conversion, is implemented with two popular D/A converter circuits, namely weighted resistor and *R-2R* ladder techniques.

16.3.1 Weighted Resistor D/A Converter

The circuit of this converter is shown in Figure 16.1. The working principle is same as that of an op-amp adder (Chapter 11). The specialty is that the number (*n*) of resistors at the input port is equal to the number of input bits and the resistance values from MSB to LSB increase as power of 2. The absolute values of the resistances need not be fixed but this weighted ratio must be maintained. The phrase 'weighted resistor' originates from this stipulation. This type of converter is also known as *resistive divider* or *binary weighted* converter. Figure 16.1 has built up the circuit using a certain arrangement of weighted resistors. Instead, the circuit could work well with the resistors arranged in decreasing powers of 2 from MSB to LSB and the theory would have to be modified accordingly.

Each resistor is connected either to the voltage source (+*V*) representing logical 1 or to ground signifying logical 0 through the switches S_0 (LSB), S_1, ... and S_{n-1} (MSB). Actually these are not mechanical switches but the on–off conditions are realized by transistors acting as switches. However, the purpose is, as expected, to provide logical 0s or 1s to the input resistors. If any switch is turned on, the corresponding resistor carries a current and if the switch is turned off, no current flows through that resistor. Considering all the switches, the total current at the inverting input is expressed as

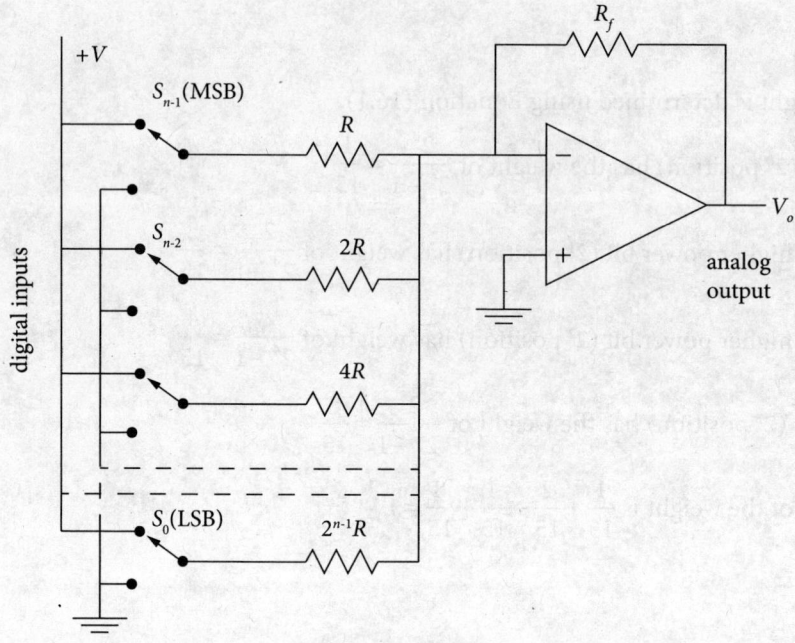

Figure 16.1 Weighted resistor D/A converter for n-bit digital input

$$I = \frac{V}{R}S_{n-1} + \frac{V}{2R}S_{n-2} + \ldots + \frac{V}{2^{n-2}R}S_1 + \frac{V}{2^{n-1}R}S_0$$
$$\Rightarrow I = \frac{V}{2^{n-1}R}(2^{n-1}S_{n-1} + 2^{n-2}S_{n-2} + \ldots + 2^1 S_1 + 2^0 S_0) \tag{16.4}$$

The output voltage (V_o) of the amplifier is given by

$$V_o = -IR_f = -\frac{VR_f}{2^{n-1}R}(2^{n-1}S_{n-1} + 2^{n-2}S_{n-2} + \ldots + 2^1 S_1 + 2^0 S_0) \tag{16.5}$$

Equation (16.5) represents the analog equivalent voltage produced by the weighted resistor D/A converter for the *n*-bit digital input given in terms of two voltage levels.

Comparing with Equation (16.3) it is understood that the proportionality constant for this weighted resistor D/A converter circuit is $-\frac{VR_f}{2^{n-1}R}$, which can be adjusted by changing three parameters:

- the input voltage level (V),
- the feedback resistance (R_f) value and
- the common multiple (R) of the input resistance value.

The negative sign of the output is due to the property of the inverting amplifier. It can be ignored, if only the magnitude is considered.

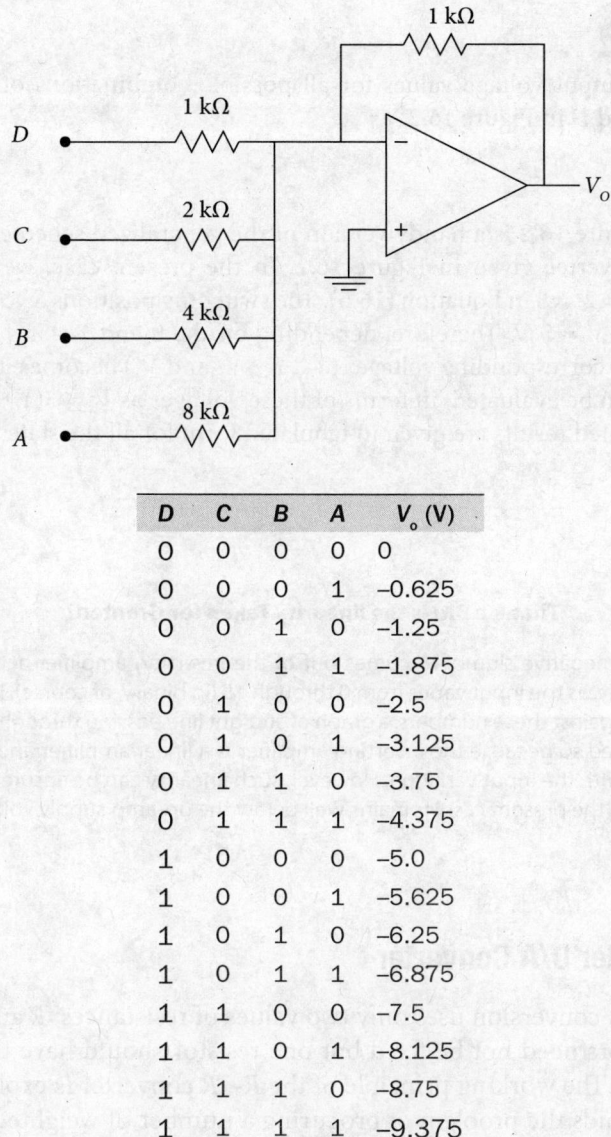

Figure 16.2 4-bit weighted resistor D/A converter and its output voltage calculation for different input states

The weighted resistor D/A converter circuit is simple but it suffers from several shortcomings. The accuracy of the analog output voltage depends on the precision of the resistance values. It is very difficult to manufacture such exact values of a set of resistor over such a wide range. Also the great mismatch between the currents flowing through the maximum and the minimum resistors limit the accuracy. Slight fluctuation in the largest resistance value may cause severe deviation in the smallest resistance and the related parameters. Therefore, D/A conversion with more than four bits generally follows another method, known as R-2R ladder method.

> **Example 16.2**
>
> Determine the output voltage values for all possible combinations of the four digital inputs: A, B, C and D in Figure 16.2.
>
> **Solution**
>
> The circuit of Figure 16.2 is a handy version of the generalized scheme of the weighted resistor D/A converter given in Figure 16.1. In the present case, we put $R_f = 1$ kΩ, $R = 1$ kΩ and $2^{n-1} = 2^3 = 8$ in Equation (16.5). The switching positions $S_o, S_1, ..., S_{n-1}$ denote either 0 or 1 and $V = 5$ V. Therefore, depending on the 0 and 1 states of the terminals D, C, B and A, the corresponding voltages (V_D, B_C, V_B and V_A) become either 0 or 5 V. The output voltage can be evaluated, in terms of these voltages as $V_o = 5(V_D + V_C/2 + V_B/4 + V_A/8)$. The calculated results are given in tabulated form for all the states.

> **Thank a Bit: Is the linearity Taken for Granted?**
>
> In Example 16.2, the negative sign of V_o comes out of the inverting amplifier action. The magnitude increases continuously, as the input varies from 0 through 15 (in binary, of course). If the output voltage values were plotted against these numbers, a graph of straight line passing through the origin would be obtained. It is expected so because the inverting amplifier is a linear amplifier and the output voltage should vary linearly with the input variation. However, such linearity can be ensured so long as the full-scale output (9.375 in the present case) remains well below the op-amp supply voltage.

16.3.2 *R-2R* ladder D/A Converter

This method of D/A conversion uses only two values of resistances, R and $2R$. The absolute values of the resistors need not be fixed but one resistor should have the resistance value double of the other. The working principle of the R-$2R$ converter is explained with Figures 16.3 and 16.4. It avoids the problem of procuring a number of weighted resistances over a wide range.

Figure 16.3 shows the circuit diagram for n-bit digital input, each connected through a switch and a resistor. The arrangement of the resistors up to the point P looks like a ladder, hence the name 'R-$2R$ ladder is coined'.

The switches $S_0, S_1, ..., S_{n-1}$, similar to those in the weighted resistor circuit represent the 0 and 1 conditions of the n-bit digital input in terms of voltage $+V$ (logical 1) or ground (logical 0) connection. The proportional analog output voltage (V_o) obtained at the op-amp output is a weighted sum of the digital input. To deduce the proper expression for V_o, the fraction of the output contributed by the 1 state of each switch is to be estimated separately and the process is interpreted with Figure 16.4.

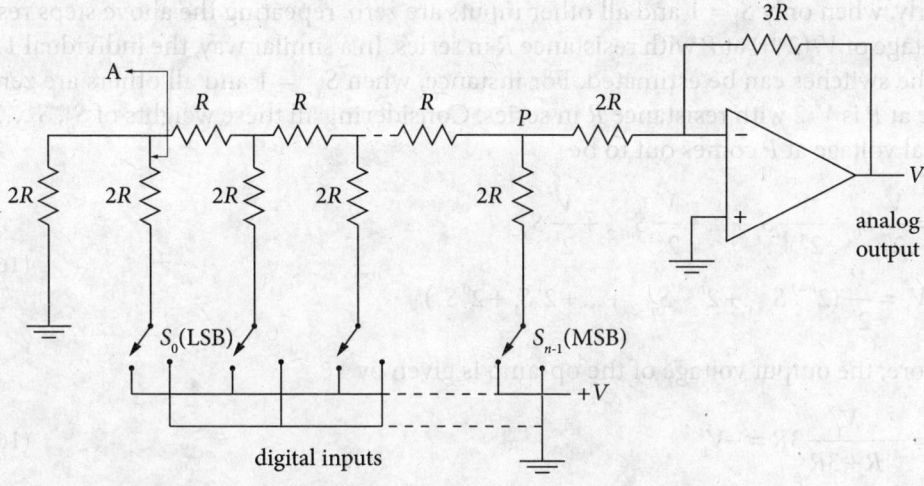

Figure 16.3 R-2R ladder D/A converter for n-bit digital input

Let only S_0 be at 1 state and all other switches ($S_1, \ldots, S_{n-1}$) be at 0 state. Following Thevenin's theorem, the A-arrowed portion of the circuit in Figure 16.3 can be reduced to that of Figure 16.4(a). Proceeding in the same way with the circuit segments denoted by B, C etc., the equivalent circuit can be reduced to that of Figure 16.4(b). Each time the equivalent voltage is halved and the equivalent series resistance remains R. Ultimately a voltage source of $V/2^n$ in series with a resistance R is posed to point P.

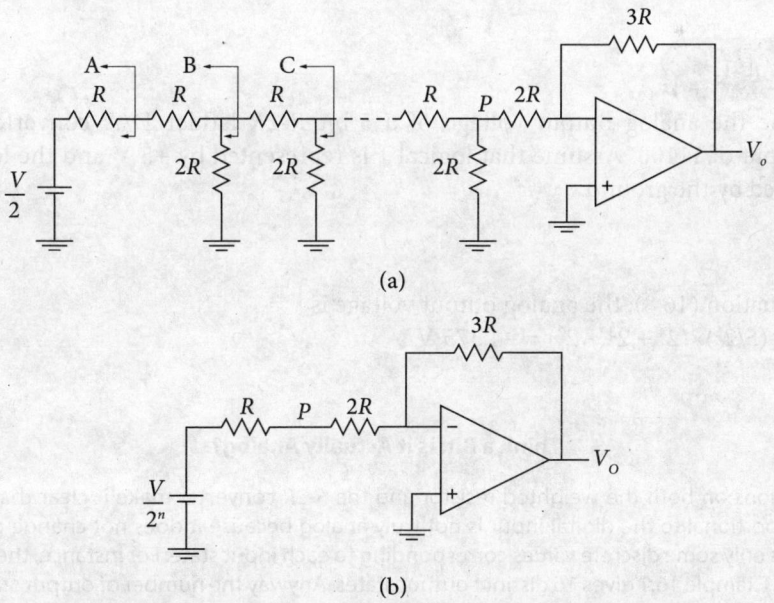

Figure 16.4 Equivalent circuit for R-2R ladder D/A converter (Figure 16.3) when $S_0 = 1$ and other switches are at zero state: (a) circuit simplification using Thevenin's theorem, (b) circuit simplification completed

Similarly, when only $S_1 = 1$ and all other inputs are zero, repeating the above steps result in the voltage of $V/(2^{n-1})$ at P with resistance R in series. In a similar way, the individual 1 states of all the switches can be estimated. For instance, when $S_{n-1} = 1$ and all others are zero, the voltage at P is $V/2$ with resistance R in series. Considering all these weights of $S_0, S_1, \ldots, S_{n-1}$, the total voltage at P comes out to be

$$V_1 = \frac{V}{2^n} S_0 + \frac{V}{2^{n-1}} S_1 + \ldots + \frac{V}{2^2} S_{n-2} + \frac{V}{2^1} S_{n-1}$$
$$\Rightarrow V_1 = \frac{V}{2^n}(2^{n-1} S_{n-1} + 2^{n-2} S_{n-2} + \ldots + 2^1 S_1 + 2^0 S_0) \quad (16.6)$$

Therefore, the output voltage of the op-amp is given by

$$V_o = -\frac{V_1}{R+3R} 3R = -V_1 \quad (16.7)$$

The feedback resistance value of $3R$ is chosen deliberately to simplify the expression. From Equations (16.6) and (16.7),

$$V_o = -\frac{V}{2^n}(2^{n-1} S_{n-1} + 2^{n-2} S_{n-2} + \ldots + 2^1 S_1 + 2^0 S_0) \quad (16.8)$$

Equation (16.8) represents the analog equivalent voltage produced by the n-bit R-$2R$ ladder D/A converter of Figure 16.3. Comparing with the general expression of Equation (16.3) and the voltage output of weighted resistor converter [Equation (16.5)], the constant of proportionality is found to be different.

> ### Example 16.3
>
> Determine the analog output voltage of a 5-bit R-$2R$ ladder D/A converter for the digital input of 11100. Assume that logical 1 is represented by +5 V and the logical 0 is represented by the ground.
>
> **Solution**
>
> Using Equation (16.8), the analog output voltage is
> $-(5/2^5) \times (2^4 + 2^3 + 2^2) = -4.375$ V

> **Think a Bit: Is it Actually Analog?**
>
> The illustrations on both the weighted resistor and the R-$2R$ converter make it clear that the output voltage proportional to the digital input is not truly analog because it does not change continuously but assumes only some discrete values corresponding to each input state. For instance, the 4-input D/A converter of Example 16.1 gives 16 distinct output states. Anyway the number of output state increases and the difference between the successive states decreases, if the number of inputs is increased keeping the reference voltage unchanged. Thus the conversion becomes better approximation to analog with larger number of input bits.

16.3.3 D/A Converter Performance

The two different types of D/A converters discussed above have different working principles but the similarity is that both produce an analog voltage proportional to the given digital input. In general, the quality of digital to analog conversion is estimated by the following parameters.

- *Resolution:* This is the smallest change in the analog output voltage caused by the change of one bit in the digital input. In other words, the resolution of a D/A converter is the number of states into which the full-scale output range can be resolved. Generally it is expressed in percentage and obviously it depends on the number of input bits. For an n-bit D/A converter, there are 2^n input states but 2^n-1 steps between the zero and the full scale. Therefore, the resolution (R) of n-bit D/A converter is given by

$$R = \frac{1}{2^n - 1} \times 100\% \qquad (16.9)$$

The larger is the number of bits, the smaller is the incremental step and the more the output steps are smoothened, hence the resolution is increased.

- *Accuracy*: It is a measure of how close the practical output voltage is to the expected theoretical value. The accuracy of the converter depends on the precision of the resistors and the perfection of the reference voltage. The accuracy of the D/A converter should be within ± half of its LSB.
- *Linearity*: The op-amp in the D/A converter circuit should act as a linear amplifier so that the output voltage can change regularly as a perfect staircase along with the change in the digital input.
- *Monotonicity*: Not only linearity, it should be ensured that the D/A converter does not make any reverse step throughout the entire range of input bits, either in increasing or in decreasing order.
- *Settling Time*: This is the time taken by the D/A converter circuit to settle within ±1/2 step-size of the final output value when a change is made in the digital input number.

The D/A converter has many important fields of application. A few are mentioned here.

Control System: In an automatic control system for motor speed, temperature or other nonelectrical quantities, the digital controlling signals from the computer are implemented through D/A converters.

Signal Reconstruction: In many signal processing systems, such as digital storage oscilloscope and digital recording of audio or video, the analog signal is digitized by converting the successive amplitude sequences to digital equivalents. These are converted back to analog signals by the use of D/A converters.

A/D Conversion: Sometimes the D/A converter becomes a part of the analog to digital conversion process.

Digital Amplitude Control: It is understood from Equations (16.5) and (16.8) that for a certain digital input number, the analog output of a D/A converter increases or decreases in proportion to the reference voltage (V) denoting the input logical state. If, in lieu of a fixed V, a voltage signal (e.g. a sine wave) is applied, the output becomes the same waveform but of a different amplitude depending on the assigned digital input value.

> **Example 16.4**
>
> What is the percentage resolution of a D/A converter with (i) 4 bits and 8 bits? What is the resolution in terms of voltage of the 8-bit converter, if the input reference voltage is +5 V?
>
> **Solution**
>
> Using Equation (16.9), the percentage resolution is
> (i) $1/(2^4 - 1) \times 100 = 6.67\%$
> (ii) $1/(2^8 - 1) \times 100 = 0.39\%$
>
> For the 8-bit input, the total number of output steps is $2^8 - 1 = 255$ and the resolution in terms of voltage is
> $$1/(2^8 - 1) \times 5 = 0.0196 \text{ V}$$

16.4 Analog-to-Digital (A/D) Conversion

A circuit of this category converts an analog input voltage to an equivalent digital number in terms of presence/absence of voltages as different bits. There are several principles of A/D conversion, such as

- applying the analog voltage to an array of parallel op-amp comparators with weighted reference voltages,
- successive approximation of the analog voltage by consecutively dividing the voltage range in half and trying one bit at a time beginning with the MSB and
- one-way or two-way counting of clock pulse and comparison of the resultant number converted to analog voltage with the input analog voltage.

Each category of conversion has some advantages and some limitations. Some popular D/A converters based on the any of the above principles are illustrated here.

16.4.1 Flash A/D Converter

This technique of analog to digital conversion is also known as *simultaneous method*. Though the technique calls for a complicated circuit, the conversion mechanism is simple and it provides with very fast conversion process. The nickname 'Flash' is related to its speedy operation. A set of analog op-amp comparators (Chapter 11) is an indispensable part of the flash A/D converter.

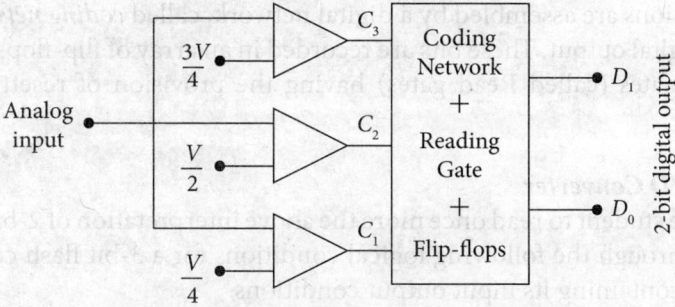

Analog input	Comparator outputs			Digital outputs	
	C_1	C_2	C_3	D_1	D_0
0 to V/4	0	0	0	0	0
V/4 to V/2	1	0	0	0	1
V/2 to 3V/4	1	1	0	1	0
3V/4 to V	1	1	1	1	1

Figure 16.5 Flash A/D converter, block diagram (top) and input–output conditions (bottom)

A. 2-Bit Flash A/D Converter

Figure 16.5 outlines the working principle of a 2-bit flash A/D converter. Both the block diagram (top) and the table (bottom) should be looked at simultaneously. Only a 2-bit output with three (= 2^2-1) comparators is considered for the sake of easy understanding but it interprets the general rules. The analog input signal to be digitized is applied to one of the inputs of each comparator. The other input of each comparator is a pre-determined fraction of a reference voltage arranged in increasing order. Depending on the reference voltage value ($+V$), the converter is capable of accepting any analog input voltage of 0 to $+V$ and depending on the number of comparators (2^n-1), it can produce n-bit digital output. In the present case, the 2-bit binary number can represent up to four distinct states. Hence, the range of 0 to V is divided into four levels and the first three of these are applied in increasing order to comparators C_1, C_2 and C_3, respectively. When the analog input voltage exceeds any of these reference values, the corresponding comparator output goes high. Comparing the digital output bits D_0 and D_1 with the output states of the comparators C_1, C_2 and C_3, the following inferences can be made.

D_0 first turns high when $C_1 = 1$ and $C_2 = 0$. The next higher input level will obviously turn on these comparators and those conditions are redundant. Also D_0 turns on when $C_3 = 1$. Therefore,

$$D_0 = C_1 \overline{C_2} + C_3$$

D_1 goes high for the first time when $C_2 = 1$ and $C_3 = 0$. Hence,

$$D_1 = C_2 \overline{C_3}$$

The above conditions are assembled by a digital network, called *coding network*, to produce each bit of the digital output. These bits are recorded in an array of flip-flops through proper combination of gates (called Read gates) having the provision of resetting the flip-flop contents.

B. 3-Bit Flash A/D Converter

Now I request the student to read once more the above interpretation of 2-bit flash converter and then to go through the following logical conditions for a 3-bit flash converter derived from Table 16.1 containing its input output conditions.

D_2 is high whenever C_4 is high.

D_1 is high whenever C_6 is high or C_2 and $\overline{C}_4$ is high. Hence, $D_1 = C_2\overline{C}_4 + C_6$

Similarly, $D_0 = C_1\overline{C}_2 + C_3\overline{C}_4 + C_5\overline{C}_6 + C_7$

Table 16.1 All combinations of the eight analog input levels, the seven comparator outputs ($C_1, C_2, ..., C_7$) and the three bits (D_2, D_1 and D_0) of the digital output for a 3-bit flash A/D converter

Analog input	Comparator outputs							Digital output bits		
	C_1	C_2	C_3	C_4	C_5	C_6	C_7	D_2	D_1	D_0
0 to V/8	0	0	0	0	0	0	0	0	0	0
V/8 to V/4	1	0	0	0	0	0	0	0	0	1
V/4 to 3V/8	1	1	0	0	0	0	0	0	1	0
3V/8 to V/2	1	1	1	0	0	0	0	0	1	1
V/2 to 5V/8	1	1	1	1	0	0	0	1	0	0
5V/8 to 3V/4	1	1	1	1	1	0	0	1	0	1
3V/4 to 7V/8	1	1	1	1	1	1	0	1	1	0
7V/8 to V	1	1	1	1	1	1	1	1	1	1

Because of the high speed of conversion, the flash A/D converter is used in radar signal processing, high-speed instrumentation and television broadcasting. However, the circuit is complicated. The *n*-bit flash A/D conversion system requires $2^n - 1$ comparators. For instance, an 8-bit converter would require 255 comparators.

16.4.2 Counter-Type A/D Converter

The simplified block diagram of this system is given in Figure 16.6. Interestingly a D/A converter is an essential component of this A/D converter. A counter (Chapter 15) counts the number of clock pulses applied to it. How the feeding of the clock pulses is controlled will be discussed later. As the counter goes on counting, the number at its output increases gradually. This output is applied to a D/A converter, may be of *R-2R* type. As the counter output goes on increasing, the output of the D/A converter produces a stepwise increasing analog voltage. Because of this ramp-like increasing voltage, this category of converter is also known as *digital-ramp* A/D converter. This voltage is applied to an op-amp comparator. The other input of the comparator is the analog voltage (V) that is to be converted to the digital equivalent.

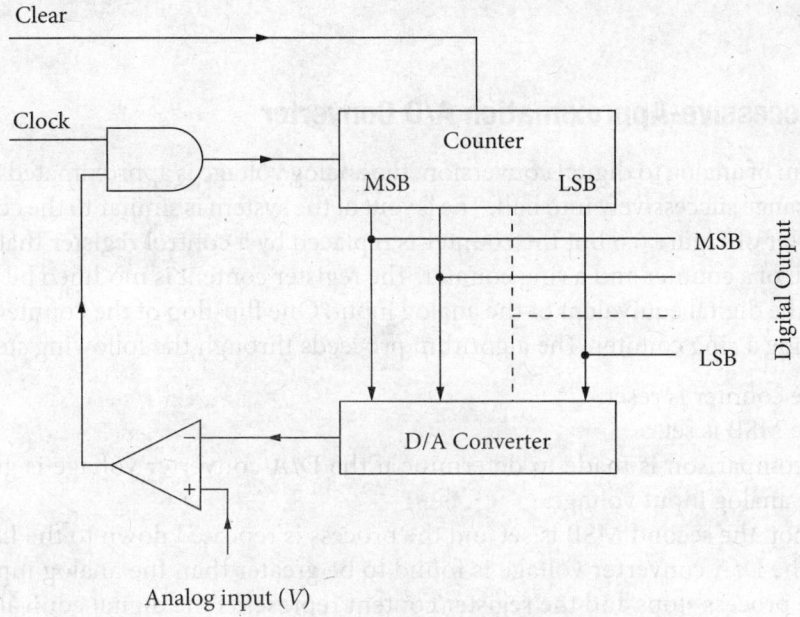

Figure 16.6 Block diagram of counter-type A/D converter

Initially the counter is reset and the D/A converter output is zero. So long as it remains less than the analog input voltage (V), the comparator output remains high, which allows the clock pulses to reach the counter through the 2-input AND gate. When the stepwise increasing D/A converter output voltage becomes slightly larger than V, the comparator output goes low and disables the AND gate. The counter no more counts pulses and the counter output becomes fixed. At this situation, the counter output represents the digital equivalent of the analog input. After the conversion is complete, the counter is reset with the Clear terminal for beginning another count sequence.

The counter-type A/D converter has a simple circuit with only one comparator and it is suitable for digitizing to high resolution. However, it requires a longer conversion time because the counting starts from zero and proceeds through normal binary sequence. The counting system can be improved by the use of up-down counters that eliminates the requirement of resetting every time and the counting begins upward or downward at the last converted point of the analog input thereby providing a continuous A/D conversion.

> **Think a Bit: Is the A/D Output Equal to the Analog Input?**
>
> The D/A converter output is not exactly equal to the analog input because of the finite increment in steps. As a result, the counter output is not exactly the digital equivalent of the analog input. Also the resolution of the internal D/A converter is an inherent source of error. So the A/D converter cannot produce the exact digital equivalent but can approximate it. The difference between the actual analog quantity and converted the digital value assigned to it is referred to as *quantization error*. It can be reduced by increasing the number of bits in the counter and the D/A converter.

16.4.3 Successive-Approximation A/D Converter

In this system of analog to digital conversion, the analog voltage is approximated by dividing the voltage range successively into half. The layout of the system is similar to the counter-type A/D converter of Figure 6.6 but the counter is replaced by a control register that contains a combination of a counter and a ring counter. The register content is modified bit by bit until it becomes the digital equivalent of the analog input. One flip-flop of the counter is selected at a time using a ring counter. The algorithm proceeds through the following steps.

- The counter is reset.
- The MSB is set.
- A comparison is made to determine if the D/A converter voltage is greater than the analog input voltage.
- If not, the second MSB is set and the process is repeated down to the LSB.
- If the D/A converter voltage is found to be greater than the analog input voltage, the process stops and the register content represents the digital equivalent.

The main advantage of the successive-approximation A/D converter is its short conversion time. Each conversion corresponds to each bit and takes the same time. The total conversion time is equal to the number of bits. Also, as inferred from the above working principle, the conversion time is independent of the analog input value. Assuming each conversion corresponds to one clock pulse, the total conversion time is almost equal to the number of bits times the clock time period. Another advantageous feature is that the successive-approximation A/D converter is most suitable for multiplexing inputs.

The limitation of successive-approximation A/D converter is that the operation must be matched to the full-scale input range, which may call for an offset from the ground so that a small positive voltage also may produce zero output. Also the circuit is vulnerable to noise.

16.5 Memory

A digital system possesses a unit referred to as *memory* that has the capability of storing information in the form of binary data. The memory is an essential component of a computer and is also part of other digital systems ranging from calculator to mobile phone and camera to photocopier. The memory unit is comprised of devices and circuits meant for storing and accessing bits in some physical form, such as two-state voltage variation, orientation of magnetized particle, or optical reflectance of a surface.

The *main memory* or the *working memory* of a computer, which has direct interaction with the CPU is essentially made of semiconductor devices. It holds the program and the related data while the computer remains in running state executing a program. The computer needs another type of memory, called *auxiliary memory* or *secondary storage*, for storing a large amount of digital information that are not in current use. This type of memory elements are made of magnetic or optical devices. The tree of Figure 16.7 presents the relative positions of different types of memory elements. The memory is generally constituted of either of the following materials.

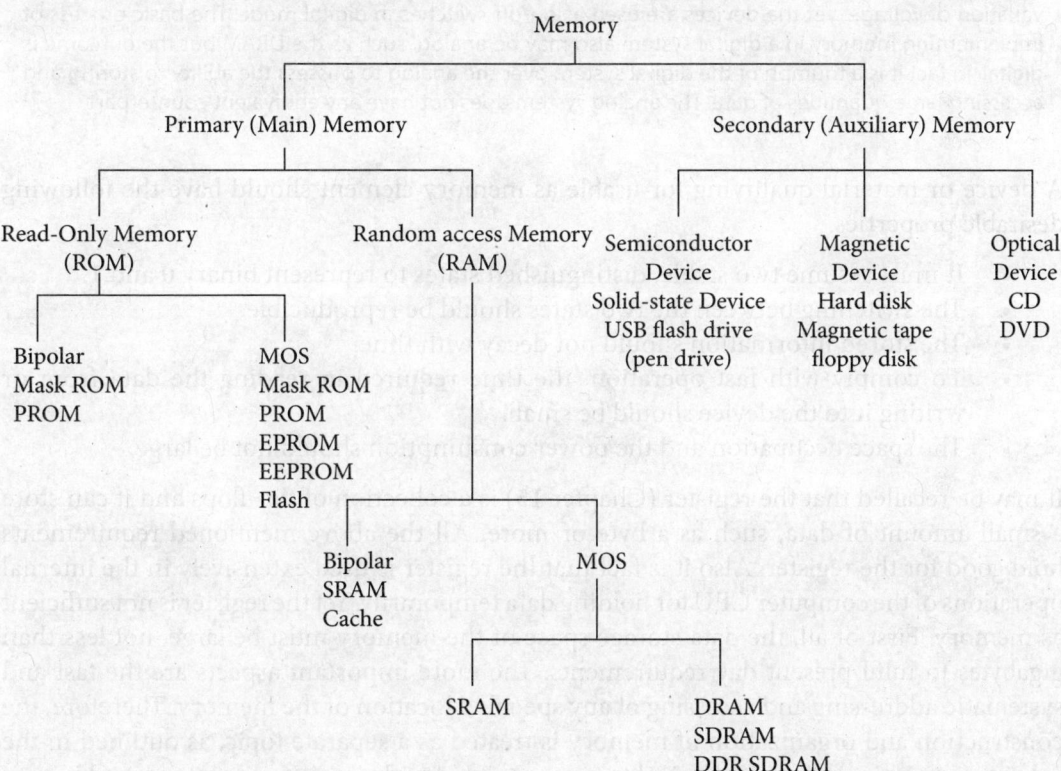

Figure 16.7 Tree-structure representation of different categories of memory

Semiconductor Memory: It is made of bipolar or MOS transistors and the digital information is stored as presence or absence of voltage. The flip-flop (Chapter 15) capable of storing a single bit is of this category. The RAM and the ROM of a computer are made of semiconductor memory where memory cells are fabricated in the form of VLSI chips.

Magnetic Memory: This is based on the principle of orientation of magnetic particles. The memory elements are obtained with magnetization of small spots in a thin film of magnetic material. The hard disk of a computer and the previously used floppy disk and magnetic tape are of this category.

Optical Memory: It uses either organic dye or metal-semimetal alloy whose reflectivity is changed by heating with laser. The intensity of reflected laser from two different surfaces informs of the binary data. The CD and DVD are of this group of memory.

> **Think a Bit: Is there Analog Memory Too?**
>
> It is interesting to recall that the fundamental units of a digital system, such as logic gates are made of bipolar transistors or MOSFETs that are analog in nature. Their properties change with continuous variation of voltage, yet the devices are used as on/off switches in digital mode. The basic process of implementing memory in a digital system also may be analog, such as the DRAM but the outcome is digital. In fact it is a triumph of the digital system over the analog to possess the ability to storing and accessing large quantities of data. The analog system does not have any equivalent counterpart.

A device or material qualifying for usable as memory element should have the following desirable properties.

- It must assume two stable, distinguished states to represent binary 0 and 1.
- The switching between the two states should be reproducible.
- The stored information should not decay with time.
- To comply with fast operation, the time required in reading the data from or writing it to the device should be small.
- The space occupation and the power consumption should not be large.

It may be recalled that the register (Chapter 15) is a collection of flip-flops and it can store a small amount of data, such as a byte or more. All the above mentioned requirements hold good for the register. Also it is fact that the register is used extensively in the internal operations of the computer CPU for holding data temporarily. Yet the register is not sufficient as memory. First of all, the data storage space of the memory must be large, not less than gigabytes to fulfil present day requirements. The more important aspects are the fast and systematic addressing and accessing of any specified location of the memory. Therefore, the construction and organization of memory is treated as a separate topic, as outlined in the subsequent sections. A few technical terms associated with memory are introduced here.

Memory Cell: It is the device that stores a single bit, 0 or 1 of the memory. A flip-flop or a single spot on the magnetic disk is a memory cell.

Memory Word: It is a group of bits in a memory representing data or instruction.

Capacity: It is the total number of bits that can be stored in a particular memory system.

Address: It is the unique binary number assigned to each cell of the memory. It identifies the location of the word stored in it for reading or writing purpose.

Read or Fetch: It is the operation of sensing the binary word stored in a specific memory address and transferring the same to another device, such as a register of the CPU. It is a data output from the cell.

Write or Store: It is the operation of replacing the word of a certain address by a new word. It is a data input to the cell.

Line or Bus: It refers to the conducting wires for connecting the memory cells to other devices. An n-bit word requires n number of input and output lines. For a memory capacity of 2^n words, n number of address lines are needed. For instance, a memory system of $2^4 = 16$ words, the address varies from 0000 through 1111 and 4-bit address lines are needed.

16.5.1 Read-Only Memory (ROM)

This type of memory comprises permanently stored data, which can be read at any time but cannot be changed. In some specific cases, such as EPROM discussed later, the data can be changed only with some special technique. Thus the ROM is either permanent or semi-permanent data storage. This is also termed as nonvolatile memory because the data bits are retained with the device even when the power supply is switched off. Some popular varieties of ROM are introduced below.

A. Mask ROM

It may be constructed with either bipolar or MOS transistors. A permanent information is incorporated during the manufacturing process, the 1s and 0s being represented by the presence or absence of transistor connection. Figure 16.8 (a) and (b) explain the formation of a single such ROM cell using bipolar transistor. In the case of MOSFETs, similar diagonal connections between the row and column wires are made with the gate and the source, respectively and the supply voltage $(+V_{CC})$ is replaced by $+V_{DD}$. The arrangement of such cells in an array is outlined in Figure 16.8(c) where each square represents a cell containing a transistor like that of Figure 16.8(a) or (b). The shaded and the unshaded square represents storage of 1 and 0, respectively. The n-bit output word yields the information contained in each row. The address input for both the row and the column is put through an address decoder discussed in section 16.5.3.

B. Programmable ROM (PROM)

The mask ROM mentioned above has fixed information specified by the manufacturer and it cannot be changed. A programmable version of the mask ROM, known as PROM allows the user to incorporate data for one time. The structure of PROM is similar to that of mask ROM with row-column diagonal connections. However, the connecting links between the emitters of the BJTs (or the sources of the MOSFETs) and the column lines are fusible. Initially all are connected, representing all 1s. During the programming, the desired links are burnt open by sending high current to create 0s. Thus a desired array of 1s and 0s can be created. Once programmed, the PROM acts like a mask ROM.

C. Erasable PROM (EPROM)

The PROM mentioned above can be programmed only once and cannot be changed further. A more flexible version is the EPROM, which can be reprogrammed by erasing an existing set of data in the memory. The EPROM is obtained with MOS devices. It comprises an array of MOSFETs with no gate connection. The isolated transistor gate can store an electric charge for indefinite time. A current pulse stores a fixed charge in a particular memory cell that represents logical 1. The absence of charge represents a logical 0. The stored memory can be erased by removing the charge. The EPROM IC contains a transparent quartz window through which ultraviolet radiation can be made incident to neutralize the stored charge. A fresh set of information can be incorporated by charging the desired cells with current pulse.

D. Electrically Erasable PROM (EEPROM)

This is similar to EPROM but the erasing is done by removing the charge with an electric pulse of opposite polarity. The advantage of EEPROM is the in-circuit erasing and reprogramming facility. However, the circuit complexity makes it expensive and less number of elements can be fabricated in IC chips. A specific type of EEPROM is called *flash memory* where different voltage levels obtained from a single supply are applied for writing and erasing purpose. It is called 'flash' because of rapid erasing and writing property.

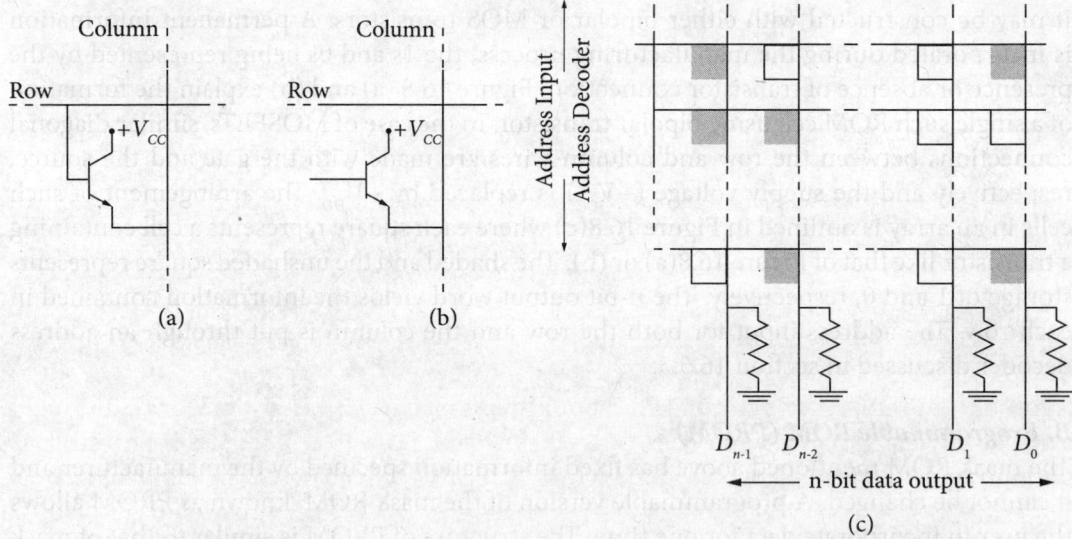

Figure 16.8 Mask ROM element constructed with bipolar transistor representing (a) storage of 1, (b) storage of 0, and (c) an array of information in mask ROM where the shaded and the unshaded squares represent 1 and 0, respectively

Whenever a pre-determined or routine work is executed, the permanent storage of information in ROM can be useful. The followings are some examples.

Look-Up Tables: Routine calculations, such as trigonometric and logarithmic functions are stored in ROM.

Code Conversion: The transformation techniques for binary-to-BCD and vice versa and other such popular code conversions are stored in ROM for repeated use.

Sequence Generator: It can generate a set of pulse trains of definite sequence in order to perform some control process. The sequence can be stored in ROM.

Decoder Driver: A popular and widely used example of ROM is BCD-to-seven-segment decoder for LED or LCD display. Alphanumeric characters are generated on video display in a similar way.

Computer Booting: The starting instructions, after switching on a computer to load the necessary software from secondary memory to main memory are obtained from a ROM chip.

16.5.2 Random Access Memory (RAM)

In this type of memory, information can be written into or read out of any address selected at random. The following two classes of RAM are widely available.

A. Static RAM (SRAM)

The basic memory cell is a flip-flop having bistable state. Once a bit is stored, it remains there so long as the power is on. That is why it is called *static*. The SRAM can be fabricated with both bipolar and MOS technologies. A small-size, high-speed and rapid-access memory unit of SRAM is referred to as *cache memory*, which stores a selected data momentarily in order to increase the operational speed of computation. It has read/write options.

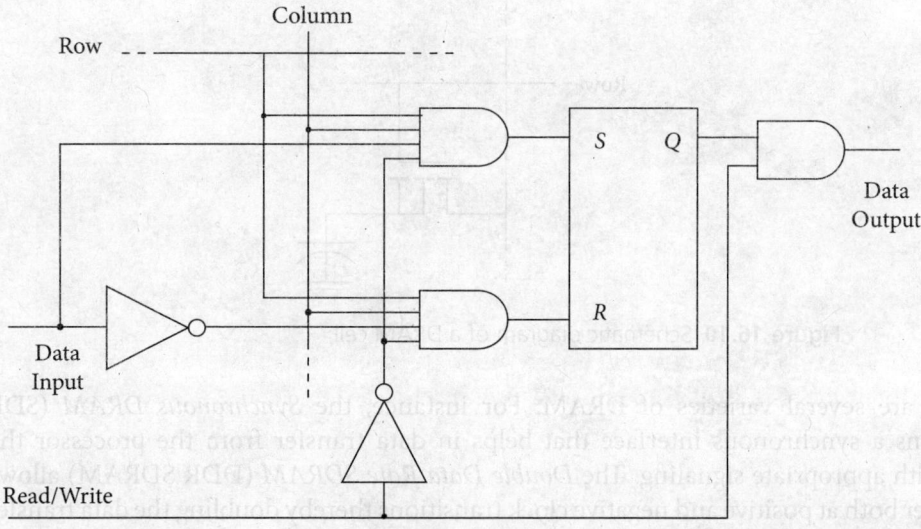

Figure 16.9 Circuit model for explaining the operation of SRAM cell

The model circuit of Figure 16.9 explains the operation of SRAM cell. An individual cell or a group cells can be selected by making the row and column lines (corresponding to S and R, respectively) high. Making these lines low disables the cells. The writing and reading processes are executed in the following way.

Write Operation: When the row and column lines are high and the read/write line is low, the input data bit and its complement are transferred through the AND gates. The flip-flop is set or reset according to 1 or 0 data bit input and the output data bit represents the input bit.

Read Operation: When the row and column lines are high and the read/write line is high, the AND gates yield 0 output state and the S and R inputs of the flip-flop get disabled. No new data bit can reach the output but the previously stored Q value remains unchanged, which appears at the output.

B. Dynamic RAM (DRAM)

This type of memory cannot retain the data indefinitely. Any stored data is to be stored again periodically at a time interval of the order of millisecond. This action is referred to as refreshing. DRAMs are made of MOS transistors. High storage density within small space is possible with it and therefore DRAMs are preferred in the case of large bulk of memory. The DRAM cell stores a data bit in the form of charge in a capacitor formed with the MOS structure. Since the charge cannot be stored indefinitely, the capacitors are recharged periodically in order to retain the data. The working principle of a DRAM cell is explained with Figure 16.10. The MOSFET acts as a switch. When the row and column lines go high, the MOSFET conducts and charges the capacitor. When the row and column lines become low, the MOSFET is cutoff and the charge is withheld within the capacitor.

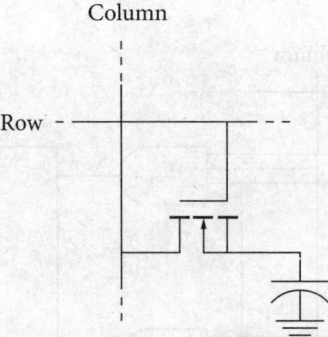

Figure 16.10 Schematic diagram of a DRAM cell

There are several varieties of DRAM. For instance, the *Synchronous DRAM* (SDRAM) contains a synchronous interface that helps in data transfer from the processor through bus with appropriate signaling. The *Double Data Rate SDRAM* (DDR SDRAM) allows data transfer both at positive and negative clock transitions thereby doubling the data transfer rate. Since the SRAM or the DRAM cell is after all a BJT or MOSFET circuit, switching off the dc power supply wipes out all the contents of the cells. Therefore, RAM is called *volatile memory*.

16.5.3 Memory Addressing

Whether the memory be ROM or RAM, there exists the requirement of selecting a memory cell for reading data bits from or writing data bits into it. Such selection of a cell in a memory array is referred to as *memory addressing*. Each location of memory element is specified by a unique binary code and only one row and one column is activated at a time in order to specify a cell. The minimizing of address lines can be understood with the following simple example of 16 memory cells.

Arranging 16 cells in 16 rows and 1 column calls for 16 address lines. Likewise, arranging 16 cells in 1 row and 16 columns would require 16 address lines. Arrangement in 8 rows and 2 columns (8×2) or 2 rows and 8 columns (2×8) needs 8+2 or 2+8, total 10 address lines. Arranging in 4 rows and 4 columns (4×4) can do with 8 address lines only. Generalizing this concept, it becomes clear that the most convenient method of addressing the memory is to organize the memory cells in a square array of n rows and n columns because that manages with the minimum number of address lines.

Figure 16.11 sketches the arrangement of $n \times n$ array where 1-of-n binary-to-decimal decoders are needed. The address decoders can reduce the number of address lines to half in order to uniquely locate a memory cell. An address line of $n/2$ bits is applied to the binary-to-decimal decoder, which produces $2^{n/2}$ output lines. So a memory array of $2^{n/2} \times 2^{n/2} = 2^n$ cells can be addressed with $n/2$ lines for rows and $n/2$ lines for columns.

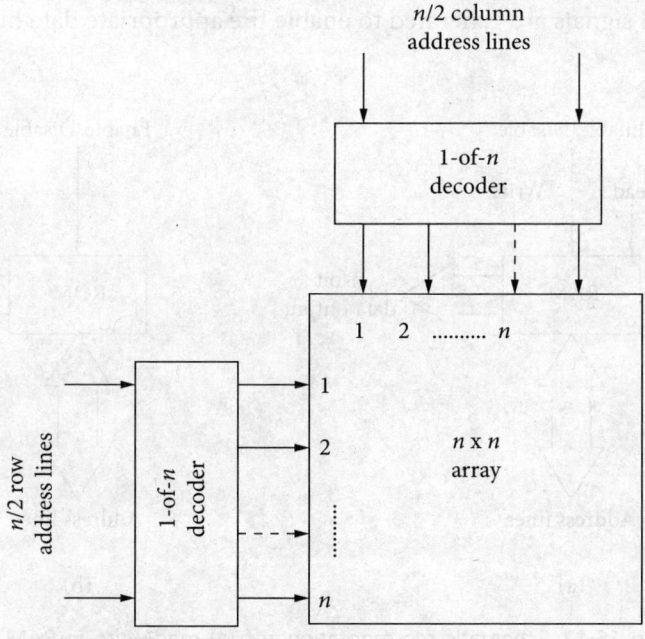

Figure 16.11 Schematic representation of $n \times n$ memory addressing

16.5.4 Memory Read/Write

To 'write' means to store a bit to the memory cell and to detect whether 0 or 1 is stored in a cell is called 'read' operation. Figure 16.12(a) outlines the read/write process for a RAM. A control signal is used to enable or disable the cells. In the read operation, data from the selected memory cell is made available to the output. In the write operation, the bit information from the input is stored in the selected cell. The cell for reading from or writing into is selected by the address lines. ROMs can be accessed randomly through address lines but those have no writing option, as indicated in Figure 16.12(b). The enable terminal allows to detect data from the selected cell. The following terms are useful in this connection.

Memory Map: As mentioned earlier, the memory address provides the location of each memory device in the system of memory unit. The pictorial representation of the locations or address of the memory devices for the entire range of access is referred to as memory map.

Memory Interface: The CPU requires frequent access to the memory unit for reading data from and writing data to it. The memory interfacing circuit provides the access through the following processes.

(i) The required address lines of the address bus are connected to the corresponding address lines of the memory.
(ii) The remaining address lines of the address bus are decoded to generate chip select signals.
(iii) Control signals are generated to enable the appropriate data buffer.

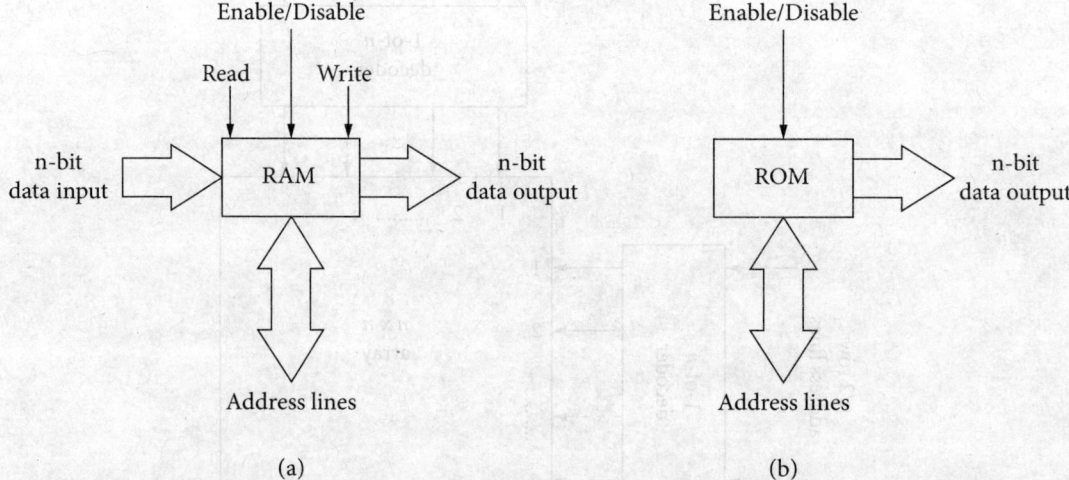

Figure 16.12 Schematic representations of (a) read/write in RAM and (b) read process in ROM

Example 16.5

A certain memory has the capacity of 1K×8. Determine (a) the number of bits that can be accommodated in the memory, (b) the required number of data input and data output lines and (c) the number of address lines.

Solution

The mentioned capacity indicates that there is provision for 1K = 1024 words, each of 8 bits. Therefore,

(a) The number of bits is 1024×8 = 8192
(b) To read and write 8-bit words, 8 data input lines and 8 data output lines are required.
(c) $1024 = 2^{10}$, therefore 10 bit address line is required to specify any one of these 1024 addresses.

Note: Matrix and Linear Addressing

The square memory configuration of *n* rows and *n* columns requires the minimum address lines and is referred to as *matrix addressing*. This is quite popular technique for memory organization. In contrast, there also exists a *linear addressing* with *n* rows and a single column (16×1 array). In this arrangement, selecting a cell means simply selecting the corresponding row and the column is always in use.

Multiple Choice-Type Questions and Answers

16.1 A digital-to-analog converter can accept a digital code in .... and convert it to a proportional ....
   (a) binary, voltage
   (b) binary or BCD, current
   (c) binary, voltage or current
   (d) binary or BCD, voltage or current

16.2 Statement-I: For each input digital code, the output voltage of a D/A converter must have a unique value.

   Statement-II: The output analog value may be greater than, less than or equal to the input digital value.
   (a) Only Statement-II is correct
   (b) Only Statement-I is correct
   (c) Both Statement-I and Statement-II are correct
   (d) None of statements I and II is correct

16.3 In a D/A converter, the weight of each bit is .... the previous bit, beginning with the LSB.
   (a) double of
   (b) 2^2 times
   (c) half of
   (d) $1/2^2$ times

16.4 The resolution of a D/A converter is .... the weight of the LSB.
   (a) less than
   (c) equal to
   (b) greater than
   (d) double of

16.5 The resolution and the accuracy of a digital-analog conversion system ....
 (a) should be compatible
 (b) are completely independent of each other
 (c) are such that one is much less than the other
 (d) are such that one is less than half of the other

16.6 The necessary resolution of a digital-analog conversion system is of eight bits. An enthusiast designer has planned for 12-bit conversion. If the circuit were constructed and implemented to the mentioned system, its performance would be of ....
 (a) better resolution (b) better accuracy
 (c) both better resolution and accuracy (d) more expenditure only

16.7 A 10-bit flash A/D converter requires .... comparators and .... reference voltage levels.
 (a) 9, 9 (b) 9, 10 (c) 10, 10 (d) 10, 9

16.8 The conversion time of a counter-type A/D converter becomes almost .... for each bit added to the counter.
 (a) double (b) 2^2 times (c) half (d) $1/2^2$ times

16.9 The memory arranged in 64 rows and 64 columns contains a total .... bits and .... address bits are required to select any one of the rows.
 (a) 128, 8 (b) 2^{12}, 64 (c) 2^8, 8 (d) 4096, 6

16.10 An address of at least .... bits is needed to locate a square memory of 4096 cells.
 (a) 12 (b) 6 (c) 8 (d) 64

16.11 A *volatile memory* means loss of stored data on switching off the power supply. Example:
 (a) ROM (b) DRAM only
 (c) SRAM and DRAM (d) SRAM only

16.12 There are .... words and .... bits in a 2K×8 EPROM chip.
 (a) 2048, 16384 (b) 2^{11}, 1
 (c) 2047, 16383 (d) 2^{14}, 2^{11}

16.13 An array with *n*-bit address and *m*-bit data has .... rows and .... columns.
 (a) n, m (b) 2^n, m (c) 2^m, n (d) m, n

16.14 Which of the following(s) memory element(s) is/are volatile? (I) flip-flop, (II) SRAM, (III) DRAM
 (a) I only (b) I and II (c) I, II and III (d) II and III

16.15 A four bit A/D converter is used to convert an analog voltage of 8V. The maximum error is ....
 (a) 0.5 V (b) 1.0 V (c) 2.0 V (d) 0.25 V

Answers

10.1 (d)   10.2 (c)   10.3 (a)   10.4 (c)
10.5 (a)   10.6 (d)   10.7 (b)   10.8 (a)
10.9 (d)   10.10 (a) 10.11 (c) 10.12 (a)
10.13 (b) 10.14 (c) 10.15 (a)

Reasoning-Type Questions and Answers

16.1 It is observed that the binary equivalent weight in both the weighted resistor and the R-2R ladder D/A converter is produced by the resistor network and not by the op-amp. So can we replace the op-amp by a resistor for obtaining the analog output voltage across it?

Ans. It might be possible ideally, if the resistor were much larger than those of the network. In reality, however, the resistor would spoil the perfection of the calculated network of resistors. The op-amp not only converts the input current to an output voltage, it also isolates the input and output ports. This permits the input currents generated at different levels proportional to the binary weights of different bits to be added without disturbing the relative weights of the resistances.

16.2 Why is the flash A/D converter so fast in operation?

Ans. The comparator output changes instantaneously with the change in the analog input level. The process is direct and not through any clock or timer. Thus the conversion time depends only on the propagation delays of the converters and that of the encoder gates while producing a change in the output state. Thus the conversion becomes very fast, of the order of nanoseconds.

16.3 What decides the mean conversion time for an n-bit counter-type A/D converter?

Ans. Since the counting starts from zero in usual binary sequence, the n-bit conversion may require up to 2^n counts. This is, however, the case of the maximum count and on the average, there are $2^n/2$ counts. So the mean conversion time is that taken for 2^{n-1} counts.

Solved Numerical Problems

16.1 A 4-bit D/A converter has an output current of 10 mA for the digital input of 1010. What would be the output current for the input of 1110?

Soln. The digital input is $(1010)_2 = (10)_{10}$. Therefore, the proportionality constant is (10 mA)/10 = 1 mA.

The digital input 1110 is equal to decimal 14 so that the analog output current is 14×1 = 14 mA.

16.2 The constant +V in the circuit of Figure 16.3 is replaced by a low frequency sine wave voltage of amplitude 5 V and let the D/A converter be of 8 bits. What will be the output, if the digital input is made fixed at (i) 11111111 and (ii) 00001000?

Soln. It is assumed that the settling time of the D/A converter is less than the time period of the low frequency sine wave voltage. Then the output will be a sine wave of the same frequency as the original with all cases of fixed digital input. However, the amplitude will be different with the digital inputs and the phase will differ by 180° with respect to the original. The amplitude of the output waveform can be calculated by using Equation (16.8).

For the input of 11111111, the output amplitude is

$(5/2^8) \times (2^7 + 2^6 + 2^5 + 2^4 + 2^3 + 2^2 + 2^1 + 2^0) = 4.98$ V

For the input of 00001000, the output amplitude is

$(5/2^8) \times (0 + 0 + 0 + 0 + 2^3 + 0 + 0 + 0) = 0.156$ V

16.3 A 16-bit D/A converter has step size of 0.1 mV. Determine the full-scale output voltage and the percentage resolution of the converter.

Soln. There are $2^{16} - 1 = 65535$ steps, each of 0.1 mV so that the full-scale output voltage is

$0.1 \times 65535 \approx 6.55$ V

The percentage resolution is

$(0.1 \text{ mV})/(6.55 \text{ V}) \times 100 \approx 0.0015$ %

16.4 A counter-type 8-bit A/D converter is driven by 100 kHz clock pulse. Calculate (i) the maximum conversion time, (ii) the average conversion time and (iii) the average conversion rate.

Soln. The clock has time period of 10 μs, hence each count advances at the rate of 10 μs.

(i) The maximum count is $2^8 = 256$ so that the maximum conversion time is 256×10 μs = 2.56 ms.

(ii) The average conversion time is $(256 \times 10 \text{ μs})/2 = 1.28$ ms.

(iii) Considering the average conversion time, the converter is capable of making $1/(1.28 \times 10^{-3}) \equiv$ approximately 781 conversions per second.

16.5 If four bits are to appear at the output, how should a 1 K ROM be organized?

Soln. The ROM has 1024 bits that should be organized as

$1024 \div 4 = 256$ words, each of 4 bits.

The 1024 memory cells are to be arranged in $2^5 \times 2^5$ square array. Therefore, 5 address bits are used to select any one of the 2^5 (= 32) rows and the 32 columns are divided into 8 groups of 4 bits.

Any one of the 8 (= 2^3) groups can be selected with 3 address lines.

Exercise

Subjective Questions

1. Draw the circuit diagram of a 4-bit *R-2R* ladder-type D/A converter and derive the expression for the output analog voltage produced by it. What is the advantage of *R-2R* ladder D/A converter over the D/A converter constructed with binary weighted resistors?
2. Construct an n-bit D/A converter of weighted resistors R, R/2, ..., $R/2^{n-1}$ and derive the expression for its analog output voltage.
3. Draw an explanatory block diagram of 3-bit flash A/D converter and express its input–output conditions in tabular form.
4. Draw an explanatory block diagram of counter-type A/D converter. How does the conversion time change with the analog input voltage value?
5. Make a comparison of the merits and limitations of flash and successive-approximation techniques of A/D conversion.
6. Put up a model circuit to explain the operation of SRAM cell. How DRAM is similar to and different from SRAM?
7. Present a brief sketch for addressing a 64×64 memory array.

Conceptual Test

1. If an 8-bit D/A converter is converted to 9-bit keeping the other parameters same, what is about the full-scale output voltage?
2. How many different output voltages can be produced with a 16-bit D/A converter?
3. If the input reference of a D/A converter is increased, what will happen to (i) the resolution and (ii) the full-scale output?
4. It is fact that a magnetic tape provides inexpensive storage of large amount of data. Also it can be read or written easily. Can it be used, instead of RAM in a computer?

Numerical Problems

1. An 8-bit D/A converter produces an output voltage of 863 mV for a digital input of 00101100. What is the full-scale value of the output voltage for this converter?
2. Estimate the maximum improvement of conversion time, if an 8-bit analog to digital conversion with 1 MHz clock pulse is shifted from counter-type to successive-approximation type system?
3. An 8-bit A/D converter produces full-scale output for an input of 5.10 V. If the converter system has an inherent fixed error of 0.2%, estimate the total possible error.
4. A range of 0 to 10 mA of current is produced by a D/A converter. The current is proportionally amplified by a perfect amplifier for driving a motor. The speed of the motor varies from 0 to 1000 rpm. If the motor speed is to be accurate within 2 rpm of the desired speed, how many bit should the D/A converter have?
5. The reference voltage of an analog-to-digital converter is 1 V. The smallest voltage step that the converter can record using a 12-bit converter is
   (a) 0.24 V (b) 0.24 mV (c) 0.24 µV (d) 0.24 nV
   [JEST 2015]
6. Specify the address lines and decoder output lines, if a square memory of 1 K is to be addressed.

Project Work on Chapter 16

Simulate the weighted resistor D/A converter circuit of Figure 16.2 with LTspice and study its output analog voltages for different combinations of the four input bits.

The LTspice simulated circuit is shown in Figure P16.1. The op-amp is chosen arbitrarily and the op-amp voltage supply is constructed by joining two equal voltage sources of 12 V. The digital inputs A (LSB), B, C and D (MSB) are either 0 (representing logical 0) or 5 V (representing logical 1). Each input combination from 0000 through 1111 is to be connected separately to the 5 V source or to the ground. For example, the figure shows the condition when D = 0 and C = B = A = 1.

From the menu, 'Simulate' and 'Edit Simulation Command' and then 'DC Sweep option is selected. The type of sweep is linear and both the start and stop value are 5 V. Right clicking on the viewer and selecting 'File' and 'Export data as text' options, the analog output voltage (V_o) values are recorded for each combination of input bits.

Since it is a simulation, the results are expected to be almost the same as those calculated with Example 16.2 thereby yielding 100% accuracy. The study should be extended with the following investigations.

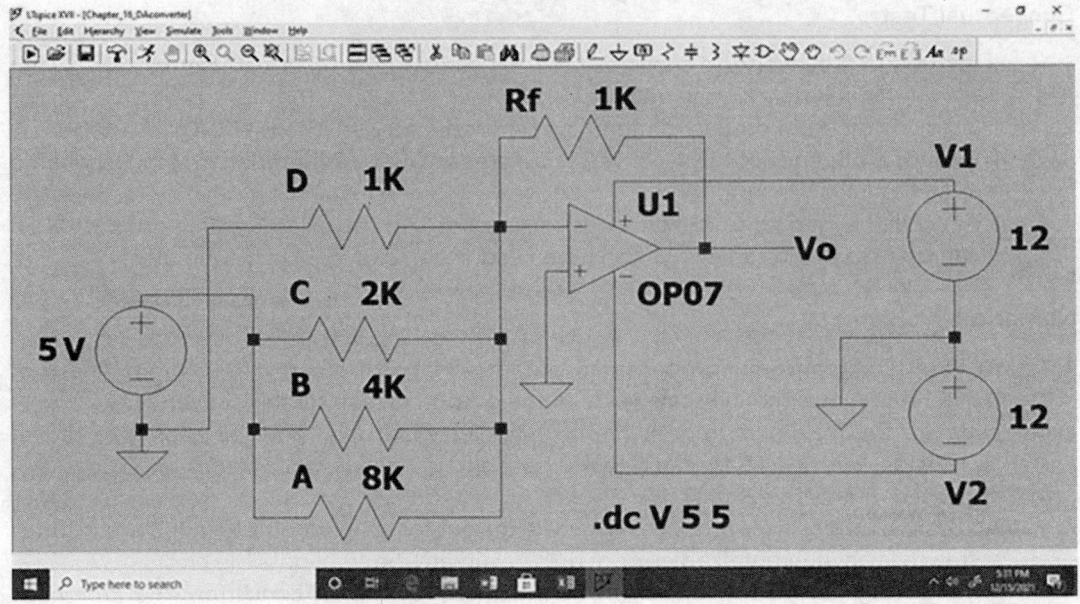

Figure P16.1 LTspice simulated 4-bit D/A converter

- Plot a graph of input digital numbers 0, 1, ..., 15 versus the corresponding analog output voltage and ensure the linearity.
- Repeat the experiment with another op-amp.
- Repeat the experiment with another set of input resistances maintaining the same ratio of R, 2R, 4R and 8R.
- Repeat the experiment with another value of the digital input, such as 3 V or 4 V for representing logical 1.

17
Microcomputer and Microprocessor

The modern computer is such as electronic gadget where all of the analog and digital electronic devices mentioned in this book are included somewhere or else. This chapter presents a brief idea on the basic principles and architectural blocks of a computer. The role of microprocessor in computer and the fundamentals of programming the microprocessor are interpreted using 8085 as a model.

17.1 Evolution of Computer

The modern computer evolved as an electronic instrument and got so much involved in scientific, technical, commercial, and household applications that it gave birth to *computer science* as a separate subject. Anyway the first computer was not electronic but a mechanical instrument. Indeed performing mathematical calculations with instruments is quite an ancient concept. The word 'calculate' is derived from the Latin word *calculos*, which means *stone*. Pieces of stone, pebbles, digits of fingers, knot of ropes, nodes of sticks and many other diversified equipment have been employed in arithmetic calculations in different ages of human civilization.

17.1.1 Historical Background

The first known calculating machine is abacus, which was popular in ancient China, Japan, Egypt, and also different countries of Europe. The earlier version of abacus consisted of slotted earthen flats over which pebbles were moved to indicate number positions. The later versions comprised of beads penetrated into framed wires. Table 17.1 outlines a brief sketch of the development of computational concepts and instruments till the middle of the twentieth century.

Charles Babbage, as mentioned in the table, was the first to incorporate all the basic features of a modern computer in his instrumentation. He constructed a large mechanical system named *Difference Engine* that possessed the following components.

- *Mill*: It was capable of executing arithmetic processors. It may be compared to the arithmetic logic unit (ALU) of a modern computer.
- *Store*: It could store numbers, analogous to the memory of today's computer.
- *In/Out Ports*: There were separate ports for giving instructions and obtaining results, similar to the input and output terminals of a modern computer.

Babbage's computer could accept instructions from the user, keep those in its memory and execute self-sustained operation with making necessary decisions on its own. Babbage had the plan for developing an improved machine named *Analytical Engine* that he could not complete despite his lifelong dedication. Monetary deficiency and lack of improved technology were the major barricades. In fact the contemporary society could not realize whether such a large and expensive machine was at all necessary for the purpose of mere calculation. George Boole was another tragic hero whose path breaking reasoning on two-state logic remained abstruse at his own time. It appears that Babbage and Boole were too advanced with references to their era.

Table 17.1 A few milestones along the pathway to the modern computational era

Name, life span and the country born in	Contribution to computation in brief
John Napier (1550–1617), Scotland	Discovered the theory of logarithm and invented the calculating instrument known as *slide rule*, which was operated by sliding and aligning one scale on the other graduated in logarithmic values.
Blaise Pascal (1623–1662), France	Invented a calculating machine at young age and incorporated the following features that are used in modern computers also. (i) Performing the carry operation. (ii) Subtraction by negative addition. (iii) Multiplication by repeated addition. Not the least, he invented the machine to help his father in calculating tax accounts.
Gottfried W. Leibnitz (1646–1716), France	Improved mechanical calculations and refined the binary number system, which is the foundation of computer.
Joseph Jacquard (1752–1834), France	Invented cards with punched holes for controlling the threads in weaving cloths. A programmable loom could be designed with such cards with different arrangements of holes. Babbage applied the concept of punch card for giving input codes to his computer.
Charles Babbage (1791–1871), England	He is honored as the 'Father of computer' because he was the first to propound the concept of digital programmable computer including the features of *memory* and *decision making*.
George Boole (1815–1864), England	Propounded a two-state logic involving objects of two classes, each with a given property. The other class was characterized by the absence of that property. The real importance of his ideas, now famous as Boolean logic was not at all realized at his time. Indeed the term *Boolean algebra* was coined after his death.

Herman Hollerith (1860–1929), America	Invented electromechanical punched cards as the input medium for computer programs where digital information is represented by presence or absence of holes in paper card for electrical connection. He also developed a methodical coding system for computer instructions.
Augustus De Morgan (1806–1871), India Maurice Karnaugh (1924 – living), America	Their work on mathematics and logic enriched Boolean logic and contributed significantly to its maturity as a full-fledged algebra.
Claude Shannon (1916–2001), America	Pioneered the application of Boolean logic through electronic circuits.
John W. Mauchly (1907–1980), America J. P. Eckert (1919–1995), America	Designed the first general purpose electronic digital computer known as Electronic Numerical Integrator and Computer (ENIAC).

> **Think a Bit: Babbage's Work Was Truly 'Interdisciplinary'**
>
> Babbage's computer was a mechanical system but digital in operation. He applied, as input device, the perforated cards invented by Joseph Jacquard originally meant for weaving of cloths with different designs. The presence or absence of holes in the card were analogous to the 1 and 0 in modern computers. It is worthwhile to mention that the computers of even a few decades ago used punched cards as input devices that used similar principle of perforation for electrical connection. Lady Augusta Ada (1815–1852), the daughter of poet Lord Byron and the Countess of Lovelace was the first to realize the significance of Babbage's machine and she composed the first algorithm for the machine. Ada is recognized as the first computer programmer of the world and is remembered for her unique comment, "The Analytical Engine weaves algebraic patterns just as the Jacquard loom weaves flowers and leaves."

17.1.2 Modern Computer

The resurrection of Babbage's dream and Boole's reasoning took place almost a century later when the following developments were achieved.

(i) The initiation of two-state logic matured into Boolean algebra through the theories propounded by De Morgan, Karnaugh, Turin, et al.

(ii) The experimental techniques put forward by Shannon, Hollerith, Newmann, et al. made it possible to implement Boolean algebra in the form of electronic circuits.

(iii) The implementation of improved algorithm and other technical accomplishments were behind the success.

The first digital computer was built up contemporary to the Second World War. Since then, the science and technology of computation has passed through many stages of development, which can be divided in several generations, as follows.

First Generation (1946–55): Considering the creation of ENIAC in 1946, the commencement of the first generation of computers can be specified with the following features.

(i) These were electronic computers made of vacuum tubes.
(ii) The instruments worked on binary system.
(iii) Instructions were given in machine language.
(iv) The whole program could be stored within the memory during the program execution.

Second Generation (1956–63): The Nobel prize for the invention of transistor was awarded in 1956. The second generations were made of transistors and other semiconductor devices and had the following features.

(i) Smaller size, lower power consumption and higher speed and efficiency.
(ii) Use of magnetic core as memory element.
(iii) Introduction to two high level languages: FORTRAN and COBOL.
(iv) Beginning of commercial use of computers.

Third Generation (1964–1970): This generation is specified with the use of integrated circuit (IC). In 1964, International Business Machine (IBM) announced the first miniature computer system for all ranges of applications having the following properties.

(i) Semiconductor ROM and RAM were used.
(ii) Higher versions of high level languages were developed.
(iii) The *byte* unit for memory and the use of *virtual memory* were introduced.
(iv) Many peripheral devices, such as optical scanner, magnetic ink character reader and graph plotter came into picture.
(v) Computers became popular among general people, academic institutions and business concerns.

Fourth Generation (1971–?): This generation of computer evolved with the invention of microprocessor, which encapsulated the whole arithmetic and logical functional blocks within a single IC. The first microprocessor was designed in 1971. The fourth generation computers have the following features.

(i) Higher speed and more data handling capability compared to the previous computers.
(ii) The operating system was developed.
(iii) The term *microcomputer* came forth. The computer became affordable to all and the phrase 'personal computer' became popular.
(iv) The idea of incorporating the mini-sized controlling device in IC form for an instrument originated during this period.

Fifth Generation (?–): It is quite difficult to demarcate the fourth and the fifth generation of computer because the fourth generation started with the implementation of microprocessors and still the computers of common use are made of improved microprocessors. Yet we must feel that so many diversified applications of computers were not available in the early days of the fourth generation.

The term 'fifth generation' was coined by Japan's Ministry of International Trade and Industry in 1982 when they started an initiative to create computers of massive parallel processing. During the last 3–4 decades, the following remarkable achievements in the field of computation are noted.

(i) High performance hardware.
(ii) Varieties of improved operating systems.
(iii) Machine intelligence, such as recognizing human voice and fingerprint through computer programming.
(iv) Diversified application software for virtually everything.
(v) Global computer networking, such as Internet.
(vi) Emerging of computer network and cyberspace as new media for data storage, information source and social interaction.

17.2 Computer, Microprocessor and Microcontroller

A common computer, whether it be a laptop or desktop, contains all its electronic circuits for arithmetic and logic operations confined within a single unit referred to as *central processing unit* (CPU). The high capacity supercomputers have a number of such units where the processes are executed simultaneously. Such a system is termed as *parallel processing system*. Examples of supercomputer in India, owned by Government organizations are Param Siddhi-AI, Pratyush, and Mihir used for meteorology, weather forecasting, health care systems, agriculture and other tasks. The personal computers that we are familiar with contain only one CPU fabricated on a single IC chip, known as microprocessor. Intel introduced (1971) the first microprocessor named 4004, which was of 4-bit operation. In today's computers, 64-bit operation is very common. Microprocessors are widely used in all digital systems including desktop and laptop computers, smartphones and many automated controlling systems.

17.2.1 Computer Organization

Figure 17.1 depicts a simplified block diagram for the architecture of a single CPU computer. It also indicates the workflow directions. It has the following main components.

Input Device: The user enters the instructions to the computer through input devices, such as keyboard or mouse. Also the input information may come from the secondary memory, such as hard disk or from other data sources like Internet or data acquisition systems. Whatever may be the input signal, electrical or nonelectrical, that must first be converted to an analog electric signal of specified amplitude. Then it is converted to the digital equivalent (A/D converter of Chapter 16) and is assigned some code with an encoder. This digital version of the user's input is the actual input to the CPU.

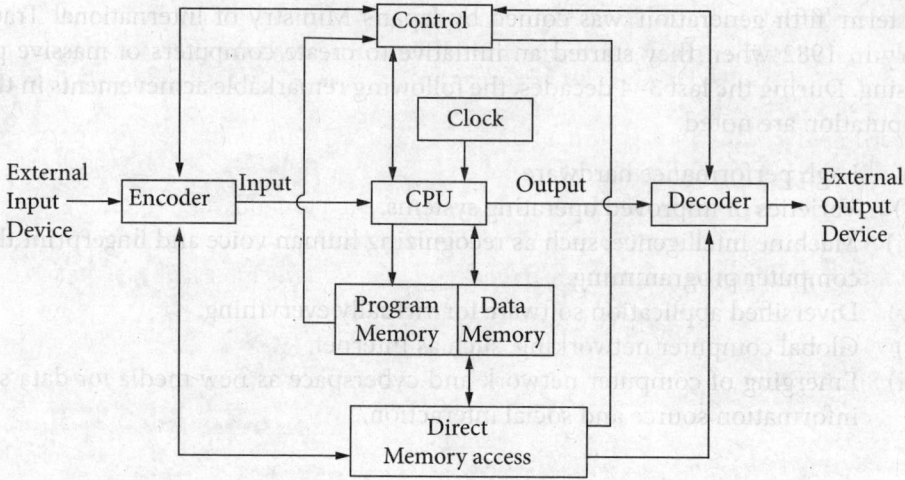

Figure 17.1 Block diagram of computer organization

Output Device: The results of the program execution are supplied to the user through output devices, such as monitor screen or printer. However, before reaching these devices, the actual output from the CPU is converted to the proper output format with a decoder and is converted to analog equivalent (D/A converter of Chapter 16).

Central Processing Unit (CPU): This is the main working unit of the computer where the mathematical and logical operations are executed. It is comprised of digital circuits of the following categories.

- *Arithmetic Circuit*: The adder (Chapter 14) can perform binary addition. The same adder can perform subtraction operation by negative addition, multiplication by repeated addition and division as repeated negative addition.
- *Logical Circuit*: The comparator circuit (Chapter 14) outlines the working principle of comparing two digital numbers in terms of voltage.
- *Dataprocessors*: Multiplexing and demultiplexing (Chapter 14) coordinate the flow of data through limited path in a methodical way. The data transfer occurs through a set of wires called *bus*. The number of wires in a bus is in power of 2.
- *Register and Counter*: These sequential logic circuits, similar to those mentioned in Chapter 15 store the instructions under process and account for the number of instructions executed in a particular sequence.

Thus the CPU contains both combinational and sequential logic circuits. It fetches instructions and necessary data from the memory, decodes those and executes by acting through the above mentioned digital circuits.

Memory: This unit stores digital information in an organized way with the provision for easy accession. The *main memory*, also termed as *program memory* stores the program or series of instructions and maintains a continuous interaction with the CPU. It is essentially semiconductor memory RAM (Chapter 16) and gets wiped out as soon as the power supply is switched off. The other part is *secondary memory* or *auxiliary storage*, which is made of magnetic (e.g. hard disk) or optical (e.g. CD) devices. The secondary memory is actually a storage of data. It simply holds a large amount of data or the necessary software that can be accessed by the CPU on requirement. So this is also called *data memory*. It is a permanent storage and does not get erased unless the user deliberately instructs for the removal.

Think a Bit: How Does It Execute Mathematics?

The computer is, of course, capable of executing various mathematical operations, such as three dimensional integration or handling complex numbers. How does the electronic circuit carry out such complicated processes? Actually the digital circuits within the computer can perform only arithmetic and comparison operations in terms of voltage. It is the user's program that translates the intricate mathematics in that way. It is pleasant to realize that virtually any mathematical process ranging from definite integral to trigonometric series can be pulled down to the arithmetic level. There is a separate subject known as *Numerical Methods* where the solution of algebra and calculus are performed in terms of only numerical processes. It is of venerable surprise that scientists like Newton, Gauss and others established powerful numerical techniques far ahead the advent of computer, which are now found to be very useful to computer programming for mathematical operations.

Note: Analog and Digital Arithmetic Circuits

The op-amp adder circuit (Chapter 11) can add two numbers in terms of voltages. The comparator or Schmitt trigger (Chapter 11) can compare two voltage levels. These circuits involve continuous variation of voltage and the voltage level itself represents the number. Today's computer and microprocessor do not use such analog circuits for computation. Instead, digital adder and comparator (Chapter 14) are used. However, analog computation is still used in some specific cases, for instance flight computer in aviation and industrial control systems. The analog circuit is straightforward and avoids the extra complexity and conversion time required for analog-digital conversions. An analog computer, custom-designed for a specific task is supposed to be energy efficient.

17.2.2 Use of Microprocessor

The common desktop and laptop computers have the CPU confined within a single IC chip referred to as microprocessor. The name implies the provision for arithmetic and logical processes within a micro-sized workplace. Intel (USA) and Busicom (Japan) developed the first microprocessors. Now many varieties of microprocessors produced by different companies are available on market. The improved versions of microprocessors of the same make involve the following qualities.

- Higher clock speed
- Larger memory handling capability
- Capability of carrying larger amount of data:16 bits, 32 bits or 64 bits at a time

Table 17.2 Comparative view of the Intel microprocessor products starting from the early days

Processor	Year of Introduction	Initial clock speed	Address bus	Data bus	Addressable memory
8085	1976	5 MHz	16-bit	8-bit	64 KB
8086	1978	5 MHz	20-bit	16-bit	1 MB
8088	1979	5 MHz	20-bit	8/16-bit	1 MB
80286	1982	8 MHz	24-bit	16-bit	16 MB
80386	1985	16 MHz	32-bit	32-bit	4 GB
80486	1989	25 MHz	32-bit	32-bit	4 GB
Pentium	1993	60 MHz	32/64-bit	32/64-bit	4 GB
Pentium II	1997	233 MHz	36-bit	64-bit	64 GB
Pentium 4	2000	1.4 GHz	36-bit	64-bit	64 GB

Table 17.2 outlines a development history by comparing several Intel products in increasing order. An improved microprocessor should have the following facilities as well.

- *Multitasking*: executing more than a single task, such as data counting, device driving etc. simultaneously.
- *Virtual Memory*: to operate by keeping some part of the program in the secondary memory that gives the facility of enhancing main memory resource.
- *Pipelining*: mutual interconnections of the given instructions.

This chapter illustrates the architecture of Intel 8085 microprocessor and demonstrates the programming process with it. This is used as a prototype for understanding the principles of computer hardware-software interaction.

17.2.3 Use of Microcontroller

It is true that the microprocessor is the 'heart' of the computer. Being within the computer, the microprocessor can do two types of jobs: controlling the computer hardware and meeting the need of software. The improvement of microprocessor upgraded the computer in terms of speed, memory size and data bits thereby enhancing the software handling capability to a great extent. Yet there is another significant field of engineering applications, such as control systems where these merits are not relevant at all. An 8-bit CPU with limited memory and in/out capacity can serve the purpose efficiently and an expensive 64-bit microprocessor is quite unnecessary in that case. Also the microprocessor is only a CPU, it does not have input and output provisions of its own. Therefore, an alternative technology is developed for digital instrumental systems where the input/output ports, timers, limited memory and other peripheral units are embedded to the single-chip IC thereby making it a miniature computer suitable for the instrumental purpose only. Such a digital IC is called *microcontroller*. Since the different working units are embedded to the CPU, a digital system comprised of microcontroller is called an *embedded system*. Table 7.3 outlines the comparative features of microprocessor and microcontroller.

Table 17.3 Comparison of microprocessor and microcontroller

Microprocessor	Microcontroller
This contains CPU, working registers, a clock timing circuit and interrupt circuits. However, it does not contain memory and input/output devices.	In addition to the CPU, registers and other components of a microprocessor, the microcontroller contains ROM, RAM, a clock circuit and parallel and serial ports for input/output operations. Thus it may be imagined as a little computer on a chip.
The prime job of a microprocessor is to fetch data, perform extensive calculations and yield the results. These can be executed only with the support of other units.	It fetches data, performs limited calculations and based on that, it communicates signals for controlling some parameter, such as temperature or flow of liquid. All the units for these tasks are present on the chip.
Microprocessors have many operational codes for moving data from memory to CPU and only one or two types of bit handling instructions.	Microcontrollers have one or two operational code but many types of bit handling instructions.
It is concerned with flow of code and data from external memory to CPU.	It is concerned with flow of bits within the chip for a specific task.
It is a general purpose device having prime use as the CPU of a general purpose computer where it can meet the requirements of various types of software instructions. Control systems can be constructed with a microprocessor with the support of other digital units but that is not convenient.	This is also general purpose device but a particular set up is dedicated to a specific application, such as controlling speed, temperature or light intensity.
Microprocessors are suitable when the main objective is executing huge amount of complex calculations and data handling at a high speed, such as that in a computer.	Microcontrollers are suitable when the amount of data is not so high but the precisely controlled functioning of a machine, such as an automated instrumentation is the major requirement.
Examples of Microprocessor: 8085 marketed by Intel Corporation in 1970s and Intel Core i7 or i9 of today's computers.	Examples of microcontroller: 8051 developed by Intel in 1970s and Atmega328p, developed by Atmel Corporation in 2016. It may be mentioned that Arduino, an open source electronics platform has developed, using microcontroller (e.g. Atmega328p), a combination of hardware and programming language that can be run by common computers. A few words on Arduino are stated in Section 17.8.

17.3 Hardware and Software

The digital circuits and the related mechanical and instrumental systems of a computer are known altogether as *hardware*. The physical elements of a computer including the keyboard, the mouse, the monitor and all the ICs within the casing are parts of the hardware. However, the equipment is lifeless until suitable binary instructions in the form of voltage levels activate it. Such instructions that operate the computer in a systematic manner are referred to as *software*.

The hardware and the software of a computer are interdependent. The software directs the hardware to execute the required task. Without the software, the hardware has no function. Similarly, the software has nothing to do in absence of the hardware. Each hardware component executes some specified task. For example, an adder cannot replace a register. The keyboard cannot do the work of the video screen. However, varieties of software can be implemented to the same set of hardware to accomplish many different tasks. Consequently, the same computer can perform mathematical operations, process images, plot graphs and play music with the help of different software, which can be classified in two broad categories.

System Software: It helps run the hardware and provides with a platform to other software. Operating system, loader, linker, computer language translator and antivirus are examples of system software.

Application Software: It enables the user to perform some specific tasks, such as word processing, designing graphics, working on spreadsheets and databases and even playing games.

Windows and Android are two popular operating systems, hence examples of system software. Microsoft Office under Windows operating system is a popular set of application software. Sometimes the application software is abbreviated as 'App'. Today there exist plethora of apps for virtually every task.

> **Note: Firmware**
>
> The firmware is something at the boundary of hardware and software. It is a machine-level computer program stored permanently or semi-permanently for running specific hardware. A prominent example of firmware is ROM BIOS (Basic Input–Output System), a computer program stored in flash ROM that executes the elementary functions for hardware initialization. A printer, a scanner and the remote control of a television contains a firmware of its own.

17.3.1 Operating System

It is the most remarkable system software of a computer. It is a set of programs that act a medium between the user and the machine. It executes the basic functions, such as

- interaction between the hardware and the software,
- the process of storing information to and accessing information from the secondary memory, and
- communication between the CPU and the peripherals.

Depending on the computer hardware, the operating systems are also available in diversified forms. The upgrading of the hardware of a certain manufacturer calls for simultaneous improvement of the operating system to a higher version. Some popular operating systems are mentioned briefly.

Disk Operating System (DOS): This single user operating system was developed by IBM and was upgraded by Microsoft. The home computers during 80s and 90s were largely operated with DOS. Many upgraded versions of DOS were marketed. Today it is not so popular but still some specific tasks, such as

- prototyping in programming
- updating firmware
- partitioning disks and
- testing embedded systems

are executed with DOS. It is a command-based operating system. There are assigned words, known as *command* for each task, such as copying files, listing directory contents or printing documents. These words are actually the names of the executable program or part of the operating system itself. To execute a command, the word is to be entered with keyboard.

Windows: This single user, multitasking operating system was developed by Microsoft, the first version being published in 1985. Passing through the evolutionary developments of many versions, such as 3.1, 3.11, 95, 2000, XP, 10, etc., the operating system is still very popular in personal computers for varieties of applications. Windows added some new dimensions to the use of computer. For instance:

- Computer networking including Internet
- Easy access to playing music, video, television, etc.
- Diversified application software
- Executing several programs at a time

Instead of command typing with keyboard, Windows popularized the technique of *Graphic User Interface* (GUI) that lets the user to instruct by clicking mouse pointer on icons, buttons and menus on the screen.

Android: This single user operating system is developed by Google (originally by Android Inc. purchased by Google) mainly for mobile devices, such as smart phone and tablet computers having on-screen objects, virtual keyboard and Bluetooth (a wireless technology for exchanging data over short distance using radio wave of frequency $\approx$ 2.4 GHz. Android devices are usually battery operated. Therefore, the system operation complies with minimum consumption of power. A lot of application programs, nicknamed as 'Apps' extend the functionality. An application when not in use, does not use CPU resources. Third-party applications can be downloaded and installed by the user.

Unix: This is a multiuser, multitasking operating system originated at Bell Labs. Many versions are developed so far, some being optimized for mainframe computers. Unix has the following features.

- It is portable, usable in different computing platforms.
- It is of modular design. A set of software tools, each capable of performing limited functions remain at core level and another software combines the tools to perform complex tasks.
- There is a master control program called *kernel*, which (i) handles files, (ii) controls hardware interaction and (iii) allocates time and memory slots to programs.
- Another set of programs, known as *shell* acts as an interface between the user and the kernel. For example, it interprets the command that the user types at the command line and arranges those for execution.

Linux: This operating system is quite flexible and is becoming popular now a days in both personal and mainframe computers and in servers. Unlike Windows etc., which are proprietary software, Linux is open source, free to use software. It was developed by a volunteer group and is always available for modification by the user.

Linux is actually the core component of the operating system. It is referred to as kernel, which manages the input–output interaction, the processing and the memory handling. The same kernel can be used in different distributions, such as Fedora (Red Hat), Ubuntu, Debian, etc. Thus the Linux operating system is a collection of software based around the kernel, available for a wide variety of systems: PC, mainframe and server.

17.3.2 Computer Languages

All the instructions for the computer including operating systems, application software and source programs are composed with some or other language. The following types of computer languages are available.

Machine Language: All instructions to the computer ultimately touches the hardware in the form of presence/absence of voltage representing binary codes. The machine can understand the language comprised of 0s and 1s only and it is termed as *machine language*.

Assembly Language: In the earlier days of computer, instructions used to be given in machine language directly. Because of the inconvenience in managing innumerable 0s and 1s, some specific combinations of 0 and 1 are assembled to coin several linguistic codes referred to as mnemonics. These are small words like MOV (to copy data), ADD (to add) or HLT (to stop). The language composed of such symbolic codes is called *assembly language*. It facilitates the user to instruct the computer. The hardware cannot understand it directly. Either manually or using a software named *assembler*, the assembly language code is converted to the equivalent machine code.

Both the machine language and the assembly language are called *low-level language* because these are closer to the machine level and these depend on the hardware. Computers of different make have different low-level languages.

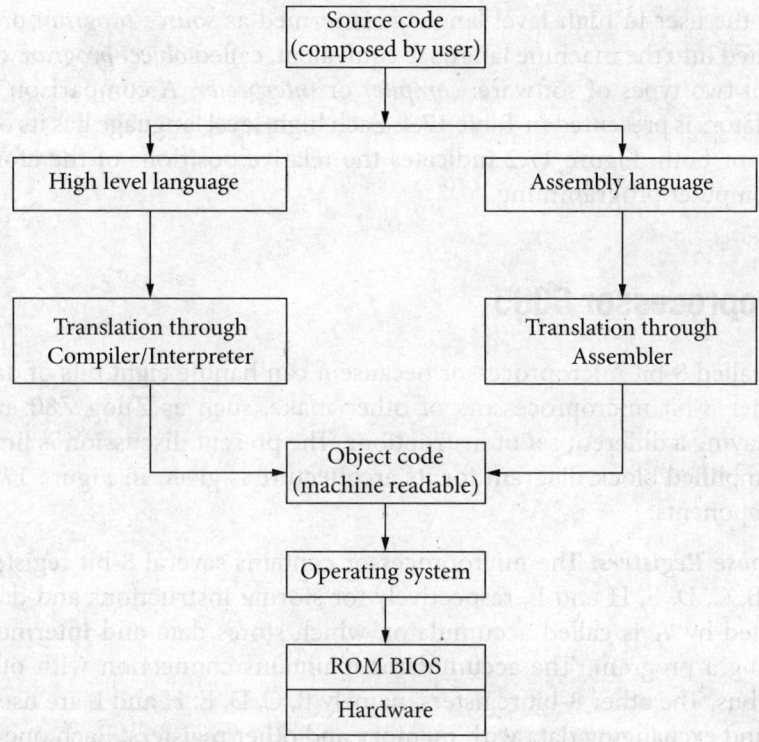

Figure 17.2 Relative positions of different types of software between the user and the hardware

Table 17.4 Comparison of compiler and interpreter

Compiler	Interpreter
It is a software for translating the source code to the corresponding object code.	It is also a software for similar purpose.
It reads the entire source code and generates the equivalent machine code but does not execute until the instruction is given.	It translates instruction by instruction into the object code and also executes the same before fetching the next one.
Compiler works faster. The time taken to analyze the source code is large but once that is completed, the overall execution time is less.	The time taken in analyzing a single statement of the source code is small but the total execution time is large.
Compiler is suitable when the structure of the program has become ready.	Interpreter is easy to debug, hence suitable for building a new program by trials.
Once the program is compiled, the source code is no longer required unless some change is made.	The source code is necessary all the way while running the program.

High-Level Language: It is further development of machine level codes and the text of a high-level language looks almost the same as a human language. Our familiar languages, such as BASIC, FORTRAN, C, C_{++}, JAVA and Python are high-level languages. The program composed by the user in high-level language is termed as *source program* or *source code*. This is translated into the machine language equivalent, called *object program* or *object code* using either of two types of software: *compiler* or *interpreter*. A comparison of these two types of translators is presented in Table 17.4. Each high-level language has its own compiler or interpreter or both. Figure 17.2 indicates the relative positions of the aforementioned software in computer programming.

17.4 Microprocessor 8085

Intel 8085 is called 8-bit microprocessor because it can handle eight bits of data at a time. There are other 8-bit microprocessors of other make, such as Zilog Z80 and Motorola 68008, each having a different set of instructions. The present discussion is limited to 8085 only and a simplified block diagram for its architecture is given in Figure 17.3. It has the following components.

General Purpose Registers: The microprocessor contains several 8-bit registers indicated by letters A, B, C, D, E, H and L, respectively for storing instructions and data. The 8-bit register denoted by 'A' is called accumulator, which stores data and intermediate results while executing a program. The accumulator maintains connection with other registers through data bus. The other 8-bit registers, namely B, C, D, E, H and L are used for storing instructions and exchanging data with memory and other registers. Each one is capable of storing one byte but can be cascaded as B-C, D-E and H-L to perform 16-bit operations. These seven registers can be accessed by the user for storage of instructions and data while running a program.

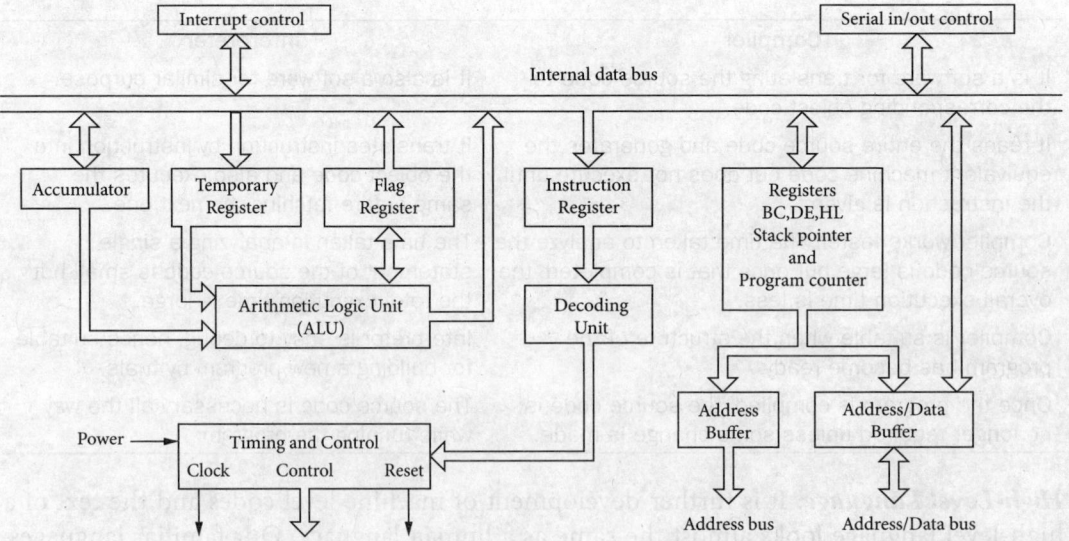

Figure 17.3 Simplified block diagram of the architecture of microprocessor 8085

Internal Registers: In addition to the above registers, there are some other registers that are not accessible to the user. These are meant for some pre-determined functions only. The followings are these internal registers.

Temporary Register: This 8-bit register stores the operands, such as multiplicand, divisor and addend. Also call-subroutine operations, mentioned later are done via this register.

Instruction Register: This 8-bit register stores the instructions fetched one by one from the memory. The decoder decodes it and establishes the sequence to follow.

Flag Register: The ALU includes five flip-flops, which are set or reset after arithmetic or logical operations. These are represented by five bits of an 8-bit register, named *flag register*, as specified in Figure 17.4. The bits, as mentioned below, work as flags to indicate some specific conditions.

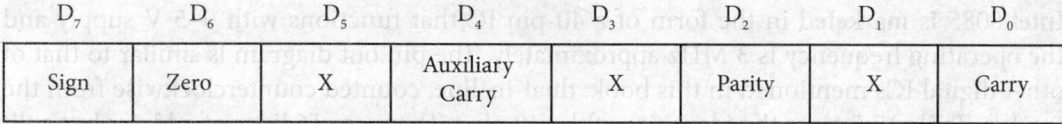

D_7	D_6	D_5	D_4	D_3	D_2	D_1	D_0
Sign	Zero	X	Auxiliary Carry	X	Parity	X	Carry

Figure 17.4 Bit positions in flag register

- *Sign*: This flag is set, if the MSB (D_7) of the result of an operation becomes 1.
- *Zero*: This flag is set when the result of an operation is zero.
- *Auxiliary Carry*: If a carry is generated from the lower nibble (generated by D_3), the auxiliary carry flag is set. This flag is used internally for BCD operations only.
- *Parity*: This flag is set when the result has even number of 1s.
- *Carry*: If an arithmetic operation results in a carry, this flag is set.

program Counter: This is a 16-bit register that points to the memory address from which the next byte is to be fetched. When a byte for machine code is being fetched, the program counter is incremented by one so as to point at the next address.

Stack Pointer: The 8085 microprocessor has the provision for utilizing a portion of the read/write memory specified by the programmer for the purpose of storing data temporarily during the execution of a program. A 16-bit register named *stack pointer* points to the memory location in the stack, i.e. an array of addresses such that the data written last is accessed first. The stack pointer tracks the storage and retrieval of data, two bytes at a time. Therefore, each time the data exchange takes place, the 16-bit memory address in the stack pointer is incremented or decremented by 2.

Timing and Control: This unit synchronizes all the operations with the clock and generates the control signals necessary for communications between the microprocessor and the peripherals. The *system clock* generates a voltage pulse of fixed frequency that acts as a time reference when specific operations are executed. The *control* unit takes timing information from the clock and provides the necessary information to the specific hardware components for data flow among CPU, memory and peripherals. The control unit decides the function that the ALU executes at a particular instant of time.

Buses: The bus, also mentioned in Chapter 16, is a group of connecting wires providing with a common path for data transfer among different components of the CPU, memory and input/output ports. The buses are of two types: the *internal data bus* carrying instructions and data between ALU and registers and the *external data bus* connected between memory and input/output devices. These are of the following types.

- 16-bit unidirectional address bus for accessing the memory addresses and identifying peripherals
- 8-bit bidirectional data bus to transfer data among ALU, memory and peripherals
- Control bus of individual single, unidirectional lines for carrying timing and synchronizing signals.

17.4.1 Pin Configuration

Intel 8085 is marketed in the form of a 40-pin IC that functions with 0–5 V supply and the operating frequency is 3 MHz approximately. The pin out diagram is similar to that of other digital ICs mentioned in this book: dual-in-line, counted counterclockwise from the top left. Table 17.5 gives the identities of the 40 pins. There are 16 lines of address bus split into two segments. The lower eight lines are bidirectional and are used for dual purpose: data bus as well as address bus through multiplexing technique. The upper eight lines are unidirectional and are used for addressing only.

Table 17.5 Pin connections of microprocessor 8085

Pin No.	Function
1 and 2	Crystal oscillator connecting points
3	Signal to indicate reset condition
4 and 5	Serial output and input data ports
6	Input to interrupt a program execution
7–9	Restart interrupt to transfer the program control to specific memory location
10	Interrupt request input
11	The processor responds to interrupt by this signal
12–19	Address/Data lines
20	Ground terminal of power supply
21–28	Address lines
29	Status signal to identify the nature of operation
30	Address latch enable, a positive going pulse generated to indicate the starting of a program
31	Memory write
32	Memory read
33	Status signal (also Pin 29)
34	Status signal (also pins 29 and 33) to identify the nature of operation

35	Input signal to delay the read/write cycle until the peripheral device is ready to send/receive data
36	A signal to set the program counter to zero
37	System clock output signal
38	Hold acknowledge signal that responds to HOLD request (pin 39)
39	Input signal indicating that the peripheral device is requesting the use of address and data
40	+5 V terminal of power supply

17.4.2 Operational Code (Opcode)

Whether it be a source code or an object code, the basic features of a computer program remains the same. The computer program ultimately reaching the hardware is a set of binary numbers that states the machine what to do at what condition and how to do. The instructions in a program get stored in sequence in the main memory. The CPU fetches one instruction at a time in the given order and executes the same. The data, result etc. are stored at data memory and are accessed by the CPU when needed. The instructions of the source code are converted, either manually or through software like compiler or interpreter, to binary operation codes, abbreviated as *opcode*. It accounts for

- the operation,
- the register(s) involved, and
- the status flag.

Each opcode is carried out in the following three different phases.

(i) *Fetch Cycle*: When the CPU brings the machine version of the given instruction (i.e. the opcode) from the memory.
(ii) *Decoding*: When it decodes the opcode and
(iii) *Execute Cycle*: When the instruction is executed.

The opcode is formed by the combination of the binary codes corresponding to the register and the instruction. The general purpose registers have the following codes.

A	111
B	000
C	001
D	010
E	011
H	100
L	101

Each instruction has certain binary code. For instance, addition is 10000 and to move (copy) register content is 01. Combining the above two types of codes, the opcode is constructed. Examples:

(i) The opcode for adding ($\equiv$ 10000) the content of register B ($\equiv$ 000) to the accumulator content is: 10000 000 = 80 (Hex).

The assembly language mnemonic is ADD B and the corresponding opcode, as coined above, is 80 H.

(ii) The opcode for moving (actually copying, $\equiv$ 01) the content of accumulator ($\equiv$ 111) to register C ($\equiv$ 001) is: 01 001 111 = 4F (Hex).

The mnemonic is MOV C, A and the corresponding opcode is 4F H.

The clock period or the inverse of clock frequency is referred to as *T-state*. Each instruction requires the duration of several T-states because it takes finite time in completing the

- fetching of the opcode from the memory using the address & data bus and the control signal by
- placing the memory address from the program counter on the address bus,
- enabling the memory to place the byte on data bus, and
- decoding and executing the instruction.

17.5 8085 Programming

In this section we will learn to write programs for 8085 in assembly language using the mnemonics given in Table 17.6. These are valid for 8085 only. For other microprocessors, the logic of programming remains the same but the prescribed mnemonics are different because the low-level machine languages are machine dependent, as stated earlier.

The assembly language is not meaningful to the hardware. This must be converted to the equivalent binary opcodes. In an actual computer, the conversion is done by compiler, interpreter or assembler software. In the present case, the user have to hand-assemble the program using the opcodes corresponding to each instruction in Table 17.6. These are given in hexadecimal for the user's convenience but actually it enters the microprocessor in binary form, the 1s and 0s being determined by the voltage levels at different pins.

For the sake of easy programming, a commercially available 8085 practice kit is supplied in the form of a 'semi-computer' by adding other supporting ICs for memory and input/output interface. The user enters the hexadecimal opcodes (Table 17.6) with a key pad and the ALU outputs are displayed with seven-segment displays (Chapter 14). The manufacturer stores permanently the built-in monitor programs in the ROM for converting the user's hexadecimal input to binary, displaying the memory address and data output and interfacing the memory and input/output ports. The range of RAM available for the user is also specified.

A. The First Program

The instruction set in Table 17.6 is not that complicated as it looks like. The list of instructions is so lengthy because often the same job is repeated for each register with a separate instruction of the same category. We will not be able to discuss each instruction

- JNZ *yyyy* : Jump to memory address *yyyy*, if the content of the register mentioned immediately before this instruction is not zero.
- JMP *zzzz* : Jump to address *zzzz* without any condition.

Example 17.2

Add the contents of memory addresses 8100 and 8150 and store the result to memory address 9100.

Solution

The assembly language program and the opcode are given below against each address. It is important to note that LDA and STA operations require three addresses in the following sequence: first the opcode, then the lower byte and then the upper byte.

8000	LDA 8100	3A
8001		00
8002		81
8003	MOV B, A	47
8004	LDA 8150	3A
8005		50
8006		81
8007	ADD B	80
8008	STA 9100	32
8009		00
800A		91
800B	HLT	76

Example: If the address 8100 and 8150 contain the numbers 4A and 8B, respectively, then the result of 4A + 8B = D5 is stored at address 9100. Also the consecutive address numbering in hexadecimal may be noted.

but an overall idea on the implementation of different types of instructions in programming will be experienced. Before starting the first program for 8085, let us choose some relevant instructions from Table 17.6 and understand the meaning.

- MVI X, xy : Move immediately the byte *xy* to register X. The byte may be 00 through FF and the register is any one of A, B, C, D, E, H and L.
- ADD X : Add the content of register X to the accumulator content.
- ADI *xy* : Add immediately the byte *xy* to accumulator content.
- SUB X : Subtract the content of register X from the accumulator content.
- SUI *xy* : Subtract immediately the byte *xy* from the accumulator content.
- RST1 : Restart-1, a method of ending the given program.
- HLT : Halt, stop program execution.

It should be noted that HLT is a machine control instruction for stopping the processing and wait whereas RST (restart) passes the control of the running program to the monitor program of the trainer kit.

Table 17.6 Alphabetic list of microprocessor 8085 instructions with the corresponding opcodes (hexadecimal)

Mnemonic	Opcode	Mnemonic	Opcode	Mnemonic	Opcode	Mnemonic	Opcode
ACI	CE	DAD B	09	LXI SP	31	MOV H, M	66
ADC A	8F	DAD D	19	MOV A, A	7F	MOV L, A	6F
ADC B	88	DAD H	29	MOV A, B	78	MOV L, B	68
ADC C	89	DAD SP	39	MOV A, C	79	MOV L, C	69
ADC D	8A	DCR A	3D	MOV A, D	7A	MOV L, D	6A
ADC E	8B	DCR B	05	MOV A, E	7B	MOV L, E	6B
ADC H	8C	DCR C	0D	MOV A, H	7C	MOV L, H	6C
ADC L	8D	DCR D	15	MOV A, L	7D	MOV L, L	6D
ADC M	8E	DCR E	1D	MOV A, M	7E	MOV L, M	6E
ADD A	87	DCR H	25	MOV B, A	47	MOV M, A	77
ADD B	80	DCR L	2D	MOV B, B	40	MOV M, B	70
ADD C	81	DCR M	35	MOV B, C	41	MOV M, C	71
ADD D	82	DCX B	0B	MOV B, D	42	MOV M, D	72
ADD E	83	DCX D	1B	MOV B, E	43	MOV M, E	73
ADD H	84	DCX H	2B	MOV B, H	44	MOV M, H	74
ADD L	85	DCX SP	3B	MOV B, L	45	MOV M, L	75
ADD M	86	DI	F3	MOV B, M	46	MVI A	3E
ADI	C6	EI	FB	MOV C, A	4F	MVI B	06
ANA A	A7	HLT	76	MOV C, B	48	MVI C	0E
ANA B	A0	IN	DB	MOV C, C	49	MVI D	16
ANA C	A1	INR A	3C	MOV C, D	4A	MVI E	1E
ANA D	A2	INR B	04	MOV C, E	4B	MVI H	26
ANA E	A3	INR C	0C	MOV C, H	4C	MVI L	2E
ANA H	A4	INR D	14	MOV C, L	4D	MVI M	36
ANA L	A5	INR E	1C	MOV C, M	4E	NOP	00
ANA M	A6	INR H	24	MOV D, A	57	ORA A	B7
ANI	E6	INR L	2C	MOV D, B	50	ORA B	B0
CALL	CD	INR M	34	MOV D, C	51	ORA C	B1

Mnemonic	Opcode	Mnemonic	Opcode	Mnemonic	Opcode	Mnemonic	Opcode
CC	DC	INX B	03	MOV D, D	52	ORA D	B2
CM	FC	INX D	13	MOV D, E	53	ORA E	B3
CMA	2F	INX H	23	MOV D, H	54	ORA H	B4
CMC	3F	INX SP	33	MOV D, L	55	ORA L	B5
CMP A	BF	JC	DA	MOV D, M	56	ORA M	B6
CMP B	B8	JM	FA	MOV E, A	5F	ORI	F6
CMP C	B9	JMP	C3	MOV E, B	58	OUT	D3
CMP D	BA	JNC	D2	MOV E, C	59	PCHL	E9
CMP E	BB	JNZ	C2	MOV E, D	5A	POP B	C1
CMP H	BC	JP	F2	MOV E, E	5B	POP D	D1
CMP L	BD	JPE	EA	MOV E, H	5C	POP H	E1
CMP M	BE	JPO	E2	MOV E, L	5D	POP PSW	F1
CNC	D4	JZ	CA	MOV E, M	5E	PUSH B	C5
CNZ	C4	LDA	3A	MOV H, A	67	PUSH D	D5
CP	F4	LDA XB	0A	MOV H, B	60	PUSH H	E5
CPE	EC	LDA XD	1A	MOV H, C	61	PUSH PSW	F5
CPI	FE	LHLD	2A	MOV H, D	62	RAL	17
CPO	E4	LXI B	01	MOV H, E	63	RAR	1F
CZ	CC	LXI D	11	MOV H, H	64	RC	D8
DAA	27	LXI H	21	MOV H, L	65	RET	C9
RIM	20	RST 5	EF	SIM	30	SUI	D6
RLC	07	RST 6	F7	SPHL	F9	XCHG	EB
RM	F8	RST 7	FF	STA	32	XRA A	AF
RNC	D0	RZ	C8	STAX B	02	XRA B	A8
RNZ	C0	SBB A	9F	STAX D	12	XRA C	A9
RP	F0	SBB B	98	STC	37	XRA D	AA
RPE	E8	SBB C	99	SUB A	97	XRA E	AB
RPO	E0	SBB D	9A	SUB B	90	XRA H	AC
RRC	0F	SBB E	9B	SUB C	91	XRA L	AD
RST 0	C7	SBB H	9C	CUB D	92	XRA M	AE
RST 1	CF	SBB L	9D	SUB E	93	XRI	EE
RST 2	D7	SBB M	9E	SUB H	94	XTHL	E3
RST 3	DF	SBI	DE	SUB L	95		
RST 4	E7	SHLD	22	SUB M	96		

Example 17.1

Write a program for 8085 to calculate 6 + 7 − 3.

Solution

This apparently simple work calls for the following tasks.
(i) To store the given three numbers (in hexadecimal form) in three registers
(ii) To instruct to add the first two
(iii) To instruct to subtract the third number from the above sum and
(iv) To instruct that the program has come to its end point.

The above instructions are given in terms of 8085 mnemonics at successive memory addresses, as given below. The corresponding opcodes may be verified with Table 17.6.

Address	Instruction	Opcode	Meaning
8000	MVI A, 06	3E	Move the number 06 to accumulator
8001		06	The number 06 is stored at address 8001
8002	MVI B, 07	06	Move the number 07 to register B
8003		07	The number 07 is stored at address 8003
8004	MVI C, 03	0E	Move the number 03 to register C
8005		03	The number 03 is stored at address 8005
8006	Add B	80	Add register B content to accumulator content
8007	SUB C	91	Subtract Register C content from accumulator
8008	RST1	CF	Stop

The register contents after the program execution will be:

Accumulator $\equiv$ 0A, B $\equiv$ 07 and C $\equiv$ 03

The above program can be simplified further with the use of other instructions, as given below.

Address	Instruction	Opcode
8000	MVI A, 06	3E
8001		06
8002	ADI 07	C6
8003		07
8004	SUI 03	D6
8005		03
8006	HLT	76

[Please note that the address need not start always from 8000. It can start from any range of the available RAM but must be in consecutive sequence.]

B. Some More Instructions

The previous instructions can execute arithmetic operations. However, interchange of data with memory addresses and repetition in program execution require some other instructions from Table 17.6, as mentioned below.

- LDA $xxxx$: To load the data of address $xxxx$ to accumulator.
- STA $yyyy$: To store the accumulator content to address $yyyy$.
- MOV Y, X : To move the content of register X to register Y.
- INR X : Increase the content of register X by 1.
- DCR Y : Decrease the content of register Y by 1.
- JZ $xxxx$: Jump (transfer access) to memory address $xxxx$, if the content of the register mentioned immediately before this instruction becomes zero.

C. Flowchart

It is a pictorial format of the thinking process for developing a program. It is a scientific way to start the program with a flowchart. Some common symbols used in flowcharting are given in Figure 17.5.

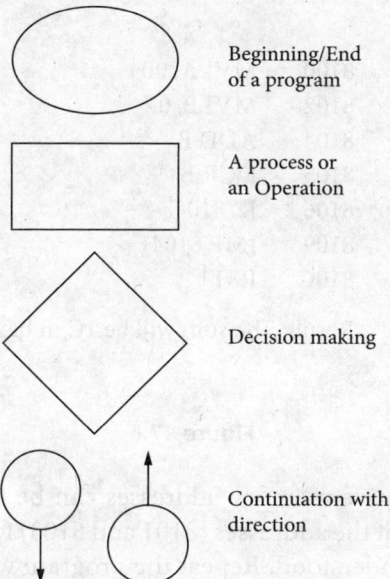

Figure 17.5 Common symbols used in flowchart

Example 17.3

Perform the addition $1 + 2 + 3 + 4 + 5 + 6 + 7$.

Solution

The 8085 program is given below with the flowchart in Figure 17.6. [Henceforth the opcodes will not be given with the program. It is the exercise of the student to hand-compile the program by finding out the opcodes from Table 17.6.]

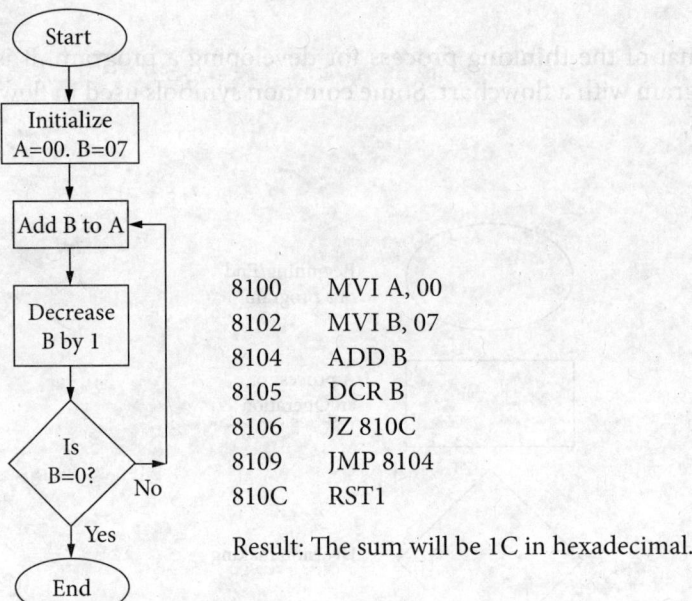

8100	MVI A, 00
8102	MVI B, 07
8104	ADD B
8105	DCR B
8106	JZ 810C
8109	JMP 8104
810C	RST1

Result: The sum will be 1C in hexadecimal.

Figure 17.6

[N.B. Note that the consecutive memory addresses can be chosen arbitrarily within the specified range. Also note that the addresses (8101 and 8103) for storing data are not written in the program. These are understood. Repeat the program with another upper limit. Can you state what the maximum sum obtainable by this process is?]

Example 17.4

Multiply decimal 10 by 7.

Solution

The principle is to add 0A (hex-equivalent of 10) seven times. The flowchart (Figure 17.7) and the program are given below.

D. More Instructions

It was mentioned earlier that except for the accumulator, the other registers can be used in pair as B with C, D with E and H with L. Now we will practice with the instructions related to such register pairs.

- LXI H *xxyy* : To load the number *xxyy* immediately to the extended register pair HL with *xx* at H and *yy* at L. Similarly,
- LXI B   : for BC pair and
- LXI D   : for DE pair.

Microcomputer and Microprocessor

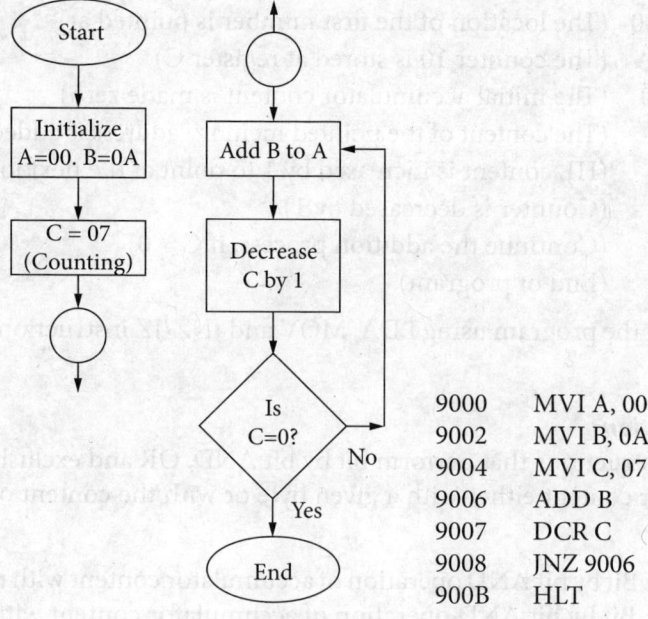

Figure 17.7

The symbol 'M' used after the LXI instruction indicates the address *xxyy* to be used in the program.

- INX H : Increase the content of HL pair by 1.
- DCX H: Decrease the content of HL pair by 1.

Similarly,

- INX B and DCX B for BC pair and
- INX D and DCX D for DE pair.

Example 17.5

Ten hexadecimal numbers are stored at consecutive memory locations starting from F900. Write a program to add these numbers.

Solution

An assembly language program is given with explanation. The student is suggested to sketch the flowchart and develop the program independently and then compare with the present one.

[Note that the address sequence is different from that of the previous example.]

F800	LXI H F900	(The location of the first number is pointed at)
F803	MVI C, 0A	(The counter 10 is stored at register C)
F805	MVI A, 00	(The initial accumulator content is made zero)
F807	ADD M	(The content of the pointed memory address is added to accumulator)
F808	INX H	(HL content is increased by 1 to point at the next location)
F809	DCR C	(Counter is decreased by 1)
F80A	JNZ F807	(Continue the addition process, if $C \neq 0$)
F80D	HLT	(End of program)

[N.B. Try to repeat the program using LDA, MOV and JNZ/JZ instructions and infer which process is better.]

E. Logical Instructions

There are some instructions that perform bit by bit AND, OR and exclusive OR operations of the accumulator content either with a given byte or with the content of another register R (= B, C, D etc.).

- ANA R : Bit by bit AND operation of accumulator content with register R content.
- ANI mn : Bit by bit AND operation of accumulator content with the byte mn.

Similar operations are:

- ORA R and ORI mn representing OR operation and
- XRA R and XRI mn representing OR operation.

Example 17.6

Let the accumulator content be CC and B register contains the byte DD. Determine the result of ANA B.

Solution

$A \equiv CC = 1100\ 1100$
$B \equiv DD = \underline{1101\ 1101}$
$ANA\ B \equiv 1100\ 1100 = CC$

Example 17.7

If the accumulator content is 3F and the content of register C is 5B, then determine ORA C.

Solution

$A \equiv 3F = 0011\ 1111$
$C \equiv 5B = \underline{0101\ 1011}$
$ORA\ C \equiv 0111\ 1111 = 7F$

Example 17.8

In the above case, what would be the result of ORI 24?

Solution

$$A \equiv 3F = 0011\ 1111$$
$$24 = \underline{0010\ 0100}$$
$$ORI\ 24 \equiv 0011\ 1111 = 3F$$

Example 17.9

If the accumulator content is D7 and the register D contains 6E, find the result of the operation XRA D.

Solution

$$A \equiv D7 = 1101\ 0111$$
$$D \equiv 6E = \underline{0110\ 1110}$$
$$XRA\ D \equiv 1011\ 1001 = B9$$

Example 17.10

Check the fourth bit from the left of the number stored at memory address 9100. If it is 1, write FF at address 9101 and if it is 0, write 00 at address 9101.

Solution

A flowchart (Figure 17.8) and the corresponding program for the above task is given here. The key idea is that the instruction ANI 10 can check for the 4th bit from the left. It that is 0, the result also becomes zero and the JZ instruction works.

[Note that no specific address is given in the program. Instead, two symbolic labels are used to indicate the required transfer of program control. This is often practiced while doing the paper work for a programming. This program has missed the position of Label-2. Can you put it properly?]

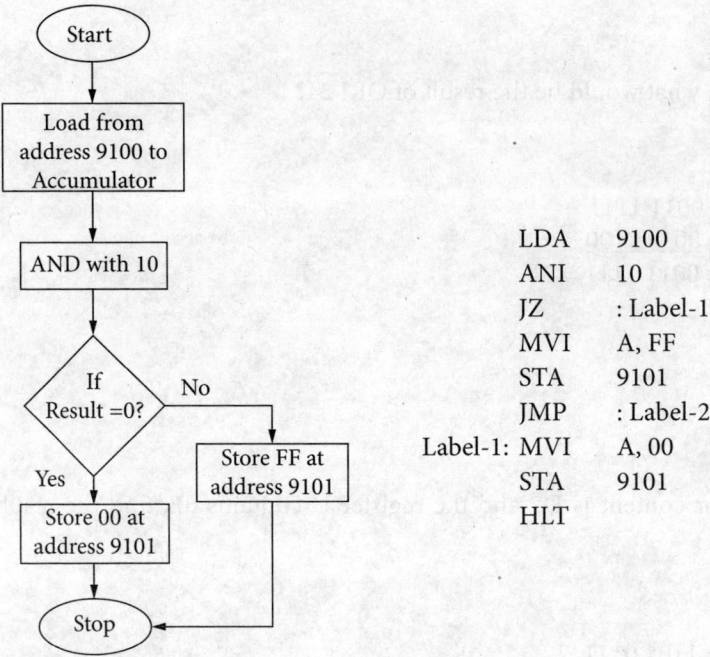

Figure 17.8

[N.B. Please note that this is not the unique program. A certain task can be programmed in several ways. For instance, try to repeat the above program using the instruction JNZ. Also repeat the program for any other bit position with suitable AND operation.]

F. Double-Byte Instructions

These instructions involve HL and other register pairs for two types of task. Either the two register contents together indicate a memory address or two data bytes are simultaneously exchanged between two registers at a time.

- LHLD *xxyy* : The content of address *xxyy* is loaded to register L and the content of address (*xxyy* + 1) is loaded to register H.
- SHLD *mnpq* : The content of register L is stored to the address *mnpq* and the content of register H is stored to the address (*mnpq* + 1).
- DAD B : The double byte stored in BC pair is added to the content of HL pair.

Similarly DAD D instruction acts for DE and HL pair.

- XCHG : This implied instruction exchanges the contents of HL and DE pair.
- LDAX D : The content of the address denoted by the double byte stored in the DE pair is loaded to accumulator.
- STAX D : It stores the accumulator content to the address indicated by the double byte stored in DE pair.

Similarly LDAX B and STAX B holds for BC pair. These instructions are similar to LDA and STA but here a pair of register acts as a pointer to the memory address.

> **Example 17.11**
>
> A 2-byte number is stored occupying memory addresses 8100 and 8101. Another 2-byte number is stored occupying addresses 8200 and 8201. Add these two numbers and store the result at addresses 8300 and 8301.
>
> **Solution**
>
> The LHLD instruction is used to load the numbers to the register pairs and the DAD instruction is used for double byte addition.
>   LHLD 8100
>   XCHG
>   LHLD 8200
>   DAD D
>   SHLD 8300
>   RST1

G. Implied Instructions

These instructions have only a mnemonic without referring to any register, memory address or data byte. Anyway the implied operation is executed properly.

- RLC : The MSB of the accumulator content shifts to the LSB position and simultaneously takes the position of carry flag. The LSB and the other bits of the accumulator content shift to the left by one place. The process can be viewed as an anticlockwise rotation of the accumulator content by one bit.
- RRC : This is the complement of the previous instruction. The LSB of the accumulator content takes the place of MSB and also that of the carry flag. The MSB and the other bits of the accumulator shift to the right by one place. The mechanism can be regarded as a clockwise rotation of the accumulator content by one bit.
- RAL : This is similar to RLC but the anticlockwise rotation of bits is executed including the carry bit. The MSB of accumulator content moves to the carry flag and the other bits shift to the left by one bit. The carry bit moves to LSB.
- RAR : This is similar to RRC but the clockwise rotation of bits takes place including the carry bit. The LSB of accumulator content transfers to the carry flag and the other bits shift to the right by one position. The carry bit takes the position of MSB.
- CMA : It complements the bits of the accumulator content. No flag is affected. It is actually of logic instruction category.
- DAA : The full form is *decimal adjust accumulator*. It converts the hexadecimal number stored in the accumulator to BCD so that the result appears to be in decimal form. All flags are affected here.

Example 17.12

Write a program to detect how many 1s are there in the given byte.

Solution

This program uses the RAL instruction and another access transfer instruction JNC (jump if no carry). The given byte in register B is indicated by X in the program. The register C contains the count for 8, the number of bits in a byte and register D counts the number of 1s, starting from zero.

```
         MVI B, X
         MVI C, 08
         MVI D, 00
         MOV A, B
Label-1: RAL
         JNC    : Label-2
         INR D
Label-2: DCR C
         JNZ    : Label-1
         HLT
```

[N.B. Try to build up the program using RLC instead of RAL.]

Example 17.13

Write a program to determine the 2's complement of the number stored in memory location 9200 and store the result in address 9300.

Solution

The program uses CMA to make the 1's complement and INR A to make the 2's complement.

```
         LDA 9200
         CMA
         INR A
         STA 9300
         HLT
```

[N.B. Convert the program to machine readable form by using the opcodes from Table 17.6 and including consecutive addresses for each instruction. Repeat with several different accumulator contents. Verify, if ADI 01 could be used for adding 1 to the 1's complement.]

Example 17.14

Write a program to add decimal 15 and decimal 25 and to show the result in decimal.

Solution

The program considers the two numbers as hexadecimal and the ADD instruction yields the result of 15 + 25 = 3A. The DAA instruction converts that to BCD and the result appears as 40, equivalent to decimal 40.

       MVI A, 15
       MVI B, 25
       ADD B
       DAA
       RST1

[N.B. DAA instruction holds for addition only.]

H. Stack Instructions

As mentioned previously, the stack is a set of memory locations that can be used to store data temporarily during the program execution. There is a dedicated 16-bit register for this purpose, known as stack pointer (SP) that contains a 16-bit memory address to point at the beginning of the stack. The related instructions are given below.

- LXI SP *xxyy*   : The memory address *xxyy* is pointed by the stack pointer.
- PUSH B   : This is to copy the contents of BC pair to two consecutive memory addresses immediately before that one initialized by the stack pointer (*xxyy* in the present case). On getting the PUSH instruction, the stack pointer is decremented by 1 and the content of register B is copied to address (*xxyy* − 1). Then the stack pointer is again decremented by 1 and the content of register C is copied to the address (*xxyy* − 2). The stack pointer then denotes the address (*xxyy* − 3).

Similar operations are PUSH D and PUSH H for DE pair and HL pair, respectively.

- POP B : This is to copy the contents of two indicated consecutive addresses to BC pair. If the POP instruction is given after LXI SP *xxyy*, the content of address *xxyy* is copied to register C and then the stack pointer is incremented by 1. The content of location *xxyy* + 1 is copied to register B. The stack pointer then points to the address (*xxyy* + 2).

Similar operations are POP D and POP H for DE pair and HL pair, respectively. The following examples will clarify the stack pointer operation.

Example 1: Let register B and C contain the numbers 5A and 13, respectively and the stack pointer instruction is given as LXI SP 8100. Then the instruction PUSH B copies 5A to address 80FF (just before 8100) and 13 to address 80FE (the next ahead).

Example 2: Let addresses 8100 8101 contain the number 11 and 22, respectively. The instruction LXI SP 8100 is followed by POP B. Then 11 will come to register C and 22 to register B. The stack pointer will be pointing to address 8102.

Example 3: Let DE = 3344 and HL = 5566. The following instructions are given.

    LXI SP 8500
    PUSH D
    PUSH H

These execute the following operations sequentially. The stack pointer initially points at the address 8500. The first PUSH instruction decrements it by 1 and copies the number 33 from register D to address 84FF. The pointer is again decremented by 1 and the number 44 from register E is copied to address 84FE. The second PUSH instruction decrements the pointer by 1 and copies 55 from H to 84FD. The pointer is decreased again by 1 and 66 from L is copied to 84FC.

Some other instructions related to the stack are mentioned below.

- PUSH PSW   : The symbol PSW means program status word. Here the concerned registers in pair are the accumulator and the flag register. The counter instruction is POP PSW.
- PCHL   : The contents of HL pair are copied to the program counter.
- SPHL   : The contents of HL pair are copied to the stack pointer.
- XTHL   : The content of the location denoted by stack pointer is copied to register L and the next location content is copied to register H.

17.6 Types of 8085 Instructions

The above examples and related discussions imply that the 8085 instructions can be classified into different categories. Indeed the same instruction can be put into different classes.

On the basis of addressing modes, there are the following varieties.

- Direct addressing instructions: LDA, STA, etc.
- Immediate addressing instructions: MVI, ORI. etc.
- Register addressing instructions: MOV A, B; MOV C, D, etc.
- Register indirect addressing instructions: MOV A, M; MOV M, A, etc.
- Implied addressing instructions: CMA, DAA, etc.

On the basis of mode of operations, the following types are found.

- Data transfer instructions: MOV A, B; MVI 03, etc.
- Arithmetic operation instructions: ADD B; SUI 2A, etc.
- Logical operation instructions: ANA, ORI, CMA, etc.
- Branching operation instructions: JZ, JMP, JC, etc.
- Machine controlling instructions: HLT, RST1, etc.

In terms of byte occupying, the following categories are found.

- One-byte instructions: ADD B; MOV C, A, etc.
- Two-byte instructions: MVI 07; ADI 35, etc.
- Three-byte instructions: LDA 9100; STA F850, etc.

17.7 Use of Subroutine

A subroutine is a group of instructions within the main program for doing some repeated job. It is written as a separate unit starting from the address *xxxx* and ending with the instruction RET (means return). On getting the instruction CALL *xxxx* in the main program, the access transfers to the address *xxxx* and after getting the RET instruction, it returns to the main program.

Example 17.15

Write a program to determine 5!

Solution

The technique of using subroutine is applied here to execute repeated multiplication. The subroutine starts at address 9000.

8000	MVI B, 05	9000	MOV C, B
8002	MVI A, 01	9001	MOV D, A
8004	CALL 9000	9002	XRA A
8007	DCR B	9003	ADD D
8008	JNZ 8004	9004	DCR C
800B	HLT	9005	JNZ 9003
		9008	RET

17.8 Arduino Programming

Arduino is a commercially available small-size ($\approx$ 7 cm × 5 cm) microcontroller board containing a 28-pin, 8-bit processor IC and several supporting electronic components. It is operated by 16 MHz quartz crystal. There are fourteen digital input/output pins and six analog input pins for general use. An open source programming language named Arduino Integrated Development Environment (IDE) is available. The combination of the board and the software creates an Arduino platform that can be interfaced directly with a common computer through USB connection. Several versions of the board and IDE are available that can work under different operating systems. The present examples are carried out with Arduino UNO board and Windows 10 operating system. The name "Arduino" was in commemoration of a bar in Italy where the co-founders of this project used to meet.

17.8.1 Arduino Board

This is the hardware part consisting of the microcontroller IC and other active and passive components. The following connection pins are ready to use.

POWER Pins	Analog Pins	Digital Pins
Vin: to provide input supply from any external voltage source. **5V:** to supply regulated voltage to the board and the on board components. **3.3V:** to provide a supply of 3.3V generated from a voltage regulator on the board **GND:** to ground the Arduino board. **Reset:** to reset the microcontroller.	The six pins numbered A0 to A5 are used as analog inputs in the range of 0 to 5V.	The fourteen pins numbered 0 to 13 are used as digital input or output for the Arduino board. Pin 0 is receiver and Pin 1 is transmitter for communication between the Arduino board and the computer or other devices. These are known as serial pins. Pins 2 and 3 are used to produce the external interrupt. Pins 3, 5, 6, 9, 10 and 11 are pulse width modulation pins that are used to convert the digital signal into analog by varying the pulse width. Pins 10, 11, 12 and 13 maintain the serial peripheral interface communication. **LED Pin:** Pin 13 drives the on/off state of a built-in LED. **AREF Pin:** This is to provide a reference analog voltage from any external power supply.

17.8.2 Arduino IDE

It contains a text editor for writing codes and provisions for communicating with the Arduino hardware. The programs composed with IDE are called *sketches*. These are typed in the text editor and are saved with the file extension .ino. There are menu-based commands under the headings File, Edit, Tools etc. For example:

- *Verify*: checks the program for errors compiling it.
- *Upload*: compiles the program and uploads its binary version to the configured board.
- *New*: for creating a new sketch.
- *Open*: to load a sketch file browsing through the drives and folders.
- *Save*: saves the current sketch.
- *Quit*: closes all IDE windows.

A sketch (Arduino program) has the following general structure.

```
void setup() {
  // put your setup code here, to run once:
}
```

```
void loop() {
  // put your main code here, to run repeatedly:
}
```

We will be practicing several programs using this structure. The setup() function initializes the variables and pin modes and starts using libraries. It runs only once. The loop() function may run repeatedly and the codes in this section control the board.

Program 1: Blinking of LED
One end of a current limiting resistor (e.g. 220 Ω) is connected to the digital pin 13. The anode of an LED is connected to the other end of the resistor. The cathode of the LED is joined to the ground (GND). The LED can be made to blink at the desired frequency by executing the following program.

```
void setup() {
// put your setup code here, to run once:
pinMode(LED_BUILTIN,OUTPUT);
}

void loop() {
// put your main code here, to run repeatedly:
digitalWrite(LED_BUILTIN, HIGH);
delay(1000);
digitalWrite(LED_BUILTIN, LOW);
delay(1000);
}
```

The instruction "`pinMode(LED_BUILTIN, OUTPUT);`" initializes the specified pin for LED (Pin 13) as an output pin.

The instruction "`digitalWrite(LED_BUILTIN, HIGH);`" turns on the LED by supplying 5 V to the anode.

The instruction "`delay(1000);`" withholds the board for 1000 milliseconds.

The instruction "`digitalWrite(LED_BUILTIN, LOW);`" turns off the LED by supplying zero volt to pin D13.

The instruction "`delay(1000);`" again withholds the board for 1000 milliseconds.

Thus the LED glows and turns off periodically for 1 second each. The on/off duration can be changed by changing the number after the delay statement.

The sketch is typed on the IDE screen. To run the program, select from the menu: Sketch → Upload. Messages like 'Compiling sketch' and 'Done uploading' will be displayed.

For the first time, the port is to be defined by selecting from menu: Tools → Port.

Program 2: Generating a clock pulse.
A simple modification to the previous program produces a different outcome. The two delay times are changed as "delay(1);" so that each of the on and off time becomes about one millisecond. If the LED is replaced by a CRO (Chapter 12), a unidirectional steady square wave of frequency about 500 Hz and amplitude about 5 V can be observed. This can be used as the clock to a digital circuit.

Program 3: Reading an analog input voltage.
This sketch, as given below with comments of explanations can read a variable voltage (0 to less than 5 V) given at analog pin 0. The built-in A/D converter of the board converts it to a number between 0 and 1023. The program normalizes that to a value between 0 and 5.

```
void setup() {
// This initializes serial communication between board and computer
@ 9600 bits per second.
Serial.begin(9600);
}
void loop() {
// This reads the input at analog pin 0 and assigns it as integer
"numvoltage".
int numvoltage = analogRead(A0);
// This converts the number to a voltage value of 0 to 5 V.
float voltage = numvoltage * (5.0 / 1023.0);
// This prints the voltage value to Serial Monitor.
Serial.println(voltage);
}
```

Program 4: Generating a continuously varying voltage.
Pulse width modulation (PWM) is a technique of producing variable analog outputs with digital inputs. A square wave is generated whose pulse width or on time duration is changed to get a varying analog voltage. Arduino has several PWM pins, namely 3, 5, 6, 9, 10 and 11. Anyone of those can be used. The following sketch has used pin 6 to generate a varying voltage. It can be utilized to vary the brightness of an LED or to generate a waveform. The programming instructions are given with detailed explanation at each line.

```
int signal = 6; // Specifying the PWM pin number where the signal
is obtained.
int amplitude = 0; // Initializing the amplitude of the signal with
zero value.
int stepsize = 2; // Step size of increase/decrease.
```

```
void setup() {
// Pin 6 is declared to be the output.
pinMode(signal, OUTPUT);
}

void loop() {
// The signal amplitude obtained at pin 6 is specified.
analogWrite(signal, amplitude);

// This loop changes the amplitude step-by-step.
amplitude = amplitude + stepsize;

// This reverses the direction of signal amplitude change after
reaching the extreme value.
if (amplitude <= 0 || amplitude >= 255) {
stepsize = -stepsize;
}
// This delay is optional, to visually observe the effect, such as
the brightness change of an LED.
delay(20);
}
```

Other novel interfacing of instruments are possible with Arduino. For example, the 5 V output of Program 1 can drive a relay to control high current on/off switching.

Multiple Choice-Type Questions and Answers

17.1 The RAM of a computer requires a constant supply of electrical power to retain the information.
   (a) True
   (b) False
   (c) Uncertain
   (d) situation dependent

17.2 It is fact that sometimes the ROM content is copied to RAM before use. The process is known as *shadowing*.

   Statement-I: RAM can be read faster than ROM.
   Statement-II: Writing data to RAM is faster than writing data to ROM.
   (a) Only Statement-II is correct and it is the reason for shadowing.
   (b) Only Statement-I is correct and it is the reason for shadowing.
   (c) Both Statement-I and Statement-II are correct and Statement-I is the reason for shadowing.
   (d) Both Statement-I and Statement-II are correct and Statement-II is the reason for shadowing.

17.3 What is the purpose of using the lower eight address lines in 8085 as both address bus and data bus?
(a) To increase the speed of operation
(b) To handle more number of data
(c) To exhibit technical novelty
(d) To reduce the number of connecting pins

17.4 The program counter of microprocessor 8085 is capable of addressing .... locations.
(a) 64K   (b) 1K   (c) 256K   (d) 32K
Hints: 16-bit register, 2^{16} = 64K

17.5 A 32-bit microprocessor has 32-bit .... bus.
(a) address
(b) data
(c) data/address
(d) control

17.6 The maximum hexadecimal number that can be stored in a 16-bit data field is ....
(a) FF   (b) FFFF   (c) 1000   (d) 9999

17.7 In a certain computational system, each memory location contains 16-bit data. The program memory contains the 8 upper bits for opcode and the lower 8 bits for address of the memory location. The maximum number of opcodes is .... and the size of accessible memory is ....
(a) 2^{16}, 16 KB   (b) 8, 8   (c) 2^8, 2^8 byte   (d) 2^8, 2^8 bit

17.8 The .... in 8085 is a register programmable by the user's instruction.
(a) accumulator
(b) instruction register
(c) temporary register
(d) flag register

17.9 The number of address lines required for an 8 KB memory is ....
(a) 8   (b) 12   (c) 16   (d) 13
Hints. 8×1024 = 2^{13} lines for 8×1024 registers.

17.10 If the size of each memory chip is 1024, the number of such chips required to fabricate an 8 KB memory is ....
(a) 8   (b) 16   (c) 64   (d) 24
Hints. 8×1024×8 ÷ 1024 = 64

Answers
10.1 (a)   10.2 (c)   10.3 (d)   10.4 (a)
10.5 (b)   10.6 (b)   10.7 (d)   10.8 (a)
10.9 (d)   10.10 (c)

Reasoning-Type Questions and Answers

17.1 What are *single user* and *multiuser* operating systems?

Ans. A single user operating system is such that only one user can access the computer at a time. A typical home computer with Windows and a tablet with Android are such systems. A multiuser operating system can allow several users to work simultaneously and independently with the same computer having their own terminals of monitor, keyboard and mouse. Specific versions of UNIX and Linux are such systems. These are suitable for mainframe computers where a number of users work at a time with the same operating system.

17.2 Distinguish between *multitasking* and *multiprocessing*.

Ans. Multitasking means using a single processor for several tasks by switching it in time slots between tasks. Since the input/output operations are much slower than the CPU speed, more than one program can be loaded in the main memory but only one program at a time gets access to the CPU for executing the instructions. Thus the use of CPU time can be maximized by this process. Windows operating system can provide with such facility for the home computer. Multiprocessing means running many programs simultaneously using more than one processor. This is an actual parallel processing at hardware level and is suitable for mainframe computers.

17.3 Compare the features of assembly language and high-level language.

Ans. The assembly language is more efficient. It requires less memory space and has one-to-one correspondence between the mnemonics and the machine code. However, it is machine dependent. The high-level language is machine independent. It is easier to write and debug. However, it requires larger memory space because several machine codes are necessary to translate into machine code.

17.4 What is the maximum address capability of microprocessor 8085?

Ans. The 16-bit address bus of 8085 can address a maximum of 2^{16} different locations, each containing one byte of data. Therefore, the maximum address capability of 8085 is $2^{16} \times 8 = 524288$ bits, more concisely stated as 64 KB.

17.5 How does the microprocessor differentiate between hexadecimal instruction code and data?

Ans. It is done just by tracking the proper sequence of instructions. Out of the bytes fetched, it interprets the first byte as the opcode and the next byte as data. If the programmer makes any mistake in this sequence, compilation error occurs causing an abnormal result.

Solved Examples

(Some programming examples are given. The same task may be programmed in different ways.)

17.1 Write a program to exchange the data existing in BC pair and DE pair.

Soln.
```
MOV H, B
MOV L, C
XCHG
MOV B, H
MOV, C, L
```

17.2 Add two large numbers, C3 and 92 that result in a carry. Store the sum at address 9200 and the carry at address 9201.

Soln. The sum 152 is greater than FF, the maximum capacity of the accumulator. The program uses the ADC instruction, which means adding with carry. The addresses are given for a convenient look.

9000	MVI A, C3	
9002	MVI B, 92	
9004	ADD B	
9005	STA 9200	(The sum is stored here.)
9008	MVI A, 00	
900A	ADC A	(Only the carry bit is added.)
900B	STA 9201	(The carry is stored here.)
900E	HLT	

17.3 Write a program to subtract a given number from another resulting in a borrow. Store the difference at address 8050 and the borrow at 8051.

Soln. The program given below is made generalized for both the occurrence and non-occurrence of borrow. The JM (jump if minus) instruction activates when the sign bit of flag register becomes 1. The program shows the example of 7 − 10 (decimal).

8000	MVI A, 07
8002	MVI B, 0A
8004	SUB B
8005	STA 8050
8008	JM 8011
800B	MVI A, 00
800D	STA 8051
8010	RST1
8011	MVI A, 01
8013	STA 8051
8016	RST1

17.4 Write a program to find out the maximum number in a given array of ten numbers stored at consecutive addresses starting at 9000.

Soln. The program uses an instruction CMP X that compares the content of register X with accumulator content and produces carry bit = 1 when the corresponding number is greater than the accumulator content.

8000	XRA A
8001	MVI B, 0A
8003	LXI H 8FFF
8006	INX H
8007	CMP M
8008	JNC 800C
800C	DCR B
800D	JNZ 8006
8010	HLT

[N.B. The principle of the program is to bring the numbers one by one to the accumulator and compare with the previous one. For this purpose, the accumulator content is initialized as zero, which is done by XRA A. Think over it and decide if it is better than MVI A, 00.]

17.5 Write a program to arrange a given set of ten numbers in ascending order and store those in the same locations. The data array starts from 8050.

Soln. The program, with helping comments is given below. It uses the instruction JC, which transfers the access to the specified address, if no carry occurs.

	MVI C, 09	(Provides the counter)
Label-1:	MOV D, C	(The counter is copied to D)
	LXI H 8050	(The first address is pointed at)
Label-2:	MOV A, M	(The first number is accessed)
	INX H	(The pointer position is increased by 1)
	MOV B, M	(The second number is accessed)
	CMP B	(Comparison with accumulator content)
	JC : Label-3	(If A < B, go to Label-3)
	MOV M, A	(Otherwise interchange the numbers)
	DCX H	
	MOV M, B	
	INX H	
Label-3:	DCR D	
	JNZ : Label-2	(Continue till count ≠ 0)
	DCR C	
	JNZ : Label-1	(Continue till count ≠ 0)
	RST1	

Note: Physical Computing

The computational processes executed by the microprocessor are 'numerical' only where digital data and digital electronic circuits are involved. The microcontroller, along with the interfaced sensors, actuators or transducers, can develop an interactive system capable of responding to an external electrical or non-electrical signal and making decisions accordingly. Such a calculation process involving physical objects and physical phenomena is called *physical computing*. The given Arduino programs in this chapter are some such examples. Calculations like counting the number of packets produced in an industry and taking the decision of switching on/off an electrical connection on crossing a certain level of temperature are other examples of this category. Such computations executed by the microcontroller for decision making, controlling and maintaining a specific situation combine the hardware and the software and establishes an integrated use of the input and output devices.

Exercise

Subjective Questions

1. What is operating system and what are its functions? State briefly, with a sketch, how does the operating system mediates between the digital circuits and the human user.
2. Distinguish between high-level and low-level languages. Compare the features of compiler and interpreter.
3. Compare the construction and functioning of a microprocessor with those of a microcontroller.
4. Draw a simplified block diagram of microprocessor 8085 and briefly explain the process of executing the given instruction with its various units.
5. What is opcode? Explain the process of coining an opcode with example.
6. What is flag register and what does it do?
7. What is stack and how it is used?
8. Discuss the different types of addressing modes in 8085 with examples.

Conceptual Test

1. In your opinion, what is the present generation of computer?
2. Suggest how can one test the (i) 'all bits zero' and (ii) 'all bits one' conditions of a number with 8085?
3. Justify that the microprocessor contains both combinational and sequential digital circuits.

Programming Problems

1. Exchange the locations of two data bytes denoted by N1 and N2 stored in two consecutive locations 9100 and 9101.
2. Two numbers, C3 and 92 are stored at consecutive memory locations 9050 and 9051, respectively. Write a program that can add these two, store the sum at address 9052 and the carry at address 9053.
3. Write a program to interchange the lower nibble and the higher nibble of the given byte F5.
4. Write a program to calculate the mean of ten numbers stored at consecutive addresses starting from 8050.

Bibliography

Bell, D. A. (1980). *Electronic Devices and Circuits*, Virginia: Reston Pub. Co.

Boylestad, R. L. and Nashelsky, L. (2013). *Electronic Devices and Circuit Theory*, New Jersey: Pearson.

Cathey, J. J. (2002). *Schaum's Outline of Theory and Problems of Electronic Devices and Circuits*, New York: McGraw-Hill.

Floyd, T. L. (2006). *Digital Fundamentals*, New Jersey: Pearson.

Floyd, T. L. and Buchla, D. A. (2014). *Electronics Fundamentals: Circuits, Devices and Applications*, New Jersey: Pearson.

Harris, D. M. and Harris, S. L. (2013). *Digital Design and Computer Architecture*, New York: Morgan Kaufmann, Elsevier.

Helfrick, A. D. and Cooper, W. D. (2008). *Modern Electronic Instrumentation and Measurement Techniques*, New Delhi: PHI.

Horowitz, P. and Hill, W. (2015). *The Art of Electronics*, New York: Cambridge University Press.

Karris, S. T. (Ed), (2005). *Electronic Devices and Amplifier Circuits with MATLAB® Applications*, San Antonio: Orchard.

Keith, L. J. and Brogue, A. (Eds.) (1950). *Father of Radio: The Autobiography of Lee de Forest*, Chicago: Wilcox & Follett Co.

Leach, D. P., Malvino, A. P., and Saha, G. (2011). *Digital Principles and Applications*, New Delhi: Tata McGraw Hill.

Malvino, A. P. and Bates, D. J. (2016). *Electronic Principles*, New York: McGraw-Hill Education.

Millman, J. and Halkias, C. C. (1972). *Integrated Electronics: Analog and Digital Circuits and Systems*, Tokyo: McGraw-Hill, Inc.

Millman, J. and Taub, H. (1965). *Pulse, Digital and Switching Waveforms*, Tokyo: McGraw-Hill, Inc.

Morrison, R. (2002). *The Fields of Electronics*, New Jersey: John Wiley & Sons Inc.

Roberge, J. K. (1975). *Operational Amplifiers: Theory and Practice*, New Jersey: John Wiley & Sons Inc.

Scherz, P. and Monk, S. (2016). *Practical Electronics for Inventors*, New York: McGraw-Hill Education.

Schultz, M. E. (2011). *Grob's Basic Electronics*, New York: McGraw-Hill.

Streetman, B. G. and Banerjee, S. K. (2009). *Solid State Electronic Devices*, New Delhi: PHI Learning Pvt. Ltd.

Sze, S. M. and Ng, K. K. (2007). *Physics of Semiconductor Devices*, New Jersey: Wiley Interscience, John Wiley & Sons, Inc.

Tocci, R. J., Widmer, N. S., and Moss, G. L. (2007). *Digital Principles and Applications*, New Jersey: Pearson.

Tokheim, R. L. (1994). *Schaum's Outline of Theory and Problems of Digital Principles*, New York: McGraw-Hill.

Index

ac beta, 180
ac emitter resistance, 180–181, 215, 366
ac equivalent circuit, 179, 218, 300–302, 330–332, 339, 366–367
ac load line, 176, 250, 257, 263
acceptor ion, 32, 62–63, 74, 85
acceptor level, 39–40
active device, 14, 70, 117, 327, 340, 402, 523
active filter, 11, 389–392
active pull up, 11, 525
active region, 13, 147–152
adder (analog), 381, 384–386, 396, 403–404, 611, 643
adder (digital), 510–512, 527–529, 531–532, 642–643, 646
address bus, 630, 644, 651–652, 654
all-pass filter, 415
alpha, 147
amorphous, 21, 27
amplification factor, 281, 283
amplifier classes, 245
amplitude distortion, 243
analog, 460
analog computation, 365, 384, 643
analog-to-digital (A/D) conversion, 618–622
AND gate, 435, 513–516
AND operation, 480–484, 488, 490, 513, 518, 527, 662
aquadag, 444
Arduino
arithmetic logic unit (ALU), 527, 588, 638, 650
associative law, 481
astable multivibrator, 345–348, 407, 560–562

asynchronous counter, 389–391
attenuator, 363
avalanche breakdown, 77–79

band gap, 25–26, 32–38, 41–42, 67, 79, 83
band-stop filter, 390–391, 395–396
band-pass filter, 390, 395
bandwidth, 226–227, 324–326, 370, 374, 380–381
Barkhausen criterion, 315, 328, 331, 336
base, 141–155
base bias, 167, 175, 247, 297
base clipper, 120–122
base current, 144–150, 153–154, 164–183, 208, 242, 258, 260, 372
base–width modulation, 150, 223
basis, 21–22
basis vector, 22–23
BCD (binary coded decimal), 471–472
beta, 147
binary equivalent weight, 610
binary number, 463–481, 489–491
bipolar junction transistor (BJT), 141–155, 176–179, 182, 185, 223, 239–241, 264–265, 276–283, 294–302, 523
bistable multivibrator, 345, 350, 564–565
bit, 463, 464
Bode plot, 328
Boolean algebra, 460, 479–499
Boolean simplification, 485–488, 493, 519
bootstrapping, 169
Bravais lattice, 22–23
breakdown, 71–72, 76–81, 108, 113, 150, 279–281, 406–407

bridge rectifier, 114–116
bubbled gates, 518–519
buffer, 184, 186, 383, 391, 516
built-in potential, 63–65, 69–70, 85, 144, 264
bypass capacitor, 179, 217–218, 221, 331, 381
byte, 464, 624, 640, 650–651, 655, 662, 664

capacitor filter, 117–118
cathode ray oscilloscope (CRO), 126, 405, 442–445
central processing unit (CPU), 471, 598, 609, 641–642
characteristic equation, 579–580
clamper, 123–125
class A amplifier, 245–247, 251
class AB amplifier, 259–260
class B amplifier, 245, 252–254, 258
class C amplifier, 245, 262–264
clear, 580–581, 586, 596–597
clipper, 120–122
clock, 455, 557–559, 553, 564, 569–580, 582, 586, 589–5928
CMRR (Common mode rejection ratio), 368–370, 387–388
collector, 141–155
collector current, 143, 145–155, 164–184, 208, 223, 226, 241–263
collector–feedback bias, 171–172
Colpitts oscillator, 327, 330–332
combinational logic, 511
common mode gain, 367–369
common-base amplifier, 184–186
common-base configuration, 151, 210
common-collector amplifier, 182–184
common-collector configuration, 153–154, 210
common-drain amplifier, 301–302
common-emitter amplifier, 178–181, 211–213
common-emitter configuration, 143–144, 148–151, 211
common-gate amplifier, 302–303
common-source amplifier, 300–301
comparator (analog), 400, 407–408, 559, 562, 618–619
comparator (digital), 510, 532–533
compensating capacitor, 364, 373
compound semiconductor, 27, 32, 83

conduction band, 25–26, 29, 33–42
conductivity, 4, 7–8, 10, 20, 25–27, 30, 39, 43–44, 49, 61, 67
consensus theorem, 481, 487
continuity equation, 47
counter, 588–598
counter modulus, 589, 592
counter-type A/D converter, 620–621
coupling capacitor, 179, 218–221
crossover distortion, 254, 258–261, 264
crystal, 5, 20–21, 23, 29–30, 33–34, 61–62, 65, 341–345
crystal momentum, 33
crystal oscillator, 341–345
current amplifier, 142, 176, 239, 316, 320–322
current gain, 148, 165–166, 176–178, 180, 183–184, 186, 208–209, 211, 227, 240, 261, 264–265, 316, 320, 332, 340, 370, 440
current source, 12–14, 50, 186, 200, 202, 215, 217, 222, 280, 293–294, 340, 366, 380
current-series feedback, 319–320
current-shunt feedback, 320–321
cut-in voltage, 67, 71, 74, 83, 101, 105, 123, 148, 258, 405, 513
cutoff region, 148, 152, 246, 290

D flip-flop, 570, 579–580, 583
Darlington pair, 265, 370
dc beta, 180
dc load line, 169, 176, 250, 257, 263
De Morgan's theorem, 484, 493, 517–518
decade counter, 595–597
decibel, 221, 226–227, 267, 321, 328, 368
decoder, 510, 536
delay line, 445
demultiplexer, 510, 536–537
depletion capacitance, 74–75
depletion region, 63–64, 67–68, 74–78, 84, 150, 279–281, 437, 43
depletion-type MOSFET, 286–288, 291, 293–294, 299
differential amplifier, 364, 365–369, 386–387
differential gain, 367, 370
differentiator, 396, 398–399
diffusion, 30, 43, 45, 62–64, 68–69, 284, 437
diffusion capacitance, 75

diffusion current, 43, 45, 63–64, 68–69
digital, 460–461
digital-to-analog (D/A) conversion, 611–618
diode, 4–6, 61
direct coupling, 240, 261, 267
distortion, 242–244
distributive law, 481
divide-by-n counter, 590, 592
don't care conditions, 498, 534, 569
donor ion, 32, 62–63, 74
donor level, 38–39
doping, 39, 41, 61–65, 76–77, 83, 87, 436–438
drain, 277–278
drain current, 277–285, 287, 289–297, 299, 302
drain resistance, 282–283, 285, 293, 296, 301, 303
drain–feedback bias, 296–297
drift current, 43–44, 68–69
drift velocity, 7–10, 43, 49, 280
duty cycle, 348, 558, 562
dynamic resistance, 71–74, 181, 215, 222

8085 instructions, 655, 668
Early effect, 150, 152, 223, 284
edge triggering, 573–574, 577
effective mas, 28, 33–35, 40–41, 43
Einstein relation, 46–47
electron, 2, 6–8, 10, 21, 24–49
emitter, 141–154
emitter current, 143–147, 151–154, 176, 180–184, 284, 294, 366–367
emitter follower, 13, 153, 182, 261, 265, 301, 316, 366, 383, 441, 516, 525
emitter–feedback bias, 168–170, 174
encoder, 510, 539–541, 642
energy band, 24–26, 28–29, 33–40, 62, 65–69, 76, 82
enhancement-type MOSFET, 289–294, 296–297, 299, 525
EPROM, 623, 625–626
exclusive-OR gate, 519–521, 531–532, 541
extrinsic semiconductor, 30, 32, 40–41, 45

555 timer IC, 434, 559–560
feedback, 11, 165, 168–174, 223, 294, 296–297, 299, 313–350
Fermi energy, 10, 84

Fermi level, 35–40, 42–43, 46, 65–67, 69, 76–77, 83, 85
Fermi–Dirac distribution, 26, 36, 39–40
FET (field-effect transistor), 141, 145, 276–303, 370, 525, 628
FET model, 293–294
field emission, 10
field ionization, 76, 78
filter, 5, 11, 117–119, 207, 389–396
firmware, 646–647
flash A/D converter, 618–620
flip-flop, 350, 512, 559–580, 622–624, 627–628
forward bias, 68–70, 72–75, 82–86, 102–115, 144–152, 245, 259, 264, 346, 513, 523–524
four-probe technique, 49–50
full-adder, 528–532
full-subtractor, 530–531
full-wave rectifier, 111–114

gain–bandwidth product, 226–227, 325–326
gate, 277–303
gate bias, 278–297, 299
ground, 363–364, 366, 370, 375–376, 378, 381–382

half-adder, 527–529
half-power frequency, 227
half-subtractor, 530–531
half-wave rectifier, 105–110
Hall coefficient, 49
Hall effect, 25, 48–49
Hall field, 48–49
hardware, 341, 641, 644–649, 670
harmonic distortion, 243–244, 255
Hartley oscillator, 327, 330, 332–333
Heavy hole, 35, 54
hexadecimal number, 469–471, 665
high-pass filter, 390, 394–395
hole, 8, 10, 26–49
hybrid parameters, 205–210, 214, 223, 338
hybrid pi (Π) model, 221–223

IC (integrated circuit), 30, 61, 363, 432–439
impact ionization, 77–79
inductor filter, 119
input impedance, 50, 178, 180, 183–186, 209, 212, 215, 223, 316, 322, 365, 370, 381, 387, 391

instrumentation amplifier, 374, 387–389
integrator, 396–399, 404–405, 407–408
intermodulation distortion, 244–245
intrinsic semiconductor, 29–30, 32, 36–39, 41–42, 61, 67
inverting amplifier, 223, 316–317, 335, 374–377, 381, 384, 392, 396, 402, 405
ion implantation, 30, 62

JFET (junction field-effect transistor), 276–285, 292, 294, 297–298, 300
JK flip-flop, 571–573, 575, 577, 579–582, 590
Johnson counter, 587–588

Karnaugh map, 471, 493–499

latch, 479, 563–565, 567, 569, 601
lattice, 21–23, 26, 30, 33–34, 43, 61, 77, 83, 341, 436
lead-lag network, 334, 337
LED (light-emitting diode), 6, 27, 67, 82–83, 87, 537
light hole, 35, 54
Lissajous figure, 452–454
load line, 102–104, 163–166, 169, 175–177, 245
logarithmic amplifier, 396, 402–403
logic families, 510, 522–526
logic gate, 20, 434, 510, 513, 516, 522–523, 565, 572, 624
loop gain, 315, 320, 327–331, 334–335
low-pass filter, 390–394
lower 3dB frequency, 227

majority carrier, 31, 68–69, 276–277, 284, 292
mass-action law, 41
master–slave flip-flop, 574–577
maximum power transfer theorem, 204
memory addressing, 629
memory read/write, 630
microcontroller, 641, 644–645, 669–670
microprocessor, 434, 637, 640–641, 643–645, 650–669
Miller effect, 219, 223–225, 373
Miller indices, 22–24
minority carrier, 31, 65, 68, 75–77, 85, 146, 150, 152, 222, 277, 284

mobility, 7–10, 28–29, 43–44, 48, 280, 290
monostable multivibrator, 345, 348–349, 559, 562–563
MOS logic, 525–526
MOSFET (metal-oxide-semiconductor field effect transistor), 276–278, 282–284, 286–293, 296–300, 525–526, 625, 628
multiplexer, 510, 533–535, 539
multistage amplifier, 239, 264–267, 370
mutual conductance, 283, 293, 316–317

9's complement, 477–478
NAND gate, 435, 484, 517–520, 523, 525–526, 529, 567–568, 573, 597
narrowband amplifier, 262
negative feedback, 165, 314–315, 321–328, 334, 364, 374–375, 378, 380, 440
negative logic, 512
nibble, 464, 471, 651
Niquist Criterion, 328
noninverting amplifier, 186, 316, 335, 374, 377–383, 385
nonlinear distortion, 326, 374
NOR gate, 484, 516–520, 524–526, 565–566, 574
Norton's theorem, 201–203, 216–217
NOT gate, 484, 515–518, 526, 536, 541, 567–568, 570, 575
NOT operation, 480, 483–484, 515, 526

1's complement, 472–474, 476–479, 531, 666
octal number, 467
offset parameters, 370
ohmic contact, 86–87, 141
ohmic region, 279–281, 286, 288, 290
op-amp adder (inverting), 384
op-amp adder (noninverting), 385–386
op-amp comparator, 396, 400–401, 407–408
opcode, 653–656
open-loop gain, 315, 322, 400, 406
operating point, 103, 163, 165, 176, 226, 241, 246–247, 294–296, 299
operating system, 640–641, 646–649, 669
operational amplifier (op-amp), 186, 264, 267, 313, 333, 337, 362–409, 434, 442, 460. 559, 611, 614, 616–618, 643

OR gate, 435, 482, 484, 513–514, 516–518, 522, 540, 584
OR operation, 481–482, 484, 488, 490, 513, 662
oscillator, 5, 313, 315, 327–349
output impedance, 142, 178, 181, 183–186, 212–213, 316, 322–323, 365, 381, 391, 516

parity, 541, 651
parity checker, 471, 520, 541–542
parity generator, 541
passive device, 14–15, 70
passive filter, 389–392
peak inverse voltage, 108, 113, 115–116, 118
phase-shift oscillator, 327, 338–340, 409
photodiode, 6, 27, 67, 84–85, 87
photoelectric emission, 10
photolithography, 436–438
piezoelectric effect, 341–342, 598
pinchoff voltage, 280–282
positive feedback, 314–315, 327–328, 331, 334–335, 364
positive logic, 512–513
potential barrier, 2, 63, 65, 69, 76, 87
power amplifier, 181, 186, 239–263
power bandwidth, 374
power gain, 180–181, 227, 267
precision rectifier, 396, 405
preset, 580–581, 585–586, 596–597
product-of-sums (POS), 488, 490, 493, 498, 519, 535
propagation delay, 488, 558, 572–574, 577, 591
pulse triggering, 573
punch through, 150
push–pull amplifier: Class A, 254–255
push–pull amplifier: Class B, 255–259

Q-point (quiescent point), 102–104, 163–165, 167, 175–178, 210, 240–241, 245–257, 262–264, 297, 299
quartz, 21, 341–342, 598, 626, 669

R-2R ladder D/A converter, 611, 614–616
racing, 564, 572–573, 575, 577
radix, 462, 477
RAM (random access memory), 623, 627
RC coupled amplifier, 266

rectifier, 2, 11, 76, 79, 86–87, 104–120
regenerative feedback, 346
register, 434, 455, 511, 581–588, 622, 624, 642, 650
regulated power supply, 12, 439–442
relaxation oscillator, 327, 345, 406
relaxation time, 8
resistivity, 7, 9–10, 26–27, 38, 43, 48–50
reverse bias, 67–69, 70–72, 75–79, 84–87, 102, 108, 111, 113–115, 123, 125, 143, 145, 148, 150, 152, 181, 262, 264, 276–277, 279–280, 287, 290, 292, 514, 523
reverse saturation current, 71, 76, 146, 152, 165, 284, 402
ring counter, 586–588, 622
ripple counter, 589, 591
ripple factor, 108–109, 113–114, 116
RMS current, 108, 113–114, 116
ROM (read-only memory), 623, 625
RS flip-flop, 565–569, 571–572, 579

saturation region (BJT), 148, 152, 154
saturation region (FET), 13, 279–282, 286, 288, 290–291, 294, 299
Schmitt trigger, 396, 401–402, 643
secondary emission, 10
self-bias, 174, 254, 294–296, 299
semiconductor, 2–5, 10, 20, 26–28, 50
sequential logic, 511, 557, 569, 579–581, 588, 642
series clipper, 120–122
seven-segment display, 537–538, 654
shift register, 455, 581–588
shunt clipper, 120–122
siemens, 283
signed binary number, 478–479
simplified hybrid model, 214–216, 218–220
sine wave generator, 409
slew rate, 370, 373–374
software, 473, 627, 641, 643–650, 653–654, 669
solar cell, 6, 67, 84–87
source, 276–302
source current, 318, 320
space–charge region, 63
square wave generator, 406–407
stability factor, 165–167, 169–170, 172, 174
state diagram, 379–380

static resistance, 72
successive-approximation A/D converter, 618, 622
sum-of-products (SOP), 488, 493, 498, 519, 535
superposition theorem, 177, 203
synchronous counter, 589, 591–592

2's complement, 472–475, 477–479, 531, 666
T flip-flop, 577–580, 590
thermionic emission, 10, 443–444
Thevenin's theorem, 200–203, 217, 224, 298, 615
timer, 407, 434, 557–560, 644
timing diagram, 522, 586, 589–591, 593–597
transconductance, 145, 222, 283
transconductance amplifier, 316–317, 319–320, 322
transfer characteristic, 149–150, 282, 286, 376, 378, 395, 402
transformer coupled amplifier, 174, 249–251
transformer utilization factor, 110, 113, 116
transistor–transistor Logic (TTL), 523–525
transition region, 63, 74
transresistance amplifier, 316–319, 322, 377
triangular wave generator, 407–408
triode, 1–4, 7, 10, 142, 145, 241, 281
tuned amplifier, 239, 245, 262, 264, 266
two-port model, 205–207

unit cell, 21–22, 341
universal gate, 484, 510, 516–517, 519–520, 522–523, 525
upper 3dB frequency, 227

valence band, 25–26, 28–29, 33–39, 41–42, 65–67, 76
valve, 4, 61, 70
virtual ground, 375–378, 381–382, 384, 396
virtual short, 378, 382–383, 385–386, 388, 392, 394
voltage amplifier, 177, 185–186, 223, 240–241, 246–247, 281
voltage doubler, 125, 126
voltage follower, 381, 383, 387, 389, 405
voltage gain, 178, 180, 183–184, 212, 219, 223, 225, 227, 240, 265, 267, 277, 300–302, 316, 336, 365, 367, 370, 376, 378, 383
voltage multiplier, 125–126
voltage regulation, 5, 78, 81, 107, 112, 440
voltage regulator, 79–81, 87, 434–435, 440, 670
voltage source, 12–14, 76, 102, 152, 200, 212, 214, 224, 377, 380, 611
voltage–divider bias, 172–174, 182, 217–218, 254, 298–299, 331
voltage–series feedback, 317–318, 323, 378, 380–381
voltage–shunt feedback, 318–319, 375, 381, 384

weighted resistor D/A converter, 611–614
wideband amplifier, 239, 270
Wien-bridge oscillator, 239, 270

Zener breakdown, 76–78, 406
Zener diode, 76, 79–82, 292, 406, 441